ENVIRONMENTAL ENGINEERING III

Environmental Engineering III

Edited by

Lucjan Pawłowski, Marzenna R. Dudzińska & Artur Pawłowski
Institute of Environmental Protection Engineering,
Lublin University of Technology, Lublin, Poland

CRC Press
Taylor & Francis Group
Boca Raton London New York Leiden

CRC Press is an imprint of the
Taylor & Francis Group, an **informa** business

A BALKEMA BOOK

CRC Press/Balkema is an imprint of the Taylor & Francis Group, an informa business

© 2010 Taylor & Francis Group, London, UK

Typeset by MPS Ltd. (A Macmillan Company), Chennai, India

Published by: CRC Press/Balkema
P.O. Box 447, 2300 AK Leiden, The Netherlands
e-mail: Pub.NL@taylorandfrancis.com
www.crcpress.com – www.taylorandfrancis.co.uk – www.balkema.nl

ISBN: 978-0-415-54882-3 (Hardback)
ISBN: 978-0-203-84666-7 (eBook)

Environmental Engineering III – Pawłowski, Dudzińska & Pawłowski (eds)
© 2010 Taylor & Francis Group, London, ISBN 978-0-415-54882-3

Table of Contents

Indoor air pollution control

Neutralization of sewage sludge and wastewater

Neutralization of solid wastes and sludges

Remediation of polluted sites

Water quality and supply

Energy saving and recovery

Environmental Engineering III – Pawłowski, Dudzińska & Pawłowski (eds)
© *2010 Taylor & Francis Group, London, ISBN 978-0-415-54882-3*

Preface

The central goals of the book *Environmental Engineering III* are to summarize research carried out in Poland, and to improve technology transfer and scientific dialogue in this time of economical transformation from a planned to a free market economy, thereby leading to a better comprehension of solutions to a broad spectrum of environmentally related problems.

Poland, like other post-communist countries, is undergoing transformation into a capitalist system. This transformation brings many problems – economical, social, psychological and also ecological. Ecological problems are strongly connected with the political, economic and psychological inheritance of the past as well as with changes in the post-communist society.

To understand these problems it is necessary to consider the following issues:

- The geographic situation of Poland,
- The political transformations that occurred after World War II – forced development of heavy industry combined with neglect of its effects on the environment, and
- The economic problems

Its geographical position in the European lowland, with mountains in the south and the Baltic Sea to the north, gives Poland some advantages such as trading and transportation opportunities. On the other hand, Poland's geography creates excellent conditions for pollution migration. Since 85% of the winds are from the west or south-west, about 50% of the sulphur dioxide in Poland comes from former East Germany and Czechoslovakia. Therefore, the western parts of Poland are much more heavily polluted by sulphur than are the eastern ones. The largest Polish rivers – the 1047-km Vistula (Wisła) and the 845-km Oder (Odra) – originate in the mountains of the highly industrialized southern part of Poland, and flow to the northern lowlands where rural areas and beautiful lakes prevail. Most of the Polish water supplies in the north are highly affected by contamination of the upper rivers.

Deposits of coal, cooper, zinc and other metals are found in the southern and southwestern parts of Poland. As a result of these raw materials, heavy industry developed in the region and caused significant degradation of air, water and soil in that area.

After the Second World War, Poland came under communist rule and heavy industrial development was forced for political reasons – to form a labour class. Most decisions on the localization of new enterprises were based on purely political reasoning, irrespective of economics or environmental health. The most typical examples are the steel works near the historical capital of Poland – Cracow, a city where the intelligentsia had previously had a strong position. Along with industrialization, a policy of neglecting ecological and psycho-social factors was developed. The ecological or human costs of living in the degraded environment were never taken into account.

Ongoing political and social changes in Poland have caused some environmental improvements, but also some new problems, both expected and unpredicted. We have observed "ecological fashion". This "fashion" for environmental protection and "ecology" has resulted in a plethora of information in the media. This situation causes social pressure on pro-ecological behavior. However, there are also new conflicts, often associated with job losses that accompany the closing of polluting industries.

Money at the local level is now distributed by local, democratically elected councils. Because of the "ecological fashion" it is easier to make the decision of spending funds on protection of the environment. Such decisions are popular among the local populace and this is a positive result of democracy. Democratic mechanisms are less satisfactory when considering the possibility of convincing people about the necessity of locating a landfill in their neighborhood or building a waste incinerator.

Increased use of motor vehicles is one of the most serious problems in Poland today. No incentives or economic stimulation for buying pro-ecological cars have yet been introduced.

Nevertheless, due to EU pro-ecological programs, a lot of very important environmentally oriented projects are realized in Poland in which also international companies are participating.

The number of multinational consortia with participation of Polish partners is steadily growing.

Therefore, a presentation of the scientific findings and technical solutions created by the Polish research community ought to be of interest not only for Polish institutions, but also for international specialists, seeking solutions for environmental problems in emerging new democracies, especially those who plan to participate in numerous projects sponsored by the European Union.

Finally, I would like to express my appreciation to all who have helped to prepare this book. Dr Sandy Williams of the Scriptoria performed a herculean task working with great patience, aiding many authors to improve the linguistic side of their papers. Anonymous reviewers who not only evaluated papers, but very often made valuable suggestion helping authors and editors to improve the scientific standard of this book. And finally, last but definitely not least Ms Katarzyna Wójcik – Oliveira and Ms Justyna Kujawska for her invaluable help in preparing a lay out of all papers.

Lublin, January 2010

Lucjan Pawłowski

Environmental Engineering III – Pawłowski, Dudzińska & Pawłowski (eds)
© 2010 Taylor & Francis Group, London, ISBN 978-0-415-54882-3

About the editors

Lucjan Pawłowski, was born in Poland, 1946. Dean of Faculty of Environmental Engineering and Director of the Institute of Environmental Protection Engineering of the Lublin University of Technology, Member of the European Academy of Science and Arts, honorary professor of China Academy of Science. He got his Ph.D. in 1976, and D.Sc. (habilitation in 1980 both at the Wrocław University of Technology). He started research on the application of ion exchange for water and wastewater treatment. As a result he together with B. Bolto from CSIRO Australia, has published a book "Wastewater Treatment by Ion Exchange" in which they summarized their own results and experience of the ion exchange area. In 1980 L. Pawłowski was elected President of International Committee "Chemistry for Protection of the Environment". He was Chairman of the Environmental Chemistry Division of the Polish Chemical Society from 1980–1984. In 1994 he was elected the Deputy President of the Polish Chemical Society and in the same year, the Deputy President of the Presidium Polish Academy of Science Committee "Men and Biosphere". In 1999 he was elected a President of the Committee "Environmental Engineering" of the Polish Academy of Science. In 1991 he was elected the Deputy Reactor of the Lublin University of Technology, and this post he held for two terms (1991–1996). He has published 22 books, over 168 papers, and authored 88 patents, and is a member of the editorial board of numerous international and national scientific and technical journals.

Marzenna R. Dudzińska received M.Sc. in physical chemistry in 1983 from Marie Curie-Skłodowska University in Lublin, Poland. She got a Fulbright Scholarship in 1989, and performed pre-doctoral research at University of Houston, USA. She received Ph.D. in environmental chemistry from Marie Curie-Skłodowska University (1992) and D.Sc. (habilitation) in 2004 from Warsaw University of Technology in environmental engineering. She is an associate professor at the Institute of Environmental Protection Engineering, Lublin University of Technology, head of Indoor Environment Engineering Division. She authored and co-authored 2 books and 85 papers and co-edited 8 books in the area of POPs in the environment, VOC and SVOC in indoor air. She is a member of Polish Chemical Society and Committee of Environmental Engineering of Polish Academy of Sciences.

Artur Pawłowski, Ph.D., D.Sc. (habilitation), was born in 1969 in Poland. In 1993 he received M.Sc. of the philosophy of nature and protection of the environment at the Catholic University of Lublin. Since that time he has been working in the Lublin University of Technology in the Faculty of Environmental Protection Engineering. In 1999 he defended Ph.D. thesis "Human's Responsibility for Nature" in the University of Card. Stefan Wyszyński in Warsaw. Also at this University in 2009 he defendend D.Sc. thesis "Sustainable Development – Idea, Philosophy and Practice". Now he works on problems connected with multidimensional nature of sustainable development. Member of International Association for Environmental Philosophy and Lublin Voivodship Board for Protection of Nature. Editor-in-chief of scientific journal "Problems of Sustainable Development". He has published 40 articles (in Polish, English and Chinese), 6 books, and has been an editor of further 13 books.

General problems

Environmental Engineering III – Pawłowski, Dudzińska & Pawłowski (eds)
© *2010 Taylor & Francis Group, London, ISBN 978-0-415-54882-3*

Environmental engineering as a tool for managing the human environment

L. Pawłowski & A. Pawłowski

Faculty of Environmental Engineering, Lublin University of Technology, Lublin, Poland

ABSTRACT: Taking into account, the current situation in Poland, comprehensive research is needed to develop:

- strategies for the management of waste and sewage sludge,
- strategies for the short- and long-term utilisation of different elements of the environment (energy supply, the role of alternative energy sources, water management, land management, management of resources)
- a better description of anthropogenic and natural sources of pollutants, as well as their transformations, pathways and dispersion through geo-ecosystems
- the means to shape and manage socioeconomic relationships through appropriate legal regulation of the use of the environment (i.e. through rationalised consumption of resources and land use and the minimisation of anthropopressure).

It is stressed that environmental engineering may be one of the most important tools in the implementation of a concept for sustainable development in the country.

Keywords: Environmental engineering, sustainable development, waste, water and wastewater management.

Information bombarding modern man suggests that the world is on the way to an ecological catastrophe.

While we do not disregard the dangers the world is now facing, it is necessary to recall that since the beginning of its existence mankind has been facing numerous threats of an ecological character. First, there were those caused by natural phenomena – huge forest fires, floods and earthquakes. Then, later on, there were those caused by the development of our civilisation. Mankind, which was becoming more and more powerful in its abilities, started creating new, anthropogenic threats.

We may look pessimistically at the development of our civilisation, having in mind the catastrophes caused by man's activities, but against that we must look at the development of knowledge and the skills derived from it, which made it possible to eliminate some of the threats and, at the same time, make people's lives richer.

It is not possible to make an in-depth analysis of the phenomena mentioned above in a short opening speech to Congress. Nevertheless, we would like to share with you an optimistic reflection.

We think that we can observe two trends in the development of our civilisation – good alternating with evil, environmental threats alternating with the hopes for their defeat. Events swing from one side to the other like the pendulum of a clock.

Environmental engineering has a leading role in the elimination of ecological threats. It has an interdisciplinary character that can deal with a wide range of technical and technological problems. It uses the knowledge of the basic sciences – biology, chemistry, biochemistry and physics – to neutralise pollution in all the elements of the environment, i.e. the hydrosphere, atmosphere and lithosphere.

Moreover, environmental engineering deals with the design and maintenance of systems of water supply, sewage disposal, heating, ventilation and air-conditioning in buildings. In brief, it deals with securing, technically, the conditions which create a safe environment for mankind to live in.

History shows that in every period of its existence mankind has been plagued by phenomena of extreme character, for example, the rise of bigger settlements led to the development of epidemics. In the second half of the 14th century the Black Death epidemics killed one-third of the European population of that time. Epidemics, called plagues, haunted whole continents not so long ago. They began in cities, which did not have adequate sanitary conditions – sufficient healthy drinking water supplies and suitable sewage disposal systems. We may, therefore, say that the existing settlements suffered from the underdevelopment of sanitary engineering, which is an important part of environmental engineering nowadays.

The improvement in the quality of water through its treatment reaches back into pre-historical times. The first information on the subject, coming from the period around 2000 BC, was found in ancient Egypt, India, Palestine, Persia and China. A Sanskrit document ordered people to boil water, as well as to heat it in the sun's rays and filter it through sand, gravel or even charcoal.

The Chinese recommended adding dried leaves from bushes in order to improve the taste of water, and in this way they discovered tea.

Also in the Bible, in Genesis 15, 'the march from the sea to the mountains of Sinai', we find information about Moses' activities in the field of environmental engineering. He led the Jews across the Shur desert and encountered water springs which were undrinkable. Moses cut bush branches and threw them into the water, which made it drinkable. Contemporary research showed that in that desert there are water springs containing excessive amounts of calcium and magnesium salts. Also a bush was found, which contains large amounts of oxalic acid in its sap. The addition of the oxalic acid from the bush branches precipitated calcium and magnesium ions in the form of oxalates with low-solubility. In light of this information we can say that Moses was the first person to use the technology of desalination of salty water. So, we may find the origin of our discipline in pre-historic times.

The quality of water influences human health in a significant way. Contrary to the common belief, it was not the development of medicine, but the development of sanitary engineering, which contributed to the sudden improvement in the health of the human population, eliminating epidemics caused by (inadequate) bad-quality water in significant areas of the globe. It did so through the improvement of the quality of the water supplied and in sewage disposal. Unfortunately, according to UN and WHO data reports, three-quarters of the people who live on Earth do not have access to water clean enough to be considered healthy. The same sources say that every year 15 million children under the age of 5 die because of diseases caused by drinking bad-quality water. It should be noted that it happens for political and economic reasons, since modern environmental engineering provides knowledge of how to effectively purify water. Unfortunately, contaminated water sources occur most frequently in poor and overpopulated areas. People who live there cannot afford to install proper facilities for water purification. Also in Poland, inhabitants of rural regions often have bad-quality water for their use.

The development of science was always affected by the twin factors of a desire on the part of the learned to better understand nature, and a need for solutions to be found for the problems considered important to the ongoing progress of civilisation. Bearing in mind the fact that the pursuit of contemporary science requires ever greater resources, attempts are being made to set priority research objectives that reflect a need to forecast directions for the development of civilisation.

A turning point as regards approaches to environmental matters was the famous U. Thant report of 1969, which spelled out the threats attendant upon environmental degradation. There was a long period of time during which environmental protection was mainly understood in terms of nature conservation and this approach remains the prevalent one in Poland. Meanwhile, for all that this approach showed that it slowed down the further degradation of the natural environment over major areas of the globe, no such success has been possible where socioeconomic relations are concerned. For this reason, the concept, hitherto understood as the protection of the natural environment, would need to be replaced by the concept of the protection of the environment for human existence. Environmental protection understood in this way takes in, not only the well-known issues geared primarily towards nature conservation, but also the whole matter of the management of the earth's resources. It is also imperative that reference be made to the social context, with account being taken of the fact that it is socioeconomic relations that exert such a major influence on the quality of life. Unemployment has just as destructive an effect upon a human being as does life in a degraded environment.

Environmental protection understood in this way encourages unavoidable changes on our planet – unavoidable since there is no alternative. This allows for a somewhat more optimistic look into the future, since the ongoing changes are becoming irrevocably linked with the need to ensure environmental conditions sufficient to allow people to live with at least minimal human dignity.

The overriding aim in protecting the environment for human life is to ensure that present and future generations enjoy healthy conditions for their existence. Similar objectives are set for the development of techniques and technologies. Furthermore, the development of the latter supplies new tools which, if used in the right way, may exert a significant influence on improving the state of the environment. Simplifying somewhat, we may say that the techniques and technologies are tools facilitating the transformation of raw materials into utilisable products as human civilisation operates. Their abrupt and accelerated development (particularly in the 20th century), resulting from a geometric increase in humankind's capacity to produce goods, led to a marked increase in living standards across large parts of the world. However, the open question arising in this context concerns whether or not the encroachment upon this of a marketing system stirring up a constant demand for new goods through slick advertising is actually raising the quality of life further.

This question is made sensible enough by the fact that the growth in output is accompanied by the accelerated utilisation of resources, itself linked on the one hand with the possibility of these being used up sooner or later, and, on the other, with an undesirable ongoing increase in the level of pollution of the environment. It is also certain that humankind's future will be very much dependent upon the way in which the flows of the Earth's resources within the human environment are managed.

The regulation of the flows in question is determined by adopting the concepts of the socioeconomic functioning of civilisation. These concepts should be shaped by knowledge of the Earth's resources and their availability, as well as of the influence exerted on the environment for human existence by the methods used to convert resources into products. The functioning of the entire biosphere – and its human component in particular – is mainly decided by how the Earth is utilised as a whole.

It was a growing awareness of these issues that led to the formulation of the sustainable development concept set out in the 1987 'Bruntland Report', officially entitled *Our Common Future*. According to this document, sustainable development is that kind of development which guarantees the meeting of the current generation's needs, without limiting the possibilities for future generations to satisfy their needs. The proper management of the Earth assumes key importance in this context, since it is upon the rational utilisation of resources that the guaranteed meeting of future generation's needs will depend.

In going beyond the questions of a purely nature-related character, attention will also need to be paid to many problems of a general kind, such as, the growing disparities between the rich and poor nations, or the increase in numbers of people going hungry and lacking access to clean water.

In our opinion, there is a need to carry out comprehensive research in the following areas:

- on developing a strategy for the management of wastes and sewage sludge.

Waste management exerts a significant influence on the degradation of the environment on the one hand (through land degradation and the generation of secondary pollutants to the soil-water environment and the air) and on the functioning of the economy on the other. Too liberal a policy will lead to excessive environmental degradation, while too restrictive a policy may stand in the way of economic development.

There is a need to better understand the consequences of dumping waste. What are the consequences of abandoning mines, pathways and the quantities of pollutants that are generated through dumping? We need to gain greater insight into the mechanisms by which these pollutants are transferred from landfills to the different components of the environment, and to determine the effects that migrating pollutants impose upon different elements of geo-ecosystems.

- on devising a strategy for the short- and long-term use of different elements of the environment (energy supply, the role of alternative energy sources, water management, land management, management of resources).

The objective of this research should be to gain a better understanding of the environmental conditioning underpinning Poland's development, taking into account our own natural resources. It is a common belief that importing primary energy resources, such as gas, is an ecological undertaking. However, no attention is paid to the fact that the gas can only be purchased if it is paid for by exporting other products whose manufacture may considerably increase environmental degradation.

Principles for the protection of natural resources need to be set out, with particular attention to underground and surface waters.

- on developing a better description of anthropogenic and natural sources of pollutants, as well as their transformations in pathways and movements through geo-ecosystems.

The negative effect of each pollutant is manifested when it passes from the place where it is generated into a living organism, wherein it is able to affect life processes.

The development of civilisation is associated with the mining of raw materials from geo-ecosystems and processing them into usable products. Following use, these return to geo-ecosystems in the form of pollution. The process is inseparable from the exertion of anthropopressure via pollutants introduced into the environment. Some of these are chemical compounds already present in nature. However, as chemistry has developed, a whole array of new chemical compounds, unknown to nature, have appeared, these sometimes displaying an exceptionally high level of biological activity. Chemical Abstracts listed in excess of 5 million chemical compounds, with around 50,000 new ones being registered annually.

In this situation, it becomes impossible to understand precisely the behaviour in the environment of all known chemical compounds. Hence, there is the need to better understand the transformations of different groups of chemical compounds and their pathways through the environment, as well the influence these exert – most especially on the biosphere and via the different food chains. Such information is indispensable if remedial actions are to be taken to limit the negative effects of chemicals introduced into the environment.

To simplify analyses it would be helpful to define the most important pathways of chemicals in geo-ecosystems.

- on the search for means to shape and manage socioeconomic relationships through appropriate legal regulation leading to a rationalised use of the environment (*i.e.* through rationalised consumption of resources and land use and the minimisation of anthropopressure).

It would seem of importance to obtain a better understanding of the attitudes representative of Polish society, and to look for means by which to shape such attitudes in order to favour implementation of the concept of sustainable development.

The objective of this work should be to better understand the socioeconomic and legal mechanisms shaping relationships between humankind and the environment.

It can be seen from all this that taking rational action to ensure sustainable development depends first and foremost on knowledge of the functioning of geo-ecosystems, as well as the skill to limit the negative effects that civilisation exerts on their functioning. Assuming particular significance in this context is the scientific research being conducted in environmental engineering on methods indispensable to the protection and appropriate shaping of the environment for human life.

Air pollution control

Environmental Engineering III – Pawłowski, Dudzińska & Pawłowski (eds)
© 2010 Taylor & Francis Group, London, ISBN 978-0-415-54882-3

Evaluation of gas emissions from graphitising of carbon products

M. Bogacki, R. Oleniacz & M. Mazur

AGH University of Science and Technology (AGH-UST), Faculty of Mining Surveying and Environmental Engineering, Department of Management and Protection of Environment, Krakow, Poland

ABSTRACT: Graphitising of carbon products emits many gaseous substances into the air, with the following having the highest emissions: CO, CO_2, SO_2, H_2S, and CS_2; aliphatic hydrocarbons (CH_4, C_2H_4, C_2H_6 and C_3H_8); and benzene, toluene, ethylbenzene and xylenes (BTEX). These emissions are of a time-variable nature and depend on technological parameters, e.g. weight and assortment of charge, furnace heating curve, type and amount of insulation packing, and efficiency of air pollution control devices. Presented are concentrations of selected gaseous substances emitted from Castner graphitising furnaces (equipped with installations of catalytic afterburning and flue-gas desulphurisation), and corresponding mass flow rates and emission factors related to carbon charge weight.

Keywords: Carbon products, graphite production, gastner furnace, gaseous pollutants, emission factors.

1 INTRODUCTION

Graphitising is a high-temperature heat treatment of amorphous carbon materials where there is rearrangement and reconstruction of the apparently amorphous structure of the carbon charge into the crystalline graphite structure. The process is gradual heating of the charge to 2500–2800°C. As a result of physical and chemical changes in the carbon material during the graphitisation cycle, gases are carried away and their composition changes dynamically with process duration. In the first stage (up to $\approx$1500°C), hydrogen and sulphur are removed. In the second stage (1500–1800°C) the majority of semi-products made from petroleum coke and pitch binder increase in volume (swelling) by $\approx$0.2–0.6%. In the third stage (>2000°C) the gradual graphitisation of carbon material and distilling of ash components begins (Lebiedziejewski 1984).

Graphitising of carbon products is usually carried out in electric-resistance Acheson or Castner furnaces (EC 2001, 2009). The Castner graphitisation method is more advanced and promising. It is based on supplying power to the graphitised preforms by direct passage of electric current, so energetic efficiency is higher than that of the Acheson method, and the time required to complete the graphitisation process is shorter (10–25 h in Castner vs. 45–80 h in Acheson furnaces). Thus the specific consumption of electric power in the Castner method can be decreased by 15–25% compared with the Acheson method (Kuznetsov 2000, Kuznetsov & Korobov 2001).

During the process, post-graphitising gases form, and are emitted into the air. The composition of the emitted gases depends on, among other things, chemical composition of the carbon mixture (semi-products), assortment of products, type of insulation packing, graphitising furnace heating-curve, and pollutant-reduction efficiency of air pollution control devices.

There are few scientific reports that examine compositions of post-graphitising gases, both those carried away from the furnace and those emitted into the air, despite them being highly noxious to the environment. The reasons for this may be, on one hand, the niche nature of the graphite industry in the world and, on the other hand, the high difficulty in sampling and chemical analysis of these gases due to high variability of their concentrations versus duration of the graphitisation process, and a high content of organic compounds in the gases. Most research in this field, covering the manufacturing process in Acheson or Castner furnaces, is found in Mazur (1995) and Mazur et al. (1990, 2004, 2005, 2006a, b). These studies generally showed the changeability of air pollutant concentrations in post-graphitising gases and hourly emission values without emission factors. The emission factors were determined in Mazur et al. (2006b), but were only concerned with the graphitising of small carbon products in an Acheson furnace. In European Commission IPPC reference documents (EC 2001, 2009) there is no information about gaseous emission from the graphitisation process, except for the range of total hydrocarbon concentrations.

The present work presents measurements of selected gaseous substance emissions from six lengthwise graphitisation (LWG) Castner furnaces equipped with catalytic afterburning and flue-gas desulphurisation installations. The main purpose of the research was to assess concentrations of the following

Table 1. Technological parameters for the Castner (LWG type) graphitising furnace.

Parameter	Unit	Value
Feedstock capacity	t	100
Insulation packing	t	220–270
Furnace long	m	3×25
Cross-section furnace area	m^2	4.8
Maximum voltage	V	280
Maximum current intensity*	kA	35
Maximum transformer power	MW	6

* one transformer.

substances emitted into the air during the graphitising cycle (when furnaces were current-operated): CO, SO_2, H_2S, CS_2, CH_4, C_2H_4, C_2H_6, C_3H_8 and BTEX (i.e. total sum of benzene, toluene, ethylbenzene and xylenes). An additional aim was to determine the gas volume flow for each graphitising furnace and enumerate the emission factors of analysed substances related to the carbon charge weight.

2 MATERIALS AND METHODS

The research covered the graphitisation of carbon products (electrodes and shapes) using six Castner electric resistance furnaces (LWG type) operated in the Graphitising Plant of SGL Carbon Polska S.A. in Nowy Sacz, Poland (Table 1).

The gases generated during graphitising were cleaned in two installations:

- catalytic afterburner, comprised of four Swingtherm-Kormoran 30.0 reactors with KERATERM ceramic filling and platinum catalyst with increased resistance to sulphur compounds (working temperature 340–425°C);
- flue gas desulphurisation installation (wet scrubber) using the double-alkaline method, which sprays the gases with NaOH solution (Na_2CO_3 batching) and post-absorption solution regeneration using $Ca(OH)_2$.

As it was necessary to provide bypassing (used in emergencies and when gases carried away do not require a specific cleaning method), each of the installations was equipped with a by-pass. This allows a proportion of the post-graphitising gases to bypass the first, second or both stages of gas cleaning, moving directly to the stack flue.

The presented results of selected air pollutant emissions are measurements taken in years 2002–2008 and cover the whole production process of graphitising of carbon and graphite products. The examinations were conducted for individual graphitising furnaces with diverse technological parameters: different charge weights, different assortments of graphitised electrodes/carbon shapes, different durations of graphitising, as well as changing temperature characteristics of furnace operation. Sampling was conducted

at the measurement point located within the stack, i.e. in flue gases upon cleaning in the catalytic afterburning and desulphurisation installations.

Each measuring series was started upon switching on the furnace power, when the post-graphitising gases were being carried away to the atmosphere in a controlled manner through the installed flue-gas hood and finished ≈2–3 h after switching the furnace power off. The times of current operation of furnaces (heating phase) for individual measuring series were 16–23 h.

As a part of the measuring series, among other things, the concentrations of the following substances in the post-graphitising gases were measured:

- furnace 4, measuring series 1 and 2: CO, SO_2, NO_2, H_2S, CS_2, CH_4, C_2H_4, C_2H_6, C_3H_8 and BTEX;
- furnaces 1–6, other measuring series: CO, SO_2 and NO_2.

The concentrations of gaseous substances, CO, SO_2, NO_2 and H_2S within measuring series 1 and 2 (furnace 4), were determined several times per hour using the automatic gas analyser Lancom Series II (Land Combustion). For the other measuring series the CO, SO_2 and NO_2 concentrations were determined using a Horiba gas analyser, type PG-250 with PSS-5 gas conditioning set; measurements and records of concentrations were continuous (sampling frequency of 1 s) with averaging interval of 1 h.

The concentrations of CH_4, C_2H_4, C_2H_6 and C_3H_8 in the post-graphitising gases were determined by sampling the gas into 0.5-dm^3 gas pipettes at a 1-h interval. Gases in samples were then identified by gas chromatography using a HP5890 chromatograph with FID detector (2-m steel column with internal diameter of 3 mm, containing phenyl isocyanate on Porasil B; column operating temperature: 40°C; carrier gas was argon at 30 cm^3/min).

The CS_2 concentration in the post-graphitising gases was determined using the manual aspiration method by taking at least one gas sample per hour (sampling time 10–55 min, depending on expected concentration). Gas samples were retained in relevant absorption solutions and then determined colorimetrically using a HACH DR/2000 spectrometer.

Gas samples for determination of BTEX content were taken at least once an hour. BTEXs were adsorbed on active carbon, extracted with carbon disulphide and determined by gas chromatography in the extract using a Pye Unicam chromatograph with FID detector (separation on two glass columns: one 2.8 m long with internal diameter 4 mm, containing 15% of tri-p-cresyl phosphate on W-AW DMCS Chromosorb, and the second 2.5 m long with internal diameter 4 mm, containing 15% of SE-30 on W-AW DMCS Chromosorb; column temperature: 120°C; injector temperature: 160°C; detector temperature: 220°C; carrier gas was argon at 30 cm^3/min).

All the measurements were made according to applicable standards and procedures. The concentration measurements of gaseous substances were taken in accordance with PN-ISO 10396: 2001.

Table 2. Selected parameters connected with the measurement runs for graphitising of carbon products.

Furnace no.	Series no.	Graphitising duration[a], h	Measuring series duration[b], h	Carbon charge (feedstock) weight, t	Average flue-gas flow rate[c], m^3_N/h
1	1	25	27	170.8	81,912
	2	22	24	125.0	75,323
2	1	21	23	155.1	77,983
	2	24	26	171.0	83,568
3	1	21	23	130.0	84,365
4	1	16	19	110.9	50,441
	2	16	19	110.9	50,376
	3	21	23	121.0	75,786
5	1	22	24	164.0	81,994
	2	19	21	148.0	87,914
6	1	23	25	168.0	81,919

[a] the furnace heating phase (in power operation).
[b] the furnace heating and ventilation phases (including 2–3 h after switching the power off).
[c] corrected to dry gas and normal conditions (pressure 101.3 kPa and temperature 273 K).

Table 3. CO, SO_2 and NO_2 concentrations in stack flue-gases from graphitising of carbon products.

		Concentration in dry gas, mg/m^3_N					
		CO		SO_2		NO_2	
Furnace no.	Series no.	Mean	Range	Mean	Range	Mean	Range
1	1	1197	164–2864	9.14	0–47	0.11	0–1.53
	2	2591	420–5987	1.84	0–18	0.61	0–3.06
2	1	2165	489–3750	7.13	0–47	1.79	0–7.65
	2	1361	39–2647	13.8	0–61	16.5	0–53.1
3	1	1820	501–3414	1.83	0–15	–	ND*
4	1	94	0–177	20.7	0–58	2.50	0–8.1
	2	161	0–353	22.0	0–54	1.20	0–8.1
	3	2417	890–4416	5.21	0–39	2.12	0–7.65
5	1	1193	355–3174	25.1	0–55	0.06	0–1.53
	2	1284	140–3034	3.91	0–38	6.99	0–35.2
6	1	920	21–2584	13.4	0–40	1.29	0–18.4
Total		1382 ± 630	0–5987	11.3 ± 7.0	0–61	3.32 ± 3.37	0–53.1

*ND = Not determined.

Simultaneously with the concentration measurements, the volume flow of gases emitted into the air was measured. The whole measuring equipment was calibrated before the measurements and checked for correct readouts using certified standard gases.

The summary of technological parameters (graphitising time, carbon charge weight and average gas-volume flow), as well as durations of individual measuring series, are provided in Table 2. This data showed that duration of graphitising (heating phase) was positively correlated with carbon charge weight ($R^2 = 0.6309$).

3 RESULTS AND DISCUSSION

The average values and the range of variability in CO, SO_2 and NO_2 concentrations in the post-graphitising gases are shown in Table 3. There was a wide range of values of measured concentrations, within the individual measuring series that reflect the emission variability during the single graphitising cycle, as well as high variability in average concentrations calculated for different graphitising cycles in furnaces 1–6. There were no correlations between the range of measured concentrations and the amount of carbon charge or LWG furnace number. The mean deviations for average concentrations during the measuring series were 45.6, 62.2 and 101.5% for CO, SO_2 and NO_2, respectively, in furnaces 1–6.

Statistical analyses on the measurements showed that CO concentration in the emitted gases changed dynamically according to duration of graphitisation. At the beginning of the heating phase, the concentrations were at their minimum levels, and in a relatively short time (1–2 h) reached their maxima; then, within another 2–3 h, concentrations dropped on average to 25% of maxima. This level was usually maintained until the end of the furnace heating phase.

Table 4.   CO, SO_2 and NO_2 mass flow rate for stack flue-gases from graphitising of carbon products.

		Mass flow rate (emission), kg/h					
		CO		SO_2		NO_2	
Furnace no.	Series no.	Mean	Range	Mean	Range	Mean	Range
1	1	98.1	13.4–234.6	0.749	0–3.850	0.009	0–0.125
	2	195.2	31.6–451.0	0.139	0–1.356	0.023	0–0.115
2	1	168.8	33.1–292.4	0.556	0–3.665	0.139	0–0.597
	2	113.1	3.3–221.2	1.150	0–5.098	1.370	0–4.437
3	1	153.6	42.3–288.0	0.155	0–1.265	–	ND*
4	1	30.0	0.2–55.2	1.040	0–2.900	0.126	0–0.408
	2	52.1	0.3–105.0	1.110	0–2.530	0.191	0–0.382
	3	183.1	67.5–334.7	0.448	0–2.956	0.141	0–0.580
5	1	97.8	29.1–260.2	2.057	0–4.510	0.005	0–0.125
	2	112.9	12.3–266.7	0.344	0–3.341	0.614	0–3.094
6	1	75.4	1.7–211.7	1.100	0–3.277	0.106	0–1.504
Total		116.4 ± 42.8	0.2–451	0.805 ± 0.443	0–5.10	0.272 ± 0.288	0–4.44

*ND = Not determined.

The maximum concentration occurrence time (after the graphitising furnace was started) differed for individual measuring series and on average were 4 h for furnace 4 (series 1 and 2); 10 h for furnaces 4 (series 3) and 2 (series 1); 14 h for furnaces 5 (series 1 and 2) and 1 (series 2); 17 h for furnace 1 (series 1); and 20 h for furnaces 3 (series 1), 2 (series 2) and 6 (series 1). Such large differences in the time of occurrence of maximum CO concentrations were weakly correlated with the carbon charge weight ($R^2 = 0.5038$). They could be supposed to be more strongly correlated with furnace heating rate, given the assortment of carbon products being the charge as well as the insulation packing used in a specific graphitising cycle. However, these technological parameters were very difficult to describe explicitly and thus were not examined. By the time of raising the graphitising furnace cover (i.e. furnace ventilation phase) the CO concentration was reduced drastically, compared with the final furnace heating phase, as a result of waste gases mixing with the air drawn in from the production hall.

The SO_2 concentration in stack flue-gases changed with graphitising process duration similarly to CO concentration; however, the difference was that SO_2 concentration was sometimes bimodal or multimodal. The concentration peaks that usually occur after the first highest peak already have the absolutely lower amplitude. The additional small SO_2 concentration peaks in gases emitted into air may be affected by both the contents of various sulphur compounds (with different decomposition times) in the charge material and periodic instability of operation of the flue-gas desulphurisation installation. Large variability in the SO_2 concentration of post-graphitising gases from different graphitising cycles (i.e. different measuring series) may also indicate influences of the assortment of graphitised products (different chemical compositions of the charge material used for preparation of products), as well as of the type and amount of insulation packing used.

NO_2 had the highest recorded variability in concentrations, both versus duration of graphitising of carbon products and depending on the measuring series. Elevated NO_2 concentrations in flue gases usually occurred in the first few hours of the furnace heating phase, upon switching the furnace power off, and the transition from heating to ventilation phases. The furnace unsealing due to the raising of the cover then took place; the exposure of the kindled graphitised products to air resulted in oxidation of nitrogen contained in the air and produced nitrogen oxides (mainly NO) in flue gases.

Large variability in average concentrations of analysed substances in different measuring series, as well as the lack of strong correlations between these concentrations and the recorded technological parameters (carbon charge weight and duration of graphitising), indicates the need for further research. This would determine the factors affecting concentrations of analysed substances in the post-graphitising gases.

In addition to the concentration of the analysed substances, the values of hourly emissions (Table 4) and emission factors (Table 5) were determined for each measuring series. To determine these values, the average flow of gases emitted into the air and the feedstock weight for all the analysed series was used (Table 2). The summary for each of Tables 4 and 5 includes the average emission values, the emission factor and the mean deviation value calculated for all measuring series. The deviations from the average of determined emissions and emission factors were correlated with the deviations of average concentrations in the stack flue-gases determined for the same substances.

The concentrations, hourly emission values, and average emission factors of H_2S, CS_2, CH_4, C_2H_6, C_2H_4, C_3H_8 and BTEX are presented in Table 6. These substances were only determined during two measuring series for furnace 4 (series 1 and 2) characterised by similar processing conditions (i.e. the same carbon charge assortment and weight, the same weight

Table 5. CO, SO$_2$ and NO$_2$ emission factors for graphitising of carbon products.

		Emission factor, kg/t-feedstock		
Furnace no.	Series no.	CO	SO$_2$	NO$_2$
1	1	15.5	0.118	0.0014
	2	37.5	0.027	0.0044
2	1	25.0	0.082	0.0206
	2	17.2	0.175	0.2083
3	1	27.2	0.027	ND*
4	1	5.1	0.178	0.0216
	2	8.9	0.190	0.0327
	3	34.8	0.085	0.0269
5	1	14.3	0.301	0.0007
	2	16.0	0.049	0.0871
6	1	11.2	0.164	0.0158
Total	Mean	19.3 ± 8.6	0.127 ± 0.068	0.038 ± 0.040
	Range	5.1–37.5	0.027–0.301	0–0.208

*ND = Not determined.

Table 6. H$_2$S, CS$_2$, aliphatic hydrocarbons and BTEX emission rate from graphitising of carbon products (furnace 4, measuring series 1 and 2).

Parameter	Substance	Series 1 Mean	Series 1 Range	Series 2 Mean	Series 2 Range	Average value
Concentration in dry gas, mg/m^{3}N	H$_2$S	3.80	0–9.6	3.90	0–11.3	3.85 ± 0.05
	CS$_2$	7.10	2.1–14.3	1.60	0–3.8	4.35 ± 2.75
	CH$_4$	47.60	1.7–495	49.10	1.6–377	48.35 ± 0.75
	C$_2$H$_6$	0.59	0.15–2.14	0.50	0–1.12	0.545 ± 0.045
	C$_2$H$_4$	0.41	0.10–1.62	0.11	0–0.46	0.26 ± 0.15
	C$_3$H$_8$	0.74	0.15–2.66	0.33	0–0.98	0.535 ± 0.205
	Benzene	0.32	0.10–1.01	0.19	0.08–0.62	0.255 ± 0.065
	Toluene + Ethylbenzene	0.27	0–0.42	0.17	0.12–0.29	0.22 ± 0.05
	Xylenes	0.26	0–0.41	0.13	0–0.17	0.195 ± 0.065
Mass flow rate (emission), kg/h	H$_2$S	0.191	0–0.484	0.193	0–0.532	0.192 ± 0.001
	CS$_2$	0.356	0.1–0.724	0.084	0–0.206	0.22 ± 0.14
	CH$_4$	2.31	0.08–23.8	2.35	0.08–19.0	2.33 ± 0.02
	C$_2$H$_6$	0.0293	0.0078–0.109	0.0253	0–0.0526	0.0273 ± 0.0020
	C$_2$H$_4$	0.0205	0.0052–0.0824	0.0053	0–0.0216	0.0129 ± 0.0020
	C$_3$H$_8$	0.0366	0.0078–0.128	0.0162	0–0.0460	0.0264 ± 0.0102
	Benzene	0.0155	0.0053–0.0483	0.0092	0.0042–0.0293	0.0124 ± 0.0032
	Toluene + Ethylbenzene	0.0134	0–0.0202	0.0084	0.0059–0.0134	0.0109 ± 0.0025
	Xylenes	0.0130	0–0.0198	0.0066	0–0.0085	0.0098 ± 0.0032
Emission factor, g/t- feedstock	H$_2$S	32.7	–	33.1	–	32.9 ± 0.2
	CS$_2$	61.0	–	14.4	–	37.7 ± 23.3
	CH$_4$	395.8	–	402.6	–	399.2 ± 3.4
	C$_2$H$_6$	5.02	–	4.33	–	4.68 ± 0.34
	C$_2$H$_4$	3.51	–	0.91	–	2.21 ± 1.30
	C$_3$H$_8$	6.27	–	2.78	–	4.52 ± 1.75
	Benzene	2.66	–	1.58	–	2.12 ± 0.54
	Toluene + Ethylbenzene	2.30	–	1.44	–	1.87 ± 0.43
	Xylenes	2.23	–	1.13	–	1.68 ± 0.55

of insulating packing, and the same furnace operation time). The only difference was new insulation packing during series 2. In both cases, the insulation packing (≈ 270 t) was a mixture of metallurgic coke and brown coal coke in a 2:3 ratio. Carrying out two measuring series under similar process conditions resulted in significantly lower deviations from average concentrations for both series.

Of the aliphatic hydrocarbons, CH$_4$ was predominant in gases emitted into the air and, in addition, only C_2H_6, C_2H_4 and C_3H_8 were also identified within the range of determinability of the applied analytical methods (Table 6). The high concentrations of substances such as CH_4 and H_2S most often occurred after 3–4 h of the furnace heating phase, reaching a maximum up from nearly zero very quickly, i.e. ≤ 1 h. After their maximum concentration in waste gases was reached there was a decrease, in the beginning very quick and then progressively more slowly (with slight variations). There was an opposite situation for CS_2; the maximum concentrations were in the later phase of the process (10–15 h of the cycle), with low concentrations in the initial phase.

The concentrations of such aliphatic hydrocarbons as C_2H_6, C_2H_4 and C_3H_8 were below the determinability limit until ≈ 2 h of the heating phase, while in the later part of the process they fluctuated at a not very high level, i.e. ≤ 2.7 mg/m^3 N. There were much lower concentrations of BTEX emitted for the whole duration of graphitisation, with maximum concentrations in the initial phase of the process.

The measured rates of all analysed substances, as hourly emission values and emission factors, are shown in Table 6.

4 CONCLUSIONS

The research revealed that, of the gaseous substances introduced into the air as a result of graphitising, the substance with the highest share of emissions was CO, followed in descending order by CH_4, SO_2, NO_2, CS_2 and H_2S (average emissions ≈ 116, 2.3, 0.8, 0.27, 0.22 and 0.19 kg/h, respectively). The other gaseous substances were emitted in small amounts.

The characteristic feature of the emission of gaseous substances was their high variability depending on the process duration, conditioned mainly by chemical composition of the carbon raw materials and the insulation packing, as well as the technological regime used in graphitising (i.e. heating curve). Moreover, the examinations confirmed that the concentration of gaseous substances emitted was also affected by the weight and assortment of the graphitised products.

ACKNOWLEDGEMENTS

The work was completed within the scope of AGH-UST statutory research for the Department of Management and Protection of Environment No. 11.11.150.008 and the contract No. 5.5.150.611. The authors would like to thank Jerzy Gałda for kind assistance in the experimental tests on the full-scale plant.

REFERENCES

EC (European Commission) 2001. IPPC Reference Document on Best Available Techniques in the Non Ferrous Metals Industries, http://eippcb.jrc.es/reference/nfm.html.
EC (European Commission) 2009. IPPC Draft Reference Document on Best Available Techniques for the Non-Ferrous Metals Industries, http://eippcb.jrc.es/reference.
Kuznetsov, D.M., 2000. Shrinkage phenomena in graphitization of preforms in Castner furnaces. *Refractories and Industrial Ceramics* 41(7–8): 279–282.
Kuznetsov, D.M. & Korobov, V.K. 2001. A comparison of properties of electrodes graphitized by the Acheson and Castner methods. *Refractories and Industrial Ceramics* 42(9–10): 355–359.
Lebiedziejewski, M. 1984. *Elektrody grafitowe w hutnictwie.* Katowice: Wyd. Śląsk.
Mazur, M., Westfal, M., Sadowska-Janusz, D. & Lipowski, J. 1990. Problemy emisji zanieczyszczeń z procesu grafityzacji elektrod. *Ochrona Powietrza* 24(6): 129–131.
Mazur, M. 1995. *Emisja związków siarki z procesów hutnictwa żelaza.* Seria: Rozprawy Monografie nr 37. Kraków: AGH-UST.
Mazur, M., Szczygłowski, P., Oleniacz, R. & Bogacki, M. 2004. Emisja BTX towarzysząca produkcji wyrobów węglowych i grafitowych. In J. Konieczyński & R. Zarzycki (eds), *Ochrona powietrza w teorii i praktyce*: 249–256. Zabrze: IPIŚ PAN.
Mazur, M., Oleniacz, R., Bogacki, M. & Szczygłowski, P. 2005. Emisja zanieczyszczeń z procesu grafityzacji elektrod węglowych w piecach LWG (Castnera). Cz. 1. Wybrane substancje gazowe, *Inżynieria Środowiska* 10(2): 149–160. Kraków: AGH-UST.
Mazur, M., Oleniacz, R., Bogacki, M. & Szczygłowski, P. 2006a. Emisja zanieczyszczeń z procesu grafityzacji elektrod węglowych w piecach LWG (Castnera). Cz. 2. Wybrane substancje pyłowe. *Inżynieria Środowiska* 11(1): 27–38. Kraków: AGH-UST.
Mazur, M., Oleniacz, R., Bogacki, M. & Szczygłowski, P. 2006b. Emisja zanieczyszczeń z pieca Achesona do grafityzacji wyrobów drobnych. *Inżynieria Środowiska* 11(2): 145–159. Kraków: AGH-UST.

Environmental Engineering III – Pawłowski, Dudzińska & Pawłowski (eds)
© 2010 Taylor & Francis Group, London, ISBN 978-0-415-54882-3

The influence of traffic-related air pollution on the ventilation efficiency of persons living in the proximity of main roads

A.J. Badyda & A. Kraszewski

Department of Informatics and Environmental Quality Research, Warsaw University of Technology, Warsaw, Poland

ABSTRACT: The influence of traffic-related air pollutants on inhabitants of Warsaw living in the vicinity of busy roads has been investigated. Residents of a rural area were taken as a control group. Pulmonary function tests were carried out on local residents to assess the risk of obstruction to their airways. The results show that, for Warsaw inhabitants, the most important spirometric parameters are significantly lower compared to the rural area group. Logistic regression models demonstrate that living in the proximity of a main road in Warsaw increases the risk of pulmonary obstruction among non-smokers more than fourfold compared with persons residing in the country.

Keywords: Traffic-related air pollutants, influence of traffic on municipal environment, influence of air pollution on health.

1 INTRODUCTION

Within urbanized areas the direct proximity of busy main roads is characterized by higher levels of air pollutants compared to areas remote from busy roads, and particularly to rural areas. As a result inhabitants living close to the busiest traffic arteries are likely to be more exposed to the harmful influence of traffic than those living in other areas. Generally, as shown by MacNee & Donaldson (2000), air pollutants have been recognized as factors of chronic obstructive pulmonary disease (COPD) for over 50 years. This recognition led to the implementation of air quality standards, which, in turn, resulted in significant decrease in the levels of air pollutants from fossil fuel combustion, in particular dust and sulphur dioxide. However, dynamic rise of road traffic has led to increased levels of other pollutants, such as ozone, particulate matter with diameters $<10\,\mu m$ (PM_{10}), or nitrogen oxides.

Numerous epidemiological results demonstrate that a relation exists between reported air pollution levels and harmful health effects, including a higher intensity of respiratory system ailments, and even a higher mortality caused by respiratory and cardio-vascular system disease. Several publications report a higher intensity of symptoms of pulmonary disease among secondary school children living in the proximity of main roads with heavy traffic. Venn et al. (2001) show that, although there is no clear evidence to prove that air pollutants lead to increased asthma incidence, a higher risk of respiratory problems can be noticed among children. The risk is inversely proportional to the distance from busy road traffic arteries. The

risk is at its highest among the inhabitants who live within a distance of 90 m of busy main roads. Having reported the results of research performed with a group of almost 1000 children, based on odds ratio calculations, the authors have clearly shown that, for children living within 150 m of main roads, each 30 m closer to the road increases the risk of asthma symptoms (specifically: wheezing breath) by 1.08 and 1.16 for age groups $4 \div 11$ and $11 \div 16$ respectively.

Research results for a group of 1129 children coming from different areas (with a low or a high ambient level of SO_2 and PM_{10}) reported by Jędrychowski & Flak (1998) show a strong relation between air pollution levels and some respiratory system symptoms (amount of secretion). Much more frequent cough and wheeze were reported (asthma symptoms) among children who had not had allergy symptoms, who lived in areas with higher levels of air pollutants. In the case of children diagnosed as having allergy coming to such conclusions seems to be more difficult, because asthma symptoms can result from numerous factors, including air pollution. Kim et al. (2004) could show, based on results of research carried out on a group of 1109 children, that there is a slight, though statistically significant, increase in bronchitis symptoms and asthma among children living in areas with higher levels of traffic-related air pollutants. The research was carried out in urban areas with relatively clean air compared to the rest of the region, where local air pollutants originated mainly from road traffic. For particular pollutants ($NO, NO_2, NO_x, PM_{2.5}, PM_{10}$) odds ratios were calculated with reference to concentration increase by the value equal to interquartile range of distribution of particular pollutant concentration. Depending on the kind of pollutant, OR amounted to 1.02–1.06 for bronchitis, and 1.01–1.08 for asthma.

Schikowski et al. (2005) reports one of the earliest results of research on the long-term influence of road traffic air pollutants on the progress of chronic obstructive pulmonary disease. Not much other research of this type has been carried out so far. The tests were performed for 10 years among a group of 4757 women, living in the Ruhr Area (in Germany). They indicate that an increase of average concentration of PM_{10} (interquartile range) by 7 $\mu g/m^3$ within 5 years caused a noticeable decrease of spirometric indicators. The forced Expiratory Volume during the First Second of Expiration (FEV_1) decreased by 5.1% (95% confidence level: 2.5%–7.7%), while the Forced Vital Capacity (FVC) decreased by 3.7% (95% confidence level: 1.8%–5.5%). At the same time the author points out that the women investigated in the test who lived closer than 100 m to busy roads showed significantly lower spirometric parameters, and the risk of COPD incidence was 1.79 times higher (odds ratio with 95% confidence level: 1.06–3.02) compared to the inhabitants of areas located further from main roads.

Another kind of study was carried out by Maheswaren & Elliott (2003), reporting the results which seem to prove the relation of death rate – deaths caused by cardio-vascular system disease – and the level of traffic-related pollutants within almost 200 000 inhabitants of England and Wales aged over 45. The results show that the percentage of deaths increases with the age of the persons and tends to increase when the distance from the place of living to main communications route shortens. Among inhabitants living within the distance shorter than 200 m from main roads the risk of death caused by cardio-vascular system diseases is higher by 5% (7% for men and 4% for women) compared to those who live within a longer distance from communications routes. The study does not show any reference to absolute levels of pollution in the studied area, making it impossible to assess the relation between exposure to pollutants and increased risk of mortality.

Grynkiewicz-Bylina et al. (2005) report the decrease in respiratory system activity, an increase in respiratory system disease symptoms, as well as more hospital admissions and higher death rate caused by higher exposure to dust pollutants. The results of the research (for a selected communications route in Gliwice, Poland) prove that average daily PM_{10} concentrations in street canyons are over 70% higher, and for polycyclic aromatic hydrocarbons over 60% higher, compared to the concentration values of these pollutants within 100 m distance from the road. The authors expect a 10% higher rate of respiratory system disease within the population exposed to the higher level of the PM_{10}, on account of its concentration, which remains at a level of over 90 $\mu g/m^3$ in the road canyon.

According to Lubiński et al. (2003), concentrations of PM_{10} higher than 10 $\mu g/m^3$ lead to a 3% higher mortality rate caused by respiratory system disease,

but also to 3% growth in the frequency of bronchial asthma attacks and to more than 12% higher use of bronchodilators (drugs used to facilitate the functioning of the respiratory system) in patients suffering from asthma and chronic obstructive pulmonary disease (COPD). Schwartz et al. (1996) reported that the daily percentage of deaths is related to changes in $PM_{2.5}$ concentrations, but not to changes in PM_{10} concentration. The test results quoted by the author show that each time growth of 2-day average concentration of $PM_{2.5}$ by 10 $\mu g/m^3$ is related to 1.5% (95% confidence level: 1.1 ÷ 1.9%) growth of daily death rate. In further study, carried out in six cities in the USA, Schwartz et al. (2002) reported that the test results prove the linear relation between $PM_{2.5}$ concentration and the death rate. Moreover, the author demonstrated that there is no minimum value of concentration of this pollutant that could be considered safe. It was shown that the above given relation between the change in $PM_{2.5}$ concentration and the mortality rate is true also for the concentrations below the minimum permissible levels, according to the United States Environmental Protection Agency, by quoting the results showing that daily mortality rate grows by 1.5% for each 10 $\mu g/m^3$ increase of $PM_{2.5}$ concentration. This indicator grows by 3% for each 10 $\mu g/m^3$ increase of solid particles arising from road transport emissions.

In work presented by Docekry et al. (1993) an influence of various risk factors on mortality rate was analysed among a group of 8111 adult persons in six cities of the United States. It was concluded that the mortality level is highly dependent on tobacco smoking. After making corrections, taking into account this and other risk factors, a statistically significant and noticeable relation was observed between air pollution caused by respiratory dust and the death rate. Interpretation of survival analysis results proves that the mortality risk coefficient for deaths caused by lung cancer and respiratory and circulatory disease among inhabitants of the most polluted cities amounts to 1.26% (95% confidence level: 1.08 ÷ 1.47) compared to the inhabitants of the cities characterized by the lowest level of air pollutants.

Hoek et al. (2002) reports the results of long-term tests (lasting 8 years), carried out in Holland on a group of 5000 persons aged 55 ÷ 69. The tests were based on the assessment of a correlation between road pollution level and mortality resulting from different causes – measurement of air pollutants was conducted (dust and nitrogen dioxide) in urban areas as well as the level of regional background, introducing a descriptor reflecting the fact of living along the main roads. The resulting analysis shows that the mortality risk coefficient among persons living along main roads amounts to 1.41 (95% confidence level: 0.94 ÷ 2.12) for all causes of death and 1.95 (95% confidence level: 1.09 ÷ 3.52) for the deaths caused by respiratory and circulatory system diseases. The authors came to the conclusion that long term exposure to traffic-related pollutants may shorten life expectancy, proving that mortality caused by disease not related to respiratory and circulatory system as well as by tumours other than those of the lung are not related to air pollution levels – the risk coefficient amounted to 1.03 (95% confidence level: $0.54 \div 1.96$).

Test results reported by Pope et al. (1995) are also worth considering. They seem to be concentrated on the estimation of the influence of pollutants (mainly S compounds; solid particles) resulting from the combustion of fossil fuel on the mortality coefficient. However, their value cannot be underestimated, as they were carried out over a vast region (151 metropolitan areas of the USA) and the investigated group was numerous (over 550 thousand). The study also took into consideration individual risks. The results prove that there is a relation between the level of respirable particulate matter and the incidence of lung tumours and circulatory and respiratory system diseases. It could also be shown that the increased risk of mortality for the most polluted areas compared to the least polluted areas was 1.15 (95% confidence level: $1.09 \div 1.22$) for sulphur compounds and 1.17 (95% confidence level: $1.09 \div 1.26$) for $PM_{2.5}$.

In the current study a review, one of the first in this part of Europe, describes the influence of traffic-related air pollutants on the ventilation efficiency of inhabitants of large urban areas. The research was conducted among inhabitants of Warsaw living along major thoroughfares. The rationale for the study was that other studies on this subject, especially those from the United States, cannot be directly applied to the corresponding situation in Warsaw. The characteristics of the exposure to traffic-related pollutants typical for inhabitants of Warsaw are different from those in other big cities, because of different traffic structure, higher average age of vehicles, different climatic and meteorological conditions etc.

In this study in the context of ventilation efficiency chronic obstructive pulmonary disease (COPD) will be referred to, as well as the term obstruction. In accordance with the definition quoted by Zieliński et al. (1998) COPD is a chronic disease distinguished by permanent impairment of bronchial patency. The airflow in the respiratory tracts is limited. The disease is not fully reversible. It is usually progressive and connected with improper inflammatory reaction of lung tissue to the presence of harmful particles or gases. Obstruction refers to disproportionate decrease in Forced Expiratory Volume during the First Second of Expiration in relation to the present Vital Capacity.

2 MATERIALS AND METHODS

2.1 Materials

In 2005, according to a previously planned schedule, pulmonary function tests were carried out among 823 persons, living along Niepodległości Avenue, a major thoroughfare crossing the centre of the city with an average daily traffic load of ca. 70,000 vehicles. The tests were conducted in the neighbourhood of an air quality monitoring station belonging to the Regional Environmental Protection Inspectorate. The tests were carried out in cooperation with lung disease specialists from the Military Institute of Medicine in Warsaw. The results of the tests from patients who were already being treated for chronic obstructive pulmonary disease (COPD) or bronchial asthma as well as those who did not cooperate with the investigating doctor during the examination (the results of these patients were difficult to assess) have been excluded from further analysis (overall of 73 tests). 750 examination results were tested, including 333 women and 417 men. The tests involved 512 non-smokers and 238 smokers aged 14 to 90 (on average 50.9 ± 19.7). Anthropometric features in each analysed group showed normal distribution.

The control group consisted of 756 persons (423 women and 333 men), inhabitants of non-urban areas (29 locations in various regions of Poland). The study was conducted using the same methodology and the same devices in corresponding months (summer period) from 2003 to 2004. 445 non-smokers participated in the examination and 311 smokers aged 18 to 85 (on average $47.8 \pm 14,3$). Anthropometric features in each analysed group showed normal distribution.

In order to exclude or at least minimize the influence of conditions from previous places of residence, only results from persons living no less than 10 years in the indicated place of residence were taken into account

According to the data provided by Regional Environmental Protection Inspectorates in all locations where control group tests were carried out, concentrations (established for the purpose of human health protection) of carbon oxide, nitrogen dioxide, PM_{10} and benzene (average annual concentrations in 2004) were lower than those detected by the monitoring station in.Niepodległości Avenue in Warsaw. In accordance with reports on environments conditions in the province from 2004, 23 places (out of 29 which were tested) belonged to A class – none of the pollutants was above the permitted levels. The other six locations belonged to the B class – PM_{10} concentrations were above the permissible level, however they did not exceed the permissible level increased by a margin of tolerance. For comparison, Niepodległości Avenue in Warsaw is classified as C class – PM_{10} and NO_2 concentrations are above the permissible values increased by the margin of tolerance.

2.2 Methods

Tests carried out from May to September, were conducted in a mobile laboratory equipped for pulmonary function tests, with the application of a Lung Test 1000 MES spirometer. The period of research resulted from the necessity of reducing the influence of pollutants from sources other than road traffic.

An outline of the research goals is as follows:

- presenting aims of the test to the examined person and informing them that the test will not have any harmful impact on their health condition,

- subject research – a questionnaire was carried out, taking into consideration anthropometric features, load of respiratory system disease, exposure to harmful factors in the workplace and place of living, presence of symptoms that might prove to be respiratory system disease, allergies,
- object research – a spirometric test carried out in the sitting position, in a specially prepared research environment, after a few minutes given to adapt to the new breathing conditions. Subsequently several flow-volume curves were recorded, to attain reproducible results in accordance with the Polish Lung Disease Association. The test results included the following variables:

- FVC (*Forced Vital Capacity*) – capacity of air, which is exhaled by a tested person during a forced exhalation after maximum slow inhalation;
- FEV_1 (*Forced Expiratory Volume during the First Second of Expiration*) – capacity of air, which is exhaled by a tested person within the first second of expiration;
- PEF (*Peak Expiratory Flow*) – maximum velocity of flow measured during forced exhalation;
- FEF_{50} (*Forced Expiratory Flow at 50% of FVC*) – velocity of air flow in middle phase of exhalation;
- $FEV_1\%FVC$ – percentage indicator of FEV_1 capacity, in its relation to the present forced vital capacity (so-called pseudo-Tiffeneau indicator).

According to the guidelines of American Thoracic Society (1991 and 1994) and the Polish Lung Disease Association, measurements should be carried out until at least three repeatable results have been obtained, i.e. results for which the values of indicators for particular measurements do not vary by more than 5%.

The values of the particular indicators (expressed in litres) were converted (taking into account any necessary additional data, such as: gender, age, height) into proper values according to European Coal and Steel Community, quoted by Quanjer et al. (1993). These standards are commonly used to assess the results of pulmonary test results in Poland.

Persons taking part in the test were divided into two groups (smokers and non-smokers). The test results were analysed in accordance with this division i.e. smoking city residents vs. smoking rural residents etc.

3 THE ASSESSMENT OF THE INFLUENCE OF AIR POLLUTANTS ON SPIROMETRIC FACTORS

3.1 *The assessment in the non – smoking group*

The non-smoking group included in total 512 inhabitants of Warsaw (249 women and 263 men) and 445 residents of non-urban areas (including 274 women and 171 men). The investigated persons showed proper average mean values for particular spirometric indicators, both among the rural area inhabitants and for inhabitants of Warsaw. However, several significant

Table 1. Mean values and standard deviations of the proper values of selected spirometric parameters among non-smoking Warsaw residents (investigated group) and inhabitants of rural areas (control group).

Parameter	Investigated group	Control group	Significance level
	Percentage of proper values		
FEV_1	100.43 ± 17.95	103.49 ± 15.06	$p < 0.05$
FVC	109.51 ± 15.82	102.85 ± 15.37	$p < 0.05$
PEF	100.58 ± 21.47	96.25 ± 18.13	$p < 0.05$
FEF_{50}	88.62 ± 36.67	100.56 ± 28.87	$p < 0.05$
$FEV_1\%FVC$	97.59 ± 10.95	106.36 ± 11.13	$p < 0.05$

differences between the groups were noticed. In the investigated group the Forced Expiratory Volume during the First Second of Expiration (FEV_1), the Forced Expiratory Flow at 50% of FVC (FEF_{50}) and the pseudo-Tiffeneau indicator ($FEV_1\%FVC$) were statistically significantly lower than in the control group, which was confirmed by t-Student test (hypothesis on equality of average values of indicators was rejected at the level of $p < 0.05$). The Forced Vital Capacity (FVC) as well as the Peak Expiratory Flow (PEF) showed lower values in the control group compared to the group of inhabitants of Warsaw. Values of FEV_1, FEF_{50} (as indicators of bronchial patency) and $FEV_1\%FVC$ are particularly important to assess potential pathological changes. The decrease in values of these parameters reveals some disorders in airflow in the respiratory tracts, and thereby – bronchial obstruction. On the other hand, lower values of FVC and PEF do not unequivocally prove pathological changes, and the situation may result from lack of sufficient cooperation with the doctor or the subject not understanding the instructions about which activities should be performed during the test. A situation is without clinical significance. The differences between values of particular breathing process indicators are shown in Table 1.

To demonstrate the influence of pollutants on individual persons in each group, the percentage of persons with disorders in the airflow in their respiratory systems was calculated, taking into account the state of advancement of the disorders. Bronchial stricture is diagnosed, when the $FEV_1\%FVC$ factor is lower than 70%, and the obstruction is considered mild if $FEV_1 \geq 80\%$ of its mean value. If the mean value of FEV_1 remains between 50% and 79% moderate obstruction is diagnosed, while a value lower than 50% proves a severe form of bronchial stricture. The results of calculations are demonstrated in Table 2.

Among non-smokers living along a main road with significant traffic intensity:

- the percentage of investigated persons with respiratory disorders is almost five times higher in comparison with rural inhabitants;
- the most significant difference is noted in the case of moderate obstruction, which was diagnosed almost

Table 2. Ratios of non-smokers with pulmonary disturbance in the investigated and control group, considering the
degree of severity of airflow obstruction.

Degree of obstruction severity	Investigated group	Control group
	Percentage of persons with airflow obstruction	
Mild	12.1%	4.0%
Moderate	9.6%	0.5%
Severe	0.2%	0.0%
All	21.9%	4.5%

20 times more frequently among inhabitants of
Warsaw than in the control group;
- mild obstruction was diagnosed 3 times more often
  among persons living along main roads in comparison with rural area inhabitants;
- severe obstruction was diagnosed in the case of
  0.2% of city inhabitants.

Because the investigated indicators demonstrated
normal distribution, the significance of differences
between the percentage of persons with obstructions of
airflow in their respiratory tracts was analyzed applying the difference between two proportions test, on the
basis of which it was proved that the hypothesis on
equality of percentages should be rejected at the level
of $p < 0.05$ for mild and moderate obstruction. For
severe cases no reasons to reject the hypothesis were
observed, which most probably resulted from the fact
that the number of diagnosed cases was not significant.

3.2 The assessment in the smoking gropup

Among the investigated persons 238 inhabitants of
Warsaw were smokers (including 84 women and 154
men) as well as 311 rural inhabitants (including 149
women and 162 men). Similarly to the non-smokers,
among the smokers average mean values of particular spirometric indicators remained within the norm,
both for the rural area residents and the inhabitants of
Warsaw.

For persons living in the proximity of Niepodległości Avenue the average mean values of FEV_1,
FEF_{50} and $FEV_1\%FVC$ were lower than for the control group. It must be noted that the t-Student test
showed that there is no basis to reject the hypothesis
on the equality of FEV_1 indicator in the investigated
and control groups. The differences of values of this
indicator in both groups are not statistically significant. Smoking and inhalation of pollutants occurring
in the cigarette glow zone are much more burdening
factors than air pollutants. Therefore, in the smoking group the influence of pollutants on health may
be not unequivocal, although it must be pointed out
that the significance level in the test is close to 0.05
(it amounts to 0.08). Lower values of FVC and PEF
compared to inhabitants of Warsaw are not clinically
significant. The differences between average values of
the discussed indicators are shown in Table 3.

Table 3. Mean values and standard deviations of the proper
values of selected spirometric parameters among smoking
Warsaw residents (investigated group) and inhabitants of
rural areas (control group). Statistically significant values
are marked in red.

Parameter	Investigated group	Control group	Significance level
	Percentage of proper values		
FEV_1	96.71 ± 18.08	99.06 ± 13.39	$p = 0.08$
FVC	107.19 ± 16.55	102.04 ± 13.47	$p < 0.05$
PEF	96.62 ± 20.48	92.70 ± 17.11	$p < 0.05$
FEF_{50}	84.49 ± 34.59	95.02 ± 28.02	$p < 0.05$
$FEV_1\%FVC$	95.18 ± 9.29	102.10 ± 11.07	$p < 0.05$

Table 4. Ratios of smokers with pulmonary disturbance in
the investigated and control group, considering the degree of
severity of airflow obstruction.

Degree of obstruction severity	Investigated group	Control group
	Percentage of persons with airflow obstruction	
Mild	9.2%	8.3%
Moderate	11.8%	2.6%
Severe	0.4%	0.0%
All	21.4%	10.9%

In both groups the percentage of persons with disorders of airflow in respiratory tracts was calculated,
taking into consideration severity of obstruction. The
results are shown in Table 4.

Similar to the non-smoking group, in the group of
smokers the percentage of persons having symptoms
of airways stricture is higher among the inhabitants of
Warsaw than respondents from rural areas. However,
the difference is much less significant and differences in frequency of diagnosing particular stages of
bronchial stricture are also smaller:

- the frequency of obstruction symptoms among persons living in the proximity of busy main roads
  is nearly twice as high as that among non-urban
  inhabitants
- similarly to the group of non-smokers, among smokers the most significant variations are noticed in case
  of moderate obstruction – it is diagnosed nearly 5
  times more often among citizens of Warsaw than
  amid rural area inhabitants
- in case of mild obstruction the difference is slight
  and comes to 1.1
- severe obstruction was diagnosed for 0.4% of investigated persons.

Testing the difference between two proportions
proves that the hypothesis on equality of percentages
of persons with obstruction in both groups should be
rejected at the significance level $p < 0.05$. For persons
with moderate obstruction, the percentage in the group

of inhabitants of Warsaw is statistically higher as well (for $p < 0.05$) compared with rural area inhabitants. Such a hypothesis can not be rejected in the case of mild and severe forms of the disease.

4 THE ASSESSMENT OF RISK OF AIRWAYS OBSTRUCTION AS A RESULTS OF EXTERMAL FACTORS

Pulmonary function indicators allow individual cases of airways obstruction to be identified. As mentioned above, the most important indicator to demonstrate disorders of the pulmonary process is a value of the pseudo-Tiffeneau indicator ($FEV_1\%FVC$) of lower than 70%. Therefore, persons with $FEV_1\%FVC < 70\%$ are persons showing symptoms of obstruction, and their FEV_1 is disproportionately lower compared with their present vital capacity. Persons having $FEV_1\%FVC \geq 70\%$ are considered to be free of symptoms of obstruction. Therefore, each subject can be described by a dichotomic variable. Further in the study a person for which the value of this variable is '0' will be considered healthy, while a result of '1' will refer to an ill person. The aim is to determine a relation, similar to a regression function, of a probability of an appearance of a obstruction, with a group of independent variables such as age, gender, smoking habit burden and place of living. In this type of analysis it is not possible to apply multiple regression, therefore logistic regression is used.

Logistic regression is a mathematical model which can be used to describe the influence of independent variables on a dichotomic dependent variable (Stanisz 2000). Let Y refer to a dependent variable with the values 0 and 1: 0 – shows no symptoms of COPD ($FEV_1\%FVC \geq 70\%$), 1 – shows symptoms of COPD ($FEV_1\%FVC < 70\%$). The model of logistic regression for such a dichotomic variable has a form shown by equation 1:

$$P\left(Y = 1 | x_1, x_2, ..., x_k\right) = \frac{e^{\left(a_0 + \sum_{i=1}^{k} a_i x_i\right)}}{1 + e^{\left(a_0 + \sum_{i=1}^{k} a_i x_i\right)}} \quad (1)$$

where: $P(Y = 1 | x_1, x_2, ..., x_k) =$ conditional probability that the Y variable will equal 1 for independent values x_1, x_2, ..., x_k; a_i, $i = 0$, ...; $k =$ regression coefficients; x_1, x_2, ..., x_k – independent values.

Apart from assessing regression coefficients and their statistical significance, values of an additional parameter – the so-called odds ratio – were calculated. It uses the term 'odds', which is defined as the ratio of the probability that a phenomenon will occur (A), e.g. a disease, to the probability that it will not occur The definition can be shown as in the equation 2:

$$S\left(A\right) = \frac{p(A)}{p(notA)} = \frac{p(A)}{1 - p(A)} \quad (2)$$

where: $S(A) =$ odds of A phenomenon appearing; $p(A) =$ probability of the absence of phenomenon A; $p(notA) =$ probability of A phenomenon non-appearance.

The logistic models presented and discussed below were worked out on the basis of pulmonary function test results and values of selected anthropometric indicators, smoking burden and place of living. The estimation of the models' parameters was carried out separately for the investigated group as a whole, as well as taking into account the division into smokers and non-smokers. The presented logistic models include only the independent variables that proved statistically significant (for $p < 0.05$). For each case the mean square error estimator and the quasi-Newton method was used. It must be mentioned that applying any other available estimator or estimation method generally caused no noticeable changes in the forms of models or values of regression coefficients. The ratio of the product of properly classified cases to the product of improperly classified ones significantly exceeded 1, which proves that the classification was much better than the one expected to occur by coincidence. Moreover, for each model odds ratios were calculated for single variation of analyzed parameters. They are presented in tables with estimation results.

The model calculated for all the considered cases is shown by equation 3:

$$P\left(X\right) = \frac{e^{-6,608 + 0,075AGE - 0,528GEN + 0,780SMK + 1,014LIV}}{1 + e^{-6,608 + 0,075AGE - 0,528GEN + 0,780SMK + 1,014LIV}} \quad (3)$$

where: AGE = age of investigated person [years]; GEN – gender = dichotomic variable: man (GEN = 0), woman (GEN = 1); SMK-smoking habit = dichotomic variable: non-smokers (SMK = 0), smokers (SMK = 1); LIV – place of living = dichotomic variable: control group (rural area inhabitants, LIV = 0), investigated group (Warsaw inhabitants, LIV = 1).

A quality factor test for chi-square fitting (χ^2) for the variation between the presented model and the one with only one absolute term shows high statistical significance ($p < 0.0001$). This shows that independent variables in the model influence the possibility of disease development. Values of model parameter estimators are also statistically significant ($p < 0.05$).

The model proves that the probability of appearance of bronchial stricture increases when values of variables such as 'AGE', 'SMK' and 'LIV' rise. For the latter two variables the increase shall be understood as the change of value from '0' to '1'. In this case the probability of developing the disease rises among smokers and urban area inhabitants. In analogy, the probability decreases when 'GEN' variable increases, which proves the lower probability of appearance of disease among women.

Table 5 shows selected values of parameter estimation for the model as well as odds ratios for single variations of particular parameters.

The test results prove that smoking increases the risk of appearance of bronchial airflow disorders more than

Table 5. Estimation of parameters of the logistic regression model of the entire investigated group.

	Variable			
	AGE	GEN	SMK	LIV
Estimated parameter values	0.075	−0.528	0.780	1.014
Significance level	<0.05			
95% confidence interval for parameters	0.063 ÷ 0.087	−0.862 ÷ −0.193	0.412 ÷ 1.142	0.659 ÷ 1.369
Odds ratio for unit change of parameter	1.07	0.59	2.18	2.76
95% confidence interval for odds ratios	1.06 ÷ 1.09	0.42 ÷ 0.82	1.52 ÷ 3.13	1.93 ÷ 3.93

Table 6. Estimation of parameters of the logistic regression model for the non-smoking group.

	Variable		
	AGE	GEN	LIV
Estimated parameter values	0.081	−0.764	1.470
Significance level	<0.05		
95% confidence interval for parameters	0.065 ÷ 0.098	−1.204 ÷ 0.324	0.945 ÷ 1.994
Odds ratio for unit change of parameter	1.08	0.47	4.35
95% confidence interval for odds ratios	1.07 ÷ 1.10	0.30 ÷ 0.72	2.57 ÷ 7.35

Table 7. Estimation of parameters of the logistic regression model for the smoking group.

	Variable		
	AGE	BM	LIV
Estimated parameter values	0.071	0.017	0.463
Significance level	<0.05		0.08
95% confidence interval for parameters	0.052 ÷ 0.091	0.001 ÷ 0.032	−0.056 ÷ 0.982
Odds ratio for unit change of parameter	1.07	1.02	1.59
95% confidence interval for odds ratios	1.05 ÷ 1.10	1.00 ÷ 1.03	0.95 ÷ 2.67

twofold, which is equivalent to developing chronic obstructive pulmonary disease (COPD). Living along a busy highway (compared to a rural area) causes an increase of the risk by nearly three times. Men are 1.7 times (0.59^{-1}) more at risk than women.

Analogous models were created based on a division between smokers and non-smokers, in order to analyze an influence of their place of living on probability of COPD appearance. The model for non-smokers is shown in equation 4:

$$P(X) = \frac{e^{-7,249+0,081AGE-0,764GEN+1,470LIV}}{1+e^{-7,249+0,081AGE-0,764GEN+1,470LIV}} \qquad (4)$$

According to the model presented above, the probability of developing the disease increases with age and is higher in the investigated group than in the control group, and is lower for women. Similarly to the previous model, the χ^2 statistical value and values of estimators of model parameters are highly statistically significant. The values are presented in Table 6.

Among non-smokers, the risk of COPD increases with age, similarly to the entire investigated group.

Men are over twice $(0,47^{-1})$ more exposed to it than women. Persons living in the proximity of main roads are more than four times more exposed to risk of obstruction than those from non-urban areas. It must be emphasised that the range of 95% confidence level is relatively high both for estimated parameters and for single variation of a parameter.

For smokers the model is presented in equation 5:

$$P(X) = \frac{e^{-6,793+0,071AGE+0,017BM+0,463LIV}}{1+e^{-6,793+0,071AGE+0,017BM+0,463LIV}} \qquad (5)$$

where: BM – body mass of investigated person [kg].

Within the group of smokers in the influence of gender on the probability of appearance of disease turned out to be not relevant. However, body mass had a slight, but statistically significant (p = 0.04) positive effect. The values of quality factor test for χ^2 fitting model is highly statistically significant. It is interesting that, among model estimators, the variable describing place of residence is statistically insignificant. A detailed list of estimation results is presented in Table 7.

The model including the statistically insignificant estimator of 'LIV' variable was presented intentionally, to show that among smokers the influence of their place of living on the possibility of COPD appearance is not highly relevant. It confirms results presented previously, according to which the difference in average values of one of the most important pulmonary function indicators (FEV_1) between urban and rural area inhabitants also appeared to be statistically insignificant The collected results prove that the increased risk of appearance of obstruction for smokers results mainly from the fact of smoking itself. The risk increases with age and is slightly higher for persons with high body mass. On the one hand, the single odds ratio of 'LIV' variable shows a 1.6 higher risk for city inhabitants however, on the other, the lower limit of 95% confidence range shows the value lower than unity, and the estimated parameter itself is not statistically significant, as previously mentioned. The logistic model calculated for the group of smokers has the form presented in equation 6:

$$P(X) = \frac{e^{-6,858+0,075AGE+0,018BM}}{1+e^{-6,858+0,075AGE+0,018BM}} \qquad (6)$$

A questionnaire, which was carried out at the same time adds the spirometric test, included information about the period of living in a particular place, workplace and external factors potentially harmful to health, connected with the present job. However, the fact that they were gathered only from the inhabitants of Warsaw, made it impossible to use them as input data for the presented models, since they included test results for inhabitants of both rural and urban areas. Moreover, the size of the investigated group was not large, and only a small number of persons confirmed any harmful factors in their workplace, making the results worked out here difficult to interpret explicitly.

5 CONCLUSIONS

To summarise the above analysis, a statistically significant increase in COPD cases is evident among non-smokers living close to busy roads compared to those of the control group living in rural areas. The results quoted in this study indicate a possibly significant role of air pollutants in the development of diseases causing bronchial stricture (mainly COPD). Statistically significant differences in average values and percentages of persons with bronchial obstruction between rural area inhabitants and those living in the proximity of main roads indicate a potential influence of air pollutants. This fact is partly confirmed by the logistic model, which proves that the probability of bronchial stricture appearance is higher among smokers and elderly people. Above all, it suggests the increase of COPD suffering risk with degree of exposure to traffic-related air pollutants – the probability of bronchial obstruction appearance is almost 3 times higher for inhabitants of urban areas, located in direct proximity of main roads with heavy traffic, than for residents of rural area.

Logistic models do not classify cases of disease in the expected way (though the cases of lack of COPD symptoms are classified very well by it), which may suggest that there are other factors which influence the incidence of obstruction which were not taken into consideration in our analysis. The low-level in the classification of disease cases proves that the size of the sample was probably too small. Therefore, for the currently available database the models could not precisely diagnose the influence of pathogenic factors, or work out the expected high level of correct classification, similarly to those for healthy persons.

Despite this, among the factors taken into account, the mere fact of living in the proximity of a busy main road in Warsaw increases the risk of disease the highest, by almost 3-fold (for non-smokers – 4 times). In conclusion, it must be pointed out that the test results based on the above presented models do not reflect the reality completely satisfactorily, although they do show a relevant and statistically significant influence. of living in proximity of busy roads on increased risk of COPD appearance.

We cannot fail to mention that other risk factors, not included in the study, may also influence the increased number of COPD among persons living along highly frequented roads. Last year, having considered the results of this study, the authors continued the research, including other factors, that potentially increase the risk of obstruction. The investigation, continued as a follow-up project financed by the Polish Ministry of Science and Higher Education, will cover 5000 persons from Warsaw and rural area. First results seem to correspond with those obtained in the above-mentioned research. The programme aims to be the basis of a further, long-term project, intended to be carried out periodically among persons living in the proximity of busy main roads in cities, in areas isolated from direct influence of traffic-related air pollutants, as well as in rural areas. It will enable tendencies in the appearance of COPD to be identified, showing how air pollutants contribute towards changes in the number of cases of respiratory system disease.

ACKNOWLEDGEMENTS

Scientific work financed as research projects, from budget funds for science for years 2005–2006 and 2008–2011.
    The part of work of Artur Badyda has been supported by the European Union in the framework of European Social Fund through the Warsaw University of Technology Development Programme.

REFERENCES

Dockery, D.W., Pope, C.A., Xu, X., Spengler, J.D., Ware, J.H., Fay, M.E., Ferris, B.G. Jr, & Speizer, F.E.1993. An association between air pollution and mortality in six

U.S. cities. *The New England Journal of Medicine* 329: 1753–1759.

GrynkiewiczBylina, B., Rakwic, B. & Pastuszka, J.S.2005. Assessment of exposure to traffic-related aerosol and to particle-associated PAHs in Gliwice, Poland. *Polish Journal of Environmental Studies* 14(1): 117–123.

Hoek, G., Brunekreef, B., Goldbohm, S., Fischer, P. & van den Brandt, P.A. 2002. Association between mortality and indicators of traffic-related air pollution in the Netherlands: a cohort study. *Lancet* 360: 1203–1209.

Jędrychowski, W.A. & Flak, E. 1998. Effect of air quality on chronic respiratory symptoms adjusted for allergy among preadolescent children. *European Respiratory Journal* 11: 1312–1318.

Kim, J.J., Smorodinsky, S., Lipsett, M., Singer, B.C., Hodgson, A.T. & Ostro, B. 2004. Traffic-related air pollution near busy roads. *American Journal of Respiratory and Critical Care Medicine* 170: 520–526.

Lubiński, W. & Chciałowski, A. 2003. Zanieczyszczenia powietrza a czynność układu oddechowego. *Polskie Archiwum Medycyny Wewnêtrznej* 110(1): 783–788

MacNee, W. & Donaldson, K. 2000. Exacerbations of COPD – environmental mechanism. *Chest* 117: 390–397

Maheswaran, R. & Elliott P. 2003. Stroke mortality associated with living near main roads in England and Wales. *Stroke* 34: 2776–2780.

Pope, C.A., Thun, M.J., Namboodiri, M.M., Dockery, D.W., Evans, J.S., Speizer, F.E. & Heath, C.W. Jr. 1995. Particulate air pollution as a predictor of mortality in a prospective study of U.S. adults. *American Journal of Respiratory and Critical Care Medicine* 151: 669–674

Quanjer, P.H., Tammeling, G.J., Cotes, J.E., Pedersen, O.F., Peslin, R. & Yernault, J.C. 1993. Lung volumes and forced ventilatory flows. Report working party: Standardization of lung function tests. European Community for Steel and Coal. Official statement of the European Respiratory Society. *The European Respiratory Journal* 16: 5–40.

Schikowski, T., Sugiri, D., Ranft, U., Gehring, U., Heinrich, J., Wichmann, E.H. & Kraemer, U.2005. Long-term air pollution and living close to busy roads are associated with COPD in women. *Respiratory Research* 6: 152–177.

Schwartz J., Dockery D.W., Neas L.M.: „Is daily mortality associated specifically with fine particles?"; *Journal of the Air & Waste Management Association*, 1996, 46: 927–939.

Schwartz, J., Laden, F. & Zanobetti, A. 2002. The concentration-response relation between PM2,5 and daily deaths. *Environmental Health Perspectives* 110(10): 1025–1029.

Stanisz. A. 2000. *Przystępny kurs statystyki z wykorzystaniem programu STATISTICA PL na przykładach z medycyny*. Tom II. StatSoft Polska: Kraków.

Venn, A.J., Lewis, S.A., Cooper, M., Hubbard, R. & Britton, J. 2001. Living near a main road and the risk of wheezing illness in children. *American Journal of Respiratory and Critical Care Medicine* 164: 2177–2180

Zieliński, J., Górecka, D. & Śliwiński, P. 1998. *Przewlekła obturacyjna choroba płuc*. Wydawnictwo Lekarskie PZWL: Warszawa.

American Thoracic Society. Medical Section of the American Lung Association. 1991. Lung function testing: selection of reference values and interpretative strategies. *The American Review of Respiratory Disease* 144 :1202–1218.

American Thoracic Society. Medical Section of the American Lung Association. 1995. Standardisation of spirometry. 1994 Update. *American Journal of Respiratory and Critical Care Medicine* 152: 1107–1136.

Environmental Engineering III – Pawłowski, Dudzińska & Pawłowski (eds)
© *2010 Taylor & Francis Group, London, ISBN 978-0-415-54882-3*

Constructions and investigation of a wet air deduster

A. Heim, M. Tomalczyk & Z. Bartczak

Faculty of Process and Environmental Engineering, Technical University of Lodz, Lodz, Poland

ABSTRACT: Sugar obtained from centrifuges in sugar production lines should be dried and then cooled. Those operations frequently occur in one apparatus: a dryer–cooler with a fluid bed. The air flowing through the dryer captures considerable amounts of sugar. A deduster combining a spray deduster (with sprinklers), a cyclone and a fixed-bed deduster was developed for separation of the sugar. Such an industrial-scale apparatus was constructed and installed in a sugar factory and its efficiency investigated. Dedusting efficiency was calculated using theoretical dependencies, giving a result of $\eta_t = 0.85$, considerably lower than the experimental result of 98–99%.

Keywords: Air dedusting, wet dedusters.

1   INTRODUCTION

Many processing technologies used for granular materials generate dust. There is a need for separation of solid particles from gas (most frequently air). This often requires the elimination or reduction of losses of valuable material and, additionally, preventing atmospheric pollution. The presence of fine particles is hazardous to organisms, including human beings. In general, the degree of hazard depends on the properties of the dust material, and its concentration and particle size. Very fine dusts of size 0.2–0.5 μm are the most dangerous for humans.

To separate solid particles from gas a variety of methods and devices which are connected with those methods are applied (Valdberg 2007). The choice of dedusting depends mainly on the dust particle size, its concentration, the solid and gas density difference and the required degree of dedusting. Generally, gas dedusting is categorised into dry and wet methods. Compared to dry dedusting, wet dedusting is characterised by high efficiency, particularly for fine dust particles at relatively low resistance to gas flow through the apparatus (Lee & Jung & Park 2008, Dubinskaya 2001, Strauss 1975).

Basic kinds of wet gas dedusters (Figure 1) are:

a) deduster with sprinklers
b) deduster with packing
c) foam deduster
d) moving - bed deduster
e) impact – inertial deduster
f) Venturi deduster
g) dynamic deduster

Dedusters with liquid sprinkler devices (Figure 1a) are vertical tanks furnished with a system of sprinklers which homogenously spray the inside of the deduster with large liquid drops. The non-uniformity of droplet distribution in the working space affects the dedusting efficiency in this type of deduster. Sprinklers are located on one or several levels but their number on any particular level depends on their type and diameter of the apparatus and can be one to a dozen or so. In case of apparatuses of large diameters, a hexagonal distribution of sprinklers in horizontal planes is recommended.

The main advantage of a deduster with sprinklers is the relatively small gas flow resistance and the consequent saving of energy in the dedusting process. The drawback of this type of deduster is the lifting of liquid droplets by gas flowing out of the deduster.

Dedusters with packing contain, in their bottom part, a grate on which Rashig rings or other elements are laid or poured. In contemporary constructions there are applied modern, self-supporting beds made of wire or steel ribbons, which are characterised by very good process parameters (e.g. good unfolding of the interface and low gas-flow resistance). Over the packing there are sprinklers to homogenously wet its surface.

In dedusters with packing, the dedusting efficiency depends on the type of packing and the effectiveness of wetting of its surface. Modern packing is produced by specialised companies and is costly and with higher gas flow resistance than in dedusters with sprinklers (Shilyaev 2005).

2   THE AIM OF STUDY

Design studies on sugar production lines are being carried out at the Technical University of Lodz. In the classical technology, sugar from centrifuges is directed to drying and cooling and, frequently, both operations occur in one apparatus which is a dryer–cooler with a

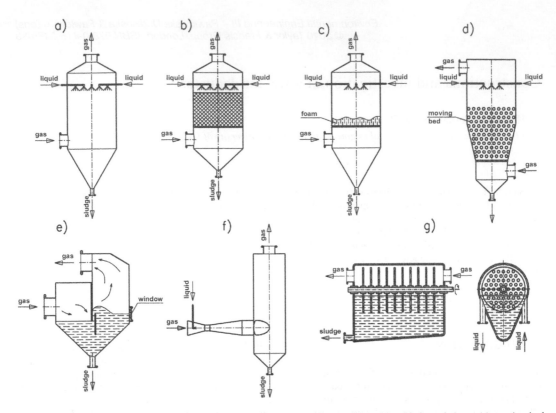

Figure 1. Schemes of wet gas dedusters: deduster with sprinklers (a), deduster with packing (b), foam deduster (c), moving-bed deduster (d), impact–inertial deduster (e), Venturi deduster (f), dynamic deduster (g).

fluid bed. The air flowing through the dryer carries away considerable amounts of sugar. For economic reasons (i.e. sugar losses) and environmental protection it is necessary to separate the sugar particles from the air. The size of particles is 0–0.3 mm, and the amount of sugar carried away in an average-sized installation is of the order of 0.2 kg/s (or 700 kg/h).

Research into wet dedusters has been carried out for many years at the Technical University of Lodz. The 'Multiwir'-type structural packing has been used for wet dedusters, and for very fine dusts this has an efficiency up to 85%. In the aforementioned sugar-production installations such efficiency would be inadequate. Therefore, a specially constructed deduster composed of the known apparatuses has been developed.

3 THE CONSTRUCTION OF THE DEDUSTER

The construction of the deduster (Figure 2) is a combination of the three known solutions: a spray deduster (with sprinklers), a cyclone and a fixed-bed deduster

The tank of the deduster comprises: a jacket 1, the bottom part in the form of a cone 2 and a cover 3. The circular smooth collars 4 are welded to the upper edge of the jacket 1 and the bottom edge of the cover 3.

The collars with screws 5 and a gasket 6 constitute a combination of the cover 3 with the jacket 1 in the form of a collar.

Inside the deduster along its axis main sprinklers 7 are installed and fed with pipes. The self-supporting packing 11 rests above the sprinklers, and is supported on the ring 10 welded to the jacket 1. Above and under the packing 11 there are located moving nozzles 12 which periodically wash the packing 11 with hot water or with hot water vapour.

The nozzles are mounted on the ends of pipes 13 which are used for feeding the nozzles 12 and their movement is in a horizontal plane. Movement of the pipes 13 and nozzles 12 is possible due to a special construction of stubs 14 on which the pipes 13 are mounted.

The inlet stub of the dusted air 15 is introduced tangentially to the bottom part of the jacket 1. In the upper part of the cover 3 the outlet stub of the air 16 is located coaxially with the vertical axis of the deduster.

The outlet stub of the liquid spray 17 and the blow-down connection 18 are mounted on the bottom part in the form of a cone 2.

In most cases, a deduster is located at the technological line in front of an exhaust fan, meaning that there is a negative pressure of 3–10 kPa. Thus, it is recommended that jacket 1 is supported with rings welded on its external surface, which are made of a bent flat-bar

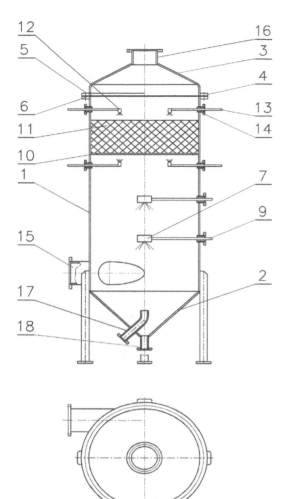

12
5

16
3
4

6
11
10
1

13
14

7
9

15

2

17
18

Figure 2.   Construction of a wet air deduster.

or two bent angle-plates. The support is indispensable for material-saving construction of a deduster made of nickel chrome steel (Figure 3).

The construction details of the main nozzles and self-supporting packing are protected as an industrial secret.

The dusted air flows into the apparatus by the stub 16 and in a circular motion is directed upwards and contacts the liquid spray droplets which absorb average and small-sized dust. The largest dust particles, under influence of centrifugal force, sediment on the internal surface of the jacket 1 and carried by water they reach the bottom part of the deduster 2.

The liquid spray absorbed from the deduster with the stub 17 flows in a closed circulation in which it may be filtrated outside the deduster.

Sugar dust particles separated from the air dissolve in water and thus the wetting liquid becomes a sugar solution. After attaining the appropriate concentration the solution is recycled to the sugar processing line.

Liquid loss caused by sugar solution recirculation as well as by evaporation is constantly supplemented with fresh water.

Small droplets of liquid containing dust settle on the surface of packing 11 and form large drops which fall down or flow down the jacket wall 1. The dedusted air flows from the apparatus through the stub 15.

A series of dedusters, constructed as described above, is produced in Poland by the OMNIKON company and have high dedusting efficiency at air flow of 8–18 m^3/s. Their characteristic parameters are:

– deducted air rate at inlet of 12–16 m/s
– gas flow resistance of 600–1000 Pa
– liquid spray demand of 0.5×10^{-3} to 1×10^{-3} m^3 of liquid / m^3 of gas;
– wetting density of 1.4×10^{-3} to 2.8×10^{-3} m^3 of liquid/s m^2 of deduster cross-section.

4   CALCULATION OF THEORETICAL EFICCIENCY

The investigated deduster may be treated as two classical dedusters connected in series as follows:

– a countercurrent deduster with sprinklers (Figure 1a),
– a countercurrent deduster with packing (Figure 1b).

The total efficiency of the deduster may be calculated from the following formula:

$$\eta_t = \eta_1 + (1 - \eta_1)\eta_2 \qquad (1)$$

where η_1 and η_2 are dedusting efficiencies at the first and second stage, respectively.

The formulae from the monograph (Warych 1979) were applied for the efficiency calculations. Some of the magnitudes needed for the calculations had to be determined experimentally:

– the size of dust particles at the inlet of the deduster,
– the size of dust particles after the first stage of dedusting,
– the size of water droplets in the sprinkler.

The granulometric analysis of dust was used to investigate the samples collected at the appropriate locations of the apparatus and the characteristics of liquid sprinklers were determined.

The dedusting efficiency at the first stage (deduster with sprinklers) was calculated from the following equation:

$$\eta_1 = 1 - \exp\left(-\frac{3V_l\, u_g\, \eta_d\, h}{2V_g\, u_d\, d_d}\right) \qquad (2)$$

where V_l = volumetric flow-rate of liquid = 10.9×10^{-3} m^3/s; V_g = volumetric flow-rate of gas = 12.0 m^3/s; u_g = gas linear velocity = 2.44 m/s; u_d = droplet sedimentation velocity, calculated for droplets of average diameter; 2.4×10^{-3} m, falling with turbulent flow $= 5.7$ m/s; $d_d =$ average diameter of droplets $= 2.4 \times 10^{-3}$ m; h $=$ height of the interface of liquid and gas contact $= 3.5$ m; $\eta_d =$ dust particle sedimentation efficiency on droplets, dependent on Stk number as expressed by the following equation:

$$Stk_1 = \frac{r_s^2 \cdot \rho_s \cdot u_g}{9 \cdot \mu_g \cdot r_d} \qquad (3)$$

where: $r_s =$ average radius of dust particles $= 0.01 \times 10^{-3}$ m (from the granulometric distribution of dust at the inlet of the deduster the average diameter of dust particles is equal to 0.02×10^{-3} m); $\rho_s =$ sugar density $= 1860$ kg/m³; $\mu_g =$ gas viscosity (of air) at temperature of $35°C = 19 \times 10^{-6}$ Pa s; $r_d =$ radius of liquid droplets $= 1.2 \times 10^{-3}$ m (from the granulometric distribution of droplets).

The value of Stoke's number calculated from Equation (3) was $Stk_1 = 2.21$.

Assuming that only the inertia mechanism occurs during dust particles sedimentation on droplets, the potential flow takes place for such a value of Stoke's number. The sedimentation efficiency of dust particles on droplets is attributed to this flow and is:

$$\eta_d = \frac{Stk_1^2}{(Stk_1 + 0.25)^2} \qquad (4)$$

Substituting the value of Stk_1 into equation gives $\eta_d = 0.807$. Hence, the dedusting efficiency at the first stage is equal to:

$$\eta_1 = 1 - \exp\left(-\frac{3 \cdot 10.9 \cdot 10^{-3} \cdot 2.44 \cdot 0.807 \cdot 3.5}{2 \cdot 12 \cdot 5.7 \cdot 2.4 \cdot 10^{-3}} \right) = 0.50$$

The dedusting efficiency at the second stage (after packing) is expressed by the following equation:

$$\eta_2 = 1 - \exp\left(-\pi \cdot n \frac{d_p \cdot u_g}{b \cdot u_g'} \cdot 2 Stk_2 \right) \qquad (5)$$

where: n $=$ the number of arches at the packing height $= 4$, $d_p =$ the characteristic size of the packing elements $= 60 \times 10^{-3}$ m; b $=$ the width of the channel for the gas flow $= 54 \times 10^{-3}$ m; $u_g =$ gas linear velocity calculated on the free cross-section $= 2.44$ m/s; $u_g' =$ gas linear velocity in packing channels $= 2.58$ m/s.

Stoke's number for the flow through the packing is calculated as follows:

$$Stk_2 = \frac{d_s^2 \cdot u_g \cdot \rho_s}{9 \cdot \mu_g \cdot d_p} \qquad (6)$$

where: $d_s =$ dust particle diameter $= 0.01 \times 10^{-3}$ m (determined from the granulometric analysis of dust after the first stage of deducting). The remaining

parameters were present in the aforementioned equations. Therefore,

$$Stk_2 = \frac{(0.01 \cdot 10^{-3})^2 \cdot 2.44 \cdot 1860}{9 \cdot 19 \cdot 10^{-6} \cdot 60 \cdot 10^{-3}} = 0.044$$

whereas:

$$\eta_2 = 1 - \exp\left(-\pi \cdot 4 \frac{60 \cdot 10^{-3} \cdot 2.44}{54 \cdot 10^{-3} \cdot 2.58} \cdot 2 \cdot 0.044 \right) = 0.69$$

thus total efficiency is:

$$\eta_t = 0,50 + (1 - 0.5)\,0.69 = 0.85$$

5 INVESTIGATIONS OF THE DEDUSTER

The investigations of dedusting efficiency used the outlet air from an industrial-scale fluid dryer–cooler (SCFM – 500 produced by the Polish company OMNIKON) with a diameter of 2.5 m.

The investigations were conducted in the course of normal operation of the sugar cooling and drying installation. The quantity of dust stopped in the deduster was calculated on the basis of the sugar concentration increase in the liquid spray flowing in the closed circulation. The amount of dust contained in the inlet air was defined on the basis of sugar mass balance in the dryer–cooler. Before beginning the test the tank of liquid spray was filled with clean water to a volume of 1 m³. During the test the mass of sugar supplied and directed away from the dryer–cooler, as well as its moisture content in the front of and behind the tank, were measured. On completion of the test the volume and concentration of the sugar in the liquid spray tank was determined.

The dedusting efficiency was derived from the following equation:

$$\eta = \frac{\frac{C_{Bx}}{100} \cdot V \cdot \rho}{(m_1 - m_2)\left(1 - \frac{w_1 - w_2}{100}\right)} \qquad (7)$$

where: $\eta =$ dedusting efficiency; C_{Bx} is the concentration of sugar in the liquid spray the moment the test is completed, in Brix degrees; V $=$ volume (m³) of the liquid spray in a closed circulation; $m_1 =$ sugar mass (kg) supplied to the fluid dryer–cooler during the test; $m_2 =$ sugar mass (kg) taken from the fluid dryer–cooler during the test; $w_1 =$ moisture content (%) in sugar at the inlet of the dryer–cooler; $w_2 =$ moisture content (%) at the outlet of the dryer–cooler.

The results of ten example tests are summarised in Table 1. The present investigations were carried out during the final technical acceptance procedure of the deduster after the optimisation of wetting intensity.

Table 1. Measurements and calculations of dedusting efficiency.

	Measurements							Calculations		
Test no.	Final concentration of syrup C_{Bx}, (Bx)	Syrup volume (m³)	Sugar mass m_1, (kg)	Sugar mass m_2, (kg)	Moisture content w_1, (%)	Moisture content w_2, (%)	Time of duration of the test τ, (min)	Wetting intensity m³/(h · m²)	Syrup density (kg/m³)	Dedusting efficiency
1	25.1	1.05	7814.5	7518.8	1.01	0.028	25	7.95	1101	0.991
2	22.0	1.06	7512.2	7251.6	0.98	0.029	23	8.20	1088	0.983
3	23.4	1.05	7923.1	7647.2	1.02	0.028	24	8.12	1094	0.984
4	24.1	1.04	7853.1	7571.2	1.01	0.030	24	8.08	1097	0.985
5	24.2	1.05	7624.8	7338.3	0.97	0.031	24	8.14	1098	0.983
6	22.1	1.06	7639.9	7379.4	1.00	0.032	24	8.22	1088	0.988
7	23.4	1.05	7789.1	7512.8	1.05	0.030	24	8.18	1094	0.983
8	22.2	1.05	7634.9	7375.9	1.03	0.029	24	8.17	1089	0.980
9	24.1	1.06	7721.2	7433.2	1.04	0.030	24	8.02	1097	0.983
10	24.8	1.04	7889.1	7599.6	1.05	0.033	24	7.99	1100	0.990

6 CONCLUSIONS

1. The slight changes observed in sugar quantity flowing through the dryer–cooler, and oscillations of wetting intensity considered in the measurements (Table 1), had little influence on dedusting efficiency.
2. The tests confirmed high dedusting efficiency in the tested deduster of around 0.985.
3. The calculated efficiency of $\eta_t = 0.85$ was considerably lower than experimental results and may be explained by beneficial construction solutions such as follows:

   – tangential gas supply to the apparatus which causes its whirling (i.e. cyclonic) motion,
   – the original construction of the packing.

REFERENCES

Lee, K. B, Jung, R. K. & Park, H. S. 2008. Development and application of a novel swirl cyclone scrubber. *Journal of Aerosol Science* 39: 1079–1088.

Dubinskaya, F.E. 2001, A priori estimation for coefficients in the energy theory of wet trapping. *Journal of Aerosol Science* 37: 525–529.

Gluba, T., Heim, A. & Kwaœniak, J. 2005. Wet dedusting efficiency with the use of "Multiwir"-type packing. *Environmental Engineering Committee Monograph PAN33*: 39–47 (in polish).

Shilyaev, M.I. & Shilyaev, A.M. 2005. Relationship between the energy and fractional methods for designing wet-type dust collectors. *Theoretical Foundations of Chemical Engineering* 39: 555–560.

Strauss, W. 1975. Industrial Gas Cleaning. *Pergamon Press. London.*

Valdberg, A.Y.U. 2007. Modern tendencies in the development of dust collecting theory and practice. *Chemical and Petroleum Engineering* 43: 423–426.

Warych, J. 1979, Dedusting of gases using wet methods. *WNT. Warsaw*(in polish).

Environmental Engineering III – Pawłowski, Dudzińska & Pawłowski (eds)
© 2010 Taylor & Francis Group, London, ISBN 978-0-415-54882-3

Pressure swing absorption of carbon dioxide in physical solvents

H. Kierzkowska-Pawlak & A. Chacuk
Faculty of Process and Environmental Engineering, Technical University of Lodz, Łódź, Poland

ABSTRACT: The mass transfer rates during CO_2 absorption and desorption from supersaturated propylene carbonate (PC) and dimethyl ether of polyethylene glycol (DMEPEG) solutions were measured at the temperature range 293.15–323.15 K in a baffled agitated reactor with a flat gas–liquid interface. Isothermal batch desorption was initiated by a sudden pressure reduction in a gas saturated system. The volumetric mass transfer coefficients for the bubbling desorption were determined from the measured overall and diffusive desorption rates and correlated by a power relationship of supersaturation, Reynolds and Weber numbers. The influence of inert solid particles suspended in the liquid phase on the mass transfer rate during desorption was investigated.

Keywords: CO_2 separation, gas–liquid mass transfer, bubbling desorption, supersaturation, volumetric mass transfer coefficient, solid particles.

1 INTRODUCTION

The removal of acid gases, such as CO_2 and H_2S, from natural, refinery and synthesis gas streams by absorption using different solvents is a significant operation in gas processing. Another important application of absorption-based technologies is CO_2 separation from flue gases in many industries – such as fossil-fuel power plants, steel and cement production. According to the Intergovernmental Panel on Climate Change (IPCC), carbon dioxide is recognized as a major man-made greenhouse gas contributing to global warming. The problem of CO_2 increasing emissions has led to international commitment to regulate the CO_2 emissions by all developed countries. The idea of carbon dioxide sequestration, which includes its capture and storage in underground rock formations, has progressed steadily over the past 10 years. It is claimed that this solution could play an important role in solving the problem of increasing greenhouse gas emissions. The physical or chemical absorption of CO_2 are generally recognized as the most efficient CO_2 separation technologies at present. The most commonly used physical solvents are methanol at low temperatures (Rectisol, Lurgi GmbH), propylene carbonate (Fluor-Solvent), N-methyl-2-pyrrolidone (Purisol, Lurgi GmbH) and dimethyl ether of polyethylene glycol (Selexol, Norton Chem) (Kohl & Nielsen 1997). The common feature of these processes is that they are used in absorber-stripper mode, requiring two separate steps in CO_2 separation. This technology is known as a pressure swing absorption process, in which low pressure is used to desorb CO_2 and to regenerate the solvent. Although, in many practical situations, the operational and capital costs of the desorption column may be greater than the costs of the absorption column,

studies devoted to desorption are not as numerous as those concerning absorption and there is little information in the literature on the design aspects of desorption columns. The problem of predicting the desorption rates arises when the process is accompanied by bubble nucleation in the liquid bulk. This phenomenon completely changes the hydrodynamic conditions in the liquid phase and the diffusive mass transfer equations cannot describe the process rate in terms of analogues to absorption.

Research on measuring and calculation of the desorption rate under dynamic conditions accompanied in the typical mass transfer equipment is very limited compared with the theoretical considerations of bubble nucleation from quiescent supersaturated liquid solutions (Lubetkin 2003). Only a few references report quantitative results on the rates of physical gas desorption accompanied by bubble nucleation from agitated liquid. Weiland et al. (1977) investigated the desorption of CO_2 from supersaturated water solutions using a stirred cell apparatus. Hikita and Konishi (1984) studied the same gas–liquid system in a baffled agitated vessel that was operated in a continuous manner. Recently, Kierzkowska-Pawlak (2007) published the preliminary results of the mass transfer rates during CO_2 desorption from saturated N-methyl-2-pyrrolidone solutions at 293.15 K in a baffled agitated reactor.

The main scope of the present paper is therefore to study CO_2 desorption from supersaturated different solvents initialized by the pressure release under dynamic conditions in a stirred cell reactor. The purpose of these measurements is to determine the influence of temperature, stirring speed and supersaturation of the solution on the CO_2 desorption rate from common physical solvents, including propylene carbonate

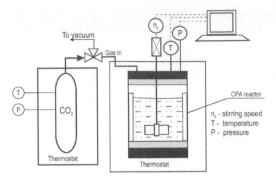

Figure 1. The schematic of the experimental set-up.

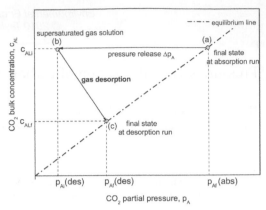

Figure 2. The schematic of the experimental procedure.

and dimethyl ether of polyethylene glycol. Propylene carbonate ($C_4H_6O_3$) is becoming increasingly popular as a gas-treating solvent (Fluor Solvent), especially for coal gasification applications. DMEPEG 250 ($CH_3(CH_2CH_2O)_nCH_3 - 250\,kg \cdot kmol^{-1}$ is an average molecular weight) is commonly known as a Selexol solvent and is an important absorbent in both natural gas and syngas purification (Kohl & Nielsen 1997). The aim of the next part of the present research work was to investigate the effect of the presence of inert solid particles on mass transfer during gas desorption from a supersaturated liquid in a mechanically stirred gas–solid–liquid system.

2 MATERIAL AND METHODS

The measurements were performed in a heat-flow reaction calorimeter (Chemical Process Analyser, ChemiSens AB, Sweden), which is a fully automated and computer-controlled stirred reactor vessel with the possibility of an on-line measurement of thermal power developed by the process. Its heart is a mechanically agitated stainless steel/glass reactor with an effective volume of $250\,cm^3$. The schematic diagram of an experimental set-up is shown in Figure 1. The reactor was equipped with four stainless steel baffles and an impeller stirrer. While in use, it is submerged in the thermostating liquid bath. A Peltier element mounted inside the bottom of the reactor serves as an efficient heating and cooling device and keeps the temperature constant at ±0.1 K. The absolute pressure was measured by a pressure transducer mounted on the reactor flange. The measurement accuracy of the digital pressure transducer was 0.1% of the full range (0–1.5 MPa). A separate tube on the top flange allows either evacuation of the cell or introduction of a gas into the reactor.

Each experiment consisted of two steps, including the absorption and subsequent desorption initialized by the pressure release, which is schematically shown in Figure 2. The solvent ($100\,cm^3$) was placed into the reactor. The gas feed lines and the content of reactor were then vacuumed. When the vacuum was shut off, under stirring conditions the system was allowed

to come to vapour liquid-equilibrium at a given temperature. At this equilibrium, the stirring was stopped for a moment. Following this, a controlled amount of acid gas was introduced into the upper part of the cell from the thermostated high-pressure gas reservoir through insulated tubing. The total pressure at this state was recorded as the initial pressure (P_i). Then, the process was initiated by switching on the stirrer at the desired mixing speed. Pressure decay versus time was recorded as the result of the CO_2 absorption through the horizontal gas–liquid interface until the equilibrium state was reached (P_f). This total pressure change (P_i–P_f) was necessary for calculating the equilibrium CO_2 solubility in the liquid expressed by Henry's law constant k_H ($MPa \cdot m^3 \cdot kmol^{-1}$). The final state at absorption run is marked by a letter (a) in Figure 2.

The desorption run was conducted in a similar manner. The stirring was stopped for a moment while the system was rapidly decompressed via a release valve in order to impose the required supersaturation of the solution (b). The resulting supersaturation of the system subsequently led CO_2 to be released from the liquid. The corresponding pressure increase was recorded, up to another gas–liquid equilibrium state (c).

During the experiment, the CO_2 concentration in the solution decreased as bubbles were formed, thus resulting in a decrease of the driving force and consequently the supersaturation of the solution. The relative supersaturation, σ, was defined as:

$$\sigma = \frac{c_{AL} - c_{AL}^*}{c_{AL}^*} \tag{1}$$

where $c_{AL}^* = CO_2$ concentration at the gas–liquid interface; $c_{AL} =$ the gas bulk concentration.

The equilibrium concentration c_{AL}^* is linearly related to CO_2 partial pressure (p_A) through Henry's law, as follows:

$$c_{AL}^* = \frac{p_A}{k_H} \tag{2}$$

By the CO_2 mass balance in the reactor, the continuous measurement of the pressure changes was related to the instantaneous rate of desorption, both CO_2 concentrations (c^*_{AL} and c_{AL}) and supersaturation σ. In these calculations, it was assumed that, in the desorption run, the initial CO_2 liquid concentration was known and equal to the final (equilibrium) CO_2 concentration that was reached in the absorption run.

The experiments were carried out at three temperatures: 293.15, 313.15 and 323.15 K. Various stirring speeds were applied: 100, 150, 200 and 250 min^{-1}. The initial pressure in the absorption run was in the range of 0.5–1 MPa. Different pressure reduction values were applied in order to impose different supersaturations of the solution. The chemicals used – CO_2 (99.995 vol % pure), PC (Fluka, 99 mass % pure) and DMEPEG 250 (Fluka, 98.5 mass % pure) – were of analytical reagent grade and used without any further purification.

The mass balance equation for the gas component entering the liquid phase predicts a linear dependence of the gas absorption rate ($N_{A,abs}$) and the concentration driving force, and has the following form:

$$N_{A,abs} = (k_L a) V_L \left(c^*_{AL} - c_{AL} \right) \qquad (3)$$

where $k_L a$ = the volumetric mass transfer coefficient in the liquid phase; V_L = the liquid volume. For gas desorption, the similar equation can be written as:

$$N_{A,des} = (k_L a)_{des} V_L \left(c_{AL} - c^*_{AL} \right) \qquad (4)$$

where $(k_L a)_{des}$ = the volumetric mass transfer coefficient for gas desorption. For relatively low supersaturations occurring at a low bubbling region, the coefficient $(k_L a)_{des}$ can be divided into the contributions of the diffusive mass transfer through the flat gas–liquid interface and the interface of bubbles generated in the liquid according to:

$$(k_L a)_{des} = k_L a + k_L a_b \qquad (5)$$

The volumetric mass transfer coefficient $k_L a$ in the Equation (5) expresses the diffusive mass transfer during desorption and is equal to the $k_L a$ determined at the absorption step of the same experiment. The volumetric mass transfer coefficient $k_L a_b$ expressing the contribution of generated bubbles into the process rate could be thus determined from the experimental absorption rate and overall desorption rate by combining Equations 3–5. During the course of every desorption run, several instantaneous values of $k_L a_b$ were evaluated for decreasing values of supersaturation of the solution as the process was realized under unsteady conditions.

For the analysis and correlation of experimental data, the physicochemical properties of the system under the applied temperature range were required. Most of these were collected from the open literature, while viscosity and surface tension of the solvent were independently measured.

In the next part of the present research work, the effect of inert solid particles at low concentrations on gas–liquid mass transfer during a gas desorption was investigated for both liquids at a temperature of 293.15 K and a stirring speed of 150 min^{-1}. Two types of solids were used:

1) PVC, $d_p = 110 \, \mu m$, $\rho = 1376 \, kg/m^3$
2) Polypropylene, $d_p = 560 \, \mu m$, $\rho = 867 \, kg/m^3$

All solid particles used were approximately spherical in shape. The solids holdup in the liquid were the same in each experiment and equaled approximately 1 vol. %.

3 RESULTS AND DISCUSSIONS

A typical example of mass transfer rates for CO_2 desorption from supersaturated propylene carbonate solutions for different stirring speeds is shown in Figure 3, where the process rate is plotted against the concentration driving force. For the comparison, the corresponding linear relation for the absorption process is plotted by dashed and solid lines. It can be seen that higher desorption rates were achieved with higher stirring speeds. A higher stirring rate produces greater turbulence and promotes detachment of the gas bubbles from the heterogeneous surface, which are contributing factors to the solvent regeneration rate. As can be seen, there exist two different regions with respect to effect of driving force. In the initial stage of desorption run, which corresponds to high values of supersaturation, the desorption rate deviates upward from the absorption rate. In this region, the bubble nucleation was visually observed through the glass walls of the reactor. The bubble nucleation in the liquid phase increases the turbulence in that phase and results in a subsequent increase of mass transfer coefficient and interfacial area. The observed rate of desorption under bubbling conditions is significantly greater than the absorption rate for the same driving force. This observation is in agreement with the previous findings of Hikita and Konishi (1984) and Kierzkowska-Pawlak (2007). As the driving force and the supersaturation substantially decrease while approaching the equilibrium state, the CO_2 desorption rate clearly decreases and becomes well represented as a linear function of driving force in a similar way to the absorption process. In this region, which is designated as diffusive or quiescent desorption, the mass transfer takes place by a diffusive mechanism and the absorption model can be successfully used to predict the desorption rates. Consequently, the volumetric mass transfer coefficient $(k_L a)_{des}$ in the quiescent desorption is equal to the relevant $k_L a$ coefficient in the absorption process. The similar trends can be observed in Figure 4, where the desorption rates of CO_2 from saturated DMEPEG solutions are plotted against the concentration driving force at different temperatures.

The transition point from bubbling to quiescent desorption can be determined from Figs. 3–4 as the value of mass transfer driving force at which desorption

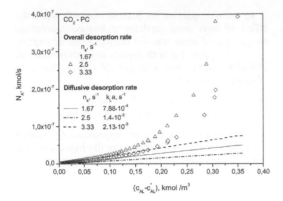

Figure 3. Effect of stirring speed on CO_2 desorption rate at 323.15 K from saturated PC solutions.

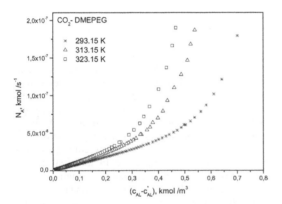

Figure 4. Effect of temperature on CO_2 desorption rate from saturated DMEPEG solutions at stirring speed of $2.5\,\mathrm{s}^{-1}$.

rate approaches absorption rate. As seen by the presented results, the inception of bubbling is not sharp and takes place smoothly. The transition concentration driving force between bubbling and diffusive desorption decreased slightly with an increase of stirring speed. The increased turbulence in the system due to higher stirring speeds enhances the desorption rate and causes lowering of the threshold for bubbling desorption. For higher stirring speeds, the bubbles find it easier to detach themselves from the walls of the vessel and its bottom. As discussed above, the effect of stirring speed is thus to promote bubble heterogeneous nucleation.

Figure 5 shows the selected data of volumetric mass transfer coefficients $k_L a_b$ determined for CO_2-PC and CO_2-DMEPEG systems over a wide range of operational conditions, including different temperatures and stirring speeds versus supersaturation values. Experimental results show that $k_L a_b$ increases with an increase of supersaturation of the solution, temperature and stirring speed. According to the analysis of experimental data, an empirical correlation of $k_L a_b$ data was derived using this method of dimensional

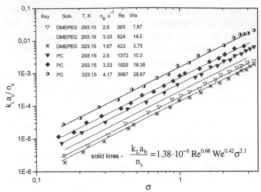

Figure 5. Volumetric mass transfer coefficient $k_L a_b$ at different operational conditions as a function of supersaturation for CO_2-PC and CO_2-DMEPEG systems.

analysis. The mass transfer coefficient can be related to other variables by the following functional equation:

$$k_L a_b = f\left(\sigma, n_s, d_s, \rho_L, \eta_L, \gamma_L\right) \tag{6}$$

where n_s = the stirring speed; d_s = stirrer diameter; and ρ_L, η_L and γ_L represent the liquid density, viscosity and surface tension, respectively. Dimensional analysis leads to the following equation:

$$\frac{k_L a_b}{n_s} = B\,\mathrm{Re}^a\,\mathrm{We}^b\sigma^c \tag{7}$$

where Weber and Reynolds numbers are defined by:

$$\mathrm{Re} = \frac{n_s d_s^2 \rho_L}{\eta_L} \tag{8}$$

$$\mathrm{We} = \frac{\rho_L n_s^2 d_s^3}{\gamma_L} \tag{9}$$

A non-linear regression of the experimental data to the Equation (7) gave the following results: $B = 1.38 \cdot 10^{-6}$, $a = 0.68$, $b = 0.42$, $c = 2.1$. The straight lines in Figure 5 represent the fitted relations for $k_L a_b$. The empirical correlation developed in this paper corresponds reasonably well with the relatively wide variety of experimental data, with an average deviation of $\pm 18\%$. The ranges of dimensionless numbers in the above equation are:

$$3.50 < \mathrm{We} < 28.67$$
$$262 < \mathrm{Re} < 3987$$
$$0 < \sigma < 4$$

The correlation can predict the values of the volumetric mass transfer coefficient for different physico-chemical properties of the liquid and under different operational conditions. The capability of $k_L a_b$ prediction is being further examined using different gas–liquid systems. It should be noted, however, that the proposed correlation is specific for the similar reactor types used in this study.

Much research has been carried out on the influence of solids on $k_L a$ in mechanically agitated reactors. Despite this, there is no universally applicable relationship that describes the influence of all types of particles in any liquid over a wide range of parameters. It is clear, however, that the effect of the particles on the system is a function of particle properties, vessel and impellor dimensions, and arrangement and operating conditions (Littlejohns & Daugulis 2007). The general trend for the effect of solid concentration on $k_L a$ is that a small content of up to 2–3 vol. % of relatively small solids has little effect on $k_L a$. As more solids are added, the value of $k_L a$ increases. The possible mechanisms that can account for this phenomenon include boundary-layer mixing and changes in interfacial area. At higher solid volumes (above 10%), $k_L a$ decreases significantly due to an increase in apparent viscosity (Beenackers & Van Swaaij 1993, Ozkan et al., 2000).

In view of the above considerations, for studying the effect of inert solid particles on gas–liquid mass transfer during bubbling gas desorption, a small solids holdup of 1 vol % was applied. In the range of the measurements conducted, it can be assumed that the solids do not affect the diffusive mass transfer rate.

The results from several absorption/desorption runs in CO_2-PC and CO_2-DMEPEG with the presence of both kinds of solids revealed that absorption rate and diffusive desorption rate were the same as in the case of pure solvents. The experimental results obtained further confirm the previous conclusions – that for a low solids concentration, the particles do not effect the value of $k_L a$, and the small amount of solids does not contribute to an increased effective viscosity of the solution.

During the occurrence of bubbling desorption, it was observed that the bubble production in the supersaturated liquid generally occurred more intensively than in the absence of solids. It is evident from the visual observations that particles initiated a significantly greater number of gas bubbles. It is remarkable, however, that the effect of polyethylene solids is rather moderate relative to PVC solids, which have a much more pronounced effect. In the region of bubbling desorption, the inert solid particles increased overall desorption rate compared to the case with pure solvent only. The influence of PVC particles on the volumetric mass transfer coefficient is plotted against supersaturation of the solution for PC and DMEPEG solvents in Fig. 6. Suspended solids give higher values of $k_L a_b$ up to 30% for the system, with PVC particles relative to the system without particles for both solvents. In contrast, larger particles of polyethylene showed a small increase of about 10% for the $k_L a_b$.

The results show that the solid particles can promote heterogeneous nucleation. These studies clearly demonstrated that the solid particles enhanced the desorption rate under bubbling conditions by the additional heterogeneous surface of solid particles. The higher enhancement effect of PVC particles can probably be attributed to the higher specific area of these particles. However, due to a limited amount of experimental data with solids, it was not possible to quantify the observed effects.

The observed significant changes in the bubble desorption rate with PVC particles relative to solution without solids could thus be attributed to the additional heterogeneous surface of solid particles, which is a contributing factor to the bubble nucleation rate.

4 CONCLUSIONS

The rates of CO_2 desorption from supersaturated PC and DMEPEG solutions were investigated in an agitated reactor under batch-wise conditions. The obtained results show that the CO_2 desorption rate increases with increasing stirring speed and supersaturation of the solution. In the non-bubbling region of the process, the desorption process can be regarded as opposite to the absorption process. The desorption rate under bubbling conditions was greater than the diffusive desorption. The measured overall desorption rate was divided into the contributions of the interface of the free surface of the liquid (diffusive desorption) and the interface of generated bubbles. The volumetric mass transfer coefficients for the bubbles, $k_L a_b$, were determined in the wide range of operational conditions and correlated as a function of supersaturation of the solution, Reynolds and Weber numbers. The proposed correlation is found to be consistent with the available experimental results within ±18%.

The effect of the presence of inert solid particles on the mass transfer rate during a gas desorption in the bubbling region was investigated. It was found that PVC particles cause an enhancement of the volumetric mass transfer coefficient $k_L a_b$ up to 30%, whereas particles of polyethylene showed only a small effect on $k_L a_b$ relative to a system without a solid phase. The observed changes in the bubble desorption rate relative to a solution without solids could thus be explained by the additional heterogeneous surface of solid particles, which is a contributing factor to the bubble nucleation rate.

The present study has provided an improved understanding of the desorption process and its quantitative description, which would be helpful for design considerations of the regeneration step in several industrial processes for separating CO_2 based on physical solvents.

ACKNOWLEDGEMETS

This work was funded by the Ministry of Science and Higher Education of Poland (Project No. 1 T09C 018 30).

REFERENCES

Beenackers, A. A. C. M. & Van Swaaij, W. P. M. 1993. Mass transfer in gas–liquid slurry reactors. *Chemical Engineering Science* 48 (18): 3109–3139.

Hikita, H. & Konishi, Y. 1984. Desorption of carbon dioxide from supersaturated water in an agitated vessel. *AIChE Journal* 30 (6): 945–951.

Kierzkowska-Pawlak, H. 2007. Pressure swing absorption of carbon dioxide in n-methyl-2-pyrrolidone solutions. *Polish Journal of Chemical Technology* 9 (2): 106–109.

Kohl, A. & Nielsen R. 1997. *Gas purification*, 5th ed., Gulf Publishing Co.: Houston, TX.

Littlejohns, ,J. & Daugulis, A. 2007. Oxygen transfer in a gas-liquid system containing solids of varying oxygen affinity. *Chemical Engineering Journal* 129: 67–74.

Lubetkin, S. D. 2003. Why is it much easier to nucleate gas bubbles than theory predicts? *Langmuir* 19: 2575–2587.

Ozkan, O., Calimli, A., Berber, R. & Oguz, H. 2000. Effect of inert particles at low concentrations on gas-liquid mass transfer in mechanically agitated reactors. *Chemical Engineering Science* 55 (14/7): 2737–2740.

Weiland, R. H., Thuy, L. T. & Liveris, A. N. 1977. Transition from bubbling to quiescent desorption of dissolved gases. *Industrial And Engineering Chemical Fundamentals* 16 (3): 332–335.

Environmental Engineering III – Pawłowski, Dudzińska & Pawłowski (eds)
© 2010 Taylor & Francis Group, London, ISBN 978-0-415-54882-3

Mass size distribution of total suspended particulates in Zabrze (Poland)

K. Klejnowski, A. Krasa & W. Roguła-Kozłowska
Institute of Environmental Engineering, Polish Academy of Science, Zabrze, Poland

ABSTRACT: From 20 April 2007 to 4 January 2008, in Zabrze (Poland), total suspended particulates (TSP) were sampled by using a 13-stage low-pressure impactor, DEKATI. The sampler worked in periods that were separated by several-hour maintenance breaks. It segregated ambient particles by their aerodynamic diameters into 13 fractions. In all periods, the mass distribution of TSP with respect to particle size (mass size distribution) functions had a maximum at a diameter not greater than $0.65\,\mu m$, and in all periods but two at a diameter not greater than $0.4\,\mu m$. In summer bimodal, in winter unimodal distributions prevailed. The shift of the maxima towards 0 expressed the relative quality of atmospheric aerosol; the variability of their number and place showed that changes in the activity of pollution sources occurred. Concentrations of particles with diameters of $0.26–0.4\,\mu m$, $0.4–0.65\,\mu m$ and $0.65–1.0\,\mu m$ contributed the most to TSP concentration.

Keywords: Ambient particles, atmospheric aerosol, cascade impactor DEKATI, particle, particulate matter, TSP.

1 INTRODUCTION

The relationships between air pollutant concentrations (also their rapid leaps or levels that are commonly considered safe) and malfunctions of the respiratory and cardiovascular systems, both of which increase human mortality and hospitalizations, are well known and widely documented (Pope et al. 1995, Pope & Schwartz 1996, Wojtyniak & Piekarski 1996, Schwela 2000, Wichmann et al. 2000, Atkinson et al. 2001, Le Tertre et al. 2002). Generally, airborne particles affect the human body through the respiratory system and their effects depend on the exact location that they reach in the lungs and also their chemical composition – both of which depend on particle size. Particles with aerodynamic diameters greater than $5.8\,\mu m$ remain in the upper part of the respiratory system. Particles with diameters between 4.7 and $5.8\,\mu m$ pass to the larynx. Particles with diameters between 3.3 and $4.7\,\mu m$ get into the larynx and primary bronchi, those with a diameter between 2.1 and $3.3\,\mu m$ get into secondary bronchi, and those with a diameter between 1.1 and $2.1\,\mu m$ get into tertiary bronchi. The finest particles, with diameters not greater than $1.1\,\mu m$, reach the alveoli. The time needed for their removal elongates with the depth of penetration into the lungs – that is, with their size. Different size fractions of the ambient aerosol affect human health in different ways. The most adverse health effects are linked with the finest particles. The more fine particles that an aerosol contains, the more harmful it is. Therefore, the size distribution of ambient particles at a site of interest is important from the point of view of human health.

The value $f(a)$ of a density distribution function f at a point a expresses a growth speed of the cumulative distribution function $F(x) = \int_{-\infty}^{x} f(z)dz$ at this point. Here, the functions describe the mass size distribution of TSP: f presented in a chart is a density function for the cumulative function F expressing PM_x mass contribution to TSP. Let a be the diameter such that f assumes local maximum (mode) $f(a)$ at a and $0 < \varepsilon$ is a small real number. Then, for any b in some neighbourhood of a, $F(a + \varepsilon) - F(a - \varepsilon)$ is not less than $F(b + \varepsilon) - F(b - \varepsilon)$. This means that $PM_{(a+\varepsilon)-(a-\varepsilon)}$ is such a particle fraction that its mass contribution $F(a + \varepsilon) - F(a - \varepsilon)$ to TSP is greater than the mass contribution $F(b + \varepsilon) - F(b - \varepsilon)$ of $PM_{(b+\varepsilon)-(b-\varepsilon)}$.

Of all fractions, $PM_{(b+\varepsilon)-(b-\varepsilon)}$, where $0 < \varepsilon$ and b is in some neighbourhood of a, the fraction $PM_{(a+\varepsilon)-(a-\varepsilon)}$ has the greatest mass contribution to TSP. This relationship between a and b translates into terms of the density f as:

$$\lim_{\varepsilon \to 0} \frac{F(b+\varepsilon)-F(b-\varepsilon)}{\varepsilon} = f(b) \le f(a) = \lim_{\varepsilon \to 0} \frac{F(a+\varepsilon)-F(a-\varepsilon)}{\varepsilon}$$

In an instant or short interval of time, among all short-interval TSP fractions, the ones whose intervals of diameters comprise a maximum of the density function have the greatest mass. If, during the time of exposure, the extrema remain within the intervals of small diameters of particles, then the health hazard from TSP is high because very fine particles stay in the lungs for a very long time (in the order of months or even a year) and their adverse effects grow due to fast particle mass accumulation. Therefore, the arrangement of maxima of a mass size distribution density

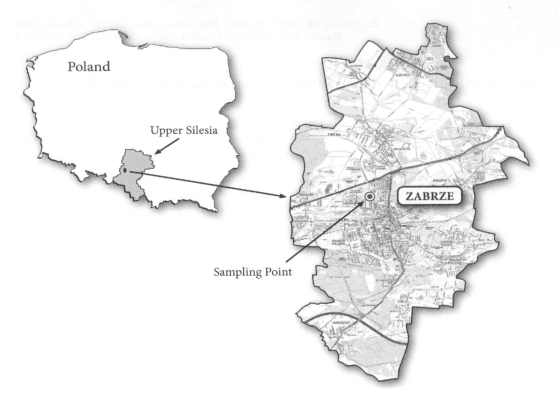

Figure 1. Location of the measuring point.

function brings information about the hazard from exposure to TSP and their shift towards 0 may be some measure of the hazard.

Measurements of mass size distribution of ambient particles were carried out in Zabrze, Poland, from April 2007 to January 2008. Like other cites in Upper Silesia, Zabrze is still greatly polluted by municipal and industrial sources (Pastuszka & Okada 1995, Houthuijs et al. 2001, Jabłońska et al. 2001, Wawroś et al. 2001), whereas in most Western European countries, vehicular emission dominates (Pakkanen et al. 2003, Slezakova et al. 2007, http://reports.eea. europa.eu/technical_report_2008_7/en). Zabrze lies in the western part of the Upper Silesian agglomeration. The sampling point (50°18'58"N, 18°46'18"E) was located in the central part of Zabrze (Figure 1). The site is surrounded by blocks of flats, detached houses and a few supermarkets. In winter, these are heated by domestic heating systems that burn hard coal or from the central heating system. About 500 m north and west, there are moderately busy roads. The measuring point meets conditions for the urban background monitoring station.

2 MATERIAL AND METHODS

Ambient dust was sampled by using a typical inertial (cascade) 13-stage low-pressure impactor, DECATI (DLPI). It separates particles into 13 size fractions with diameters in consecutive intervals: 0.03–0.06, 0.06–0.108, 0.108–0.17, 0.17–0.26, 0.26–0.4, 0.4–0.65, 0.65–1.0, 1.0–1.6, 1.6–2.5, 2.5–4.4, 4.4–6.8, 6.8–10.0 and >10 μm. Particles with diameters <0.03 μm were not measured; the mass of particles with aerodynamic diameters greater than 20 μm, collected at the 13th stage, was assumed to be negligible.

Dust was collected on aluminium or polycarbonate substrates. Masses of the fractions were determined gravimetrically. A Mettler Toledo microbalance (accuracy 2 μm) and a gate that neutralized electric charges were used. Ambient concentrations of particular fractions were computed from their masses and volumes of passed-through air (aerosol).

In the period from 20 April 2007 to 4 January 2008, 28 measurements were carried out. These covered all of this time period – except for short breaks used for changing substrates and one longer maintenance break between 21 and 23 May 2007. The shortest measuring period was 95.75 h; the longest was 357.17 h. Each measurement yielded a distribution of TSP mass among 13 aerodynamic diameter intervals, defined by the cut-off diameters of the impactor, for one measuring period.

3 RESULTS AND DISCUSSION

All observed extrema of all density functions f are shown in Table 1; approximate charts of some fs are

Table 1. Extrema of distribution functions for all periods.

	Max. I [μm]	Max. II [μm]	Min. [μm]
20–27 April 2007	0.26–0.4	–	–
27 April–7 May 2007	0.03–0.06	0.26–0.4	0.06–0.108
7–14 May 2007	0.06–0.108	0.4–0.65	0.108–0.17
14–21 May 2007	0.17–0.26	6.8–10	4.4–6.8
23–31 May 2007	0.26–0.4	–	–
31 May–6 June 2007	0.108–0.17	2.5–4.4	1.6–2.5
6–15 June 2007	0.108–0.17	0.4–0.65	0.26–0.4
15–22 June 2007	0.108–0.17	–	–
22 June–2 July 2007	0.108–0.17	0.26–0.4	0.17–0.26
2–12 July 2007	0.17–0.26	0.4–0.65	0.26–0.4
12–23 July 2007	0.17–0.26	–	–
23 July–2 August 2007	0.26–0.4	2.5–4.4	1.6–2.5
2–16 August 2007	0.26–0.4	–	–
16–31 August 2007	0.06–0.108	0.26–0.4	0.108–0.17
Summer			
20 April–31 August 2007	0.26–0.4	–	–
31 August–14 September 2007	0.4–0.65	6.8–10.0	4.4–6.8
14–28 September 2007	0.17–0.26	–	–
28 September–12 October 2007	0.26–0.4	–	–
12–26 October 2007	0.17–0.26	0.4–0.65	0.26–0.4
26 October–5 November 2007	0.26–0.4	0.65–1.0	0.4–0.65
5–12 November 2007	0.26–0.4	–	–
12–19 November 2007	0.17–0.26	–	–
19–26 November 2007	0.26–0.4	10.0–20.0	6.8–10.0
26–30 November 2007	0.03–0.06	0.17–0.26	0.06–0.108
30 November–6 December	0.17–0.26	–	–
6–14 December 2007	0.17–0.26	–	–
14–19 December 2007	0.4–0.65	–	–
19–27 December 2007	0.26–0.4	–	–
27 December 2007–4 January 2008	0.26–0.4	–	–
Winter			
31 August 2007–4 January 2008	0.17–0.26	–	–
Whole period			
20 April 2007–4 January 2008	0.17–0.26	–	–

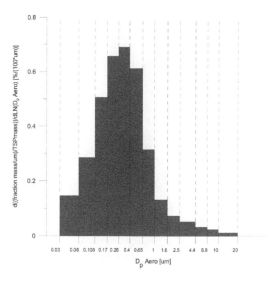

Figure 2. PM mass distribution with respect to particle size in 20 April–31 August 2007 (summer).

shown in Figures 2–10. All observed $F(a) - F(b)$ for longer periods (summer, winter and the entire measuring period), where $b < a$ are consecutive cut-off diameters of the impactor, are listed in Table 2 as 'Contributions' to TSP concentrations.

The distribution of mass of TSP among the 13 contiguous intervals of particle aerodynamic diameters varied depending on the measuring period. This indicates short-period fluctuations in the characteristics of emissions that affect the measuring site. In summer, there were 9 periods that have a bimodal distribution of particles and 5 periods that have unimodal ones. In winter, the situation was inverted: 5 distributions were bimodal and 9 were unimodal (Table 1). Unimodality of the distributions in winter may be due to the start of the heating season – the site is affected by hard-coal-fuelled heating stoves. In addition, the shift in the maximum from the interval 0.26–0.4 μm in summer to 0.17–0.26 μm in both winter and the whole measuring period is also due to the effects of coal combustion. Periodical changes in the modality of functions reflects such changes in the activity of sources.

For all the longer periods – summer, winter and the whole measuring period – the distributions were unimodal (Table 2; Figures 2, 3 and 4).

Examples of unimodal mass distributions for shorter periods are shown in Figures 5 (summer) and 6 (winter); some bimodal distributions are presented in Figures 7, 8, 9 (summer) and 10 (winter). Figures 7 (23–31 May 2007) and 10 (31 May–6 June 2007) present distributions of mass particles in two consecutive periods. In the first, the distribution is unimodal; in the second, it is bimodal. Maxima of the second distribution are far from each other. In almost all periods – including longer ones – all distribution functions had maxima at a diameter not greater than 0.4 μm (exceptions are 31 August–14 September 2007 and 14–19 December 2007, with maxima between 0.4 and 0.65 μm). All the maxima were reached at diameters close to 0 – that is, in the whole measuring period, the site was relatively abundant in very fine particles. The TSP was toxic – most of the TSP was able to reach the alveoli when inhaled.

The actual inhaled dose of fractions of TSP depends on their ambient mass concentrations. In all periods, three fractions of particles – particles with aerodynamic diameters of 0.26–0.4 μm, 0.4–0.65 μm and

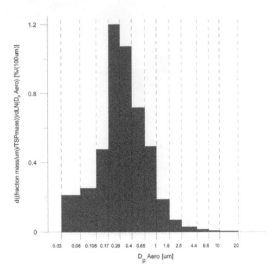

Figure 3. PM mass distribution with respect to particle size in 31 August 2007–4 January 2008 (winter).

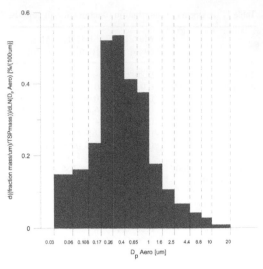

Figure 5. PM mass distribution with respect to particle size in 23–31 May 2007.

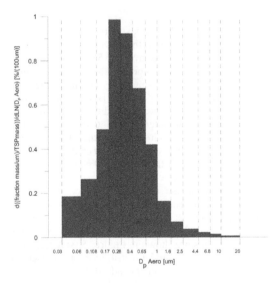

Figure 4. PM mass distribution with respect to particle size in 20 April 2007–4 January 2008 (whole measuring period).

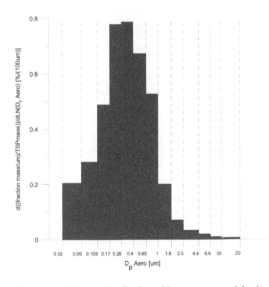

Figure 6. PM mass distribution with respect to particle size in 28 September–12 October 2007.

0.65–1.0 μm – had the greatest ambient mass concentrations.

Fractions with diameters of 0.17–0.26 μm and 1.0–1.6 μm had the next greatest ambient mass concentrations in winter and the whole measuring period; fractions with diameters of 2.5–4.4 μm and greater than 10 μm had the next greatest ambient mass concentrations in summer (Table 2).

The diameter intervals most frequently comprising a maximum were 0.108–0.17 μm, 0.17–0.26 μm and 0.26–0.4 μm in summer and the whole measuring period; and 0.17–0.26 μm and 0.26–0.4 μm in winter (Table 1). The average concentrations of the three fractions are shown in Table 2.

Summing these up for each period (the volume of passed-through air was the same for each impactor stage), we receive concentrations of the most toxic fractions at the sampling point: 4.2 μg/m³ for $PM_{0.4-0.108}$ in summer, 11.9 μg/m³ for $PM_{0.4-0.17}$ in winter and 8.0 μg/m³ for $PM_{0.4-0.108}$ in the whole measuring period.

The fraction PM_1, comprising all these fractions, had the whole period average mass concentration of 18.9 μg/m³. In summer, this was 10.5 μg/m³ and in winter it was 30.1 μg/m³ (Table 2). The distribution function reached its maximum between 2.5 and 10 μm – that is, for $PM_{10-2.5}$ – three times in summer, once in winter and always as a second maximum.

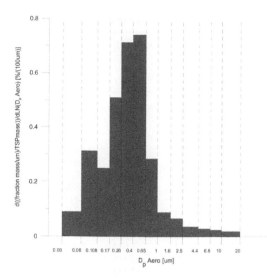

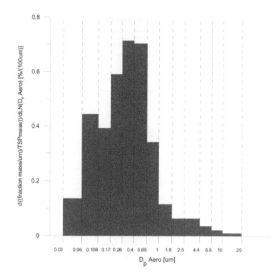

Figure 7.   PM mass distribution with respect to particle size in 7–14 May 2007.

Figure 9.   PM mass distribution with respect to particle size in 16–31 August 2007.

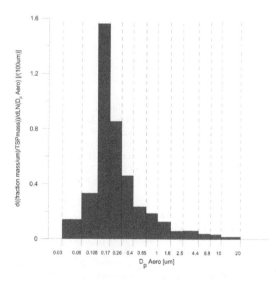

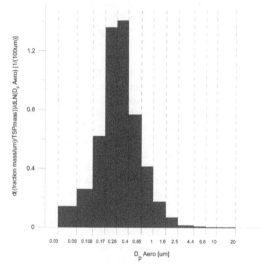

Figure 8.   PM mass distribution with respect to particle size in 31 May–6 June 2007.

Figure 10.   PM mass distribution with respect to particle size in 19–26 November 2007.

The only maximum at a diameter greater than $10 \,\mu m$ (19–26 November 2007) was reached also as a second maximum (Table 2). The whole period concentrations of the most often measured and important for environmental legislation TSP fractions, $PM_{2.5}$ and PM_{10}, are also given in Table 2.

4   CONCLUSIONS

At the sampling point in Zabrze, the mass distribution of TSP with respect to particle size changed from one measuring period to another. The shorter period mass distribution functions differed in numbers of their extrema (Table 1). This proves short term (several days, 2 weeks) variability of the characteristics of dust emission in Zabrze.

The functions for shorter periods had one or two maxima in the interval 0.03–20 μm (Figures 7–10, Table 1). In summer, a greater number of bimodal and, in winter, a greater number of unimodal distributions occurred. All the distribution functions had maxima at diameters not greater than 0.65 μm, and almost all at diameters not greater than 0.4 μm. In longer periods – summer, winter and the whole measuring period – the functions had a single maximum that was reached at a

Table 2. Average ambient concentrations of particular dust fractions and their contributions to TSP concentrations.

D_p interval [μm]	Concentration [μg/m^3]			Contribution to TSP concentrations [%]		
	Summer[1]	Winter[2]	Whole measured period[3]	Summer	Winter	Whole measured period[3]
0.03–0.06	0.098	0.292	0.181	0.438	0.636	0.558
0.06–0.108	0.306	0.553	0.412	1.368	1.205	1.269
0.108–0.17	0.701	1.354	0.982	3.138	2.953	3.026
0.17–0.26	1.322	4.952	2.883	5.919	10.798	8.883
0.26–0.4	2.165	6.884	4.193	9.689	15.012	12.922
0.4–0.65	3.428	8.222	5.489	15.344	17.929	16.914
0.65–1.0	2.457	7.881	4.789	10.999	17.185	14.757
1.0–1.6	1.757	5.142	3.212	7.863	11.212	9.898
1.6–2.5	1.492	2.864	2.081	6.676	6.244	6.414
2.5–4.4	2.263	2.47	2.352	10.128	5.385	7.247
4.4–6.8	1.797	1.716	1.762	8.041	3.742	5.43
6.8–10.0	1.793	1.23	1.551	8.025	2.682	4.78
>10	2.764	2.3	2.564	12.371	5.015	7.902
PM$_1$	10.477	30.138	18.928	46.895	65.718	58.329
PM$_{2.5}$	13.725	38.144	24.221	61.434	83.175	74.641
PM$_{10}$	19.577	43.56	29.886	87.629	94.985	92.098
PM$_{0.108}$	0.403	0.845	0.593	1.806	1.842	1.828
Average TSP concentration	22.341	45.86	32.45	–	–	–

[1] Average in 20 April 2007–31 August 2007
[2] Average in 31 August 2007–4 January 2008
[3] Average in 20 April 2007–4 January 2008

diameter not greater than 0.4 μm (Table 1; Figures 2, 3 and 4). Switching from unimodal to bimodal distribution meant changes in the activity of emission sources. The site is affected mainly by local (domestic stoves) and industrial (power plants) coal combustion and traffic – whereas traffic stays at the same level during the whole year, the pollution from hard coal combustion increases during winter (especially during the season when heating is required). More data for consecutive years would help to establish the pattern of periodic (seasonal or yearly) behaviour of the distribution of atmospheric aerosol at the site, but this requires further investigation.

The fractions PM$_{0.65–0.26}$, PM$_{0.4–0.65}$ and PM$_{1–0.65}$ had the greatest ambient mass concentrations in all periods (Table 2). This is significant because the maxima of the density functions were at 0.108–0.17 μm, 0.17–0.26 μm and 0.26–0.4 μm most frequently. The maxima occur close to 0, at diameters not greater than 0.65 and less than 1.0 μm.

The cascade impactor that was used did not have a very good resolution – some extrema of density functions may not be noticed. Nevertheless, the idea of observing changes in emission source activities by investigating the densities of TSP mass size distribution functions with cascade impactors seems to be promising as long as the impactors have good enough resolution – that is, a large number of stages (to distinguish extrema) and a high flow rate (to shorten necessary operating time), or there are several impactors.

REFERENCES

Atkinson, R.W., Anderson, H.R., Sunyer, J., Ayres, J., Baccini, M., Vonk, J.M, Boumghar, A., Forastiere, F., Fosberg, B., Touloumi, G., Schwartz, J. & Katsouyanni, K. 2001. Acute effects of particulate air pollution on respiratory admissions: results from APHEA 2 project. Air Pollution and Health: a European Approach. *American Journal of Respiratory and Critical Care Medicine* 164: 1860–1866.

Houthuijs, D., Breugelmans, O., Hoek, G., Vaskövi, È., Micháliková, E., Pastuszka, J.S., Jirik, V., Sachelarescu, S., Lolova, D., Meliefste, K., Uzunova, E., Marinescu, K., Volf, ,J., De Leeuw, F., Van De Wiel, H., Flecher, T., Lebret, E. & Brunekreef, B. 2001. PM-10 and PM-2.5 concentrations in central and eastern Europe: Results from the CESAR study. *Atmospheric Environment* 35: 2757–2771.

http://reports.eea.europa.eu/technical_report_2008_7/en.

Jabłońska, M., Rietmeijer, F.J.M. & Janeczek, J. 2001. Fine-grained barite in coal fly ash from Upper Silesia Industrial Region. *Environmental Geology* 40: 941–948.

Le Tertre, A., Medina, S., Samoli, E., Forsberg, B., Michelozzi, P., Boumghar, A., Vonk, J.M., Bellini, A., Atkinson, R., Ayres, J.G., Sunyer, J., Schwartz, J. & Katsouyanni, K. 2002. Short-term effects of particulate air pollution on cardiovascular diseases in eight European cities. *Journal of Epidemiology & Community Health*

Pakkanen, T.A., Kerminen, V.M., Loukkola, K., Hillamo, R.E., Aarnio, P., Koskentalo, T. & Maenhaut, W. 2003. Size distributions of mass and chemical components in street-level and rooftop PM1 particles in Helsinki. *Atmospheric Environment* 37: 1673–1690.

Pastuszka, J.S. & Okada, K. 1995. Features of atmospheric aerosol particles in Katowice. Poland. *Science of the Total Environment* 175: 179–188.

Pope, C.A., Dockery, D. & Schwartz, J. 1995. Review of epidemiological evidence of health effects of particulate air pollution. *Inhalation Toxicology* 7: 1–18.

Pope, C.A. & Schwartz, J. 1996. Time series for analysis of pulmonary health data. *American Journal of Respiratory and Critical Care Medicine* 154: 229–33.

Schwela, D. 2000. Air pollution and health in urban areas. *Reviews on Environmental Health* 15(1–2): 13–42.

Slezakova, K., Pereira, M.C., Reis, M.A. & Alvim-Ferraz, M.C. 2007. Influence of traffic emissions on the composition of atmospheric particles of different sizes – Part 1: Concentrations and elemental characterization. *Journal of Atmospheric Chemistry* 58: 55–68.

Wawroś, A., Talik, E. & Pastuszka, J.S. 2001. Investigations of aerosols from Świętochłowice, Pszczyna and Kielce by XPS method. *Journal of Alloys and Compounds* 328: 171–174.

Wichmann, H.E., Spix, C., Tuch, T., Wolke, G., Peters, A., Heinrich, J., Kreyling, W.G. & Heyder, J. 2000. Daily mortality and fine and ultrafine particles in Erfurt, Germany Part I: Role of particle number and particle mass. In: *HEI Report 98*.

Wojtyniak, B. & Piekarski, T. 1996. Short-term effects o fair pollution on mortality in Polish urban populations – what is different? *Journal of Epidemiology & Community Health* 1: 36–41.

Environmental Engineering III – Pawłowski, Dudzińska & Pawłowski (eds)
© 2010 Taylor & Francis Group, London, ISBN 978-0-415-54882-3

Microwave – assisted desorption of volatile organic compounds from activated carbon to a water phase

A. Kozioł & M. Araszkiewicz

Division of Chemical and Biochemical Process, Faculty of Chemistry, Wroclaw University of Technology, Wroclaw, Poland

ABSTRACT: In laboratory-scale experiments, microwaves were used as a heating medium for desorption of 1-butanol (as an example of volatile organic compounds) from a granular activated carbon bed to a water phase. Temperatures were examined using an infra-red camera. The efficiency of desorption was determined by concentration measurements. To better present the specific microwave effects, the microwave heating process was compared to a similar one without microwaves. Microwave process efficiency was similar to that without microwaves, but the microwave-assisted process was significantly shorter.

Keywords: Microwaves, desorption, VOC, activated carbon, 1-butanol.

1 INTRODUCTION

The increasing emission of volatile organic compounds (VOCs), which are components of greenhouse gas emissions, is now a major problem. Decreasing the total emissions of these gases requires installing devices for capturing and later decomposing them, thus preventing secondary emission. Biodegradation by bacteria of VOCs seems to be the most promising and 'ecological' of the many VOC utilisation methods. However, local emissions of VOCs are non-stable and their concentrations are relatively low. These features make biodegradation methods directly from waste gas very difficult. Therefore, VOCs should be removed from gas and transferred to a water solution. The water solution can then subsequently be decomposed in bioreactors. The method of such transfer depends on the water solubility of each VOC. Standard water absorption will be efficient for compounds with good water solubility (i.e. hydrophilic ones, Figure 1).

In the case of compounds with medium to weak solubility in water, adsorption is much more effective for the separation of the VOCs from the waste gases (Figure 2).

There is a need to add an additional process stage, where the VOCs are moved from adsorbent pores to the water solution (desorption) with simultaneous adsorbent regeneration.

This process requires a supply of energy to the interior of the adsorbent to overcome the surface forces retaining the VOC molecules on its surface. Heat energy is supplied to the adsorbent with hot steam in classical desorption processes (Paderewski, 1999). The bonding between the adsorbent and adsorbate

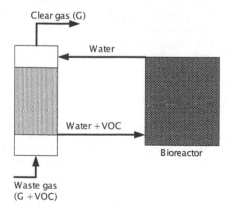

Figure 1. Absorption and biodegradation of water soluble VOC.

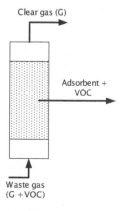

Figure 2. Gas purification by adsorption in the case of weakly water-soluble VOC.

breaks and the VOC evaporates into the steam stream. Next, the steam has to be condensed. For VOCs with weak water solubility, there are two phases in a separation funnel: almost pure VOC with a minor amount of water and a water phase with little VOC (Figure 3).

This kind of process is complicated and expensive, and in the case of bioregeneration requires further VOC dissolution in water. If the VOC could be moved directly from the adsorbent to the water phase it would be possible to dispense with the steam phase and the regeneration process would link directly to the biodegradation one (Figure 4).

The greatest difficulty is connected with the low efficiency of VOC desorption in the water phase. This difficulty arises from the need to deliver the heat energy into the adsorbent pellets. Our preliminary studies, and other research, have indicated the possibility of using microwave energy for this heating.

The main objective of this work was to evaluate experimentally the use of microwaves as an energy source in the 1-butanol desorption process from granulated activated carbon (GAC). Matisova & Skrabakova (1995) made a detailed analysis of different carbon sorbents and their use. The simulation of VOC adsorption in GAC from waste air was tested by Nastaj et al. (2006). The most significant problem in the Nastaj

study was connected with the next process step of carbon regeneration – desorption of the VOC. The primary objective of this work was to move benzene from GAC directly to the water solution. The use of steam is not desirable because of the high possibility of a secondary movement of benzene into the atmosphere. Submerging the compound in water decreases this risk. The water solution containing the VOC can be further processed to deactivate the pollution; however, there are some drawbacks. First, benzene has limited solubility in water. Second, benzene is nonpolar and do not generate heat from its interaction with the microwaves. Therefore, benzene desorption from GAC to water in a microwave environment will show clear thermal effects in the process.

The use of microwaves to increase the effectiveness of desorption processes is quite popular, and was reviewed in detail by Cherbanski & Molga (2009). Given the specific features of microwaves, they seem an almost perfect way of delivering energy to a bed of GAC containing organic compounds, in a water medium. Microwaves are a means of delivering energy to the material, with the energy later converted into heat. The conversion process depends on many different parameters, such as the dielectric factors of the material, the water content, sample dimensions and geometry and the uniformity and strength of the electric field. Microwaves offer a number of advantages over traditional convective heating. These include the simplicity of the microwave delivery setup, lack of heat losses and perfect process control. Additionally, energy reaches the core material without major losses. Microwaves can significantly increase the effectiveness of the separation processes (Eskillson et al. 2000, Lupinski et al. 2006). Heat generation is especially intensive in the carbon pellets, temperatures can reach 1000°C and that phenomenon can be successfully used to recover gold from waste active carbon (Amankwah et al. 2005). Microwaves can be used for desorption and destruction of some VOCs (for example, trichloroethylene) from GAC (Chih-Ju Jou 1998). A similar process of microwave-driven degradation of an azo dyestuff (Congo red), catalysed by activated carbon powder, was described by Zhang et al. (2007). Menendez et al. (1999) analysed changes in the GAC pellet surface during microwave irradiation. Heating carbon to a high temperature can destroy (melt) the pellet surface and close the pores, making it impossible for the carbon to return to activated state. Therefore, such high temperatures should be avoided during desorption. Microwaves are widely used in environmental engineering in processes like contaminated soil remediation, waste processing, mineral processing and activated carbon regeneration (Jones et al., 2002). Desorption of a VOC (ethylene) from mordenites with microwaves was tested successfully by Kim et al. (2005). Microwave-assisted thermal desorption of oil from North Sea drill cuttings was demonstrated by Shang et al. (2006). Ho et al. (2005) described the regeneration of activated carbon (for reuse in the process of mercury adsorption from combustion flue

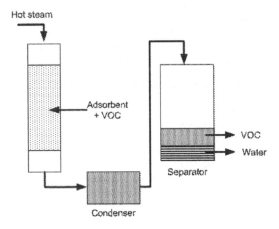

Figure 3.   Classical desorption of VOC by hot steam.

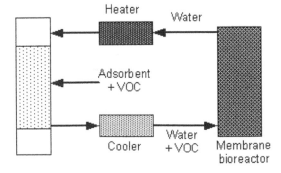

Figure 4.   Proposed method of simultaneous desorption and biodegradation of the VOC.

gas) with microwaves. The use of microwaves in the desorption of VOC processes generally results in the destruction of the VOCs due to the very high temperatures generated if the GAC is placed directly in a microwave chamber. Unfortunately, this method leads to total destruction of the GAC. Preliminary laboratory tests indicated the need to immerse the GAC in water; the water plays a buffer role and prevents the GAC from overheating.

2 MATERIALS AND METHODS

The experiments were conducted at a laboratory scale. A single sample, comprising 1 g of GAC (of which approximately 0.5 g were adsorbed 1-butanol) in 5.5 ml of water, was placed inside a microwave chamber. The microwave power supply was turned on until the sample temperature reached one of three predetermined levels, 40°C, 60°C or 80°C. The times required by the microwave irradiation to achieve these temperatures mentioned were 30, 70 and 90 s, respectively. After the given time, the microwave power was turned off and 3 dm³ of solution was taken for chromatographic analysis using a Shimadzu GC – 2014 gas chromatograph. To determine the microwave performance, four tests were conducted with traditional convective heating (thermostat without microwaves). The time of desorption in the convection experiments (without microwaves) was constant at 600 s. Thus, a long time (almost 10 times longer than for microwaves) was necessary to achieve a similar 1-butanol concentration in water. The sample temperature was observed with an IR camera (Flir Systems, ThermaCam E20; Araszkiewicz et al., 2007) during the microwave heating process. The experimental setup is presented in Figure 5.

To test the relationship between the times of processing and desorption efficiency using convection heating, additional experiments were conducted. Four identical probes were prepared with GAC and adsorbed 1-butanol. The samples were kept at a constant temperature of 20°C for 30 s, 60 s, 90 s and 600 s. Then 3 dm³ samples of the water were taken for chromatographic analysis.

3 RESULTS AND DISCUSSION

Table 1 presents the basic information and results of the basic comparative experiments of desorption efficiency.

As can be seen from Figure 6, after about 600 s the efficiency of desorption remained stable. Given the good water solubility of 1-butanol it was assumed that this length of time was necessary to achieve the determined VOC concentration in the water. Tests are compared in Figure 7. There were similar efficiencies at 40°C and 60°C, both with and without microwaves. At 80°C, the efficiency was somewhat lower using microwaves as compared with convection heating.

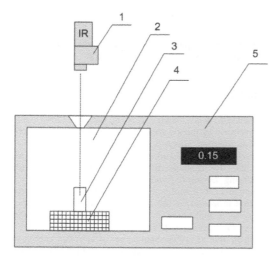

Figure 5. Experimental setup. Legend: 1 – IR camera; 2 – microwave cavity; 3 – sample material; 4 – base; 5 – microwave control panel; 0.15 – time counter.

Table 1. 1-Butanol desorption: experimental data.

Temp. (°C)	Thermostat				Microwaves		
	20	40	60	80	40	60	80
GAC weight (g)	1	1	1	1	1	1	1
GAC + butanol (g)	1.49	1.47	1.52	1.41	1.51	1.53	1.52
Butanol (g)	0.49	0.47	0.52	0.41	0.51	0.53	0.52
Butanol conc. in water (g/dm³)	26.50	27.26	29.80	30.27	27.70	30.09	31.72
Time (s)	600	600	600	600	30	70	90
C_0*		0.49	0.47	0.52	0.41	0.51	0.53	0.52
C**		0.34	0.32	0.35	0.24	0.36	0.36	0.34
C/C_0		0.70	0.68	0.68	0.59	0.70	0.69	0.66
η***		29.75	31.90	31.52	40.61	29.88	31.23	33.55

* initial butanol concentration in GAC (g/g GAC).
** butanol concentration in GAC (g/g GAC).
*** $\eta = ((C_0 - C)/C_0) \times 100\%$.

However, it should be stressed that the microwave-assisted process took less time and therefore the process rate was much higher. The difference between processes conducted with and without microwaves is shown by the efficiency of both processes when compared with time (Figure 8).

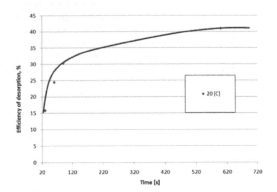

Figure 6. Efficiency of desorption compared with the time of process at low temperature (process conducted without microwaves).

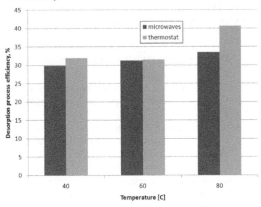

Figure 7. Comparison of the efficiency of 1-butanol desorption using microwave heating and heating with a thermostat at various temperatures.

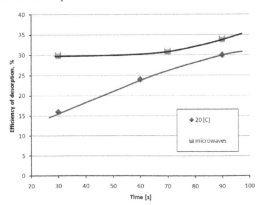

Figure 8. Comparison of the efficiency of 1-butanol desorption, with and without microwave heating, with the duration of the process.

The comparison shows that the process using microwaves was significantly more efficient. It is noteworthy that the temperature was not stable during the microwave process and rose significantly; however, efficiency up to 60°C was very weakly dependent on temperature (Figure 7).

4 CONCLUSIONS

1. Desorption using microwave heating of a mixture of water, activated carbon and 1-butanol gave almost identical results as the same process conducted without microwaves.
2. Desorption processes conducted with microwaves ran in a significantly shorter time.
3. Using microwaves may improve the intensity of the desorption process.

ACKNOWLEDGEMENTS

This paper is a part of the project Polish Ministry of Science (MNiSW) No. PBZ-3/2/2006.

REFERENCES

Amankwah, R.K., Pickles, C.A. & Yen, W-T. 2005. Gold recovery by microwave augmented ashing of waste activated carbon. *Minerals Engineering* 18: 517–526.
Araszkiewicz, M., Koziol, A., Oskwarek, A. & Lupinski, M. 2007. IR technique for studies of microwave assisted drying. *Drying Technology* 25: 569–574.
Cherbanski, R. & Molga, E. 2009. Intensification of desorption processes by use of microwaves—An overview of possible applications and industrial perspectives. *Chemical Engineering and Processing* 48: 48–58.
Chih-Ju Jou, G. 1998. Application of activated carbon in a microwave radiation field to treat trichloroethylene. *Carbon* 36(11): 1643–1648.
Ho, T.C., Lee, Y., Chu, H.W., Lin, C.J. & Hopper, J.R. 2005. Modeling of mercury desorption from activated carbon at elevated temperatures under fluidized/fixed bed operations. *Powder Technology* 151: 54–60.
Jones, D.A., Lelyveld, T.P., Mavrofidis, S.D., Kingman, S.W. & Miles, N.J. 2002. Microwave heating applications in environmental engineering – a review. *Resources, Conservation and Recycling* 34: 75–90.
Kim, S-I., Takashi, A. & Niiyama, H. 2005. Binary adsorption of very low concentration ethylene and water vapor on mordenites and desorption by microwave heating. *Separation and Purification Technology* 45: 174–182.
Lupinski, M., Koziol, A., Lupinska, A. & Araszkiewicz, M. 2006. Leaching of β-escin from chestnut seeds with microwave aid. *Chemical and Process Engineering* 27(2): 475–484.
Matisova E. & Skrabakova S. 1995. Carbon sorbents and their utilization for the preconcentration of organic pollutants in environmental samples. *Journal of Chromatography* 707: 145–179.
Menedez, J.A., Menedez, E.M., Iglesias, M.J., Garcia, A. & Pis, J.J. 1999. Modification of the surface chemistry of active carbons by means of microwave – induced treatments. *Carbon* 37: 1115–1121.

Nastaj, J.F., Ambrozek, B. & Rudnicka, J. 2006. Simulation studies of vacuum and temperature swing adsorption process for the removal of VOC from waste air streams. *International Communications in Heat and Mass Transfer* 30: 80–86.

Paderewski M.L. 1999. *Adsorption processes in chemical engineering*. WNT. Warsaw (in Polish).

Shang, H., Snape, C.E., Kingman, S.W., Robinson, J.P. 2006. Microwave treatment of oil-contaminated North Sea drill cuttings in a high power multimode cavity. *Separation and Purification Technology* 49: 84–90.

Sparr Eskilsson, C. & Bjorklund, E. 2000. Analytical-scale microwave- assisted extraction. *Journal of Chromatography A* 902: 227–250.

Zhang, Z., Shan, Y., Wang, J., Ling, H., Zang, S., Gao, W., Zhao, Z., Zhang, H. 2007. Investigation on the rapid degradation of congo red catalyzed by activated carbon powder under microwave irradiation. *Journal of Hazardous Materials* 147: 325–333.

Environmental Engineering III – Pawłowski, Dudzińska & Pawłowski (eds)
© 2010 Taylor & Francis Group, London, ISBN 978-0-415-54882-3

Application of activated carbons for the removal of volatile organic compounds in the automotive sector

A. Marecka

Faculty of Energy and Fuels, AGH – University of Science and Technology, Krakow, Poland

ABSTRACT: The current study presents a short review concerning the problem of VOCs emissions from vehicles and vehicle-related use of fuels, and the methods of their removal, with particular focus on methods involving physical adsorption.

The isotherms of argon (77.5 K) and benzene vapours (298 K) adsorbed on microporous activated carbon (Norit) were analyzed on the basis of Dubinin-Polanyi theory of adsorption. The porous characteristics (volume values of micropores, characteristic adsorption energy, and mean width of micropores filled with each adsorbate), were evaluated.

It was shown that in the case of benzene adsorption, it is possible to predict both the equilibrium isotherms and the thermodynamic parameter (Q_{st}) in a wide range of temperatures and relative pressures using the characteristic curve.

Keywords: Activated carbon, adsorption properties, carbon canisters, VOCs emissions.

1 INTRODUCTION

Sustainable development has been defined as one of the answers to the degradation of the natural environmental, especially with regard to environmental pollution. Actions aimed at limiting the impact of pollution are thus becoming one of the main objectives connected with environmental protection and human health. Unfortunately, the influence of various detrimental factors on the natural environment is not easily assessed, as some of the negative consequences may become apparent only after many years.

The emissions of Volatile Organic Compounds (VOCs), which are released into the atmosphere from the automotive sector (road transport) as the result of liquid fuel evaporation from vehicles, are considered to be a significant threat. The vapours from fuel tanks and pipes, as well as storage tanks, may also be a notable source of hydrocarbon emissions (Mellios et al. 2007, Hoshi et al. 2008, Wang et al. 2008, Huo et al. 2009). Recently, there has also been a lot of discussion about the emissions and the selection of appropriate techniques for the removal of VOCs produced during composting of different solid organic wastes (exhaust gases and exhaust vapours), from wastewaters, and during organic solvent recovery in chemical industry (Sene et al. 2002, Pagans et al. 2006).

VOCs emissions are dependent on two main factors, i.e. vehicle/engine technology and the properties of the gasoline and diesel fuels. The classification of vehicles, fuel systems, and various emission factors is also a relevant issue (Yu et al. 2000, Kawashima et al. 2006, Huo et al. 2009).

Recent years have brought about legislation and numerous directives that set limits pertaining to the ecological characteristics of engine fuels. In the European Union, VOCs emissions are controlled through the EU Directive 1999/13/EC (Pagans et al. 2006), and, in addition, the emission standards for vehicles are set according to different legislative certification tests (Mohr 1997, Mellios et al. 2007). In the USA, the institution responsible for monitoring and control of VOCs emissions is the US-EPA Environmental Protection Agency (through CAAA-Clean Air Act Amendments 1990). In particular, the most rigorous norms were introduced by CARB (California Environmental Protection Agency) in the state of California.

Volatile Organic Compounds comprise various groups of biogenic and anthropogenic chemical compounds (Kelly & Holdren 1995, Zielińska et al. 1996). Three main groups may be distinguished: aromatic compounds (e.g. benzene, toluene, xylenes), olefins, and compounds with functional groups. Their defining features are relatively high vapour pressure and low solubility in water. Among these organic compounds, benzene is a special case, since it is commonly found in emissions from exhaust vapours (Hoshi et al. 2008). In addition, it is well-known that benzene is very harmful for human health, as it paralyzes the central nervous system (Medinsky et al. 1995).

Volatile organic compounds may also be precursors of ozone. Sillman (1999) discusses the relation between tropospheric ozone and its main precursors, (including VOCs), in urban air pollution. The toxicity of these organic compounds varies and their source is anthropogenic activity. When released to the atmosphere, they may undergo incidental photochemical reactions, which lead to the formation of many new organic compounds and fine aerosol organic particles. They are hazardous to the environment and human health, and contribute to global warming.

Modelling the processes taking place in atmospheric air (i.e. the transformations and transport of pollutants) on the basis of estimated concentrations of pollutants and mathematical models has been getting increasingly popular in the recent years, especially in countries with strongly developed automotive industry. There are noticeable effects of these studies in the literature (Mellios et al. 2007), and the methods are constantly being improved. It should be noted that in this way, many attempts to evaluate the influence of relevant factors on the emissions of given groups of VOCs from automotive-related sources and into the atmosphere have been made (Harley & Cass 1995, Zielińska et al. 1996, Mensink et al. 2000, Mellios et al. 2007, Hoshi et al. 2008).

Among the various methods of VOCs removal that utilize different physico-chemical processes (thermal or catalytic combustion/oxidation, condensation, bio-filtration, absorption and membrane separation), advanced technology that combines adsorption with other methods seems to be the most promising solution. It is worth noting that on its own, adsorption is the most efficient and widely used technique for pollutant removal, especially with regard to the separation of components at low concentration levels (of the order of 10-1000 ppmv) from gaseous streams. In order to apply this technique, it is necessary to select an adsorptive material with sufficiently high adsorption capacity, appropriate chemical surface properties, favourable physico-chemical stability, and advantageous hydrodynamic characteristics (fast kinetics of adsorption/desorption).

The adsorption of organic pollutants may be carried out by employing various adsorbents, e.g. hydrophobic zeolites such as high silica faujasite (Baek et al. 2004). However, the adsorption of these pollutants onto carbonaceous adsorbents, followed by desorption for reuse or destruction, has been a more effective way of purifying atmospheric air and environmental protection. Porous materials, including granular or pelletized activated carbon, carbon molecular sieves, activated carbon fibres, or monoliths (Jankowska et al. 1991, Suzuki 1994, Luo et al. 2006), have been applied in the removal of pollutants both from the gas and liquid phases for many years now. They are currently also used to purify waste gases, and municipal and industrial waste, among others. Its low cost and efficiency are interesting for industrial applications.

Activated carbons are materials with a highly developed microporous structure and a large internal surface area. Their advantage, aside from high adsorption capacity, is the relative ease with which the parameters relevant to their application may be modified.

An interesting example of the application of activated carbons for the removal of hydrocarbons, and a comprehensive review of the related solutions in the vehicle industry, were presented by Mohr (1997). The installation of carbon canisters acting as adsorptive filters that remove toxic compounds from the fluid in the fuel system is one of the ways in which hydrocarbon emissions may be reduced (Kelli et al. 1995, Mohr 1997, Mellios 2007). According to Mohr, when the engine is inactive, fuel vapours travel to a filter containing granulated activated carbon, where they are adsorbed. While the engine is active, they are pushed out of the filter by a stream of air, they undergo desorption, and then they are transported to the chamber where combustion takes place. Presently, similar solutions in the adsorption-desorption cycle are also used in fuel storage. It should be emphasized that one of the features that makes it possible to apply activated carbons for these purposes is the hydrophobic character of their surface.

These new applications of activated carbons carry with them the need for constant improvement of their performance, including adsorption properties. Furthermore, there are efforts to develop cheaper and more efficient methods of production of activated carbons (Jankowska et al. 1991, Buczek & Marêché et al. 1998, Mianowski et al. 2007), e.g. such that would operate at their full adsorption capacity, with constant performance.

Studies on adsorption of benzene and other organic vapours have shown that both porosity and surface chemistry are factors with regard to the adsorption properties of activated carbons. As mentioned earlier, activated carbons are generally hydrophobic. However, chemically modified activated carbons with acid or bases (Kim et al. 2006), with carbon-oxygen surface groups (Moreno-Castilla et al. 2000) or carbon-nitrogen groups (Biniak et al. 1997, Bimer et al. 1998), have been described in literature. By adsorbing polar and nonpolar molecules on modified activated carbons (with unaltered porous structure), Burg and his co-workers (2002) have shown the role of their surface chemistry. The modification of carbon surface results in better selectivity towards a specific organic compound.

Activated carbons have been extensively investigated in both fundamental and practical aspects. Comprehensive reviews of adsorption methods for assessing porous structure of microporous activated carbons has been given, among others, by Dubinin et al. (1960, 1989); Bering et al., (1972); Jaroniec (1998).

Adsorption isotherms from the gaseous phase (argon, benzene vapours) were presented and analyzed using the D-R equation to obtain the structural parameters and temperature dependences of the adsorption processes taking place on the activated carbon. Attempts were made to predict isotherms from the knowledge of characteristic curve of benzene adsorption. This adsorbate was selected for both scientific and practical reasons, since it is one of the typical species of volatile non-methane organic compounds.

2 APPLICATION OF THE D-R EGUATION TO ADSORPTION ISOTHERMS

Adsorption isotherms of nitrogen or argon at low temperature are the most widely used for the determination of the basic characteristics of the porous texture of carbonaceous materials. One of the most important factors determining the adsorption capacity is the microporosity of carbon adsorbent. To evaluate the parameters of the microporous structure, the classical Dubinin-Radushkevich equation, which is frequently used for analyzing the adsorption isotherms on activated carbons, was selected. This D-R equation, based on Polanyi potential theory of adsorption, may be written as follows (Dubinin 1989):

$$W = W_0 \exp[-B(A/\beta)^n] = W_0 \exp[-(A/\beta E_0)^n] \quad (1)$$

with: $n = 2$, $E = \beta E_0$ and the adsorption potential (A):

$$A = -\Delta G = RT \ln\left(\frac{p_0}{p}\right) \quad (2)$$

where: W – volume of gas adsorbed at temperature T and relative pressure p/p_0 of the adsorbate; W_0 – maximum (limiting) micropore volume accessible to the given adsorbate; β – an affinity coefficient; $B = E_0^{-2}$ – structural constant dependent on the pore structure; E_0 – characteristic adsorption energy; G – Gibbs free energy.

The energy E_0 in the characteristic point is related to the mean size of micropores. Consequently, carbonaceous adsorbents are characterized structurally by means of three parameters: W_0, E_0, and x (the half-width of micropores in slit-shaped model). The determination of the values of these parameters and the application of the D-R equation to empirical data can be found in the literature (Dubinin 1960, McEnaney 1987, Jaroniec 1988).

The function $A = f(W)$ should be independent of temperature of the given adsorbate. For the adsorbate-adsorbent system, this relation is expressed by a characteristic adsorption curve. The characteristic curve may be obtained on the basis of the Dubinin-Radushkevich equation, by plotting empirical data as the volume of the adsorption space vs. the reduced adsorption potential, as follows:

$$W = f(A/\beta) \quad (3)$$

Another issue of interest is the temperature dependence of benzene adsorption which represent of a VOCs component. The basic thermodynamic function is the differential molar heat of adsorption. The heat of adsorption may be estimated in several ways: from direct calorimetric measurements, calculations based on the experimentally determined temperature coefficient of volumetric expansion of the adsorbate, or from adsorption isosters obtained experimentally from measurements of isotherms in at least two temperatures.

The isosteric heat of adsorption Q_{st} is obtained by inserting formula (2) in the Clausius-Clapeyron equation (Bering 1972, Do 1998), which leads to:

$$Q^{st} = -\Delta H = \frac{RT^2}{p_0}\frac{dp_0}{dT} + A - \alpha T\left(\frac{\partial A}{\partial \ln a}\right)_T \quad (4)$$

where: Q_{st} – isosteric heat of adsorption; ΔH – enthalpy of adsorption; a – amount of adsorption at the relative pressure p/p_0; R and T – the gas constant and temperature.

The expansion coefficient of liquid adsorbate may easily be calculated from the following dependence:

$$\alpha = \frac{1}{T_c - T_b}\ln\left(\frac{\rho_b}{\rho_c}\right) \quad (5)$$

where: ρ_b i ρ_c are the densities of the adsorbate in boiling temperature T_b (under normal pressure), and at the critical temperature T_c, respectively.

Equation (4) may be used to calculate the differential heat of adsorption based only on a single experimental isotherm. Its application is limited to the region of temperature invariance of a characteristic curve.

3 EXPERIMENTAL

3.1 Materials and Method

The adsorption experiments were performed by means of the volumetric method, which is suitable for determining general characteristics of adsorbents. This method is also commonly used for obtaining single adsorption isotherms.

The argon adsorption isotherm was measured at 77.5 K on one activated carbon sample using the sorption apparatus of Ciembroniewicz and Lasoń (1972) to characterize the porous structure and adsorption properties of the carbon adsorbent under investigation. The benzene sorption experiment (298 K) was performed using fluid micro-burettes equipment (Lasoń & Żyła 1963). By using this method, it was possible to measure the adsorption of benzene in a relative pressure range (p/p_0) of 10^{-4} to 1.

An activated carbon of the Norit type (0.18–0.25 mm), imported from the USA, was selected in this study, as these carbon materials are commonly used in industrial application as filters. This activated carbon is produced from peat. Details as to their manufacture were not available.

Measurements were carried out for gas-free carbon samples preheated at 393 K (degassing up to $\leq 1 \cdot 10^{-3}$ Pa).

Table 1. Selected physico-chemical properties of adsorbates (Landolt-Börnstein 1956).

Adsorbate	Molecular weight	$T_{boil.}$ K	T_c K	V_{mol} $cm^3 \cdot mol^{-1}$	Dipole moment D	Polarizibility cm^3
Argon	39.95	87.46	151.06	29.08	0	$1.63 \cdot 10^{-24}$
Benzene	78.11	353.46	561.66	89.41	0	$10.74 \cdot 10^{-24}$

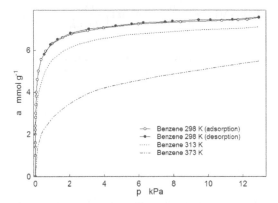

Figure 1. Adsorption of benzene on activated carbon, experimental and predicted data.

The properties of the adsorbates are presented in Table 1. All experiments were performed below the critical temperatures of each adsorbates.

4 RESULTS AND DISCUSSION

4.1 Isotherms of adsorption

The porosity of selected activated carbon was analyzed using adsorption isotherms of argon (at low temperature) and benzene (at ambient temperature). The former was preferred because of the spherical shape and small dimensions of argon (Mianowski et al. 2007). In addition, argon adsorption is dependent mainly on weak adsorbent-adsorbate interaction and, consequently, on the type of adsorbent.

Figure 1 shows the isotherm of benzene adsorption at 298 K. This isotherm is type I according to the IUPAC classification, and describes the adsorption of benzene vapours in a wide range of relative pressures. Its shape is characteristic of physical adsorption occurring in micropores. It has to be noted that at low relative pressures, which are relevant with regard to the practical application of the adsorptive method for the removal of benzene at low concentration levels, pores smaller than 1.5 nm are almost entirely filled. Activated carbon owes the ability to adsorb large amounts of benzene vapours to its strong affinity for hydrocarbons and the specific microporosity of the carbon adsorbent.

Furthermore, empirical data indicate that C_6H_6 sorption is reversible, which means it is easy to desorb molecules which had been adsorbed earlier.

Argon and nonpolar organic vapour (benzene) isotherms are further interpreted in terms of pore structure.

Table 2 gives the values of different parameters obtained from the adsorption data:

- specific surface area of the sample calculated from a_m values (in the relative pressure range of 0.1–0.3, using the BET method), assuming that the cross-sectional area (ω) of adsorbed argon atom is equal to $0.166 \cdot 10^{-18}$ m^2 (0.166 nm^2), and that of adsorbed benzene molecule to $0.41 \cdot 10^{-18}$ m^2 (0.410 nm^2),
- micropore volumes (W_0) and characteristic adsorption energy values (E_0) obtained from the D-R equation (1) for both Ar (77.5 K) and C_6H_6 (298 K)
- mean micropore width calculated from the McEnaney equation (1987).

As can be seen in Table 2, the specific surface area values, calculated from the experimental isotherms obtained at severely different temperatures of measurements, were high and relatively comparable for both argon and benzene. According to the majority of the data published on the subject, activated carbons with S_{BET} values of 1500–2000 $m^2 \cdot g^{-1}$ are the ones that find the widest application in many fields of industry (Suzuki 1994).

The total pore volumes were estimated from the amount adsorbed at a relative pressure of 0.95 (Table 2). The contribution of mesopore adsorption to the total adsorption is rather negligible, especially for benzene. Under these circumstances, the application of the Dubinin-Radushkevich equation to evaluate W_0 values is justified.

In general, the D-R plots have good linearity at high values of relative pressure. The linearity regions of these plots may also depend on the mean pore width and temperature, as suggested in literature (Ohba & Kaneko 2001). This is a result of different adsorbat-adsorbent pore wall interactions. In this study, the Dubinin-Radushkevich plots were linear in nearly the whole range of pressures used.

It is found that the characteristic energy of adsorption (E_0), derived from the experimental data, shows the increase in the case of benzene, which is typical for adsorption of strongly adsorbed compounds in microporous material. Table 2 includes also the mean micropore widths, estimated using the equation of McEnaney (1987):

$$2x = 4,691 \exp(-0,0666\, E_0) \qquad (6)$$

The results obtained from the adsorption data of argon and benzene are consistent with each other.

Table 2. Specific surface areas and structural parameters as obtained from the D-R and related equations.

Adsorbate	S_{BET} $m^2 \cdot g^{-1}$	Pore volume $cm^3 \cdot g^{-1}$			E_0 $kJ \cdot mol^{-1}$	Mean pore width nm
		W_0	V_{mes}^a	V_t^b		
Ar	1978	0.623	0.126	0.759	5.7	0.69
C_6H_6	1391	0.619	0.041	0.660	17.7	0.71

[a] Mesopore volume calculated from Kelvin equation.
[b] Total volume of pores estimated from the amount adsorbed at $p/p_0 = 0.95$.

The values of limiting pore volumes estimated from both adsorbates are very close (around $0.62 \, cm^3 \cdot g^{-1}$), which proves that the pore network is accessible to the same extent.

From the results presented in Table 2, it may be concluded that activated carbon under investigations is largely microporous (but does not feature very narrow micropores).

4.2 Characteristic curves for Ar and C_6H_6 adsorption

Another problem to be considered is the comparison between the micropore volumes inferred from the adsorption of Ar (at a very low adsorption temperature) and benzene (at ambient temperature) on the same carbon sample. In order to obtain the characteristic curves of the adsorbate – activated carbon adsorption systems, the dependence of ln of the W_0 (volume of adsorbed phase per gram of carbon in $cm^3 \cdot g^{-1}$) vs. the reducing adsorption potential was plotted according to equation (3). W_0 is the limiting adsorption which corresponds to complete filling of the micropores.

Figure 2 shows the resulting curves for both adsorbates on the same activated carbon. The characteristic curve for argon corresponds to the isotherm measured in a relative pressure range from $5 \cdot 10^{-3}$ to nearly 1. The values of the affinity coefficient selected for the calculations were $\beta = 1.00$ for benzene and $\beta = 0.31$ for argon (McEnaney 1987).

It can be seen from the empirical data that the amount adsorbed in the higher range of relative pressures and hence adsorption capacities, are essentially determined on the micropore volume of the carbon sample and it is not affected by the temperature of measurements. Moreover, the W_0 values extracted from these curves were in good agreement with those obtained using the D-R equation (Table 2). This suggests that the mechanism of adsorption is comparable for both adsorbates.

On the other hand, the adsorption potential $(A/\beta)^2$ and hence the dispersion interactions with carbon surface, is higher for benzene than for argon.

Figure 2, however, clearly illustrates that the Ar and C_6H_6 data cover different ranges of the micropore filling process. When the adsorption data are compared in the experimental conditions used, one may conclude that for the investigated activated carbon, Ar is used at a range of higher relative pressures, and

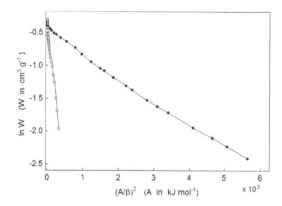

Figure 2. Characteristic curves for activated carbon (Norit): o – argon (77.5 K), and, • – benzene (298 K), plotted using their respective β values.

determines the total micropore volume (which indicates that this sample has also some amount larger microporosity), whereas benzene molecules only signify the volume of narrow micropores. In this way, the achieved results reflect the different sizes of both adsorbates (the kinetic size of argon is 0.34 nm, while that of C_6H_6 is 0.37 nm). Consequently, micropores are first filled with benzene molecules, and only then by argon. This sequence corresponds to the increase in values of relative pressure at which pores of a given width are filled with the particular adsorbate. For this reason, the values of the adsorption energy are affected by the complexity of the microporous structures in the carbon material.

A unique feature of the Dubinin-Polanyi theory is the invariance of the characteristic curve of adsorption, which may be used to predict adsorption isotherms when there is only a minimum of available data.

From the calculated characteristic curve for benzene adsorption, two isotherms were predicted for 313 K and 373 K, as shown in Figure 1. They are good representations of the empirical isotherm on the microporous activated carbon. The relative pressures at these two temperatures were calculated based on the standard vapour pressures at each temperature.

4.3 Heat of benzene adsorption

As is commonly known, the heat of adsorption depends for the most part both on the nature of the organic

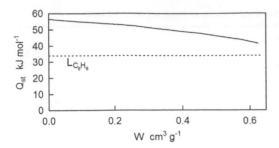

Figure 3. Isosteric heat of benzene adsorption on activated carbon calculated from equation (4).

compounds (polar/nonpolar) and on the type of activated carbon (porous texture), as well as on the VOCs concentration.

Figure 3 shows the plot of the isosteric heat of benzene adsorption process on the carbon adsorbent, calculated from the characteristic curve according to equation (4).

As can be seen, the isosteric heat of adsorption decreases continuously with the amount of adsorbed benzene. The adsorption of nonpolar vapour of benzene on the activated carbon sample under investigation is determined by nonspecific dispersion interactions.

In the case of microporous carbon, the effect of microporosity should not be underestimated, since even a slight increase in pore size results in a very large increase in potential energy of these interactions (with the sites being micropores with different adsorption energies). These factors are defined in the literature data (Ohba & Kaneko 2001). The above-mentioned fact also confirms the changes in the affinity of activated carbon for argon and benzene (see Figure 2).

As a result, the adsorption equilibrium must be affected by the distribution of the micropore size, and it is suggested that the shape of the Q_{st} plot reflects the degree of surface heterogeneity of the activated carbon. That is why micropore size distribution is a very appropriate issue for investigation in further studies, e. g., in order to optimize the benzene adsorption at very low concentrations.

5 CONCLUSIONS

The adsorption of VOCs (benzene) from the gaseous phase onto porous adsorbents has been suggested as an innovative technology to be used in environmental applications. In practice, microporous activated carbon and gas/vapour adsorption investigations of its structural properties have attracted much attention. These new applications of activated carbons carry with them the need for continuous improvement of their performance, including adsorption capacity.

As expected, that the microporous structure of the carbon material is the main parameter implicated in activated carbon – benzene adsorption system. The results of this study confirm that the investigated

activated carbon exhibits the highest adsorptive efficiency with regard to benzene at low pressures (at low concentrations). The effect of molecular sieving by very narrow micropores does not occur in this carbon material.

It is shown that based on Dubinin-Polanyi theory it is possible to predict, with reasonable accuracy, the equilibrium adsorption isotherms and the isosteric heat of adsorption in a wide range of temperatures and relative pressures from only a single isotherm, without performing laborious and time-consuming experiments. The fact, that it is not necessary to apply a particular model of adsorption, is an additional advantage.

From the practical point of view, the obtained results indicate, that in order to achieve optimal adsorption capacity, especially with regard to the adsorption of low benzene concentrations, it is preferable to use carbon adsorbents with a maximal volume of narrow micropores (≤ 0.7 nm).

ACKNOWLEDGEMENTS

The support of this work referring to the Grant AGH (No. 11.11.210.117) has been greatly appreciated.

REFERENCES

Baek, S. W., Kim, J. R. & Ihm S. K. 2004. Design of dual functional adsorbent/catalyst system for the control of VOC's by using metal-loaded hydrophobic Y-zeolites. *Catalysis Today*. 93–95: 575–581.

Bering, B.P., Dubinin, M. & Serpinsky, V.V. 1972. On thermodynamics of adsorption in micropores. *Journal of Colloid and Interface Science* 38(1): 185–194.

Bimer, J., Sałbut, P.D., Berłożecki, S., Boudou, J.P., Broniek, E. & Siemieniewska, T. 1998. Modified active carbons from precursors enriched with nitrogen functions: sulfur removal capabilities. *Fuel* 77(6): 519–525.

Biniak, S., Szymański, G., Siedlewski, J. & Świątkowski, A. 1997. The characterization of activated carbons with oxygen and nitrogen surface groups. *Carbon* 35(12): 1799–1810.

Buczek, B., Ziętkiewicz, J., Marecka, A., Albiniak, A., Siemieniewska, T., Marêché, J.F., Payot, F. & Bégin, D. 1998. A comparison of adsorption properties of industrial activated carbons and activated anthracite. *Proceedings of Annual Conference* Zakopane, 27–30 September 1998: 90–91 (in Polish).

Burg, P., Fydrych, P., Bimer, J., Salbut, P.D. & Jankowska, A. 2002. Comparison of three active carbons using LSER modeling: prediction of their selectivity towards pairs of volatile organic compounds (VOCS). *Carbon* 40(1): 73–80.

Ciembroniewicz, A. & Lasoń, M. 1972. Sorptional Manostat a Semi-Automatic Device for Sorption Investigations. *Polish Journal of Chemistry* 46(4): 703–710 (in Polish).

Do, D.D. 1998. Adsorption Analysis: Equilibria and Kinetics, Imperial College Press. London. Series on Chemical Engineering 2: 168–170.

Dubinin, M.M. 1960. The Potential Theory of Adsorption of Gases and Vapors for Adsorbents with Energetically Nonuniform Surfaces. *Chemical Review* 60(2): 235–239.

Dubinin, M.M. 1989. Fundamentals of the theory of adsorption in micropores of carbon adsorbents: characteristics of their adsorption properties and microporous structures. *Carbon* 27(3): 457–467.

McEnaney, B. 1987. Estimation of the dimensions of micropores in active carbons using the Dubinin-Radushkevich equation. *Carbon* 25(1): 69–75.

Harley, R.A. & Cass, G.R. 1995. Modeling the atmospheric concentrations of individual volatile organic concentration. *Atmospheric Environment* 29(8): 905–922.

Huo, H., Wu, Y. & Wang, M. 2009. Total versus Urban: Well-to-wheels assessment of criteria pollutant emissions from various vehicle/fuel systems. *Atmospheric Environment* 43(10): 1796–1804.

Hoshi, J., Amano, S., Sasaki, Y. & Korenaga, T. 2008. Investigation and estimation of emission sources of 54 volatile organic compounds in ambient air in Tokyo. *Atmospheric Environment* 42(10): 2383–2393.

Jankowska, H., Świątkowski, A. & Choma, J. 1991. Active Carbon. Chichester: Ellis Horwood Ltd.

Jaroniec, M. & Madey, R. 1988. Physical adsorption on heterogeneous solids. Elsevier. Amsterdam. Chapter 2.

Kawashima, H., Minami, S., Hanai, Y. & Fushimi, A. 2006. Volatile organic compound emission factors from roadside measurements. *Atmospheric Environment* 40(13): 2301–2312.

Kelly, T.J. & Holdren M.W. 1995. Applicability of canisters for sample storage in the determination of hazardous air pollutants. *Atmospheric Environment* 29(19): 2595–2608.

Kim, K.J., Kang, Ch.S., You, Y.J., Chung, M.Ch., Woo, M.W., Jeong, W.J., Park, N.C. & Ahn, H.G. 2006. Adsorption-desorption characteristics of VOCs over impregnated activated carbons. *Catalysis Today* 111(3–4): 223–228.

Landolt-Börnstein, N. 1956. Physico-Chemical Constants. Springer Verlang: Berin 1.

Lasoń, M. & Żyła, M. 1963. Apparatus for determination of vapour sorption and desorption isotherms by means of microburetes. *Analytical Chemistry* 8(2): 279–287 (in Polish).

Luo, L., Ramirez, D., Rood, M.J., Grevillot, G., Hay, K.J. & Thurston, D.L. 2006. Adsorption and electrothermal desorption of organic vapors using activated carbon adsorbents with novel morphologies. *Carbon* 44(13): 2715–1723.

Medinsky, M.A., Kenyon, E.M. & Schlosser, P.M. 1995. Benzene: a case study in parent chemical and metabolite interactions 105(2–3): 225–233.

Mellios, G. & Samaras, Z. 2007. An empirical model for estimating evaporative hydrocarbon emissions from canister-equipped vehicles. *Fuel* 86(15): 2254–2261.

Mensink, C., De Vlieger, I. & Nys, J. 2000. An urban transport emission model for the Antwerp area. *Atmospheric Environment*. 34(27): 4595-4602.

Mianowski, A., Owczarek, M. & Marecka, A. 2007. Surface Area of Activated Carbon Determined by the Iodine Adsorption Number. *Energy Sources, Part A* 92: 839–850.

Mohr, U. 1997. Activated Carbon Canisters for Automobiles. *Filtration and Separation* 34(10): 1018–1020.

Moreno-Castilla, C., López-Ramón, M.V. & Carrasco-Marin, F. 2000. Changes in surface chemistry of activated carbons by wet oxidation. *Carbon* 39(14): 1995–2001.

Ohba, T. & Kaneko, K. 2001. GCMC Study on relationship between DR plot and micropore width distribution of carbon. *Langmuir* 17(12): 3666–3670.

Pagans, E., Font, X. & Sanchez, A. 2006. Emission of volatile organic compounds from composting of different solid wastes: Abatement by biofiltration. *Journal of Hazardous Materials* 131(1–3): 179–186.

Sene, L., Converti, A., Felipe, M.G.A. & Zilli, M. 2002. Sugarcane bagasse as alternative packing material for biofiltration of benzene polluted gaseous streams: a preliminary study. *Bioresource Technology* 83(2): 153–157.

Sillman, S. 1999. The relation between ozone, NO_x and hydrocarbons in urban and polluted rural environments. *Atmospheric Environment* 33(12): 1821–1845.

Suzuki, M. 1994. Activated carbon fiber: Fundamentals and applications. *Carbon* 32(4): 577–586.

Wang, P. & Zhao, W. 2008. Assessment of ambient volatile organic compounds (VOCs) near major roads in urban Nanjing. China. *Atmospheric Research* 89(3): 289–297.

Yu, T. Y., Lin, Y.Ch. & Chang L.F.W. 2000. Optimized combinations of abatement strategies for urban mobile sources. *Chemosphere* 41(3): 399–407.

Zielińska, B., Sagebiel, J.C., Harshfield, G., Gertler, A. & Pierson, W.R. 1996. Volatile Organic Compounds up to C_{20} Emitted from Motor Vehicles; Measurement Methods. *Atmospheric Environment* 30(12): 2269–2286.

Environmental Engineering III – Pawłowski, Dudzińska & Pawłowski (eds)
© 2010 Taylor & Francis Group, London, ISBN 978-0-415-54882-3

Emission of polycyclic aromatic hydrocarbons (PAHs) during the production of carbon and graphite electrodes

M. Mazur, R. Oleniacz, M. Bogacki & P. Szczygłowski

AGH University of Science and Technology (AGH-UST), Faculty of Mining Surveying and Environmental Engineering, Department of Management and Protection of Environment, Krakow, Poland

ABSTRACT: Many organic contaminants, including polycyclic aromatic hydrocarbons (PAHs), are released during the production of carbon and graphite electrodes. We studied the concentration and mass streams of 16 PAHs released into the air via flue gases during the processes involved in the storage, handling, calcination and mixing of raw materials, as well as the forming, baking, impregnation and graphitising of electrodes. Individual emission sources and production cycles differ significantly in PAH emission profiles. The biggest single source of PAH emissions is the baking of carbon electrodes, pressing of the solid raw materials and pitch, and storage or handling of liquid raw materials.

Keywords: Carbon and graphite industry, PAHs, stack gas, emission, air pollution control.

1 INTRODUCTION

Polycyclic aromatic hydrocarbon (PAHs) compounds are a class of complex organic chemicals, containing carbon and hydrogen with a fused ring structure containing at least two benzene rings (Ravindraa et al. 2008). PAHs are widely distributed in the atmosphere due to the incomplete combustion of carbonaceous materials, and they were one of the first atmospheric pollutants to be identified as suspected carcinogens. They have relatively low vapour pressure and are resistant to chemical reaction (Ferreira 2001), so they persist in the environment and tend to accumulate in biota, soils and sediments. They are included in the Protocol on Persistent Organic Pollutants of the Convention on Long-Range Transboundary Air Pollution (UNECE 1998) and measures should be put in place to control their emission (Denier van der Gon et al. 2007).

The major anthropogenic sources of atmospheric PAH emission include biomass (biofuel) burning, coal and petroleum combustion, waste incineration and coke and metal production (Dyke et al. 2003, Li et al. 1999, Pacyna et al. 2003, Ravindraa et al. 2008, Xu et al. 2006, Zhang & Tao 2009). The global PAH emission inventory for 2004 shows that major industrial activities contribute less than 10% of the total global PAH emission, with coke production contributing the most (Zhang & Tao 2009). Industrial sources including the use of solvents and other products, production of iron, steel and non-ferrous metals and other industrial processes contribute much more to the total PAH emission in Europe (Breivik et al. 2006). Further emission data from these activities are needed.

PAHs are also emitted during the manufacture of carbon products and graphite electrodes (EC 2001, 2009, Hagen 2007, Mazur et al. 2006a, b, 2008). The carbon and graphite industry supplies basic elements for smelting technology such as the graphite electrodes used in steel making and anodes and cathodes for aluminium industry. Steel production consumes electrodes at the rate 1.5 to 3 kg per tonne of steel. Production of inert solid carbon and graphite is based mainly on petroleum coke and coal and a highly annealed coke derived from coal tar, with petroleum pitch and coal tar-pitch used as binders (EC 2001, 2009).

Manufacture of carbon products usually involves the following consecutive stages:

- calcining, grinding (crushing/milling), classifying and mixing the raw materials,
- forming (pressing/moulding/extruding) and baking of green shapes, and
- machining and finishing of the final products.
- Graphite electrode production is more complicated and consists of the following stages:
- grinding and classifying the solid raw materials,
- mixing the solid and liquid materials,
- forming and baking (baking I) of green shapes,
- impregnation of prebaked electrodes,
- re-baking of impregnated electrodes (baking II),
- graphitising of baked (re-baked) electrodes, and
- machining and finishing of the final products.

Some of these stages result in significant release of PAHs into the atmosphere with total-PAH concentrations in flue gases similar to those found in coke drying and calcination (Bayram et al. 1999), carbon

Table 1. Description of emission sources, air pollution control devices (APCDs) and test conditions.

Installation (process)	Stack no.	Emission sources	APCDs	Operating characteristics	Number of samples
Calcination plant (raw material calcining)	1	Calcining furnaces with high calorific natural gas heating	Afterburning in waste heat boilers with gas burner and calcining flue gas recirculation	Operation of one retort furnace and one or two waste heat boilers (gas consumption: 82–220 m³/h, steam production: 1.4–3.2 t/h)	4
Liquid raw material store (liquid material storage, heating and handling)	2	Heavy tar tank, 100 m³	Dephlegmator	Heavy tar storage (47 t, temp. 86°C)	2
	3	Pitch tank, 200 m³	Dephlegmator	Pitch storage (185 t, temp. 188°C)	2
	4	Tar tank, 100 m³	Dephlegmator	Tar storage (39 t, temp. 89°C)	2
	5	Carbomass tank, 21.5 m³	Dephlegmator	Carbomass storage (15 t, temp. 66°C)	2
	6	Impregnant tank, 100 m³, and tanker	One-section coke filter	Impregnant pumping over the tanker (22 t)	2
Pressing plant (solid material grinding; solid and liquid material handling; paste production; product packaging)	7	Jeffrey crusher, conveyors, dry mixers	Two fabric filters (efficiency 98%)	Solid material grinding and handling (5–7.5 t/h)	4
	8	Wet mixers, 40 MN extrusion press, Eirich mixer, pressing and conveyors, packers	Five-section coke filter (efficiency 50%)	Mixing and pressing of materials: coke - 75%, heavy tar, carbomass, pitch and tar - 25% (7.5–11 t/h)	4
	9	62 MN extrusion press, belt conveyor, paste cooler	Two-section coke filter (efficiency 50%)	Solid material and pitch moulding and paste handling (5–7.5 t/h)	4
	10	40 MN extruder outlet	Swingtherm reactor (efficiency 97.6%)	Green shape extrusion and outlet heating: coke - 75%, pitch and tar - 25% (3–5 t/h)	4
Impregnation plant (baked or graphitized electrode impregnation)	11	Impregnation unit (autoclave), 19.6 m long, with high calorific natural gas heating	Afterburner (internal gas burner)	Electrode impregnation with DEZA impregnant (DEZA consumption: 0.5–0.7 t/cycle, duration: 30–60 minutes)	4
Baking plant (carbon electrode baking)	12	20-section multiple chamber open type (horizontal) ring furnaces and a closed type (vertical) ring pit furnace, natural gas fuelled	Two-stage electrostatic precipitators with pipe coolers (Hosokawa-Mikropul) and two-section coke filters (total efficiency: tars 98%, dust 99%)	Carbon shape baking in three open ring furnaces (feedstock: 350–450 t per furnace) and one closed ring pit furnace (feedstock: 160–340 t); total feedstock: 1412–1514 t (30 % coal, 70 % graphite)	4
Graphitising plant (baked electrode graphitising)	13	Castner (LWG) electric-resistance furnaces, 25 m long	Swingtherm-Kormoran type catalytic thermo-reactor and FGD (flue-gas desulphurisation) wet-scrubber	Electrode graphitising in single furnace (heating and ventilation phases); total feedstock: 111 t - electrodes, 270 t - insulation packing	14

black manufacture (Tsai et al. 2001), coke production in the iron and steel industry (Yang 2002) and asphalt mixing (Lee et al. 2004). The measurement of PAHs in stack emissions from major sources in the carbon and graphite electrode industry is the main purpose of this study. The research was carried out at SGL CARBON Polska S.A. in Nowy Sacz (Poland).

2 MATERIALS AND METHODS

The characteristics of emission sources, air pollution control devices (APCDs) and test conditions are shown in Table 1. At least two samples were collected from each stack flue gas using a suitable sampling system. The flue gas was sampled iso-kinetically using an automatic gravimetric dust monitor (Emiotest 2592) according to Polish Standard PN-Z-04030-7:1994 Gravimetric Method. The particle-bound phase of PAHs was collected using glass microfiber filters. The gaseous phase was collected using a glass cartridge (an adsorption tube) packed with resin (XAD-2).

Sampling points after the wet-scrubbing and cooling systems used only the low-temperature particle phase sampling without the XAD resin collector. In cases with very low flue gas flow (e.g. heating

Table 2. Polyaromatic hydrocarbon (PAH) concentrations in stack flue gases in the carbon and graphite industry ($\mu g/m_N^3$).

PAHs	Installation											
	Calcination plant (stack 1)		Liquid raw material store (stacks 2–6)		Pressing plant (stacks 7-10)		Impregnation plant (stack 11)		Baking plant (stack 12)		Graphitising plant (stack 13)	
	Mean	Range	Mean	Range	Mean	Range	Mean	Range	Mean	Range	Mean	Range
Nap	0.75	0–2.4	2919	0–5221	11.2	0–176	0.67	0–1.8	–	ND*	0.19	0–0.43
AcPy	–	ND	–	ND	1.14	0–16	0.45	0–1.8	0.22	0–0.49	0.03	0–0.16
Acp	1.59	0–5.5	125	0–207	38.6	0–240	0.91	0–3.6	0.28	0–0.64	0.45	0–1.64
Flu	0.65	0–1.6	62.1	0–241	22.1	0–180	0.35	0–1.1	1.06	0.33–1.7	0.44	0.08–0.98
PA	2.74	0.47–4.7	145	0–747	188	0.1–1570	1.91	0.5–5.8	81.5	46.8–122	0.25	0–0.53
Ant	0.54	0–1.1	33.6	0–194	16.9	0–115	0.21	0–0.83	5.08	3.1–9.3	0.62	0–1.72
FL	5.33	1.7–10.7	122	0–406	121	0–898	2.98	2.0–3.6	138	76–210	0.26	0.03–0.51
Pyr	3.56	0.6–8.0	48.4	0–326	88.6	0.1–646	0.89	0–1.5	99.8	57–138	0.23	0.11–0.42
BaA	1.67	0–5.2	57.9	0–212	9.19	0–70	1.09	0–2.7	20.3	9.2–38	0.91	0–2.36
CHR	1.98	0–5.6	0.6	0–4.7	9.31	0–67	0.77	0–2.3	55.3	22–103	0.04	0–0.26
BbF	0.77	0–1.6	28.4	0–282	2.89	0–23	0.67	0–1.5	39.2	21–56	0.03	0–0.34
BkF	0.57	0–2.3	24.9	0–249	2.49	0–38	1.21	0–3.3	6.79	0–27	0.26	0–0.59
BaP	0.22	0–0.89	–	ND	3.92	0–57	0.41	0–1.7	6.73	2.3–11	0.02	0–0.12
IND	0.21	0–0.43	110	0–719	0.06	0–0.53	0.47	0–1.3	3.60	1.7–6.2	0.12	0–0.90
DBA	–	ND	96.0	0–956	0.69	0–6.3	–	ND	2.81	1.3–5.7	0.01	0–0.22
BghiP	–	ND	32.5	0–325	–	ND	–	ND	3.64	1.6–7.9	0.09	0–0.58
Total – 16 PAHs	20.6	3.3–49.3	3805	72–6297	515	0.33–4103	13.0	4.2–24.6	464	261–666	4.0	1.9–6.6

*ND = Not determined.

Table 3. Averaged polyaromatic hydrocarbon (PAH) emissions for selected stack flue gases.

Process* Stack no.	Calc. 1	Storage and handling 2	3	4	5	6	Grinding, mixing and forming 7	8	9	10	Impr. 11	Baking 12	Graph. 13
Average dry flue-gas flow (m^3_N/h)	9425	46.4	40.2	47.0	47.9	324	24,012	21,231	15,323	831	6543	119,718	50,375
Average mass flow rate (mg/h):													
Nap	5.7	130.7	158.0	155.3	217.6	0.3	2.8	8.6	548	—	4.1	—	9.8
AcPy	—	—	—	—	—	—	8.9	—	52	—	2.9	28	1.5
Acp	12.3	5.3	6.9	6.6	7.2	16.9	6.8	81.4	2452	2.0	5.7	36	22.6
Flu	4.9	—	—	5.1	5.7	29.7	5.7	11.7	1120	0.4	2.2	123	22.6
PA	21.9	1.5	15.0	4.4	0.0	81.0	37.8	31.0	9652	4.0	12.2	9547	12.9
Ant	4.0	4.5	—	1.4	0.0	14.6	8.9	28.9	930	1.3	1.3	613	30.5
FL	48.1	—	6.7	9.5	9.1	16.6	12.8	120.5	6360	5.0	19.6	16,192	13.0
Pyr	31.1	—	6.5	2.2	0.0	11.2	17.4	87.5	4653	3.6	5.6	11,613	11.7
BaA	14.5	—	2.5	6.8	3.6	2.7	0.3	17.1	488	0.7	6.8	2414	45.9
CHR	17.7	—	—	—	—	1.0	4.2	15.7	504	0.7	4.6	6210	2.2
BbF	7.4	—	—	6.6	—	0.4	4.9	—	148	—	4.3	4552	1.5
BkF	4.5	—	—	5.9	—	—	—	—	119	0.3	7.7	637	13.0
BaP	1.7	—	—	—	—	—	—	—	191	0.5	2.6	767	1.0
IND	2.4	—	—	16.9	9.1	0.2	—	3.0	—	0.1	3.0	409	6.3
DBA	—	—	—	22.5	—	0.7	—	6.3	33	—	—	312	0.7
BghiP	—	—	—	7.6	—	—	—	—	—	—	—	400	4.4
Total 16 PAHs	176.2	142.1	195.6	250.8	252.4	175.3	110.4	411.6	27,248	18.7	82.5	53,852	199.5

*Raw material calcining (Calc.); Storage and handling of liquid raw materials (i.e. tanks and tankers); Grinding and mixing of raw materials and green shape forming (pressing/moulding); Prebaked electrode impregnation (Impr.); Green shape baking or impregnated electrode re-baking; and Baked electrode graphitising (Graph).

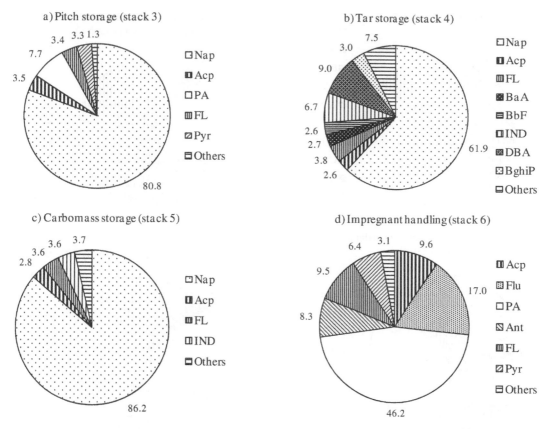

Figure 1. Averaged percentage of emissions of 16 polyaromatic hydrocarbons during: a) pitch, b) tar and c) carbomass storage, and d) impregnant handling.

tank emissions), vapours were collected with an oil dephlegmator and adsorption tubes only. Sampling times varied from 10 to 180 minutes depending on the expected PAH levels, sampling method and installation operating conditions.

PAH samples collected by the filters and XAD resin were analysed by the gravimetric method according to the AEERL/12-9/2/86 Procedure (1986). Each sample was extracted with dichloromethane in a Soxhlet apparatus for 8 hours. The extracts were condensed to a volume of 10 ml by means of solvent evaporation in a Kuderna-Danish concentrator.

Gas chromatography was used to determine the PAH content in the extracts. An HP6890 gas chromatograph with flame ionisation detector (FID) and a 30 m long HP DB5 capillary column was used, working with a programmed temperature rise of 8°C/min, from 40°C to 280°C. The concentrations of 16 EPA priority PAHs were determined: naphthalene (Nap), acenaphthylene (AcPy), acenaphthene (Acp), fluorene (Flu), phenanthrene (PA), anthracene (Ant), fluoranthene (FL), pyrene (Pyr), benzo(a)anthracene (BaA), chrysene (CHR), benzo(b)fluoranthene (BbF), benzo(k)fluoranthene (BkF), benzo(a)pyrene (BaP), indeno(1,2,3-cd)pyrene

(IND), dibenzo(a,h)anthracene (DBA) and benzo(ghi) perylene (BghiP).

3 RESULTS AND DISCUSSION

Mean, minimum and maximum values of PAH levels in stack flue gases for each installation (manufacturing process) are shown in Table 2. The averaged mass flow rates for selected stack emissions (determined using individual PAH concentrations and flue gas flow-rate measurements) are shown in Table 3.

The results obtained confirm that the major source of PAH emissions in the carbon and graphite industry is the baking of carbon electrodes (EC 2001, 2008, Hagen 2007). This is because of both the relatively high concentration of PAHs in the gases released (0.26–$0.67\,mg/m_N^3$ for the total of 16 PAHs studied) and the very large volume stream of the flue gases (up to $130{,}000\,m^3/h$). These two factors result in emissions of 33–85 g/h from a single furnace for the total of the 16 PAHs studied, including high emissions of carcinogenic substances: BaP (0.3–1.3 g/h), BaA (1.2–4.9 g/h), IND (0.2–0.6 g/h), BkF (0–2.5 g/h), and BbF (2.6–7.0 g/h).

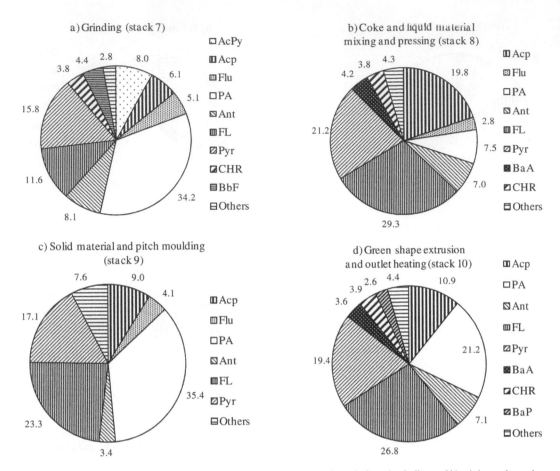

Figure 2.   Averaged percentage of emissions of 16 polyaromatic hydrocarbons during: a) grinding and b) mixing and pressing of raw materials, and c), d) forming of green electrode shapes.

Turning to the level of PAH emissions, the processes of forming (moulding) electrodes from solid raw materials and pitch (stack 9) are also important. Here, the total levels of all 16 PAHs studied range from 0.7 to 4.1 mg/m$_N^3$, with emissions of 13 to 51 g/h. Although there are even higher PAH levels in the vapours released into the air from liquid raw materials storage and handling (usually a total of a few mg/m$_N^3$ for all 16 PAHs studied), these operations are less important for total emissions (mass flow rate of 0.1–0.3 g/h) due to the relatively low level of gases carried away. Calcining, grinding, mixing and handling of solid raw materials and impregnation and graphitising of electrodes are characterised by much lower concentrations of PAHs in the stack flue gases (1–70 μg/m$_N^3$), resulting in lower emissions into the air (0.001-1.4 g/h).

The individual processes also differ in their profiles of PAHs released (Fig. 1–3). When storing or handling heavy tar, pitch, tar and carbomass, Nap is the predominant PAH present, represents about 60–90% of the total. When storing or handling the saturant used in this plant, large amounts of PA, Flu,

Acp and FL, and sometimes Pyr and BaA, are released. The main share in the total release of PAHs from grinding and mixing of raw materials and from shaping of electrodes is due to PA, FL, Pyr and Acp, and in some cases, Ant and AcPy. During the calcination of raw materials emissions of FL and Pyr predominate (altogether approximately 45% of all 16 PAHs studied), although the share of PA, CHR, BaA and Acp is also significant. PAH emissions during impregnation of pre-baked electrodes distributes rather uniformly among various PAHs, with FL and PA predominating. In turn, when baking electrodes the emissions of FL, Pyr and PA predominate (altogether almost 70% of all 16 PAHs studied), whereas BaA, Ant, Acp and Flu make up ≈60% of all 16 PAHs studied in emissions during graphitising.

The profiles of PAH emissions during the production of carbon and graphite electrodes are different from those seen in various medium- and high-temperature industrial processes (Li et al. 1999, Tsai et al. 2001, Yang 2002). Only Bayram et al. (1999) obtained similar PAH emission profiles for drying and calcination of coke. The concentration of total

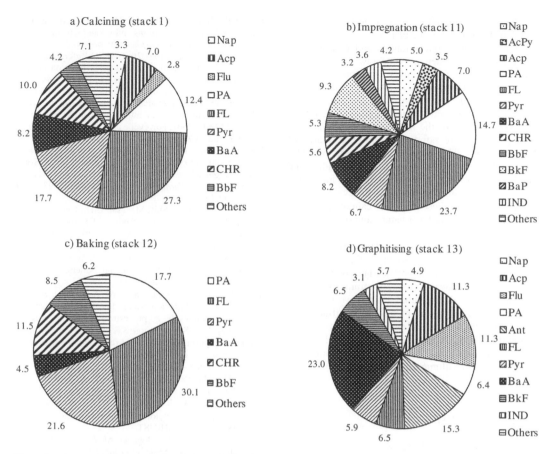

Figure 3. Averaged percentage of emissions of 16 polyaromatic hydrocarbons during: a) calcining of raw materials, and b) impregnation, c) baking and d) graphitising of carbon electrodes.

PAHs in waste gases from different industrial boilers and furnaces is subject to a wide range of fluctuations and remains at approximately $0.01–3.5 \text{ mg/m}_N^3$, depending on the type and source of emissions and the type of APCDs used. In our study of the carbon and graphite industry, even more diversity in concentration was seen. Higher total levels of all 16 PAHs studied ($3.5–5.0 \text{ mg/m}_N^3$) were seen, for example, in the vapours released from tanks used for the storage of liquid raw materials such as pitch and carbomass and, sporadically from solid raw material and pitch moulding (electrode forming). Conversely, lower concentrations (below 0.01 mg/m_N^3) occurred in the stack flue gases from graphitising of electrodes. However, all these emissions are within the range of variability as defined for the production of carbon and graphite products in the best available techniques (BAT) reference documents for the non-ferrous metals industries (EC 2001, 2009).

The type and levels of PAH release is significantly affected by the type of APCDs used. Dry and wet scrubbers, adsorbers, coolers and fabric filters retain significant amounts of PAHs, e.g., special electrostatic precipitators (with pipe coolers) and coke

filters such as those used for cleaning waste gases from carbon electrode baking, turn out to be useful here. On the other hand, afterburners or thermocatalytic reactors, often used in afterburners for carbon monoxide and hydrocarbons, may cause rearrangement and formation of some PAHs (Mazur et al. 2003, 2006a).

4 CONCLUSIONS

The manufacturing processes used in the graphite electrode industry are a significant source of PAH release into the atmosphere, and therefore efficient preventative methods should be used. The highest emissions of PAHs, including BaP, BaA, IND, BkF and BbF, which are known to be potentially carcinogenic, occur during the baking of carbon electrodes. The average total content of all 16 PAHs studied in waste gases cleaned by electrostatic precipitators with pipe coolers and coke filters is under 0.5 mg/m_N^3. However, because of the large amounts of combustion gas this results in significant emissions, usually of more than 50 g/h, from one ring furnace (including, on average, 0.8 g/h of BaP, 2.4 g/h of BaA and 4.5 of BbF).

Relatively high concentrations and emissions of PAHs also occur when pressing raw materials and shaping electrodes from solid raw materials and pitch. The highest total levels of all 16 PAHs studied occurred in the waste gases accompanying the storage and handling of liquid raw materials (including pitch). However, because of the low flow of these vapours, these processes are not the predominant source of PAH emissions. The lowest concentrations of PAHs occur in the waste gases from calcining, grinding and handling solid raw materials and during the impregnation and graphitising of electrodes. However, the emission of PAHs from these processes cannot be ignored because of their long-term nature and/or significant flow of waste gases.

Diversity in the profiles of PAHs released is very high, for both the various sources of emissions and specific processes. These profiles largely depend on the type of raw materials, process parameters and air pollution control devices used. Further work needs to be done to identify more precisely the influence of these factors on the type and amount of PAHs released. It would also be advisable to make known the emission factors and relate them to the consumption of raw materials or volume of production for individual processes.

ACKNOWLEDGEMENTS

The work was completed within the scope of AGH-UST statutory research for the Department of Management and Protection of Environment No. 11.11.150.008, and contracts No. 5.5.150.611 and 5.5.150.687.

REFERENCES

AEERL/12-9/2/86 Procedure 1986. Standard procedure for gravimetric analysis of organic extracts (based on: Haris J.C. et al., *Laboratory Evaluation, Level 1 Organic Analysis Procedure. EPA-600/S7-82-048, NTIS PB 82-239, pp. 30-36, March 1982* and Lentzen, D.E., Wagoner, D.E., Estes, E.D. & Gutknecht W.F., *IERL-RTP Procedures Manual, Level I Environment Assessments, Second Edition, EPA 6000/7-78/201, NTIS No. PB293-795, pp. 140–142, October 1978*).

Bayram, A., Müezzinoğlu, A. & Seyfioğlu, R. 1999. Presence and control of polycyclic aromatic hydrocarbons in petroleum coke drying and calcination plants. *Fuel Processing Technology* 60: 111–118.

Breivik, K., Vestreng, V., Rozovskaya, O. & Pacyna, J.M. 2006. Atmospheric emissions of some POPs in Europe: a discussion of existing inventories and data needs. *Environmental Science & Policy* 9: 663–674.

Denier van der Gon, H., van het Bolscher, M., Visschedijk, A. & Zandveld, P. 2007. Emissions of persistent organic pollutants and eight candidate POPs from UNECE–Europe in 2000, 2010 and 2020 and the emission reduction resulting from the implementation of the UNECE POP protocol. *Atmospheric Environment* 41: 9245–9261.

Dyke, P.H., Foan, C. & Fiedler, H. 2003. PCB and PAH releases from power stations and waste incineration processes in the UK. *Chemosphere* 50: 469–480.

EC (European Commission) 2001. IPPC Reference Document on Best Available Techniques in the Non Ferrous Metals Industries, http://eippcb.jrc.es/reference/nfm.html.

EC (European Commission) 2009. IPPC Draft Reference Document on Best Available Techniques for the Non-Ferrous Metals Industries, http://eippcb.jrc.es/reference.

Ferreira, M.M.C. 2001. Polycyclic aromatic hydrocarbons: a QSPR study. *Chemosphere* 44: 125–146.

Hagen, M. 2007. Results of operating a new RTO based fume treatment system at a baking furnace. In: M. Sorlie (ed.), *Light Metals 2007 Volume 4: Electrode Technology Symposium (formerly Carbon Technology)*: 977–980. Warrendale: TMS.

Lee, W.J., Chao, W.H., Shif, M., Tsai, C.H., Chen, T.J.H. & Tsai, P.J. 2004. Emissions of polycyclic aromatic hydrocarbons from batch hot mix asphalt plants. *Environmental Science & Technology* 38(20): 5274–5280.

Li, C.T., Mi, H.H., Lee, W.J., You, W.C. & Wang, Y.F. 1999. PAH emission from the industrial boilers. *Journal of Hazardous Materials* A69: 1–11.

Mazur, M., Oleniacz, R., Bogacki, M. & Szczygłowski, P., 2003, Ocena funkcjonowania instalacji oczyszczania gazów odlotowych z procesu grafityzacji elektrod węglowych w SGL Carbon S.A. w Nowym Sączu. In J. Konieczyński & R. Zarzycki (eds), *Problemy ochrony powietrza w aglomeracjach miejsko-przemysłowych*: 141–150. Łódź - Gliwice: PAN.

Mazur, M., Oleniacz, R., Bogacki, M. & Szczygłowski, P. 2006a. Emisja zanieczyszczeń z procesu grafityzacji elektrod węglowych w piecach LWG (Castnera). Cz. 2. Wybrane substancje pyłowe. *Inżynieria Środowiska* 11(1): 27–38. Kraków: AGH-UST.

Mazur, M., Oleniacz, R., Bogacki, M. & Szczygłowski, P. 2006b. Emisja zanieczyszczeń z pieca Achesona do grafityzacji wyrobów drobnych. *Inżynieria Środowiska* 11(2): 145–159. Kraków: AGH-UST.

Mazur, M., Bogacki, M., Oleniacz, R. & Szczygłowski, P. 2008. Air pollutant emissions from process of mixing materials used for manufacturing small products from carbon and graphite. *Environment Protection Engineering* 34(4): 119–127.

Pacyna, J.M., Breivika, K., Münch, J. & Fudala, J. 2003. European atmospheric emissions of selected persistent organic pollutants, 1970–1995. *Atmospheric Environment* 37 (Supplement no. 1): S119–S131.

Ravindraa, K., Sokhia, R. & van Grieken, R. 2008. Atmospheric polycyclic aromatic hydrocarbons: Source attribution, emission factors and regulation. *Atmospheric Environment* 42: 2895–2921.

Tsai, P.J., Shieh, H.Y., Hsieh, L.T. & Lee, W.J. 2001. The fate of PAHs in the carbon black manufacturing process. *Atmospheric Environment* 35: 3495–3501.

UNECE 1998. Convention on long-range transboundary air pollution: The 1998 Aarhus Protocol on Persistent Organic Pollutants, http://www.unece.org/env/lrtap/pops_h1.htm.

Xu, S., Liu, W. & Tao, S. 2006. Emission of Polycyclic Aromatic Hydrocarbons in China. *Environmental Science & Technology* 40(3): 702–708.

Yang, H.H., Lai, S.O., Hsieh, L.T., Hsueh, H.J. & Chi, T.W. 2002. Profiles of PAH emission from steel and iron industries. *Chemosphere* 48: 1061–1074.

Zhang, Y. & Tao, S. 2009. Global atmospheric emission inventory of polycyclic aromatic hydrocarbons (PAHs) for 2004. *Atmospheric Environment* 43: 812–819.

Environmental Engineering III – Pawłowski, Dudzińska & Pawłowski (eds)
© 2010 Taylor & Francis Group, London, ISBN 978-0-415-54882-3

Mutagenic properties of PM_{10} and $PM_{2.5}$ air pollution in Wroclaw (Poland)

K. Piekarska

Institute of Environment Protection Engineering, Wroclaw University of Technology, Wroclaw, Poland

ABSTRACT: Particulate matter below $10\,\mu m$ (PM_{10}) and $2.5\,\mu m$ ($PM_{2.5}$) in size were collected during summer and autumn using a high performance Staplex air aspirator, and dichloromethane-extracted in a Soxhlet apparatus. Two *Salmonella typhimurium* strains, TA98 and YG1041, were used in assays in two experiment types: with and without metabolic activation. There were different levels of mutagens in PM_{10} and $PM_{2.5}$ fractions. Mutagenicity of $PM_{2.5}$ was higher than of PM_{10}. The highest mutagenic ratios were in autumn-collected samples. There were both types of pollutants in samples, with indirect and direct effects on genetic material. The YG1041 strain was highly sensitive to mutagenicity of nitro-aromatic compounds in extracts.

Keywords: Air pollutants, suspended PM_{10} and $PM_{2.5}$, mutagenicity, *Salmonella* assay, polycyclic aromatic compounds (PAH), nitro-PAH.

1 INTRODUCTION

Air pollution is a very complex mixture of gases and particles with condensed organic matter. So far, >500 compounds in ambient air have been identified as mutagenic (Claxton et al. 2004). Coke plants, gas works, steelworks, petroleum refineries, plants using fossil fuels, coal tar, asphalt, as well as chemical works are among substantial sources of air pollutant emission. Household coal-fired stoves and road transport also make considerable contributions to total emissions (De Kok et al. 2006). The mutagenic and carcinogenic effects of polycyclic aromatic hydrocarbons (PAH) are widely known. PAH air concentration results from numerous factors, such as: quantity of particulate fall-out generated by emissions from industrial plants, heating techniques, intensity of road transport, implemented town planning solutions that would facilitate or impede air change, meteorological and climatic conditions (Nisbet & LaGoy 1992). A major source of PAH content in the atmosphere is from incomplete combustion of organic compounds. Fuel combustion and chemical reactions that proceed in atmospheric air and involve organic pollutants also lead to formation of other genotoxic compounds: polar aromatic compounds, heterocyclic compounds and phenols. Nitro- and amino-PAH derivatives are also considered highly mutagenic compounds (Bamford et al. 2003).

Particulate matter (PM) is a mixture of small solid particles suspended in air. PM can be in the form of submicroscopic aerosols, and even visible dust particles. Depending on particle size, PM is classified as PM_{10} (particles of diameter $<10\,\mu m$), PM_5 (particles $<5\,\mu m$), $PM_{2.5}$ ($<2.5\,\mu m$), and PM_1 ($<1\,\mu m$) (Škarek et al., 2007). The health effects of PM in humans depend on both size and concentration of particulates.

Biological effects of respirable particles ($<10\,\mu m$) are significant, as their relatively long half-life periods in atmospheric air mean that they are more likely to penetrate into human respiratory systems. Because of the large active surface of lungs, the initial resorption of chemicals adsorbed on dust particles proceeds intensively in pulmonary alveoli. This thin and delicate tissue is susceptible to absorption of mutagenic substances and the resulting harm. Pollutants deposited on the surface of alveoli can penetrate directly into the circulatory system and cause other tissues to be exposed to activity of mutagenic compounds (Cohen 2000, Brunekreef & Holgate 2002, Risom et al. 2005).

Total analysis of air pollution is rather impracticable, considering the complexity of pollutant composition and interactions between the individual pollutants. Therefore chemical analysis should not be used for prognosis of biological effects from air pollution in humans and animals; and similarly for the mutagenic activities thereof. Thus, besides analytical methods, implementation of bioindicative assays in ambient air quality control is necessary (Claxton et al. 2004, Claxton & Woodall 2007).

The purpose of this work was to compare the mutagenicity of organic pollutants adsorbed on suspended PM_{10} and $PM_{2.5}$ particulates from samples collected over the Wrocław city area during summer and autumn of 2007. A *Salmonella* plate assay (Ames assay) was used in the investigations for its high prognostic usefulness for mutagenic and potentially carcinogenic factors. It was recognised as the first short-term assay within the methods used in genetic toxicology (Mortelmans & Zeiger 2000), and is the most commonly used bioassay in studies on mutagenicity of particulate air pollution (Claxton et al. 2004, Claxton & Woodall 2007). The assay consists in checking whether the investigated material causes reverse mutation in special histidine-dependent (his⁻) strains of *Salmonella typhimurium* LT2 bacteria. The prognosed value from the *Salmonella* assay for potentially carcinogenic chemicals is near 90% (Maron & Ames 1983).

2 MATERIALS AND METHODS

The material used in investigations was PM samples collected using a high performance PM_{10} and $PM_{2.5}$ Staplex air aspirator. The air was aspired on sintered-glass TFAGF810 filters (size 20 cm × 25 cm) at a rate of 71.7–82.08 m^3/h. The aspirator filters were replaced every 24 h. The sampling site was on the roof of a two-storey building on the campus of the Wrocław University of Technology (at the corner of the Norwida and Wybrzeże Wyspiańskiego Streets).

The filters were weighed immediately after sampling to determine PM mass, and kept at −18°C until extraction. The particulates collected on eight filters were combined into one pool, and put into a Soxhlet extractor. Then extraction by dichloromethane was performed in darkness for 8 h plus a 15-min reflux. The obtained extract was divided into two portions, for chemical analyses and biological assays. Then the extracts were thickened until dry in a vacuum evaporator, and finally blown through with nitrogen (Lenicek et al. 2000, Binková et al. 2003).

PAH content was determined by high performance liquid chromatography using fluorescence detection, whereas the nitro-PAH content was by gas chromatography using mass detection (Bamford et al. 2003, Zaciera 2006).

The dry particulate extract used for biological examinations was dissolved in dimethyl sulfoxide (DMSO) so that 1 cm^3 of stock solution contained pollutants from 1000 m^3 of atmospheric air. A bacterial test in *Salmonella* (Ames assay) carried out according to Maron and Ames (1983) was the basis for determination of the PM_{10} and $PM_{2.5}$ fractions. Two test strains: *S. typhimurium* TA98 and its derivative strain YG1041 were used. The TA98 strain is useful in detection of mutagens of the reading frame-shift type. The *S. typhimurium* YG1041 strain shows increased sensitivity to nitro-, amino- and hydroxylamino-PAH derivatives, due to the presence of plasmids in its cells, which relate to overproduction of nitroreductase and O-acetyltransferase (Table 1) (Watanabe et al. 1989, Hagiwara et al. 1993). *Salmonella* test strains were obtained from Dr. T. Nohmi, Division of Genetics and Mutagenesis, National Institute of Hygienic Sciences, Tokyo, Japan.

Experiments were carried out in the presence of a microsomal S9 fraction obtained from rat liver and activated with Aroclor 1254, as well as in the absence of the fraction. The S9 fraction was used in the assay for metabolic activation of promutagens. Protein content in the microsomal fraction (determined by Lowry's method) was 64.44 mg/cm^3. S9 content in the S9-mix used in experiments was 4% (v/v).

Table 1. *Salmonella typhimurium* strains used.

Strain	Description
TA 98	TA 1538 his D3052 (pKM101)
YG 1041	TA 98 (pYG233): a nitroreductase and O-acetyltransferase-overproducing strain

Table 2. Collection data of study samples.

Type of sample	Time of collection (h)	Capacity of air sample (m^3)	Mass of pollutants ($\mu g/m^3$)
PM10 Summer	197.11	13364.058	53.00
PM2.5 Summer	238.76	20772.120	31.20
PM10 Autumn	239.31	16225.218	92.39
PM2.5 Autumn	244.03	21230.610	66.24

Initially obtained particulate extracts (1000 m^3/ 1 cm^3) were dissolved in DMSO and used in experiments such that the quantity of extract per plate was: 50, 25, 12.5, 6.25, 3.125, 1.56, 0.78, 0.39, 0.195, 0.098, 0.049 and 0.024 m^3 of the tested air. All analyses were repeated five times. Cultures were incubated for 48 h (TA98) or 72 h (YG1041) at 37°C. After that time the number of (his⁺)-revertant colonies growing on Petri dishes was determined. Before starting each experiment, genetic markers of the test strains and their degree of spontaneous reversion were checked (negative check), as well as the test strain sensitivity to diagnostic mutagens (positive check – without S9: 0.2 μg of 2,4,7-trinitro-9-fluorenone per plate for the TA 98 strain, and 50 μg of 2,6-dinitrotoluene per plate for the YG 1041 strain; with S9 addition: 5 μg of 2-aminofluorene per plate for both test strains). The test results were presented as a mutagenic ratio (MR), a quotient of the mean number of revertants grown in the presence of the tested sample and the mean number of spontaneous revertants. Samples were considered mutagenic if MR ≥2.

3 RESULTS AND DISCUSSION

PM_{10} and $PM_{2.5}$ samples were collected in summer and autumn of 2007 in Wrocław city (Table 2). The amount of PM collected in autumn was twice that collected from the same test point in summer. The ratio of PM_{10} to $PM_{2.5}$ was about 1.5. Concentrations of collected PM in air ranged from 31.20 μg/m^3 ($PM_{2.5}$, in summer) to 92.39 μg/m^3 (PM_{10}, in autumn). In most cases the concentrations were below the acceptable concentration limits of US EPA regulations for airborne PM in a 24-h period (Juda-Rezler 2000), i.e. 150 μg/m^3 for PM_{10} and 65 μg/m^3 for PM2.5.

Distribution of annual average Total Suspended Particulate values in Europe is evidence of high health

Table 3. PAH concentration in organic extract of air pollutants (ng/m^3).

Examined PAH	PM_{10} Summer	$PM_{2.5}$ Summer	PM_{10} Autumn	$PM_{2.5}$ Autumn
Phenanthrene	0.183	0.035	0.700	0.435
Anthracene	0.165	0.042	0.334	0.390
Fluoranthene	0.506	0.086	1.149	1.850
Pyrene	0.340	0.071	1.350	2.100
Benzo[a]anthracene	0.214	0.029	2.200	2.900
Chrysene	0.162	0.048	1.700	1.450
Benzo[b]fluoranthene	0.489	0.145	8.650	8.550
Benzo[k]fluoranthene	0.232	0.043	2.400	2.250
Benzo[a]pyrene	1.049	0.079	9.400	10.000
Dibenzo[a,h]anthracene	0.262	0.072	2.700	2.250
Benzo[g,h,i]perylene	0.476	0.076	6.250	1.800
Indeno[1,2,3-c,d]pyrene	0.464	0.104	8.500	8.000
TOTAL	4.542	0.83	45.333	41.975

hazard to people in central–south Europe (Gilli et al. 2007). Concentrations of suspended PM collected in summer and autumn over the Wrocław area were similar to those measured in spring–summer and autumn–winter in various European cities. For example, PM_{10} and $PM_{2.5}$ autumn–winter concentrations in downtown Parma, Italy (Buschini et al. 2001) had ranges 43.81–75.77 and 33.04–54.76 $\mu g/m^3$, respectively. In Antwerp (Du Four et al., 2004), PM_{10} concentrations were 25.4–45.3 $\mu g/m^3$ in winter, and 31.4–41.6 $\mu g/m^3$ in summer. In such towns as Prague (Czech Republic), Košice (Slovak Republic) and Sofia (Bulgaria) (Gábelová et al. 2004, 2007) the summer and winter PM_{10} concentrations were 24.30–36.91 and 57.99–89.88 $\mu g/m^3$, respectively. PM_{10} concentrations in previous measurements in Wrocław (Piekarska & Karpińska-Smulikowska 2007a, 2007b) were 38.33–87.32 $\mu g/m^3$. Higher PM_{10} concentrations were noted in Upper Silesia (Motykiewicz et al. 1991, Chorąży et al. 1994), where in summer the range was 22.5–101.9 $\mu g/m^3$ and the $PM_{2.5}$ was comparable to observations in Italy; $PM_{2.5}$ concentrations in six Italian towns (Cassoni et al. 2004) had range 10.45–60.38 $\mu g/m^3$. Gilli et al. (2007) found concentrations in the Turin area were 20–60 $\mu g/m^3$.

In particulate extracts, concentrations were determined of 12 PAHs and eight nitro-PAH derivatives, among the most frequent of atmospheric air pollutants.

The total PAH was 4.542 ng/m^3 in the PM_{10} summer-sample and 45.333 ng/m^3 in autumn; in the $PM_{2.5}$ sample there was, respectively, 0.83 and 41.975 ng/m^3 (Table 3). The total concentration of PAH in the summer-collected PM_{10} extract was 5.5 times that of the $PM_{2.5}$ extract, whereas for PM collected in autumn the total PAH concentrations in both PM_{10} and $PM_{2.5}$ were comparable.

Among PAHs present in the extracts, three (benzo[a]anthracene, benzo[a]pyrene, and dibenzo [a,h]anthracene) were classified by IARC in 2A group as hydrocarbons probably carcinogenic to humans. Another three (beno[b]fluoranthene, benzo[k] fluoranthene], and indeno[1,2,3-c,d]pyrene) were

classified in 2B group, as possibly carcinogenic to humans (Nisbet & LaGoy 1992). Polyaromatic hydrocarbons with 3–6 condensed rings, and adsorbed on particulates, particularly show carcinogenic activity (De Kok et al. 2006). The concentrations of individual PAHs were within limits reported in the literature, very close to those in urban areas of Belgium, the Czech Republic and Italy (Černa et al. 2000, Buschini et al. 2001, Binková et al. 2003, Brits et al. 2004, Du Four et al. 2005). Organic extracts contained high percentages of benzo[a]pyrene, concentrations were 1.049–9.4 ng/m^3 (for PM_{10}) and 0.079–10.0 ng/m^3 (for $PM_{2.5}$). Benzo[a]pyrene is an indicator compound, to which carcinogenic activities of other PAHs are compared. A relative carcinogenicity index (k) for benzo[a]pyrene (B[a]P) is assumed to be 1 (Nisbet & LaGoy 1992). In Poland the B[a]P 1-h-reference value is 12 ng/m^3, while the relevant value for a calendar year is 1 ng/m^3 (Directive of the Ministry of Environment Protection, 2002).

There are usually high concentrations of benzo[a] pyrene in autumn–winter (Claxton et al. 2004, Claxton & Woodall 2007). The tested samples also contained, at lower concentrations, dibenzo[a,h] anthracene – a compound of higher carcinogenic potential (k = 5), and benzo[g,h,i]perilene – a compound of lower carcinogenic potential (k = 0.01). The latter compound is an indicator of exposure to aromatic hydrocarbons emitted from exhaust of diesel engines (Nisbet & LaGoy 1992, Claxton et al. 2004).

No tested dinitro-PAHs were found in the extracts, within their determination limits. However, five mononitro-PAH derivatives were noted, among others 2-nitrofluorene and 1-nitropyrene, both classified in the 2B group (possibly carcinogenic to humans). The compound 3-nitrofluoranthene was also present in the extracts. Both 1-nitropyrene and 3-nitrofluoranthene are typical emissions from diesel engines. Those compounds are not observed in any other reactions proceeding in a gaseous phase (Bamford et al. 2003). The total concentration of mononitro-PAHs (Table 4) in the summer-collected PM_{10} sample was about 1.5 times

Table 4. Nitro-PAH concentration in organic extract of air pollutants (ng/m^3).

Examined nitro-PAH	PM$_{10}$ Summer	PM$_{2.5}$ Summer	PM$_{10}$ Autumn	PM$_{2.5}$ Autumn
1-nitronaphthalene	0.024	0.017	0.020	0.020
2-nitrofluorene	0.005	0.002	0.026	0.039
9-nitroanthracene	0.044	0.030	0.084	0.158
3-nitrofluoranthene	0.043	0.029	0.047	0.278
1-nitropyrene	0.021	0.014	0.020	0.046
1,3-dinitropyrene	n.d.	n.d.	n.d.	n.d.
1,6-dinitropyrene	n.d.	n.d.	n.d.	n.d.
1,8-dinitropyrene	n.d.	n.d.	n.d.	n.d.
Total	0.137	0.092	0.197	0.541

n.d. – not detected.

that in the PM$_{2.5}$. However, for samples collected in autumn, the total concentration of mononitro-PAHs adsorbed on suspended particles of the PM$_{2.5}$ fraction was 2.7 times that adsorbed on PM$_{10}$ particulates. The nitro-PAH concentrations in organic extracts of air pollutants are usually considerably lower than the PAH content. Such compounds are highly stable in a solid phase; however, some show higher mutagenicity (2×10^5 times) and carcinogenicity (10 times) than PAHs (Bamford et al. 2003, Zaciera 2006).

Dichloromethane PM$_{10}$ and PM$_{2.5}$ extracts were mutagenic to *S. typhimurium* TA98 and YG1041 strains used in experiments (Tables 5, 6). In all assays the highest MR were for samples collected in autumn. The same seasonality has been observed in other countries. Mutagenic power of air-polluting particulates

Table 5. MR values for summer extracts of particulate pollutants, determined using TA98 and YG1041 strains in the absence (−F) and presence of S9 (+F).

Sample concentration (m^3/plate)	TA98 PM$_{10}$ −F	+F	PM$_{2.5}$ −F	+F	YG1041 PM$_{10}$ −F	+F	PM$_{2.5}$ −F	+F
50	3.97	2.97	4.48	3.67	4.31	3.69	0.86	5.90
25	2.10	2.54	3.17	1.98	4.99	3.86	0.98	4.70
12.5	1.82	2.10	2.17	1.26	4.54	4.03	4.98	4.00
6.25	1.69	1.50	1.87	1.22	3.89	2.54	4.15	3.10
3.125	1.54	1.42	1.43	1.14	3.81	1.44	3.26	2.50
1.56	1.26	1.34	1.23	0.95	2.93	1.07	2.57	2.20
0.78	1.12	1.24	1.13	1.00	1.79	0.93	1.83	1.96
0.39	1.09	1.21	1.00	1.09	1.49	0.85	1.58	1.87
0.195	0.98	1.16	0.98	1.14	1.29	1.05	1.57	1.64
0.098	1.02	1.10	0.91	0.86	1.03	0.99	1.18	1.38

Table 6. MR values for autumn extracts of particulate pollutants, determined using TA98 and YG1041 strains in the absence (−F) and presence of S9 (+F).

Sample concentration (m^3/plate)	TA98 PM$_{10}$ −F	+F	PM$_{2.5}$ −F	+F	YG1041 PM$_{10}$ −F	+F	PM$_{2.5}$ −F	+F
50	17.80	7.47	20.98	15.60	2.96	1.22	1.12	1.75
25	10.67	6.73	18.87	12.78	3.96	1.98	2.10	1.94
12.5	8.18	5.56	14.02	9.87	5.52	5.06	2.90	2.40
6.25	5.40	3.03	10.71	9.10	6.61	5.71	3.59	3.61
3.125	4.97	2.99	8.56	6.84	6.21	4.58	5.04	3.86
1.56	3.02	2.71	6.60	6.04	5.36	2.45	4.49	3.02
0.78	2.70	2.05	5.56	5.23	4.40	1.72	4.15	2.29
0.39	1.78	1.56	4.50	4.44	2.68	1.24	3.04	1.18
0.195	1.67	1.49	4.31	3.99	1.74	1.12	2.46	1.10
0.098	1.43	1.43	3.90	2.01	1.57	0.91	1.78	1.05
0.049	1.23	1.37	2.24	1.84	1.16	1.02	1.52	0.99
0.024	1.02	1.19	1.57	1.23	0.98	0.99	1.15	1.10

increases in a heating season, and decreases in summer (Claxton et al. 2004, Claxton & Woodall 2007).

High MR in assays conducted with and without metabolic activation, were evidence that there were pollutants able to indirectly (promutagens) as well as directly (direct mutagens) affect the genetic material in tested samples. Indirect impact of pollution on genetic material is linked to the presence of non-substituted PAHs, whereas the direct impact is linked with nitro-, amino- and oxy-PAH derivatives (Binková et al. 2003, Claxton et al. 2004).

Dichloromethane PM_{10} and $PM_{2.5}$ extracts were mutagenic to *S. typhimurium* TA98 and YG1041 strains used in experiments (Tables 5 and 6). In all assays the highest MR were for samples collected in autumn. The same seasonality has been observed in other countries. Mutagenic power of air-polluting particulates increases in a heating season, and decreases in summer (Claxton et al. 2004, Claxton & Woodall 2007).

High MR in assays conducted with and without metabolic activation, were evidence that there were pollutants able to indirectly (promutagens) as well as directly (direct mutagens) affect the genetic material in tested samples. Indirect impact of pollution on genetic material is linked to the presence of non-substituted PAHs, whereas the direct impact is linked with nitro-, amino- and oxy-PAH derivatives (Binková et al. 2003, Claxton et al. 2004).

In assays using the TA98 strain, the highest MR were for PM_{10} and $PM_{2.5}$ extracts from autumn-collected samples (Table 6). In assays using this strain, all tested extracts evoked greater response when there was no applied metabolic activation by S9 fraction (prevalence of direct-acting mutagens). The smallest air volumes able to produce mutagenic effects in the TA98 strain were in the presence of pollutants from PM2.5, collected in autumn. The volumes were: $0.049\,m^3$ for assays without the S9 fraction, and $0.098\,m^3$ with the S9 fraction. The lowest PM_{10} extract concentration that produced mutagenic effects in assays, with or without metabolic activation, was from $0.78\,m^3$ of air. To evoke mutagenic effects by extracts of summer-collected particulates, pollutants were derived from greater volumes of tested air. For the PM2.5 fraction, the mutagenic effect in assays with metabolic activation required pollutants derived from $50\,m^3$ of air.

In assays using the YG1041 strain (sensitive to nitro-aromatic compounds), similar revertant numbers were obtained, both in the presence and absence of the S9 fraction. Within the examined range of air pollution concentrations, there was a clear dose–response relationship for this strain, fully explaining the biological effect of the chemical compounds in the samples versus their concentrations. The lowest air volume to produce mutagenic effects in the YG1041 strain was for the PM2.5 fraction. For PM collected in autumn, for assays without metabolic activation, the volume was $0.195\,m^3$. For the PM_{10} fraction, there was a mutagenic effect in assays without S9 addition for pollutants derived from $0.39\,m^3$ of air, sampled in autumn.

Although the tested samples differed in total content and percentage share of individual PAHs and nitro-PAHs, their extracts produced mutagenic effects in both strains. Mutagens in the PM2.5 fractions of airborne particulates evoked greater response from both test strains; the air volume that produced mutagenic effects was also smaller. As previously mentioned, the size of airborne particulates is very important for biological effects, including mutagenic effects. The airborne particles initially enter the body through the respiratory tract. Many studies have confirmed that mutagenicity increases as the aerodynamic diameter of suspended particles decreases. Thus, the PM2.5 fraction contains more adsorbed organic compounds than the PM_{10} fraction. This is due to the finer particles having a larger surface area and so can adsorb more organic pollutants in relation to mass of airborne particulates (Brunekreef & Holgate 2002, De Kok et al. 2006).

The studies confirmed the high sensitivity of the YG1041 strain to mutagenic compounds in atmospheric air (Claxton et al. 2004, Claxton & Woodall 2007). Assays with the YG1041 strain confirmed the presence of mutagens in small volumes of sampled air, and provided evidence for occurrence of nitroaromatic compounds in tested extracts. Most studies on the sensitivity of the YG1041 strain to detect nitroaromatic compounds indicate that compounds of moderate–high polar classes in particular are responsible for the mutagenic effects of suspended particles. In more polar fractions the concentrations of nitro-PAH derivatives are usually higher than the concentrations of non-substituted PAHs (Bamford et al. 2003, Claxton et al. 2004, Zaciera 2006).

Therefore, using the YG1041 strain enabled detection of nitro-PAH derivatives in extracts.

Usefulness of the YG1041 strain in air pollution monitoring was reported in studies in North Bohemia (Černa et al. 2000), Prague (Binková et al. 2003), Upper Silesia (Motykiewicz et al. 1991) and Wrocław city (Piekarska & Karpińska-Smulikowska 2007a, Smulikowska 2007b).

Atmospheric air pollution indicators used in standard monitoring (determination of suspended particulate concentration and concentrations of EPA-listed PAHs) give only approximate information on health hazards. Atmospheric air pollution monitoring should be supplemented with testing the mutagenic activity of organic pollutants by the *Salmonella* assay.

4 CONCLUSIONS

1. The *Salmonella* assay provided evidence for mutagenicity of organic air pollutants adsorbed on suspended particles of the PM_{10} and $PM_{2.5}$ fractions, sampled in summer and autumn in Wrocław city.
2. PAHs and their nitro derivatives in tested extracts, even at low concentrations, were jointly responsible for mutagenic activity of tested fractions of airborne particulates.

3. Mutagens in extracts of the PM2.5 fraction evoked greater responses from the tested strains.
4. The studies confirmed the high sensitivity of the YG1041 strain to mutagenicity of aromatic compounds adsorbed on airborne particulates.
5. MR from assays with and without S9 addition showed that pollutants in tested samples affected genetic material indirectly or directly.
6. The *Salmonella* assay is a reliable and effective test, suitable for monitoring atmospheric air quality.

ACKNOWLEDGEMENTS

The project was carried out within the N305 096 31/3476 grant awarded by the Polish Ministry of Science and Higher Education

REFERENCES

Bamford, H.A., Bezabeh, D.Z., Schantz, M.M., Wise, S.A. & Baker, J.E. 2003. Determination and comparison of nitrated-polycyclic aromatic hydrocarbons measured in air and diesel particulate reference materials. *Chemosphere* 50: 575–587.

Binková, B., Černa, B., Pastorková, A., Jelìnek, R., Beneš, I., Novák, J. & Šrám, R.J. 2003. Biological activities of organic compounds adsorbed onto ambient air particles: comparison between the cities of Teplice and Prague during the summer winter seasons 2000–2001. *Mutation Research* 525: 43–59.

Brits, E., Schoeters, G. & Verschaeve, L. 2004. Genotoxicity of PM 10 and extracted organic collected in an industrial, urban and rural area in Flanders, Belgium. *Environmental Research* 96: 109–118.

Brunekreef, B. & Holgate, S.T. 2002. Air pollution and health. *Lancet* 360: 1233–1242.

Buschini, A., Cassoni, F., Anceschi, E., Pasini, L., Poli, P. & Rossi, C. 2001. Urban airborne particulate: genotoxicity evaluation of different size fractions by mutagenesis tests on microorganisms and comet assay. *Chemosphere* 44: 1723–1736.

Cassoni, F., Bocchi, C., Martino, A., Pinto, G., Fontana, F. & Buschini, A. 2004. The Salmonella mutagenicity of urban airborne particulate matter (PM 2,5) from eight sites of the Emilia- Romagna regional monitoring network (Italy). *Science of the Total Environment* 324: 79–90.

Černa, M., Pochamanova, D., Pastorková, Beneš, I., Leniček, J., Topinka, J. & Binková, B. 2000. Genotoxicity of urban air pollutants in Czech Republic. Part I. Bacterial mutagenic potencies of organic compounds adsorbed on PM 10 particulates. *Mutation Research* 469: 71–82.

Chorazy, M., Szeliga, J., Strozyk, M., Cimander, B. 1994. Ambient air pollutants in Upper Silesia: partial chemical composition and biological activity. *Environmental Health Perspectives* 102. Supplement 4: 61–66.

Claxton, L.D., Matthews, P.P. & Warren, S.H. 2004. The genotoxicity of ambient outdoor air, a review: Salmonella mutagenicity. *Mutation Research* 567: 347–399.

Claxton, L.D. & Woodall, G.M. JR. 2007. A review of the mutagenicity and rodent carcinogenicity of ambient air. *Mutation Research* 636: 36–94.

Cohen, A.J. 2000. Outdoor air pollution and lung cancer. *Environmental Health Perspectives* 108(4): 743–750.

De Kok, T.M.C.M., Driece, H.A.L., Hogervorst, J.G.F. & Briede, J.J. 2006. Toxicological assessment of ambient and traffic-related particulate matter: A review of recent studies. *Mutation Research: Reviews in Mutation Research* 613(2–3): 103–122.

Ministry Of Environment Protection (Poland). 2002. Directive of the Ministry of Environment Protection; in: *December 5th 2002 year in matter of value of reference for some substances in air (Dz. U. z 2003 r. Nr 1, poz. 12)*. Ministry of Environment Protection (Poland).

Du Four, V.A., Van Larebeke, N. & Janssen, C.R. 2004. Genotoxic and mutagenic activity of environmental air samples in Flanders, Belgium. *Mutation Research* 558: 155–167.

Du Four, V.A., Janssen, C.R., Brits, E. & Larebeke, N.V. 2005. Genotoxic and mutagenic activity of environmental air samples from different rural, urban and industrial sites in Flanders, Belgium. *Mutation research – Genetic toxicology and environmental mutagenesis* 588(2): 106–117.

Gábelová, A., Valovicová, Z., Bacová, G., Lábaj, J., Binková, B., Topinka, J., Sevastyanova, O, Šrám, R.J., Kalina, I., Habalová, V., Popov, T.A., Panev, T. & Farmer, P.B. 2007, Sensitivity of different endpoints for in vitro measurement of genotoxicity of extractable organic matter associated with ambient airborne particles (PM$_{10}$). *Mutation Research* 620: 103–13.

Gábelová, A., Valovicová, Z., Horváthová, E., Slameòová, D., Binkova, B., Šrám, R.J., Farmer, P.B. 2004. Genotoxicity of environmental air pollution in three European cities: Prague, Košice and Sofia. *Mutation Research* 563: 49–59.

Gilli, G., Pignata, C., Schiliro, T., Bono, R., La Rosa, A. & Traversi, D. 2007. The mutagenic hazards of environmental PM2.5 in Turin. *Environmental Research* 103(2): 168–175.

Hagiwara, Y., Watanabe, M., Oda, Y., Sofuni, T. & Nohmi, T. 1993. Specificity and sensitivity of Salmonella typhimurium YG1041 and YG1042 strains possessing elevated levels of both nitroreductase and acetyltransferase activity. *Mutation Research* 291: 171–180.

Juda-Rezler, K. 2000. *Environmental impact of air pollution*. Oficyna Wydawnicza Politechniki Warszawskiej: Warszawa.

Lenicek, J., Sekyra, M., Badnarkova, K., Benes, I. & Sipek, F. 2000. Fractionation and chemical analysis of urban air particulate extracts. *International Journal of Environmental Analytical Chemistry* 77(4): 269–288.

Maron, D.M. & Ames, B.N. 1983. Revised methods for the Salmonella mutagenicity test. *Mutation Research* 113: 173–215.

Mortelmans, K. & Zeiger, E. 2000. The Ames Salmonella/ microsome mutagenicity test. *Mutation Research* 455: 29–60.

Motykiewicz, G., Cimander, B., Szeliga, J., Tkocz, A. & Chorazy, M. 1991. Mutagenic activity of complex air pollutants in Silesia. In: H. Vainio, M. Sorsa, A. Mc Michael & J. Lyon (eds); *Complex Mixtures and Cancer Risk*. IARC Scientific Publications.

Nisbet, I.C.T. & Lagoy, P.K. 1992. Toxic (TETs) for polycyclic aromatic hydrocarbons (PAHs). *Regulatory Toxicology and Pharmacology* 16: 290–300.

Piekarska, K. & Karpińska-Smulikowska, J. 2007a. Seasonal variability in mutagenicity of airborne particulate pollution in Wrocław urban area. *Polish Journal of Environmental Studies* 16(3B): 408—413.

Piekarska, K. & Karpińska-Smulikowska, J. 2007b. Mutagenic activity of environmental air samples from the area of Wrocław. *Polish Journal of Environmental Studies* 16(5): 757–764.

Risom, L., Møller, P. & Loft, S. 2005. Oxidative stress-induced DNA damage by particulate air pollution. *Mutation Research* 592: 119–137.

Škarek, M., Janošek, J., Čupr, P., Kohoutek, J., Novotnă-Rychetskă, A. & Holoubek, I. 2007. Evaluation of genotoxic and non-genotoxic effects of organic air pollution using in vitro bioassays. *Environment International* 33: 859–866.

Watanabe, M., Ishidate, M. Jr. & Nohmi, T. 1989. A sensitive method for the detection of mutagenic nitroarenes: construction of nitroreductase-overproducing derivatives of *Salmonella typhimurium* strains TA 98 i TA 100. *Mutation Research* 216: 211–220.

Zacieram, M. 2006. Metoda oznaczania nitrowych pochodnych WWA w powietrzu [Nitro-PAH determination in air]. In:J. Konieczynski (ed.); *Ochrona powietrza w teorii i praktyce*. Instytut Podstaw Inżynierii Środowiska Polskiej Akademii Nauk: Zabrze.

Environmental Engineering III – Pawłowski, Dudzińska & Pawłowski (eds)
© 2010 Taylor & Francis Group, London, ISBN 978-0-415-54882-3

Concentration and elemental composition of atmospheric fine aerosol particles in Silesia Province, Poland

W. Roguła-Kozłowska, K. Klejnowski, A. Krasa & S. Szopa
Institute of Environmental Engineering, Polish Academy of Sciences, Zabrze, Poland

ABSTRACT: In Silesia Province, airborne dust concentration limits are constantly exceeded. This paper presents the results of gravimetric measurements of $PM_{2.5}$ concentrations in three cities – Katowice, Zabrze and Czestochowa – for the period 2001–2007. The elemental composition of $PM_{2.5}$ from selected diurnal samples was determined using the energy dispersive X-ray fluorescence (EDXRF) technique. Diurnal ambient concentrations and the masses of 40 selected elements in $PM_{2.5}$ were determined in Zabrze and Katowice. The results and analysis of Pearson correlations between these factors confirmed that energy production, traffic and industry are likely to be the main sources of $PM_{2.5}$ in both these cities.

Keywords: Atmospheric aerosol, $PM_{2.5}$, elemental composition, trace elements, heavy metals, X-Ray Fluorescence Spectroscopy (X-RFS), Energy Dispersive X-Ray Fluorescence Spectroscopy (EDXRF), Poland.

1 INTRODUCTION

Although the past two decades have brought improvements in atmospheric air quality in Poland, the problem of exceeding PM_{10} standards has remained unsolved (Pastuszka et al. 2003, Klejnowski 2008). In 2002, amendments to Polish environmental law according to European Union regulations divided Poland into 362 monitoring zones. Atmospheric air has been monitored for dust pollution in each zone since then. Depending on the results of yearly assessments of PM_{10} concentrations, air quality improvement programs have been implemented in most, but not all, zones. Because in the Upper Silesian and the Częstochowa agglomeration zones ambient air quality has always been poor, yearly improvement programs have been in place in these areas since 2003. Directive 2008/50/EC of the European Parliament and the directive of the Council on 21 May 2008 on ambient air quality and cleaner air for Europe, plus air quality standards for PM_{10}, introduced regulations concerning $PM_{2.5}$. It should also be emphasized that the European Community emission inventory report states that, although the estimated $PM_{2.5}$ emissions in Poland fell by 5% during 2001–2007, it is still 10% of the total $PM_{2.5}$ emissions from the 27 EU member-states (4th place after France, Spain and Italy, EEA 2009).

An unavoidable challenge preceding any attempt to lower concentrations of PM_{10}, $PM_{2.5}$ and related toxic metal compounds (Schroeder et al. 1987) and hydrocarbons (Sheu et al. 1997) in ambient air is to achieve a precise identification of their sources in the analyzed area.

Measurements at a site permit the determination of bulk concentrations of dust that arise from many sources. The contributions of specific groups of sources and shares of individual sources can be determined and estimated, respectively, by the application of statistical methods (mainly receptor models; Hopke 1991). Typical data from air monitoring systems functioning in Poland over the last 15 years cover daily concentrations of PM_{10}, but rarely concentrations of $PM_{2.5}$. Usually, these data lack information on the chemical composition of both dust fractions. The data provided by the Inspection for Environmental Protection (IEP) in Warsaw show that the monitoring of $PM_{2.5}$ dust concentrations has been carried out in Poland since 2003 only in Cracow. Since 2004, there have also been functioning $PM_{2.5}$ concentration monitoring stations in Warsaw and Łódź, and occasionally in Lublin and Slupsk, and in 2006 a $PM_{2.5}$ monitoring system was set-up in Szczecin (IEP 2009). Apart from the results of measurements in Zabrze, Katowice and Częstochowa presented in this work (financed from the Regional Fund for Environmental Protection in Katowice since 2005), up until 2007 there was no other data available concerning the concentrations and chemical composition of $PM_{2.5}$ in Silesia Province.

The goal of this work is to analyze the ambient concentrations of $PM_{2.5}$ and selected $PM_{2.5}$-related elements, as well as the relationships between the daily contributions of these elements to $PM_{2.5}$, based on data collected during the last 7 years in Silesia Province. From an environmental point of view, the area is one of the most interesting sites in Europe. Industry has been developing here since the Middle Ages and manifests its past and present influence on the environment in many ways. An increasing traffic load, a well-developed yet traffic-jam-prone road network (especially in the city centres) and a central

Figure 1. Location of measuring sites.

heating network (which, for the most part, is based on the combustion of hard coal) contribute to high concentrations of ambient dust across the region.

2 MATERIALS AND METHODS

In the second half of 2001, from 2002 to 2003 and from January to October 2004, diurnal samples of dust were taken once every 3 days. In the period 2005–2007, diurnal samples of dust were taken daily (apart from a 3-month maintenance break in 2006). During the whole of 2005, the sampler worked in Czestochowa, every even month of 2007 in Katowice, and during the rest of the time in Zabrze (Figure 1).

Katowice and Zabrze are big cities – the former in the central and the latter in the western part of the Upper Silesian agglomeration. Częstochowa lies in the northern part of Silesia Province. In all three cities, the measuring points were located at sites with representative urban backgrounds (according to the definition from Directive 2008/50/EC – sites in urban areas where levels of pollutants are representative of exposure of the urban population to pollutants). Each measuring point was representative for an area of a few square kilometres and was located in such a way that dust concentrations could be influenced by emissions from all sources upwind of the measuring station. The estates surrounding the measuring points are blocks of flats and houses heated from the central network or by combusting hard coal. Each site lies within several hundred meters of a busy road, which might also be a source of suspended dust.

A Rupprecht & Patashnick Dichotomous Partisol® Plus model 2025 was used to sample $PM_{2.5}$. $PM_{2.5}$ was collected on Teflon filters. The sampler recorded meteorological parameters and the volume of air passing through the system. $PM_{2.5}$ concentrations were determined gravimetrically (Mettler Toledo microbalance, electric charge neutralizer). Air humidity and temperature in the weighing room were kept at about 50%

and 20°C, respectively. Filters were conditioned for 48 hours in the weighing room before weighing.

For 2002–2004, from two to four samples of $PM_{2.5}$ from each month (beginning, middle and end of a month, the same number of samples from summer and winter) were analysed chemically. In 2007, $PM_{2.5}$ was measured alternately: in odd months in Zabrze, in even months in Katowice, and six or seven samples of $PM_{2.5}$ from each month were analysed. Because all samples from 2005 and 2006 were used for other purposes, there were no samples from these years left for elemental analyses.

Trace elements in $PM_{2.5}$ were determined with the use of a Panalytical EPSILON 5 energy-dispersive X-ray fluorescence spectrometer (EDXRF). It was equipped with an X-Ray water-cooled tube with side window (gadolinium anode, voltage range 25–100 kV, 150 μm thick beryllium window), a system of nine secondary targets (Al, Ti, Fe, Ge, Zr, Mo, Ag, Ce_2O_3, Al_2O_3) and a Ge detector (resolution 140 eV, range of energy 0.7–100 keV, 30 mm^2 of work surface, 8 μm thick beryllium window).

Thin-layer single-element Micromatter standards were used to calibrate the apparatus according to the methodology of US EPA (US Environmental Protection Agency, 1999). Daily energy calibrations (performed automatically by the instrument), weekly measurements of the NIST, SRM2873 samples (recovery of each element was between 80 and 120% of the certified value) and monthly measurements of the monitor were routinely performed to control the quality of the analytical procedure (Szopa & Rogula-Kozlowska, 2008). The detection limits for the procedure were determined by using two blanks (Teflon filters). Each of the blanks underwent the entire EDXRF procedure, designed for actual samples, 30 times. The detection limit for each element was computed as a standard deviation from the 60 results received for this element by measuring the blanks (Table 1).

3 RESULTS AND DISCUSSION

Annual concentrations of $PM_{2.5}$ were similar at all three sampling sites for the period 2001–2007 (Figure 2). The statistical parameters of the yearly series of diurnal $PM_{2.5}$ concentrations from Katowice (K, 2007) and Czestochowa (C, 2005) were close to the respective parameters of the yearly series from Zabrze (Z, 2001–2004, 2006–2007).

Annual concentrations (arithmetic means from all diurnal $PM_{2.5}$ concentrations for a given calendar year) were not lower than 25 μg/m^3 – the Directive's proposed future limit for $PM_{2.5}$, which is valid from 2015 onwards (Directive 2008/50/EC). However, in most cases, the median of the yearly series of diurnal $PM_{2.5}$ concentrations were lower than the arithmetic mean, which was probably heightened by winter $PM_{2.5}$ concentrations (Figure 2).

In Poland, especially due to the temperature drop, winter $PM_{2.5}$ concentrations increase in urban areas

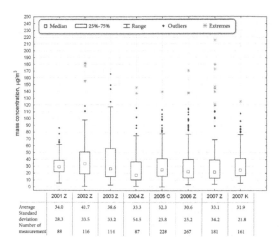

	2001 Z	2002 Z	2003 Z	2004 Z	2005 C	2006 Z	2007 Z	2007 K
Average	34.0	41.7	38.6	33.3	32.3	30.6	33.1	31.9
Standard deviation	28.3	33.5	33.2	54.5	23.8	25.2	34.2	21.8
Number of measurement	88	116	114	87	228	267	181	161

Figure 2. Comparison of ranges (minimum – maximum, percentile 25% – percentile 75%) of diurnal $PM_{2.5}$ concentrations, numbers of outliers and extremes, medians, arithmetic means, standard deviations of sets of diurnal $PM_{2.5}$ concentrations. The symbols 2001Z, 2002Z, 2003Z, 2004Z, 2006Z, 2007Z stand for Zabrze sets of data from the respective years; 2005C and 2007K denote the sets of data for Czestochowa from 2005 and Katowice from 2007, respectively.

like in other European urban areas (e.g. Artiñano et al. 2001, Manoli et al. 2002). However, the effect is more distinct in Poland due to combustion of hard coal in inefficient house furnaces.

This phenomenon was particularly well observed across Silesia Province in the winter of 2006 – in January – when two smog episodes were observed (the mean diurnal air temperature dropped below $-20°C$, there was a shallow (max. 200 m) inversion layer and mean diurnal SO_2 concentrations above $250\,\mu g/m^3$; Pastuszka et al. 2009); and in Zabrze in the urban background site (Z, 2006), the mean diurnal concentrations of $PM_{2.5}$ were as high as $170\,\mu g/m^3$ (Figure 2).

However, in other measurement periods, great differences were observed between minimum (usually mean diurnal concentrations during summer days) and maximum (during winter days) concentration values of $PM_{2.5}$ and great single outliers and extrema – the latter reaching $220\,\mu g/m^3$ (Z, 2007). Nevertheless, 75% of diurnal $PM_{2.5}$ concentrations reached $40\,\mu g/m^3$ in each year, and two of them (Z, 2002 and 2003) exceeded $50\,\mu g/m^3$. This, and the $PM_{2.5}/PM_{10}$ average for the period 2005–2007 – equal to 0.7 (Klejnowski et al. 2009) – prove that the limit for daily or annual PM_{10} concentration was exceeded due to high $PM_{2.5}$ concentrations.

Comparison of the results from this study with historical concentrations may be interesting. In the second half of the 1990s (1995–1996), the mean concentration of $PM_{2.5}$ in the three towns – Świetochłowice, Pszczyna, Kędzierzyn-Koźle – was $59.7\,\mu g/m^3$ (Houthuijs et al. 2001). The mean concentration in Zabrze, Katowice and Czestochowa in 2001–2007 was $34.5\,\mu g/m^3$ (Figure 2). Therefore, in Silesia Province, the drop in $PM_{2.5}$ concentration

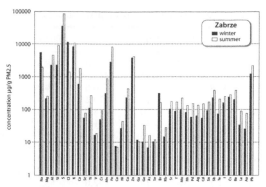

Figure 3. Comparison of winter and summer daily mass contributions of elements to $PM_{2.5}$ in Zabrze.

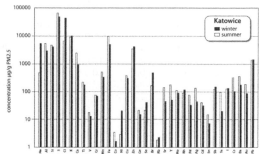

Figure 4. Comparison of winter and summer daily mass contributions of elements to $PM_{2.5}$ in Katowice (contributions of Mg, Sc, Ga, As and La are not shown because their averages equal 0 in either summer or winter).

through the decade was 42%. This drop was certainly due to reducing emissions of dust and its gas precursors by restructuring industry and closing lots of factories across the Province. It seems, therefore, that the main causes of achieving air quality that is compliant with standards in the three cities are municipal emissions – highest in winter – and, as across Europe, emissions from the transport sector (EMEP, 2009). The same problem (high municipal emissions) also applies to other regions of Poland. In the provinces, where $PM_{2.5}$ concentrations were measured at urban background sites (i.e. Pomerania, Lubelskie, Łódzkie, Małopolskie and Masovia), high diurnal concentrations have also been observed in winters. However, in these provinces at almost all measuring points, the mean annual concentrations of $PM_{2.5}$ were below $25\,\mu g/m^3$ (IEP 2009). Exceptions were concentrations measured in street canyons in Cracow and Warsaw, where the road traffic raised mean annual concentrations to above $50\,\mu g/m^3$ (IEP 2009).

These conclusions are also confirmed through an analysis of the latest data published by EMEP (2009). The data show that, in Eastern Europe, emissions of both $PM_{2.5}$ and PM_{10} from Public Electricity and Heat Production definitely make up a greater share of total dust emissions (19% to PM_{10}, 18% $PM_{2.5}$) compared with Western Europe (less then 4% to PM_{10} and

Table 1. Ambient concentrations of $PM_{2.5}$-related elements, ng/m^3.

El.[1]	DL[2]	DL[3]	2002 Z		2003 Z		2004 Z		2007 Z		2007 K	
Na	2.7	1.3	181.2[4]	0.0[6]	198.6	0.0	95.8	0.0	60.3	0.0	115.0	0.0
			171.3[5]	664.2[7]	240.1	922.4	124.1	441.5	117.3	428.9	175.2	564.2
Mg	41.8	20.8	11.0	0.0	9.4	0.0	9.3	0.0	2.3	0.0	6.7	0.0
			19.4	70.2	15.5	3.7	21.4	74.5	11.4	68.7	25.5	129.4
Al	16.6	8.3	60.2	8.0	61.6	0.0	39.0	0.0	77.0	0.0	75.2	0.0
			32.1	161.7	38.1	163.2	27.9	120.4	65.5	296.5	45.5	168.3
Si	21.3	10.6	106.3	0.0	100.7	0.0	114.1	0.0	142.3	0.0	101.2	0.0
			105.4	3723.7	138.4	650.3	164.9	818.3	186.2	713.5	131.2	435.3
S	1.5	0.7	1390.9	418.4	1744.4	236.3	1044.7	357.2	811.1	321.9	1365.0	253.3
			888.0	3916.6	1277.7	4331.5	803.4	4183.3	457	2148.8	737.6	3157.3
Cl	1.3	0.6	33.7	2.0	63.1	0.0	141.7	0.7	655.9	4.5	904.4	0.0
			40.3	166.2	107.2	362.7	366.7	1519.0	1009.2	3187.7	1156.5	4830.5
K	1.1	0.5	291.7	64.7	279.4	36.1	206.5	20.1	174.4	25.4	233.0	36.3
			188.2	809.0	256.6	979.4	211.1	988.9	188.4	1102.5	124.8	534.8
Ca	1.2	0.6	24.6	5.0	18.4	5.0	18.4	2.8	25.6	4.5	30.6	5.0
			21.2	96	14.7	78.6	18.0	84.4	21.8	109.0	20.1	100.5
Sc	0.9	0.4	2.8	1.0	0.0	0.0	0.9	0.0	0.0	0.0	0.0	0.0
			0.8	4.0	0.1	0.5	1.3	4.2	0.0	0.0	0.0	0.0
Ti	0.8	0.4	4.5	1.0	3.3	0.5	2.4	0.3	3.4	0.0	3.9	1.0
			2.1	10.0	2.2	11.4	1.5	6.2	2.6	11.9	5.4	36.8
V	0.3	0.1	0.5	0.0	0.8	0.0	0.4	0.0	0.1	0.0	0.4	0.0
			0.6	2.5	1.2	5.0	0.6	2.5	0.3	1.0	0.8	3.5
Cr	0.9	0.4	2.4	0.5	1.2	0.0	1.4	0.0	1.3	0.0	1.4	0.0
			1.6	7.0	1.2	4.7	1.4	6.3	1.9	9.0	1.4	7.0
Mn	3	1.5	18.7	5.0	13.5	0.2	11.1	0.0	14.2	0.0	8.8	0.0
			20	102.0	13.5	59.0	11.9	39.3	40.2	244.3	8.4	39.8
Fe	0.7	0.3	177.4	33.3	131.7	12.4	103.4	12.3	150.8	20.4	147.7	52.2
			196.3	760.2	115.1	455.7	89.1	321.4	292.7	1797.6	67.1	353.2
Co	0.4	0.2	0.4	0.0	0.2	0.0	0.2	0.0	0.1	0.0	0.1	0.0
			0.4	1.5	0.4	1.7	0.3	1.6	0.3	1.4	0.2	1.0
Ni	0.6	0.3	1.4	0.5	1.1	0.0	0.7	0.0	0.6	0.0	0.3	0.0
			1.0	5.0	0.9	3.7	0.7	3.1	1.8	10.4	0.4	1.7
Cu	1.4	0.7	10.7	4.0	7.9	2.5	6.4	0.2	6.0	0.0	8.1	2.5
			10.9	63.2	5.7	27.4	5.6	26.1	6.7	39.8	4.1	20.9
Zn	0.9	0.4	117.4	17.4	145.4	12.2	113.5	8.0	70.5	8.0	94.3	18.4
			87.6	365.7	151.1	492.3	129.3	562.5	62.9	328.4	53.4	242.8
Ga	1.4	0.7	0.7	0.0	0.4	0.0	0.3	0.0	0.0	0.0	0.0	0.0
			1.4	5.5	0.7	2.7	0.8	3.8	0.1	0.5	0.2	1.5
Ge	1.0	0.5	0.3	0.0	0.6	0.0	0.2	0.0	0.5	0.0	0.5	0.0
			0.6	2.5	0.8	2.5	0.4	1.3	1.0	3.5	0.8	2.5
Number of samples			30 (17 s[8], 13 w[9])		26 (16 s, 10 w)		30 (16 s, 14 w)		39 (21 s, 18 w)		43 (21 s, 22 w)	

[1]Element, [2]ng/cm^2, [3]ng/m^3 (flow rate $21.6\ m^3$), [4]Average, [5]Standard deviation, [6]Minimum, [7]Maximum), [8]Summer, [9]Winter

$PM_{2.5}$); $PM_{2.5}$ and PM_{10} emissions from the transport sector contribute significantly to PM emissions in both Western and Eastern Europe - PM_{10} by 7.2 % and 7%, and $PM_{2.5}$ by 16.5% and 8.3% (EMEP 2009).

Thirty $PM_{2.5}$ samples from Zabrze 2002, 26 samples from Zabrze 2003, 30 samples from Zabrze 2004, 39 samples from Zabrze 2007 and 43 samples from Katowice 2007 were analysed for $PM_{2.5}$ elemental composition.

Average ambient concentrations of 40 determined elements for these periods and sites are presented in Table 1.

The winter versus summer averages of diurnal mass contributions for the determined elements in $PM_{2.5}$ in Zabrze are presented in Figure 3. Figure 4 compares the averages for Katowice.

The diurnal ambient concentrations of most $PM_{2.5}$-related elements were greater than their detection limits (Table 1). Because concentrations lower than the detection limit entered the average as zeros, values lower than the detection limits appear in Table 1.

Average concentrations of crustal elements (Al, Si, Ca, Fe, Ti) were between 2.4 ± 1.5 (Z, 2004) for Ti and $177.4 \pm 196.3\ ng/m^3$ (Z, 2002) for Fe. In Zabrze and in Katowice, the summer averages of daily mass contributions to $PM_{2.5}$ were higher than the winter ones (Figures 3 and 4). In Zabrze, daily mass contributions of all five elements to $PM_{2.5}$ were pair-wise significantly correlated. The contributions of Ca and Ti were correlated most closely (Pearson correlation coefficient $R = 0.88$, $p = 0.05$). Contributions of Al and Ti ($R = 0.83$), Ca and Al ($R = 0.73$) and Ca and

78

Table 2. Ambient concentrations of PM$_{2.5}$–related elements, ng/m^3 - continued.

El.[1]	DL[2]	DL[3]	2002 Z		2003 Z		2004 Z		2007 Z		2007 K	
As	11.6	5.8	0.1[4]	0.0[6]	0.7	0.0	0.3	0.0	0.2	0.0	0.8	0.0
			0.5[5]	3.0[7]	1.6	5.5	0.9	4.7	1.2	7.5	1.2	4.5
Se	0.2	0.1	0.1	0.0	0.3	0.0	0.4	0.0	0.4	0.0	10	0.0
			0.7	3.5	0.6	2.5	1.0	4.4	0.7	3	9.1	34.3
Br	0.4	0.2	7.5	0.5	10.1	0.0	7.8	0.0	6.3	0.0	0.1	0.0
			7.9	27.4	15.4	55	12.1	60.6	7.4	29.4	0.3	1.5
Rb	0.7	0.3	0.7	0.0	0.6	0.0	0.3	0.0	0	0.0	1.6	0.0
			0.8	3	0.7	2	0.7	3.1	0.2	1.0	2.6	11.4
Sr	1.0	0.5	3.9	0.0	0.5	0.0	1.4	0.0	2.5	0.0	2.2	0.0
			2.2	9.5	0.9	3	1.9	9	3.6	17	2.4	10.4
Y	4.3	2.1	2.2	0.0	2.7	0.0	1.7	0.0	2.2	0.0	1.8	0.0
			2.3	6.5	2.5	9	2.6	13.7	2.3	7.5	1.9	6.7
Mo	4.9	2.4	5.8	0.5	1.1	0.0	2.4	0.0	1.7	0.0	1.5	0.0
			3.1	13.9	1.4	5.2	2.6	10	2.5	11.4	2.1	7.5
Rh	4.1	2	2.5	0.0	0.6	0.0	1.6	0.0	2.1	0.0	1.0	0.0
			2.2	7	1.2	4	1.5	5.2	2.6	12.7	1.6	6.0
Pd	3.9	1.9	1.8	0.0	1.2	0.0	1.2	0.0	1.5	0.0	1.7	0.0
			1.6	5	1.4	5.2	0.9	3.2	2	8.5	1.8	7.0
Ag	3.3	1.6	2.7	0.0	0.7	0.0	1.5	0.0	1.2	0.0	0.6	0.0
			1.8	7	1.1	3.7	1.3	5.2	1.9	6.5	1.2	5.5
Cd	4.5	2.2	3.9	0.0	1.4	0.0	1.9	0.0	0.7	0.0	0.3	0.0
			2.9	12.9	1.7	5.2	2.6	11	1.2	5	1.0	5.5
Sn	4.4	2.2	6.5	0.0	3.2	0.0	2.1	0.0	0.3	0.0	3.4	0.0
			3.3	11.4	2.6	9	2.5	10.5	0.8	4	3.8	12.9
Sb	5.7	2.8	10.1	3	6.4	0.0	6.2	0.0	3.4	0.0	1.4	0.0
			6.2	28.4	7.5	35.3	6.6	25.8	6.4	38.3	2.2	8.0
Te	5.1	2.5	3.3	0.0	1.3	0.0	1.3	0.0	1.4	0.0	3.4	0.0
			2.4	8.5	1.9	6.2	1.7	6.8	2.2	7	3.3	11.9
I	4.7	2.3	3.9	0.0	3.9	0.0	3.1	0.0	3.2	0.0	3.9	0.0
			3.3	11.4	4.7	16.9	3.6	16.3	3.8	17	3.3	11.4
Cs	7.4	3.7	7.1	1.5	0.6	0.0	2.8	0.0	5	0.0	4.5	0.0
			3.3	14.4	1.2	4.5	±2.3	10	5.2	29.7	3.9	17.4
Ba	7.6	3.8	8	0.0	2.5	0.0	4.6	0.0	6.7	0.0	0	0.0
			3.3	14.9	3.6	13.9	3.6	13.1	10	62.2	0.0	0.0
La	14.3	7.1	3	0.0	3.3	0.0	0.9	0.0	0	0.0	7.1	0.0
			5.8	17.9	5.7	17.9	2.8	14.2	0.0	0	5.9	20.9
Au	8.6	4.3	0.1	0.0	0.1	0.0	0.9	0.0	1.6	0.0	2.3	0.0
			0.6	3.5	0.4	2	1.5	4.7	2.4	9.9	3.4	12.4
Pb	0.3	0.1	53.5	6	49.4	3.5	34.4	0.5	23.5	0.0	35.6	3.5
			45.4	208.5	46.5	187.6	37.3	180.1	20.5	112.4	21.8	90.1
Number of samples			30 (17 s[8], 13 w[9])		26 (16 s, 10 w)		30 (16 s, 14 w)		39 (21 s, 18 w)		43 (21 s, 22 w)	

[1]Element, [2]ng/cm^2, [3]ng/m^3 (flow rate 21.6 m^3), [4]Average, [5]Standard deviation, [6]Minimum, [7]Maximum), [8]Summer, [9]Winter

Fe (R = 0.71) were also positively correlated, to a lesser extent. In Katowice, the daily mass contributions of Al and Ca to PM$_{2.5}$ were correlated most closely (R = 0.84), and the contributions of Al and Ti (R = 0.75) and Ca and Ti (R = 0.72) were correlated, to a lesser extent. The contributions of Ca and Si were not correlated significantly.

Let C_{Al}, C_{Si}, C_{Ca}, C_{Fe}, C_{Ti} be mean mass concentrations of Al, Si, Ca, Fe, Ti in ambient air, respectively. Then $C_{soil} = 2.2C_{Al} + 2.49C_{Si} + 1.63C_{Ca} + 2.42C_{Fe} + 1.94C_{Ti}$ is the mass concentration of soil dust in ambient air (Malm et al., 1994). For C_{Al}, C_{Si}, C_{Ca}, C_{Fe}, C_{Ti} being respective mean concentrations for 2002–2004 and 2007 in Zabrze or for 2007 in Katowice (Table 1), the mass concentration C_{soil} of soil dust both in Zabrze and Katowice is not greater than 0.85 μg/m^3. The mean PM$_{2.5}$ concentrations in these periods were 37 μg/m^3 in Zabrze and 32 μg/m^3 in Katowice (Figure 2) and the contribution of soil dust to PM$_{2.5}$ concentrations was not greater than 2.2% for Zabrze and 2.6% for Katowice.

Of all PM$_{2.5}$-related elements, S had the greatest ambient concentration. Its lowest yearly concentration was 811.1 ± 457 ng/m^3 (Z, 2007), and the highest was 1744.4 ± 1277.7 ng/m^3 (Z, 2003) (see Table 1). In Zabrze, the daily mass contributions of S to PM$_{2.5}$, arising from the combustion of fossil fuels (liquid in diesels and coal in furnaces) and the transformation of various compounds of sulphur in the atmosphere (Seinfeld, 1986; van Loon and Duffy, 2008; Lough et al., 2005) was primarily correlated with the contributions of K, Cu, Cd and Pb the most (R = 0.93), and also with the contributions of Cu (R = 0.91) and Zn (R = 0.84). Contributions of S were also significantly correlated with contributions of Sc, Ti, Cr, Mn, Ni, Rb, Sr, Mo, Ag and Pd. In Katowice, they were highly correlated with contributions of K ($R = 0.78$), Se ($R = 0.75$), Pb ($R = 0.74$), and also with contributions of Br and Zn ($R = 0.66$), Cd ($R = 0.64$), Cr and Cu ($R = 0.61$). There were also significant correlations between S contributions and contributions of Na, Al, Si, Cl, Ca, Ti, Mn, Fe, Sr, Mo, Rh, Sb and Ba.

Daily mass contributions of S to $PM_{2.5}$ were greater in summer than in winter in Zabrze and in Katowice (see Figures 3, 4). A high correlation between the contributions of S and K in Katowice suggests biomass burning as one of the sources of ambient $PM_{2.5}$.

Recognized in the literature as traffic-related trace elements, Mn, Ni, Zn, Pb, Cd, Sb, Cu, Sn, Pd, Rh (Lough et al. 2005, Sternbeck et al. 2002, Gomez et al. 2001, Gandhi et al. 2003, Morcelli et al. 2005, Park & Kim 2005, Birmili et al. 2006 Bocca et al. 2006, Wahlin et al. 2006) had ambient concentrations from $0.3 \pm 1.0 \, \text{ng/m}^3$ for Cd (K, 2007) through $145.4 \pm 151.1 \, \text{ng/m}^3$ for Zn (Z, 2003) (see Table 1). In Zabrze, the daily mass contributions of these elements to $PM_{2.5}$ were greater in summer than in winter (Figure 3); in Katowice, contributions of Zn, Ni, Pb, Sb and Rh to $PM_{2.5}$ were greater in winter.

The daily mass contributions of these elements to $PM_{2.5}$ were pair-wise significantly correlated in Zabrze. Significant correlations between the contributions of all these elements and the contributions of Ti and Mo confirm that road traffic is their primary source.

In Katowice, the contributions of Pd, Rh, Sn and Cd were not significantly correlated with contributions of any other traffic-related trace elements. The contributions of Mn were significantly correlated with contributions of Fe ($R = 0.81$), contributions of Mn with contributions of Zn ($R = 0.61$), and contributions of Cu with contributions of Zn ($R = 0.61$). Therefore, sources other than traffic were also responsible for the presence of these elements in $PM_{2.5}$ in Katowice. On the other hand, the contributions of all these elements, except Rh and Pd in Zabrze and Ni, Sn and Pd in Katowice, were significantly correlated with the contributions of S (i.e. they might have come from coal combustion as well).

Daily Pb mass contributions to $PM_{2.5}$ were correlated with the contributions of Zn (in Zabrze and Katowice $R = 0.87$), Cu, Cd ($R = 0.94$ for both in Zabrze), Cl and K ($R = 0.77$ and 0.82, respectively, in Katowice) the most, which may confirm that industry – zinc and lead smelters or steelworks – is a Pb, Zn, Cu and Cd source. This is more so as the Pb contributions, despite lead-content reductions in gasoline, are higher in Katowice and Zabrze than in other parts of Poland and Europe (Hueglin et al. 2005, Vallius et al. 2005, Krzeminska-Flowers et al. 2006, Salvador et al. 2007, Ragosta et al. 2008, Central Statistical Office 2009).

Daily mass contributions of alkali metals (Mg, Na, K) and halogens (Cl) to $PM_{2.5}$ are not correlated with each other significantly in Zabrze. In Katowice,

strong correlations are seen between contributions of K and Na ($R = 0.77$), Cl and Br ($R = 0.95$), Cl and Na ($R = 0.94$), Cl and K ($R = 0.8$), and Cl and Zn ($R = 0.78$).

Higher winter than summer daily mass contributions of Cl and Na to $PM_{2.5}$ (Figures 3 and 4) and a significant correlation between the contributions of S and K ($R = 0.52$) indicate that $PM_{2.5}$-related Cl and Na arises from the combustion of fuels (especially of coal). Neither daily mass contributions of Na to $PM_{2.5}$ nor of Cl are correlated with contributions of S. The contributions of Na and Cl are significantly correlated with contributions of Fe, Mn, Ni, Cu, Cr, Zn, Br, Rb, Sb and Pb in Katowice. In Zabrze, the contributions of these metals are correlated with the contribution of K. Some of these correlations may indicate the effects of (nonferrous or ferrous) metal-processing plants, which are numerous in Silesia. Sodium and potassium hydroxides, inorganic (sulphuric or boric) acids and salts of Na and K are used in almost all galvanic processes together with compounds of all these metals.

The rest of the determined elements (Table 1) had ambient concentrations not greater than the detection limits. Therefore, their yearly concentrations (Table 1) were lower than the detection limits. In general, their daily mass contributions to $PM_{2.5}$ were greater in summer than in winter. The strongest correlations were between the contributions of Mo and Sc ($R = 0.64$) and Sc and Cs ($R = 0.51$) in Zabrze, and Cs and Sr ($R = 0.49$), Cd and Se, Pd and Au (both $R = 0.41$) in Katowice.

4 CONCUSIONS

The future limit for yearly $PM_{2.5}$ concentrations, $25 \, \mu\text{g/m}^3$, is constantly exceeded in Silesia Province. The yearly averages of daily $PM_{2.5}$ concentrations are exceeded by peak-like daily concentrations occurring in winter. In general, the problem of high $PM_{2.5}$ concentrations in Silesia Province is connected with high emissions of dust from hard coal combustion in the power sector and inefficient domestic furnaces.

The basic sources of elements related to $PM_{2.5}$ are probably municipal emissions and road traffic, but the industrial sector may also be a large contributor to the emissions of certain elements in Silesia. However, in order to confirm these assumptions, a comprehensive analysis of obtained results and the results of analyses (determinations of contents of sulphates, nitrates and organic and elementary carbon in $PM_{2.5}$) currently being conducted is necessary.

ACKNOWLEDGEMENTS

The work was subsidised by the Regional Fund for Environmental Protection in Katowice according to the National Environmental Monitoring program realized by the Regional Inspectorate of Environmental Protection in Katowice.

REFERENCES

Artiñano, B., Querol, X., Salvador, P., Rodríguez, S., Alonso, D.G. & Alastuey, A. 2001. Assessment of airborne particulate levels in Spain in relation to the new EU-directive. *Atmospheric Environment* 34: 543–553.

Birmili, W.A., Allen, G., Bary, F. & Harrison, R.M. 2006. Trace metal concentrations and water solubility in size-fractionated atmospheric particles and influence of road traffic. *Environmental Science & Technology* 40: 1144–1153.

Bocca, B., Caimi, S., Smichowski, P., Gomez, D. & Caroli, S. 2006. Monitoring Pt and Rh in urban aerosols from Buenos Aires. Argentina. *Science of the Total Environment* 358: pp. 255–264.

Central Statistical Office. 2009. *Ochrona Środowiska 2008. Informacje i opracowania.* Warszawa.

Directive 2008/50/EC of the European Parliament and of the Council of 21 May 2008 on ambient air quality and cleaner air for Europe.

EEA 2009. European Community emission inventory report 1990–2007 under the UNECE Convention on Long-range Transboundary Air Pollution (LRTAP). Technical report, no 8.

EMEP 2009. *Transboundary particulate matter in Europe. Status report,* no 4.

Gandhi, H.S., Graham, G.W. & Mccabe, R.W. 2003. Automotive exhaust catalysis. *Journal of Catalysis* 216: 433–442.

Gomez, B., Gomez, M., Sanchez, J.L., Fernandez, R. & Palacios, M.A. 2001. Platinum and rhodium distribution in airborne particulate matter and road dust. *Science of the Total Environment* 269: 131–144.

Hopke, P.K. 1991. *Receptor modelling for air quality management.* Elsevier Science Publishing Company. Inc. New York.

Houthuijs, D., Breugelmans, O., Hoek, G., Vaskövi, E., Micháliková, E., Pastuszka, J.S., Jirik, V., Sachelarescu, S., Lolova, D., Meliefste, K., Uzunova, E., Marinescu, K., Volf, J., De Leeuw, F., Van De Wiel, H., Flecher, T., Lebret, E. & Brunekreef, B. 2001. PM-10 and PM-2.5 concentrations in central and eastern Europe: results from the CESAR study. *Atmospheric Environment* 35: 2757-2771.

Hueglin, C., Gehrig, R., Baltensperger, U., Gysel, M., Monn, C. & Vonmon, H. 2005. Chemical characterization of PM2.5, PM10 and coarse particles at urban, near-city and rural sites in Switzerland. *Atmospheric Environment* 39: 637–651.

Inspection for Environmental Protection. 2009. *Unpublished data provided by IEP.*

Klejnowski, K. 2008. Stan formalnoprawny zagadnien związanych z ochroną i monitoringiem powietrza. In *Ocena jakości powietrza w województwie śłaskim w latach 2002-2006.* WIOŚ: Katowice.

Klejnowski, K., Rogula-Kozłowska, W. & Krasa A. 2009. Structure of atmospheric aerosol in the Upper Silesian Agglomeration (Poland). Contribution of PM2.5 to PM10 in Zabrze, Katowice, Częstochowa in 2005-2007. *Archives of Environmental Protection,* in press.

Krzemińska-Flowers, M., Bem, H. & Górecka, H. 2006. Trace metal concentration in size-fractioned urban air particulate matter in Lodz. *Polish Journal of Environmental Studies* 15(5): 759–767.

Lough, G.C., Schauer, J.J., Park, J.S., Shafer, M.M., Deminter, J.T. & Weinstein, J.P. 2005. Emission of metals associated with motor vehicles roadways. *Environmental Science and Technology* 39: 826–836.

Malm, W.C., Sisler, J.F., Huffman, D., Eldred, R.A. & Cahill, T.A. 1994. Spatial and seasonal trends in particle concentration and optical extinction in the United States. *Journal of Geophysical Research* 99: 1347–1370.

Manoli, E., Voutsa, D. & Samara, C. 2002. Chemical characterization and source identification/apportionment of fine and coarse air particles in the Thessaloniki, Greece. *Atmospheric Environment* 36: 949–961.

Morcelli, C.P., Figueiredo, A.M., Sarkis, J.E., Enzweiler, J., Kakazu, M. & Sigolo, J.B. 2005. PGEs and other traffic-related elements in roadside soils from Sao Paulo, Brazil. *Science of the Total Environment* 345: 81–91.

Park, S.S. & Kim, Y.J. 2005. Source contributions to fine particulate matter in an urban atmosphere. *Chemosphere* 59: 217–226.

Pastuszka, J.S., Rogula-Kozlowska, W. & Zajusz-Zubek, E. 2009. Characterization of PM10 and PM2.5 and associated heavy metals at the crossroads and urban background site in Zabrze, Upper Silesia, Poland, during the smog episodes. *Environmental Monitoring and Assessment,* in press.

Pastuszka, J.S., Wawroś, A., Talik, E. & Paw, U.K.T. 2003. Optical and chemical characteristics of the atmospheric aerosol in four towns in southern Poland. *Science of the Total Environment* 309: 237–251.

Ragosta, M., Caggiano, R., Macchiato, M., Sabia, S. & Trippetta, S. 2008. Trace elements in daily collected aerosol: Level characterization and source identification in a four-year study, *Atmospheric Research* 89: 206–217.

Salvador, P., Artíñano, B., Querol, X., Alastuey, A. & Costoya, M. 2007. Characterisation of local and external contributions of atmospheric particulate matter at a background coastal site. *Atmospheric Environment* 41: 1–17.

Schroeder, W.H., Dobson, M., Kane, D.M. & Johnson, N.D. 1987. Toxic trace elements associated with airborne particulate matter: a review. *Journal of Air Pollution Control Association* 37: 1267–1285.

Seinfeld, J.H. 1986. *Air pollution: physical and chemical fundamentals,* 2nd ed. McGraw-Hill, New York.

Sheu, H.L., Lee, W.J., Lin, S.J., Fang, G.C., Chang, H.C. & You, W.C. 1997. Particle-bound PAH content in ambient air. *Environmental Pollution* 96: 369–382.

Sternbeck, J., Sjödin, A. & Andreasson, K. 2002. Metal emissions from road traffic and the influence of resuspension – results from two tunnel studies. *Atmospheric Environment* 36: 4735–4744.

Szopa, S. & Rogula-Kozlowska, W. 2008. Wykorzystanie techniki ED-XRF do oznaczania wybranych 42 pierwiastków, w tym pierwiastków śladowych, we frakcji PM2,5 pyłu zawieszonego. *Materiały XVII Poznanskiego Konwersatorium Analitycznego*: 72.

U.S. Environmental Protection Agency. June 1999. *Compendium of methods for the determination of inorganic compounds in ambient air. Compendium method IO-3.3 Determination of metals in ambient particulate matter using X-ray fluorescence (XRF) spectroscopy.* Cincinnati.

Vallius, M., Janssen, N.A.H., Heinrich, J., Hoek, G., Ruuskanen, J., Cyrys, J., Van Grieken, R., De Hartog, J.J., Kreyling, W.G. & Pekkanen, J. 2005. Sources and elemental composition of PM2.5 in three European cities. *Science of the Total Environment* 337: 147–162.

Van Loon, G. W. & Duffy, S.J. 2008. *Chemia srodowiska.* Panstwowe Wydawnictwo Naukowe: Warszawa.

Wahlin, P., Berkowicz, R. & Palmgren, F. 2006. Characterisation of trafficgenerated particulate matter in Copenhagen. *Atmospheric Environment* 40: 2151–2159.

Environmental Engineering III – Pawłowski, Dudzińska & Pawłowski (eds)
© 2010 Taylor & Francis Group, London, ISBN 978-0-415-54882-3

Modification of the gas flow in electrostatic precipitators and their influence on the ESP efficiency

B. Sładkowska-Rybka & M. Sarna
Institute of Engineering and Environmental Protection, University of Bielsko-Biala, Bielsko-Biała, Poland

ABSTRACT: Electrostatic precipitators (ESP) are the most commonly used devices for gas cleaning in the power industry. From the beginning of ESP usage on a commercial scale, it has been said that all swirls and turbulences should be eliminated from the gas flow, approaching uniform gas distribution in an ESP chamber. Application of CFD (Computer Fluid Dynamics) methods in electrostatic precipitation caused radical changes in views on the role of the gas flow. Series of non-uniform gas flows was then indicated, causing an increase in ESP efficiency. This paper is a review of the gas flow distributions used in ESP and their influence on ESP efficiency. Authors studied the efficiency of multi-zonal electrostatic precipitators with different combinations of skewed gas flow patterns. The results of computer analysis presented in this paper show that diversification of gas velocity in the ESP chamber leads to efficiency improvement.

Keywords: Gas cleaning, electrostatic precipitator (ESP), skewed gas flow, computer simulation, efficiency improvement.

1 INTRODUCTION

It is more than 100 years after the construction of the first electrostatic precipitator in industrial scale by Cottrel. During that period a significant development and improvement of these devices was observed. ESP are used for gas de-dusting from various manufacturing processes. The main area of the ESP applications are power industry, cement industry, metallurgical industry, chemical industry and waste incineration (Parker 1997). ESP is a device complicated in operation and its description requires the involvement of many scientific disciplines, including physics, chemistry, aerosol theory, chemical engineering, electronics, automatics, aerodynamics, building engineering and mechanics. The key property of the ESP is its de-dusting efficiency. Until recently dust concentrations at the ESP outlet were in the level of $100–200\,mg/m^3$, nowadays it is even $5\,mg/m^3$ or less. Such a considerable progress in the ESP technology was possible thanks to the investigation led in many scientific centers in the world.

In this paper the results of computer simulation studies are presented, associated with the ESP performance improvement in relation to the gas flow distribution in the ESP chamber.

2 INDUSTRIAL SCALE GAS DE – DUSTING

A standard ESP is a building with dimensions of approximately 25 m of length, 16 m of width and 22 m in height. A typical installation in the power industry treats about $500,000\,m^3$ of gas per hour and precipitates about 8–10 tones of dust per hour (Kabsch 1992, Lutynski 1965, White 1963).

The main element in the ESP construction is a chamber with two types of electrodes placed inside. The collecting plates, vary in size and design, are placed in parallel rows, forming ducts. The second type – corona electrodes (discharge, emitter electrodes), are in the form of wires or brands on a frame, evenly spaced between the plates. Discharge electrodes are connected to the negative end of the voltage source and provide necessary corona current, while collecting plates are grounded. Typical values of rectified DC voltages are of the order of 60–70 kV. In modern ESP a new model of power supply is used – negative DC voltage with nano-impulses of high voltage. High voltage equipment determines and controls the strength of the electric filed generated between the electrodes. The power system maintains the voltage at the highest level without causing excess sparkover between electrodes. As traditional wire-to-plate ESPs are characterized by non-homogenous current distribution, that is disadvantageous for small particle removal, new types of emitter electrodes are used i.a. barbed plates, barbed tapes, pipes with double-spice, pipes with square tabs or even some nonconventional as moving electrode (Jaworek et al. 2009, Jędrusik 2008, Jędrusik & Świerczok 2008).

The basic process taking place in the ESP chamber is particle charging by means of the electric filed created in the inter-electrode area. Waste gases flow into the chamber having velocity about 1 m/s. Dust particles contained in the gas are rapidly (in matter of millisecond) charged by electron and ion current and attain their saturation charge. Charged dust particles are strongly attracted to the collection plates. They move mainly towards the collecting plates, settle on them creating a dust layer. The charge of the particle is slowly leaked to the grounded electrodes. A portion of the charge is retained and contributes to the intermolecular adhesive and cohesive forces that hold the particles on the plates. The degree of particle discharge depends on the resistivity of the dust layer. The layer is allowed to built up on the collecting electrodes to an appropriate thickness, and then the dust is dislodged from the plates by mechanical impulses, sent to the electrodes (the rapping process). Internal forces resulting from the mechanical oscillation of the plates allows the dust layer to fall of them as large agglomerates. Dislodged dust falls into the hoppers placed bellow the ESP chamber. Unfortunately, during the rapping process, a part of dust particles is re-entrained into the gas flow, but in can be captured again (Riehle 1997, Benítez 1993).

The proper progress of gas de-dusting in the ESP depends on many factors, inter alia, on the suitable gas flow distribution across the ESP. To achieve an adequate gas flow profile, different types of diffusers and perforated screens are used (Jędrusik 2008, Parker 2003). The concept of gas flow distribution has radically changed over the last 15 years (Boyd 2001a, Frank 1996, Gibson et al. 1998, Hein 1989, Lind 1986, Schmitz et al. 1998). Until the 80's of XX century, the model of laminar and even gas flow in the ESP chamber was assumed (Lutyński 1965, White 1963). An even gas flow distribution achieved at the inlet of the ESP should be maintained, almost unchanged throughout the ESP. The basic model of the ESP efficiency was given by Deutsch and Anderson (Deutsch 1922):

$$\eta = 1 - \exp\left(-\frac{L \cdot w}{s \cdot v_s}\right) \tag{1}$$

where L = electric filed length; w = migration velocity of a dust particle; s = wire to plate distance; v_s = mean value of gas flow velocity in a cross section of an ESP chamber.

More accurate estimates of the collection efficiency were obtained by White (1963), who modified the Deutsch-Andsrson equation (1), taking into consideration uneven gas flow distribution. White introduced to the model a gas flow distribution function (White 1963):

$$\eta = 1 - \int_0^{v_{max}} \exp\left(-\frac{L \cdot w}{s \cdot v_s}\right) \cdot f(v) \cdot dv \tag{2}$$

where f(v) = gas flow distribution function in a cross section perpendicular to the gas flow.

As a typical function f(v) a Gaussian distribution function was assumed.

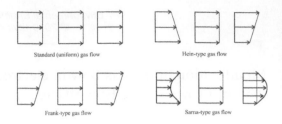

Figure 1. Gas flow distribution profiles in the ESP.

Another modification to the Deutsch-Anderson equation was given by Idelcik and Aleksandrov (1974):

$$\eta = 1 - \exp\left(-\frac{L \cdot w}{s \cdot v_s} \cdot \frac{1}{M_k}\right) \tag{3}$$

$$M_k = \frac{1}{A} \cdot \int_A \left(\frac{v}{v_s}\right)^2 \cdot dA \tag{4}$$

where M_k = coefficient defining the non-uniformity of the gas velocity profile in the cross section of the ESP; A = cross section area of the ESP chamber; v = local gas velocity in the cross section of the chamber.

The literature data shows, that both equations (2) and (3) were verified experimentally with satisfactory results (Self et al. 1990).

Another more complicated models of gas flow were proposed by Mc Donald and Dean (1982).

Owing to the rapid development of computer technology in 80's of XX century, discrete computer models of ESP chamber were introduced (Haque et al. 2009a, Haque et al. 2009b, Hein 1989, Lind 1986, Varonos et al. 2002). The ESP chamber was subdivided into separate cells. Each cell (element) is a miniaturized ESP. Acceptance of such models allowed to discover new possibilities of the ESP performance improvement. The efficiency may be increased only by an adequate gas flow distribution in the ESP chamber (Boyd 2001b, Frank 1996, Hein & Gibson 1996, Jedrusik et al. 2002, Lockhart & Weiss 2001).

In literature 3 types of gas flow distribution that caused a significant improvement in ESP performance can be found (so called skewed flows). The first, suggested by G. Hein (Gibson et al. 1998, Hein 1989, Hein & Gibson 1996), where on the inlet of the ESP gas velocity is the highest on the bottom of the ESP and linearly decreases to the bottom, whereas on the outlet of the ESP gas flow distribution is reversed. Gas distribution proposed by W. Frank (1996), is characterized with velocities decreasing linearly from the top to the bottom both on the inlet and the outlet cross-sections of the ESP. Sarna (1998) suggested a new type of gas flow pattern: concave-convex flow, where gas velocities on the inlet and outlet of the ESP are formed as concave and convex "surfaces". Figure 1 shows gas flow distribution profiles in the ESP.

Skewed gas flows are identified using the skew coefficient "q" that is the quotient of the extreme

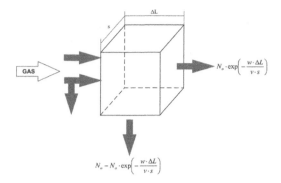

$$N_o \cdot \exp\left(-\frac{w \cdot \Delta L}{v \cdot s}\right)$$

$$N_o - N_o \cdot \exp\left(-\frac{w \cdot \Delta L}{v \cdot s}\right)$$

Figure 2.   Single cell in the ESP model.

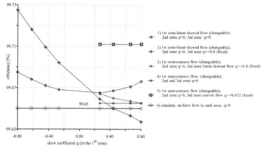

Figure 3.   Efficiency of the ESP as a function of different gas flow types.

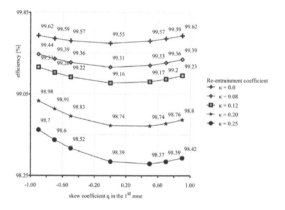

Figure 4.   ESP efficiency as a function of the re-entrainment coefficient for various gas flow distributions.

velocity values difference to the mean value of the velocity in the profile.

Experimental research led by Technology Research and Investigation ESKOM in the Republic of South Africa on four industrial electrostatic precipitators, confirmed efficiency improvement by means of skewed gas flow technology (SGFT), especially by Hein type of flow (Frank 1996, Hein 1996). Achieved results show that dust concentrations are 24–80% lower comparing with the ESP with standard, uniform gas flow. Required gas flow distribution can be attained using suitable diffusion perforated screens and gas turning vanes installed in the diffuser and in the confusor.

3   GAS FLOW AND THE ESP EFFICIENCY

Authors of that paper in their own investigation of gas flow in the ESP also assumed discrete computational model. Each zone of the ESP was divided into individual cells ($10 \times 10 \times 10$ elements or $10 \times 10 \times 15$ elements). In a single cell (Figure 2) a partial dust precipitation occurs, the remains move with the gas to the next cell for further precipitation. The process of gas de-dusting in a single cell is described by the Deutsch-Anderson equation (1).

Dust precipitated in a cell move down to the hoppers but during that movement it is partially re-entrained by gas flow and can be captured in cells placed below. Using that model of the ESP, a software module was created, run in the Mathcad® environment. Computer programs enabled to carry on a series of simulation tests for different gas flow distributions. Three- and multi-zonal ESP were included in the investigation, with Hein-type, Frank-type and Sarna-type of flow. Also dust re-entrainment and particle size distribution were studied. Obtained results are consistent with literature information reports, regarding the advantage of skewed gas flow over the standard uniform flow.

Let's follow some of this analysis.

3.1   Linear and concave – convex gas flows

A three-zonal ESP chamber was assumed. Different combinations of the gas flow types were used in the ESP chamber. Results of that analysis are shown on Figure 3. Standard, uniform flow enabled to achieve 99.65% efficiency. The greatest efficiency improvements were acquired for Hein type of gas flow (curve no. 2) and concave-convex flow (curve no. 3).

3.2   Dust re – entrainment

Collected dust is removed from the plates by mechanical vibrations of the electrodes. The dust is dislodged as large aggregate sheets and move (due to gravity) to the dust hoppers. Unfortunately during that movement a part of the dust is re-entrained by the gas flow. That process reduces the ESP efficiency. The ratio of re-entrained dust mass to mass of the dust collected on the plates is called the re-entrainment coefficient (κ). Figure 4 shows simulation results of the ESP efficiency of 4-zonal ESP as a function of the re-entrainment coefficient κ.

Obtained results confirm that re-entrainment reduces the ESP efficiency, but skewed gas flows are able to diminish that effect. The efficiency

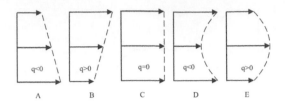

Figure 5.   Gas flow profiles.

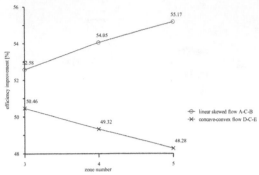

Figure 7.   ESP efficiency improvement in comparison to the ESP with uniform gas flow ($\kappa = 0.08$).

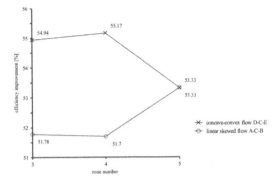

Figure 6.   ESP efficiency improvement in comparison to the ESP with uniform gas flow ($\kappa = 0.00$).

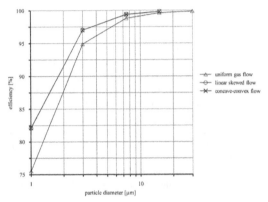

Figure 8.   Fractional efficiency of the ESP.

improvement is the higher, the greater the dust re-entrainment is.

The relationship between the re-entrainment, type of the gas flow and number of the ESP zones was published by authors of that paper. The efficiency analysis was led for 3, 4 and 5-zonal ESP. Various combinations of the gas flow distribution shown on Figure 5 were assumed, with the mean value of gas velocity $v_s = 1$ m/s and skew coefficient $q = \pm 1.5$.

Two levels of re-entrainment were assumed: no re-entrainment ($\kappa = 0.00$) and 8% ($\kappa = 0.08$). Obtained results of the ESP performance were compared with results for the ESP with standard uniform gas flow on each zone. Figure 6 and Figure 7 show acquired results.

Achieved results show that properly formed gas flow distributions enable apparent efficiency improvement. For example: dust concentrations at the outlet of the ESP are more than 60% lower for linear-skewed (A-C-B) and concave-convex (D-C-E) flows and for gas flow type A-A-C-B-B dust concentration is 73% lower.

Surprising is the impact of the dust re-entrainment shown at Figure 6 and Figure 7. If there is no re-entrainment in the ESP linear-skewed gas flow is more effective in internal zones, but on the outlet of the ESP both types of gas flow enable the same efficiency improvement. In case of re-entrainment at the level of 8% linear-skewed flow is more effective in each of the ESP zones. Efficiency improvement increases in following zones, while for concave-convex flow, it decreases.

3.3 *Particle size distribution and gas flow configuration*

The skewed distribution of the gas flow have a significant impact on the fine dust fraction (smaller than $10 \, \mu m$) removal in the ESP. Authors of that paper made studies for 3-zonal ESP with uniform, linear-skewed and concave-convex flows. Two types of dust with different size distribution were investigated. The dust re-entrainment was established in terms of 8% ($\kappa = 0.08$). Figure 8 shows calculated values of fractional efficiency for particles up to $30 \, \mu m$. For larger particles obtained results of efficiency improvement are presented in Table 1.

Figure 8 and Table 1 show that both skewed flows enable to achieve higher efficiencies of de-dusting for smaller dust fractions, than standard uniform flow.

4 CONCLUSIONS

The numerical analysis of the dust removal in the ESP leads to the following observations and conclusions:

– linear-skewed (especially Hein-type) and concave-convex flows lead to ESP efficiency improvement;

Table 1. Fractional efficiency improvement for linear – skewed and concave – convex flows.

Particle diameter (μm)	Efficiency improvement (%)	
	Gas flow type	
	Linear-skewed	Concave-convex
1	6.57	6.62
3	2.12	2.1
7.5	0.6	0.58
15	0.21	0.19
30	0.07	0.06
45	0.04	0.03
60	0.02	0.02
80	0.01	0.01
100	0.01	0.01

- the efficiency improvement increases with the level of the dust re-entrainment;
- both types of flows: linear-skewed and concave-convex enable efficiency improvement, but concave-convex flow appears to be adequate in case of no re-entrainment, while linear-skewed flow – when re-entrainment in the ESP occurs;
- skewed gas flow distributions are advantageous also for fine dust fractions.

Selection of the proper gas flow distribution each time requires an analysis of physicochemical gas properties and ESP construction. Modifications of the gas flow are a proper and easy way to reduce the air pollution.

REFERENCES

Benítez J. 1993. Process engineering and design for air pollution control. Englewood Cliffs: Prentice Hall.
Boyd M. 2001a. Skewed gas flow technology offers antidote to opacity derates. Power Engineering 105(6): 47–50.
Boyd M. 2001b. The evaluation of skewed gas flow technology. In: Proc VIII Int. Conf. on Electrostatic Precipitation, Birmingham AL, 14–17 May 2001.
Deutsch W. 1922. Bewegung und Landung der Electricitätstränger im Zylinder Kondensator. Annalen der Physik 68(335).
Frank W.J. 1996. Aspects of ESP upgrading. In: Proc VI Int. Conf. on Electrostatic Precipitation, Budapest, 17–22 June 1996.
Gibson D., Schmitz W., Hein A.G. 1998. Electrostatic precipitator skew gas flow technology application and quantification. In: Proc VII Int. Conf. on Electrostatic Precipitation, Kyongju, 20–25 September 1998.
Haque M.E., Rasul M.G., Khan M. M. K., Deev A.V., Subaschandar N. 2009a. Influence of the inlet velocity profile on the prediction of velocity distribution inside electrostatic precipitator. Experimental Thermal and Fluid Science 33(2): 322–328.
Haque M.E., Rasul M.G., Khan M. M. K., Deev A.V., Subaschandar N. 2009b. Flow simulation in an electrostatic precipitator of a thermal power plant. Applied Thermal Engineering 29(10): 2037–2042.
Hein A.G. 1989. Dust re-entrainment, gas distribution and electrostatic precipitator performance. Control Technology 39(5): 766–771.
Hein A., Gibson D. 1996. Electrostatic precipitator skewed gas flow technology – ESKOM experience in South Africa. In: Proc. 86th Annual Meeting and Exhibition, Nashville, 1996.
Idelcik I.E., Aleksandrov V.P. 1974. The effect of non-uniformity of the gas flow on efficiency of electrostatic precipitators. Teploenergetika 21(8).
Jaworek A., Krupa A., Czech T. 2007. Modern electrostatic devices and methods for Exhaust gas clearing. A brief review. Journal of Electrostatics 65(3): 133–155.
Jędrusik M. 2008. Electrostatic precipitators. Wrocław: OWPWr.
Jędrusik M., Świerczok A. 2008. Structure of collected dust layer on collecting electrode. Polish Journal of Environmental Studies 17(3A): 259–263.
Jędrusik M., Świerczok A., Nowaczewski A., Sarna M. 2002. Application to the asymmetric profiles of gas velocity to an electrostatic precipitator. Archives of Environmental Protection 28(2): 5–16.
Kabsch P.1992. Dedusting and dust collectors. Warszawa: WNT.
Lind L.1986. Influence of gas distribution on precipitator performance. In: Proc. Six Joint EPA/EPRI Symposium on Transfer and Utilization of Particulate Control Technology, New Orleans, 1986.
Lockhart J., Weiss O. 2001. The application of skewed gas flow technology at the Israel Electric Corporation MD-A Station. In: Proc VIII Int. Conf. on Electrostatic Precipitation, Birmingham AL, 14–17 May 2001.
Lutyński J. 1965. Electrostatic gas dedusting. Warszawa WNT.
Mc Donald J.R., Dean A.H. 1982. Electrostatic precipitator manual, New York: Noyes Data Corporation.
Parker K.R. 1997, Milestones in the history of electrostatic precipitation. In: K.R. Parker (ed.),Applied Electrostatic Precipitation: 11–24. London: Chapman and Hall.
Parker K.R. 2003. Electrical operation of electrostatic precipitators. London: Institution of Engineering and Technology.
Riehle C. 1997. Basic and theoretical operation of ESPs. In: K.R. Parker (ed.),Applied Electrostatic Precipitation: 25–88. London: Chapman and Hall.
Sarna M. 1998. Some aspects of flow skew technology in ESP performance improvements. In: Proc VII Int. Conf. on Electrostatic Precipitation, Kyongju, 20–25 September 1998.
Schmitz W., Gibson D., Pretorius L. 1998. ESP performance prediction and flow optimization using CFD modeling. In: Proc VII Int. Conf. on Electrostatic Precipitation, Kyongju, 20–25 September 1998.
Self S.A., Choi D.H., Mitchner M., LEACH R. 1990. Experimental study of collector plate rapping and re-entrainment in electrostatic precipitators. Proc IV Int. Conf. on Electrostatic Precipitation, Beijing, 1990.
Varonos A.A., Anagnostopoulon J.S., Bergeles G.C. 2002. Prediction of the cleaning efficiency of an electrostatic precipitator. Journal of Electrostatics 55(2): 111–133.
White H.J. 1963. Electrostatic Precipitation. New York: Addison Wesley (Reprint by ISESP).

Environmental Engineering III – Pawłowski, Dudzińska & Pawłowski (eds)
© *2010 Taylor & Francis Group, London, ISBN 978-0-415-54882-3*

A method to quantify light pollution and results from Poland 1994–2008

T. Ściężor & M. Kubala
Cracow University of Technology, Faculty of Environmental Engineering, Cracow, Poland

T.Z. Dworak
AGH University of Science and Technology, Faculty of Mining Surveying and Environmental Engineering, Cracow, Poland

W. Kaszowski
Institute of Meteorology and Water Management, Cracow Branch, Cracow, Poland

ABSTRACT: This paper describes the artificial night sky glow, known as "light pollution", which has recently become appreciable in Poland. The history of the problem, causes of the light pollution, its types and its effects on the environment are described. An innovative, simple, inexpensive and accessible method of quantifying light pollution based on the observation of dim comets is presented. An assessment of the accuracy of the method has been carried out and results are presented for Poland for the period 1994–2008.

Keywords: Light pollution, photopollution, comets, surface brightness, surface magnitude, sky glow.

1 INTRODUCTION

The negative influence of humankind on the Earth's atmosphere manifests itself not only in widely-known phenomena such as industrial pollution. We should not forget the emission of electromagnetic waves across a broad spectrum that interfere with the natural electromagnetic background. The danger of some kinds of electromagnetic radiation has been known for a long time (e.g. x-ray radiation), while in other cases, problems have only recently been discovered (e.g. microwave radiation) or are still being investigated (e.g. radio waves). Radiation from another area of the electromagnetic spectrum must also be taken into account. This is visible light in the range from infrared to ultraviolet. This so-called light pollution is the subject of the present publication.

1.1 *The light pollution problem: The history of research*

Humans have long exploited the attraction of other species to light, or their repulsion by it (Rich & Longcore 2006b). Initial concerns about the adverse impact of such attraction were raised by studies of the mortality of birds at lighthouses and light ships (Gauthreaux & Belser 2006), such as that by Allen in New England (Allen 1880) and the Californian ornithologist Carlos Lastreto on the Pacific coast (Squires & Hanson 1918). Other concerns were echoed during the early days of the electrification of street lighting in the United Kingdom, where concern was expressed about the loss of insects as food for songbirds (Los Angeles Times 1897).

In the 1950s, the Dutch ecologist F. J. Verheijen reviewed the attraction of animals to light as reported in the European and Japanese literature (Verheijen 1958). McFarlane (1963) described the risk to turtles from coastal lighting and Verheijen formally described "photopollution" in the 1980s (Verheijen 1985).

Meanwhile, astronomers had also noticed the negative effect of the growth in artificial illumination on the environment. In the 1950s, astronomers from the Kitt Peak National Observatory, south of Tucson (Arizona, USA) called attention to the fast growing brightness of the night sky. Moreover, the wide spectral range of visible light made it very difficult to filter out (Massey & Foltz 2000).

In 1988 in the USA, following the suggestion of David Crawford, the International Dark-Sky Association (IDA) was founded (IDA, accessed 2009). The purpose of IDA is to stop the influence of light pollution on the environment through informing public opinion about light pollution and how to avoid it. Sections of IDA are currently active in 25 US states, as well as in Australia, Austria, Chile, China, Czech Republic, Greece, Ireland, Israel, Canada, Malta, Slovenia, Switzerland and Italy (IDA Sections, accessed 2009).

In the 1990s, the astronomical aspects of the light pollution problem were widely known. At the 1992 UNESCO Congress in Paris attention was called to the phenomenon of the excessive emission of light. At the same time it was said that as much as 30% of the electrical energy used to light cities is wasted in the form of light pollution – uncontrolled light emission to the sky (Laulainen 1994).

Astronomers were still the main animators of the night sky protection movement. In 1997, the Resolution for the "Protection of the Night Sky" agreed at the XXIII General Assembly of the International Astronomical Union in Kyoto (Japan) included the following statement:

"Considering that proposals have been made repeatedly to place luminous objects in orbit around the earth to carry messages of various kind…and that the night sky is the heritage of all humanity, which should therefore be preserved untouched…to take steps with the appropriate authorities to ensure that the night sky receive no less protection than has been given to the world heritage sites on earth" (IAU, 1998).

But it was not only astronomers that were concerned about light pollution. Various aspects of this problem, including ecological ones, have been covered at the European Symposium for the Protection of the Night Sky that has been organised annually since 2000 (IDA Archives, accessed 2009) and at other conferences (Raevel & Lamiot 1998, Schmiedel 2001).

The ecological effects of light pollution were summarised at the conference on the Ecological Consequences of Artificial Night Lighting (Los Angeles, California) (Rich & Longcore 2006c).

At this conference, attention was drawn to the following points (Rich & Longcore 2006a):

At the IDA/New England Light Pollution Advisory Group (NELPAG) Fall Meeting held in October 2002 in Boston and Cambridge, astronomers were in the minority for the first time. Astronomers and many other specialists – environmentalists, lawmakers, criminologists, lighting engineers, city planners, medical researchers and wildlife conservationists – agreed on the importance of the light pollution problem (Poulsen 2003). The following definition of light pollution was also drawn up at this meeting:

"Light pollution is any adverse effect of artificial light, including sky glow, glare, light trespass, light clutter, decreased visibility at night, and energy waste. Light pollution wastes energy, affects astronomers and scientists, disrupts global wildlife and ecological balance, and has been linked to negative consequences in human health" (IDA FAQ, accessed 2009).

The health consequences of circadian disruption have also received scientific attention (see Navara & Nelson 2007 for a review). In particular, Kloog and colleagues have shown convincing correlations between exposure to elevated nocturnal illumination and breast (Kloog et al. 2008a, Kloog et al. 2008b) and prostate cancers (Kloog et al. 2009).

1.2 *Forms of light pollution*

The adverse effects of artificial night lighting on ecosystems and on astronomical observation have been called "ecological light pollution" and "astronomical light pollution" (Longcore & Rich 2004). In addition, lighting can cause a range of other undesirable effects on human environments. For the most part, astronomical light pollution is concerned with sky glow, where the combination of all light escaping up into the sky is scattered (redirected) by the atmosphere back towards the ground. Disturbance of human environments and that of other species (Longcore & Rich 2004) can result from sky glow, but also from "light trespass", when unwanted light enters an area, for instance, by shining over a neighbour's fence, "over-illumination" as in the excessive use of light, and "glare", often the result of excessive contrast between bright and dark areas in the field of view.

Most publications are devoted to these problems, and technical proposals for their solutions are proposed. The Light Pollution Science and Technology Institute in Thiene (Italy) researches sky glow, as does the Night Sky Team of the US National Park Service, which has measured the brightness of the night sky at 80 locations, predominantly in US National Parks, by the use of CCD cameras (Duriscoe et al. 2007).

1.3 *Artificial sky glow in the world and in Poland*

Satellite photos of the night hemisphere of the Earth are the best illustrations of the extent of the light pollution problem (Cinzano et al. 2001a, Elvidge et al. 1997). They clearly show that it is very hard to find places on our planet that have maintained their original, night-time darkness. There are certainly areas that are not inhabited or not industrialised, e.g. areas of the Sahara desert and central Africa, Amazonia in South America, the interior of Australia and the northern parts of Asia and North America. Three areas, with the highest population density, present the main sources of light pollution. These are the eastern states of the USA, western and central Europe and Japan.

In Europe, the area with the most evident increase in sky glow spreads from England to eastern and central Europe, including the Benelux countries, Germany, Italy, Hungary, the Czech Republic, southern Scandinavia and southern Poland. Dark skies can still be found in the interior of the Iberian Peninsula, the Balkan Peninsula, northern Poland and in the whole of eastern Europe (Cinzano et al. 2001b).

Cinzano (2001a) used satellite data to model the distribution of night sky brightness in Europe, with calculations of the expected values and the brightness of the dimmest stars visible to the naked-eye. The results show the area subject to high light-pollution includes east England, the Benelux countries, southern Germany, northern Czech Republic and Upper Silesia and (separately) northern Italy. In Poland, the main light sources are the Katowice – Cracow region and Warsaw. A dark night sky can still be seen in north-eastern and north-western Poland (Cinzano et al. 2001a).

Light pollution levels are of large practical significance. They are correlated to not only the quality and quantity of the light sources, but also to the quantity of aerosols and industrial dusts suspended in the atmosphere. Consequently, measurements of the brightness of the night sky can facilitate estimates of, among other things, industrial atmospheric pollution. Currently, systematic surveys of night sky brightness in urban areas have only been made in a few countries of the world – Netherlands, Japan, Italy, Venezuela and the United States (Narisada & Schreuder 2004, Duriscoe et al. 2007, Kosai & Isobe 1991, Isobe & Kosai 1998, Della Prugna 1999, Schreuder 2001).

Based on DMSP (Defense Meteorological Satellite Program) data, Cinzano (2001a) modelled expected night sky brightness and the brightness of the dimmest stars visible to the naked-eye from any site. This method allows for the brightness of the night sky to be quantified but depends on many subjective factors not directly connected to sky glow, e.g. a decrease in eye pupil diameter because of bright light sources in the neighbourhood. Moreover, this is a comparison of the point light sources (stars) to the extended one (the sky).

The aim of our research was to make ground-based measurements of night sky brightness in Poland and to compare the results obtained with earlier models. The original method was worked up, allowing for the use of astronomical observations for valuations in the light pollution level. The positive verification of values obtained with scientific instruments will allow quantification of the level of light pollution in the past, when the phenomenon itself had not been noticed.

2 MATERIALS AND METHODS

2.1 *Use of existing astronomical observations to quantify light pollution*

The direct measurement of the brightness of the night sky is possible only with sensitive instruments, limited to the people or institutions having access to them (Duriscoe et al. 2007). The problem is much simpler, however if the whole sky is treated as the astronomical object for which brightness is to be determined. In astronomy, the measurement of an object's brightness is always made by comparison to a model object.

The brightness of celestial objects in astronomy (magnitude) is measured using the unit magnitudo (mag, ^{m}), where a 0^m star is 100 times brighter than a 5^m star, and the dimmest stars visible to the naked-eye in ideal circumstances are 6^m. This means that the magnitudo scale is logarithmic, and the proportion of light intensity of object A to object B can be described by the formula (Green 1997):

$$\frac{I_A}{I_B} = 2.512^{m_B - m_A} \qquad (1)$$

where I_A = light intensity of object A (brighter one), I_B = light intensity of object B (fainter one), m_A = magnitude of the object A, m_B = magnitude of the object B.

The luminous sky must, however, be treated as a diffuse object, having some area.

For diffuse objects with an area A, the so-called surface magnitude S_a was introduced, defined as:

$$S_a = m_1 - 2.5 \log A^{-1} = m_1 + 2.5 \log A \qquad (2)$$

where S_a = surface magnitude of the object, m_1 = magnitude of the object, A = surface area of the object.

The total magnitude m_1 of the diffuse object is determined by the comparison of this object, in a special way, with the out-of-focus telescopic images of stars with known catalogue brightness (Green 1997).

The unit of surface brightness in the International System of Units (SI) is candela per square meter (cd/m^2), while for the surface magnitude S_a the commonly used unit is magnitudo per square arcsecond $(mag/arcsec^2)$. These units can be converted by the formula $[cd/m^2] = 10.8 \cdot 10^4 \cdot 10^{\wedge}(0.4 \cdot [mag/arcsec2])$.

Because of natural phenomena, mainly the northern lights, the darkest night sky can be found in the near-equator regions and becomes lighter and lighter in the direction of the poles. This difference is minimal in the solar minimum period and greatest in the solar maximum period (solar activity changes in an 11-year cycle). A widely accepted value of S_a for the night sky in zenith, at a site lacking any artificial lights, at the time of the solar minimum (e.g. the years 1994–1998 and from 2005) is equal to about 22 mag/arcsec² (Garstang 1989). During the solar maximum period (e.g. the years 1989–1992, 1999–2003), S_a is equal to about 21 mag/arcsec² (Krisciunas et al. 2007).

The model map of night sky brightness in Europe presented by Cinzano and colleagues (2001a) is calibrated in mag/arcsec². This unit is also used here, because of the specificity of the measurements and because of the aim of our study – the comparison of measured values with model ones.

Our original method of quantifying night sky brightness is based on the measurement of the surface magnitude of the faintest visible diffuse objects, the value of which should be only slightly lower than the surface magnitude of the night sky. However, these must be objects with flat brightness distribution, without any marked variations.

Comets are an example of such generally observed objects (Green 1997). A comet is a diffuse object, where surface brightness decreases systematically from the centre outwards until it disappears into the sky background. The value DC (degree of condensation) was introduced in astronomy, to describe the degree to which the comet diffuses into the sky background (in fact it describes the brightness gradient between the centre of the comet envelope and its edge). A comet with a DC=9 looks like a star, whereas the surface brightness of a comet with a DC = 0 is uniformly distributed and almost equal to the sky surface brightness. A comet with a DC = 5 has a distinct, brighter core and an envelope diffusing into the sky. A comet observer reports the total magnitude of the comet, as well as the maximum noticeable diameter of the envelope (up to point where it diffuses into the sky background). The surface magnitude of the comet can be easily computed on the base of these two values. This means that for the faintest comets still visible through the telescope that have a very low DC (0 to 2), the surface brightness approximates the surface brightness of the sky (more precisely, it marks the lower limit of the surface brightness of the sky).

The main advantage of this method is its simplicity. There are many comet observers (several dozen regular ones in Poland) and each tries to make as thorough observations as possible, so providing a sample size sufficient for further analysis. A dozen or so comets are observed each year, which allows for analysis of changes in sky brightness during the year. It is also possible to take advantage of archival observation of comets from the start of the 20th century to determine long-term changes in sky glow.

To verify this astronomical method, we used the rich archive of the Comet Observers Section (SOK) of the Polish Amateur Astronomical Society (PTMA), including observations for the period 1994–2004 and additionally some later observations taken from the Comet Observations Centre (COK), both including tens of thousands of comet observations. For further analysis, observations of only the faintest visible comets were used. For urban observers, comets fainter than 9^m were included, for those elsewhere, fainter than 11^m. We further limited observations to those comets with a low DC (0 to 2), observed at high elevations above the horizon, at about midnight local time, and in a moonless sky. In this way, a sample of over 700 observations was obtained that was optimised for our research purpose.

It must be noted that the estimation of comet magnitude and envelope diameter is subjective and consequently can vary between observers. It is also impossible to take into consideration the variability in weather conditions at the time of the measurements. This means that the method is a statistical one, requiring the consideration of many observations made by many observers. Single observations, or even a series of observations made by just one observer, cannot be used to estimate the light pollution level for a given area.

2.2 Qualitative testing of the astronomical method

Two tests were conducted to confirm that the S_a value determined by the astronomical method actually reflects the surface brightness of the sky. Their aim was to see if the changes in S_a reflect changes in the brightness of the night sky that can be accounted for by natural processes.

2.2.1 1st test – changes in the night sky brightness related to the phase of the moon

The sun's light reflected off the moon illuminates the night sky. The degree of illumination is correlated to the phase of the moon – there is no such effect in the new moon phase and it is the strongest at the full moon. We compared the surface magnitude of the faintest observed comets (S_a), determined by the astronomical method, against the phases of the moon for the years 1994–2004.

2.2.2 2nd test – changes in the night sky brightness related to the season of the year (depth of the sun below the horizon at night)

The changes in the surface magnitude of the faintest visible comets (S_a) during the year were investigated at those sites in Poland that are considered practically free of light pollution (north-eastern and eastern Poland). These sites were chosen to investigate the natural changes in the night sky related to the depth of the sun below the horizon at midnight local time. Is well known that the sun achieves its most negative altitude (below the horizon) during the winter (the sky is darkest), and its least negative altitude during the summer (the sky is brightest), when one even speaks of "astronomical white nights".

2.3 Quantitative testing of the astronomical method

We used specialist sky quality meters (SQMs) to verify that the average value of S_a obtained by the astronomical method approximates the real night sky surface magnitude value. These SQMs measure the sky surface magnitude in a solid angle of 40° at a precision of 0.01 mag/arcsec^2.

SQM measurements were taken at localities chosen from among the observations for the period 1994–2004 analysed by the astronomical method under consistent, and possibly the most conducive, circumstances for observation. Parallel observations of the faint comets were also made.

3 RESULTS AND DISCUSSION

3.1 Qualitative tests

During the full moon (−1.0 and +1.0 values on the horizontal axis; Figure 1), the surface magnitude of the faintest visible comets is about 1 mag/arcsec^2 lower

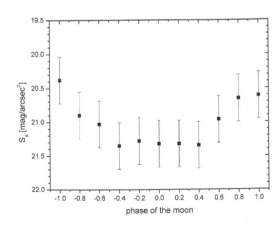

Figure 1. Changes in the value of S_a during the lunar month, averaged for the period 1994–2004.

than during the new moon (moon phase 0.0). Keeping in mind the nature of the magnitudo scale, this indicates a greater than two-fold linear increase in surface brightness detected by comet observations during the full moon. It must be noted that the moon near the full phase spends the majority of the night in the sky, which is maybe even more important than its brightness. This could explain the constant value of S_a in the phase range -0.4 to $+0.4$, during which the moon sets in the evening or rises in the morning.

The averaged graph for the period 1994–2004 shows that seasonal changes are distinctly visible (Figure 2). The differences between December and June are equal to about 0.7 mag/arcsec^2, which is a nearly two-fold linear increase in sky surface brightness in the summer compared with the winter. To show the dependence of the determined value on the sun depth below the horizon, that curve is also drawn, representing the predicted trend of the sky surface magnitude resulting from this value.

Based on these two qualitative tests, the average surface magnitude of the faintest observed comets (S_a) is in fact correlated with the surface magnitude of the sky as it varies according to known natural patterns, so this method can also be used to measure light pollution.

3.2 Quantitative tests

The night sky surface magnitude measured in the field by the SQM was slightly lower (at maximum 1 mag/arcsec^2) than the value of S_a determined by the astronomical method (Table 1).

This quantitative validation gives us confidence that our method provides realistic values that are close to and highly correlated with actual night sky surface magnitude.

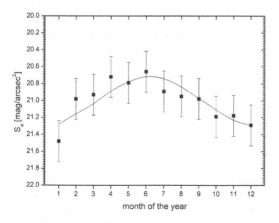

Figure 2. Annual variations in the value of S_a for the north-eastern and eastern regions of Poland averaged for the period 1994–2004. The dotted line represents the predicted surface magnitude of the night sky due to changes in sun depth below the horizon at midnight local time during the year.

3.3 Geographic and temporal variability in S_a

We calculated mean values of S_a for areas with a range of population sizes and industrialisation to compare the light pollution of these environments. This value was determined for two very large, industrialised cities (Cracow, Warsaw), three medium to small cities located in agglomerations (Lublin, Niepolomice, Czestochowa), three average cities (Sanok, Szczecin, Wałbrzych) or suburbs of large cities (Jerzmanowice, 21 km NW of Cracow; Izabelin, 18 km N of Warsaw) and two rural towns sited far away from agglomerations (Krosno, Białystok).

As we expected, the average value of S_a achieved its highest level of 19 mag/arcsec^2 in large cities in highly industrialised areas, such as Warsaw and Cracow. For smaller cities in highly industrialised areas, such as Lublin or Częstochowa, it was 20 mag/arcsec^2. The highest value (representing the darkest skies), 22 mag/arcsec^2, was found in medium-sized and small cities in less industrialised areas located in north-eastern and south-eastern Poland, such as Krosno or Białystok (Figure 3). These values agree with data from satellite images of the night sky in Poland and, most importantly, they are consistent with the night sky surface magnitude predicted by the satellite data model worked out for the specified regions (Cinzano et al. 2001a). The highest values obtained agree with the values in the literature for ideal observational circumstances (Garstang 1989).

We described changes in sky surface magnitude over time for specific sites where a series of observations were made by more than one observer.

Table 1. Results of comparative measurements for selected localities in Poland. The maximum error of the astronomical method (AM) data is equal to 1.4 mag/arcsec^2 while the maximum error for the SQM data is equal to 0.1 mag/arcsec^2.

Locality & coordinates	S_a [mag/arcsec^2]			
	AM	SQM	Difference	Sample size
Cracow 50°04′N 19°54′E	19.7	18.7	1.0	26 (AM) 206 (SQM)
Lublin 51°14′N 22°33′E	19.4	18.6	0.8	28 (AM) 4 (SQM)
Jerzmanowice, near Cracow 50°12′N 19°45′E	21.4	20.6	0.8	17 (AM) 120 (SQM)
Lubomir mountain 49°45′N 20°02′E	21.5	20.9	0.6	24 (AM) 5 (SQM)
Bieszczady mountains 49°09′N 22°19′E	22.1	21.7	0.4	16 (AM) 15 (SQM)

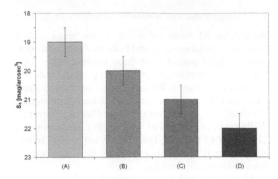

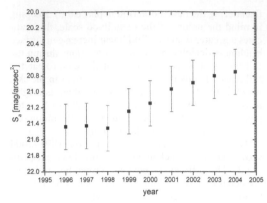

Figure 3. The value of S_a for selected sites in Poland averaged for the period 1994–2004. (A) large cities in highly industrialised areas; (B) medium-sized and small cities in highly industrialised areas; (C) medium-sized and small cities in moderately industrialised areas or distant suburbs of big cities in highly industrialised areas; and (D) medium-sized and small cities in less industrialised areas.

Figure 5. Changes in the value of S_a for the whole territory of Poland for the period 1996–2004.

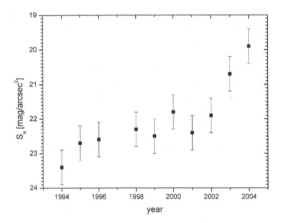

Figure 4. Changes in the value of S_a for the Lubomir and Łysina Range in the Beskid Makowski mountains for the period 1994–2004.

First, we determined the changes in the brightness of the night sky for the Lubomir and Łysina mountain range in the Beskid Makowski mountains (near Myślenice city) during the period 1994–2004. These measurements are important for further research into the ecological aspects of light pollution because the location is not far from the Dobczyce reservoir, the main water supply for Cracow. These observations were made mostly at the astronomical camps of the Polish Amateur Astronomical Society that were held annually in this area until 2004.

A systematic decrease in S_a is clear, from about 23 mag/arcsec^2 in 1994 to 20 mag/arcsec^2 in 2004 (Figure 4). A distinct decrease in sky surface magnitude can be seen beginning in 2002, coinciding with the observable growth of the nearest villages. These calculations show an almost 16-fold linear increase in the brightness of the night sky over the past 10 years. Unfortunately, in 2005 observations of comets

from this site were ended, mainly because of the increasing light pollution. Direct measurements by the SQM, made at this site in analogous circumstances in 2008, showed a night sky surface magnitude equal to 20.9 mag/arcsec^2.

A similar analysis made for areas of Poland traditionally considered only slightly affected by urbanisation, i.e. north-eastern Poland, Podlasie and Bieszczady, showed no measurable changes in the average value of S_a. Over the period 2001–2008, the S_a value was constant and equal to only 21.8 mag/arcsec^2.

Taking advantage of all observations made by the PTMA SOK during 1994–2004, granting the conditions mentioned earlier, we calculated the country-wide average value of S_a (Figure 5). Up to 1998, these values remained practically constant, but starting from 1999, we can see a systematic decrease. The difference is equal to about 0.7 mag/arcsec^2, which is exactly as much as between the summer and winter months in the absence of light pollution. This means that human activity on average across the country results in the extension of the "white nights" period to the whole year, which without doubt must be treated as a serious infringement of the natural environment.

4 CONCLUSIONS

This paper introduces a novel method of estimating the brightness of the night sky. The method has been verified by instrumental measurements and provisional conclusions have been presented.

Our results show that it is possible to treat the average surface magnitude of the faintest observable comets as very good approximations of the surface magnitude of the night sky. The advantage of this method is its simplicity and the large number of widely distributed reliable comet observers, who have been making observations for many years. These observations were not made to measure light pollution, so they will show no bias. The new method allows a very thorough investigation of the problem without requiring expensive means such as obtaining satellite photographs of the earth's night hemisphere and the associated computer processing.

The main advantage of the method is that it allows us to describe the light pollution level at all locations for which we have the proper observational database. Such observations often start early in the 20th century (and, in some cases, even earlier), and so include periods when light pollution was not researched or even noticed.

Because observable light pollution is highly dependent on the content of aerosols and suspended dust in the atmosphere it is possible to indirectly research this problem as well, especially for areas or periods of time where direct measurements were not, or could not be made. This would require, however, the creation and verification of an appropriate model for light diffusion in the atmosphere.

Our next goal is to make a current map of light pollution in Cracow and its vicinity and then in the mountain regions, mainly using the SQMs together with comet observations, to directly connect instrumental measurements to the observational data. In the future, we also plan to determine the long-term changes in sky brightness in individual regions of Poland using the archive of comet observations. We will also attempt to determinate the character of changes in sky surface brightness in large cities over a year, and to compare it with the content of aerosols and suspended dust in the atmosphere known from the other measurements.

REFERENCES

Allen, J.A. 1880. Destruction of birds by light-houses. *Bulletin of the Nuttall Ornithological Club* 5: 131–138.

Cinzano, P., Falchi, F., Elvidge, C.D. 2001a. Naked eye star visibility and limiting magnitude mapped from DMSP-OLS satellite data. *Monthly Notices of the Royal Astronomical Society* 323: 34–46.

Cinzano, P., Falchi, F., Elvidge, C.D. 2001b. The first World Atlas of the artificial night sky brightness. *Monthly Notices of the Royal Astronomical Society* 328: 689–707.

Della Prugna, F. 1999. Visual measurements and spectral survey of night sky brightness in Venezuela and Italy, *Astronomy and Astrophysics Supplement Series* 140(3): 345–349.

Duriscoe, D.M., Luginbuhl, C.B., Moore, C. 2007. Measuring night-sky brightness with a wide-field CCD camera. *Publications of the Astronomical Society of the Pacific* 119: 192–213.

Elvidge, C., Baugh, K.E., Kihn, E.A., Davis, E.R. 1997. Mapping city lights with nighttime data from the DMSP Operational Linescan System. *Photogrammetric Engineering & Remote Sensing* 63: 727–734.

Garstang, R.H. 1989. Night-sky brightness at observatories and sites. *Publications of the Astronomical Society of the Pacific* 101: 306–329.

Gauthreaux, S.A., Belser, C.G. 2006. Effects of artificial night lighting on migratory birds. In C. Rich & T. Longcore (eds), *Ecological Consequences of Artificial Night Lighting*: 67–93. Washington DC: Island Press.

Green, W.E. 1997. Guide to observing comets. *International Comet Quarterly* Special issue: 1–216.

[IAU] International Astronomical Union, 1998. Resolution A1: Protection of the night sky. *Information Bulletin* 81: 20. http://www.iau.org/static/publications/IB81.pdf (accessed March 12, 2009).

[IDA] International Dark-Sky Association. http://www.darksky.org (accessed Mar 12, 2009).

[IDA Archives] International Dark-Sky Association. Calendar Archives. http://www.darksky.org/mc/page.do?sitePageId=56434&orgId=idsa (accessed March 12, 2009).

[IDA FAQ] International Dark-Sky Association. Frequently Asked Questions. http://www.darksky.org/mc/page.do?sitePageId=61045&orgId=idsa (accessed March 12, 2009).

[IDA Sections] International Dark-Sky Association. IDA Sections. http://www.darksky.org/mc/page.do?sitePageId=56418&orgId=idsa (accessed March 12, 2009).

Isobe, S., Kosai, H. 1998. Star Watching Observations to Measure Night Sky Brightness. *ASP Conference Series* 139:175–184.

Kloog, I., Haim, A., Portnov, B. 2008a. Using kernel density function as an urban analysis tool: Investigating the association between nightlight exposure and the incidence of breast cancer in Haifa, Israel. *Computers, Environment and Urban Systems* 33(1): 55–63.

Kloog, I., Haim, A., Stevens, R., Barchana, M., Portnov, B. 2008b. Light at night co-distributes with incident breast but not lung cancer in the female population of Israel. *Chronobiology International* 25(1): 65–81.

Kloog, I., Haim, A., Stevens, R., Portnov, B. 2009. Global co-distribution of Light at Night (LAN) and cancers of prostate, colon, and lung in men. *Chronobiology International* 26(1): 108–125.

Kosai, H., Isobe, S. 1991. Organised Observations of Night-sky Brightness in Japan during 1987–1989. *Proceedings of the Astronomical Society of Australia* 9(1): 180–183.

Krisciunas, K., Semler, D.L., Richards, J., Schwartz, H., Suntzeff, N.B., Vera, S., Sanhueza, P. 2007. Optical Sky Brightness at Cerro Tololo Inter-American Observatory from 1992 to 2006. *Publications of the Astronomical Society of the Pacific* 119(856): 687–696.

Laulainen, N. 1994. Light Extinction in the Atmosphere. In D. McNally (ed.), *The Vanishing Universe. Adverse Environmental Impacts on Astronomy*: 48–58. New York: Cambridge University Press.

Longcore, T., Rich, C. 2004. Ecological light pollution. *Frontiers in Ecology and the Environment* 2(4): 191–198.

Los Angeles Times 1897. Electricity and English song birds. September 14:7.

Massey, P., Foltz, C.B. 2000. The Spectrum of the Night Sky over Mount Hopkins and Kitt Peak: Changes after a Decade. *Publications of the Astronomical Society of the Pacific* 112(770): 566–573.

McFarlane, R.W. 1963. Disorientation of loggerhead hatchlings by artificial road lighting. *Copeia* 1963: 153.

Moore, M.V., Pierce, S.M., Walsh, H.M., Kvalvik, S.K., Lim, J.D. 2000. Urban light pollution alters the diel vertical migration of *Daphnia*. *Verhandlungen des Internationalen Verein Limnologie* 27: 1–4.

Narisada, K., Schreuder, D. 2004. *Light pollution handbook*. New York: Springer Verlag.

Navara, K.J., Nelson, R.J. 2007. The dark side of light at night: physiological, epidemiological, and ecological consequences. *Journal of Pineal Research* 43: 215–224.

Poulsen, T. 2003. The Light-Pollution Fight Goes Mainstream. *Sky and Telescope* 105(2): 89–90.

Raevel, P., Lamiot, F., 1998. Impacts écologiques de l'éclairage nocturne [Ecological impacts of night lighting]. In *Premier Congrès Européen sur la Protection du Ciel Nocturne*. Paris–La Villette: Cité des Sciences.

Rich, C., Longcore, T. 2006a. *Ecological Consequences of Artificial Night Lighting.* Washington DC: Island Press.

Rich C., Longcore, T. 2006b. Introduction. In C. Rich, T. Longcore (eds), *Ecological Consequences of Artificial Night Lighting*: 1–13. Washington DC: Island Press.

Rich C., Longcore T. 2006c. Preface. In C. Rich, T. Longcore (eds), *Ecological Consequences of Artificial Night Lighting*: xv–xx. Washington DC: Island Press.

Schmiedel, J. 2001. Auswirkungen künstlicher Beleuchtung auf die Tierwelt – ein Überblick [Effects of artificial lighting on the animal world – an overview]. *Schriftenreihe für Landschaftspflege und Naturschutz* 67: 19–51.

Schreuder, D.A. 2001. Sky Glow Measurements in the Netherlands. *IAU Symposium* 196: 130-133.

Squires, W.A., Hanson, H.E. 1918. The destruction of birds at the lighthouses on the coast of California. *Condor* 20: 6–10.

Verheijen, F.J. 1958. The mechanisms of the trapping effect of artificial light sources upon animals. *Archives Néerlandaises de Zoologie* 13: 1–107.

Verheijen, F.J. 1985. Photopollution: Artificial light optic spatial control systems fail to cope with. Incidents, causations, remedies. *Experimental Biology* 1985: 1–18.

Environmental Engineering III – Pawłowski, Dudzińska & Pawłowski (eds)
© *2010 Taylor & Francis Group, London, ISBN 978-0-415-54882-3*

Modeling of PM_{10} and $PM_{2.5}$ particulate matter air pollution in Poland

W. Trapp, M. Paciorek & M.K. Paciorek
Air Protection Unit of Ekometria, Gdansk, Poland

K. Juda-Rezler, A. Warchałowski, M. Reizer
Faculty of Environmental Engineering, Warsaw University of Technology, Warsaw, Poland

ABSTRACT: A three-dimensional chemical transport model, CAMx, is used for the first time to simulate particulate matter (PM) mass concentrations in Poland during the year 2005. The model is coupled to the mesometeorological model WRF and the original emission model EMIL. The resulting maps of PM_{10} and $PM_{2.5}$ concentrations are presented; these concentrations exceed the EU limit values. The WRF–EMIL–CAMx modelling system predictions are evaluated against PM_{10} measurements extracted from the European monitoring database. The performance of the system is good for annual means and autumn–winter diurnal levels and patterns. During spring and summer, the measured values are underpredicted.

Keywords: Air quality modelling, particulate matter, Poland, CAMx, emission model, model evaluation.

1 INTRODUCTION

The third-generation Eulerian Grid Models (EGMs), commonly referred to as Chemical Transport Models (CTMs), are at present the most complex three-dimensional numerical tools capable for representing majority of the atmospheric processes. Recently, a number of worldwide studies have focused on particulate modelling with CTMs, such as: the Unified EMEP model (Fagerli et al. 2004), CMAQ, Community Multiscale Air Quality Model (Byun & Ching 1999), CAMx, Comprehensive Air Quality Model with eXtensions (ENVIRON, 2008), CHIMERE (Bessagnet et al. 2004) and REM-CALGRID (Beekmann et al. 2007). Studies have mainly been performed for the USA (Tesche et al. 2006, Gaydos et al. 2007, Pun et al. 2009, Smyth et al. 2009), but also for Europe (Vautard et al. 2007, Stern et al. 2008). However, application of the CTM models for particulate matter (PM) (aerosols) dispersion simulations is highly demanding, as the PM in the atmosphere has both primary and secondary origins and differs widely in physical (see Van Dingenen et al. 2004) and chemical (see Putaud et al. 2004) properties. Primary PM is emitted from a large variety of anthropogenic, biogenic and natural sources, whereas secondary particles are formed in the atmosphere by chemical and physical processes from gaseous precursors such as nitrogen dioxide (NO_2), sulphur dioxide (SO_2), ammonia (NH_3) and non-methane volatile organic compounds (NMVOCs).

Due to serious health problems associated with the exposure to PM (Pope et al. 2004, WHO 2006), the particulate pollution is, at present, one of the priorities of the European Thematic Strategy on Air Pollution. The PM_{10} (particulate matter with an aerodynamic diameter less than $10\,\mu m$) limit values of the new European Union Directive on Ambient Air Quality and Cleaner Air for Europe (EC, 2008) are currently exceeded over a large part of Europe, especially in urban areas. The PM_{10} short-term limit value (max 35 days above $50\,\mu g/m^3$) is usually exceeded to a much larger extent than the annual mean limit value ($40\,\mu g/m^3$). In 2003–2005, the highest urban background concentrations were measured in cities in Central, Eastern and Southern European countries (EEA 2009, EMEP 2007). Moreover, according to the latest European Environment Agency report (EEA 2009), measured PM concentrations in Europe have not shown, in general, any downward tendencies over the period from 2000 to 2005.

Poland, which is home to more than 38 million people, is associated with some of the highest PM_{10} levels measured in the European Union. Thus, modelling efforts are necessary in order to design cost-effective emission control strategies to attain or make progress towards the attainment of air quality limit values.

The present work is a first attempt at applying the complex CTM model to PM simulation in Poland. It focuses on assessing the ability of our chosen modelling system to reproduce long-term and short-term PM_{10} and $PM_{2.5}$ concentrations. The main objectives of this study are to gain experience in PM modelling and to try to understand a specific "PM climate" in Poland.

The year 2005 was selected as the reference year of the study. During 2005, short-term as well as long-term limit values were exceeded over relatively large parts of Southern Poland. On several winter days, observed PM_{10} daily mean levels reached values well above $150\,\mu g/m^3$ (three times the EU limit value).

Figure 1. The modelling domain with the locations of rural background PM$_{10}$ monitoring sites, with annual data coverage of at least 75% in 2005 (blue squares).

The Eulerian grid CTM model, CAMx, was used to simulate the temporal and spatial PM concentration distribution. The model was coupled to a WRF (Weather and Research Forecasting) mesometeorological model of the NCAR (National Center for Atmospheric Research). For the purpose of this study, the original emission model EMIL was developed and coupled with meteorological and CAMx models. The results of the WRF–EMIL–CAMx modelling system simulations, as well as model performance evaluation, are presented.

Section 2 gives an overview of the modelling tools applied, as well as of the established modelling domain. Section 3 presents the results and discussion, and Section 4 contains the conclusions.

2 MATERIALS AND METHODS

2.1 Modelling domain

The modelling domain established for the study is presented in Figure 1. It covers a part of Central-Eastern Europe, centred over Poland (52.00°N, 19.30°E). The Lambert conformal projection is applied to the modelling grid of 70 grid cells in both east–west and north–south directions, with a spatial resolution of 10 km. The domain's vertical profile contains 18 layers of varying thickness, extending up to 200 hPa. The PM dispersion simulations were performed for the whole of the year 2005.

2.2 Model description

The Comprehensive Air Quality Model with eXtensions (CAMx) is a complex third-generation Eulerian Grid Model, developed at ENVIRON International Corporation (Novato, California). CAMx is a publicly available software for the assessment of gaseous and particulate air pollution (ozone, PM$_{2.5}$, PM$_{10}$, air toxics) over many scales ranging from sub-urban to continental. CAMx simulates the emission, dispersion, chemical reactions and removal of pollutants in the lower troposphere by solving the pollutant

Table 1. Summary of major CAMx components.

Component	CAMx
Gas phase chemistry	CBM-IV (Gery et al. 1989)
Gas phase chemistry solver	CMC fast solver
Horizontal advection	PPM (Colella & Woodward 1984)
Heterogeneous chemistry	N$_2$O$_5$ hydrolysis
Aerosol size distribution	CF (Coarse/Fine)
Inorganic aerosols	Thermodynamic equilibrium with ISORROPIA (Nenes et al. 1998)
Organic aerosols	SOAP (Strader et al. 1999)
Aqueous phase chemistry	RADM (Chang et al. 1987)
Dry deposition/sedimentation	Included
Sea salt	Not included
Meteorological driver	WRF mesoscale model
Boundary conditions	Monthly means from observations

Table 2. Aerosol species included in the PM calculations.

CAMx NAME	Description	PM fraction
PSO4	Sulphate	Fine/coarse
PNO3	Particulate nitrate	Fine/coarse
PNH4	Particulate ammonium	Fine/coarse
POA	Primary Organic Aerosol	Fine/coarse
SOA1-2	Anthropogenic Secondary Organic Aerosols	Fine/coarse
SOA 3-7	Biogenic Secondary Organic Aerosols	Fine/coarse
SOPA	Polymerized anthropogenic SOA	Fine/coarse
SOPB	Polymerized biogenic SOA	Fine/coarse
FPRM	Fine Other Primary (diameter ≤2.5 μm)	Fine/coarse
FCRS	Fine Crustal (diameter ≤2.5 μm)	Fine/coarse
PEC	Primary Elemental Carbon	Fine/coarse
CPRM	Coarse Other Primary	Coarse
CCRS	Coarse Crustal	Coarse

continuity equation for each chemical species on a system of nested three-dimensional grids (ENVIRON 2008).

In this study, air quality simulations have been carried out using CAMx v. 4.51. The Carbon Bond Mechanism IV (Gery et al. 1989) with extensions for aerosol modelling, including 117 reactions – 11 of which are photolytic – and up to 67 species (37 state gases, up to 18 state particulates and 12 radicals) was used. A descriptive summary of CAMx model is presented in Table 1.

Eighteen aerosol species (listed in Table 2) are included in the PM calculations. Several of these species – primary organic aerosols, primary elemental carbon and crustal materials – are present only in the aerosol phase. The secondary organic species – ammonium, nitrate and sulphate – partition between the gas and aerosol phases.

All meteorological fields required by CAMx were calculated by a WRF–CAMx pre-processor, which was developed by ENVIRON International Corporation.

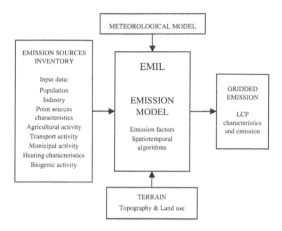

Figure 2. Schematic representation of the EMIL emission model for Poland.

2.3 Emissions

Primary PM (PM_{10} and $PM_{2.5}$) is emitted from a large variety of anthropogenic (energy production, transport, domestic heating and a wide range of industries), biogenic and natural sources. The secondary particles are formed in the atmosphere by chemical and physical processes from gaseous precursors such as nitrogen dioxide (NO_2), sulphur dioxide (SO_2), ammonia (NH_3) and non-methane volatile organic compounds (NMVOCs).

For the purpose of this study, the original emission model EMIL (EMIssion modeL) was developed for Poland and coupled with meteorological and CAMx models. The model generates PM_{10}, $PM_{2.5}$, SO_2, NO_x, NH_3 and NMVOCs emissions from Large Combustion Plants (LCPs), as well as gridded area emissions with 10 km resolution. A schematic representation of the EMIL model is shown in Figure 2.

For the quantification of NO_x, NMVOC and PM emissions, the method proposed by Krüger et al. (2008) was used: the emissions were split sector specifically into the speciation of the chemical compounds that corresponds to the chemical mechanisms of the CAMx model. For NO_x ($NO + NO_2$), it was assumed that 10% (in molar percentages) is emitted directly as NO_2 and 90% as NO (for all emission sectors). The emission of NMVOCs was disaggregated into specific emissions of the species defined in CBM-IV mechanism: PAR (alkane groups), ETH (ethene), OLE (alkene groups), TOL and XYL (aromatics of different reactivity), FORM (formaldehyde) and ALD2 (aldehydes, ketones). The emission of $PM_{2.5}$ was split into organic aerosol (POA), elementary carbon (PEC), and remaining aerosol following the split suggested in the EMEP model (Simpson et al. 2003).

For a LCP with a stack that has a height that is equal or above 100 m ($h \geq 100$ m), a detailed emission and emission parameters database was developed. In Poland, air emissions are strongly connected with the national composition of energy supplies. Hard and brown coal consumption is still the main energy

source, contributing 60% to the total consumption of primary energetic raw materials in the national economy. This results in a significant share of LCP sources in the total national emission of SO_2, NO_x and PM. The created point sources database for reference year 2005 contains data for 445 stacks and consists of the following information: name of the LCP, number of stack, geographical coordinates, installed technological units, stack height and diameter, temperature of exhaust gases, velocity of exhaust gases (yearly average, seasonal averages) and emission amount (yearly average, seasonal averages).

For area (municipal, agricultural, biogenic-forests), mobile (traffic) and point sources with a stack height below 100 m ($h < 100$ m), the model generates emissions for each cell of computational grid (10×10 km^2). This is based on a detailed emission sources inventory composed for reference year 2005 in 1 km $\times$ 1 km resolution, meteorological data and terrain characteristics. Data on population density, sector-specific activity, fuel demands and characteristics, and sector-dependent Polish specific emission factors were gathered. For temporal distribution of emissions, the EMIL model applies sector-specific monthly, daily and hourly emission factors.

Anthropogenic emissions for the parts of the modelling domain not belonging to Poland were prepared based on the UNECE/EMEP database (http://webdab.emep.int/) and treated by CAMx as area sources only. For temporal distribution of these emissions, the method proposed by Krüger et al. (2008) was applied.

2.4 Meteorological fields

In our modelling system, CAMx is forced by the WRF model, which was developed by the collaboration of several research centres (http://www.wrf-model.org). The WRF is a next-generation mesoscale numerical weather prediction system designed to serve both operational forecasting and atmospheric research needs. The WRF software is freely available and can be used for downscaling of weather and climate, ranging from a kilometre to thousands of kilometres, as well as for providing meteorological parameters required for air quality models.

In this study, the WRF version 3.0.1.1 was applied. Microphysical processes are represented by the WRF Single-Moment 3-class (WSM3) scheme (Hong et al. 2004). The Rapid Radiative Transfer Model (RRTM) – based on Mlawer et al. (1997) and the Dudhia scheme (Dudhia 1989) – are used to represent long-wave and short-wave radiation processes, respectively. For Cumulus parameterization, the modified Kain–Fritsch scheme (Kain 2004) with deep and shallow convection is applied. The parameterization of the Planetary Boundary Layer (PBL) follows the non-local Yonsei University scheme (Hong et al., 2006). The similarity based on Monin–Obukhov with a Carslon–Boland viscous sub-layer and standard similarity functions from look-up tables similarity theory and Noah Land

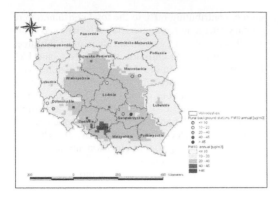

Figure 3. Modelled annual mean PM_{10} concentrations in Poland (2005). Results are from the WRF–EMIL–CAMx modelling system. Dots indicate measured values at rural background monitoring sites. The EU limit value is 40 μg/m³.

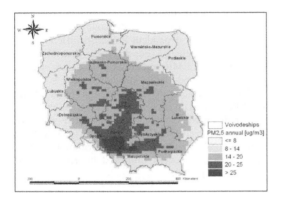

Figure 4. Modelled annual mean $PM_{2.5}$ concentrations in Poland (2005). Results are from the WRF–EMIL–CAMx modelling system. The EU standard is 25 μg/m³.

Surface Model (Chen and Dudhia 2001) are used to represent surface layer and land surface processes, respectively. Meteorological simulations were driven by NCEP FNL (Final) Operational Global Analysis data, with 1.0 degree spatial resolution and temporal resolution of 6 h.

3 RESULTS AND DISCUSSION

3.1 Predicted PM_{10} and $PM_{2.5}$ levels

The EU air quality legislation sets two legally binding limit values for PM_{10} mass concentrations (EC 2008). Exposures to annual mean PM_{10} levels exceeding 40 μg/m³ and PM_{10} concentrations exceeding 50 μg/m³ for more than 35 days per year (36th maximum daily average) are acknowledged as hazardous for European citizens. In addition, the so-called CAFE Directive (EC 2008) sets a non-binding annual $PM_{2.5}$ standard of 25 μg/m³.

Figures 3–5 show the distribution of calculated annual mean PM_{10} and $PM_{2.5}$ levels, as well as the 36th daily maximum of PM_{10}, respectively. High

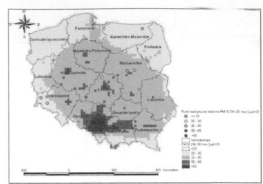

Figure 5. Modelled daily PM_{10} concentrations in Poland – 36th highest daily value (2005). Results are from the WRF–EMIL–CAMx modelling system. Dots indicate measured values at rural background monitoring sites. The EU limit value is 50 μg/m³.

PM_{10} annual mean levels, exceeding the EU limit value, were obtained for the urban areas of Upper and Lower Silesia. High concentrations were also found in Southern-Central Poland. The lowest PM_{10} levels occurred in the north and west, as well as in the east of Poland. The distribution of $PM_{2.5}$ levels (Figure 4) is very similar to that of PM_{10}, however the area with exceedances is larger (Upper and Lower Silesia, Cracow region, Poznań and Łódź from Central Poland). The area with "safe" $PM_{2.5}$ levels is much smaller than in the case of PM_{10}. The 36th daily maximum of PM_{10} (Figure 5) had an analogous distribution to both previous parameters. The highest 36th daily maximum values with exceedances of EU limit value are seen in Southern-Central Poland, as well as in the urban areas in Upper and Lower Silesia, Cracow, Warsaw, Łódź, Poznań, Rzeszów, Gdańsk, Lublin and Bydgoszcz. The areas with high PM concentrations are those with high emissions of PM and PM precursors mainly from combustion processes in energy production, as well as from industrial and residential sectors. Burning coal, frequently of low quality, in domestic stoves is, at present, the biggest PM pollution problem to be solved in Poland.

3.2 Model evaluation

The model results obtained for the reference year 2005 were compared to observations. Measured PM_{10} data were extracted from the European monitoring database AirBase, supplemented by EMEP stations that are not reported on AirBase. Only stations classified by AirBase and/or EMEP as rural background were considered. Other station types (suburban, urban background, traffic and industrial) were not taken into account, as they represent local-scale concentration levels that are not applicable to the regional-scale model evaluation. In the reference year 2005, only 10 rural background stations in Poland had required an annual data coverage of at least 75%. The location of stations used for model evaluation is presented in Figure 1. Measured values of annual mean and of the 36th highest daily PM_{10} levels are represented by coloured dots, respectively, in Figures 3 and 5.

$PM_{2.5}$ was measured only at one rural background station; however, the data coverage was poor. The data regarding chemical speciation of $PM_{10}/PM_{2.5}$ was not available as well. Therefore, CAMx performance evaluation was carried out only for PM_{10}.

For model evaluation, a number of statistical measures can be applied (see, for example, Borrego et al. 2008). For this study, a subset of parameters proposed by Juda–Rezler (2009), which characterise the general uncertainties estimation, was applied. The subset consists of the following measures: NMB (Normalized Mean Bias), RMSE (Root Mean Square Error), with its systematic ($RMSE_s$) and unsystematic ($RMSE_u$) part, NMSE (Normalized Mean Square Error), correlation coefficient (r) and IA (Index of Agreement). The formulas are given below (Juda–Rezler 2009):

$$NMB = \frac{\sum_{i=1}^{N}\left(C_{oi} - C_{pi}\right)}{\sum_{i=1}^{N}\left(C_{oi}\right)} \quad (1)$$

$$RMSE = \sqrt{\frac{1}{N}\sum_{i=1}^{N}\left(C_{oi} - C_{pi}\right)^2} \quad (2)$$

$$RMSE_s = \sqrt{\frac{1}{N}\sum_{i=1}^{N}\left(C_{oi} - \hat{C}_{pi}\right)^2} \quad (3)$$

$$RMSE_u = \sqrt{\frac{1}{N}\sum_{i=1}^{N}\left(C_{pi} - \hat{C}_{pi}\right)^2} \quad (4)$$

$$NMSE = \frac{\frac{1}{N}\cdot\sum_{i=1}^{N}\left(C_{oi} - C_{pi}\right)^2}{\overline{C}_o \cdot \overline{C}_p} \quad (5)$$

$$r = \frac{\sum_{i=1}^{N}\left(C_{oi} - \overline{C}_o\right)\cdot\left(C_{pi} - \overline{C}_p\right)}{\sqrt{\sum_{i=1}^{N}\left(C_{oi} - \overline{C}_o\right)^2 \cdot \sum_{i=1}^{N}\left(C_{pi} - \overline{C}_p\right)^2}} \quad (6)$$

$$IA = 1 - \frac{\sum_{i=1}^{N}\left(C_{oi} - C_{pi}\right)^2}{\sum_{i=1}^{N}\left(\left|C_{pi} - \overline{C}_o\right| + \left|C_{oi} - \overline{C}_o\right|\right)^2} \quad (7)$$

where N = the total number of monitoring stations; C_o and C_p = the concentrations observed and predicted; $\hat{C}_{pi}$ = the conditional mean value of C_p on a given C_o.

The scatter plot of the predicted versus observed annual mean PM_{10} concentrations in 2005 is presented in Figure 6, whereas the statistical indices are given in Table 3. The results indicate a satisfactory model performance; however, in general, the model tends to underestimate long-term observations. The predicted

Table 3. Evaluation results of the WRF–EMIL–CAMx modelling system for 2005.

Statistical measures Number of stations, N = 10	WRF–EMIL–CAMx
NMB – Normalised Mean Bias	0.20
RMSE – Root Mean Square Error [$\mu g/m^3$]	8.48
$RMSE_s$ – Systematic Root Mean Square Error [$\mu g/m^3$]	5.44
$RMSE_u$ – Unsystematic Root Mean Square Error [$\mu g/m^3$]	6.51
NMSE – Normalised Mean Square Error	0.18
r – Correlation coefficient	0.66
IA – Index of Agreement	0.74
Predictions within a factor of 2 of the observations [%]	90.00

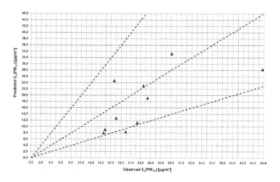

Figure 6. The scatter plot of the predicted and observed annual mean PM_{10} levels in Poland, for the year 2005. Dashed lines indicate perfect agreement (middle line) and a difference of a factor of 2.

spatial mean concentration is very close to the measured mean and the NMB bias is equal to 20%. The NMSE is small and equals 18%. The RMSE is also quite low; however, its systematic part is substantial, which means that there is a systematic error in the PM predictions. The correlation measures (r, IA) of 0.66 and 0.74, respectively, indicate a good correlation between measurements and predictions. Finally, nine from ten compared pairs lie inside a factor 2 agreement area.

The time series of daily mean values are presented for two stations from Central Poland Tłuszcz (Figure 7) and Legionowo (Figure 8). During winter, the patterns of measured levels are fairly well simulated by the model; moreover, high (peak) values are well captured. The ability of the model to capture high PM_{10} concentrations is promising and is probably due to detailed emission data provided by the emission model EMIL, which was developed specifically for the country. The good MM5–CAMx modelling system performance for winter PM levels has also been obtained for the New Zealand city of Christchurch.

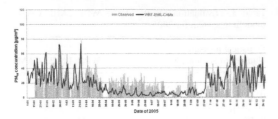

Figure 7. Observed and calculated PM$_{10}$ time series of daily mean concentrations at Tłuszcz (Poland) – for the year 2005.

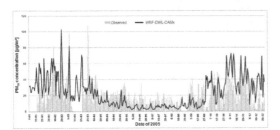

Figure 8. Observed and calculated PM$_{10}$ time series of daily mean concentrations at Legionowo (Poland) – for the year 2005.

Titov et al. (2007) simulated PM distribution for June–August 2000 (winter time in New Zealand), for area dominated by fuel combustion in domestic, transport and industrial sources. The authors obtained reasonably good diurnal variation in ground-level PM$_{10}$, as well as an IA between 0.90 and 0.96 and an r between 0.85 and 0.89, depending on the number of vertical layers used in the calculations. However, in a recent PM modelling study performed for Northern Germany by Stern et al. (2008), an increasing underestimation of the concentrations obtained with increasing observed PM$_{10}$ was obtained for five three-dimensional CTM models applied for simulation of an 80-day, winter–spring period during 2003.

Surprisingly, during the spring–summer seasons, the modelling system applied in the present study performed worse than in winter (see Fig. 7 and 8). The system reproduces diurnal patterns quite well; however, it underpredicts the measured PM$_{10}$ levels. A significant production of PM by processes, which are not included in the CAMx model, might be responsible for that situation. We suspect that, in Poland, especially during the warm seasons, there are several processes that produce PM, such as: grass burning, forest fires, as well as windblown or resuspended dust.

The underprediction of PM (PM$_{2.5}$) levels in summer has also been found for long-term simulations (entire 2001 year) performed by Appel et al. (2008) for the Eastern USA (MM5–CMAQ modelling system). At the same time, they obtained overpredictions in winter. Two CTM models – CMAQ and CAMx – driven by the MM5 meteorological model have been used for a full-year simulation (entire 2002 year), performed for the Eastern USA by Tesche et al. (2006). During all seasons, both models tended to overpredict

PM$_{10}$, whereas PM$_{2.5}$ was overpredicted by CAMx and slightly underpredicted by CMAQ.

It is clear from the above discussion that the complexity of the particulate problem makes PM modelling very difficult. The current state of PM modelling in Europe was recently summarized in the EMEP Particulate Matter Assessment Report (2007). The report states that, currently, all models underestimate the total PM$_{10}$ and PM$_{2.5}$, and therefore do not achieve mass closure. Similar conclusions were drawn from the results of the CityDelta and EuroDelta modelling experiments for Europe (Vautard et al. 2007, Stern et al. 2008). One of the proposed methods for improving CTM performance is to adjust the model values with observations through the use of data assimilation techniques. Such methods have been applied with promising results by Denby et al. (2008) to estimate the exceedances of the limit values for PM$_{10}$ on a regional scale in Europe.

4 CONCLUSIONS

In this study, the high-resolution modelling system based on the mesometeorological model WRF, the original emission model EMIL and the chemical transport model CAMx was implemented for the first time for the Central-Eastern European domain, centred over Poland.

The highest concentrations were calculated for the Upper Silesia and Malopolskie regions of southern Poland. The resulting maps suggest that the EU PM$_{10}$ and PM$_{2.5}$ annual air quality standards are exceeded for several urban areas of Poland, with the PM$_{2.5}$ standard being exceeded to a larger extent than the PM$_{10}$ standard, and the short-term standard being exceeded to a much larger extent than the long-term ones. In the present study, we obtained a reasonably good WRF–EMIL–CAMx modelling system performance for the annual mean PM$_{10}$ levels in Poland. The system reproduces the measured PM$_{10}$ concentrations with a normalised mean square error (NMSE) of less than 20% and with a correlation coefficient (r) of 0.66. The predictions are within a factor of 2 of the measurements for 90% of the data points. A clear seasonal cycle appears both in observations and simulations. Much larger levels are observed in the cold seasons than in warm seasons, and seasonal variation is well captured by the modelling system. However, for short-term calculations, we obtained mixed system performance. During the cold season, the system is able to reproduce the measured daily mean PM$_{10}$ levels and time patterns quite well. It is also able (in contrast to the CTMs tested in a recent study by Stern et al., 2008) to capture the high concentrations in moderately polluted sites in central Poland. This is probably a result of a detailed emission inventory and country-specific emission factors used in the emission model EMIL. On the other hand, the system performance during the warm seasons is worse: the time pattern of PM$_{10}$ levels is simulated quite well, but measured values are substantially underpredicted.

The relatively poor modelling system performance during the spring and summer can be partly explained by the following model features:

- Absence of some natural (as sea-salt) and anthropogenic PM sources.
- Absence of some PM production processes, such as grass burning, forest fires, as well as windblown and resuspended dust.
- Uncertainties in the prediction of the meteorological input data.
- Poor representation of the physical and chemical processes that lead to the formation of secondary inorganic (SIA) and organic (SOA) aerosols.

As an overall conclusion, in our opinion, this study has helped to understand the specific "PM climate" in Poland and to gain experience in PM modelling with complex CTM model. However, it has also opened a huge set of questions about particulate pollution in Poland and accompanying modelling complexities. Future research is needed and will focus on the formation mechanisms, and physical and chemical transformations of atmospheric aerosols. There is also a necessity for model evaluation of $PM_{2.5}$ mass and composition (sulphate, nitrate, ammonium, elemental carbon and organic carbon). As there are no such measurements in Poland yet, appropriate German and Czech stations could be used for this end. Thus, as a next step, it would be interesting to use the WRF–EMIL–CAMx modelling system for extended domains and for a broader evaluation exercise. The application of data assimilation techniques, as a promising method for improving PM model performance, is also taken into consideration.

ACKNOWLEDGEMENTS

This research was partially supported by the Central and Eastern Europe Climate Change Impact and Vulnerability Assessment Project (CECILIA), financed by EU 6. FP Contract GOCE 037005 to Warsaw University of Technology (Warsaw, Poland) and is partially the product of work carried out for the Polish Main Environment Protection Inspectorate in Warsaw, Poland.

REFERENCES

Appel, K.W., Bhave, P.V. & Gilliland, A.B., Sarwar G. & Roselle S.J. 2008. Evaluation of the community multiscale air quality (CMAQ) model version 4.5: Sensitivities impacting model performance; Part II—particulate matter. *Atmospheric Environment* 42: 6057–6066.

Beekmann, M., Kerschbaumer, A., Reimer E., Stern R. & Moller D. 2007. PM measurement campaign HOVERT in the Greater Berlin area: model evaluation with chemically specified observations for a one year period. *Atmospheric Chemistry and Physics* 7: 55–68.

Bessagnet, B., Hodzic, A., Vautard, R., Beekmann, M., Cheinet, S., Honore, C., Liousse, C. & Rouil C. 2004.

Aerosol modeling with CHIMERE—preliminary evaluation at the continental scale. *Atmospheric Environment* 38: 2803–2817.

Borrego, C., Monteiro, A., Ferreira, J., Miranda, A.I., Costa, A.M., Carvalho, A.C. & Lopes M. 2008. Procedures for estimation of modelling uncertainty in air quality assessment. *Environment International* 34: 613–620.

Byun, D.W. & Ching, J.K.S. 1999. *Science Algorithms of the EPA Models-3 Community Multiscale Air Quality (CMAQ) Modelling System.* U.S. EPA Report EPA/600/R-99/030.

Chang, J.S., Brost, R.A., Isaksen, I.S.A., Madronich, S., Middleton, P., Stockwell, W.R. & Walcek C.J. 1987. A three-dimensional eulerian acid deposition model: Physical concepts and formulation. *Journal of Geophysical Research* 92: 14681–14700.

Chen, F. & Dudhia J. 2001. Coupling an advanced land-surface/hydrology model with the Penn State/NCAR MM5 modeling system. Part I: Model description and implementation. *Monthly Weather Review* 129: 569–585.

Colella, P. & Woodwar, D.P.R. 1984. The Piecewise Parabolic Method (PPM) for gas dynamical simulations. *Journal of Computational Physics* 54: 174–201.

Denby, B., Schaap, M., Segers, A., Builtjes, P. & Horalek, J. 2008. Comparison of two data assimilation methods for assessing PM_{10} exceedances on the European scale. *Atmospheric Environment* 42: 7122–7134.

Dudhia, J. 1989. Numerical study of convection observed during the winter monsoon experiment using a mesoscale two-dimensional model. *Journal of the Atmospheric Sciences* 46: 3077–3107.

EC 2008. Directive 2008/50/EC of the European Parliament and of the Council of 21 May 2008 on ambient air quality and cleaner air for Europe.

EEA 2009. Spatial assessment of PM10 and ozone concentrations in Europe (2005). *Technical Report 1/2009.* Copenhagen: EEA.

EMEP 2007. EMEP Particulate Matter Assessment Report. *EMEP/CCC-Report 8/2007.* Oslo: EMEP.

ENVIRON 2008. CAMx Users' Guide, version 4.50.

Fagerli, H., Simpson, D. & Tsyro, S. 2004. Unified EMEP model: Updates. *EMEP/MSC-W Status Report 1/2004.* Oslo: EMEP.

Gaydos, T.M., Pinder, R., Koo, B., Fahey, K.M., Yarwood, G. & Pandis, S.N. 2007. Development and application of a three-dimensional aerosol chemical transport model, PMCAMx. *Atmospheric Environment* 41: 2594–2611.

Gery, M.W., Whitten, G.Z., Killus, J.P. & Dodge M.C. 1989. A photochemical kinetics mechanism for urban and regional scale computer modelling. *Journal of Geophysical Research* 94: 12925–12956.

Hong, S.Y., Dudhia, J. & Chen, S.H. 2004. A revised approach to ice microphysical processes for the bulk parameterization of clouds and precipitation. *Monthly Weather Review* 132: 103–120.

Hong, S.Y., Noh, Y. & Dudhia, J. 2006. A new vertical diffusion package with an explicit treatment of entrainment processes. *Monthly Weather Review* 134: 2318–2341.

Juda-Rezler, K. 2009. New Challenges in Air Quality And Climate Modelling. *Archives of Environmental Protection,* in print.

Kain, J.S. 2004. The Kain-Fritsch convective parameterization: An update. *Journal of Applied Meteorology* 43: 170–181.

Krüger, B.C., Katragkou, E., Tegoulias, I., Zanis, P., Melas, D., Coppola, E., Rauscher, S., Huszar, P. & Halenka, T. 2008. Regional photochemical model calculations for Europe concerning ozone levels in a changing climate.

Quarterly Journal of the Hungarian Meteorological Service 112 (3–4): 285–300.

Mlawer, E.J., Taubman, S.J., Brown, P.D., Iacono, M.J. & Clough, S.A. 1997. Radiative transfer for inhomogeneous atmosphere: RRTM, a validated correlated-k model for the longwave. *Journal of Geophysical Research* 102: 16663–16682.

Nenes, A., Pilinis, C. & Pandis, S.N. 1998. ISORROPIA: A new thermodynamic model for multiphase multicomponent inorganic aerosols. *Aquatic Geochemistry* 4: 123–152.

Pope, C.A. III, Burnett, R.T., Thurston, G.D., Thun, M.J., Calle, E.E., Krewski, D. & Godleski, J.J. 2004. Cardiovascular mortality and long term exposure to fine particulate air pollution: Epidemiological evidence of general pathophysiological pathways of disease. *Circulation* 109: 71–77.

Pun, B.K., Balmori, R.T.F. & Seigneur, C. 2009. Modeling wintertime particulate matter formation in central California. *Atmospheric Environment* 43: 402–409.

Putaud, J.P., Raes, F., Van Dingenen, R., Brüggemann, E., Facchini, M.-C., Decesari, S., Fuzzi, S., Gehrig, R., Hüglin, C., Laj, P., Lorbeer, G., Maenhautg, W., Mihalopoulos, N., Müller, K., Querol, X., Rodriguez, S., Schneider, J., Spindler, G., ten Brink, H., Tørseth, K. & Wiedensohler A. 2004. A European aerosol phenomenology – 2: chemical characteristics of particulate matter at kerbside, urban, rural and background sites in Europe. *Atmospheric Environment* 38: 2579–2595.

Simpson, D., Fagerli, H., Jonson, J.E., Tsyro, S., Wind, P. & Tuovinen J.-P. 2003. Unified EMEP Model Description. *EMEP/MSC-W Status Report 1/2003 Part I*. Oslo: EMEP.

Smyth, S.C., Jiang, W., Roth, H., Moran, M.D., Makar, P.A., Yang, F., Bouchet, V.S. & Landry, H. 2009. A comparative performance evaluation of the AURAMS and CMAQ air-quality modelling systems. *Atmospheric Environment* 43: 1059–1070.

Strader, R., Lurmann, F. & Pandis, S.N. 1999. Evaluation of secondary organic aerosol formation in winter. *Atmospheric Environment* 33: 4849–4863.

Stern, R., Builtjes, P., Schaap, M., Timmermans, R., Vautard, R., Hodzic, A., Memmesheimer, M., Feldmann, H., Renner, E., Wolke, R. & Kerschbaumer, A. 2008. A model inter-comparison study focussing on episodes with elevated PM10 concentrations. *Atmospheric Environment* 42: 4567–4588.

Tesche, T.W., Morris, R., Tonnesen, G., McNally, D., Boylan, J. & Brewer P. 2006. CMAQ/CAMx annual 2002 performance evaluation over the eastern US. *Atmospheric Environment* 40: 4906–4919.

Environmental Engineering III – Pawłowski, Dudzińska & Pawłowski (eds)
© 2010 Taylor & Francis Group, London, ISBN 978-0-415-54882-3

Lead and zinc in the street dust of Zielona Gora, Poland

B. Walczak

Department of Land Protection and Reclamation, Institute of Environmental Engineering, University of Zielona Gora, Zielona Gora, Poland

ABSTRACT: The lead and zinc contents (total and readily soluble) of street dust from arterial roads in Zielona Gora, Poland were studied. Dust samples were collected twice, in 56 places within the city. Street-dust concentrations of these metals were found to be higher than those in soils adjacent to the streets sampled. Lead and zinc levels in street dust met Polish soil and ground standards (Regulation of the Minister of Environment 9.09.2002) for industrial land and communication routes, but not for urban areas. However, levels in street dust were similar to those reported for other cities in the world.

Keywords: Street dust, lead, zinc.

1 INTRODUCTION

The worsening health conditions of the inhabitants of large cities, and especially the increasing occurrence of allergic diseases and asthma in children, has led to a new approach towards street dust as an aspect of the urban environment, which has a significant impact on the health of city dwellers. We must also take into account the fact that streets form a large percentage of urban areas. We are not able to eliminate street dust; we can only reduce its emission or remove it regularly by cleaning streets. The presence of heavy metals in suspended dust and soils of the urban environment is a well-investigated subject both in Poland and all over the world. The problem of street dust (loose material deposited on streets and squares in cities) is very poorly investigated in the world, and almost hardly at all in Poland. Street dust is common in cities. The composition of street dust may include soil erosion from land, industrial emissions, construction and automotive exhausts, debris from the surface of streets and car tires. Research scientists from Norway have shown that street dust and suspended matter PM_{10} depend on each other.

Metals present in an urbanized area may come from many different sources. One of the most important heavy metal sources is vehicle emissions. Three main factors are known to influence the level of heavy metals in dust samples: traffic, industry and weathered materials, particularly house and street dust (Arslan 2001, Al-Khashman 2004). Road paint degradation, vehicle wear, and lubricating oil and particulate emissions are also known to have a great impact on heavy metal content (Al-Khashman 2007).

The main source of lead in street dust is fuel burned by cars (Lagerwerff & Specht 1970). The introduction of unleaded gasoline all over Europe has resulted in a reduction in lead emissions and a loss of Pb in street dust. The effect of lead on plants causes shortening of roots, dark-green chlorosis, wilting foliage and necrosis; in plant tissues, oxidation processes are affected (photosynthesis, lipid phosphorylation and breakdown processes) (Zablocki et al. 1998).

Lead in large concentrations affects human beings and causes nervous system disorders, anaemia, infertility, impaired foetal development, impaired mental development and lower levels of intelligence. Toxic concentrations may occur in bone marrow in which red blood cells are produced. Lead can affect the course of at least five stages of the formation of haemoglobin, which can lead to anaemia. Lead also irritates the human nervous system, causing deterioration of the mental system (depression, aggression, weakness of memory), and damages heart muscle, the kidneys and the cerebral cortex.

Zinc is a component of diesel fuel and, as such, also contributes to the formation of pollution arising from industrial activities (Zablocki et al. 1998). Zinc leads to disorders in plant development, chlorosis and deterioration of photosynthesis. Tissues become horny and shrink. The critical zinc concentration is 400 ppm. The negative impact of zinc on human beings can be illustrated by swelling of the lungs, blood poisoning and cancer.

The aim of this paper is to present the concentrations of lead and zinc in the street dust of a city in western Poland. The monitoring of metal content assists us in assessing the distribution of pollution in an urban area.

2 MATERIALS AND METHODS

2.1 *Location and characteristics of the research area*

The research area was located in Zielona Góra, a town in the western part of Poland in the Lubuski voivodeship (province).

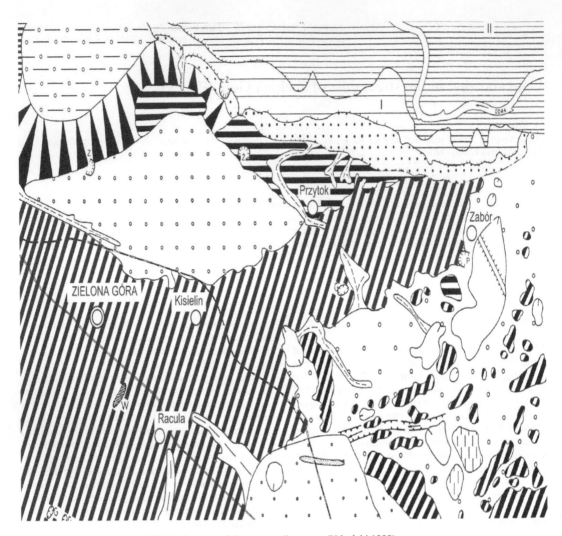

Figure 1. Geomorphology of the study area and the surrounding areas (Urbański 1998).

The area of Zielona Góra can be characterized by high geological diversity. This is a result of the activity of glaciers (moraine deposits and ice), as well as wind erosion and the accumulation of river sediments. The central and northern part of the town is located within the Highlands. Figure 1 shows the geomorphological structures of this region. Zielona Gora has 119,000 inhabitants (as reported in 2006). The city is mainly a service city. Until 1989, however, the town was heavily industrialized. The environmental conditions in Zielona Góra are affected by municipal heating plants and power stations, as well as by heating of individual houses by coal. The main threat to the city is large-vehicle traffic, which is increasingly common. Dustiness of the city is also, to some extent, a result of the fact that many ex-building sites are left uncovered with plants or pavements. Zielona Góra is also subject to a large loss of soil due to erosion and the soil falls onto the streets.

2.2 Location of sampling sites of street dust and soil

Street dust samples were collected from the streets within Zielona Gora city, at 56 points; soil samples were collected from 10 different places. The samples obtained of industrial, park, rural and residential origin. The locations of sampling sites of street dust and soil are shown in Figure 2.

2.3 Sample collection

Samples of street dust were collected on two separate occasions: first in February 2001, and then in May 2002 from the same places. The material was taken from the lane of the road adjacent to the edge of the road, up to 0.5 m from the edge, over a length of about 10 m. Street dust was collected using a brush and was then placed in boxes. In order to detect and recognize one of the potentially significant sources of

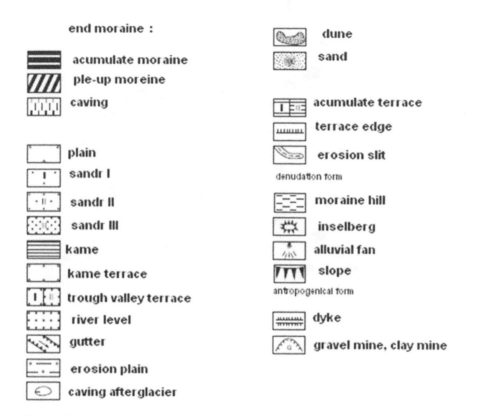

Geomorfology sketch

Skala 1:100 000

Glacier form

end moraine :

- acumulate moraine
- ple-up moreine
- caving

- plain
- sandr I
- sandr II
- sandr III
- kame
- kame terrace
- trough valley terrace
- river level
- gutter
- erosion plain
- caving after glacier

- dune
- sand

- acumulate terrace
- terrace edge
- erosion slit

denudation form

- moraine hill
- inselberg
- alluvial fan
- slope

antropogenical form

- dyke
- gravel mine, clay mine

Figure 1. (*Continued*)

the material occurring on road surfaces, the averaged soil samples were collected from squares near the road dust collection sites. Each time, these samples were obtained by collecting unit samples from a soil surface layer (0–10 cm) from an area of approximately 10 m² and by averaging them.

2.4 *Analytical methods*

The content of total lead and zinc in street dust and soils was determined by means of atomic absorption spectrometry AAS FL; extracts were obtained after igniting the samples in a muffle furnace at a temperature of 550°C and after dissolving them in aqua regia. The content of lead and zinc in street dust available to living organisms was assessed using atomic absorption spectrometry AAS FL in 0.1 M of hydrochloric acid. Multiple repetitions of the results were performed, and the results were found to be reproducible.

3 RESULTS AND DISCUSSION

The content of lead, zinc, organic carbon and the soil pH are presented in Table 1. Soils showed a low pH level and a low organic carbon content. The total and readily soluble content of lead values are shown in Table 2. The total and readily soluble content of zinc is presented in Table 3. Organic carbon and pH values in street dust samples were found to be higher than those in soil samples. Table 4 illustrates the organic carbon content and pH values of 56 samples of street dust.

Analyzing the reaction of urban dust led us to conclude that this is different from the reaction of urban soils. The reaction of road dusts in Zielona Góra has a basic tendency, showing a higher pH level than soil. In Zielona Góra, there were no significant differences in the reaction of road dusts collected in the winter and summer months. The reaction did not show a specific tendency to show a higher pH level either in winter or in summer. The basic reaction of Zielona Góra road dusts does not differ from road dusts in other cities all over the world. The road dusts in Hong Kong showed a reaction of pH 8.5–9.20 (Wong et al. 1984). Road dust from Lancaster in the UK showed a reaction of pH 6.9–8.4; in the majority of cases, it was around pH 7.5 (Harrison et al. 1981).

The organic matter content of Zielona Góra street dust is higher than that in the city's soils; this may be

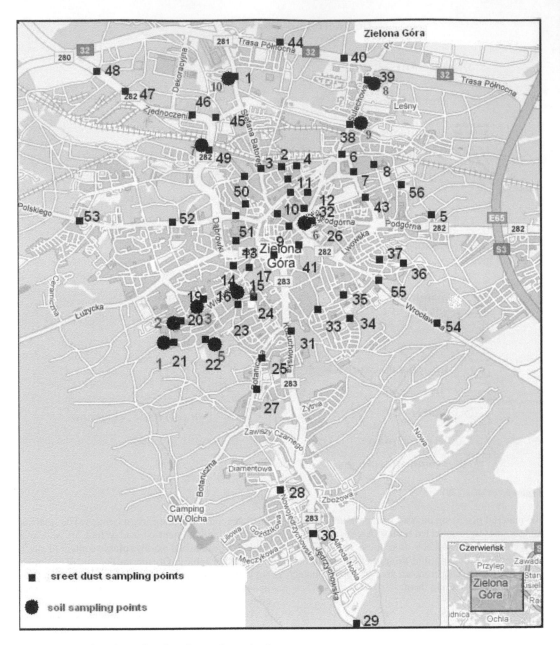

Figure 2.   Location of sampling sites of street dust and soil.

caused by migration of the materials from the surface layer of the road shoulder soil, which is (in general) richer in organic matter. Xie et al. (2000) suggest in their research work that street dust is an important carrier of organic matter. Organic compounds arising from street dust have a great impact on the development of asthmatic diseases, which is a result of lead bonding by organic matter. This process is referred to in the research of Al-Chalabi and Hawker (1996). If it is assumed that the high pH is responsible for an increased lead retention in street dusts and soil, this leads to an increased amount of lead being present in street dust where pH is high. Static analysis of Zielona Góra street dusts does not show a significant correlation between the content of organic matter and lead, which can to some extent be explained by a major difference in results.

This fact was inevitably influenced by a diversity of composition of liquids from versatile mineral and organic matter with a diversified anthropopression.

The lead and zinc contents of street dusts in Zielona Góra are higher than those of lead and zinc in soils

Table 1. Organic carbon, pH level and concentrations of lead and zinc in soil.

Sample	pH	Organic carbon %	Lead (Pb) mg·kg^{-1}		Zinc (Zn) mg·kg^{-1}	
			0.1 M HCl	Aqua regia	0.1 M HCl	Aqua regia
1	4.73	0.05	24.53	30.80	11.42	59.00
2	6.05	0.09	4.91	16.40	12.54	51.60
3	6.74	0.10	14.75	16.60	24.30	78.60
4	5.10	0.12	19.38	22.37	13.21	52.80
5	6.15	0.13	40.98	47.20	46.68	126.80
6	6.98	0.39	5.93	65.40	59.60	245.40
7	6.52	0.14	21.84	30.60	26.94	85.60
8	6.72	0.09	36.92	40.40	39.01	100.60
9	6.65	0.15	18.20	19.67	54.22	104.00
10	5.01	0.03	9.40	13.61	7.84	36.00

Table 2. Concentrations of lead in street samples.

Sample number	Lead (Pb) [mg·kg^{-1}]				Sample number	Lead (Pb) [mg·kg^{-1}]			
	Aqua regia I series	0.1 M HCl I series	Aqua regia II series	0.1 M HCl II series		Aqua regia I series	0.1 M HCl I series	Aqua regia II series	0.1 M HCl II series
1	151.46	77.09	49.14	40.99	29	262.4	29.83	49.72	29.34
2	278.48	178.3	67.94	53.11	30	27.02	20.31	92.74	78.01
3	129.62	64.58	70.8	53.46	31	40.96	31.97	68.7	54.68
4	30.88	27.25	35.5	32.52	32	42.08	34.65	30.04	20.69
5	45.02	42.85	81.18	43.59	33	36.32	23.94	33.74	24.48
6	37.84	28.83	145.3	53.39	34	55.4	37.59	35.86	25.37
7	28.86	25.19	59.66	20.84	35	34.6	23.36	53.02	37.35
8	41.56	35.11	48.04	36.3	36	21.28	18.08	112.14	75.19
9	32.82	29.6	94.94	36.6	37	26.38	19.87	130	54.12
10	28.42	24.58	60.06	23.96	38	46	34.82	39.4	28.72
11	16.78	15.26	137.14	21.84	39	102.86	45.89	39.22	34.61
12	58.54	43.3	150.96	31.5	40	69.72	47.31	56.06	22.14
13	23.76	18.9	42.4	28.85	41	75.24	39.65	37.78	13.97
14	57.6	52.3	44.9	29	42	55.4	23.3	31.94	18.89
15	17.38	13.25	109.88	53.19	43	28.62	22.32	31.62	22.73
16	21.2	19.7	72.28	48.99	44	104.1	52.63	38.66	24.73
17	8.7	6.7	39.02	30.2	45	97	78.74	92.38	33.9
18	51.54	47.53	56.96	47.03	46	56.4	51.29	46.52	40.16
19	20.24	18.2	34.5	29.25	47	41.92	21.47	43.24	21.83
20	14.06	12.1	50	10.81	48	57.66	21.82	46.08	25.47
21	16.62	14.32	60.5	30.1	49	148.5	75.91	88.98	52.29
22	37.42	18.29	124.02	20.13	50	36	34.75	56.3	35.65
23	26.5	14.58	158.14	29.26	51	47.66	27.17	83.12	42.02
24	38.5	30.81	93.72	25.79	52	44.08	34.61	130.06	19.74
25	30.6	28.75	172.78	38.42	53	40.34	27.99	119.3	93.01
26	111.5	32.43	27.56	18.81	54	49.94	30.17	71.7	37.43
27	30.6	26.68	36.14	24.77	55	34.46	23.07	51.76	34.06
28	18.5	17.8	62.54	38.95	56	80.86	46.87	120.56	65.14
Minimum						8.7	6.7	27.56	10.81
Maximum						278.28	178.3	172.78	93.01
Mean						56.57	34.7	71.72	36.02

located near roads. Zinc is presented in series I and II of studies in similar areas, but the content of these elements is repeated at the same level of series I and II. The content of Pb in road dust in series I in 80% of cases is also higher than in series II.

There are no qualifying standards for street dust on city streets, but it is a material that is similar in its composition to soil. By comparing the content of pollutants in street dust of Zielona Góra to the Polish Regulation of the Minister of Environment of 9.09.2002 regarding standards of soil and ground quality (Table 5), we can conclude that all of the ingredients in the street dust in Zielona Góra meet soil and land quality standards

Table 3. Concentrations of zinc in street dust samples.

| | Zinc (Zn) [mg · kg^{-1}] | | | | | Zinc (Zn) [mg · kg^{-1}] | | | |
| | Aqua regia | 0.1 M HCl | Aqua regia | 0.1 M HCl | | Aqua regia | 0.1 M HCl | Aqua regia | 0.1 M HCl |
Sample number	I series	I series	II series	II series	Sample number	I series	I series	II series	II series
1	409.86	123	77.82	74.37	29	68.4	45.87	154.74	103.62
2	405	128	138.4	110.37	30	62.96	39.59	135.8	119.6
3	292.6	96.49	138.4	123.5	31	98.46	73.87	160.78	127.89
4	103.84	68.39	85.92	74.12	32	143.62	97.9	80.46	76.74
5	148.56	84.05	102.78	93.2	33	91.54	54.92	68.24	57.26
6	123.58	59.02	183.9	91.41	34	145.68	109.82	157.16	71.22
7	138.62	89.9	69	52.1	35	79.14	51.09	156.92	124.36
8	105	65.67	111.32	108.41	36	33	22.11	172.66	117.56
9	185.46	115	113.04	88.82	37	43.08	23.23	272	161.58
10	124.14	80.76	106.22	70.26	38	102.78	76.83	72.08	70.1
11	84.02	50.35	58.34	45.75	39	162.66	100.51	109.04	106.2
12	314.2	135	75.24	66.89	40	154.96	109.62	75.1	68.39
13	146.28	87.24	85.28	73.24	41	264.2	138	32.04	29.37
14	189.32	103.63	115.66	94.77	42	106.94	84.66	77.12	71.25
15	103.62	35.76	242.4	116.8	43	95.36	64.6	63.68	51.56
16	82.58	59.28	128.48	117.3	44	174.34	112	83.76	72.55
17	70.84	54.62	96.7	93.2	45	280.6	125	176.16	124.87
18	190.52	102.76	128.24	111.3	46	112.3	86.33	127.08	118.36
19	100.7	47.75	57	52.54	47	98.9	72.07	65.7	61.23
20	94.48	56.71	28.24	21.38	48	92.64	59.32	82.98	78.8
21	95.02	42.87	63.88	48.07	49	193.16	109	208.42	144.6
22	71.64	29.01	41.24	39.96	50	102.84	100.24	124.1	96.4
23	62.88	30.43	121.28	97.88	51	113.06	70.74	220.2	156.87
24	115.02	44.44	132.92	119.5	52	98.96	78.01	87.98	49.84
25	85.4	38.47	133.86	109.26	53	82.16	66.91	250	155.87
26	84.56	38.46	39.44	32.13	54	110.06	67.83	87.56	73.24
27	95.94	46.35	51.6	46.92	55	75.14	55.15	133.34	125.45
	40.58	31.92	251	160.58	56	20.39	18.93	154.36	131.25
Minimum						20.39	18.93	28.24	21.38
Maximum						409.86	138.00	272.00	161.58
Mean						129.84	74.09	117.26	90.71

for group C; group A and B standards are exceeded (Table 5).

In Zielona Góra, no significant relationship between the high level of pollutants and heavy traffic was noted, but in some of the city's streets, pollution (especially lead and zinc) is more prevalent than in others. After careful consideration of these points based on the city's topography, it appears that increased quantities of street dust occur on roads which are located on low ground and also on streets that are generally flat. On inclined roads, there are considerably less pollutants.

Many of the pollutants are present in readily soluble forms, which facilitates their transport in water with rain over long distances. Many streets in Zielona Góra take the form of long, straight stretches of steadily declining gradients, which results in water from heavy rainfall flowing there as easily as in water courses. If water does not flow into sewers, it stops only in depressions or on flat areas. This also explains why, in the areas adjacent to the places from which street dust samples were collected, there are more heavy

metals, particularly lead, in street dust than in soils (Jabeen et al. 2001, Kusinska et al. 2005, Yongming et al. 2006). Grigalaviciene et al. (2005) and many other researchers confirm the impact of increased road concentrations of heavy metals, particularly lead and zinc, in street dust.

The relatively high rainfall in Zielona Góra, due to the climatic zone in which it is located (humid climate), has a positive impact on the washing of pollutants compared with urban areas in the dry climate zone, which is referred to in works of researchers such as Al-Khashman (2007), in Amman, Jordan.

Knowing both the topography of the city, as well as the gradients of the streets and the amount of rainfall, we can model the pollution spread following this path. We are able to predict in which locations on the streets the concentration of pollutants may be high. Sometimes, a cloud of road dust can be easily spotted on a street, which may be a sign that this is the place where the water always flows, and there may be large quantities of pollutants. A comparison of heavy metals in street dusts of Zielona Góra to the content of these

Table 4. Organic carbon and pH level in street dusts.

Sample number	pH I series	pH II series	Organic carbon % I series	Organic carbon % II series	Sample number	pH I series	pH II series	Organic carbon % I series	Organic carbon % II series
1	7.9	7.2	3.41	3.72	29	7.5	7.6	3.58	0.74
2	7.7	7.9	6.50	2.08	30	7.3	7.3	2.38	1.26
3	7.9	7.5	4.06	2.52	31	7.4	6.7	2.16	2.32
4	7.6	7.5	3.73	4.88	32	7.3	6.7	2.66	2.29
5	7.5	7.7	1.79	2.18	33	7.1	7.3	3.12	1.70
6	7.5	7.8	2.06	1.71	34	7.3	7.5	11.2	1.92
7	7.2	7.3	8.00	2.69	35	7.3	7.1	1.77	1.67
8	7.1	6.8	3.26	1.73	36	7.3	7.0	2.14	2.06
9	7.1	7.3	7.75	3.60	37	7.3	6.2	2.20	2.75
10	7.1	7.1	2.93	2.31	38	7.3	7.1	1.92	0.83
11	7.3	7.7	2.63	3.77	39	7.3	6.8	1.88	1.56
12	7.2	7.6	11.54	2.01	40	7.5	7.1	3.43	2.56
13	7.7	7.6	1.50	7.08	41	7.3	7.4	2.94	1.86
14	7.1	6.6	3.01	3.12	42	7.3	6.9	2.08	3.14
15	7.2	7.6	1.19	2.72	43	7.2	6.4	2.81	1.06
16	7.0	7.4	3.06	5.91	44	7.1	7.5	3.63	2.74
17	7.4	7.9	1.81	2.70	45	7.1	7.5	4.16	3.77
18	7.3	8.0	2.66	1.26	46	7.5	7.4	1.87	1.28
19	7.1	7.9	3.61	1.62	47	7.3	7.8	4.09	1.45
20	7.2	7.6	4.43	1.47	48	7.3	7.2	4.28	1.17
21	7.3	7.2	3.25	4.14	49	7.3	6.9	2.99	2.91
22	7.1	7.0	1.11	1.55	50	7.3	7.3	4.19	1.58
23	7.1	6.9	1.67	3.90	51	7.2	7.4	1.82	1.93
24	7.2	7.1	3.21	1.56	52	7.1	7.5	2.97	2.03
25	6.9	7.4	2.67	2.42	53	7.3	7.7	3.35	2.49
26	7.4	8.0	1.63	2.40	54	7.2	7.4	2.49	2.97
27	7.2	8.3	2.28	0.90	55	7.3	7.2	1.85	3.21
28	7.4	7.5	1.36	1.08	56	6.0	7.6	5.75	1.63

Table 5. Lead and zinc in street dusts were compared with the instructions in the Regulation of the Minister of Environment of 9.09.2002 concerning standard quality for soil and ground.

Contents		Group Aa	Group Bb	Group Cc
I series	II series		0.0–0.3 m under ground	0.0–2.0 m under ground
Heavy metals			mg · kg^{-1}	
Zinc (Zn) 20.4-409.9	28.2-272.0	100	300	1000
Lead (Pb) 8.7-278.5	27.5-172.8	50	100	600

[a]Land forming part of an area subject to protection under the Water Act and the areas subject to protection under the provisions on the protection of nature.
[b]Land classified as agricultural land, excluding land under ponds and ditches, forest land and trees and bushes, barren land and built without urban areas and industrial use of fossil sites and communication.
[c]Industrial areas, fossil sites, communication sites.

elements in street dusts of other cities in the world is shown in Table 6. It may be concluded that the amount of heavy metals in road dust is not proportional to the size of the city, and that it depends on a number of other factors.

In many cities, the amount of different heavy metals in road dust was recorded in similar areas. Aviles in Spain (80,000 inhabitants), as well as Hong Kong and Istanbul (a few million inhabitants), show ranges of heavy metals that are higher than those in Zielona Góra. Oslo, Madrid, Hawaii, Ontario – the capitals of countries and regions that have a much greater number of residents – represent areas in which there is a lower content of heavy metals in road dust than in the dust of Zielona Góra. In the road dusts from Halifax, Christchurch, Kingston and Seoul, concentrations of lead and zinc were higher than in the dust collected in Zielona Góra.

The content of pollutants in street dust is higher than in the soils adjacent to roads. The content of lead and

Table 6. Lead and zinc in street dust in cities all over the world and in Zielona Góra.

City	Authora	Lead (Pb) $mg \cdot kg^{-1}$	Zinc (Zn)	Solution
Amman	1	219–373	ndb	50% HNO_3
London	2	413–2241	nd	HNO_3
London	3	3030	1174	4 M HNO_3
New York	3	2583	1811	4 M HNO_3
Halifax	3	1297	468	4 M HNO_3
Christchurch	3	1090	548	4 M HNO_3
Kingston	3	863	765	4 M HNO_3
Seoul	4	245	296	HNO_3 + HCl
Oslo	5	180	412	HNO_3
Madrid	5	193	476	$HClO_4$ + HF
Bahrain	6	697	152	HNO_3 + HCl
Manchester	7	970	nd	2 M HNO_3
Cincinnati 1990	8	662	nd	2 M HNO_3
Cincinnati 1998	8	650	nd	2 M HNO_3
Ontario	9	90	227	$HClO_4$ + HF
Hawaii	10	106	434	$HClO_4$ + HF
Bursa, Turkey	11	210	57	HNO_3 + HCl
Karak, Jordan	12	1.4–609	1.8–123	HNO_3 + HCl
Hong Kong	13	233–2006	571–2372	HNO_3 + HCl
Istanbul	14	61–382	1852	HNO_3 + HCl
Aviles Spain	15	330–964	2422–23400	HNO_3 + HCl
Zielona Góra	16	8–278	20–409	HNO_3 + HCl

Authors: 1. Jires et al. 2001, 2. Leharne et al. 1992, 3. Fergusson & Ryan 1984, 4. Chon et al. 1995, 5. de Miguel et al. 1997, 6. Akheter & Madany 1993, 7. Day et al. 1975, 8. Tong 1998, 9. Stone & Marsalek 1996, 10. Sutherland & Tolosa 2000, 11. Arslan 2001, 12. Al.-Khashman 2004, 13. Wang et al. 1998, 14. Sezgin et al. 2005, 15. Ordonez et al. 2003, 16. Author Walczak B., own research, 2001
nd, no data available

zinc in street dusts exceed the allowable standards as proposed by the Regulation of the Minister of Environment of 9.09.2002 in terms of standard quality of soil and ground for groups A and B. The content of lead and zinc in street dust is similar to that of other cities in the world.

REFERENCES

Al-Chalabi, A.S. & Hawker, D. 1996. Retention and exchange behaviour of vehicular lead in street dusts from major roads. The Science of the Total Environment 187: 105–119.
Al-Khashman, O.A. 2004. Heavy metal distribution in dust, street dust and soil from the work place in Karak Industrial Estate, Jordan. Atmospheric Environment 38: 6803–6812.
Al-Khashman, O.A. 2007. The investigation of metal concentrations in street dust samples in Aqaba city, Jordan. Environmental Geochemistry and Health 29: 197–207.
Arslan, H. 2001. Heavy metals in street dust in Bursa, Turkey. Journal of Trace and Microprobe Techniques 19: 439–445.
Chon, H.T., Kim, K.W. & Kim, J.Y., 1995, Metal contamination of soil land dust in Seoul metropolitan city, Korea. Environmental Geochemistry and Health 17: 134–136.
Day, J.P., Hart, M. & Robinson, S.M. 1975. Lead in urban street dust. Nature 253: 343–345.
de Miguel, E., Liamas, J.F., Chacon, E., Berg, T., Larssen, S, Royset, O. & Vadset, M. 1997. Origin and patterns of distribution of trace elements in street dust: unleaded

petrol and urban lead. Atmospheric Environment 31(17): 2733–2740.
Fergusson, J.E. & Ryan, D.E. 1984. The elemental composition of street dust from large and small urban areas related to city type, source and particle size. The Science of the Total Environment 34: 101–116.
Grigalaviciene, I., Rutkoviene, V. & Marozas, V. 2005. The accumulations of heavy metals Pb, Cu and Cd in roadside forest soil. Polish Journal of Environmental Studies 14(1): 109–115.
Harrison, R.M., Laxen, D.P.H. & Wilson, S.J. 1981. Chemical associations of lead, cadmium, copper, and zinc in street dusts and roadside soils. Environmental Science & Technology 15(11): 1378–1383.
Jabeen, N., Ahmet, S., Hassan, T. & Alam, N.M. 2001. Levels and sources of heavy metals in house dust. Journal of Radioanalytical and Nuclear Chemistry 247(1): 145–149.
Lagerwerff, J.V. & Specht, A.W. 1970. Contamination of roadside soil and vegetation with cadmium, nickel, lead, and zinc. Environmental Science & Technology 4(7): 583–586.
Leharne, S., Charlesworth, D. & Chowdry, B., 1992, A survey of metal levels in street dusts in an inner London neighbourhood. Environmental International 18: 263–270.
Kusińska, A., Bauman-Kaszubska, H. & Dziegielewska-Sitko, A. 2005. Soil environment contamination in the Plock agglomeration. Ecological Chemistry and Engineering 12(3): 251–259.
Ordonez, A., Loredo, J., De Miguel, E. & Charlesworth, S. 2003. Distribution of heavy metals in street dusts and

soils of an industrial city in northern Spain. *Environmental Contamination and Toxicology* 44: 160–170.

Sezgin, N., Ozcan, H. K., Demir, G., Nemiloglu, S. & Bayat, C. 2005. Determination of heavy metal concentrations in street dusts in Istanbul E-5 highway. *Environmental International,* 29: 979–985.

Stone, M. & Marsalek, J. 1996. Trace metal composition and speciation in street sediment: Sault Ste. Marie, Canada. *Water, Air and Soil Pollution* 87: 149–169.

Sutherland, T. A. & Tolosa, C. 2000. Multielement analysis of road-deposited sediment in an urban drainage basin, Honolulu, Hawaii. *Environmental Pollution* 110: 483–495.

Tong, S. 1998. Indoor and outdoor household dust contamination in Cincinnati, Ohio, USA. *Environmental Geochemistry and Health* 20: 123–133.

Urbański, K. 1998. *Szczególowa mapa geologiczna Polski.* Panstwowy Instytut Geologiczny: Wroclaw.

Wang, W.H., Wong, M.H., Leharne, S. & Fisher, B. 1998. Fractionation and biotoxicity of heavy metals in urban dusts collected from Hong Kong and London. *Environmental Geochemistry and Health* 20: 185–198.

Wong M.H., Cheung L.C. & Wong W.C. 1984. Effects of roadside dust on seed germination and root growth of *Brassica chinensis* and *B. parachinensis. The Science of the Total Environment* 33: 87–102.

Yongming, H., Peixuan, D., Jungi, C. & Posmentier, E.S. 2006. Multivariate analysis of heavy metal contamination in urban dusts of Xi'an, Central China. *The Science of the Total Environment* 355(1–3): 176–186.

Xie, S., Dearing, J.A. & Bloemendal, J. 2000. The organic matter content of street dust in Liverpool, UK, and its association with magnetic properties. *Atmospheric Environment* 34: 269–275.

Zabłocki, Z., Fudali, E., Podlasinska, J. & Kipas- Kokot, A. 1998. *Pozarolnicze obciazenia srodowiska.* AR Szczecin.

Environmental Engineering III – Pawłowski, Dudzińska & Pawłowski (eds)
© 2010 Taylor & Francis Group, London, ISBN 978-0-415-54882-3

NOx emission control technologies in sludge pyrolysis and combustion

X. Yang, X. Wang, Y. Cai & L. Wang

Department of Thermal Energy Engineering, Beijing University, Beijing, China

ABSTRACT: The characteristic of NOx emission in sludge pyrolysis and sludge combustion has been carried out in the fixed bed reactor. The influence factors including final temperatures, the heating rates, types of sludge were studied on the effects of the NOx emissions. The results showed that CaO, MgO can lead to the reduction of NO. Staged combustion and injection of NH_3 have been performed to study the effects on NOx release.

Keywords: Sludge, incineration, pyrolysis, pollution, nitrogen oxide.

1 INTRODUCTION

The main formation of nitrogen element in sewage sludge is organic nitrogen and nitrogen pollutant is an inevitable product whether in sludge pyrolysis or sludge incineration. The emission and controlling of nitrogen pollutant is one of problems in thermo-chemical treatment. Therefore, it is necessary to further research for control of nitrogen pollution in sludge pyrolysis and combustion.

NO_x is the sum of NO and NO_2 formed during thermal-chemical sludge treatment. NO is usually the dominant species with NO_2 being 5–10%. In generally, NO can be oxidized to NO_2 at low temperature. There are mainly three routes of NO_x formation in thermal-chemical sludge treatment. (1). Thermal-NO_x. The nitrogen in gas phase is oxidized into NO_x, which is the main route of formation of NO_x at high temperature. The yield rate of thermal-NO_x formation has an exponential growth with the rise of temperature. The percent of thermal-NO_x formation is close to zero at 800°, and the percent of thermal-NO_x is up to 12% at the temperature of 1000°; (2) Prompt-NO_x. The prompt-NO_x formation org from the radical CH, which is formed as an intermediate at the flame front only. The radical CH reacts with the nitrogen in air, forming HCN which is oxidized further to NO. Comparing with thermal-NO_x and fuel-NO_x, the emission of prompt-NO_x can be often neglected. (3) Fuel-NO_x. The nitrogen element in sludge is oxidized to NO_x in thermal-chemical treatment. The formation mechanism of fuel-NO_x is very complex. NO_x comes from the oxygenation of the nitrogen in volatile matter or residence char (Volatile-N and char-N). Firstly, Volatile-N pyrolyzes into the intermediate radicals including NH_3, CN and HCN, which are formed by the quick reaction of fuel-N with the intermediate product from hydrocarbon decomposition. The char-N plays an important role in controlling of NO_x release and it

supplies the contact surface of gas-solid reaction. The main process of reducing NO is the catalytic reaction of NO with CO on the surface of char.

The present research work mainly focused on the NO_x emission in sludge incineration but lack of the comparison of the NO_x emission in sludge pyrolysis and sludge combustion. In this paper, NO_x emissions in sludge pyrolysis and sludge combustion were detected and the control technologies were also discussed.

2 MATERIALS AND METHODS

The experiments were conducted in a fixed bed. The temperature and heating rate were controlled automatically. The flow gas can be applied by air compressor for sludge combustion experiment and nitrogen gas cylinder for sludge pyrolysis experiment. The gas emission is tested by flue gas analyzer (testo-360, Germany) and the data is collected by a computer connected with the flue gas analyzer.

Two types of sludge were sampled from two different wastewater treatment plants. Petrochemical waste activated sludge (sludge A) was obtained from a petrochemical wastewater treatment plant that treats wastewater discharged from alkenes, plastic and chloral-alkali industries using a pure oxygen aerobic process. Municipal waste activated sludge (sludge B) was obtained from a municipal wastewater treatment plant.

3 RESULTS AND DISCUSSION

From Figure 1, four peaks of NO_x release from two types of sludge combustion were observed. The first peak of NO_x release is in the temperature range of

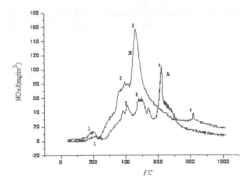

Figure 1. NOx concentration in different types of sludge combustion.

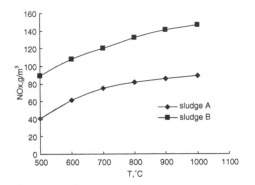

Figure 2. Effects of combustion heating rate on NOx emission.

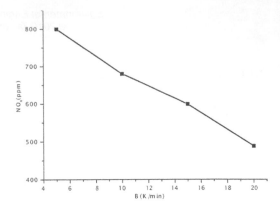

Figure 3. Effect of heating rate on NOx emission in sludge combustion.

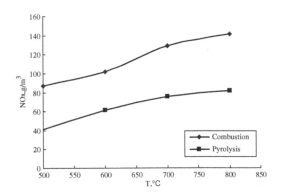

Figure 4. Final temperatures on NO_x emission in sludge combustion and pyrolysis.

150~250°C, which could be the release of volatile-N leading to the formation of NO_x. The second and third peaks are in the range of 400~550°C. Protein compounds, which is the main form of nitrogen element in the sludge, could combust at this temperature to produce nitric oxide. The four peak of petrochemical sludge is in the temperature range of 600~700°C. Integrating of the NO_x emission curves, the total emission of NO_x is 89.0 g/m^3 in sludge A and 146.3 g/m^3 in sludge B.

The NO_x emission concentration increases with the increasing of final combustion temperature, given in Figure 2. That could be the reason that NO increase with the rise of temperature and the decrease of carbon load results in the decrease of NO_x reduction reaction. NO_x emission decreases with the rise of heating rate, shown in Figure 3. With the increase of heating rate, the residence time becomes shorter at the low temperature stage and results in the decrease of NOx emission.

The routes of controlling the nitrogen oxides includes decreasing the combustion temperature, reducing the concentration of oxygen, shortening the residence time at high temperature stage. Stage combustion and selective non-catalytic reduction are useful for NO_x emission control during the sludge combustion. NO_x emissions could be decreased from 146.3 g/m^3 to 78.2 g/m^3 by stage combustion. In the injection of both NH_3 and urea during the combustion,

NO_x reduced from 146.3 g/m^3 to 54.3 g/m^3. Further research should be conducted to explain the results and mechanism.

4 CONCLUSIONS

NO_x emission concentration of petrochemical sludge decreases with the increase of heating rate in a combustion process. NO_x emission of pyrolysis process was lowere than that of combustion process. NOx emissions decreased from 146.3 g/m^3 to 78.2 g/m^3 by stage combustion, while NO_x reduced from 146.3 g/m^3 to 54.3 g/m^3 by the injection both NH_3 and urea during the combustion. Further research should be conducted to explain the results and mechanism.

REFERENCES

Leckner, B., Amand, L.-E., Luck, K. & Werther, J. 2004. Gaseous emissions from co-combustion of sewage sludge and coal/wood in a fluidized bed. *Fuel* 83: 477–486.
Nadziakiewicz, J. & Kozio, M. 2003. Co-combustion of sludge with coal. *Applied Energy* 75: 239–248.

Werther, J. & Ogada, T. 1999. Sewage sludge combustion. *Progress in Energy and Combustion Science* 25: 55–116

Sanger, K., Werther, J. & Ogada, T. 2001. NO_x and N_2O emission characteristics from fluidised bed combustion of semi-dried municipal sewage sludge. *Fuel* 80: 167–177.

Bin, L., Yong, C. & Tinghua, Z. 1998. A study of NOx and SO_2 emission of sewage sludge and paper mill sludge in fluidized bed incineration. *Journal of Engineering Thermophysics* 19(6): 776–779.

Shimizu, T., Toyono, M. & Ohsawa, H. 2007. Emissions of NOx and N_2O during co-combustion of dried sewage sludge with coal in a bubbling fluidized bed combustor. *Fuel* 86: 957–964.

Shimizu, T. & Toyono, M. 2007. Emissions of NOx and N_2O during co-combustion of dried sewage sludge with coal in a circulating fluidized bed combustor. *Fuel* 86: 2308–2315.

Xiaoping, C., Lifeng, G., Changsui, Z. & Xi, S. NO_x and N_2O emission characteristics during co-combustion of municipal sewage sludge and coal. *Journal of Southeast University Natural Science Edition* 35(1): 22–24, 2005.

Yu, Z. & Zeng, T. 1998. A study on NOx emission in sludge incinerating fluidization bed. *Environmental Protection* 2: 36–38.

Zhu, G., Zhao, C., Chen, X., Lin, L. & Zhou, W. 2008. Characteristics of pollutant emissions from co-combustion of petrochemical sludge with coal in fluidized bed. *Journal of Chemical Industry and Engineering (China)* 59(10): 2627–2633.

Indoor air pollution control

Environmental Engineering III – Pawłowski, Dudzińska & Pawłowski (eds)
© 2010 Taylor & Francis Group, London, ISBN 978-0-415-54882-3

Application of the thermal manikin for ventilation and air-conditioning system assessment

M. Chludzińska & B. Mizieliński
Faculty of Environmental Engineering, Warsaw University of Technology, Warsaw, Poland

A. Bogdan
Central Institute for Labour Protection – National Research Institute, Department of Ergonomics, Poland

ABSTRACT: This paper presents the history of the use of thermal manikins for experimental purposes and shows the modifications and upgrading that were made to the manikins in order to fine-tune the functions and data required. Furthermore, the paper shows and compares the results of thermal plumes developing above a seated person. The discrepancy analysis of the results obtained in tests with humans and thermal manikins confirms the occurrence of some differences; these differences are, however, so insignificant that the devices prove to be feasible for indoor environmental research.

Keywords: Thermal manikin, Thermal comfort, Personalised ventilation, Indoor air quality.

1 INTRODUCTION

New technological advances in the HVAC systems dedicated to permanently inhabited structures are usually verified before implementation in filed study or laboratory conditions. However, due to the cost of such research under real-life conditions, and the availability and repeatability issues, most tests are performed in a laboratory, often with the participation of volunteers or substitutes in the form of human simulators, e.g. thermal manikins. The thermal manikin is a model of the human being, and can imitate heat exchange between the body and the surrounding environment in terms of convection, conduction and radiation. Such devices have been produced for more than 60 years. Particular applications enable both evaluation of the heating conditions in enclosed environments (rooms, cars) and modelling thermal insulation of clothes or sleeping bags.

Performing tests with volunteers requires a few conditions to be met. An adequate number of people have to be tested to render the results reliable, so that statistical analysis can follow. It is physiological and psychological differences that make the measurements non-repeatable. Seasonal acclimation of the human organism is also connected with human physiology. Thus, tests have to be performed in specified months and the current season of the year needs to be matched with the simulated one. Furthermore, there appear to be some difficulties related to fatigue, involuntary movement during a measurement and other circumstances.

Replacing humans with a mechanical device has led to an improvement in end acceleration of the whole measurement process by way of eliminating the abovementioned factors.

1.1 Thermal manikin history

The first one-segment thermal manikin was made of copper in the 1940s and was commissioned by the US Army. As the development of thermal manikin construction stemmed from the necessity to adjust the device to the specific research conditions, the successive models have been fitted with considerable improvements and new solutions – for example, breathing and perspiration simulation. New solutions in terms of resources have been sought in order to improve heat-exchange conditions and also a segmented structure was introduced.

The first female manikin appeared in laboratories in 1989. This 'gender change' was due to the fact that female manikins are shorter and lighter than the male ones. Researchers also hoped that, as women usually have more layers of clothing, the manikins would be used more efficiently.

We have recently observed two directions in thermal manikin development.

The first focuses on designing 'complex manikins', which are simultaneously furnished with many functions (possibility of movement, sweating), which are used, for instance, in specialist research of the indoor environment. At the same time, the branch of industry releasing smaller manikins has developed – manikins that are provided only with the basic heating function and are specialised in thermoinsulation measurements of garments. These manikins are utilised in companies that manufacture sports and protective clothing. A significant step in thermal manikin development was the application of computer control and adjustment methods in the manikin's automatic system, which has enabled more accurate testing by being able to precisely administer temperature values on the surface of particular segments.

At first, the manikins could only work in the standing posture. In the course of development, seated manikins were designed; further improvements enabled movement and walking with the adequate number of steps in the defined time.

Ever-wider use of manikins in indoor air quality assessment led to a new device being designed, one that could also simulate human perspiration.

1.2 Applications of thermal manikins in ventilation research

Being complex instruments, manikins can imitate heat exchange with the environment by means of convection, conduction and radiation in all directions from the whole surface or part of it only. This allows imitation of the processes of heat exchange for particular segments, as well as the whole surface of the thermal manikin. Measurements are possible in a three-dimensional system, allowing heat-exchange simulation in real-life human environment conditions. What is more, the time spent on testing has become considerably less due to the increased accuracy and repeatability.

The software of thermal manikins enables us to operate in three different modes depending on what kind of measurement is taken; this is dependent on:

- Maintenance of constant surface temperature segments. In this mode, various levels of power are supplied to each segment to provide a constant temperature on the surface. Usually, the adjusted temperature is 34°C, which is the mean temperature of human skin. This mode is used for thermal insulation measurements of clothing or sleeping bags.
- Maintenance of constant heat loss. Constant power is supplied to each segment. In this mode, depending on the environmental conditions (air velocity, temperature), skin temperature varies and is measured.
- The third mode is "COMFORT", which is a physiological mode. This setting allows either comfort-controlled skin temperature or control by a physiological model – both skin temperature and heat-loss changes. This mode helps to simulate heat exchange between humans and the environment, and it is used in ventilation research.

Moreover, manikins have a function ("no heat"), which allows use as a thermometer. In this case, power is not supplied to segments, but air temperature is measured by wiring that is located in segments.

In ventilation or air-conditioning system research, thermal manikins are used to assess the impact of the thermal environment on the human body; by reading local temperature values on a manikin's surface, it is possible to carry out analysis of the impact of the microclimate on the operation of ventilation and air-conditioning systems, the influence of temperature asymmetry, heterogeneous environment conditions, and also in small enclosed spaces such as cars. Owing to this, the manikins are used in many research programmes that aim to provide construction guidelines and tools to predict the behaviour of the environment in a building, to design controlling and operating strategies for ventilation and air-conditioning systems, and also to analyse the energy needs of a building and to plan respective energy expenditure savings.

The thermal manikin has been used in research of the thermal environment under conditions of displacement ventilation (Cheong et al 2006). The manikin used in this experiment was a female of 1.68 m standing height. It is divided into 26 thermal segments that can be independently measured. Using a measuring apparatus, the manikin was operating in a 'comfort' mode. In this mode, surface temperatures of various body segments follow a comfort equation and are allowed to change to adapt to the environment. So, the manikin is always in the thermal-neutral state. Sustaining the comfort state depends on the adjustment of surface temperatures on each and every segment to indoor environment conditions. Maintaining a fixed temperature value on the surface is possible due to heat-loss compensation brought about by heating elements. A nickel wire evenly covering the surface of the manikin provides the necessary power. Measurement of the resistance on individual wires translates to the surface temperature measurement in each segment.

The tests were carried out in a special room, which was 11.12 m long, 7.53 m wide and 2.6 m high, where displacement ventilation supply units were installed in the corners in order to maintain the given temperature parameters of the room. Each variant lasted 3 h and, within this time, office work was simulated. Measurements were made at three values of temperature of surrounding air, taken at the height of 0.6 m and for different variants of temperature gradient at the height of 0.1 m to 1.1 m.

The measurements using the manikin showed that the average temperature of the manikin's surface increased in parallel to the rise of air temperature in the room. Temperature on the surface of particular segments was invariable, regardless of the value of temperature gradient; however, as the air temperature in the room was rising, the temperature difference between particular segments decreased. Assuming that higher temperature differences between individual segments increases draft rating, and consequently leads to a feeling of discomfort, it was proven that the air temperature value in a room has a greater impact on thermal comfort than the temperature gradient.

A similar, 16-segment human simulator was also used in thermal plume measurements appearing above a person (Bogdan & Chludzinska 2007). This simulator was a female manikin of 1.63 m tall and with a surface area of 1.46 m². It has movable hip joints,

Table 1. Development phases of thermal manikins (Holmer 2004).

1	one-segment	copper	analogue	–	USA 1945
2	multiple-segment	aluminium	analogue	–	UK 1964
3	radiation manikin	aluminium	analogue	–	France 1972
4	multiple-segment	plastic	analogue	movement possibility	Denmark 1973
5	multiple-segment	plastic	analogue	movement possibility	Germany 1978
6	multiple-segment	plastic	digital	movement possibility	Sweden 1980
7	multiple-segment	plastic	digital	movement possibility	Sweden 1984
8	immersible	aluminium	digital	movement possibility	Canada 1988
9	sweating manikin	aluminium	digital	–	Japan 1988
		plastic	digital	movement possibility	Finland 1988
		aluminium	digital	movement possibility	USA 1996
10	female manikin	plastic, single wire providing energy	digital, possibility of operation in the 'comfort' mode	movement possibility	Denmark 1989
11	breathing manikin	plastic, single wire providing energy	digital, possibility of operation in the 'comfort' mode	movement possibility, breathing simulation	Denmark 1996
12	sweating manikin	plastic	digital, 30 dry and 125 'sweating' zones	simulation of real human movement	Switzerland 2001
13	compact sweating manikin	metal	digital, 126 zones	articulated	USA 2003
14	numerical thermal manikin	geometric model	simulation of heat and water vapour transmission	articulated	China 2000 UK 2001
15	one-segment sweating manikin	air-permeable fabric	digital, water-heated	movement possibility	China 2001
16	one-segment sweating manikin	air-permeable fabric	digital, air-heated	movement possibility	USA 2003

knee joints and arm joints. The test was performed in the Thermal Loads Laboratory of Central Institute for Labour Protection – National Research Institute (CIOP-PIB), equipped with a climatic chamber (having the following parameters: height 3 m, width 2.5 m, depth 3 m). During the tests, the manikin was seated on a chair and the temperature on its surface equalled the average temperature of humans (34°C) (Parsons 2003; ISO 15831, 2004); power delivered to the manikin was 108 W and air temperature was 23°C ± 0.2 W.

In order to verify the data obtained from the manikin, measurements were also taken from volunteers. The results of both variants are presented further in the paper.

Introducing the possibility of breathing simulation in thermal manikins considerably enhances their scope of operation (Figure 1). The manikins are fitted with artificial lungs, which inhale and exhale air with adequate amounts of pollution, appropriate temperature levels and humidity.

Intensiveness of breathing varies between 5 and 30 litres/min, and frequency of inhalation varies between 5 and 30 inhales/min. The air breathed through the nose of the manikin is sent to the quality assessment instrument. Breathing is facilitated by circular holes in the nose and mouth of the manikin.

Additional application of tracer gases (i.e. SF_6), which mark the amount of gas getting through to the artificial lungs, enabled verification of indoor air quality assessment and performance of new ventilation systems.

Melikov and Kaczmarczyk (2007) applied the thermal manikin equipped with the breathing function in indoor air quality assessment under the conditions of personalised ventilation. A desk with installed personalised ventilation supply units and a computer to simulate office work were set up in a $5 \times 6 \times 2.5$ m climatic chamber. Apart from the personalised ventilation installation, the climate chamber was furnished with a general ventilation system providing air through a perforated floor. The installation enabled continuous air flux in the chamber. A female manikin with nostrils and mouth was set up at the desk. The nostrils, each 38.5 mm², were tilted upwards at an angle of 40°. The slightly open mouth covered 158 mm². Together with the air blown through the double floor into the climate chamber, a tracer gas SF_6 was provided. The personalised ventilation system supplied clean air without SF_6. The gas concentration measurement was taken for the air supply into the climate chamber, for the air supply by the PV system and for the air exhaled by the breathing manikin. The continuous gas concentration measurement was carried out by applying the methods of photo-acoustic detection and infra-red radiation. These tests facilitated performance assessment of different types of air supply units. None of the analysed variants provided the conditions that would lead to inhaling clean air only, as provided by the supply unit at the desk. However, the effectiveness of the individual supply units was rising in parallel to the increase in the amount of air provided. The highest efficiency of the personalised ventilation installation was achieved at a level of 80%, which means that in the air inhaled by the manikin, 80% of the air was clean. This value was attained at an air flow rate of 83 m³/h of fresh air. Although air usage met the hygienic standards, the

Figure 1. Visualization of the thermal manikin's breathing process (Melikov & Kaczmarczyk et al. 2007).

effectiveness of the tested supply units remained at the level of 30–50%.

2 MATERIALS AND METHODS

2.1 Comparison of results obtained with the thermal manikin and volunteers

In order to verify the reliability of obtained data and the possibility of using human simulators in ventilation and indoor air distribution research, the thermal plume measurements (Bogdan & Chludzinska 2007) were carried out with the thermal manikin and also with volunteers. The individual versions of the same test were performed both with the manikin and volunteers, and then compared.

2.2 Experimental facilities

During the tests, air temperature was 23°C ± 0.2, air velocity was below 0.05 m/s, and both the manikin and the volunteers were seated on a chair with 0.06 clo.

Measurements of temperature and air velocity in the thermal plume were made with the use of a HT-400 system by Sensor Electrics. The system is composed of a set of five thermoanemometric, spherical sensors used for measurements of air velocity and air temperature. Measurement sensors were distributed on a movable stand at the following heights above the top of the manikin's/human's head: 0.1 m, 0.2 m, 0.4 m, 0.7 m, 0.9 m.

Measurements of air velocity and temperature covered an area of 0.6 × 0.6 m, with subsequent measurement points creating a square grid with a 0.1 m pitch. The centre of the chair's seat was localised in the axis of the central point of the grid. The tests performed at five different height levels in this area resulted in a total of 175 measurement points.

The tests were conducted in two series:

Series I: with the use of thermal manikin "Diana".
Series II – the manikin was replaced by volunteers of similar height and body surface area, wearing

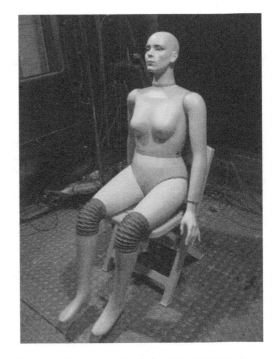

Figure 2. Thermal manikin during measurements.

underwear that was insulated at 0.03 clo (ISO 7730, 2005), and seated on the same chair. During the test, the subjects were in a state of thermal comfort. Volunteers had temperature measured in 14 skin points in accordance with the ISO 9886 standard (2004).

3 RESULTS AND DISCUSSION

The results with the thermal manikin and volunteers were alike. The thermal plume generated by the humans was found to be much more unstable than the thermal plume that is produced above the manikin.

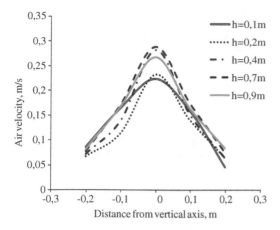

Figure 3. Vertical air velocities at different heights 'h' above seated manikin head.

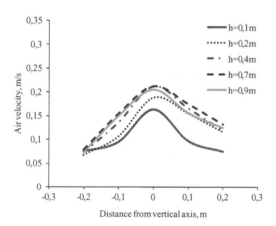

Figure 4. Vertical air velocities at different heights 'h' above seated vol. Head.

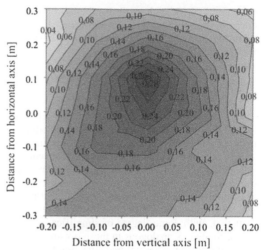

Figure 5. Horizontal air velocities at height 0.7 m above seated manikin head.

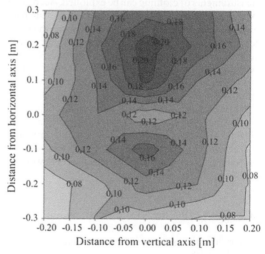

Figure 6. Horizontal air velocities at height 0.7 m above seated vol. head.

In all test variants, the highest values of central air velocity were obtained at a height of 0.7 m above the top of the manikin's/human's head (Figures 3–4).

Figures 5–6 present air velocity values in the thermal plume at a height of 0.7 m above the top of the manikin's/human's head. It is clear from the graphs that air velocity values are higher in the case of the plume generated above the manikin than above human subjects.

At the same time, in the nose and mouth area of the human subjects, it was easy to notice a "warm cloud" forming a separate thermal plume (Figure 6). The "warm cloud" did not have any effect on air velocity above the head; however, in cases of low-temperature sources of heat (e.g. a human being), the cloud should be treated as a separate thermal plume.

Apparent lack of symmetry (along the vertical axis) of the thermal plume resulted from a seated body position of both the manikin and human subjects.

On the basis of these tests and a literature review, the following conclusions have been drawn:

Tests with volunteers confirmed formation of the second plume ("warm cloud"). This phenomenon was not noted in test sessions carried out with a thermal manikin.

Tests have shown differences in air velocity values in thermal plumes above the head of a manikin/human. These differences are attributable to the thermal heterogeneity of the human skin surface. In tests performed with a thermal manikin, a constant temperature value of all segments is assumed, whereas tests should also take into account the physiological distribution of temperatures on the surface of human skin.

From this, it follows that tests with thermal manikins should be preceded by tests with volunteers and should consider the real distribution of temperatures or relevant corrective coefficients based on physiological models – e.g. Fiala et al. (2001) in a programme controlling the manikin.

4 CONCLUSIONS

Using thermal manikins for ventilation and air-conditioning tests is a convenient solution facilitating analyses that would be difficult while working with people, as in the case of verifying the effectiveness of personalised ventilation. The measurements presented above show the occurrence of certain discrepancies in comparison with the results obtained during the tests with volunteers. They are, however, so insignificant that, with some degree of simplification, they can be treated as analogous to those obtained during tests with people. Thermal manikins are, by far, the closest to human simulators, and their use facilitates swift testing and repeatability. There is a particularly important advantage – that is, the thermal insulation measurements of clothing can be exploited. According to Maria Konarska et al. (2007), carrying out such an accurate test would demand examining more than 600 people. Among other things, it assesses the differences in metabolism and skin surface temperature. What is more, relying on the thermal manikin eliminates factors such as human acclimation to external conditions or fatigue.

Nevertheless, it has to be remembered that even the thermal manikins that are equipped with a range of functions (i.e. breathing, perspiration) are only machines and, in the case of a very detailed phenomena analysis, physiological parameter measures at work or in a room are still necessary.

ACKNOWLEDGEMENTS

Research work is supported from the funds for science in 2008–2010.

REFERENCES

Bogdan, A. & Chludzinska, M. 2007. Comparative evaluation of thermal plumes formed above a thermal manikin and humans – the pilot study results. *The 11th International Conference on Indoor Air Quality and Climate.*

Cheong, K.W.D., Yu, W.J., Kosonen, R., Tham, K.W. & Sekhar, S.C. 2006. Assessment of thermal environment using a thermal manikin in a field environment chamber served by displacement ventilation system. *Building and Environment* 41: 1661–1670.

Fiala, D., Lomas K. J., Stohrer M. 2000. Computer prediction of human thermoregulatory and temperature responses to a wide range of environmental conditions. *Int. J. Biometeorol* 45: 143–159.

Holmer, I. 2004. Thermal manikin history and applications. *European Journal of Applied Physiology* 92: 614–618.

ISO 15831.2004. *Clothing. Physiological effects. Measurement of thermal insulation by means of a thermal manikin.*

ISO 7730. 2005. *Ergonomics of the thermal environment. Analytical determination and interpretation of thermal comfort using calculation of the PMV and PPD indices and local thermal comfort criteria.*

ISO 9886. 2004. *Ergonomics. Evaluation of thermal strain by physiological measurements.*

Konarska, M., Soltynski, K., Sudol-Szopinska, I. & Chojnacka, A. 2007. Comparative evaluation of clothing thermal insulation measured on a thermal manikin and on volunteers. *Fibres & Textiles in Eastern Europe* 15(2) (61): 73–79.

Melikov, A. & Kaczmarczyk, J. 2007. Measurement and prediction of indoor air quality using a breathing thermal manikin. *Indoor Air* 17(1): 50–59.

Parsons, K.C. 2003. *Human thermal environments.* The effects of hot, moderate, and cold environments on human health, comfort and performance. Second edition. Taylor & Francis: London and New York.

Environmental Engineering III – Pawłowski, Dudzińska & Pawłowski (eds)
© 2010 Taylor & Francis Group, London, ISBN 978-0-415-54882-3

Bioaerosolos and carbonyl pollutants in small, naturally ventilated office spaces in Lublin Poland – a case study

M.R. Dudzińska & U. Gąska-Jędruch

Faculty of Environmental Engineering, Lublin University of Technology, Lublin, Poland

ABSTRACT: Carbonyl compounds have been known as major eye irritants in indoor environments for many years. Bioaerosols are another common form of pollutant found in indoor environments and are thought to be involved in "sick building syndrome". Concentrations of selected aldehydes, ketones and bioaerosols in office spaces at the Lublin University of Technology, Poland have been measured. Six carbonyl pollutants, as well as fungus and bacteria were detected in all the collected samples. Concentrations vary in rooms with carpeting compared to those with wood-based floor panels. Higher levels of carbonyl compounds were found in rooms which had been recently painted prior to the measurements. No similar findings were observed for bioaerosols, but higher concentrations of them were found in rooms with north-facing windows.

Keywords: Indoor air quality, work efficiency, office environment, SBS, VOCs, bioaerosols.

1   INTRODUCTION

Due to the increasing time spent indoors by many people, indoor air quality has greater impact on human health and work performance than outdoor air. The length of time spent in offices and classrooms often exceeds 8 hours per day and therefore indoor air quality (IAQ) and pollutant levels became important factors influencing our effectiveness and well being.

Building materials (gypsum wallboard, ceiling tiles, fiberglass insulation), indoor finishing materials (wallpapers, paints, ceiling tiles), furniture and textiles, as well as air-conditioning systems and even indoor occupants themselves release numerous substances, which influence air quality. Indoor concentrations of some pollutants sometimes exceed their outdoor levels.

There are two main syndromes connected with indoor air quality – sick building syndrome (SBS) and building related illnesses (BRI), both described widely in the literature. According to the World Health Organization (WHO 1982, WHO 1983), the symptoms and signs reported by more than 20% of the occupants of air conditioned premises, that may, or not, disappear when they leave the place, are classified as "sick building syndrome". These unspecific symptoms and signs related to sick building syndrome involve, headaches, dizziness, nausea, apathy, drowsiness, tiredness, weakness, difficulty in concentrating, urticaria, irritation and dryness of the skin, shortage of breath, chest noise, nasal catarrh, nose and throat irritation, and sore-throat, irritated, burning and watering eyes.

Sick building syndrome (SBS) it generally has a multifactorial etiology, including chemical, physical, biological and psychological risk factors (Engelhart

et al. 999). WHO reported (1983) that sick building syndrome may be associated with 10 to 30% office working area

Symptoms of SBS are well recognized, but our knowledge about its direct causes is still patchy. Two main types of pollutant have to be distinguished in indoor premises i) chemical pollutants, mainly volatile organic compounds (VOCs) and semivolatile organic compounds (SVOCs), and ii) biological pollutants - microorganisms, particularly fungi and bacteria. No single chemical or biological pollutant can be seen as the origin of "sick building syndrome", because many factors and substances, usually under measurement thresholds, act synergistically to cause this effect. However, examination of spaces in which occupants suffer from SBS included VOC, aldehydes and bioaerosols (moulds).

Knowledge of the impact of indoor air quality on occupants grew rapidly by the end of the last century and considerable research was undertaken to minimize such problems. However, substances causing eye irritation and bioaerosols are still found in indoor spaces and can cause lack of concentration and reduced work efficiency of occupants. At the level of satisfying user IAQ, the effectiveness of work grows by about 6.5% (Fanger et al. 2003).

There are three main streams of research on IAQ in offices. Some researchers focus on physical parameters (temperature, humidity), others on bioaerosols and moulds and a third group on chemical pollutants, including the main eye irritants such as formaldehyde and other carbonyl compounds.

Chemical pollutant levels in office premises were reported from survey in European Community countries (ECA report, 1990), Mexico (Baez et al.

2003), Brazil (Calvacante et al. 2006), Canada (Seaman et al. 2007), France (Marchand et al. 2008) and Poland (Dudzinska et al. 2009).

Eye irritants commonly occurring in office environments include a number of aldehydes which should be of special concern (Wolkoff & Kjaergaard 2007). Due to the analytical methods some aldehydes are measured, although levels of only a few of them are regulated, due to known hazardous effects. NIOSH (2005) recommends limiting the values in the work environment for some of them: formaldehyde (methanal), acetaldehyde (ethanal), valeraldehyde (pentanal) and acrolein. Only formaldehyde is regulated in some countries (including Poland) in residential areas.

Formaldehyde alone or in combination with other chemicals, serves a number of purposes in manufactured products. It is used to add permanent-press qualities to clothing and draperies, as a component of glues and adhesives, and as a preservative in some paints and coating products (Niu & Burnett 2001). Therefore, sources of formaldehyde in indoor air include building materials, smoking, household products, and the use of unvented, fuel-burning appliances, like gas stoves or kerosene space heaters. Formaldehyde has been found to be a major secondary pollutant from cleaning products in the presence of ozone (Singer et al. 2006, Wolkoff et al. 2006).

Acrolein is another aldehyde, ubiquitous in the environment, produced by the incomplete combustion of organic materials as well as the oxidation of other atmospheric pollutants such as 1,3-butadiene, a component of motor vehicle exhaust. Less is known about indoor air sources, although heated cooking oil, cigarette smoke, as well as candles, wood burning and incenses are sources under suspicion (Ho & Yu 2002). Acrolein might be formed in many oxidation processes involving oxidation of indoor VOCs, however the formation of unsaturated aldehydes from other VOCs is not yet well understood (Morrison & Nazaroff 2002).

Acetaldehyde (ethanal), is a recognized eye and lung irritant. Nasal cancer connected to acetaldehyde exposure was confirmed in animals (NIOSH 2005) therefore OSHA recommends that acetaldehyde be considered a potential occupational carcinogen. The main indoor source of acetaldehyde is tobacco smoke, although particle board furniture and carpeting are also recognized sources (Dassonville 2009).

Compared with residential indoor areas, less data is available on aldehyde levels in office environments and other public spaces (Chuah et al. 1997, Wolkoff 1995, Cavalcante et al. 2006, Hutter 2006, Miu et al. 2008). Most current research was carried out in large space offices with several dozens of occupants and an air conditioning system (Hutter 2006, Miu et al. 2008).

Acrolein concentrations were mainly studied for dwellings (Ho & Yu 2002, Gilbert et al. 2005, Seaman et al. 2007). Indoor concentrations were 4–30 times higher compared to the outdoor levels. Additionally, concentrations registered in the morning in houses were 2.5 times lower than those registered in the evenings, which confirms the connection with cooking activities. Information about levels of acrolein in offices is scarce (Cavalcante et al. 2006).

Although bioaerosols in office premises were mainly studied in countries with hot climates, such as Hong Kong (Law et al. 2001), Italy (Sessa et al. 2002, Bonetta et al. 2009), Taiwan (Wu et al. 2005) and the USA (Zhu et al. 2003, Tsai & Macher 2005), data for France (Parat et al. 1996) and Denmark (Kildeso et al. 1998) are also available.

Similarly to chemical pollutants, most research was concentrated on air-conditioned offices.

In Hong Kong, the average bioaerosols exposure during office hours inside both offices was found to be less than $1000\,CFU/m^3$, which was considered to be lower than the recommended guideline. As the highest value found exceeded $3800\,CFU/m^3$ during the morning the occupants should be aware of the situation and avoid working in a non-ventilated environment (Law 2001). In Europe (Parat et al. 1996, Kildeso et al. 1998, Sessa et al. 2002, Bonetta et al. 2009) concentrations were lower (bacteria within the range $50–500\,CFU/m^3$ and fungi within the range $100–2000\,CFU/m^3$ was reported).

Three main types of fungi dominated: *Penicillium spp., Aspergillus spp. i Cladosporium spp.* The main types of bacteria were *Bacillus, Pseudomonas, Enterobacter, Flavobacterium, Alcaligenes, Micrococcus and Streptomyces.* about 70% of all microflora of air in offices are bacteria. However, the main health problems were related to exposure to fungi.

In some earlier work carried out in Poland, measurements of bioaerosols were made inside fifteen offices. In the office environment the concentration of airborne bacteria was approximately $300\,CFU/m^3$ (Pastuszka et al. 2000, Gołofit-Szymczak et al. 2005).

Reports about air quality which refer to both chemical and biological contaminants are very scare (Nilsson et al. 2004). Another important gap in the research is the lack of information about IAQ in small, naturally ventilated offices. Such rooms still prevail in many countries, including new members of EU or developing countries in Asia.

In Poland a typical office environment consists of small office rooms (up to $20\,m^2$) occupied by 1–3 people. Such offices have gravity ventilation and a central heating system operating in the winter. Window opening is not very popular, because most of these types of offices are situated in noisy city centers. Until now there has been no detailed information about pollutant levels, including eye irritants and bioaerosols in offices in Poland. Therefore, in this work measurements of indoor pollutants in this kind of office environment has been undertaken together with an attempt to assign possible sources of pollutants.

2 MATERIALS AND METHODS

Samples for measurements performed in this study were collected in offices located at the Lublin University of Technology, Faculty of Environmental

Table 1. Characteristics of individual offices in which measurements of bioaerosols were conducted.(winter and spring 2009).

Office	O-325	O-328	O-324	O-217	O-219	O-216
Orientation of office	south	north	south	south	north	south
Storey	3 storey	3 storey	3 storey	2 storey	2 storey	2 storey
Surface	13.9 m²	13.9 m²	14.2 m²	13.9 m²	14.2 m²	14.2 m²
Height	3.25 m	3.25 m	3.25 m	3.25 m	3.25 m	3.25 m
Equipment	Plywood furniture:	Plywood furniture:	Plywood furniture:	Plywood furniture:	Plywood furniture:	Plywood furniture:
	2 desks	4 desks	2 desks	2 desks	3 desks	4 desks
	Book shelves	Book shelves	Book shelves	Book shelves	Book shelves	Book shelves
	2 computers	1 computer	2 computers	2 computers	2 computers	one computer
	One printer		One printer	One printer	One printer	
Cover of wall	Paint	Paint	Paint	Paint	Paint	Paint
Cover of floor	Panels	panels	panels	carpet	carpet	carpet
Number of occupants	2	0	2	3	2	1

Engineering – in winter and spring 2009. Lublin has 360,000 inhabitants and is located in the eastern part of Poland in a rural area. The building in question has been in use since 2002. The rooms investigated (identical in volume ca. 400 m³, naturally gravitationally ventilated) were situated on the third and fourth floors of the five storey building. The building has a central heating system. The rooms were occupied by university employees – teachers and researchers who performed daily activities – preparing teaching materials, evaluating research data and writing reports and proposals. Most occupants spent more then 40 hours per week at the university amenities for the most part in such activities. Especially in their roles as members of the environmental engineering department they complained persistently about the IAQ of their working environment.

During measurements, the temperature in the examined rooms was kept at a stable level ($20 \pm 1°C$) due to weather conditions (winter – spring) and the heating season. The average indoor relative humidity was recorded to be between 39.5 to 45.1 %. Every room has one window (directed to the north or south) and one door which opens to the same corridor. Rooms were equipped with typical office furniture of variable age.

The rooms on the third floor had carpeting, while rooms on the fourth floor had wood based panels. Carpets were vacuumed daily, whereas panels were cleaned using special liquid products. Smoking was not permitted in the building and occupants of the examined rooms were all non-smokers. Occasional visits of smokers were reported in only one of the offices. One of rooms had not been used for six months prior to our measurements. Detailed information about the examined offices is gathered in Table 1.

2.1 VOC measurements

Samples for measurements of eye irritants were collected during seven days of normal occupation (including the weekend). Aldehydes were sorbed using a passive sampling method on Radiello dosimeters (RAD165) (exposure time: 168 hours). Adsorbed

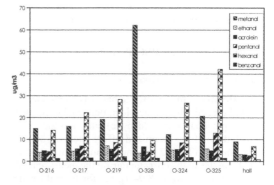

Figure 1. Carbonyl compounds in selected office premises ($\mu g/m^3$).

aldehydes were extracted with acetonitrile (HPLC grade) and analyzed using the RP HPLC method (Waters) with a Restek Allure AK column; acetonitrile/water elution and UV detection set at 365 nm. All measurements were carried out in triplicates.

The method allows measurement of about ten different carbonyl compounds; six of these were detected in all premises. The calibration of dosimeters was performed in a 5 L glass chamber under controled conditions of temperature, air flow and aldehyde concentration. Detection limits were 0.1 $\mu g/m^3$ for formaldehyde and 0.2 $\mu g/m^3$ for acrolein. Blank concentrations of carbonyl compounds were lower than detection limits.

2.2 Bioaerosol measurements

Air samples were collected using an Andersen 6-stage cascade impactor MAS-6 (Stage 6 corresponds to 0.65–1.1 µm, Stage 5:1.1–2.1 µm, Stage 4: 2.1–3.3, Stage 3: 3.3–4.7 µm, Stage 2: 4.7–7 µm and Stage 1: 7 µm or above) located in the center of the examined room at a height of 1.0-1.5 m above the floor. Windows and doors were kept closed. The sampling time was 10 min. Microorganisms were collected on nutrient media (specific to either fungi or bacteria) in Petri-dishes

Table 2. Office carbonyl concentrations ($\mu g/m^3$): dependence on floor covering and window orientation.

Carbonyl compound	With carpeting		With panels		Window to the north		Window to the south	
	range	mean	range	mean	range	mean	range	mean
HCHO	14.9–19.1	16.6	12.2–62.1	28.3	19.1–62.1	50.2	12.2–20.6	16.4
ethanal	4.0–7.0	5.2	3.7–5.6	4.8	3.7–7.0	5.7	4.0–5.7	4.8
acrolein	4.7–5.6	5.3	4.8–6.6	5.6	5.3–6.6	6.1	4.7–5.6	5.1
benzanal	1.2–2.1	1.6	1.3–1.8	1.5	1.4–2.1	1.7	1.2–1.8	1.6
pentanal	4.3–8.5	6.6	4.1–12.9	8.5	4.1–8.5	6.3	4.3–12.9	8.2
hexanal	14.1–28.8	21.5	9.5–42.1	26.1	9.5–28.8	19.1	14.1–42.1	26.3

located on all impactor stages. Agar Sabouranda with chloramphenicol (BTL) was applied for fungi, tryptone soy agar (TSA, BTL) was used for bacteria. The Petri dishes were incubated 48h at $36 \pm 1°C$ for bacteria and for 14 days at $27 \pm 1°C$ for fungi. No identification of bacteria and fungi was performed.

3 RESULTS AND DISCUSSION

3.1 Carbonyl compounds

Six carbonyl compounds: methanal (formaldehyde), ethanal (acetaldehyde), acrolein, benzanal (benzaldehyde), pentanal (valeraldehyde) and hexanal were detected in all the examined office rooms. The data are gathered in Table 2. The mean concentrations of the six aldehydes in the corridor during the measurements in office spaces were: $8.94 \mu g/m^3$ for formaldehyde, $3.07 \mu g/m^3$ for acetaldehyde, $3.0 \mu g/m^3$ for acrolein, $0.78 \mu g/m^3$ for benzanal, $2.57 \mu g/m^3$ for pentanal and $6.83 \mu g/m^3$ for hexanal. Formaldehyde concentrations measured in the examined offices showed the biggest discrepancies and varied from 12.2 to almost $62.1 \mu g/m^3$, while benzanal concentrations were relatively constant, varying only from 1.2 to 2.1 $\mu g/m^3$. Acetaldehyde (ethanal), another aldehyde common in indoor air, was found in the range 4.01–6.96 $\mu g/m^3$. Concentrations of acrolein, a known strong eye irritant were found within the range 4.8-6.6 $\mu g/m^3$. Relatively high concentrations of hexanal (9.48 to 42.10 $\mu g/m^3$) were found in the majority of the offices examined.

Formaldehyde is a product of a complex set of photochemical and radical reactions (Wolkoff et al. 2006), but as can be seen from Table 2, there were no significant differences between formaldehyde concentrations measured in the rooms with a window to the north or to the south. Hence it appears that direct light has little or no impact on the indoor formaldehyde levels we measured. However, there were significant differences between rooms with carpeting and those with wood based panels. The concentrations in the rooms with carpeting were in the range 14.9–19.1 $\mu g/m3$ (mean: 16.64 $\mu g/m3$), while in rooms with floors covered with panels they were in the range 12.2–62.1 $\mu g/m^3$ (mean: 38.1 $\mu g/m^3$). Both carpeting (by vacuuming) and panels (liquid floor cleanser) were cleaned daily.

Therefore, the higher concentrations might be caused by the plywood panels themselves, or by secondary products from the liquid cleansers. Evidence of such a reaction has been presented by Singer et al. (2006) and Wolkoff et al. (2006) in model studies.

Furniture, especially plywood based, is another typical indoor source of formaldehyde. Pressed wood products, made using adhesives that contain urea-formaldehyde (UF) resins, are used to produce furniture as well as plywood panels. However, plywood panels are made from hardwood paneling, while parts of furniture (drawer fronts, cabinets) are made of medium density fiberboard (MDF) containing a higher resin-to-wood ratio than other pressed wood products. Medium density fiberboard has been recognized as being the highest formaldehyde-emitting pressed wood product in model studies using test chambers (Hodgson et al. 2002). Unexpectedly, the highest concentration of formaldehyde was measured in the room with the oldest furniture, painted and decorated seven years prior to the experiment. Furniture in this room was even older as it was removed from another building, and was probably purchased in the late 1980s. Studies suggest that emissions decrease with time, but this applies to emissions from the panels and fiberboard covering. Other studies suggest that coatings applied on the furniture surface may reduce formaldehyde emission from wood pressed boards. However, to be effective such a coating must cover all surfaces and edges. Pieces of furniture are subject to damage during daily use and the edges of the "old" desks and cabinets in this particular room had been damaged. Taking into consideration that wood-pressed products used in this furniture were manufactured 20 years ago, when no such strict production standards were in force, the emission above 50 $\mu g/m^3$ might be explained.

Formaldehyde has been known as a major eye and nose irritant in indoor environment for many years. Exposure to it is seen as a threat to human health (Hodgson et al. 2002). Some studies have suggested that people exposed to formaldehyde levels ranging from 50 to 100 $\mu g/m^3$ for long periods of time are more likely to experience asthma-related respiratory symptoms, such as coughing and wheezing (Wolkoff et al. 2006).

The National Institute for Occupational Safety and Health (NIOSH), USA has set the recommended exposure limit for formaldehyde to REL = 0.016 ppm, which might be converted to approximately 20 $\mu g/m^3$(NIOSH 2005). The levels of formaldehyde permitted in non-occupied indoor spaces in Poland are set at $100\,\mu g/m^3$, while in occupied spaces at $50\,\mu g/m^3$ is allowed (Ministry of Health 1997). The current regulations are under evaluation, and the new standard is proposed to be set at $20\,\mu g/m^3$ for both occupied (residential) and non-occupied (workplace) spaces. The new value is in compliance with exposure limits recommended by NIOSH, however NIOSH REL values apply only to workplace hazards, whereby the term "workplace" is still understand in the traditional sense as, for example industrial sites with chemical emissions from processes and materials used. In the future it will become more relevant to consider offices, schools, shops and similar indoor spaces as workplaces, especially as the impact of formaldehyde on memory skills has recently been confirmed by Lu et al. (2008).

High concentrations were found for hexanal (see Table 2). Higher levels of hexanal were reported for the rooms with panels, due to its release from wood pellets. Hexanal is also used as masking agent in cosmetics and fragrances. The majority of the examined offices were occupied by females, which may indicate that fragrances are the source of considerably high concentration of hexanal. This finding could be confirmed by interviewing the occupants of the room, in which the highest concentrations of hexanal were measured. Office O-325 has wood based panels on the floor and female occupants confirmed regular application of strong fragrances. The lowest concentration of hexanal ($9.48\,\mu g/m^3$) was measured in the room occasionally occupied by male subjects.

Acrolein was found in all the examined office rooms. Except for one site, acrolein levels were very similar: 4.7–$6.6\,\mu g/m^3$. No simple correlation between different room finishing was observed as in the case of formaldehyde, although slightly lower concentrations were measured in rooms with carpeting compared to those with plywood panels (Table 2). This is in compliance with findings of Seaman et al. (2007), who reported no detectable emission from carpeting, but measurable emission from particle board and latex paints.

Acrolein is mainly found in kitchens (with gas stoves). Another source of acrolein is ETS (environmental tobacco smoke). Smoking was completely prohibited in the FEE building, both in public areas, and in individual office rooms. But in some rooms occasional visits of smokers were reported during interviewing.

Acrolein concentrations were significantly lower then the levels of formaldehyde, although higher than found in the literature concerning residential buildings (Gilbert et al., 2005). NIOSH has set the exposure limit of acrolein to REL $= 0.01$ ppm ($250\,\mu g/m^3$) (NIOSH, 2005). Acrolein is not listed in the ordinance of the Polish Ministry of Health.

Other detected carbonyl compounds were measured at levels similar to that of acrolein. Acetaldehyde (ethanal) was ubiquitous in the examined offices. This is in agreement with findings of other author (Baez et al. 2003, Cavalcante et al. 2006, Hutter 2006, Miu et al. 2008), as acetaldehyde is commonly detected in all indoor environments. In agreement with the findings of Dassonville et al. (2009) for homes, slightly higher concentrations were found in offices with wall-to-wall carpeting than those with panels.

Exposure to acetaldehyde has produced nasal tumors in rats and laryngeal tumors in hamsters, NIOSH therefore recommends that acetaldehyde be considered potential occupational carcinogen in conformance with the OSHA carcinogen policy (NIOSH, 2005). The detected levels were far below the recommendations of OSHA exposure limits.

One room (visiting professors room – O-328) was occasionally occupied by male individuals. Because of random usage and airing (window opening), concentrations of the examined chemicals were higher in this room, compared with other studied premises.

3.2 Bioaerosols

Bacteria and fungi were detected in all the examined office rooms. The concentrations of bacterial and fungal bioaerosols in the indoor environment are shown in Table 3. The results are expressed in colony forming unit per cubic meter (CFU/m^3). The total concentrations of fungi and bacteria with respect to window facing and floor covering are gathered in Table 4.

The total number of bacteria was higher in all the examined offices than the total number of fungi. These results are in agreement with finding of Jo (2005). Other authors (Wang-Kun Chen & Lee S. C. 1999, Meklin et al 2002a, Godwin C. S., 2007) found that in the schools they investigated the count number of fungi was greater than the bacteria.

The total number of bacteria was the lowest in the office which was permanently unoccupied during the measurements. The level of airborne bacteria was highest in one particular office, often visited by coworkers of the room's normal occupants. According to Sessa et al. (2002), concentrations of bioaerosols are higher in the presence of humans. However, no significant differences were observed in rooms occupied by one or by two persons

Concentrations of bacteria in one office used for storage of books on open shelves were significantly higher compared to other examined office rooms. This could indicate that books (paper and dust) potentially increase bacteria concentrations, due to their role in providing good environmental conditions for multiplying bacteria.

The total number of bacteria in the corridor was lower than in the offices, but the total number of fungi measured in the hall was comparable to the number of fungi in the offices examined.

The results show that the both airborne bacteria and fungi remained at their highest level in the offices with carpeting. The mean concentration of bacterial bioaerosols was $496\,CFU/m^3$ in offices with carpeting and $334\,CFU/m^3$ in offices with panels. As the highest fungi value found in an office with carpeting, the average of the concentrations of fungi was $185\,CFU/m^3$.

Table 3. Concentrations (cfu/m³) of bacteria and fungi in the indoor air of the examined offices.

Offices	328	325	324	217	219	216	Hall
Total number of bacteria (cfu/m³)							
1. (>7.0 μm)	10	163	13	76	90	20	20
2. (7.0–4.7 μm)	33	40	16	36	83	33	23
3. (4.7–3.3 μm)	26	93	56	70	86	73	40
4. (3.3–2.1 μm)	30	126	33	50	153	93	43
5. (21–1 μm)	83	113	120	126	336	96	73
6. (1.1–0.65 μm)	10	23	16	20	33	14	3
Total	192	558	254	378	781	329	202
Total number of fungi (cfu/m³)							
1. (>7.0 μm)	16	3	13	16	6	20	10
2. (7.0–4.7 μm)	23	6	13	13	10	16	10
3. (4.7–3.3 μm)	16	43	33	46	20	20	23
4. (3.3–2.1 μm)	33	43	63	116	103	46	36
5. (2.1–1.1 μm)	26	6	16	26	26	56	50
6. (1.1–0.65 μm)	3	0	6	6	6	3	6
Total	117	101	144	223	171	161	135
Number of occupants	0	2	2	3	2	1	–

Table 4. Concentrations (cfu/m³) bacteria and fungi in the indoor air of the examined offices.

Bioaerosols	With carpeting		With panels		Window to the north		Window to the south	
	range	mean	range	mean	range	mean	range	mean
Total number of bacteria	329–781	496	192–558	334	192–781	486	254–558	379
Total number of fungi	161–223	185	101–144	120	117–171	144	101–223	157

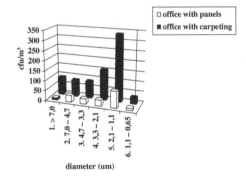

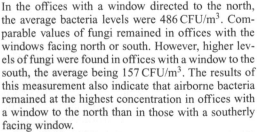

Figure 2. Bacteria concentrations in offices with N-facing window.

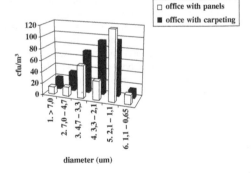

Figure 3. Bacteria concentrations in offices with S-facing window.

In the offices with a window directed to the north, the average bacteria levels were 486 CFU/m³. Comparable values of fungi remained in offices with the windows facing north or south. However, higher levels of fungi were found in offices with a window to the south, the average being 157 CFU/m³. The results of this measurement also indicate that airborne bacteria remained at the highest concentration in offices with a window to the north than in those with a southerly facing window.

Figures 2–3 compare the concentrations of bacteria in the rooms with panels and carpeting in rooms facing north and south, respectively. Figures 4–5 compare the concentrations of fungi in the rooms with panels and carpeting in rooms facing north and south respectively. For all the measured fractions the concentration of bacteria was higher than fungi. The concentrations of bacteria and fungi were highest with four (3.3–2.1 μm) and five (2.1–1.1 μm) impactor stages. Direct light does not seem to have such a strong impact on the bioaerosol concentration as floor covering (see figures 2–5). Textiles, cellulose and glue used to attach wallpaper apparently generate a favorable microenvironment for bacterial and fungal growth. Additionally, the observed high concentration of bioaerosols in rooms with carpets may be connected with the difficulty of

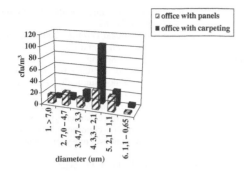

Figure 4. Fungi concentrations in offices with N-facing window.

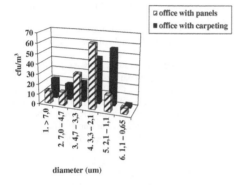

Figure 5. Fungi concentrations in offices with S-facing window.

maintaining the necessary high standard of cleanliness. The measurement of the highest concentration of bacteria in a room containing a large number of books on open shelves appears to support this conclusion.

4 CONCLUSIONS

Chemical and biological measurements carried out in a university office milieu show different dependencies between materials used in room fittings and particular pollutant levels. Six carbonyl pollutants, as well as fungus and bacteria, were detected in all the collected samples. The highest concentrations were found for methanal (formaldehyde), although hexanal concentrations were also at a high level. Concentrations vary in rooms with carpeting compared to those with wood-based panels. In particular, methanal, ethanal and hexanal concentrations were higher in the offices with wall-to-wall carpeting.

Similar findings were observed for bioaerosols, as concentrations of fungus and bacteria were higher in the rooms with carpeting compared to those with panels. The dominant fraction was 3.3–2.1 μm and 2.1–1.1 μm for both fungus and bacteria. Slightly higher concentrations of bacteria were found in rooms with windows directed to north. The total number of

bacteria was higher than the total number of fungi in all the examined offices.

ACKNOWLEDGEMENTS

This work was supported by scientific grant No. 4955/T02/2008/34 from the Ministry of Science and Higher Education, Poland.

REFERENCES

Arts, J., Rennen, M. & Hee,r C. 2006. Inhaled formaldehyde: Evaluation of sensory irritation in relation to carcinogenicity. *Regulatory Toxicology and Pharmacology* 44: 144–160.

Báez, A., Padilla, H., Garcia, R., Torres, M., Rosas, I. & Belmont, R. 2003. Carbonyl levels in indoor and outdoor air in Mex ico City and Xalapa, Mexico. *The Science of the Total Environment* 302: 211–226.

Brickus, L.S.R., Cardoso, J.N. & De Auino Neto, F.R., 1998. Distribution of indoor and outdoor air pollutants in Rio de Janeiro, Brazil: Implication to indoor air quality in Bayside offices. *Environ. Sci. Technol.* 32: 3485–3490.

Cavalcante, R.M., Seyffert, B. H., Montes, D'Oca, M.G., Nasci,Mento, R.F., Campelo, C.S., Pinto, I.S., Anjos, F.B. & Costa ,A.H.R. 2005. Exposure to assessment for formaldehyde and acetaldehyde in the workplace. *Indoor and Build Environment* 14: 165–172.

Cavalcante, R.M., Campelo, C.S., Barbosa, M.J., Silveira, E.R., Carvalho, T.V. & Nascimento, R.F. 2006. Determination of carbonyl compounds in air and cancer risk assessment in an academic institute in Fortaleza, Brazil. *Atmospheric Environment* 40: 5701–5711.

Chen, W-K., Tsao, R-Ch. & Hsu, W-Ch. 2009. Department of Property and Environmental Management Jinwen University of Sci. & Tech., Taipei, Taiwan (www.researcher.nsc.gov.tw, assessed Sept. 1, 2009)

Chuah, Y.K., Fu, Y.M., Hung, C.C. & Tseng, P.C. 1997. Concentration variations of pollutants in a work week period of an office. *Building and Environment* 32: 535–540.

Dassonville, C., Demattei, C., Laurent, A.-M., Le Moullec, Y., Seta, N., Momas, I. 2009. Assessment and predictor determination in indoor aldehyde levels in Paris newborn babies' homes. *Indoor Air* 19: 314–323.

Engelhart, S., Burchardt, H., Neumann, R., Ewers, U., Exner, M. & Kramer, M. H. 1999. Sick Building Syndrome in an Office Building Formerly Used by a Pharmaceutical Company: A Case Study. *Indoor Air* 9: 139–143.

European Concerted Action Report. 1990. Indoor Air Quality & Its Impact on Man, Cost Project 613, *Environment and Quality of Life, Report No. 7: Indoor Air Pollution by Formaldehyde in European Countries, Commission of the European Communities,* Luxembourg .

Fanger, P.O., Popiołek, Z. & Wargocki, P. 2003. *Środowisko wewnętrzne, Wpływ na zdrowie, komfort i wydajność pracy,* Gliwice.

Ghilarducci, D.P. & Tjerdema, R.S. 1995. Fate and effects of acrolein. *Rev. Environ. Contam. Toxicol.* 144: 95–146.

Gilbert, N.L., Guay, M., Miller, J.D., Judek, S., Chan, C.C. & Dales, R.E. 2005. Levels and determinants of formaldehyde, acetaldehyde, and acrolein in residential indoor air in Prince Edward Island, Canada. *Environmental Research* 99: 11–17.

Gilbert N., Gauvin D., Guay M., Heroux M., Dupuis G., Legris M., Chan C., Dietz R. & Levesque B. 2006. Housing characterristics and indoor concentrations of nitrogen dioxide and formaldehyde in Quebec City, Canada. *Environmental Research* 102: 1–8.

Gilbert N., Guay M., Gauvin D., Dietz R., Chan C. & Levesque B. 2008. Air change rate and concentration of formaldehyde in residential indoor air. *Atmospheric Environment* 42: 2424–2428.

Godwin C. & Battermann S., 2007. Indoor air quality in Michigan schools. *Indoor Air* 17: 109–121.

Gołofit-Szymczak M. & Skowron J. 2005. Zagrożenia mikrobiologiczne w pomieszczeniach biurowych. *Bezpieczeństwo Pracy* 3: 29–31 (in Polish).

Ho S.S.H. & Yu J.Z. 2002. Concentration of formaldehyde and other carbonyls in environments affected by incense burning. *J. Environ. Monitor.* 4(5) : 728–733.

Hodgson A.T., Beal D. & McIlvaine J.E.R., 2002. Sources of formaldehyde, other aldehydes and terpenes in a new manufactured house. *Indoor Air* 12: 235–242.

Hutter H.-P., Moshammer H., Wallner P., Damberger B., Tappler P. & Kundi M. 2006. Health complaints and annoyances after moving into a new office building: a multidisciplinary approach including analysis of questionnaires, air and house dust samples. *Int. J. Hyg. Environ.-Health* 209: 65–68.

IARC 2006. IARC Monographs on the Evaluation of Carcinogenic Risks to Humans, Formaldehyde, 2-Butoxyethanol and 1-tert-Butoxypropan-2-ol, Volume 88.

Jankowska E., Kondej D. & Pocńiak M. 2003. Subiektywna ocena jakości środowiska pracy w pomieszczeniach biurowych, *Medycyna Pracy* 54(5): 437–444.

Jo W. K. & Seo Y-J. 2005. Indoor and outdoor bioaerosol levels at recreation facilities, elementary schools and homes. *Chemosphere* 61: 11.

KildesØ J., Tornvig L, Skov P. & Schneider T. 1998. An Intervention Study of the Effect of Improved Cleaning Methods on the Concentration and Composition of Dust. *Indoor Air* 8: 12–22.

Law A. K. Y., Chau C.K. & Chan G Y.S. 2001. Characteristics of bioaerosol profile in office buildings in Hong Kong. *Building and Environment* 36: 527–541.

Lee S. C. & Chang M. 2000. Indoor and outdoor air quality investigation at schools in Hong Kong. *Chemosphere* 41: 109–113.

Leikauf G.D. 2002. Hazardous air pollutants and asthma. *Environ. Health Perspect.* 110: 505–526.

Lu Z., Li C., Qiao Y., Yan Y. & Yang X. 2008. Effect of inhaled formaldehyde on learning and memory of mice. *Indoor Air,* 18: 77–83.

Marchand C., Le Calve S., Mirabel Ph., Glasser N., Casset A., Schneider N. & de Blay F. 2008. Concentrations and determinants of gaseous aldehydes in 162 homes in Strasbourg (France). *Atmospheric Environment* 42: 505–516.

Melin T., Reponen T., Toivola M., Koponen V., Husman T., Hyvärinen A., Morrison G.C. & Nazaroff W.W. 2002. Ozone interactions with carpet: Secondary emissions of aldehydes. *Environ. Sci. Technol.* 36: 2185–2192.

Mui K.W., Wong L.T. & Chan W.Y. 2008. Energy impact assessment for reduction of carbon dioxide and formaldehyde exposure risk in air conditioned offices. *Energy and Building* 40: 1412–1418.

Nevalainen A. 2002. Size distributions of airborne microbes in moisture-damaged and reference school buildings of two construction type. *Atmospheric Environment* 36: 39–40.

Nilsson A., Kihlstrom E., Lagesson V., Wessen B., Szponar B., Larsson L., Tagesson C. 2004. Microorganisms and volatile organic compounds in airborne dust from damp residences. *Indoor Air* 14: 74–82.

NIOSH Pocket Guide to Chemical Hazards. 2005. National Institute for Occupational Safety and Health.

Niu J.L. & Burnett J. 2001. Setting up the criteria and credit- awarding scheme for building interior material selection to achieve better indoor air quality. *Environment International* 26: 573–580.

Parat S., Perdrix A., Fricker-Hidalgo H., Saude I., Grillot R. & Baconniers P. 1997. Multivariate analysis comparing microbal air content of an air – conditioned building and a naturally ventilated building over one year. *Atmospheric Environment* 31: 441–449.

Pastuszka J.S., Paw U. K. T., Lis D.O., Wlazło A. & Ulfig K. 2000. Bacterial and fungal in indoor environment in Upper Silesia, Poland". *Atmospheric Environment* 34: 3833–3842.

Reports Studies 78, 1983, Word Health Organization

Sakai K., Norback D., Mi Y., Shibata E., Kamijima M., Yamada T. & Takeuchi Y. 2004. A comparison of indoor air pollutants in Japan and Sweden: formaldehyde, nitrogen dioxide, and chlorinated volatile organic compounds. *Environmental Research* 94: 75–85.

Salthammer T. 1997. Emissions of volatile organic compounds from furniture coatings. *Indoor Air* 7: 189–197.

Seaman V.Y., Bennett D.H. & Cahill T.M. 2007. Origin, occurrence, and source emission rate of acrolein in residential indoor air. *Environ. Sci. Technol.* 41: 6940–6946.

Sessa R., Di P.M., Schiavoni G., Santino I., Altieri A., Pinelli S. & Del P.M. 2002. Microbiological indoor air quality in healthy buildings. *New Microbiology* Jan; 25(1): 51–56.

Singer B. C., Coleman B.K., Destaillats H., Hodgson A.T.H., Lunden M.M., Weschler C.J. & Nazaroff W.W. 2006. Indoor secondary pollutants from cleaning products and air freshener use in the presence of ozone. *Atmospheric Environment* 40: 6696–6710.

Toftum J. 2002. Human response to combined indoor environment exposures. *Energy and Buildings* 34: 601–606.

Tsai F.C. & Macher J.M. 2005. Concentrations of airborne culturable bacteria in 100 large US office buildings from the BASE study. *Indoor Air* 15: 71–81.

Wiglusz R., Sitko E., Nikel G., Jarnuszkiewicz I. & Imielska B. 2002. The effect of temperature on the emission of formaldehyde and volatile organic compounds (VOCs) from laminate flooring case study. *Building and Environment* 37: 41–44.

Wolkoff P. 1995. Volatile organic compounds – sources, measurements, emissions, and the impact on indoor air quality. *Indoor Air* (Suppl. No 3) 1–73.

Wolkoff P. & Kjaergaard S.K. 2007. The dichotomy of relative humidity on indoor air quality. *Environ. International* 33: 850–857.

Wolkoff P., Wilkins C.K., Clausen P.A. & Nielsen G. D. 2006. Organic compounds in office environments- sensory irritation, odor, measurements and the role of reactive chemistry. *Indoor Air* 16: 7–19.

Word Health Organization, Indoor air pollutants: exposure and health effects. EURO

Wu P.-C., Li Y.-Y., Chiang C.-M., Huang C.-Y., Lee C.-C., Li F.-C. & Su H.-J. 2005. Changing microbial concentrations are associated with ventilation performance in Taiwan's air-conditioned office buildings. *Indoor Air* 15: 19–26.

Zhu H., Phelan P., Duan T., Raupp G. & Fernando H. J. S. 2003. Characterizations and relationships between outdoor and indoor bioaerosols in an office building.

Environmental Engineering III – Pawłowski, Dudzińska & Pawłowski (eds)
© 2010 Taylor & Francis Group, London, ISBN 978-0-415-54882-3

Gender differences in odor perception of n-butanol neutralized by ozone

M.R. Dudzińska, B. Polednik & M. Skwarczyński
Faculty of Environmental Engineering, Lublin University of Technology, Lublin, Poland

ABSTRACT: Higher prevalence of SBS in case of women is statistically significant and the differences between men and women in the sensory perception of odorous pollutants present in indoor air may be responsible for this fact. The performed sensory assessments of the odor detection threshold of n-butanol in mixtures with indoor air confirmed these gender differences. The mean value differences for both genders amounted to 37.1%. Similar effect was observed when the deodorization by ozone was tested. Higher ozone doses increased the odor detection concentrations in the same extent for men and women.

Keywords: Indoor Air Quality, Sick Building Syndrome, Gender differences, Odor detection threshold, Ozone deodorization.

1 INTRODUCTION

Gender, apart from age, upbringing and the environment is more and more frequently considered to be one of the most important factors differentiating people's behavior. Among personal, psychological and occupational variables, gender is also of considerable importance in many epidemiological investigations (Brasche et al. 2001;, Brasche et al. 2005, Steinberg & Wall 1995). Gender effect is reported in different sensory analyses as well as e.g. in assessments of perceived indoor air quality (IAQ) (Meyer et al. 2004, Meyer et al. 2005, Skyberg et al. 2003). Because IAQ is considered as a major determinant of the sick building syndrome (SBS), one can assume that its symptoms are gender dependent. The contemporary SBS studies prove that gender is in fact a powerful impact factor (Ebbehøj et al. 2005; Muchic & Butala 2006, Runeson et al. 2006). It was namely revealed that the symptoms often reported from "sick buildings" are more prevalent among women. Stenberg and Wall (1995) found that differences in the occurrence of SBS symptoms among women and men result from factors outside the indoor environment. Similar results were presented by Brasche et al. (2001). They showed that females suffer more from SBS than men independent of personal, most work related, and building factors. Brasche et al. (2005) also proved that women are characterized by significantly increased risks of subjective perception of eye symptoms. According to the results of indoor air studies (Brasche et al. 2001, Steinberg & Wall 1995, Topp et al. 2005, Yousem et al. 1999, Lindgren & Norback 2005) women more likely to report impairments and more frequently complain of bad indoor air quality than men. It could be hypothesized that these observations are connected with the gender differences in odor sensitivity. In fact they were revealed in research. Hedge and Erickson (1996) showed that

gender differences in responses to odor are statistically significant. The extent to which gender directly affects odor sensitivity, or acts as a surrogate for the influence of other variables, remains to be clarified. Assuming that the gender effect in the SBS symptoms and odor sensitivity is significant, it should be taken into account in research and actions toward improving the perceived indoor air quality. Therefore, the gender profile of the tested persons in the SBS studies and the testing persons (gender of the sensory panel members) in odor sensitivity studies should also be considered. For example in case of office spaces in which female users outnumber male users, women should prevail among the panel members.

The SBS studies should not only be limited to monitoring its symptoms among the occupants of specific indoor environments. Generally applicable effective SBS lowering methods should be developed. They may concern a comprehensive removal of all kinds of indoor air pollutants responsible for SBS or only their reduction. Ozone treatment has become a focus of attention because ozone generating devices are still used to improve the IAQ despite the fact that this method can be considered as a controversial one as recent research has shown that in addition to the harmful effects of ozone itself, the products of indoor chemistry initiated by ozone are equally or even more hazardous (Weschler 2000, Boeniger 1995, Hubbard et al. 2005).

The paper presents various odor perception of n-butanol in the indoor air with small doses of ozone assessed by the sensory panel consisting of men and women.

2 MATERIALS AND METHODS

The odor threshold concentrations of n-butanol in indoor air and of n-butanol in defined mixtures with ozone in the indoor air were determined by means of the dynamic method using a modified Ecoma T07 olfactometer. n-butanol has been chosen because it is one of more abundant and relevant VOCs in the indoor air and due to the fact that according to the EN 13725 (2003) it is a standard compound for determining an odorous unit. The sensory analysis and sample preparations were carried out in an air quality laboratory at the Faculty of Environmental Engineering in Lublin University of Technology (LUT).

The laboratory consisted of two adjacent mechanically ventilated rooms: a main room for olfactometric assessments ($4.5 \times 3.5 \times 2.5$ m), and an anteroom ($3.5 \times 2.5 \times 2.5$) where the olfactometer control and dosing panel was located and where the samples were prepared. There was an environmental chamber with Plexiglas walls and aluminum frames (Figure 1a) in the main room. The olfactometer with diffuser tubes for sensory assessments together with the assessors' seats was located inside the environmental chamber. The dosimetric and the control part of the olfactometer was beyond the assessors' view.

Charcoal-filtered and conditioned outdoor air with the temperature of 23°C and the relative humidity of 45% and flow of 0.9 dm³/s was supplied to the environmental chamber through the inlet located on the bottom of the wall opposite the chamber's entrance. The exhaust air exited the chamber through two outlets situated over the olfactometer diffusers and was directed to the ventilation system exhaust in order to avoid the contamination of the room's air.

The evaluations of the odor detection threshold concentrations of n-butanol in mixtures with indoor air C_{th} and in mixtures with indoor air and ozone C_{tho}

were conducted for samples prepared according to the scheme shown in figure 1b. The indoor air mixtures of n-butanol with ozone were dynamically diluted with conditioned ($T = 23 \pm 0.2°C$, $RH = 45 \pm 3\%$) odorless air in the olfactometer and assessed by the sensory panel. The results of the assessments were registered by the computer. Ozone of a stable concentration of 12.6–13.7 g/m³ was obtained in a custom made corona discharge generator. A calibrated photometric ozone analyzer (Anseros Model Ozomat GM 6030) was used to monitor this concentration. Doses of ozone used in mixtures ranged from 1 to 6 ml and the concentrations obtained in diluted, sensory assessed mixtures were lower than the ones permissible by standards in the indoor air (Ashrae 2001, US EPA 1995) and ranged from 2.9 to 31.2 ppb.

The indoor air with a presumably low (not measured in the study) concentration of VOC's and ozone was sampled at one of the unoccupied, gravitationally ventilated LUT classrooms and was supplied to the vapor delivery pipette (Suppelco Inc., Bellefonte, Pa.) with a flow rate of 0.2 dm³/min. At first the required amount of n-butanol and then ozone were introduced to the air flowing through the pipette by means of a chromatographic syringe. Thorough mixing of the n-butanol vapors and ozone with the air took place in a custom stream rotating mixer. The supplied mixture was divided into two streams which collided while rotating in opposite directions. The obtained homogeneous mixture was directed to an odorless polyethylene bag. After one hour of sample conditioning the chromatographic analyses (Trace Ultra GC/Polaris Q MS – Thermo Electron Corp.) were performed.

At the same time the sample was dynamically diluted in the olfactometer and sensory evaluations were performed according to the European Standard EN 13725 (2003).

The sensory panel was recruited from among the LUT students. Male and female, trained and untrained (EN 13725, 2003), non-smoking panelists were aged from 20 to 23. Among the panelists (54% female) 16 participated in more then two odor assessments and thus were considered to be trained. All subjects passed screening for general health and they were informed about the tasks and the used agents (n-butanol and ozone). The C_{th} evaluations of n-butanol in dynamically diluted mixtures with the air as well as C_{tho} evaluations of ozone and n-butanol in such air mixtures were performed by a group of sixteen trained students and a group consisting of a variable number of untrained students. Each sample was evaluated by at least twelve students. The sessions of the C_{th}(C_{tho}) evaluations consisted of three series. Between the series there were short breaks of a few minutes duration. During each series the panelists had to assess not only samples with the given mixture, but also blank samples (clean, carbon filtrated air) in randomized order. The blank samples constituted 20% of all the assessed samples. The mixture inhalation time was set to 2.2 s.

All obtained results of sensory assessments were statistically analyzed using STATISTICA 7.1 software

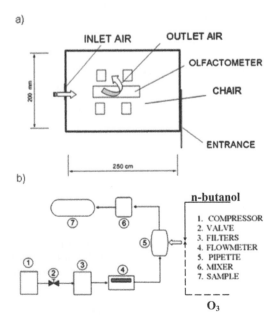

a)

INLET AIR   OUTLET AIR

OLFACTOMETER

CHAIR

ENTRANCE

200 mm

250 cm

b)

n-butanol

1. COMPRESSOR
2. VALVE
3. FILTERS
4. FLOWMETER
5. PIPETTE
6. MIXER
7. SAMPLE

O_3

Figure 1. Scheme of a – olfactometric laboratory, b – sample preparation procedure.

and appropriate statistical methods (Student's t-test and ANOVA for one way analysis of variance). The results were statistically significant for P < 0.05.

3 RESULTS AND DISCUSSION

The dependencies of odor detection threshold of n-butanol C_{th} and ozone mixed with n-butanol C_{tho} on gender and the assessing practice of the sensory panel and on the gender and dose of ozone are presented in Figure 2.

The obtained C_{th} for both trained and untrained men and women were slightly higher than the official odor threshold (123 $\mu g/m^3$) according to the European Standard EN 13725 (2003). It can be assumed that the ozone concentrations in the assessed mixtures were below or close to its detection threshold. The concentrations presented in the diagram were calculated from the ozone doses and mixture dilutions and did not take into account ozone's reactions and its concentration decline in the mixtures. According to Cain et al. (2007) ozone odor is not evident until the concentrations are slightly lower than 20 ppb. From Figure 2 it can be seen that women are characterized by a lower C_{th} than men for both sensory trained and untrained panelists. The mean value differences reached 37.1% and were statistically significant (P < 0.005). From the relation between the C_{tho} and the ozone dose in case of trained panelists it can be noticed that the detection thresholds C_{tho} for women are also lower than for men (Table 1). Besides, it can be seen that if the doses (concentrations) of ozone are higher in the assessed mixtures, n-butanol is perceived at a higher C_{tho} (a smaller dilution rate of the mixtures). Neglecting the effects of not stable agent delivery (Cain et al. 2007) and possible adsorption of agents on the sample bag inner walls (Lee et al. 2001) one could assert that higher doses of ozone in the evaluated mixtures cause a large amount of n-butanol deodorize and a smaller dilution rate of the mixtures (higher C_{tho}) is necessary for its sensory perception. The above was confirmed by a chromatographic analysis of the applied n-butanol and ozone mixture performed by the authors (Skwarczynski et al. 2007).

The influence of the ozone dose on deodorization efficiency of the n-butanol-indoor air mixture is illustrated in Figure 3. The observed deodorization effect is probable not of chemical decomposition origin, as only molecules with multiplied bonds could react with ozone at room temperature (Weschler 2000).

The deodorization efficiency was calculated from the formula:

$$\eta = (C_{tho} - C_{th})/ C_{tho} \qquad (1)$$

where: C_{th} = the odor detection concentration for n-butanol in mg/m^3; C_{tho} = the odor detection concentration for n-butanol neutralized by ozone in mg/m^3.

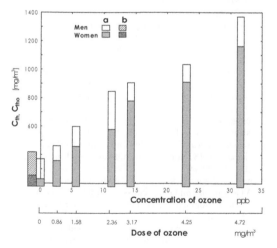

Figure 2. Odor detection threshold concentrations of n-butanol assessed by a – trained, b – untrained panelists vs. doses (concentrations) of ozone.

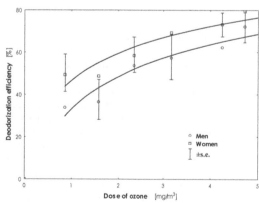

Figure 3. Deodorization efficiency of ozone doses.

Table 1. Estimated mean values and s.e. of n-butanol detection thresholds vs. gender of assessors and initial concentration of ozone in the assessed mixtures.

	Concentration of ozone [ppb]						
	0	2.9	6.8	11.8	14.7	23.2	31.2
Man	385 (45)	582 (194)	606 (68)	832 (188)	902 (211)	1021 (169)	1386 (337)
Women	242 (34)	478 (58)	472 (59)	583 (78)	788 (197)	904 (162)	1177 (303)

Analyses of variance and Student's t-test, P < 0.05.

It is visible that the deodorization efficiency increases with the ozone dose in the whole studied range. However, there are doses of ozone which, when exceeded, do not cause such an evident increase of this efficiency. This concerns both men and women from the tested sensory panel. It can also be observed that alike the occurred differences in odor sensitivity, the deodorization effect of ozone is similar for both men and women. Therefore, it could be assumed that ozone doses which allow to obtain its concentrations in the assessed mixtures below the ones permitted by standards, could reduce n-butanol odor complaints among men and women in the same degree. Taking into account a higher or lower efficiency action of ozone in relation to other odorous pollutants in the indoor air (Weschler 2000), one could conclude that such doses of ozone could improve the IAQ. Consequently, it is to be supposed that certain low levels of ozone in "sick buildings" could reduce the SBS symptoms among both men and women. However, taking into consideration that ozone directly impacts human health and that its reactions with some indoor air pollutants could be a source of hazardous secondary pollutants, mostly more irritating and more odorous than their precursors (Weschler 2000, Hubbard et al. 2005), ozone should not be applied. However, it could be exceptionally used in unoccupied spaces in buildings radically treated for SBS by a comprehensive elimination of bioaerosol and gaseous pollutants. In that case fully controlled indoor air ozone treatments could be performed and the ozone level should be carefully monitored with a possibility to check the potential byproduct concentrations.

4 CONCLUSIONS

The sensory assessed odor detection concentrations of n-butanol and n-butanol with ozone in the indoor air were lower for women from the sensory panel. Small doses of ozone, for which its obtained concentrations are many times lower than the ones that are permitted by standards, increase the odor detection concentration of n-butanol in the indoor air to the same extent for both genders. These findings confirm that gender differences in odor sensitivity exist and therefore higher doses of ozone could eliminate n-butanol odor discomfort in case of women. Considering the applicability of the indoor air ozone treatment as a remedy for SBS, the harmfulness of ozone and its reaction products should be taken into account. Generally this method should not be applied unless in specific conditions in unoccupied spaces with carefully monitored ozone concentration and a possibility to check the potentially harmful byproduct concentrations.

Further research including a detailed investigation of the sensory perception of other indoor air odorous pollutants could be essential to clarify gender-related differences in odor sensitivity and to explain why women, more often than men, suffer from the Sick Building Syndrome.

ACKNOWLEDGEMENTS

The research was carried out within the Project No 4084/T02/2007/32, financed by the Polish Ministry of Science and Higher Education.

REFERENCES

ASHRAE Standard 13-2001, ASHRAE Fundamentals Handbook: Odors. Atlanta. 2001.

Boeniger, M.F. 1995. Use of ozone generating devices to improve indoor air quality. *American Industrial Hygiene Association Journal* 56: 590–598.

Brasche, S., Bullinger, M., Morfeld, M., Gebhardt, H.J. & Bischof, W. 2001. Why do Women Suffer from Sick Building Syndrome more often than Men? – Subjective Higher Sensitivity versus Objective Causes. *Indoor Air* 11: 217–222.

Brasche, S., Bullinger, M., Petrovitch, A., Mayer, E., Gebhardt, H., Herzog, V. & Bischof, W. 2005. Self-reported eye symptoms and related diagnostic findings – comparison of risk factor profiles, *Indoor Air* 15: 56–64.

Ebbehøj, N.E., Meyer, H.W., Würtz, H., Suadicani, P., Valbjørn, O., Sigsgaard, T. & Gyntelberg, F. 2005. Molds in floor dust, building-related symptoms, and lung function among male and female schoolteachers. *Indoor Air* 15: 7–16.

Cain, W.S., Schmidt, R. & Wolkoff, P. 2007. Olfactory detection of ozone and D-limonene: reactants in indoor spaces. *Indoor Air* 17: 337–347.

EN 13725, 2003. Air quality – Determination of odour concentration by dynamic olfactometry.

Hedge, A., Erickson, W.A. & Rubin, G. 1996. Predicting sick building syndrome at the individual and aggregate levels. *Environment International* 22: 3–19.

Hubbard, H.F., Coleman, B.K., Sarwar, G. & Corsi, R.L. 2005. Effects of an ozone-generating air purifier on indoor secondary particles in three residential dwellings. *Indoor Air* 15: 432–444.

Kjærgaard, S.K., Hempel-Jørgensen, A., Mølhave, L., Andersson, K., Juto, J.-E., Stridh, G. 2004. Eye trigeminal sensitivity, tear film stability and conjunctival epithelium damage in 182 non-allergic, non-smoking Danes. *Indoor Air* 14: 200–207.

Lee, S.C., Lam, S. & Ho, K.F., 2001. Characterization of VOCs, ozone, and PM$_{10}$ emissions from office equipment in an environmental chamber. *Building and Environment* 36: 837–842.

Lindgren, T. & Norbäck, D. 2005. Health and perception of cabin air quality among Swedish commercial airline crew. *Indoor Air* 15: 65–72.

Meyer, H.W., Würtz, H., Suadicani, P., Valbjørn, O., Sigsgaard, T. & Gyntelberg, F. 2004. Molds in floor dust and building-related symptoms in adolescent school children. *Indoor Air* 14: 65–72.

Meyer, H.W., Würtz, H., Suadicani, P., Valbjørn, O., Sigsgaard, T. & Gyntelberg, F. 2005. Molds in floor dust and building-related symptoms among adolescent school children: a problem for boys only? *Indoor Air* 15: 17–24.

Muchič, S. & Butala, V. 2004. The influence of indoor environment in office buildings on their occupants: expected-unexpected. *Building and Environment* 39: 289–296.

Runeson, R., Wahlstedt, K., Wieslander, G. & NORBÄCK, D. 2006. Personal and psychosocial factors and symptoms compatible with sick building syndrome in the Swedish workforce. *Indoor Air* 16: 445–453.

Skyberg, K., Skulberg, K.R., Eduard, W., Skåret, E., Levy, F. & Kjuus, H. 2003. Symptoms prevalence among office employees and associations to building characteristics. *Indoor Air* 13: 246–252.

Skwarczynski, M., Polednik, B., Dudzinska, M. & Wronski, M. 2007. Improvement of perceived indoor air quality by ozone (based on n-butanol); *6th international conference on Indoor Climate of Buildings '07: Indoor Environment and Energy performance of buildings* vol. 5: 179–184.

Steinberg, B. & Wall, S. 1995. Why do women report "sick building symptoms" more often than men. *Social Science & Medicine* 40: 491–502.

Topp, R., Thefeld, W., Wichmann, H.-E. & Heinrich, J. 2005. The effect of environmental tobacco smoke exposure on allergic sensitization and allergic rhinitis in adults. *Indoor Air* 15: 222–227.

US EPA 1995. Ozone Generators in Indoor Air Settings, U.S. EPA (U.S. EPA.-600/R-95-154).

Weschler, C.J. 2000. Ozone in Indoor Environments: Concentration and Chemistry. *Indoor Air* 10: 269–288.

Yousem, D.M., Maldjian, J.A., Siddiqi, F., Hummel, T., Alsop, D.C., Geckle, R.J., Bilker, W.B. & Doty, R.L. 1999. Gender effects on odor-stimulated functional magnetic resonance imaging. *Brain Research* 818: 480–487.

Environmental Engineering III – Pawłowski, Dudzińska & Pawłowski (eds)
© *2010 Taylor & Francis Group, London, ISBN 978-0-415-54882-3*

Impact of systems generating local air movement on thermal environment and occupants

J. Kaczmarczyk & T. Nawrat

Department of Heating, Ventilation and Dust Removal Technology, Silesian University of Technology, Gliwice, Poland

ABSTRACT: One possible way to create more comfortable thermal conditions for occupants in warmer environments is to increase the velocity of the air and thus increase the convective heat exchange between the body and the environment. The air movement can be generated by free standing cooling fans, ceiling fans or personalized ventilation systems. Such systems generate a very non-uniform environment for the individual. This paper presents examples of such devices and documents their performances. The most important problems in evaluating the conditions created by these devices for the thermal comfort of the occupants of the space are discussed.

Keywords: Air movement, thermal comfort, draught, individually controlled environment.

1 INTRODUCTION

People spend most of their time indoors: in workplaces, homes, cars, trains or in other enclosed spaces. Creating acceptable thermal environments in such enclosures is one of the main goals set for systems design engineers. Several standards and guidelines have been developed to assist engineers in designing systems and providing solutions to assure suitable indoor thermal conditions for occupants (ISO 7730: 2005; ANSI/ASHRAE 55-2004).

The temperature indoors during summer, especially in buildings without air conditioning systems, may often exceed the upper limits recommended by the standards for thermal comfort. This may be due to intensive solar radiation and high outdoor temperatures. In many cases, however, there are indoor heat sources that substantially contribute to the heat load and thus further increase the indoor temperature. Such a situation may occur in offices, but it occurs even more frequently in industrial premises.

One way to compensate for the increased air temperature could be to install cooling panels and thus increase the radiant heat exchange between the surrounding and the occupant. Another promising and relatively easy to apply method to offset the elevated air temperature is to increase the convective heat exchange with an increased air movement. The air movement can compensate for the heat radiation from the warm surfaces, and also, as recently indicated, for the negative impact of the increased relative humidity of indoor air on thermal comfort and air quality (Melikov et al. 2008).

The air movement can be generated by free standing cooling fans, ceiling fans or by other devices. Cooling with vertical air jets has been shown to be effective in cooling workers (Melikov et al. 1994a, b). Another

method is to use localized ventilation systems, such as personalized ventilation. The idea of personalized ventilation is to provide clean, ventilation air directly to the occupant's breathing zone. Such ventilation serves both to provide clean air and to cool the occupant. Different designs and operating strategies (clean outdoor air, re-circulated room air, different air supply temperatures) have been studied (Kaczmarczyk et al. 2004, 2006) and the positive effects on thermal comfort, perceived air quality and general health (intensity of 'Sick Building Syndrome' symptoms) was proved. Also different designs of Air Terminal Devices (ATDs) for personalized ventilation have been tested. It was found that air movement towards the face, i.e. a cooling of the face, is preferred over an air movement towards the abdomen (Kaczmarczyk et al. 2006).

Cooling fans as well as personalized ventilation systems provide occupants with the ability to control the velocity level and the direction of the flow. People differ as regards their preferences for an indoor environment. Individual controls can accommodate some of these different preferences for the thermal environment. Large individual differences in the preferred air velocity were observed in several studies. Under identical environmental conditions there were subjects who preferred less, more or no change in the preferred velocity (Toftum et al. 2002). Studies of spot cooling with air jets (Melikov et al. 1994a, b) found that under identical conditions of the ambient air temperature and the air jet temperature, the minimum and maximum velocities selected by subjects differed by a factor of five. Fountain et al. (1994) studied the preferred air movement generated by three devices: a desk fan, a desk-mounted diffuser and a floor-mounted diffuser. At a temperature of approximately 25.5°C the selected velocities ranged from less than 0.1 m/s up to almost 0.9 m/s. With an increased air

temperature, a trend towards selecting higher velocities was observed. An experiment on the air velocity
preferred at elevated temperatures – from 26.0°C up
to 29.5°C – performed under laboratory conditions,
also showed large individual differences in the selected
velocities (Toftum et al. 2003). At the highest temperature tested the selected velocities varied from 0.35
to 1.7 m/s. Recently a strong preference for more air
movement, with a small risk of draughts, at a temperature of 22.5°C was observed in a field study (Arens
et al. 2009).

The positive effect of air movement on the occupants' thermal comfort was also acknowledged in
the recent updates of the standards. ANSI/ASHRAE
55-2004 and ISO 7730: 2005 specify the necessary air velocity increment to offset temperature
increases above 26°C. So far this temperature has
been recognized as the upper limit for the occupants' thermal comfort. Devices and systems used for
the convective cooling described above may create a
very non-uniform environment. Thus application of
the recommendations given in the standards may be
limited.

The purpose of this paper is to demonstrate the
non-uniformity of the local environment generated
by specially designed ATDs that can be used for
personalized ventilation and commercially available
cooling fans. Measurements of the physical parameters, primarily the air velocity, were taken and reported.
Furthermore the limitations of the methods to evaluate
the effect of locally applied air movement on people
are underlined and discussed.

2 METHODS

Several ATDs were constructed and tested. In this
paper just two types are presented, one with a circular shape and one with a rectangular shape. The ATDs
are shown in Figures 1 and 2. The ATDs were connected to a flexible air supply duct. The circular ATD
was a cone-shaped exhaust device (inlet air diameter 25 cm) modified to supply air through a circular
opening of 16 cm diameter. In order to distribute the
supplied air uniformly, a perforated plate was attached
inside the ATD at the junction with the air supply duct.
The position of the plate was tested for the best possible
performance. Furthermore, to decrease the turbulence
at the outlet a special material was placed just behind
the front net.

The rectangular ATD was made of a wall air terminal with dimensions 13.5 cm × 12 cm. The front panel
was mounted on a small metal plenum box. Inside the
box a perforated plate and a special material were introduced to make the velocity distribution uniform, in a
similar way as for the circular ATD.

In addition to the ATDs, a commercially available
cooling fan was used for the tests. The fan is shown in
Figure 3. It consisted of three rotating blades of 25 scm
diameter covered by a protective net. The fan had two
speed settings and a power rating of 31 W.

Figure 1.   Circular ATD.

Figure 2.   Rectangular ATD.

Figure 3.   Cooling fan.

The fan and the ATDs were tested in a large laboratory hall with low air velocity (<0.1 m/s) and a uniform
and constant temperature. The ATDs were tested at
three airflow rates: 8, 10 and 15 L/s and the cooling
fan at both possible speeds. The air for the ATDs was
taken from the laboratory hall, thus the temperature
of air supplied through the devices was close to the
temperature of the air in the hall. A slight increase in
temperature (up to 1°K) was observed in some of the
cases tested.

A set of eight low air velocity, thermal anemometers were used to measure the velocity profiles. The
velocity probes were mounted at a distance of 4 cm

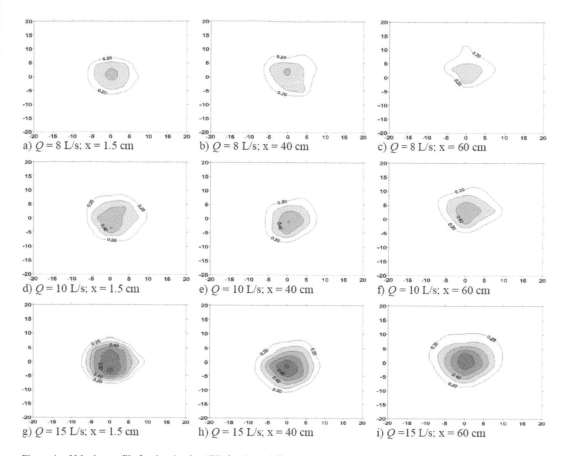

a) $Q = 8$ L/s; x = 1.5 cm   b) $Q = 8$ L/s; x = 40 cm   c) $Q = 8$ L/s; x = 60 cm

d) $Q = 10$ L/s; x = 1.5 cm   e) $Q = 10$ L/s; x = 40 cm   f) $Q = 10$ L/s; x = 60 cm

g) $Q = 15$ L/s; x = 1.5 cm   h) $Q = 15$ L/s; x = 40 cm   i) $Q = 15$ L/s; x = 60 cm

Figure 4.   Velocity profile for the circular ATD for three airflow rates and three distances x cm.

from each other on the shaft of a traversing mechanism. The shaft was attached to an arm that allowed the anemometers to be moved in a vertical direction. The arm was moved in a horizontal direction along a stationary base placed on a table. All adjustments to the anemometers' locations were made manually.

Uncertainty of the mean air velocity (v m/s) measurement was U(v) = 0.02 + 0.02 v (m/s). The procedure for determining the mean air velocity uncertainty for thermal anemometers is described in the literature (Melikov et al., 2007, Popiolek et al., 2007). The upper frequency (the highest frequency up to which the standard deviation ratio remains between the limits of 0.9 to 1.1) was 1.5 Hz. The following parameters were recorded: mean air velocity, standard deviation of the air velocity and the air temperature. The averaging time was set at 2 minutes. This was determined during the initial measurements as being sufficiently long. Velocity profiles were measured at three distances from the outlet: 1.5, 40 and 60 cm from the ATD. The velocity profiles at 1.5 cm were used to determine the initial uniformity of the velocity as well as the initial turbulence intensity. Because of the relatively high velocities generated by the cooling fans the measuring distances were set at 60 and 160 cm.

The airflow rates were measured with a thermal flowmeter. The uncertainty of the flow rate (Q L/s) determination was U(Q) = 0.9 + 0.01 Q(L/s). Additional measurements were taken to identify the velocity decay along the centrelines of the ATDs and the fans. The velocities were measured at 10 cm intervals.

3 RESULTS AND DISCUSSION

This paper reports the results for just three types of devices for generating local airflow – two ATDs and one cooling fan. The velocity profiles were measured for all combinations of distance and three settings of the airflow rate for the ATDs and the two speed settings of the fan. In order to allow direct comparisons of the velocity profiles between the devices studied, all the profiles cover an equal area of 40 × 40 cm². Figures 4 and 5 present the velocity profiles for the circular and rectangular ATDs, respectively.

The velocity fields generated by the ATDs were nearly symmetrical. For the circular outlet it was axisymmetrical, but with the rectangular ATD the profile was flattened along the horizontal axis. For the highest airflow rate studied, the highest velocities observed at a

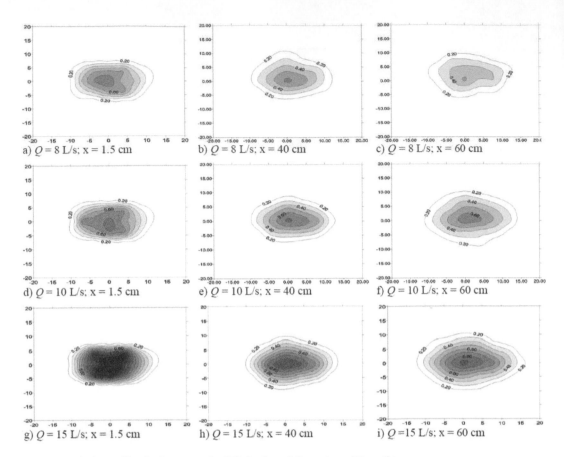

a) $Q = 8$ L/s; x = 1.5 cm   b) $Q = 8$ L/s; x = 40 cm   c) $Q = 8$ L/s; x = 60 cm

d) $Q = 10$ L/s; x = 1.5 cm   e) $Q = 10$ L/s; x = 40 cm   f) $Q = 10$ L/s; x = 60 cm

g) $Q = 15$ L/s; x = 1.5 cm   h) $Q = 15$ L/s; x = 40 cm   i) $Q = 15$ L/s; x = 60 cm

Figure 5. Velocity profiles for the rectangular ATD for three airflow rates and three distances x cm.

distance of 60 cm were in the centre of the air jet. These were: 0.7 m/s for the circular ATD and 0.9 m/s for the rectangular one. In both cases the area with the highest velocity was relatively small, less than 7 cm². The velocity measured away from the jet core decreased rapidly and had dropped to 0.1 m/s at a distance of 15 cm. For the rectangular outlet this distance was even shorter in the vertical direction.

Some of the irregularities in the observed velocity distribution resulted from a not completely uniform velocity distribution at the outlet. With increased distance from the ATD the irregularities became less evident.

For the cooling fan, the velocities recorded were much higher (Figure 6). The highest fan speed reached was up to 2.6 m/s at a distance of 60 cm. The velocity distributions for the fan were substantially less uniform than those for the ATDs.

Detailed analysis revealed that the top surface of the table, on which the traverse mechanism for holding the velocity sensors was mounted, affected the air flow pattern. This phenomenon, however, would also occur in practice, as the devices generating air movement are typically placed either on a table or directly above it.

For the ATDs the velocity remained relatively constant up to a distance of between 40 and 50 cm when it started to decrease. In the zone near the fan, the velocity changed more rapidly with a distance, but then became relatively constant up to a distance of 80 cm, as shown in Figure 7. The reason for such a velocity decay pattern was the shape of the fans' blades and the construction of the protective net (the area in the centre was covered by a plastic plate, as shown in Figure 3).

The most common approach for analyzing the impact of velocity on occupants is to compare the measured velocity to the recommendations given by the standards (ISO 7730:2005). The recommend maximum velocity depends on the occupants' activities, their clothing, and the indoor air temperature. In Category C buildings, buildings with the lowest requirements for a sedentary occupant experiencing summer conditions of operative temperature within the range 22 to 27°C, the maximum velocity recommended is 0.24 m/s. The requirements for buildings with higher expectations regarding the environment are even stricter.

v-velocity at a distance; v_o – initial velocity at 1.5 cm from the ATD and 10 cm from the cooling fan

For category B buildings the recommended maximum velocity is 0.19 m/s while for category A buildings a velocity of just 0.12 m/s is suggested. Considering these recommendations, and taking into

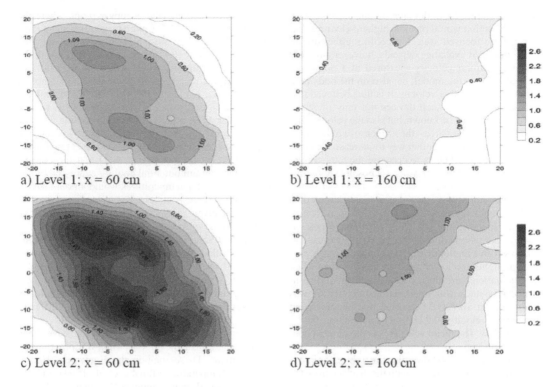

a) Level 1; x = 60 cm

b) Level 1; x = 160 cm

c) Level 2; x = 60 cm

d) Level 2; x = 160 cm

Figure 6.   Velocity profiles for the cooling fan for the two set levels and two distances x cm.

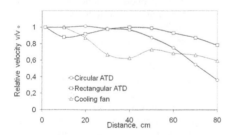

Figure 7.   Relative velocity decay measured on the center-lines of the ATDs and the fan.

account the maximum velocity measured, it should be noted that all the devices tested generated velocities which were too high and would thus cause discomfort through draughts. It should be remembered that the values given in the standards are based on the draught rate (DR) model. The DR index expresses the percentage of occupants complaining of draught discomfort. According to the model the dissatisfaction depends on the mean air velocity, v, air temperature, t_a, and the turbulence intensity, Tu. This latter is the ratio of the standard deviation of the air velocity to the mean velocity. The model was developed for a uniform environment. That is, it assumes the constant velocity and temperature of the air within the evaluated area. When discussing the maximum velocity allowed, the key issue is the possibility of controlling the velocity by the device users rather than a single velocity

value. Therefore the approach of the DR model is not applicable for the evaluation of such systems.

A more appropriate way to evaluate velocity would seem to be the percentage satisfied (PS) model proposed by Fountain et al. (1994). The PS model expresses the predicted percentage of occupants satisfied with the facially applied velocity, v_a, supplied at a particular operating temperature, T_{op}. The assumption of the model is that the occupant has the possibility to control the air velocity. It should be noted that the PS model is not connected with the DR model, although both models are based on the occupants' satisfaction/dissatisfaction. The PS is determined by equation (1):

$$PS = 1.13T_{op}^{0.5} - 0.24T_{op} + 2.7v_a^{0.5} - 0.99v_a \qquad (1)$$

The velocity, v_a, in equation (1) is determined as an average velocity measured at three heights; the maximum velocity at head height, at the lower mid-body and at foot level.

Applying the PS model to the current data, and assuming an operative temperature of 26°C, it turns out that only the cooling fan would be able to satisfy all occupants. The rectangular ATD would satisfy 80% and the circular ATD would satisfy only 70% of the users.

In order to generate the velocity of 1.6 m/s, which would assure all were satisfied according to the model (PS = 100%), either the airflow rate should be increased and/or the outlet area should be decreased. This latter action would result in higher velocities, but over a more focused area. The cooling effect due to convection heat exchange depends, however, not only on the maximum velocity measured at a point and accounted for in the model, but also on the body area that is exposed to this velocity. It is thus believed that in order to evaluate such devices not only must the maximum velocity be known, but also the velocity distribution at the target area – the surface area of the body exposed to the flow. Another way to increase the cooling effect of the ATDs is to decrease the temperature of the air supplied. It is however not possible to realize this with cooling fans.

Another important issue that is not addressed in the PS model (it was found not to be significant), but is included in the DR model for human perception, is fluctuations in the velocity. People exposed to airflows with a higher degree of turbulence would perceive draughts at a lower mean air velocity than when the fluctuations are lower. These findings were confirmed in a study when personalized ventilation was used. Yang et al. (2002) reported that people found fluctuating velocity (0.2 Hz) more distracting than a constant flow. The mean velocity was the same in both cases studied.

The current results showed that the turbulence intensity of the velocity increased with the distance from the device. It was however always from five to ten times lower for the ATDs than for the cooling fan. For the center velocity presented in Figures 4i, 5i and 6c the turbulence intensity was 5% for the circular ATD, 10% for the rectangular ATD and up to 55% for the cooling fan. Thus, it could be expected that the cooling fan would give rise to more complaints than the ATDs. In case of the personalized ventilation system the high turbulence intensity, apart from affecting thermal comfort, would have another downside. The higher the turbulence, the greater is the mixing of the supply air and the surrounding room air. This intensive mixing would thus decrease the perceived air quality by lowering the proportion of clean air in the air being inhaled.

In warmer environments people sweat in order to dissipate heat from the body. In such cases the cooling effect of the velocity of the air applied to the body will be intensified by increasing the rate of sweat evaporation from the skin. A standard effective temperature (SET) model was developed to accommodate the effect of this thermo-physiological regulation. The forthcoming Addenda d and f of ANSI/ASHRAE 55:2004 propose a SET model to evaluate more accurately the thermal environment. The SET model takes into account air temperature and velocity, radiant temperature, relative humidity and two parameters dependent on the occupants. These parameters are the individual's metabolic rate and the insulting properties of the clothing worn. The standard effective temperature can be determined by the ASHRAE Comfort Tool. Based on the SET model it is possible to evaluate the cooling effect for each of the devices studied. For people (0.5 clo and 1.2 met) who occupy a room, where the air temperature is 28°C (and equal to the mean radiant temperature), relative humidity is 50% and the air velocity is low (0.15 m/s), then the SET would be 27.4°C. Applying this model to a room with devices generating the measured air velocity at a distance of 60 cm from a person would decrease the calculated SET to 26.7°C for the circular ATD, 26.2°C for the rectangular ATDs and 24.7°C for the cooling fan. The maximum air flow for the ATDs was 15 L/s and the highest speed setting for the fan was 2. Thus for the circular ATD, the rectangular ATD and the cooling fan, the maximum potential cooling expressed by the difference in SET temperatures would amount to 0.7, 1.2 and 2.7°C, respectively. The velocity used in the model, as given in ANSI/ASHRAE 55:2004, is an average velocity measured at three levels, similar to that in the PS model. The SET takes into account only a single value of the velocity at each level (ankle, waist and head), As illustrated in Figures 5 and 6 the devices used created substantially different local velocity fields. Even with the same maximum velocity (Figures 5f and 6b) the velocity profiles for different devices may differ.

Several other methods and indexes have been defined to evaluate the effect of a non-uniform thermal environment. These include the Equivalent Homogeneous Temperature (Tanabe et al. 1994) and the 'Cooling fan efficiency' indexes (Schiavon & Melikov 2009). These indexes take into account all the parameters affecting heat exchange due to convection and radiation (including non-uniform velocity distribution) between the occupant and the environment and are determined based on measurements with thermal manikins. These indexes quantify the cooling effect, but it is rather difficult to predict the occupants' responses to these systems, as there are no models accurate enough to transform these indexes into human perceptions. Moreover, thermal manikins are an expensive item of equipment and in order to acquire accurate results calibration in a special climate chamber is required. This makes this approach difficult to apply widely in practice. In order to make appropriate comparison of different devices possible, it is recommended that comprehensive data on the systems' performances be collected and reported. Such data as the velocity profiles and turbulence intensity are easier to obtain.

The standards (ISO 7730: 2005, EN 15251:2007, and ANSI/ASHRAE 55-2004) suggest appropriate levels of increased velocity to offset an operative temperature above 26°C (resulting from high air and/or high mean radiant temperature), under conditions where the occupants can control the velocity. Such recommendations could result in possible energy savings. However, they do not directly guide engineers in the design of systems to cool occupants. Nor do they give methods for determining the velocity for systems affecting the local environment. In practice, the application of convective cooling devices could be limited by the energy used to run the fans. An analysis of different cooling strategies for office buildings showed that only under certain control scenarios would convective cooling be a cheaper alternative than mechanical cooling (Schiavon & Melikov 2008). Convective cooling is also used in industrial premises, and is achieved, in the main, by installing cooling fans. However, the energy consumption in such indoor environments has not yet been analyzed.

4 CONCLUSION

The devices used for the local, convective cooling of the occupants of rooms create a non-uniform environment. The degree of the non-uniformity depends on the design of these devices. The cooling effect depends on the location of the occupant relative to the device; this includes both distance from the device and the direction of the air jet.

Standard methods for evaluating a uniform environment created by an overall ventilation system may not be suitable for evaluating local, convective cooling devices. It is recommended that comprehensive data on the performance of the system be reported. Such data would include the velocity profiles and turbulence intensities, They would make the comparison of different devices possible.

In order to verify the cooling effect of a device, measurements with thermal manikins are very useful. However, in order to quantify the effect of people's perceptions and evaluate the overall acceptability of the system it is necessary to perform tests with human subjects.

ACKNOWLEDGEMENT

This work was supported by Polish Ministry of Science and Higher Education within research grant 2007 – 2009.

REFERENCES

Ansi/Ashrae Standard 55-2004, *Thermal environmental conditions for human occupancy*, ASHRAE, Inc., Atlanta, GA.

Arens, E., Turner, S., Zhang, H. & Paliaga, G. 2009. Moving air for comfort. *ASHRAE Journal:* 18–28.

EN 15251:2007, *Indoor environmental input parameters for design and assessment of energy performance of buildings addressing indoor air quality, thermal environment, lighting and acoustics*, European Committee for Standardization. Brussels.

Fountain, M., Arens, E., de Dear R., Bauman F. & Miura K. 1994. Locally controlled air movement preferred in warm isothermal environments. *ASHRAE Transactions* 100(2): 937–951.

ISO Standard 7730: 2005, *Ergonomics of the thermal environment – Analytical determination and interpretation of thermal comfort using calculation of the PMV and PPD indices and local thermal comfort criteria*, ISO, Geneva

Kaczmarczyk, J., Melikov, A. & Fanger., P.O. 2004. Human response to personalized ventilation and mixing ventilation. *Indoor Air* 14(8): 17–29.

Kaczmarczyk, J., Melikov, A, Bolashikov, Z., Lazarov, L. & Fanger, P.O. 2006. Human response to five designs of personalized ventilation. *HVAC&R Research* 12(2): 365–384.

Melikov, A., Arakelian, Rs., Halkjaer, L. & Fanger, P.O. 1994a. Spot cooling – Part 1: Human responses to spot cooling with air jets. *ASHRAE Transactions* 100(2): 500–510.

Melikov, A., Halkjaer, L., Arakelian, Rs. & Fanger, P.O. 1994b. Spot cooling – Part 2: Recommendations for design of spot cooling systems. *ASHRAE Transactions* 100(2): 476–499.

Melikov, A., Popiolek, Z., Silva, M., Care, I. & Sefker, T. 2007. Accuracy limitations for low-velocity measurements and draft assessment in rooms. *HVAC&R Research* 13(6): 971–986.

Melikov, A., Kaczmarczyk, J. & Sliva, D. 2008. Impact of air movement on perceived air quality at different relative humidity. In: *Proceedings of Indoor Air,*. Copenhagen: Denmark, Paper ID: 1037.

Popiolek, Z., Jørgensen, F., Melikov, A., Silva, M. & Kierat, W. 2007. Assessment of uncertainty in measurements with low velocity thermal anemometers. *International Journal of Ventilation* 6(2): 113–128.

Schiavon S. & Melikov A. 2008. Energy saving and improved comfort by increasing air movement. *Energy and Buildings* 40: 1954–1960.

Schiavon, S. & Melikov, A. 2009. Evaluation of the cooling fan efficiency index for a desk fan and a computer fan. In: *Proceedings of Roomvent,* Busan, Korea.

Tanabe, S., Arens, E.A., Bauman, P.E., Zhang, H. & Madsen, T.L. 1994. Evaluating thermal environments by using a thermal manikin with controlled skin surface temperature. *ASHRAE Transactions* 100(1): 39–48.

Toftum, J., Meliov, A., Tynel, A., Bruzda, M. & Fanger, P.O. 2002. Human preference for air movement. In: *Proceedings of Roomvent.* Copenhagen, Denmark.

Toftum, J., Meliov, A., Tynel, A., Bruzda, M., Fanger, P.O. 2003. Human response to air movement. Evaluation of ASHRAE's draft (RP-834). *HVAC&R Research* 9(2): 187–202.

Yang, J., Melikov, A., Fanger, P.O, Li, X. & Yan, Q. 2002. Impact of personalized ventilation on human response: comparison between constant and fluctuating airflows under warm condition. In: *Proceedings of Roomvent.* Copenhagen, Denmark.

Environmental Engineering III – Pawłowski, Dudzińska & Pawłowski (eds)
© 2010 Taylor & Francis Group, London, ISBN 978-0-415-54882-3

Assessment of perceived air quality for selected flat in the residential building

A. Raczkowski & A. Wywiórka

Faculty of Environmental Engineering, Lublin University of Technology, Lublin, Poland

ABSTRACT: The influence of indoor air temperature and relative humidity on the perceived air quality and thermal sensation, in the flat of residential building, was studied. The measurements of air temperature and humidity were carried out. concerning on the acceptability and sensation of temperature and relative humidity of indoor air were carried out, to determine the influence of the microclimate parameters on air quality in the flat. The study confirmed that the most important weight on assessing the indoor air quality have the air temperature. Measurements showed that the thermal acceptability increases with increasing temperature until it reaches optimal value, and then decreases with the temperature increasing. Finally, the line of acceptability of perceived air quality were correlated with the line of acceptability of temperature, in temperature over $21.2°C$.

Keywords: Perceived air quality, acceptability of indoor air, thermal comfort, thermal sensation.

1 INTRODUCTION

Health problems and poor disposition in humans are triggered by factors present in the environment that surrounds them. Both temperature and relative indoor air humidity significantly affect the health and comfort of flat residents and the durability of building structure. Therefore, the interior microclimate should be as clean as possible, and shaped reliably and at a low cost (Fanger et al. 2003, Burek et al. 2006).

The quality of indoor air can be assessed by analysing its chemical composition and comparison of measured concentrations with the maximum acceptable values of compound concentration, determined with respect to specific environments of human presence. However, in the case of general architecture, such task is very difficult, if not entirely impossible. In such interiors as nurseries, schools, office buildings, and housings, the concentrations of particular pollutants are very low, and so many compound mixes are present that the exact measurement of concentrations, emission times, and diffusion patterns is impracticable. Due to the difficulties with chemical analysis, a method was introduced in which it is the man who is the measuring instrument determining the pollutants, and assessing air quality. This is possible as during respiration, human olfactory receptors conduct a detailed analysis of air quality and freshness, reacting even to the smallest amount of pollutants which would be difficult to measure otherwise (Fanger et al. 2003).

The interior microclimate is determined by conditions existing in the room, resulting from thermal and non-thermal elements simultaneously affecting human senses. The present study concerns the conditions present in residential spaces of a multi-family building.

2 OUTDOOR AIR QUALITY

Achieving thermal balance of the human body – i.e. thermal comfort – is not the only criterion for interior climate assessment. Interior spaces often contain various kinds of pollutants which negatively affect human health and composition. The comfort and sensations of residents may be influenced by cigarette smoke, construction materials, steam, smells of prepared food, smells of furniture, and operating devices. To this day, approximately 8000 causes of pollution have been discovered, however, there is little information concerning their acceptable concentrations and their influence on the human body (Nantka 2000).

The quality of interior spaces may be shaped through control and elimination of pollution sources, through the use of ventilation, or through the processes of pollutant filtration, adsorption and absorption (Afshari et al. 2003, ASHRAE Standard 62 2007). Ventilation is definitely the best way to maintain proper air quality, though it may be difficult to implement. Despite the possibility of using low-emission finishing materials, the emission of pollutants is caused mainly by residents (Popiolek 2005). The second method increases the investment and utility cost related to the process of ventilated air conditioning in the mechanical ventilation system. Gravitational ventilation does not always work, as it is largely dependent on external conditions (Baker et al. 1993, Nantka 2007). The third method using alternative air cleaning techniques removes pollutants from the air but

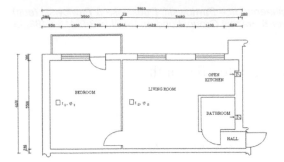

Figure 1. Projection of the flat.

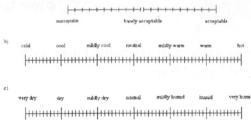

Figure 2. Sample scales used for assessment of acceptability of air quality (a), thermal sensation (b), acceptability of thermal sensation (a), assessment of relative humidity (c), and acceptability of relative air humidity (a).

still remains controversial, as it does not improve the oxygen-carbon dioxide balance inside the room. The above-described limitations mean that indoor air quality depends fundamentally on the process ventilation, with a properly adjusted airstream (Chociaj 2006, Kabza 2005).

3 METHODOLOGY

The study was conducted in the living premises of a 1984 multi-family block. The studied flat is situated on the 2nd floor of a 3-storey building. It has two external walls. The building has undergone thermal modernisation, and the thermal ratios of external walls conform to the current regulations on thermal insulation of buildings. Window frames are made of wood; they are tightly-fitted and highly insulative.

The studied flat is equipped with a gravitational ventilation system. The outside air flows into living premises through various leakage points, flows through the hall, and then gets out through gravitation ducts in the kitchen and bathroom. In the external walls of the flat there are two outlet grilles with a section area of $0.0196\,m^2$, situated $0.26\,m$ below the ceiling level. The kitchen ventilation duct is covered with a round outlet grille with a 12 mm interior diameter, whereas the bathroom duct is covered with a 14 by 14 mm square grille.

In the winter the flat is heated by a central heating system: a set of radiators supplied by a group heat transfer station, itself powered by the municipal heat and power plant.

The studied premises are inhabited by two adult people. The measurements were conducted in two rounds: fall-winter and spring.

The study aimed at measuring the temperature and relative humidity of indoor air, and conducting a survey on the acceptability of physical parameters. The measurement of particular physical parameters of interior environment was conducted with the use of LAB-EL 520 thermo-hygrometers (produced in Poland). The device uses a Pt-1000 thermoresistor to measure temperature, and a capacity sensor to measure humidity. The measurement uncertainty is $\pm0.1°C$ for air temperature, and $\pm2\%$ for relative humidity.

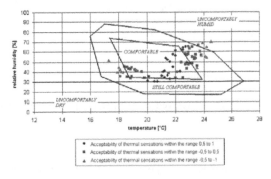

Figure 3. Measurement results of relative humidity and temperature of indoor air in living premises.

The scope of the study included:

• measuring the temperature and relative humidity in the living room and bedroom,
• a survey conducted among 8 people, investigating the assessment of acceptability of air quality, thermal sensation, acceptability of thermal sensation, assessment of relative humidity, and acceptability of relative air humidity in the living room and bedroom (Figure 2) (ASHRAE Standard 62 2007).

In order to determine the influence of microclimate parameters on indoor air quality, the survey studied the acceptability of thermal and humidity conditions. The questions in the survey concerned the perception of temperature in relative humidity, which were marked on a scale divided into 6 areas (Figs. 2b and 2c). The marks were ascribed numerical values from −3 to 3, where −3 corresponded to "cold" and "very dry", 0 corresponded to neutral sensations, and +3 to "hot" and "very humid". Acceptability of air quality, thermal sensation, and relative humidity were marked on a scale divided into 2 areas. The marks were ascribed numerical values from −1 to 1, where −1 corresponded to "unacceptable", 0 to "barely acceptable", and +1 to "acceptable" (Figure 2a) (Fanger 1973, Melikov et al. 2005, Prek 2006, Zhang et al. 2009).

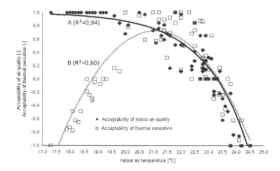
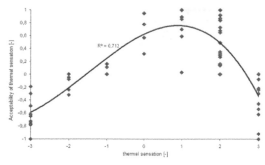

Figure 4. Acceptability of indoor air quality (A) and thermal sensation (B) depending on indoor air temperature in the studied living premises.

Figure 5. Relationship between indoor air thermal sensation acceptability and thermal sensation.

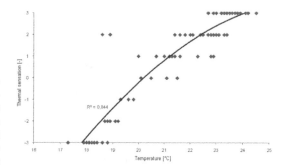

Figure 6. Relationship between indoor air temperature and thermal sensation of the whole human body.

4 RESULT ANALYSIS

The obtained values of indoor air temperature and relative humidity along with their corresponding acceptability values are shown in Figure 3. The same figure features an area of scale points representing comfortable thermal conditions, another area representing "still comfortable" conditions, and an area corresponding to uncomfortable conditions (Gazinski 2005). It was determined that part of the scale assessed as unacceptable conditions was found in the "still comfortable" area, and part of the scale assessed as acceptable was found in the "thermal comfort" area. Most results met the requirements of thermal comfort or were defined within the "still comfortable" range. Temperature reaching 24°C and relative humidity exceeding 65% was assessed as uncomfortable conditions in the "uncomfortably humid" range.

Figure 4 shows the relationship between the assessment of air quality acceptability and thermal sensation acceptability depending on indoor air temperature. Concurrence of acceptability assessments is observed for temperature above 21.2°C. As the temperature falls under 21°C, the acceptability of thermal sensation also falls, while acceptability of air quality rises.

Throughout the entire study period the temperature in the room varied from 17.3°C to 24.5°C. The analysis revealed a strict correlation between indoor air temperature and thermal sensations of persons assessing it (Figure 4). This relationship may be expressed by means of a second-degree polynomial. Analysis of survey results allowed us to determine the variation pattern of thermal sensation acceptability, and air quality acceptability depending on indoor air temperature. It was determined that the acceptability of thermal sensation is highest (amounting to 0.75) when the temperature reaches 21.2°C.

The line approximating the variation pattern of thermal sensation acceptability is a parabola ($R^2 = 0.80$). It was also discovered that with indoor air temperature above 21.2°C the line of air quality acceptability assessment converges with the line of thermal sensation acceptability. The line of air quality acceptability

assessment was presented by means of an exponential curve ($R^2 = 0.94$).

Assessment of thermal sensation acceptability depending on thermal sensation was presented as a third-degree line ($R^2 = 0.712$) which within the value range $(-3;3)$ displays an extremum at the value of 0.75 (Figure 5). The highest acceptability value was discovered for conditions defined as "mildly warm".

Thermal sensation defined as "mildly warm" corresponds to indoor air temperature of approx. 21.1°C (Figure 6). Neutral conditions correspond to 20.2°C. Bearing in mind that the study was conducted in the fall, winter and spring, the temperature corresponds to the computational indoor air temperature in the heating season. In the studied range of indoor air temperatures, i.e. between 17.3°C and 24.5°C, the curve illustrating the relationship between indoor air temperature and thermal sensation is a second-degree curve ($R^2 = 0,844$).

Assessing the influence of relative humidity on air quality is very difficult. It largely depends on air temperature. Research shows that dry and cool air is perceived as fresh, while humid and dry air is regarded as sultry, even if its chemical composition is identical (Wargocki 2004). Temperature and relative humidity, two major parameters determining air quality, are bound by enthalpy. It was shown that enthalpy influences the perceived air quality in rooms where most pollutants were human-emitted (Fang et al. 1998, Toftum et al. 2004, Burek et al. 2006). The line

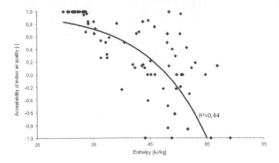

Figure 7. Relationship between air quality acceptability and indoor air enthalpy.

of air quality acceptability assessment was presented by means of an exponential curve ($R^2 = 0.74$) (Figure 7). Points of air quality acceptability in relation to enthalpy are characterised by wider spread, much wider than the points of air quality acceptability in relation to temperature, as shown in Figure 4.

5 CONCLUSIONS

The study showed that in the assessment of perceived air quality indoor air with a temperature of 23.3°C was assessed as barely acceptable, and as temperature fell, the acceptability assessment grew. A convergence was discovered between thermal sensation acceptability assessment and air quality assessment for temperature over 21.2°C. The convergence of thermal sensation acceptability with air quality acceptability proves the crucial role of temperature in the assessment of perceived indoor air quality in living premises of a multi-family building. The highest thermal sensation acceptability value occurs when the temperature in a room is perceived as mildly warm, corresponding to 21,2°C. Thermal sensation defined as cool or cold has very low acceptability, similar to sensation defined as hot. The indoor air parameter best correlated with air quality acceptability is air temperature ($R^2 = 0.94$).

REFERENCES

Ashrae Standard 62. 2007. American Society of Heating. Refrigerating and Air-Conditioning Engineers.

Afshari, A. & Bergsoe, N.C. 2003. Humidity as a Control Parameter for Ventilation. *Indoor and Built Environment* 12: 215–216.

Baker, N. & Standeven, M. 1993. Thermal comfort for free-running buildings. Energy and Buildings 23: 175–182.

Burek, R., Polednik, B. & Raczkowski, A. 2006. Study of the relationship between the perceived air quality and the specific enthalpy of air polluted by people. *Archives of Environmental Protection* 32(2): 21–26.

Chociaj, M. 2006. Indoor air in the Polish and international regulation. District Heating, Heating, Ventilation 3: 32–35.

Fang, L., Clausen, G. & Fanger, P.O. 1998. Impact of temperature and humidity on perception of indoor air quality during immediate and longer whole-body exposures. *Indoor Air* 8: 276–284.

Fanger, P.O. 1974. *Thermal comfort*. Arkady: Warsaw.

Fanger, P.O., Popiolek, Z. & Wargocki, P. 2003. *Indoor air*. Publisher Silesian University of Technology: Gliwice.

Gaziński, B. 2005. Air conditioning technique for practitioners. *Thermal comfort, principles calculations and equipment*. Systherm 2005.

Kabza, Z., Kostyrko, K. & Zator, S. 2005. Regulating the microclimate enclosure. Agenda Publishing PAK: Warsaw.

Melikov, A., Pitchurov, G., Naydenov, K. & Langkilde, G. 2005. Field study on occupant comfort and the office thermal environment in rooms with displaced ventilation. *Indoor Air* 15.

Nantka, M.B. 2007. Indoor air in buildings with natural ventilation. *District Heating, Heating, Ventilation* 3: 33–38.

Popiołek, Z. 2005. *Energy-efficient development in the indoor air*. Publisher Silesian University of Technology: Gliwice.

Prek, M. 2006. Thermodynamical analysis of human thermal comfort. *Energy* 31: 732–743.

Toftum, J., Jørgensen, A.S. & Fanger, P.O. 1998. Upper limits for air humidity to prevent warm respiratory discomfort. *Energy and Buildings* 28(1): 15–23.

Zhang, Y. & Zhao, R. 2009. Relationship between thermal sensation and comfort in non-uniform and dynamic environments. Building and Environment 44: 1386–1391.

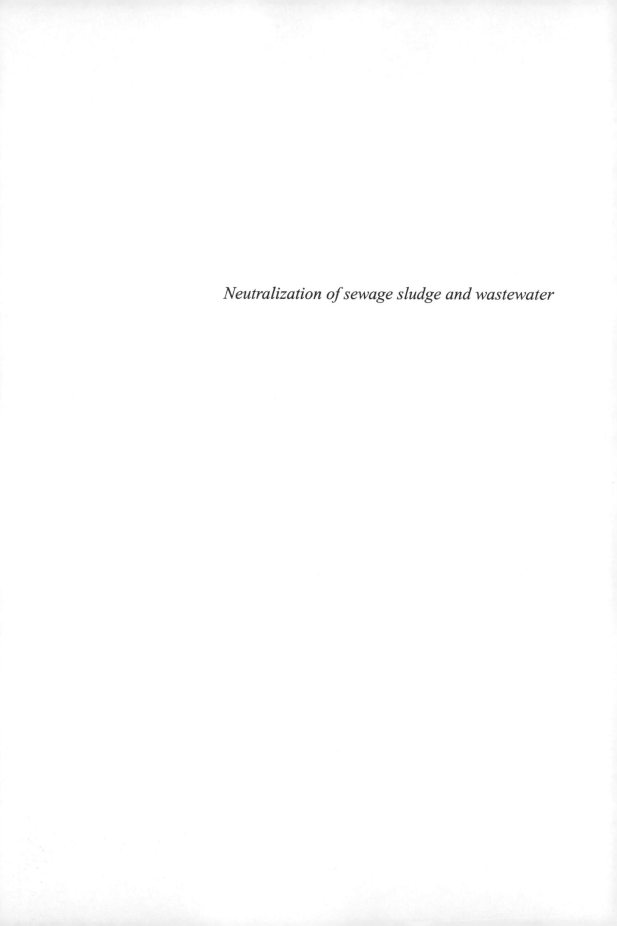

Neutralization of sewage sludge and wastewater

Environmental Engineering III – Pawłowski, Dudzińska & Pawłowski (eds)
© 2010 Taylor & Francis Group, London, ISBN 978-0-415-54882-3

Wastewater treatment with zeolites at Dygowo wastewater treatment plant

A.M. Anielak, K. Piaskowski, M. Wojnicz, M. Grzegorczuk & L. Lewandowska
Department of Water and Wastewater Technology, Koszalin University of Technology, Koszalin, Poland

ABSTRACT: The Dygowo wastewater treatment plant in Poland has been operating since 2000. Sewage treated in the plant comes from the Dygowo commune. It also treats leachate transported from the landfill in Ryman. It is a mechanical-biological wastewater treatment plant treating, on average, 300 m³/day of sewage in two sequential batch reactors (SBRs). The treated effluents are discharged into the Olszynka River. This wastewater treatment plant is trialling the use of natural zeolite to treat sewage. The SBR was dosed with zeolite in loose form during the sewage aeration phase. The research showed a positive influence of the mineral on biological sewage treatment.

Keywords: SBR system, zeolites, municipal wastewater.

1 INTRODUCTION

Public utilities, consuming approximately 2500 million cubic metres (Mm³) of water per year, are the main sources of surface water pollution. This amount of water is then discharged as sewage, although only 50% of the sewage is treated. Production plants are another source of sewage, consuming approximately 8500 Mm³ of water per year. Only 13% of this water, discharged as sewage, is treated. For the most part industrial effluents, after pre-treatment in factory wastewater treatment plants, are merged with domestic sewage and treated in municipal wastewater treatment plants. According to data from the Central Statistical Office, 1200 Mm³ of sewage per year is processed through biological wastewater treatment plants. Treatment of this sewage is often supported by coagulation. Chemical substances applied during the sewage treatment process are partly precipitated and removed with surplus sludge in a secondary settlement tank. However a certain proportion of these penetrate into the surface water along with treated effluents causing an increase in salinity. For example PIX and PAX, industrial coagulants, are added to sewage at a rate of 40 to 120 g/m³; an average consumption approaching 96,000 tonnes/year. A major part of these reagents forms organic complexes and penetrates into the surface water along with treated effluents. These complexes are found in tap water which has been oxidised and disinfected by wastewater treatment plants. The compounds are precursors of substances of high mutagenic activity, such as 3-chloro-4-(dichloromethyl)-5-hydroxy-2(5H)-furanone – also known as MX. MX and its isomers are substances that constitute between 30 and 60% of the mutagenic activity of all substances in water. Trihalomethanes, haloacetic acids (such as monochloroacetic acid, bromoacetic acid, dichloroacetic acid, trichloroacetic acid), acetonitrile,

chloral hydrate, 2-chloro-5-oxo-3-hexene diacyl chloride (COHC) and other chloroorganic compounds are hazardous substances, too. The intent of water treatment plants is to replace such reagents with natural substances.

Zeolites are ranked among those substances showing catalytic, ion exchange, sorption and molecular sieve properties. Research by Se-Jin Park et al. (2002; 2003) showed that the presence of clinoptilolite intensifies nitrification and contributes to the growth of nitrifying bacteria. The authors demonstrated that nitrification runs effectively even in the presence of metal cations, such as zinc cations, or in the presence of toxic phenol or high amounts of potassium salts and gives better results than nitrification in the presence of active carbon. The method is similar to others, reported by various authors, that function by forming a bio-film on materials of diverse shapes, like fittings, immobilised fibrous inserts or powdered substances of different granulation – where the smaller the grain size the higher the surface contact (Surampalli et al. 1997, Ling & Chen 2005).

Research that has been carried out to date has shown the influence of zeolites on the treatment of sewage both in the laboratory and under industrial conditions. That is why tests performed at a technical scale were designed to confirm these previously obtained relationships (Anielak 2006, Anielak & Smarzynska 2007, Anielak & Piaskowski 2005).

2 CHARACTERISTICS OF WASTEWATER TREATMENT PLANT IN DYGOWO

The wastewater treatment plant in Dygowo has been operating since the year 2000 and receives, on average, 300 m³/day of sewage from a commune inhabited

Figure 1. Wastewater treatment plant in Dygowo; a) general view, b) multifunctional SBR reactor.

by approximately 5500 people. Sewage is also transported to the treatment plant at a rate of approximately 49 m³/month. This sewage is, primarily, leachate from the landfill at Ryman. The wastewater treatment plant operates a mechanical-biological system with two biological sequential batch reactors (SBRs) which alternate in parallel.

The technical schema of the wastewater treatment plant is presented in Figure 2. (The numbers in parentheses in the following text refer to elements of Figure 2.) Mechanical pre-treatment of the sewage takes place in a joined system of screw sieves with grit separators. These devices are placed above a sewage retention tank (4) in which pre-treated sewage is stored. The sewage is then sent to two multifunctional SBRs. The reactors are cylindrical and divided into identical tanks with an internal closed denitrification chamber (5). When the reactor is filling, sewage is sent to the denitrification chamber and then through outlets placed at the bottom of the chamber (slow flow) to an external reactor (6) in which aeration, sedimentation and decantation take place. Horizontal agitators are installed in the denitrification chamber, whereas vertical agitators and fine bubble diffusers are installed in the multifunctional reactor. Air, from blowers located above the denitrification chamber, is sent to the diffusers.

Sewage with sludge is cyclically recirculated by a pump from the external chamber into the denitrification chamber during the aeration phase. After the sedimentation phase is completed, decantation begins with the chamber being filled with raw sewage and

the air blower being turned on. The treated effluents are sent to a neighbouring surface stream. The surplus sludge (about 8 to 10 m³/reactor) is disposed of once a day during decantation. It is then thickened, under gravitation, in a sludge tank (7).

Between 76 and 94 m³ of sludge is removed from the thickener six to seven times a month and carried away to the wastewater treatment plant in Kolobrzeg. The water above the sludge is sent back to the retention chamber. The reactors work on a 5-hour cycle; the process phases are presented in Figure 3. It takes 15 minutes to fill up the reactor and during this time the sewage collected in the retention chamber (4) is being pumped into the denitrification chamber (5). The agitators are operating in both the denitrification and multiphase chambers of the SBRs during the filling process. In the next step the sewage and sludge are aerated for 3 hours in the multiphase reactor (6). When aeration occurs, recirculation of the sludge/sewage mixture is carried out in the denitrification chamber three times for 30 minutes, with a 30 minute break during which the sewage in the denitrification chamber is mixed by the mechanical agitator. Sedimentation of the activated sludge lasts for 30 minutes then the treated effluents undergo decantation. Before the next fill up phase begins there is a stand-by phase lasting 30 minutes.

Reactor decantation is achieved using the uplift pressure of the treated effluents – filling up the denitrification chamber with the raw sewage and operating the air blower. When the treated effluent-table reaches a sufficiently high level, it flows out through apertures in the pipe placed around the circumference of the reactor and is discharged into the outflow chamber and water container.

3 MATERIALS AND METHODS

The wastewater treatment plant in Dygowo was selected for the research. It is characterised by unstable operation resulting from the quantitative and qualitative changeability of the sewage flowing in to it and by the use of two SBRs. A loose form (diameter <0.25 mm) of the zeolite (Hungarian clinoptilolite) was added to one SBR during the sewage aeration phase for a few months. The zeolite contained 55% clinoptilolite, 26% ash and volcanic glass, 6% quartz and 13% montmorillonite. A qualitative analysis of the zeolite revealed that it includes 320 mg/g silicon, 60 mg/g aluminium and elements, such as potassium, sodium, calcium, iron and magnesium in its structure (Anielak & Schmidt 2007). The research was carried out on a two-reactor system. The second reactor (SBR2), which was not dosed with zeolite, was a control for the quality of treated effluents in the presence of zeolite in the first reactor (SBR1). The amount of zeolite added to the reactor per day varied from 16 to 32 kg/day per volume of the reactor During the period of the research basic qualitative indicators (acidity (pH), total organic carbon (TOC), five-day biochemical oxygen demand (BOD_5), ammonia

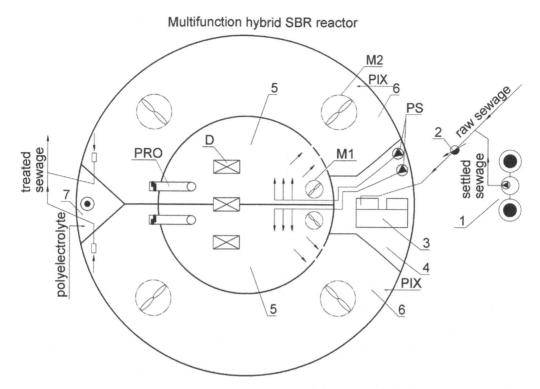

Multifunction hybrid SBR reactor

Figure 2. Technological scheme of the wastewater treatment plant in Dygowo.
Designation: 1 – sewage acceptance station, 2 – measurement chamber, 3 – screw sieve with grit separator, 4 – retention tank, pre-fermentation selector, 5 – denitrification tank, 6 – multifunction SBR, 7 – gravity sludge thickener, PS – raw sewage pumps, M1, M2 – mixers, D – air blower, PRO –sludge recirculation pump, PIX – chemical reagent $Fe_2(SO_4)_3$.

Process phase	Hour 1				Hour 2				Hour 3				Hour 4				Hour 5			
Fill																				
Mixing of multifunction reactor																				
Mixing of denitrification tank																				
Aeration																				
Internal recirculation																				
Settle																				
Draw																				
Idle																				

Figure 3. Process phases of the 5-hour working cycle of the SBR in the Dygowo wastewater treatment plant.

nitrogen (N-NH$_4$), nitrate nitrogen (N-NO$_3$) and total phosphate (P$_{total}$)) of the raw and treated sewage were determined for both reactors. The working conditions and technological parameters of both reactors were similar.

4 RESULTS AND DISCUSSION

Results for the quality of the raw and treated sewage in both reactors are presented in Figures 4 to 10 while

the influence of the zeolite on the effectiveness of the sewage treatment is shown in Table 1.

All the qualitative indicators analysed showed an improvement in the treatment of the sewage in the reactor to which the zeolite had been added. Qualitatively stabilised treated sewage was obtained; in particular with respect to organic substances. Despite increased BOD$_5$ and TOC in the raw sewage, the qualitative indicators of the zeolite-treated effluent remained at a constant level. A small improvement in the removal of phosphorus compounds and a significantly more

157

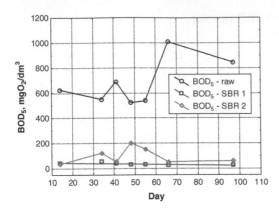

Figure 4. Changes in BOD$_5$ concentrations in raw and treated sewage in SBR1 (with zeolite) and SBR2 (control) during the research period.

effective removal of nitrogen compounds were also observed. After 40 days of dosing with zeolite the ammonia and total nitrogen concentration decreased to near zero, whereas nitrate concentration increased, but did not exceed acceptable limits. The increase in nitrate nitrogen concentration while ammonia nitrogen concentration decreased proved that nitrification was correct and intensive and the decrease of total nitrogen concentration indicated that denitrification was effective.

Changes in the concentration of nitrogen compounds in the sewage influenced the pH value due to the increase in sewage basicity during nitrification. The reaction value was at a level that did not pose a danger to the condition of the microorganisms in the activated sludge (pH > 6.5).

The influence of the zeolite on the effectiveness of sewage treatment is shown in Table 1. Evaluation of the average values of the qualitative indicators analysed shows that if the zeolite amount increases, sewage treatment is more effective. Based on the results of the qualitative analyses of the sewage, the effectiveness of using zeolite to treat municipal sewage was evaluated ($\Delta\%$). The per cent increase in treatment effectiveness of the reactor with zeolite was calculated for all the indicators analysed (Figure 11, Table 2) for the entire research period. The average increase in treatment effectiveness using zeolite ranged from 8.7 to 46.7%. The least difference in effectiveness was observed for the removal of phosphorus and organic compounds, as measured by BOD$_5$ and TOC. The greatest difference in effectiveness was found for the removal of nitrogen compounds, especially ammonia nitrogen (Levipan et al. 2004).

The greatest differences in the treatment's effectiveness were for ammonia nitrogen, 84%, and for total nitrogen, 61%. These values are high. The increase in the removal of nitrogen compounds from sewage as a consequence of using zeolite results from the structural and physiochemical properties of the mineral, which shows a high affinity for ammonia cations and intensifies the biological processes (nitrification). The zeolite

Table 1. The influence of the zeolite dose in the SBR1 on the effectiveness of sewage treatment (arithmetic average).

| Parameter | Unit | Dose of zeolite (kg/volume of reactor) | | | | | | | | |
| | | 16.0 | | | 24.0 | | | 32.0 | | |
		Raw sewage	SBR1	SBR2	Raw sewage	SBR1	SBR2	Raw sewage	SBR1	SBR2
BOD$_5$	mgO$_2$/dm^3	622	44	70	529	30	178	928	26	56
TOC	mg/dm^3	306	30	58	229	26	51	348	27	51
Total phosphorus	mgP/dm^3	16.0	0.82	2.14	16.0	0.63	1.7	18.0	0.26	2.9
Total nitrogen	mgN/dm^3	127	52	91	115	5	55	131	8	16
Ammonia nitrogen	mgN-NH$_4$/dm^3	50.0	41.0	61.0	43.0	1.3	34.0	44.0	1.5	14.0
Nitrate nitrogen	mgN-NO$_3$/dm^3	0.50	0.95	0.32	0.21	0.22	2.13	0.08	2.13	1.02

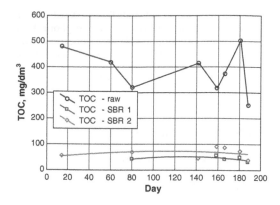

Figure 5. Changes in total organic carbon (TOC) concentrations in raw and treated sewage in SBR1 (with zeolite) and SBR2 (control) during the research period.

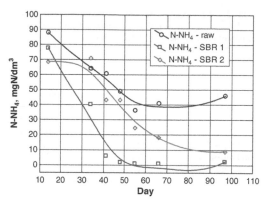

Figure 8. Changes in N-NH₄ concentrations in raw and treated sewage in SBR1 (with zeolite) and SBR2 (control) during the research period.

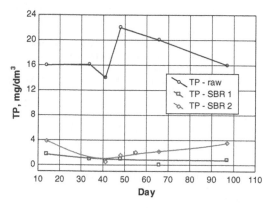

Figure 6. Changes in total phosphorus concentrations in raw and treated sewage in SBR1 (with zeolite) and SBR2 (control) during the research period.

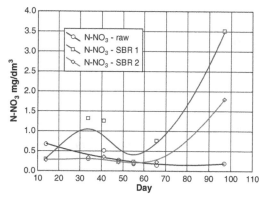

Figure 9. Changes in N-NO₃ concentrations in raw and treated sewage in SBR1 (with zeolite) and SBR2 (control) during the research period.

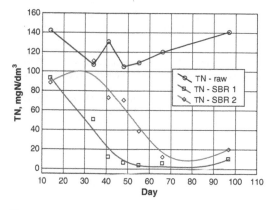

Figure 7. Changes in total nitrogen concentrations in raw and treated sewage in SBR1 (with zeolite) and SBR2 (control) during the research period.

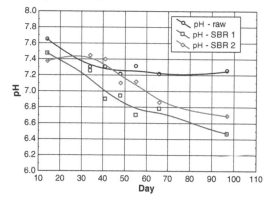

Figure 10. Changes in pH values in raw and treated sewage in SBR1 (with zeolite) and SBR2 (control) during the research period.

was a good substrate for nitrification bacteria growth. These bacteria prefer a stable substrate that influences the rate and effectiveness of ammonia nitrogen oxidation.

The ammonia nitrogen is also removed by zeolite through sorption and ion exchange. A mineral with adsorbed ammonia ions favours the growth of nitrification autotrophic bacteria, which is why nitrification runs faster and more effectively.

Table 2. The increase in sewage treatment effectiveness in SBR1 with zeolite (Δ%) as compared with the sewage treatment effectiveness in SBR2 without zeolite.

Parameter	Unit	Percentage increase in effectiveness (Δ%)		
		average	min	max
BOD$_5$	mgO$_2$/dm^3	11.1	0.5	33.1
TOC	mg/dm^3	6.4	0.1	11.4
Total phosphorus	mgP/dm^3	8.7	0.7	16.9
Total nitrogen	mgN/dm^3	30.2	3.2	61.0
Ammonia nitrogen	mgN-NH$_4$/dm^3	46.7	10.6	84.0

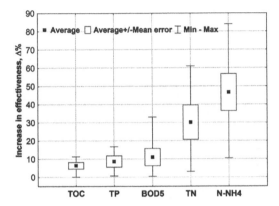

Figure 11. The increase in sewage treatment effectiveness in the reactor with zeolite (Δ%).

5 CONCLUSIONS

The following conclusions can be drawn based on this research, carried out on a technical scale, in using natural zeolite in the biological treatment of municipal sewage:

- adding zeolite to the sequencing batch reactor increases and stabilises the removal of organic compounds from sewage,
- an almost 50% increase in the effectiveness of ammonia nitrogen removal and a more than 30% reduction in total nitrogen were obtained when using natural zeolites.

- zeolites form a good substrate for the growth of nitrifying bacteria, and the removal of phosphorus from the reactor with zeolite is slightly higher than in the reactor without zeolite. In this latter instance the amount removed is consistent with the requirements for the discharge of treated effluents into surface streams.

ACKNOWLEDGEMENTS

This work was supported by the Ministry of Sciences and Higher Education (Development grant No 0628/R/T02/2007/03).

REFERENCES

Anielak, A.M. & Piaskowski, K. 2005. Influence of zeolite on kinetics and effectiveness of the process of sewage biological purification in sequencing batch reactors. *Environment Protection Engineering* 31(2): 107–133.
Anielak .M. 2006. Unconventional methods of biogenic substances removal in sequencing bioreactors, *Gaz Woda i Technika Sanitarna*, no. 2, pp. 23–27
Anielak, A.M. & Schmidt, R. 2007. Adsorption of As(III) on natural clinoptilolite modified with manganese. *Polish Journal of Environmental Studies* 16(2A) Part II: 239–245.
Anielak, A.M. & Smarzyńska, M., 2007, Wastewater treatment with zeolites in SBR system at Krokowa wastewater treatment plant, in: *Gaz Woda i Technika Sanitarna*, no. 5, pp. 30–35
Ling, J. & Chen, S. 2005. Impact of organic carbon on nitrification performance of different biofilters. *Aquacultural Engineering* 33: 150–162.
Levipan, H.A., Aspe, E. & Urrutia, H. 2004. Molecular analysis of the community structure of nitrifying bacteria in a continuous-flow bioreactor. *Environmental Technology* 25: 261–272.
Park, J.S., Lee, S.H. & Yoon, I.T. 2002, The evaluation of enhanced nitrification by immobilized biofilm on a clinoptilolite carrier. *Bioresource Technology* 82: 183–189.
Park, J.S., Oh, W.J., Yoon, I.T. 2003. The role of powdered zeolite and activated carbon carriers on nitrification in activated sludge with inhibitory materials. *Process Biochemistry* 39: 211–219.
Surampalli R.Y., Tyag I.R.D., Scheible O.K. & Heidman J.A. 1997. Nitrification, denitrification and phosphorus removal in sequential batch reactors. *Bioresource Technology* 61: 151–157.

Environmental Engineering III – Pawłowski, Dudzińska & Pawłowski (eds)
© 2010 Taylor & Francis Group, London, ISBN 978-0-415-54882-3

Structure and granulometric composition of suspensions in sewage sludge and activated sludge

E. Burszta-Adamiak, M. Kęszycka & J. Łomotowski
Department of Building and Infrastructure, Wroclaw University of Environmental and Life Sciences, Wroclaw, Poland

ABSTRACT: The characteristics of suspension structures can be represented in one-, two- and three-dimensional space using fractal dimension values. Fractal dimensions allow evaluation of the interior of solid-state aggregates in liquids and their physical properties, such as porosity, density, permeability or size, concurrently showing the degree of the object's complexity. This work presents the methods of assessing fractal dimensions and the distribution of particle sizes in raw and treated sewage and activated sludge, and three-dimensional fractal dimensions determined with a laser granulometer. The results obtained were used for the diagnosis of the state and structure of suspensions in sewage.

Keywords: Activated sludge, dimensional analysis, light scattering, flocs, fractal dimension.

1 INTRODUCTION

Solid particles in a liquid suspension constitute one of the most complicated two-phase systems encountered in nature. The shape and structure of particles suspended in sewage play an important role in the processes of separating solid particles that occur in sewage treatment plants, such as coagulation, sedimentation and sludge dehydration. Most of the suspensions found in sewage are characterized by irregular and chaotic structures (Figure 1), which excludes the application of classical Euclidian geometry in their description. The theory developed by Mandelbrot in the 1980s enables the characterization of irregular structures by using fractal geometry. The most important numerical parameter in the fractal concept is the fractal dimension, which assumes values from 1 to 3. Once the fractal dimension is known, we can estimate, among others, the "packing degree" of irregular particles in a suspension. A low value of fractal dimension shows

the presence of a large number of open spaces that can absorb many contaminants. In addition, a sludge that is characterized by a large packing degree has better dehydration properties (Smoczyński & Wardzyńska 2003).

In the literature, data on the structure of flocs in activated sludge can be found; this data is characterized by the measured fractal dimension (Table 1). However, results are lacking on the structure of suspensions in raw and treated sewage.

The present report outlines the results of studies on the distribution of particle sizes and fractal dimensions of suspensions in activated sludge and raw and treated sewage, taken from three sewage treatment plants located in the Lower Silesia province of Poland. These physical parameters were measured with a laser granulometer, the principle of operation of which is based on the method of low-angle scattering of laser light.

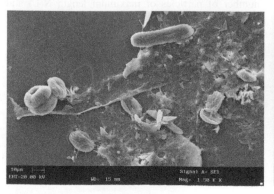 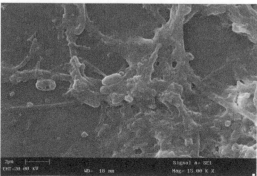

Figure 1. Differentiation of forms occurring in suspensions in activated sludge.

Table 1. Fractal dimensions of flocs from activated sludge taken from different sewage treatment plants, determined with the method of low-angle laser light scattering.

Fractal dimension	Sampling point	Source of wastewater	References
2.16 ± 0.23	Aeration tank	Domestic, industrial	Jin et al. 2004
1.96 ± 0.20	Aeration tank	Domestic	Jin et al. 2004
2.16	Aeration tank	Domestic, industrial	Wilén et al. 2003
1.96 ± 0.32	Aeration tank	Domestic	Wilén et al. 2003
1.85 ± 0.17	Activated sludge external recirculation	Domestic	Wu et al. 2002
2.26 ± 0.47	Aeration tank	Domestic	Li & Yuan, 2002
2.04 ± 0.03	Aeration tank	Domestic	Waite 1999

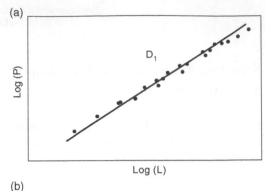

(a)

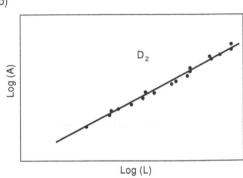

(b)

Figure 2. Methods of assessing: a) one-dimensional fractal dimension D_1, b) two-dimensional fractal dimension D_2.

2 MATERIAL AND METHODS

2.1 Theoretical basis

Fractal geometry is used to describe the morphology of irregular forms by characterizing their structure in one-, two- and three-dimensional space using the fractal dimensions D_1, D_2 and D_3 (Tang et al. 2001, Sorensen et al. 1995, Logan & Klips 1995). The one-dimensional fractal dimension D_1 is determined on the basis of measured maximum lengths of particles of suspension L and their perimeter P using the proportion:

$$P \propto L^{D_1} \tag{1}$$

The two-dimensional fractal dimension D_2 can be found from the relationship:

$$A \propto L^{D_2} \tag{2}$$

where A is the surface of particles of maximum length L.

The values of fractal dimensions D_1 and D_2 can be read from the angle of slope of the straight line in a double log frame, log (P)-log (L), when determining the one-dimensional fractal dimension (Figure 2a), and log (A)-log (L), when determining the two-dimensional fractal dimension (Figure 2b). The starting data, which is necessary for finding D_1 and D_2 (perimeter, area, maximum floc length), are usually obtained by using the technique of image analysis.

The three-dimensional fractal dimension D_3 that characterizes the volume V of particles described by the proportion:

$$V \propto L^{D_3} \tag{3}$$

can be determined by various methods depending on the method used for setting the initial data. In techniques using image analysis, the three-dimensional fractal dimension can be inferred from slopes of the lines S(L) and S(V) in the double-log reference frame log N(L) and log (L), and log N(V) and log (V), respectively (Figure 3a).

In this case, the value of fractal dimension D_3 is found from the relationship:

$$D_3 = \frac{S(L)}{S(V)} \tag{4}$$

Between the coefficients S(L) and S(V) and the number of particles of maximum length N(L) and the number of particles in volume N(V), apply the relationships:

$$N(L) = A_L L^{S(l)} \tag{5}$$

$$N(V) = A_V V^{S(V)} \tag{6}$$

where A_L and A_V are empirical constants, V is particle volume, and L is as in equations (1)–(3).

Another method of finding the three-dimensional fractal dimension is by the analysis of the intensity of laser light scattering I(Q), where Q is the wave number calculated from the equation:

$$Q = (4\pi n / \lambda)\sin(\theta / 2) \tag{7}$$

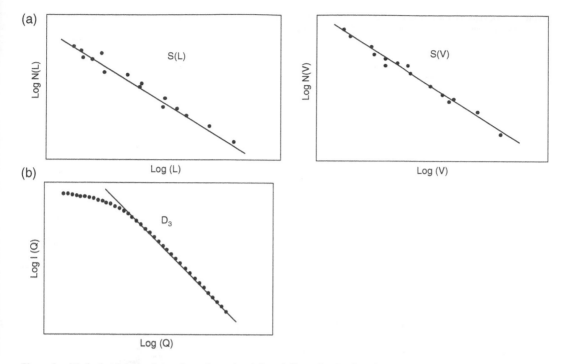

Figure 3. Methods of determining three-dimensional fractal dimension D_3 by using: a) techniques of image analysis, b) method of low-angle light scattering.

where n is the light refraction coefficient of the medium, θ is the angle of laser light scattering and λ is the laser light wave length.

Between I(Q), wave number Q applies the relationship:

$$I(Q) \propto Q^{D_3} \qquad (8)$$

Fractal number D_3 can be determined by finding the slope of a straight line in the double-log reference log I(Q) and log (Q) (Figure 3b). Such plots can be performed using data from, for example, the laser granulometer, the principle of operation of which is based on the low-angle laser light scattering. The basic result of granulometric analysis of a suspension is the percentage distribution of the volume of particles of equivalent diameters, the volume of which is the same as that of particles determined from fractal analysis.

From data reported in the literature, it follows that it is possible to determine the dimension D_3 if the relationship between one- and two-dimensional fractal analysis is known. However, studies by Lee and Kramer (2004) confirm that this method is valid only if the second fractal dimension D_2 of the suspension is less than 2 and for inorganic suspensions.

Due to the characteristics of the suspension particles analyzed and the availability of the measuring equipment, relationship (8) was applied. The measurements were performed using the laser granulometer Mastersizer 2000 of Malvern Instruments Ltd. (UK). The fractal dimension was determined on the basis of

Figure 4. Location of experimental sites.

a plot of radiation intensity I, not versus wave number Q but versus scattering angle θ. For that purpose, an Excel calculation format with a calculation procedure reserved by the Malvern company was applied.

2.2 Location of experimental sites

Samples of raw and treated sewage, and activated sludge, were taken from three sewage treatment plants, located in the Lower Silesia area (Figure 4). The plants selected are characterized by similar time in service, but differ in the amount and type of the sewage treated and kinds of technological solutions (Table 2).

Table 2. General characteristics of plants from which samples were taken for assays of granulometric composition of suspensions in raw and treated sewage and activated sludge.

Sewage treatment plant	Population equivalent	Flow capacity m³/d	Source of wastewater	Technological solution
Kąty Wrocławskie	7498	2400	Domestic, industrial, storm water	Sewage-storage tank, grid, horizontal sand trap, pump station, oxygen-free reactor, biological reactor (with activated sludge), secondary settling tank
Siechnice	10905	1800	Mainly domestic	Sewage-storage tank, scarce grid, pump station, sand trap with primary settling tank, sequencing bath reactors (SBR) A and B
Sobótka	14900	2500	Domestic, industrial, storm water	Grid, pump station, sand tank, aeration tank (nitrification), secondary clarifier tank, facultative ponds

2.3 Sample collection

The samples were collected in plastic containers of 1 dm³ volume at a point of sewage inflow to a sewage treatment plant (after the grid); samples were also collected from tanks of activated sludge and at efflux from secondary settling tanks at monthly intervals. The assays of floc size distribution in sewage and activated sludge were performed immediately after the samples arrived in the laboratory. Samples of raw sewage and activated sludge had to be diluted, as too high a concentration of suspended particles did not allow proper values of laser light obscuration (from 10 to 20%) to be obtained in the measurement cell of the granulometer. In order to avoid sedimentation during the measurement, a mechanical stirrer was applied.

Investigations were performed for the spring and summer months of 2006, 2007 and 2008.

3 RESULTS AND DISCUSSION

Representative distributions of suspension particle sizes in activated sludge and raw and treated sewage taken from a plant at Kąty Wrocławskie are shown in Figure 5.

The measurements, recorded with a laser granulometer, indicate that there is a large variation in the size distribution of particles in raw sewage, whereas there is a smaller variation in treated sewage at the efflux from the settling tank, and an even smaller variation in activated sludge samples.

Particle distributions in raw and treated sewage differed with respect to particle size and their percentage share in the volume. In raw sewage, the volume of particles of sizes in the range 12–168 μm was dominant. In treated sewage, particles were 26–250 μm in diameter. From the size distribution of particles presented in Figure 5, from Kąty Wrocławskie, it can be seen that the largest volume was taken by particles

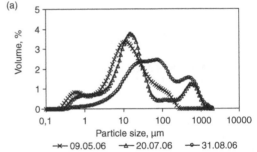

(a)

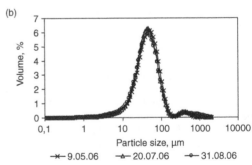

(b)

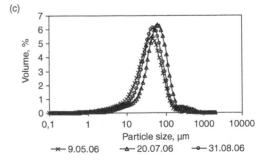

(c)

Figure 5. Particle size distributions of suspensions in: a) raw sewage, b) activated sludge, c) treated sewage at Kąty Wrocławskie sewage treatment plant.

from within the range 10–100 µm. By comparing the plots $\Delta V(d_i)$ compiled for activated sludge, and raw and treated sewage for individual months, there was no obvious difference in their time variation.

Distributions of particle sizes in raw sewage were characterized by an increased number of modes compared with the situation before treatment. In samples of treated sewage taken from plants at Kąty Wrocławskie and Siechnice, one- and two-mode systems were identified, whereas in raw sewage two- and three-mode systems were identified. In samples of sewage taken from a plant at Sobótka, the least identifiable characteristic was a change in the number of modes in sewage after treatment. In samples of activated sludge, the obtained particle size distributions were mostly of a one-mode type. It can therefore be suggested that the decrease in the number of modes occurred as a result of the sewage treatment process.

The granulometric composition of suspensions in raw sewage is a characteristic feature of the individual sewage systems. Particle size distributions in activated sludge reflect, to a lesser degree, the differentiation in composition of raw sewage conducted via an integrated sewage system to the sewage-treatment plant monitored. In spite of this, the suspensions present in raw sewage have an effect on the granulometric composition of activated sludge. To a greater degree, the granulometric composition of activated sludge affects the size of suspended particles in the sewage that leaves the plant.

The increase in size of particles in treated sewage is the effect of flocs being carried from sludge in secondary tanks.

On the basis of plots in Figure 6, using a Mastersizer 2000 instrument, the optical fractal dimension D_3 was determined. The remaining fractal dimensions found are compiled in Table 3.

Suspensions from activated sludge were characterized by a more compact structure ($D_3 = 2.04 - 2.31$) than suspensions from raw and treated sewage, the fractal dimensions of which were within the range $D_3 = 1.58 - 1.98$ and $D_3 = 1.37 - 2.09$, respectively.

In Figure 7, the results obtained are compared in a block plot system that presents the values of median, quartile 25% and 75%. The median for the set of raw sewage was 1.88 and 1.86 for treated sewage, the range scatter for deviating values being several times greater. The set median for activated sludge was 2.16, being thus markedly different from the values of fractal dimensions for treated and raw sewage. The suspension of activated sludge is characterized by a significantly greater degree of compaction. The flocs' structure is a developed feather structure, hence the fractal dimensions of this group of suspensions are greater.

The lowest values of fractal dimensions of activated sludge were found for samples taken from Sobótka and Siechnice plants, and the greatest from Kąty Wrocławskie, where the scatter of data was smallest.

Of note is the fact that the fractal dimension of suspensions in treated sewage is significantly smaller than

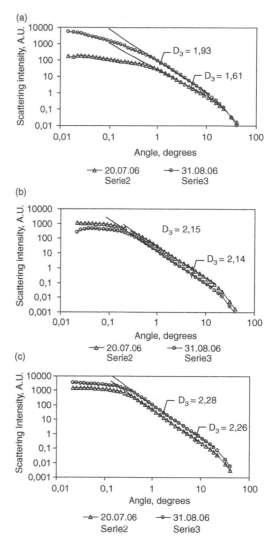

Figure 6. Plot of the function $I(\theta)$ used for determination of fractal dimension D_3 of suspension occurring in: a) raw sewage, b) activated sludge, c) treated sewage; taken in July and August of 2006 from the plant at Kąty Wrocławskie.

that of activated sludge. This means that, in treated sewage, there are not only suspensions of the flocs of activated sludge but also particles of suspensions of a different and less compact spatial structure. Results of studies on fractal dimensions of activated sludge and treated sewage can, in the future, be utilized for diagnosing the causes of low efficiency of suspension removal in secondary setting tanks.

4 CONCLUSIONS

Modern laser granulometers based on light diffraction enable quick and precise measurements of a broad range of particle sizes. With these instruments, optical

Table 3. Fractal dimensions of suspensions in activated sludge, raw and treated sewage from plants at Kąty Wrocławskie, Siechnice and Sobótka, Poland.

| | Date of sampling | | | | | |
| | 2006 year | | 2007 year | | 2008 year | |
Sample type	April	May	April	May	April	May
Kąty Wrocławskie						
Activated sludge	2.26	2.25	2.22	2.31	2.27	2.31
Raw sewage	1.58	1.57	1.89	1.66	1.88	1.76
Treated sewage	1.84	2.09	2.06	1.94	2.09	1.87
Siechnice						
Activated sludge	2.06	2.12	2.04	2.10	2.08	2.10
Raw sewage	1.88	1.86	1.70	1.93	1.94	1.98
Treated sewage	1.83	1.70	1.90	1.96	1.94	1.46
Sobótka						
Activated sludge	2.17	2.14	2.16	2.23	2.15	2.14
Raw sewage	1.80	1.98	1.78	1.90	1.92	1.93
Treated sewage	2.02	1.47	1.85	1.67	1.37	1.67

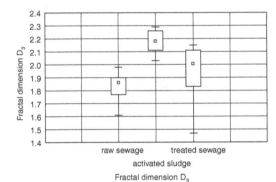

Figure 7. Comparison of fractal dimensions of suspensions in raw sewage, treated sewage and activated.

fractal dimensions can also be determined; this identifies the spatial structure of agglomerates composed of smaller particles. This is especially important when studying the coagulation and flocculation processes, and also for identifying the morphological traits of flocs of activated sludge. Hence, it seems purposeful to promote the instrumental methods of determination of sewage structure and granulometric composition of sewage sludge.

The present studies have shown that, for sewage and activated sludge taken from various sewage treatment plants, the granulometric composition of their suspensions is differentiated and exhibits spatial and temporal variation. The granulometric composition of activated sludge also depends on factors other than the nature of suspensions occurring in raw sewage, as the spatial structure characterized by the fractal dimension of activated sludge is different from that of suspensions in plant sewage influx.

REFERENCES

Jin, B., Wilén, B-M. & Lant, P. 2004. Impacts of morphological, physical and chemical properties of sludge flocs on dewaterability of activated sludge. *Chemical Engineering Journal* 98: 115–126.

Lee, Ch. & Kramer, T.A. 2004. Prediction of three-dimensional fractal dimension using the two-dimensional properties of fractal aggregates. *Advances in Colloid and Interface Science* 112: 49–57.

Li, X. & Yuan, Y. 2002. Settling velocities and perme abilities of microbial aggregates. *Water Research* 36: 3110–3120.

Logan, B.E. & Klips, J.R. 1995. Fractal dimension of aggregates for DEM in different fluid mechanical environments. *Water Research* 29(2): 443–453.

Smoczyński, L. & Wardzyńska, R. 2003. Fractal characteristics of post-coagulation sludge from sewage. *Environment Protection Engineering* 29(1): 65–72.

Sorensen, C.M., Lu, N. & Cai, J. 1995. Fractal cluster size distribution measurement using static light scattering. *Journal of Colloid and Surface Science* 174: 456–460.

Tang, S., Ma, Y. & Sebastine, I.M. 2001. The fractal nature of *Escherichia coli* biological flocs. *Colloids and Surfaces B: Biointerfaces* 20: 211–218.

Waite, T.D. 1999. Measurement and implications of flocs structure in water and wastewater treatment. *Colloids and Surfaces A: Physicochemical and Engineering Aspects* 151: 27–41.

Wilén, B-M., Jin, B. & Lant, P. 2003. Impacts of structural characteristics on activated sludge floc stability. *Water Research* 37: 3632–3645.

Wu, R.M., Lee, D.J., Waite, T.D. & Guan, J. 2002. Multilevel structure of sludge flocs. *Journal of Colloid and Interface Science* 252: 383–392.

Environmental Engineering III – Pawłowski, Dudzińska & Pawłowski (eds)
© 2010 Taylor & Francis Group, London, ISBN 978-0-415-54882-3

The structure of influent time series in wastewater treatment plants

M. Chuchro

AGH – University of Science and Technology, Krakow, Krakow, Poland

ABSTRACT: The main aim of this paper is to identify the method that gives the best results for influent time series of wastewater treatment plant (WWTP) structure recognition. Influent quantity time series of WWTPs have seasonal and other cyclical structures. Based on this, several methods of analysis were used. The wavelet model, single spectrum analysis, multiple regression and nonlinear estimation helped to recognize weekly, monthly and seasonal structures of influent time series. Correlation, Principal Component Analysis and Independent Component Analysis revealed the hidden structure of data. The afore mentioned methods demonstrate that influent time series have complicated, noisy structures that vary with periodicity, season and weather.

Keywords: Wastewater treatment plant, influent, time series, periodicity.

1 INTRODUCTION

The most important factor for wastewater treatment is the type of variability in the quantity of influent into wastewater treatment plants (WWTPs). The main aim of this paper was to establish a WWTP influent time series structure, which is necessary to create a prediction model. To do this, it was necessary to identify which methods and functions give the best results for environmental time series data.

The data from WWTPs are a sequence of data points that are measured at successive time points, spaced at uniform (1-day) time intervals (Box & Jenkins 1983). These data are typical environmental time series. Time series analyses are useful tools for the interpretation, structure recognition and prediction of repeated measurements data (Box & Jenkins 1983). Knowledge about influent wastewater structures facilitates the design and extension of WWTPs. At present, analyzed WWTPs do not have a prediction model for wastewater influent quantity. Correct information about influent time series structure helps in the selection of the optimum amount of chemicals and detention time. Information about influent structure is crucial for the accurate prediction of model realization. Three analyzed WWTPs have been used to obtain results of time series structure recognition and modelling.

Data used in this analysis were obtained from three WWTPs. The largest analyzed WWTP is situated in Warsaw; this is also the largest WWTP in Poland. The medium-sized WWTP is situated in Cracow and the smallest one in Sandomierz. The WWTPs analyzed are municipal because they intercept and purify wastewater from private houses, industry and commerce (Margel 2000). In these WWTPs, mechanical and biological treatment is applied with increased removal of nitrogen and phosphorous compounds. This

Table 1. Descriptive statistics for influent series data from studied wastewater treatment plants (WWTPs).

WWTP	Inflow average [Mm³/day]	Standard inflow deviation [Mm³/day]	Skewness	Coefficient of variation
Cracow	52.3	8.76	3.28	0.17
Sandomierz	2.86	0.63	3.34	0.22
Warszawa	179.1	31.1	0.92	0.17

is the most popular method of municipal wastewater treatment (Margel 2000).

2 MATERIALS AND METHODS

2.1 *The one – dimensional time series structure analysis*

For influent structure time series analysis, data were taken from three municipal WWTPs. The data were collected daily from January 2000 to December 2007. The smallest WWTP in Sandomierz had an average inflow capacity during the studied period equalling 2.9 m³/day and the highest right-sided skewness of distribution (around 3; Table 1). The second WWTP (Cracow) had an inflow capacity average equal to 52.3 m³/day; the value of skewness was similar to that of the Sandomierz. The largest average inflow capacity of the WWTP in Warsaw was close to 180 m³/day. The high inflow quantity in the Warsaw WWTP is probably the reason for the much smaller value (0.92) of the right-sided skewness of distribution. Parameters of basic statistics are shown in Table 1.

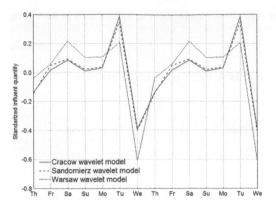

Figure 1. Influent wavelet week model (repeated twice).

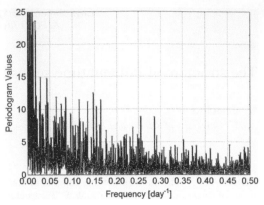

Figure 3. Single spectrum analysis.

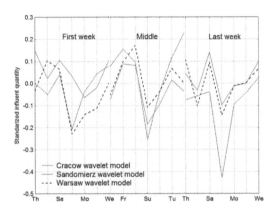

Figure 2. Wavelet month model.

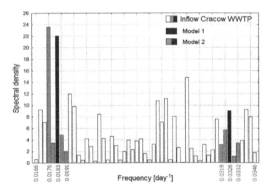

Figure 4. Graphical explanation of model 1 and 2.

The first model prepared was a week-long wavelet inflow model (Percival & Walden 2000). Data from the WWTP were standardized each week. Thursday was chosen as the first day of the week to analyze, in order to show changes between the weekdays and the weekend. Data were characterized by skewness; therefore, we normalized the data using the Box–Cox transformation (Peltier et al. 1998). A weeks' worth of data was used to compute from normalized data. As a result, we obtained untypical inflow wavelet models of the week, with characteristic changes of inflow capacity during the weekend and high values on Wednesday (Figure 1).

The second model prepared was a monthly wavelet inflow model (Figure 2). The model was realized in the same way as the previous model, with the exception that the month was divided into the first week, the middle two weeks and the last week of the month. The model consists of nine submodels, as seen in Figure 2.

Next, the method used to determine the structure of the WWTP influent time series was a Single Spectrum Analysis (Golyandina et al. 2001). The purpose of spectrum analysis is to decompose time series into underlying sine and cosine functions of different frequencies. The analyses also help to determine strong or important frequencies. In wastewater inflow time series, spectrum analysis from 1416 values of the periodogram were chosen as high-value frequencies – 0.0180, 0.0324, 0.074, 0.1482, 0.284 and 0.48 (Figure 3) (Box & Jenkins 1983). The spectrum frequencies scale in this case is 1 per day: 2 months, 1 month, 14 days, 7 days and 3.5 days' periods of time (Golyandina et al. 2001, Iacobucci 2003).

Four models were devised from data gained from spectrum analysis. From 1461 periodogram values, the highest one values were chosen, as well as other important frequencies. The first model contained sine and cosine functions of six frequencies: 0.0180, 0.0324, 0.074, 0.1482, 0.284 and 0.48 (day^{-1}) (Box & Jenkins 1983). The second model was a modification of the first. In this case, there were six frequencies chosen for the first model. Furthermore, from the nearest neighbourhood of each chosen frequency (from model 1), four frequencies were chosen. Fragments of the Single Spectrum Analyses from the Cracow WWTP, with graphical explanation of methods 1 and 2, are shown in Figure 4.

The highest periodogram values were exploited for the third model. This model is a variation of the third one, with four frequencies from the neighbourhoods. All models were compared with standardized influent WWTP data (Golyandina et al. 2001).

Standardized data were used for linear multiple regression. The general purpose of using this method was to test whether there were relationships are among several independent variables and dependent variables, and how strong these relationships were. Sine and cosine coefficients of independent variables were tested. The model constructed contains yearly and weekly sine and cosine coefficients. A model of regression, shown below (Equation 1), was developed for 2007 (Koronacki & Cwik 2005).

$$X_t = a_0 + \sum_{k=1}^{q} \left[a_k \cdot \cos(2\pi \cdot f_k \cdot t) + b_k \cdot \sin(2\pi \cdot f_k \cdot t) \right] \quad (1)$$

where: a_k = cosine regression coefficient; b_k = sine regression coefficient; $f_k = k/q$; q = number of sample; t = time variable.

Spectrum analysis does not adjust the sine functions with specified periodicity. The combination shown consists of various sine and cosine functions. In this case, the sum is equal to 2922 (Box & Jenkins 1983). The main aim of the inflow series nonlinear estimation was to match the sine function arguments for the inflow series data. Standardized, normalized and filtered data were used for estimation. Each full set of data was fitted using least squares criterion with Levenberg–Marquardt or Gauss–Newton algorithms, and a user-specified loss criterion with a quasi-Newton procedure (absolute value of difference, instead of square, in order to diminish the influence of "peaks" in data) (Borovkova 1998, Iacobucci 2003).

2.2 The time series multiple dimensional structure analysis

Factor analysis as a multivariate statistical method is used to assess a small number of factors from a data set of many correlated variables. Principal Component Analysis (PCA) is an orthogonal linear transformation that transforms the data to a new coordinate system such that the greatest variance by any projection of the data comes to lie on the first coordinate (Weron & Wójcik 2004). PCA was used for dimensionality reduction in data by retaining those characteristics of the data that contribute most to its variance. This is a non-parametric analysis and the solution is unique and independent of any hypothesis about the data's probability distribution (Weron & Wójcik 2004).

Analyses were conducted for 1 weeks' and 2 weeks' standardized, normalized data. In this case, data were assessed from each day of the week. So, Mondays are the first dimension and, Sundays are the seventh dimension. During 2 weeks of PCA, there were 14 dimensions: the first and eighth were Monday dimensions, etc. The main goal of this method was to separate important data components (Larose 2008). The principal components of lesser significance were ignored. The difference between important and ignored components depends on scree tests. If the eigenvalues are small, the loss of information is not high. Data with a smaller number of dimensions are easier to analyze

(Walanus 2000). A few analyses need to be made to assess the appropriate variation of this method for influent data series (Larose 2008).

Wastewater inflow data are non-Gaussian. This is why Independent Component Analysis (ICA) is a good complementary method to PCA (Pasztyła 2004). ICA is a method that is used to assess underlying factors – components from multivariate or multidimensional statistical data. What distinguishes this method from other methods is that components have to be *statistically independent* and *non-Gaussian*. Three time series data sets, with daily and weekly resolution, were taken for this analysis. For these data, a deflation algorithm and exponential negative entropy functions were chosen. Results of these analyses were compared to standardized and normalized data (Hyvarinen et al. 2001, Pasztyła 2004).

The Pearson product-moment correlation indicates the strength and direction of a linear relationship between two random variables. Inflow data graphs have a similar course, so a correlation coefficient is essential. The three WWTPs are situated as follows: 160 km between the Sandomierz and the middle WWTP in Cracow, 220 km between the smallest WTTP and the WWTP in Warsaw, 270 km between the largest and the Cracow WWTP. Changes in distance between WWTP should be visible in a correlation coefficient. A municipal WWTP collects household liquids from houses, industry and rainfall wastewater. A correlation between the weather and rain rate should be essential too; results of Pearson product-moment correlation are seen in Table 2. Preparation of data for this analysis included the standardization, normalization and high-pass filtering of raw data. Weather data were prepared as a weather condition (0 – rain, 0.5 – rain/sun, 1 – sun) and rain rate (mm/day).

3 RESULTS AND DISCUSSION

Changes in influent quantity are a characteristic property of municipal wastewater, as is seen in the wavelet influent model (Figure 1). Wednesdays and Saturdays have the highest flow discharge, which is correlated with public and industrial activities. The smallest flow discharge is observed on Sundays, which probably reflects the lack of industrial activities on this day of the week. Model comparison with standardized data shows similarities in direction changes, as is shown in Figure 5. A good match was observed in the middle part of the graph. The period of time chosen for this analysis is a dry January to March 2003. Only in the first part of this period we observed precipitation (first 10 days). This is visible in the graph as a period with a low match of data with the model. In the middle part of the graph, changes in standardized influent data curves are at the same points as in the wavelet week model curves. The low values in model and in standardized data are visible on Sundays. The high values in model and in data are visible on Saturdays and on Wednesdays. The wavelet model is accurate and explains the changes in inflow capacity rate into WWTPs. The wavelet month model shows the same variation of flow discharges as a wavelet week model. Variation between the first and the last week is visible in Figure 2. The month model was tested, on the assumption that the smallest absolute distance between submodels from three WWTPs will be the confirmation of the month model correctness. Models were designed as a variation of the created month submodels. For example, in the first tested model, the first (first week) submodel was swapped with the third model (last week) from the Cracow WWTP. The absolute distance month model should be lower than for other models. This condition was not realized, so this model is non-essential.

The Single Fourier Spectrum Analysis (SSA) confirms the correctness of the wavelet week model. We obtained weekly, fortnightly, monthly, half-yearly and yearly periodicity in the inflow data from WWTPs. The models created from SSA frequencies show a strong correlation with the data, especially for the second and fourth models. The multiple regression model was developed using information from the SSA. Analyses were performed for weekly, lunar monthly and yearly periodicity. Regression coefficients were essential for week and year sine and cosine functions. The model of regression explains 10% of variance.

Week and month structures are confirmed by nonlinear estimation. For the inflow data of each WWTP, five estimation criteria and algorithms were used. Satisfactory compatibility of estimation was obtained for the user-specified loss criterion with a quasi-Newton procedure. The absolute value of differences between observed and predicted values, used as a loss criterion, reduced the influence of extreme values in the data and, as a result, gives a best match of sine functions. Local extremes of 7 days' sine functions are between Saturdays and Sundays and on Wednesdays. The extremes values of month sine function are observed in the last week of the month. Results of nonlinear estimation were correlated with normalized data (Figure 6). A correlation within the moving Gaussian window

coefficient was chosen as a correlation method. The widths of the window (Gaussian sigma) were equal to 5, 10 and 30 (Figure 7). Nonlinear estimation with a loss criterion has periods that are significantly correlated with the data. There are also noted periods with negative correlations. The large correlation changes are connected with weather – after a dry period, periods with precipitation are observed.

The data structure was defined using Principal Component Analysis. As a dimension, we used weekdays. The highest score of the first factors, for 1-week analysis, was received for normalized data with a constant lambda value in Box–Cox transformation. The values of the first factor explained more than 60% of the variance in data. The second factor explained more than 10%, and the third factor more than 7%.

From 2 weeks' PCA data, 14 factors were identified and the first of these explained more than 30% of data variance; the second and the third explained more than 10% of the data. The first factor is connected to weather. The second factors reflect the variance in data connected with the time structure (Figure 8). Extreme values on Saturdays are observed in the third factors.

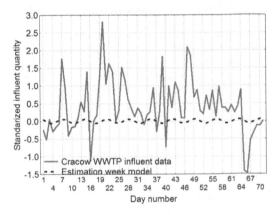

Figure 6. Comparison of nonlinear estimation measured and standardized influent data.

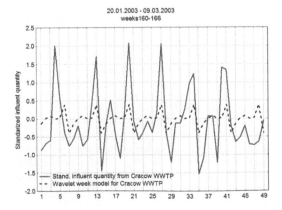

Figure 5. The comparison of wavelet week model with the standardized data from Cracow.

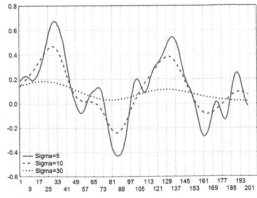

Figure 7. Correlation within the moving Gaussian window coefficient.

The rest of the factors were non-essential. These non-essential factors explain less than 10% variance of the data, probably as it is only noise.

A good complement for PCA is Independent Component Analysis (ICA). The results of this analysis show that influent data from three WWTPs have common aspects, which explain about 60% of the data variance. The second ICA factors explain nearly 25% of the variance and the third factors explain more than 16% of variance (Figure 9).

Data from three WWTPs are correlated. The strength of Pearson product-moment correlation decreases with distance between the WWTPs. The highest score of influent data correlation, which was equal to 0.51, was between the Cracow and Sandomierz WWTPs.

The smallest correlation in influent data was between the Warsaw and Cracow WWTP (0.30). Data are dependent on the weather, as is seen in Table 2. We can observe that taking the square root transformation of precipitation for daily data improves the correlation with precipitation. High values of correlation are also observed for weather conditions written only with three values: {0, 0.5, 1}. The correlation values improved for high-pass filtered data (Table 2).

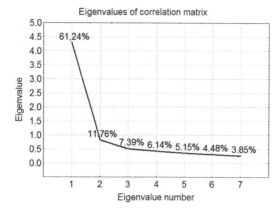

Figure 8. PCA results for week analysis.

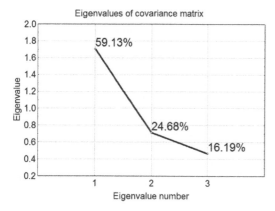

Figure 9. ICA results.

4 CONCLUSIONS

The influent data from WWTPs are noisy and depend on many factors. We observed 3.5, 7, 14, 30, 60, 180 and 365.25 days periodicity in terms of the time structure of the data. What is more, the influent capacity during a given month and year change. The lowest inflow capacities into WWTPs were observed in July and December. The highest influents capacities were observed in February and March. Changeability in inflow to WWTPs, during a month, was observed between the first and the last week.

The influent data series from three WWTPs have a similar structure, which was visible in the PCA and ICA results. A satisfactory prediction model could be created if more complete data were to be obtained. At present, we do not have any data about industrial and commercial wastewater treatment in those three WWTPs. This information is essential for improving the influent prediction model. An accurate prediction model could be achieved only if data are relatively precise. The methods presented in this paper are a good starting point for environmental time series data structure recognition. All analyzed methods provided new information about the influent WWTP structure, and they can be used in a complementary manner.

ACKNOWLEDGEMENTS

This work has been financially supported by Faculty of Geology, Geophysics and Environmental Protection, AGH University of Science and Technology, grant no 11.11.140.561.

Table 2. The correlation of influent data with weather.

Correlation	Cracow	Sandomierz	Warsaw	Cracow after filtration	Sandomierz after filtration	Warsaw after filtration
Precipitation amount Warsaw	0.189	0.208	0.306	0.210	0.207	0.406
Weather Warsaw	−0.275	−0.179	−0.385	−0.278	−0.180	−0.446
Weather Cracow	−0.577	−0.298	−0.234	−0.579	−0.320	−0.296
Square root of precipitation amount Warsaw	0.285	0.242	0.417	0.299	0.241	0.518

REFERENCES

Borovkova, S. 1998. *Estimation and prediction for nonlinear time series*. Amsterdam: University of Groningen.

Box, G.E.P. & Jenkins, G.M. 1983. *Analiza szeregów czasowych: Prognozowanie i sterowanie*. Warszawa: Państwowe Wydawnictwo Naukowe.

Golyandina, N., Nekrutkin, V., Zhigljavsky, A. 2001. *Time series structure, SSA and related techniques*, Monographs on Statistics and Applied Probability. Volume 90. Boca Raton: Chapman &Hall/CRC.

Hyvarinen, A., Karhunen, J., OJA, E. 2001. *Independent component analysis*. New York: Wiley-Interscience Publication, John Wiley and Sons.

Iacobucci, A. 2003.*Spectral analysis for economic time series*. Documents de Travail de l'OFCE 2003-07 France: Observatoire Francais des Conjonctures Economiques (OFCE).

Koronacki, J. & Ćwik,. J. 2005. *Self learning statistical systems*. [In Polish]. Warszawa: Wydawnictwa Naukowo-Techniczne.

Larose, D.T. 2008. *Data mining methods and models*. [In Polish]. Warszawa: Wydawnictwo Naukowe PWN.

Margel, L. 2000.Water treatment and wastewater treatment installations, processes, methods. [In Polish]. Bialystok: Wydawnictwo Ekonomia i Środowisko.

Pasztyła, A. 2004. *Shares rate analysis using ICA metod*. [In Polish]. Kraków: StatSoft.

Peltier, M.R., WILCOX, C. J., SHARP, D. C. 1998. Application of the Box–Cox data transformation to animal science experiments. *Journal of Animal Science* 76: 847–849.

Percival, D.B. & Walden, A.T. 2000. *Wavelet methods for time series analysis*. Cambridge: Cambridge University Press.

Walanus, A. 2000. Statistical relevance of conclusions from quantitative analysis on the Upper quaternary example research. [In Polish]. *Geologia, Kwartalnik AGH 26* (4): 1–59.

Weron, R. & Wójcik, S. 2004. *Principal components analysis in implied volatility modeling*. Research Report HSC/04/3 Hugo Steinhaus Center for Stochastic Methods, Institute of Mathematics. Wroclaw: Wroclaw University of Technology.

Environmental Engineering III – Pawłowski, Dudzińska & Pawłowski (eds)
© 2010 Taylor & Francis Group, London, ISBN 978-0-415-54882-3

Polychlorinated dibenzothiophenes (PCDTs) in leachates from landfills

J. Czerwiński, M. Pawłowska & B. Rut
Faculty of Environmental Engineering, Lublin University of Technology, Lublin, Poland

ABSTRACT: Polychlorinated dibenzothiophenes (PCDTs) in landfill leachate have been characterized. It was shown that the concentrations of PCDTs are higher in older sites having higher fractions of organic material. The increase of PCDTs concentration with the age of the landfill site suggests that these compounds are formed under anaerobic conditions from secondary products of biochemical transformations of compounds present in the landfill. The decrease of PCDTs concentration with storage time, suggests that these compounds undergo transformations in semi-aerobic conditions.

Keywords: Polychlorinated Dibenzothiophenes, Leachate, SEC-PYE clean-up, GC-MS/MS analysis.

1 INTRODUCTION

Two important reports related to the high, dioxin-like toxicity of sulphur-analogues of polychlorinated dibenzofurans have been published in recent years (Nakai et al. 2004; Hosomi 2005). In prior works, Kopponen et al. (1994) analysed the influence of three organosulphur compounds: 2,3,7,8-TeCTA (tetrachlorothianthrene) – sulphur analogue of tetrachlorodibenzodioxin, 2,3,7,8-TeCDT (tetrachlorodibenzothiophene) and 3,3',4,4'-TeCDPS (tetrachlorodiphenylthioether) – sulphur analogue of tetrachlorodiphenylether, on the induction of CYP1A1 gene expression, which was measured as an activity of aryl hydrocarbon hydroxylase (AHH) and ethoxyresorufin-O-deethylase (EROD) in liver cells of Hepa-1 mice. Significant differences between 2,3,7,8-TeCDT and analysed sulphur analogs in EROD and AHH activity were observed. They found EC (50) values as follows: 2,3,7,8-TeCTA - 700 [pM]; 2,3,7,8-TeCDT – 7500 [pM]. However, they did not observe the response for 3,3',4,4'-TeCDPS. Calculated Response Equivalency Potencies (REP) for these compounds were respectively: 0.011; 0.001 and 0. Moreover, the value of the REP factor (0.00425) was estimated by Giesy for a synthetic mixture of PCDT congeners in the H-4IIE rat liver cells (Giesy et al. 1994).

Analogous studies of Ah immunoenzymatic activity were also carried out for a number of dioxin-like congeners of polychlorinated dibenzothiophenes by a group of researchers from Tokyo University of Agriculture and Technology (Nakai et al. 2004, 2007, Hosomi 2005). The above research was carried out using a high-tech Ah-immunoassay® test (Paracelsian USA). Nakai and Hosomi found EC (50) values for 2,3,7,8-TeCDT – 111 [pM], 1,2,3,7,8-PeCDT (pentachlorodibenzothiophene) – 117 [pM]; 1,2,3,7,8,9-HxCDT (hexachlorodibenzo-thiophene) – 44 [pM];

1,2,3,4,7,8,9-HpCDT (heptachlorodibenzothiophene) – 50 [pM], and found no response for octachlorodibenzothiophene. Calculated REP factors were as follows: 2,3,7,8-TeCDT – 0.1; 1,2,3,7,8-PeCDT – 0.1; 1,2,3,7,8,9-HxCDT – 0.04; 1,2,3,4,7,8,9-HpCDT – 0.05. Similar e-TEF factors were obtained by Kobayashi (2001).

These investigations, showing much higher values of the dioxin-like toxicity of polychlorinated dibenzothiophenes have resulted in growing interest in their presence in different compartments of the environment (Sinkkonen et al. 1997, 2001, 2003). During the last fifteen years, PCDTs have been analyzed in samples from combustion and metallurgy, sediments, pulp mill effluents and in aquatic biota. Combustion and metallurgy have been considered as the major sources of PCDTs in the environment (Aittola et al. 1994, Cai et al. 1994, Koistinen et al. 1992, Nakai et al. 2004, 2007, Sinkkonen et al. 1997, 2001, 2003). They were recognized as priority contaminants by the Swedish Environmental Protection Agency in the so-called "Scandinavian Monitoring" for the year 2005 (Report 10050974).

Therefore, the established dioxin-like toxicity of PCDT, together with their known presence in leachates from municipal landfills moved us to investigate their possible migration from such leachates into the environment.

2 MATERIALS AND METHODS

2.1 *Materials*

Leachate samples were collected four times during 2006 (May - October) from three municipal landfills, different in age and kind of deposited waste.

The Landfill I wastes were deposited for 12 years without sorting. In the landfill site II sorted wastes had been deposited for over five years. The sorting

Table 1.   GC-MS system conditions.

The operation conditions of the chromatograph:

Injector:	splitless @ 320°C
Capillary column:	RTx Dioxin (Restek) 60 m × 0.25 mm df = 0.18 μm
Oven temperature programming:	60°C (0.5 min hold) ramp 15°/min to 260°C than ramp 5°/min to 320°C, 5 min hold
Carrier gas:	He (99.9996%) @ 40 cm/s

The MS operating conditions:

The ion source temperature	250°C
The transfer line temperature	275°C
Scanning mode I: Full Scan	50.0–550.0 amu
Scanning mode II: MS/MS (scan windows of ion precursors and daughter ions)	

	Ion-precursor	Daughter ions		Ion-precursor	Daughter ions		Ion-precursor	Daughter ions
TeCDF	306	237–247	TeCDD	322.0	253–263	TeCDT	322.0	284–290
PeCDF	340	272–282	PeCDD	356.0	288–298	PeCDT	356.0	319–325
HxCDF	374	306–318	HxCDD	390.0	322–332	HxCDT	390.0	352–357
HpCDF	410	340–350	HpCDD	424.0	357–367	HpCDT	424.0	387–394
OCDF	444	375–385	OCDD	460.0	391–401	OCDT	460.0	422–428

The limit of detection of the Finnigan GCQ for this screening methodology was estimated from a mass chromatogram of a 500 fg/μL injection of 2,3,7,8-TeCDT. For other chlorinated dibenzothiophenes an equimolar (to suitable PCDD) response was assumed, according to Wiedmann et al. (1998).

involved plastic (PET), aluminum and glass fractions. Landfill III had had 8-years of uninterrupted sorting. The wastes were sorted to recycle plastic (PET), metals, glass and biodegradable organic fractions.

Samples were collected from leachate collectors into teflon coated bottles, and were transported to the laboratory within one hour.

A standard solution of 2,3,7,8-TeCDT and OCDT was obtained from Promochem (Germany). Standard mixtures of other PCDTs were obtained from Prof J.T Andersson (University of Münster). Labelled (^{13}C) and native standards of polychlorinated dibenzodioxins and dibenzofurans (for determination of recoveries) were obtained from Wellington Laboratories (Ontario, Canada). All solvents were "for residue analysis" grade and were obtained from JT Baker - Germany.

2.2   Methods

Extraction of analytes from leachate (500 cm^3 of sample) was carried out using classical LLE technique with two portions of dichloromethane (200 cm^3). Afterwards they were evaporated to 3 cm^3 and filtered through Millex-FG filters (PTFE membrane, 0.45 μm pore size, 25 mm i.d. - Millipore, Bedford). The filtrate was purified in two stages: with the use of size-exclusion chromatography (SEC) and on columns filled with silica gel modified with pyrenyl (PYE) groups (Phenomenex – Japan). SEC was performed on coupled Envirogel GPC cleanup columns (19 mm I.D. × 150 mm and 19 mm I.D. × 300 mm)

with dichloromethane as a mobile phase @ 5 ml/min on Breeze 1525 system.

The fraction collected from 12.2 to 20.5 min was evaporated to dryness, under a gentle nitrogen stream, and dissolved in hexane. The volume of hexane extract was 550 μl, where 500 μl was injected to the previously described HPLC system, equipped with semipreparative COSMOSIL PYE column 10 mm I.D. × 250 mm (Phenomenex – Japan), working in the system of normal phases (hexane as the mobile phase @ 1 ml/min – 24 min and the last 5 minutes hexane/CH$_2$Cl$_2$ 80/20 v/v) to accelerate the elution of the mentioned PCDTs. Column was kept at 10°C to make the most of π – π interactions between stationary phase and planar analytes. During this fractionation, the initial 8 ml of eluate was discarded because of their content of non-planar compounds (for example PCB and PCDE). The fraction was collected from 9.5 to 24.0 min. Purified extracts, after evaporation to 100 μl, were analysed on a gas chromatograph coupled to a mass spectrometer GCQ (Finnigan) under the operating conditions shown in Table 1.

Recoveries of PCDT varied from 63 to 78 % (for tetra- and octa- substituted congeners). These results were obtained by addition of standard mixture of PCDTs to the 0.5 l of real leachate sample. The limit of detection "per sample" was estimated to be from 0.7 ng/dm^3 for 2,3,7,8-TCDT to 2.5 ng/dm^3 for OCDT. The RSD for triplicate analysis of the same samples was better than 21% for tetra- substituted congeners, 17% for penta-, 24% for hexa- and hepta- congeners, and 18% for octachlorodibenzothiophene.

Table 2. Results obtained from analysis of DX-2 Standard Reference Material (NWRI Canada).

	Certified value [ng/g]	Confidence interval [ng/g]	Obtained concentrations [ng/g]	SD [ng/g]
2,3,7,8-TeCDF	134	±61	169	±47
1,2,3,7,8-PeCDF	46	±10	54	±12
2,3,4,7,8-PeCDF	88	±28	93	±28
1,2,3,4,7,8-HxCDF	825	±348	804	±314
1,2,3,6,7,8-HxCDF	153	±61	161	±61
1,2,3,7,8,9-HxCDF	36	±45	22	±7
2,3,6,4,7,8-HxCDF	70	±47	79	±11
1,2,3,4,6,7,8-HpCDF	3064	±745	2943	±214
1,2,3,4,7,8- HpCDF	152	±84	151	±41
OCDF	7830	±3087	6760	±1231
2,3,7,8-TeCDD	262	±51	249	±32
1,2,3,7,8-PeCDD	28	±14	19	±9
1,2,3,4,7,8-HxCDD	25	±8	23	±6
1,2,3,6,7,8-HxCDD	85	±33	73	±33
1,2,3,6,7,8-HxCDD	58	±19	56	±14
1,2,3,4,6,7,8-HpCDD	757	±320	920	±141
OCDD	4402	±1257	4389	±296
2,3,7,8-TeCDT*	–	–	97	±23
2,3,6,8-TeCDT	–	–	11	±7
1,2,3,7,8-PeCDT	–	–	32	±11
1,2,3,4,7,8-HxCDT	–	–	8	±6
1,2,3,4,6,7,8-HpCDT	–	–	56	±14
OCDT*	–	–	131	±18

*spiked with ca 100 ng/g TeCDT and OCDT.

In almost all publications connected with determinations of PCDT, the clean-up procedure is based on the EN-1948 analysis of PCDD. This is a time consuming, multistep procedure based on column chromatography which is difficult to automate. Additionally, these methods (EN-1948) need GC-HRMS with a resolution of about 20 000 to prevent coelutions of PCDD and PCDTs. Our method, based on combined SEC-PYE clean-up procedure, can be applied in small laboratories equipped with quite common Ion Trap- type MS systems. Till now a procedure based on Alumina-PYE clean-up and GC-MS/MS analysis was applied for the analysis of coplanar PCB and PBDE and dioxins (Hanaari et al. 2003, 2006).

Collisionally induced fragmentation (CID) was used for selective PCDTs analysis. During the secondary fragmentation in the ion trap, polychlorinated dibenzothiophenes produced mainly $[M-Cl]^+$ ions. Polychlorinated dioxins, which have almost the same molecular mass, fragment mainly by expulsion of a COCl fragment. Hence the base peak of secondary fragmentation corresponds to $[M-COCl]^+$.

Additionally the retention windows of n-chloro-substituted dibenzothiophenes are similar to the $n+1$ chlorosubstituted dioxins (where n is the number chlorine atoms connected to the dibenzodioxin/dibenzothiophene rings).

Figure 1 shows the selectivity of the method applied in the analysis of mixtures of tetrachlorinated dibenzothiophenes and tetrachlorinated dibenzodioxins.

TOC analysis was carried out on a Shimadzu TC/TOC 5050A analyzer calibrated with phthalate/hydrocarbonate standards – POCh-Poland.

3 RESULTS AND DISCUSSION

In Table 3 the concentrations of dibenzothiophenes in each of the analysed leachate samples are shown. Results of these determinations are the averages of the three independently analysed samples. The analyses show that the highest concentrations are obtained for the 12-year-old landfill (landfill I), which collects unsorted wastes from Lublin and its surroundings. In this case, the highest concentration was for octachlorodibenzothiophene, which is the most stable and the most easily bioaccumulated (Rostkowski et al. 2004). In the sample reanalysed after 32 days much lower concentrations of chlorinated dibenzothiophenes were observed (0,7 ng/dm³ 2,3,7,8-TeCDT, 8 ng/dm³ 1,2,3,7,8-PeCDT, nd – 1,2,3,7,8,9-HxCDT, 21 ng/dm³ 1,2,3,6,7,8,9-HpCDT and 107 ng/dm³ OCDT).

An increase in the concentration of 2,3,7,8-TeCDF (to 100.7 ng/dm³) was also observed. However, the landfill II, in operation since 2001, showed only slightly lower concentrations of polychlorinated dibenzothiophenes. Moreover, different PCDT congener profiles were noted here. On this landfill, most of the plastic fraction (mainly PET), glass and metals were separated. This observation may indicate

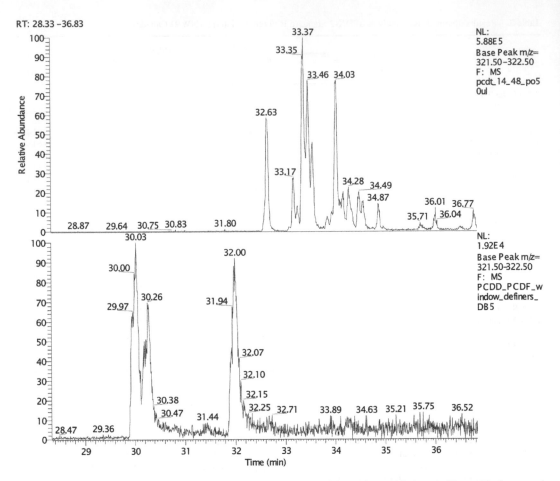

Figure 1. Selectivity of the analytical technique: example of a mixture of tetrachlorinated dibenzothiophenes and tetrachlorinated dibenzodioxins.

Table 3. PCDT concentrations [ng/dm^3] in the studied leachates. (Results shown are the averages of three independent analyses).

Congener Lechates	2,3,7,8-TeCDT [ng/dm^3]	1,2,3,7,8-PeCDT [ng/dm^3]	1,2,3,7,8,9-HxCDT [ng/dm^3]	HpCDT [ng/dm^3] HpCDT [ng/dm^3]	OCDT [ng/dm^3]	2,3,7,8-TeCDF [ng/dm^3]
samples from	43.8	11	21	nd	124	66.2
landfill I	24	nd	45	17	134	71
	21	13	55	32	145	78
	24	26	32	25	98	69
samples from	23	17	5	nd	32	45
landfill II	nd	11	nd	nd	24	51
	nd	27	14.7	nd	41	79
	13	4.2	8	nd	33	21
samples from	7.1	4.3	nd	nd	nd	24.2
landfill III	nd	nd	nd	nd	42	nd
	nd	nd	nd	21	65	18
	4.5	2.1	nd	23	67	nd

nd – not detected.

that PCDTs come from different sources or undergo various transformations within the landfill.

The lowest concentrations were observed for the leachate originating from landfill III (8 years in operation) The wastes disposed there are fully-sorted and plastics, metals, glass and biodegradable fraction are collected separately. This part of the study shows not only the lowest concentration of PCDT but also of other organic compounds measured as TOC, which for this particular landfill was from 45 to 70 mgC/dm^3. In comparison, the value of TOC for the leachate from the 5-year-old landfill (Landfill 2), where plastics, metals and glass were separated, was 170–432 mgC/dm^3. In

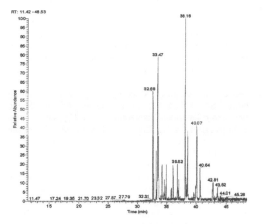

Figure 2. Chromatogram GC-MS of polychlorinated dibenzothiophenes in real sample of leachates.

this case, higher concentrations of PCDTs were also observed.

It is very difficult to explain why the PCDT concentrations increase. It may be that the increase in the proportion of organic substances influences the formation and/or elution of PCDT from the wastes in some way.

In Figure 3 the patterns of polychlorinated dibenzothiophenes from known sources are shown.

Patterns obtained in this work are neither similar to the patterns presented previously by Aittola et al. (1996) which are typical of PCDTs formation via combustion processes nor to those presented by Cai et al. (1994) whose patterns represent a "chemical" origin obtained by way of condensation of chlorinated thiophenols. It suggests that PCDTs are formed in landfills by biodegradation of organic matter via unknown bio-chemical processes.

Possible formation of PCDD/F in wastes deposited in landfills was described by Dudzinska et al. (2004). We suppose that during similar processes the formation of PCDTs also occurs. A possible source of the aromatic rings for this biosynthesis may be the degradation of polymers. Other possibility is formation of dibenzothiophenes during humification processes and then enzymatic chlorination of aromatic rings as a

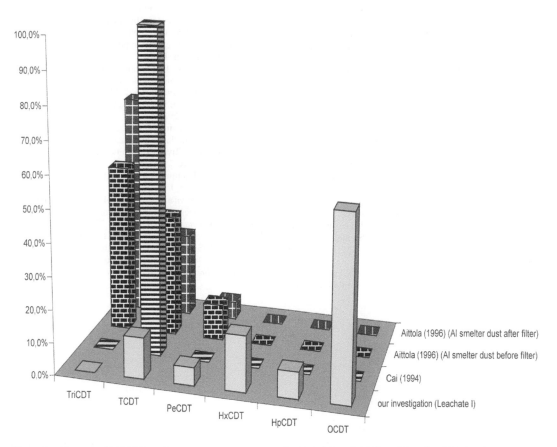

Figure 3. Patterns of PCDT from known sources (Aittola et. all 1996; Cai et al 1994) and from landfill I leachate (current work).

177

result of the activity of natural haloperoxidases. For the elucidation of plausible degradation and transformation mechanisms of PCDTs, ^{13}C standards are necessary but at this time they are not commercially available.

The concentrations of PCDTs present in landfill leachate are significant because they strongly influence the dioxin-like toxicity of the latter causing living organisms to be affected. For that reason, further investigation must be conducted with the aim of better defining their concentrations and sources.

ACKNOWLEDGEMENTS

This work was supported by scientific grant No 1 T09D 024 30 of the Polish Ministry of Science and Higher Education.

REFERENCES

Aittola, J-P., Paasivirta, J., Vattulainen, A., Sinkkonen, S., Koistinen, J. & Tarhanen, J. 1996. Formation of Chloroaromatics at a Metal Reclamation Plant and Efficiency of Stack Filter in their Removal from Emission. *Chemosphere* 32: 99–108.

Cai, Z., Giblin, D.E., Ramanujam, V.M.S., Gross, M.L. & Cristini, A. 1994. Mass-profile monitoring in trace analysis: Identification of polychlorodibenzothiophenes in crab tissues collected from the Newark/Raritan Bay system. *Environ. Sci. Technol.* 28: 1535–1538.

Chen, S.D., Liu, X.H. & Wang, Z.Y. 2007. Study of thermodynamic Properties for Polychlorinated Dibenzothiophenes by Density Functional Theory. *J. Chem. Eng. Data* 52: 1195–1202.

Dudzinska, M., Czerwiński, J. & Rut, B. 2004. Comparison of PCDD/Fs Levels and Profiles in Leachates from "New" and "Old" Municipal Landfills. *Organohalogen Compounds.* 66: 884–889

Giesy, J.P., Ludwig, J.P. & Tillit, D.E. 1994. Dioxins, Dibenzofurans, PCB's and Colonial, Fish-Eating Water Birds. In A. Schecter (ed), *Dioxins and Health,* Plenum Press: London.

Hanrai, N., Hori, Y., Taniasu, S. & Yamashita, N. 2003. Analysis of polychlorinated naphthalenes and dioxin-like compounds In pine needle leaf by high resolution GC/high resolution MS. *Bunseki Kagaku* 52: 127–138 (in Japanese).

Hanrai, N., Miyake, Y., Hori, Y., Okazawa, T. & Yamashita, N. 2006. Congener-Specific Determination of Brominated/ Chlorinated Dioxins and Related Compounds by Two-Dimensional Cleanup System. *Bunseki Kagaku* 55: 491–500 (in Japanese).

Hosomi, M. 2005. Photodegradation and dioxin-like endocrine potential of polychlorinated dibenzothiophenes, Science Council of Asia Conference, poster, May 12th, 2005, http://www.scj.go.jp/en/sca/pdf/5thposter6.pdf

Kobayashi, Y. 2001. Quick measurement of dioxins using Ah receptor. *J. Resour. Environ.* 9: 65–70 (in Japanese)

Koistinen, J., Navalainen, T. & Tarhanen, J. 1992. Identification and Level Estimation of Aromatic Coeluates of Polychlorinated Dibenzo-p-dioxins and Dibenzofurans in Pulp Mill Products and Wastes. *Environ. Sci. Technol.* 26: 2499–2507.

Kopponen, P., Sinkkonen, S., Poso, A., Gynther, J. & Kärenlampi, S. 1994. Sulfur Analogues of Polychlorinated Dibenzo-*p*-Dioxins, Dibenzofurans and Diphenyl Ethers as Inducers of CYP1A1 in Mouse Hepatoma Cell Culture and Structure-Activity Relationships. *Environ. Toxicol. Chem.* 13: 1543–1548.

Nakai, S., Kishita, S., Espino, M.P. & Hosomi, M. 2004. Detection of polychlorinated dibenzothiophenes in Japan and investigation of their dioxin like endocrine disrupting potency. *Organohalogen Compounds* 66: 1495–1499.

Nakai, S., Kishita, S., Espino, M.P. & Hosomi, M. 2007. Polychlorinated dibenzothiophenes in Japanese environmental samples and their photodegradability and dioxin-like endocrine-disruption potential. *Chemosphere* 67: 1852–1857.

Repport 10050974 (2004) Framtagning av mer information om kemiska ämnen eventuellt aktuella för screening 2005, Sweden 2004; http://www.naturvardsverket.se/dokument/mo/modok/export/screenprio.pdf.

Rostkowski, P., Puzyn, T., Świeczkowski, A. & Falandysz, J. 2004. Prediction of Environmental Partition Coefficients for 135 Congeners of Polychlorinated Dibenzothiophenes. *Organohalogen Compds.* 66: 2386–2391.

Sinkkonen, S., (1997), PCDTs in the environment. *Chemosphere* 34: 2585–2594.

Sinkkonen, S., Paasivirta, J. & Lahtiperä, M. 2001. Chlorinated and Methylated Dibenzothiophenes in Sediment Samples from a River Contaminated by Organochlorine Wastes. *J. Soils & Sediments.* 1: 9–14.

Sinkkonen, S., Lahtiperä, M., Vattulainen, A., Takhistov, V.V., Viktorovskii, I.V., Utsal, V.A. & Paasivirta, J. 2003. Analyses of known and new types of polyhalogenated aromatic substances in oven ash from recycled aluminum production. *Chemosphere* 52: 761–775.

Wiedmann, T., Schimmel, H. & Ballschmitter, K-H. 1998. Ion trap MS/MS of polychlorinated dibenzo-*p*-dioxins and dibenzofurans: confirming the concept of the molar response. *Fresenius. J. Anal. Chem.* 360: 117–119.

Environmental Engineering III – Pawłowski, Dudzińska & Pawłowski (eds)
© *2010 Taylor & Francis Group, London, ISBN 978-0-415-54882-3*

Industrial installation for integrated bioremediation of wastewater contaminated with ionic mercury

P. Głuszcz & S. Ledakowicz
Department of Bioprocess Engineering, Technical University of Lodz, Lodz, Poland

I. Wagner-Doebler
Helmoltz Zentrum fuer Infektionsforschung, Braunschweig, Germany

ABSTRACT: A novel and environmentally friendly technology for neutralization of toxic mercury in industrial wastewaters by live bacteria has been developed in Germany at HZI (former GBF). The basic principle of this process is the enzymatic reduction of Hg(II) to water-insoluble and relatively non-toxic Hg(0) by mercury resistant bacteria immobilized on porous carrier material in a fixed-bed bioreactor.

Recently the method was improved at Technical University of Lodz, Poland, by replacing the non-adsorptive carrier in a bioreactor by activated carbon which enabled integration of adsorption and bioreduction processes in one piece of apparatus. The modified technology was applied in industrial scale in Poland. The paper presents results of one-year operation of this installation.

Keywords: Mercury, bioremediation, adsorption, wastewater treatment, process integration

1 INTRODUCTION

Extensive industrial use of mercury has led to significant pollution of the environment. Mercury introduced to the natural environment, regardless of the form can be relatively easily converted into highly toxic volatile and water soluble forms, i.e. methyl- or ethyl-mercury chloride, far more bioavailable and much more toxic than other forms of mercury. Furthermore, mercury is highly retained in living organisms, and therefore becomes biomagnified. Hence, mercury is a very dangerous, global pollutant because of its extreme toxicity, global atmospheric transport and accumulation in the food chain. (Boening 2000)

Today, the major anthropogenic sources of mercury emissions to the environment include the burning of coal to produce electricity, the incineration of waste and the chlor-alkali technology. The chlor-alkali industry produces chlorine and alkali, sodium hydroxide or potassium hydroxide, by electrolysis of brine. The main technologies applied for chlorine and alkali production are mercury-, diaphragm- and membrane-cell electrolysis. The mercury-cell (amalgam) process has been in use, mainly in Europe, since 1892 and now still accounts for ca. 50% of total production of chlorine in Europe. In the EU and EFTA countries there are presently about 50 operating mercury-cell chlor-alkali installations, with a chlorine production capacity of over 5.8 million tons per year. In line with the commitment of Euro-Chlor members, these plants will be decommissioned and/or converted to an alternative mercury-free process by 2020, as will a number of mercury-cell chlor-alkali plants in the US and other countries. Considering only the European mercury-cell plants, this decommissioning activity will release ca. 12 000 tons of process mercury, and even if mercury-cell processes are phased out in the European Community they will be still operating in Eastern Europe and developing countries worldwide. (Eurochlor 2006)

Due to the amalgam chlor-alkali technology characteristics, mercury can be emitted to the environment through air, water, solid wastes and in the products. Total mercury emission from chlor-alkali plants in Europe was about 40 tons in 2006, ranging from 0.15 to 3.0 g Hg per t of chlorine produced at the individual plants. As it was mentioned mercury is a dangerous global pollutant and therefore removal of mercury from industrial emissions in all possible places is mandatory and should take into account the latest achievements in science and technology.

Several chemical processes have been utilized for the removal of mercury from mercury contaminated chlor-alkali industrial wastewaters, but it is difficult to apply chemical processes for the purification of large volumes of wastewater containing Hg of low concentration, because it requires great amounts of chemicals and can lead to a secondary pollution. Common treatment techniques for mercury removal from polluted wastewater are mostly based on the sorption onto different materials such as activated carbon or ion exchange resins. Although there are many advantages of ion-exchange resins, such as high efficiency, selectivity and insensitivity to concentration variability, high cost of using ion exchange resins is the main problem.

For mercury remediation from industrial wastewaters a unique biotechnological method based on the enzymatic reduction of Hg(II) compounds to water-insoluble Hg(0) by live mercury resistant bacteria has been developed in Helmholtz Centre for Infection Research, HZI, former German Research Centre for Biotechnology (GBF). (von Canstein et al. 1999)

The novel bioremediation technology (called *bioMER*) was tested in two chlor-alkali factories (in Germany and Czech Republic) in a pilot-plant scale. (Wagner-Doebler 2003). The core of the technological process developed at GBF was a packed-bed bioreactor with pumice stones as a carrier material for microorganisms and an activated carbon adsorber used as a "polishing" filter after the bioreactor. The experience gained during operation of the pilot plant led to the idea, that the process of bioremediation may be integrated in one piece of apparatus by combining the sorption of mercury from wastewater on the activated carbon and further biochemical reduction by bacteria. It may be obtained by immobilization of microorganisms directly on the granulated activated carbon filling the bioreactor. Such process integration should further increase efficiency of the technology and lower its costs. Certainly, at the beginning it was necessary to define several significant parameters of the activated carbon used in the bioreactor and of the adsorption process in the flow-through apparatus itself. These problems were solved in other investigations. (Gluszcz et al. 2005, 2006).

The extensive comparison of effectiveness of the original mercury bioreduction method and the integrated technique was done in a laboratory scale (Gluszcz et al. 2008). It was shown that the integrated system has higher mercury removal efficiency and ability to avoid fouling of sorption media. The bioreactor with activated carbon bed assured higher mercury reduction efficiency (97–99%) in a wider range of inflow concentration, i.e. up to $18\,mg/dm^3$ while the column containing pumice – up to $8\,mg/dm^3$. The integrated bioreactor worked more steadily and responded moderately to the changes of mercury concentration in a wider range. Hence, the laboratory investigations led to the conclusion that integration of the bioremediation and sorption processes in one piece of apparatus enabled higher efficiency of the technology. The results obtained in a laboratory-scale were so promising that the modified technology was implemented in industrial scale in one of Polish chemical companies for bioremediation of wastewater from chlor-alkali amalgam-cell installation.

2 MATERIALS AND METHODS

The installation essentially consists of $1\,m^3$ bioreactor filled with porous material (pumice stone or granulated activated carbon) as a carrier for the microorganisms and an additional tank of ca. $0.9\,m^3$ volume

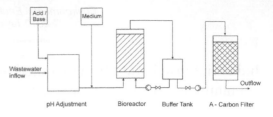

Figure 1. Flow scheme of the *bioMER* plant for continuous treatment of chloralkali electrolysis factory wastewater.

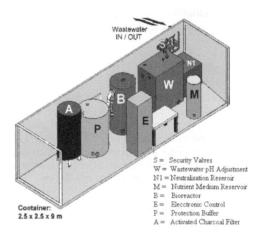

Figure 2. Scheme of the mobile standard container with the *bioMER* plant.

situated after the bioreactor, also containing activated carbon (AC). This tank may be used as an additional, "polishing" filter if the mercury concentration after the bioreduction step is too high or in case of emergency. Figure 1 shows a flowchart of the *bioMER* installation and Figure 2 is a scheme of the technical plant placed in a container.

The microorganisms used in the bioreactor are natural, non-pathogenic soil bacteria (mainly *Pseudomonas*), which possess natural mercury resistance. They convert enzymatically reactive ionic mercury to elemental mercury which remains in the packed bed of the bioreactor as almost water- insoluble metal and is no longer toxic for the bacteria. $NADPH_2$ is the biologically active electron donor within the cell, which is provided by normal metabolism of the bacteria.

The stoichiometric equation of this conversion is as follows:

$$Hg(II) + NADPH_2 \xrightarrow{\;(merA)GP\;} Hg(0) + NADPH + 2H^+ \quad (1)$$

The secreted metallic mercury which accumulates in form of small droplets is retained in the packing inside the bioreactor.

To enable the bio-transformation the microorganisms must be fed with a nutrient medium (e.g. sucrose/yeast extract) and need aerobic conditions

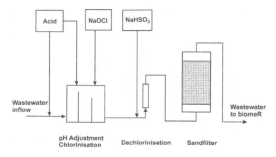

Figure 3. Schematic diagram of the pre-treatment unit.

whereby the initial oxygen saturation of the wastewater (6–8 mg O_2 dm^3) at the inlet is sufficient for the whole process and the additional aeration of the solution in the bioreactor is not necessary. An activated carbon filter situated after the bioreactor may be used as a polishing filter (as it was in the former version of the installation) to reach a final concentration of less than 50 μg/dm^3, or as a spare/emergency filter in the integrated-bioreactor version of the *bioMER*. In the first case more than 90% of the mercury removal is done by the microorganisms in the bioreactor but the rest must be adsorbed in the final filter; it was assumed that the polishing step would be not necessary if the bioreactor itself is filled with AC. The filter tank may be used then as a second bioreactor or as an emergency filter in the case of main bioreactor service necessity or its failure. If the packed bed material in the bioreactor is saturated (fouled) with droplets of mercury, packing may be removed and the metallic mercury could be recovered by distillation.

The bioreactor is protected automatically by a bypass if the inflow wastewater parameters are outside of pre-set ranges. The inflow and outflow concentration of mercury in wastewater is continuously measured by two on-line mercury analyzers PA-1 (Mercury Instruments, Germany) using the cold vapour technique. The installation is placed in a compact typical mobile container, completely automated and can be controlled via remote control through user shared process software.

Depending on the wastewater conditions a proper pre-treatment of the mercury contaminated wastewater should be applied. It should consist of pH adjustment, oxidation of mercury, removal of chlorine and filtration. The main task of the pre-treatment unit is to oxidize different forms of mercury which may be present in the wastewater (e.g. particle bound and different complexes of Hg besides Hg(II)) to ionic mercury Hg(II). In the first step the raw wastewater pH is adjusted to pH = 3. Then mercury is oxidized by adding sodium hypochlorite (NaOCl, 100 mg chlorine/dm^3). The excess of chlorine is removed by adding sodium hydrogen sulphite (NaHSO$_3$) controlled by the redox potential of 40 mV. The mercury free solid fines are removed by filtration. The flow-scheme of the pre-treatment unit is shown in Figure 3.

The *bioMER* installation was applied in one of Polish chemical companies in Tarnow, Poland, for bioremediation of wastewater coming from the mercury-cell chlor-alkali plant. At the beginning the bioreactor was filled with pumice stones, as in the original version tested previously by GBF, to check the known technique in new industrial conditions specific to the Tarnow factory and to collect data for the comparison of the old and the modified technology in industrial scale.

The bioreactor was inoculated with the mixture of 6 strains of microbes possessing mercury bioreduction ability: four strains of *Pseudomonas putida* (Spi 3, Spi 4, Kon 12, Elb 2), *Pseudomonas fulva* and *Pseudomonas stutzeri*. The strains were identified at DSMZ (Deutsche Sammlung von Mikroorganismen und Zellkulturen GmbH), Germany.

The volumetric flow rate of wastewater through the bioreactor was adjusted initially at 1 m^3/h (i.e. 1 BV/h), the inflow mercury concentration was kept at the level of 2.5–3.5 mg/dm^3 and pH in the range of 6.5–7.5.

3 RESULTS AND DISCUSSION

The *bioMER* plant undertook the operation immediately after the inoculation and within several hours the outlet concentration of mercury in wastewater decreased from ca. 3 mg/dm^3 at the inlet to 120–150 μg/dm^3 at the outlet from the bio-container. The *bioMER* plant was operated in the same conditions continuously in a time period of six month with a wastewater flow changing in the range of 1.0–2.0 m^3/h depending on the factory requirements. This was sufficient to clean the whole mercury contaminated wastewater from the chlor-alkali installation so the bioremediation process replaced the hydrazine method of mercury neutralization used in Tarnow by then.

In the Figure 4 typical Hg inflow and outflow concentrations during the initial 6-month operation is shown. The inlet Hg concentration was changing in the range of 2.5–6.0 mg/dm^3 although the shock loadings up to 15 mg/dm^3 occurred occasionally. The mercury concentration in the liquid after the bioreactor varied from 150 μg/dm^3, at the beginning of the *bioMER* operation, to 700–800 μg/dm^3 when the Hg content in the raw wastewater suddenly increased for a short time; the average steady-state value of 350–400 μg/dm^3 was reached after ca. 3 month of operation.

This mercury concentration was only an effect of the bioprocess taking place in the bioreactor packed with pumice (i.e. without of adsorption onto the activated carbon in the filter). The additional analysis showed that the 30–50% of the total mercury present in the outlet stream is in the form of Hg(0), so the observed increase of the outlet mercury concentration was partially a result of washing-out of the micro-suspension of metallic mercury from the bioreactor; this effect was visible especially when sudden fluctuations of the flow rate of wastewater through the bioreactor occurred

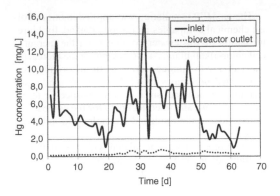

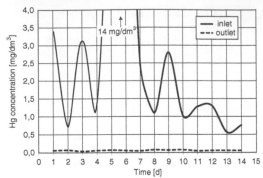

Figure 4. Concentration of mercury at the inlet and at the outlet of the *bioMER* bioreactor in the first period of the operation in Tarnow factory (bioreactor with pumice-stone packing).

Figure 5. Comparison of the inlet and outlet concentration of mercury after the exchange of the packing in the bioreactor.

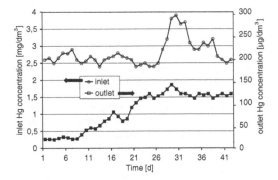

Figure 6. Comparison of the inlet and outlet Hg concentration in the integrated bioreactor after the breakthrough of mercury.

(e.g. as a result of pumps failure) or after a periodical back-flush of the packing.

After the first 6 months of the *bioMER* operation the pumice packing in the bioreactor was replaced by an activated carbon to obtain the effect of process integration. The granulated carbon type KA 2–5 (Elbar-Carbon, Poland) was used. It is a porous coke-type carbon prepared of a granulated pit-coal, used in wastewater treatment; the particle diameter is 2–5 mm and density of the packing bed 350 g/dm³. The new packing in the bioreactor was inoculated using the wastewater previously treated in *bioMER*, collected in the buffer tank after the bioreactor. It was checked that in the cleaned water the sufficient concentration of live bacteria is present (CFU in the range 10^5–10^6). It was expected that the community of microorganisms present in the bioreactor after six month of operation is fully adapted to the specific local process conditions and form the best material for the re-inoculation of the bioreactor.

After the packing exchange the concentration of mercury at the outlet from the bioreactor decreased to 20–30 μg/dm³, i.e. below the value of metallic mercury solubility in water, obviously due to almost total adsorption of mercury onto pure and "fresh" activated carbon. The same low Hg concentration was observed within the period of 95 days after the modification, irrespective from the inlet concentration changes Figure 5.

In this time period the average inlet Hg concentration was 3.1 mg/dm³ and the average flow rate of wastewater was about 1.2 m³/h, so total amount of mercury adsorbed in the packing was ca. 8.5 kg. Although such amount of mercury is still lower than the total saturation adsorption capacity of the packing the breakthrough of mercury, i.e. gradual increase of the outlet Hg concentration, was observed after that time; this effect is shown in Figure 6. It is worth to notice that only since the moment when the mercury breakthrough at the outflow occurred it makes sense comparing the efficiency of the integrated bioreactor

and its former version with pumice. After the breakthrough it is also possible to prove that the removal of mercury is not only a result of Hg adsorption onto the activated carbon: if there was no mercury-reducing bacteria in the packing the concentration of mercury at the outlet should successively increase once the breakthrough was detected, according to the typical shape of a breakthrough curve in adsorbers.

In the Figure 6 it is visible that between the 10th and the 30th days the outlet mercury concentration actually goes up (after a breakthrough) but then it stabilizes at the level of 110–120 μg/dm³, aside from the fluctuations of the Hg concentration in the inlet stream. To confirm undoubtedly the effect of the mercury-reducing activity of microorganisms immobilized on the activated carbon it would be necessary to compare – at the same time – efficiency of the integrated bioreactor containing the live bacteria and the similar tank filled with the sterile activated carbon. Such experiment was successfully done in a laboratory scale (Gluszcz et. al. 2008) but in the real, industrial conditions of the Tarnow factory it would be very difficult and costly. Nevertheless, the similar experiment was performed in the *bioMER* installation, as follows.

Before the experiment the Hg concentration after the bioreactor was steady and at the sufficiently low level for a certain time so it was assumed that the

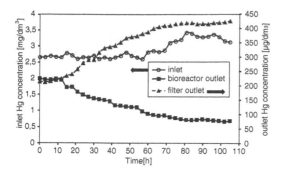

Figure 7. Comparison of the mercury removal efficiency of the integrated bioreactor and the activated carbon adsorber fed simultaneously with the same wastewater.

polishing filter was not necessary. Hence, the treated wastewater was directed from the bioreactor right to the outflow from the installation. Then the inlet stream of raw wastewater (containing ca. $3.0\,mg/dm^3$ of Hg and of pH $= 3$) was divided into two – one flowing as usual through the neutralization tank to the bioreactor and the second – flowing directly from the installation inlet to the activated carbon filter. Although during the usual operation of the *bioMER* it may be expected that the AC packing in the filter tank is also inoculated by the microorganisms from the bioreactor, the very low pH value of the raw wastewater enabled the assumption that all mercury-reducing microorganisms (if there were any) should be deactivated. The amount of activated carbon inside the filter is similar as in the bioreactor and it was assumed that the filter packing should be – at least partially – also saturated with mercury, as it was not changed for a long time of the previous operation.

The results of this experiment are shown in the Figure 7. It is interesting to compare simultaneous changes of mercury concentration at the outlet from the bioreactor and after the filter tank (working now solely as an activated carbon adsorber).

In Figure 7 one may observe the progressive increase of the Hg concentration after the filter and decreasing mercury content in the liquid flowing out of the bioreactor. The initial outflow Hg concentration presented in Figure 7 was $220\,\mu g/dm^3$, (somewhat higher than usually) because the experiment was started several hours after the back-flush of the packing in the bioreactor.

This experiment confirmed that in the integrated bioreactor containing AC the synergistic effect of two phenomena i.e. adsorption and bioreduction of mercury may be observed and the efficiency of Hg removal from the wastewater may be higher than in a standalone AC adsorber or in a fixed-bed bioreactor with a non-adsorptive packing.

In the previous laboratory investigations (Gluszcz et. al. 2008) it was also found that the integrated bioreactor is more flexible and robust than its former version in the case of sudden fluctuations and shock loadings of the mercury. In Figure 6 similar conclusion may be withdrawn comparing the inlet and outlet concentrations changes between 26th and 36th day, but actually during the discussed period of the modified bioreactor operation there were no extreme inlet fluctuations of Hg concentration observed and this feature of the integrated system should be yet confirmed. Unfortunately, in the real industrial conditions it is almost impossible to perform on purpose some experiments which are easy in a laboratory scale. Nevertheless, taking into account the presented results it may be expected that the integrated process of the mercury removal in the industrial scale would give the same results as in the laboratory, also in this aspect.

4 CONCLUSIONS

After the extensive laboratory investigations the integrated process of mercury bioreduction using activated carbon as a carrier for microorganisms immobilization and at the same time as an adsorbent for different forms of mercury was applied in industrial scale. The modified *bioMER* technology replaced the hydrazine method used previously for mercury neutralization in wastewater from the chlor-alkali plant in chemical company in Tarnow, Poland. Comparison of the results, obtained during 6-month operation of the installation before the modification of the technology and after bioreactor packing exchange, leads to the conclusion that the integration of the bioremediation and sorption processes in one piece of apparatus leads to higher efficiency of the integrated process and to higher robustness and flexibility of the bioreactor in variable conditions. In the case of Hg concentration fluctuations activated carbon may play a role of a buffer repository for the temporary excess of mercury, which may be treated by microorganisms in later time when the Hg concentration in solution goes down again. The *bioMER* installation is working at the chlor-alkali plant in Tarnow continuously and it is expected that it will prove its advantages also in more extreme conditions than prevailed during the presented period of its operation.

ACKNOWLEDGEMENT

The work was supported by the Polish Ministry of Science from funds for science development in the years 2007–2009 as a R&D project R14 006 02.

REFERENCES

Boening, D. W. 2000. Ecological effect, transport, and fate of mercury: a general review. *Chemosphere* 40: 1335–1351.
von Canstein, H., Li, Y., Timmis, K. N., Deckwer, W-D. & Wagner-Doebler, I. 1999. Removal of mercury from chlor-alkali electrolysis wastewater by a mercury resistant. *Pseudomonas putida* strain, *Appl. Environ. Microbiol.* 65: 5279–5284.

Eurochlor; 2006. Chlorine online, Information resource. www.eurochlor.org.

Gluszcz, P., Zakrzewska, K., Ledakowicz, S. & Deckwer, W-D. 2005. Modification of the microbiological method for mercury remediation of industrial wastewater. *J. Biotechnol.* 118: S1:163

Gluszcz, P., Zakrzewska, K., Ledakowicz, S., Deckwer, W.-D. & Wagner-Doebler, I.; 2006. Adsorption of mercury from aqueous solutions in a fixed-bed sorption column, Proc. of the 17th International Congress of Chemical & Process Engineering. CHISA: 27–31.

Gluszcz, P., Zakrzewska, K., Ledakowicz, S. & Wagner-Doebler, I. 2008. Bioreduction of ionic mercury from wastewater in a fixed-bed bioreactor with activated carbon, *Chem. Pap.* 62: 232–238.

Wagner-Doebler. I. 2003. Pilot plant for bioremediation of mercury-containing industrial wastewater. *Appl. Microbiol. Biotechnol.* 62: 124–133.

Environmental Engineering III – Pawłowski, Dudzińska & Pawłowski (eds)
© *2010 Taylor & Francis Group, London, ISBN 978-0-415-54882-3*

A laboratory study on toxicity removal from landfill leachate in combined treatment with municipal wastewater

J. Kalka, A. Oślislok, J. Surmacz-Górska, K. Krajewska, D. Marciocha & A. Raszka
Environmental Biotechnology Department, Silesian University of Technology, Gliwice, Poland

ABSTRACT: Combined treatment of landfill leachate and municipal wastewater were performed in order to investigate the changes of leachate toxicity during biological treatment. Three laboratory A2O models were operating under the same parameters (HRT- 1.5-1.6 d) except from the influent characteristic. The influent of model A consisted of 15% (v/v) of landfill leachate and 85% (v/v) of tap water; influent of model B consisted of 15% (v/v) landfill leachate and 85% (v/v) of municipal wastewater; model C served as a control and its influent consisted of municipal wastewater only. Toxicity of raw and treated wastewater was determinate by three acute toxicity tests with Daphnia magna, Thamnocephalus platyurus and Vibrio fischeri. Landfill leachate increased initial toxicity of wastewater. It was concluded that synergetic toxic effect in mixture of leachate and wastewater was detected. During biological treatment significant decline of acute toxicity was observed but still mixture of leachate and wastewater was harmful to all tested organisms.

Keywords: Biological treatment, landfill leachate, toxicity.

1 INTRODUCTION

Municipal solid wastes management becomes today a major environmental, economical and social concern. Municipalities are forced by the legislation to follow a minimization of wastes. Recycling and recovery of materials and energy is encouraged so as to safeguard natural resources and obviate wasteful use of land (EEC 1999). Nevertheless waste volume is growing faster than the world's population (Salem et al. 2007, Renou et al. 2008).

Landfilling is a widely accepted and used method for the ultimate disposal of solid waste material, due to its economic advantages. It is estimated that 90% of solid waste in Poland is disposed of in landfill sites (Renou et al., 2008). The internal biochemical decomposition processes taking place within a landfill play a crucial role in determining potential adverse impacts that landfills may have during and beyond its active life. Landfill leachate resulting from rainfall percolating through layers of waste may endanger aquatic environment due to uncontrolled overflow, subsidence and infiltration (Isidori et al. 2003, Bodzek et. al. 2004). Many studies have shown that landfill leachate consisted of different groups of pollutants such as organics: alkenes, aromatic hydrocarbons, acids, esters, alcohols, hydroxybenzene, amides etc.; as well as ammonia nitrogen and heavy metals (Benfenati et al. 2003, Yang & Zhou 2008). The most common practice to avoid risk of contamination is to discharge leachate into wastewater treatment plant. High load of macro- and micropollutants may disrupt biochemical processes in biological reactors. More important is that some pollutants may pass biological treatment plant unchanged and contribute to still high toxicity of the effluent. It is well known that toxicity of environmental samples (like wastewater or leachate) is a consequence of numerous contaminants, their synergistic or antagonistic effects and physical-chemical properties. Toxicity tests thus better characterize samples quality than chemical analyses alone. The aim of present study was to investigate the toxicity of landfill leachate both before and after biological treatment.

2 MATERIAL AND METHODS

2.1 Leachate

Leachate was collected from municipal solid waste landfill in Zabrze (Poland). Zabrze landfill has been receiving municipal waste since 2007. The landfill leachate is recirculated to the waste dump and excess is pumped to the sewage collection system. Samples were collected from the equalization basin. Leachate characteristic is presented in Table 1.

Wastewater was collected from wastewater treatment plant in Zabrze-Mikulczyce (Poland). The place for wastewater collection was selected to ensure lack of earlier wastewater contamination by leachate.

2.2 Chemical analysis

Ammonium nitrogen was measured with Kjeltec 1026 analyzer. Total organic carbon (TOC) was measured with a Shimadzu total organic carbon analyzer, model V_{CSH}. Chemical and biological oxygen demands

(COD and BOD) were determined by Standard Methods (PN-74 C-04578/03, PN-EN 1899-2:2002). Chemical analysis were performed two times a week during 8 weeks research period.

2.3 Treatment

The experiment was carried out in three activated sludge A2O systems – A, B and C. All systems were operated under the same technical parameters (Table 2) except for influent characteristic and load. Influent of system A consisted of 15% (v/v) of landfill leachate and 85% (v/v) of water; influent of system B consisted of 15% (v/v) of landfill leachate and 85% (v/v) of municipal wastewater. System C served as a control and was fed with municipal wastewater.

A2O activated sludge systems composed of an anaerobic/anoxic/oxide process which removed biological phosphorus with simultaneous nitrification-denitrification. Each system composed of three separate reactors with the following working volumes: anaerobic – 2 L, anoxic – 5 L, aerobic 7 L (Figure 1).

The reactors were inoculated with an activated sludge sampled from municipal wastewater treatment plant.

Activated sludge in reactors A and B was acclimated to the increasing concentration of landfill leachate in the influent. Final concentration of landfill leachate

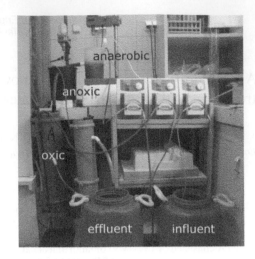

Figure 1. Photograph of A2O laboratory model.

reached 15 % (v/v) and exceeded 100 times maximum possible concentration of landfill leachate in wastewater stream. After the acclimation period, systems had been operated for 8 weeks. Samples for chemical analysis as well as toxicity testing were collected from average daily sample of influent/effluent.

2.4 Bioassays

Whole effluent toxicity tests were performed which means that the aggregate toxic effect of respectively influent or effluent was measured directly by a toxicity test.

Following tests were proposed for toxicity evaluation:

- *Vibrio fischeri* luminescence inhibition – Microtox (Azur Environmental 1991).

The test was carried out in the Microtox M500 toxicity analyzer according to the standard procedure

Table 1. Characteristic of municipal landfill leachate.

Parameter	Unit	Range	Average±SD*	Median
TOC	mg/L	1460–2300	1950 ± 350	2010
COD	mg/L	1873–3600	2560 ± 615	2330
BOD	mg/L	150–273	210 ± 55	210
N_{NH4}	mg/L	971–1250	1100 ± 92	1200

*SD – standard deviation; number of measurements n = 8; time-dependent variation of parameters value was caused by unstable composition of leachates.

Table 2. Operational parameters of activated sludge systems A, B and C.

Parameter	Unit	System	Range	Average ± SD	Median
COD	g COD	A	0.07–0.11	0.08 ± 0.02	0.08
load	gTSd	B	0.07–0.236	0.16 ± 0.07	0.15
		C	0.06–0.175	0.90 ± 0.04	0.06
		A	2.3–3.1	2.6 ± 0.5	2.3
SS	g /L	B	1.6–2.1	1.9 ± 0.2	1.9
		C	1.7–3.7	2.6 ± 0.9	2.8
		A	8.7–10.0	9.5 ± 0.5	9.7
Q	L/d	B	6.2–10.4	9.0 ± 1.9	9.7
		C	8.4–10.0	9.2 ± 0.6	9.6
		A	1.4–1.6	1.5 ± 0.1	1.4
HRT	d	B	1.3–2.2	1.6 ± 0.4	1.5
		C	1.4–1.7	1.5 ± 0.1	1.5

SD – standard deviation; number of measurements n = 16; time-dependent variation caused by unstable composition of influent; COD – chemical oxygen demand; SS – activated sludge suspended solids; Q – wastewater flow; HRT – hydraulic retention time.

Table 3. Characteristic of raw and treated wastewater.

Parameter	Unit	System	Influent			Effluent		
			Range	Average ± SD	Median	Range	Average ± SD	Median
TOC	mg/L	A	218–322	288 ± 45	317	83–100	92 ± 8	93
		B	317–517	398 ± 77	384	62–167	105 ± 45	95
		C	167–236	193 ± 34	184	19–53	27 ± 13	22
COD	mg/L	A	281–537	368 ± 85	334	167–242	192 ± 43	167
		B	340–695	525 ± 136	440	108–322	197 ± 82	150
		C	211–398	281 ± 58	310	25–127	72 ± 40	65
BOD	mg/L	A	30–40	35 ± 7	35	10	10 ± 0	10
		B	180–200	190 ± 10	190	10–20	13 ± 6	10
		C	240–260	247 ± 12	240	10–20	17 ± 6	20
N_{NH4}	mg/L	A	150–175	163 ± 13	171	2–4	3.4 ± 0.6	3.9
		B	176–236	205 ± 27	195	3–11	5.5 ± 3	7
		C	82–134	103 ± 23	87	0.5–6	2.6 ± 1.7	2.6

SD – standard deviation; number of measurements n = 16; time-dependent variation of caused by unstable composition of influent.

(Microtox, 1992), which is in accordance with ISO – DIN 38412 Part 34, 9/91. The lyophilized bacteria *Vibrio fischeri* were purchased from Azur Environmental (Carlsbad, US). As a diluent 2% NaCl was used. As the samples of wastewater were coloured – light absorbance were measured at 490 nm and colour correction procedure was applied.

Vibrio fischeri luminescence inhibition test was performed three times;

• *Daphnia magna* immobilisation test (ISO 1996) Tests were carried out with neonates (<24 h) . Five test dilutions were prepared in a 50% dilution series for each sample with three replicates of seven animals. The test volume was 20 mL. The animals were not fed during the experiment. Each test had a duration 48 h; the temperature was 20 ± 1°C. After an exposure, the number of immobile daphnids for each dilution was recorded. *Daphnia magna* immobilization test was performed five times;

• *Thamnocephalus platyurus* acute toxicity test (MicroBioTest, 1998). Tests were carried out according to MicroBioTest Standard Operational Procedure. Readily hatched organisms were used for the test. Five test dilutions were prepared in a 50% dilution series.

Each sample was with 3 replicates of 10 animals in disposable multiwall test plates. Test volume was 1 ml per well. After 24 h in a 25°C incubator in the dark, the number of dead crustaceans was recorded. *Thamnocephalus platyurus* acute toxicity test was performed five times.

3 RESULTS AND DISCUSSION

Biological treatment of leachate has been shown to be effective in removing organic and nitrogenous matter from immature effluent characterized by high BOD/COD ratio (Renou et al. 2008, Salem et al. 2008).

In present study biodegradability of influents containing 15% of leachate was low: BOD/COD ratio was 0.1 and 0.4 for system A and B respectively. In contrary BOD/COD ratio of wastewaters (system C) was 0.9.

Although removal of organic content reached 71% BOD (48% COD) – for system A and 93% BOD (63% COD) for system B, low biodegradability of influents resulted in high content of organic substances in the effluents (Table 3). The effluents of systems enriched by leachate (A and B) didn't meet the quality standards described for wastewaters introduced to surface waters or ground (EEC, 1991). In parallel treated wastewater (system C), BOD removal reached 93% (75% COD).

Combined treatment of landfill leachate and municipal wastewater was also investigated by Diamadopoulos et al. (1997) in sequencing batch reactor. Parameters of the process were similar to those in present study, except for volumetric ratio leachate: wastewater – 1:9 (1:6 in present study). The authors reported that efficiency of BOD removal was 95%, but still quality criteria were not met.

Several authors revealed the possibility of leachate treatment in combining aerobic – anaerobic conditions, which allowed to perform treatment with higher organic loading rates (Yang & Zhou 2008, He et al. 2007).

Effective ammonia nitrogen removal was observed in all three systems. Removal efficiency was as high as 98% for system A and C and 97% for system B.

The results of toxicity tests were presented in Table 4 as median effect concentrations (EC_{50}).

The results of toxicity tests were also examined for environmental relevance by calculation Toxicity Units (TU) as reported in Table 5. The toxic unit of an effluent is the inverse of its EC_{50} (or LC_{50}):

$$TU = \frac{100}{\% EC_{50}} \qquad (1)$$

Table 4. Toxicity of raw and treated wastewater.

Organism	No of tests	System	Influent – average EC_{50} [%] $\pm$ SD	Effluent – average EC_{50} [%] $\pm$ SD
D. magna	5	A	27.5 $\pm$ 1.9	0 (h.e* < 10%)
	5	B	13 $\pm$ 2.3	0 (h.e. 40%)
	5	C	39.7 $\pm$ 4.6	0
T. platyurus	5	A	15.5 $\pm$ 0.7	0 (h.e. < 10%)
	5	B	7.6 $\pm$ 0.6	34.5 $\pm$ 2.1
	5	C	18.8 $\pm$ 2.6	0
V. fischeri	3	A	14.1 $\pm$ 1.1	55.6 $\pm$ 2.4
	3	B	8.3 $\pm$ 0.9	38.5 $\pm$ 3.1
	3	C	14.9 $\pm$ 0.8	0

Table 5. Toxic Units of influents and effluents.

Organism	No of tests	System	Influent – average EC_{50} TU $\pm$ SD	Effluent – average EC_{50} TU $\pm$ SD
D. magna	5	A	3.6 $\pm$ 0.3	0
	5	B	8.4 $\pm$ 0.8	0.8 $\pm$ 0.6
	5	C	2.5 $\pm$ 0.3	0
T. platyurus	5	A	6.4 $\pm$ 0.3	0
	5	B	13.2 $\pm$ 0.9	2.9 $\pm$ 0.7
	5	C	5.4 $\pm$ 0.8	0
V. fischeri	3	A	7.1 $\pm$ 0.2	1.8 $\pm$ 0.3
	3	B	12 $\pm$ 0.2	2.6 $\pm$ 0.1
	3	C	6.7 $\pm$ 0.2	0

If the mortality in a 100% effluent concentration was between 10% and 49%, the TUs were derived as follows:

$$TU = 0.02 \times \% \text{ mortality} \qquad (2)$$

The highest initial toxicity of influent was observed for system B. After biological treatment toxicity was declined but still effluent of system B was harmful. In comparison – effluents from systems A and C were not toxic except for effluent A, which was harmful to *Vibrio fischeri*.

Relatively low toxicity of system A influent was probably connected with low bioavailability of pollutants. Influent of system A was characterized by extremely low BOD. It might be expected that this parameter influenced also initial toxicity of samples.

Thamnocephalus platyurus revealed the highest sensitivity to examined samples.

Important differences were observed between toxic response of *T. platyurus* and *Daphnia magna*. Similar effect was observed by Isidori et al. (2003), who also suggested the freshwater crustacean *Thamnocephalus platyurus* to test battery for detecting hazard and risk of landfill leachate.

Important increase of toxicity was observed in all tested bioassays while landfill leachate was mixed with municipal wastewater (system B). Toxic units calculated at the base of toxicity test results for system B were greater than expected on the basis of exposure to influents/effluents of system A and C individually:

$$TU_B > TU_A + TU_C \qquad (3)$$

It might be therefore concluded that mixture of leachate and wastewater contaminants revealed synergetic character of impact.

Acute toxicity of landfill leachate is often attributed to high ammonium nitrogen concentration (Isidori et. al. 2003, Martinen et. al. 2002, Dave & Nilsson 2005). This hypothesis was also confirmed in present study where after satisfactory reduction of N-NH$_4$ concentration in effluents of system A and C, acute toxic effects were not observed (except from system A effluent to *V. fischeri*). However, acute effects for all tested organisms were observed for system B effluent even though ammonium nitrogen was successfully removed during biological treatment. Synergetic interaction between components of landfill leachate and wastewater mixture resulted in inefficient toxicity removal during biological treatment of system B influent.

4 CONCLUSIONS

Landfill leachate significantly disrupt biological treatment of wastewater. After biological treatment wastewater enriched with leachate did not meet the water quality standards and still was harmful to aquatic organisms. The greatest share in overall toxicity of samples might be connected to the ammonium concentration. The mixture of contaminants present in wastewater and leachate revealed synergetic mechanism of toxic response. It have to be remembered that for toxicity measurement only acute tests were selected – it is planned to extent the biotest battery to chronic and reproducive tests, which should allow to detect hazard in sublethal concentrations.

ACKNOWLEDGEMENTS

The research was supported by Polish Ministry of Science and Higher Education – grant N52307732/2900.

REFERENCES

Azur Environment 1991. Microtox – Standard Operational Procedure.

Benfenati, E. Porazzi, E. Bagnati, R. Former, F. Martinex, M. Mariani & G. Fanelli R. 2003. Organic tracers identification as convenient strategy in industrial landfills monitoring. *Chemosphere* 51: 667–683.

Bodzek, M. Surmacz-Górska & J. Hung, Y. 2004. Treatment of landfill leachate. In: L. Wang, Y. Hung, C. Yapijakis (eds), *Handbook of Industrial and Hazardous Wastes Treatment*, New York: Marcel Dekker.

Dave, G. & Nilsson, E. 2005 Increased reproductive toxicity of landfill leachate after degradation was caused by nitrite. *Aquatic Toxicology* 73: 11–30.

Diamadopoulos, E. Samaras, P. Dabou, X. & Sakellaropoulos, G. 1997. Combined treatment of leachate and domestic sewage in a sequencing batch reactor. *Water Science and Technology* 36: 61–68.

EEC 1991. Urban Wastewater Treatment Directive *91/271/EEC*.

EEC 1999. Council Directive on the landfill of waste 99/31/EEC

He, R. Liu, X. Zhang, Z. & Shen, D. 2007. Characteristics of the bioreactor landfill system using anaerobic-aerobic process for nitrogen removal. *Bioresource Technology* 98: 2526–2532.

Isidori, M. Lavgorna, M. Nardelli, A. & Parrella A. 2003. Toxicity identification evaluation of leachate from municipal solid waste landfills: a multispecies approach. *Chemosphere* 52: 85–94.

ISO 6341:1996 Water quality – Determination of the inhibition of the mobility of *Daphnia magna* Straus (*Cladocera, Crustacea*) – Acute toxicity test.

MicroBioTest 1998. Thamnotoxkit Standard Operational Procedure.

Martinen, S. Kettunen, R. Sormunen, K., Soimasuo, R. & Rintala, J. 2002. Screening for physical-chemical methods of organic material, nitrogen and toxicity removal from low strength landfill leachates. *Chemosphere* 46: 851–858.

Persoone, G. Marsalek, B. Blinova, I. Törökne, A. Zarina, D. Manusadzianas, L. Nalecz-Jawecki, G. Tofan, L. Stepanova, N. Tothova, L. & Kolar, B. 2003. A practical and user-friendly toxicity classification system with microbiotests for natural waters and wastewaters. *Environmental Toxicology* 18: 395–402.

Ra, J.S. Lee, B.C. Chang, N. & Kim, S.D. 2008. Comparative Whole Effluent Toxicity Assessment of Wastewater Treatment Plant Effluents using *Daphnia magna. Journal Bulletin of Environmental Contamination and Toxicology* 80: 196–200.

Renou, S. Givaudan, J. Poulain, S. Dirrasouyan, F. & Moulin, P. 2008. Landfill leachate treatment: Review and opportunity. *Journal of Hazardous Materials* 150: 468–493.

Salem, Z. Hamouri K.. Djemaa, R. Allia, K. 2008. Evaluation of landfill leachate pollution and treatment. *Desalination* 220: 108–114.

Yang, Z. & Zhou, S. 2008. The biological treatment of landfill leachate using a simultaneous aerobic and anaerobic (SAA) bio-reactor systems. *Chemosphere* 72: 1751–1756.

Environmental Engineering III – Pawłowski, Dudzińska & Pawłowski (eds)
© *2010 Taylor & Francis Group, London, ISBN 978-0-415-54882-3*

Applying the treedendrical scheme failure method to evaluate the reliability of sewage collection draining reliability evaluation subsystems

J. Królikowska & A. Królikowski
The Technical University of Krakow, Krakow, Poland

ABSTRACT: The article presents an analysis of the reliability of functioning of a chosen sewage system, namely, the sewage draining subsystem in the town of Nowy Sącz. The analysis was based on the failure frequency of the subsystem research during field research. From the point of view of reliability testing, sewage draining subsystems are characterized by a highly complicated "dendrical" structure – i.e. a hierarchical one. Its main feature is the existence of many different states of its performance. With that in mind, the reliability evaluation of the discussed subsystem was accomplished by means of the dendrical model method. This method allow for an unambiguous evaluation of the reliability of this subsystem with more than one parameter allowed. Additionally, the reliability may be evaluated on different levels, i.e. a particular percentage of sewage drained by the system.

Keywords: Sewage draining subsystem, reliability, dendrical model method.

1 INTRODUCTION

Sewage collection and treatment systems (SUsOS) are functional sets of objects – interconnected and mutually conditional. Generally, they can be divided into sewage collection subsystems (PUs) and sewage pre-treatment subsystems (POŚ). A sewage collection subsystem is characterised by a defined technological structure, consisting of elements which maintain individual functions. The reliability of this subsystem is defined as its ability to collect continuously both the sewage from all users and rainfall, in satisfying amounts and in all working conditions during a required time. Reliability must be evaluated by means of calculable reliability parameters, i.e. quantitative characteristics of the object, such as failure frequency, maintenance capability, durability etc. The reliability parameters can be defined by (Wieczysty A. & Kapcia J. 2001):

– prognosis
– field testing (performance testing)
– reliability calculations.

From the perspective of reliability testing, a sewage subsystem has a complicated "tree structure", i.e. a hierarchical one, characterised by the existence of many possible states of operation, resulting from various combinations of working and damaged elements.

To date, only a few papers have been published on reliability parameters for sewage networks. Therefore, this paper has been prepared to present a reliability analysis of the performance of a small sewage subsystem in the town of Nowy Sacz. The analysis uses a dendrical failure model, which allows for the determination of several reliability parameters.

2 LITERATURE REVIEW

The principles of the reliability theory are used more and more often both in planning and in running tests. While much has been recently written on water supply systems, little has been published on the reliability of sewage subsystems operations (Алексеев & Ермолин 1996, Denczew & Królikowski 2002, Kapcia et. al. 2005, Wieczysty & Kapcia 2001). It should, however, be noted that these underground infrastructure systems are of importance and they should:

– guarantee collection and treatment of sewage from a given area
– respect the environment as a priority.

A sewage collection subsystem is a structurally integral unit, its specific feature being the lack of possibility of determining an idle state (Wieczysty & Kapcia 2001). Repair states are usually discrete, since subsystem checkups are performed at fixed periods. The hydraulic efficiency of the elements of the sewage subsystem does not really reflect their technical status, which is only identified after a malfunction has occurred or when the deterioration processes have already advanced. The causes of sewage system failure are usually (Kapcia et al. 2005):

– internal factors (corrosion and material wear, prolonged operation under high pressure, loosening at the joints)

- factors destructive to the elements of the system (soil subsidence towards the sewer),
- external factors (infiltration, mechanical damage, corrosiveness and subsidence).

In spite of many difficulties, in the recent years some original mathematical models and flow modelling software have been produced to examine the conditions of transient flow required for designing and modernisation of sewage subsystems (Denczew & Królikowski 2002). Technological developments have allowed the placing of television cameras in a system to monitor its status and provide its fully automatic service. Automated servicing is used increasingly frequently, and both developments have value as far as the reliable and safe functioning of sewage subsystems and their equipment are concerned.

The methods of evaluation of sewage systems functions presented in the literature concern the reliability and safety aspects, which are described by the means of a standby parameter and the probability of the occurrence of catastrophic losses (Wieczysty & Kapcia 2001). It can be stated, on the basis of a literature review, that the issue of sewage collection subsystem reliability is still largely unrecognised, and defining standards of reliability for this subsystem is still an open question.

Areas for further development in examining sewage collection subsystems will need to concentrate on determining the estimated reliability parameters applicable for local conditions, developing simulation methods of sewage system operation with respect to failure or catastrophe scenarios, as well as developing recognition of the reliability issues for individual elements of the system. Only once the reliability of these elements has been studied, will a complex evaluation of the subsystem be possible.

3 MATERIALS AND METHODS

The sewage collection subsystem in Nowy Sącz comprises a network of reinforced ducts. The oldest part of the town has a combined sewer system, while the post-war part has a separate sewer system. The length of combined pipelines is 76 km, while the separated pipelines extended for 94 km, with 95 km of drains. The network consists of concrete, ceramic and PVC or PE round-section pipes, 0,15 m to 0,60 m in diameter.

Unlike water-supply systems, it is relatively difficult to detect malfunctions in sewage systems. The time of waiting for repair is usually known, since checkups are performed at fixed dates. Moreover, the failure consequences are partially unaccountable.

A malfunction is a state where there is (Алексеев & Ермолин 1996, Dąbrowski 2004, Kapcia et al. 2005):

- clogging of the pipeline, which may lead to flooding of basement and streets
- leakage in sewage networks and/or network elements
- breakage of pipelines or network elements.

The most frequent reason for malfunctions of sewage systems is leakage and infiltration of groundwater with ground particles into the sewer. Resulting air gaps lead to the ground sinking directly above the sewer. Another characteristic event in sewage networks is silting. Silting in pipelines transporting sewage decreases their openings, increases the head losses and therefore limits the flow velocity. The intensity of silting is influenced by any type of protrusions or obstructions inside the collector, tree roots or uneven slopes, as well as a flow velocity below the self-purification level. When a system is run properly, this phenomenon does not need to lead to breaks in the sewer operation.

In order to evaluate the failure frequency for the sewage subsystem and acquire the data necessary for an estimation of reliability parameters, the test field method was employed. Information was collected on all malfunctions and damages within the network. The damage intensiveness parameter λ, which determines the number of malfunctions per square km of the network within one hour, was used to determine that the failure frequency of the conventional sewage system. The parameter varied between 0.265×10^{-4} [1/kmxh] to 3.17×10^4 [1/km $\times$ h]. These values were calculated on the basis of currently existing data on sewage subsystems.

The sewage network in Nowy Sącz revealed:

- collapsing of the stoneware and concrete pipelines
- leakage at flared connections of Vipro, concrete and stoneware pipes
- fat layers in the sewer (usually in the proximity of slaughterhouses and meat factories),
- poor sewer operation – solid wastes inside the sewer
- breakages of manholes
- breakages of concrete manhole covers
- rat infestations
- plant overgrowth in the sewers (tree roots)
- clogging of combine sewers with sand.

To determine the intensity of failures within Nowy Sącz sewer system, the study focused on those events only, which resulted in such effects as:

- exfiltration of sewage into the ground
- ground errosion
- subsidence.

The calculated singular intensity of failure was $\lambda = 2,8539 \times 10^{-4}$ [1/km $\times$ h].

4 RELIABILITY STUDY OF THE SEWAGE SUBSYSTEM DISCCUSION OF RESULTS

Malfunctions, depending on their location, can result in different types of damages. A malfunction can affect users of the system, users of neighboring systems, and/or the environment. With this in mind, the reliability of the sewage subsystem in Nowy Sącz was evaluated by using the tree failure method. This

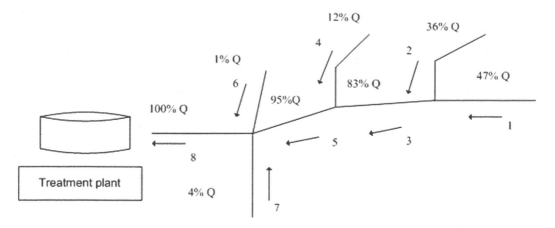

Figure 1. Simplified layout of the sewer network in the town of Nowy Sacz.

method provides an unequivocal evaluation of the system's reliability (Wieczysty & Kapcia 2001). The calculation of parameters is performed on the basis of the tree model, which represents (graphically and logically) various combinations of possible events that could happen during the system's operation and which could lead to the top (system failure).

Combinations of failures must be mutually exclusive. This condition is met when each entry of the tree failure means a failure of another element of the system.

In order to determine the reliability parameters, one must proceed from the tree structure model to a cross-section structural representation by cross-sections and failure paths. Isolating the minimal cross-sections from the collection of cross-sections, the parameters of system reliability, considering one- to three-element cross-sections, are calculated using the following relationships:

• system failure intensity Λ s:

$$\Lambda s = \sum_{i=1}^{M(1)} \lambda^{(1)}_i + \sum_{j=1}^{M(2)} \lambda_j^{(2)} + \sum_{l=1}^{M(3)} \lambda^{(3)}_l \qquad (1)$$

where: $\lambda_i^{(1)}$ = failure intensity of the one-element failure section i concerns individual elements; it is determined on the basis of statistical analysis, employing the data collected during running tests and employing the reliability parameter estimator formulae, $\lambda_j^{(2)}$ = failure intensity in the two-element failure section j, calculated from the formula: $\lambda_j^{(2)} = \lambda_{j1} \cdot \lambda_{j2} \cdot (Tn_{j1} + Tn_{j2})$; $\lambda_l^{(3)}$ – failure intensity in the three-element failure section l, calculated from the formula: $\lambda^{(3)}_l = \lambda_{l1} \cdot \lambda_{l2} \cdot \lambda_{l3}$ $(Tn_{l1} \cdot Tn_{l2} + Tn_{l2} \cdot Tn_{l3} + Tn_{l1} \cdot Tn_{l3})$
• average operating time of the Tps subsystem:

$$Tps = \frac{1}{\Lambda s} \qquad (2)$$

where: Λ_s = as defined above.
• average repair time Tns:

$$Tns = \frac{\sum_{i=1}^{M(1)} \lambda_i \cdot Tn_i + \sum_{j=1}^{M(2)} \lambda_{j1} \cdot \lambda_{j2} \cdot Tn_{j1} Tn_{j2} + \sum_{l=1}^{M(3)} \lambda_{l1} \cdot \lambda_{l2} \cdot \lambda_{l3} \cdot Tn_{l1} \cdot Tn_{l2} \cdot Tn_{l3}}{\Lambda s} \qquad (3)$$

• renewal intensity μs:

$$\mu s = \frac{1}{Tns} \qquad [1/h] \qquad (4)$$

• operation parameter K

$$K = \frac{Tps}{Tps + Tns} \qquad (5)$$

The starting point was to adopt a simplified scheme of the network (Figure 1). The basis for its construction was the layout of the sewage network of Nowy Sącz. In order to adapt this layout to the requirements of the method several simplifications have been introduced, without which the calculations would be too complicated and time-consuming. The analysis included only the main collectors, while their utilities were skipped. The calculations involved the case where pipeline failures occur in the lower nodes.

On the basis of the simplified scheme of the sewage network and the symbols used in the tree failure method, the tree scheme for the Nowy Sącz sewage system was prepared (Figure 2). The bottom of the diagram are base events, ie. failures of the basic elements of the system, in this case the collectors numbered from 1 to 8. The elements of the scheme referred to as "output" (1, 2, 3, 4) close a combination of the base events and correspond to a certain percentage of sewage that had not been collected due to system malfunction. The root node (top event; output 1) refers to an event when 100% of sewage has not been discharged to the treatment plant. The gates "And"

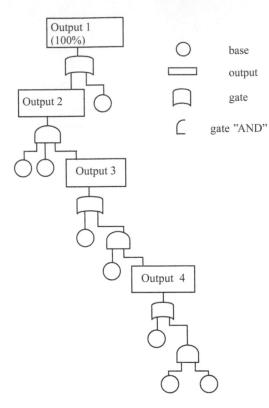

Output 1 (100%)

Output 2

Output 3

Output 4

○ base

▭ output

⊓ gate

⌐ gate "AND"

Figure 2.   Tree failure diagram.

and "Or" allow for creating a logical progression via connecting individual base events.

For the above tree diagram the following failure cross-sections have been established:

Top event: "ouput 1"
[output 2] [8]

Event: "output 2"
[7, 6, output 3]

Event: "output 3"
[7, 6, 5], [7, 6, 4, output 4]

Event: "output 4"
[7, 6, 4, 3], [7, 6, 4, 1, 2]

This way all sections have been described. However, taking into account the fact that the probability of occurrence of more than three failures at the same time is close to zero, it was decided that the final calculations will only consider the one-, two- and three-element failure cross-sections: $M(1) = 1$, $M(2) = 0$, $M(3) = 1$.

The basic values in our calculations are the failure intensity of individual sections (collectors) in the network and the times of their repair. These values were presented in Table 1. They were calculated based on the field tests and the reliability theory.

Table 1.   Reliability parameters of individual cross-sections of the studied network.

Section No.	Failure intensity λ [1/km × h]/	Section length L [km]	Failure intensity at section λ_1 [1/h]	Time of repair Tn [h]
1	2.8539×10^{-4}/	5.9	1.684×10^{-4}	39.54
2	2.8539×10^{-4}	5.1	1.455×10^{-4}	39.54
3	2.8539×10^{-4}	2.9	8.276×10^{-5}	39.54
4	2.8539×10^{-4}	6.0	1.712×10^{-4}	39.54
5	2.8539×10^{-4}	1.8	5.137×10^{-5}	39.54
6	2.8539×10^{-4}	3.6	1.027×10^{-4}	39.54
7	2.8539×10^{-4}	3.0	8.562×10^{-5}	39.54
8	2.8539×10^{-4}	1.7	4.852×10^{-5}	39.54

The calculated reliability parameters of the entire subsystem, determined by the above relationships, are as follows:

– failure intensity of the system Λs

$$\Lambda s = \sum_{i=1}^{M(1)} \lambda^{(1)}_i + \sum_{l=1}^{M(3)} \lambda^{(3)}_l = 4.852 \cdot 10^{-5} \quad [1/h]$$

– average running time of the subsystem, Tps:

$$Tps = \frac{1}{\Lambda s} = 20610 \quad [h]$$

– average time of system repair, Tns:

$$Tns = \frac{\sum_{i=1}^{M(1)} \lambda_i \cdot Tn_i + \sum_{l=1}^{M(3)} \lambda_{l1} \cdot \lambda_{l2} \cdot \lambda_{l3} \cdot Tn_{l1} \cdot Tn_{l2} \cdot Tn_{l3}}{\Lambda s} = 15$$

– renewal intensity μs:

$$\mu s = \frac{1}{Tns} = 0.00667 \quad [1/h]$$

– operation parameter K:

$$K = \frac{Tps}{Tps + Tns} = 0.999272$$

5   CONCLUSIONS

Following the field research on the sewage collection subsystem in the town of Nowy Sącz and the reliability analysis of its performance, the following can be concluded:

– most malfunctions of the studied subsystem occurred in the spring and fall
– in recent years, nearly 30% of the pipeline network was renovated or replaced, which has significantly improved their reliability
– the working reliability and safety of the system has also been considerably improved by systematic monitoring of the pipelines

194

- the calculated unit failure intensity − λ of the studied subsystem varied during the analysis from 1.4556×10^{-4} [1/km × h] to 8.562×10^{-5} [1/km × h], while the mean value was close to the above-mentioned upper value
- the application of the tree structure method in the evaluation of the level of reliability of the subsystem proved to be justified, since it made possible to establish more than one reliability parameter and allowed an unambiguous reliability evaluation.

REFERENCES

Алексеев, М,И. & Ермолин, Ю.А. 1996. Надежность канализационных сетей: цели, задачи и методология исследования. *Водоснабеинцеиние и санитарна техника – Haustechnik* 10: 2–5.

Dąbrowski, W. 2004. The Effect of Sewage Networks on the Environment. The Technical University of Krakow: Krakow (in Polish).

Denczew S. & Królikowski A. 2002. Introduction to Modern Exploitation of Water-supply and Sewage Systems. Arkady: Warsaw (in Polish).

Kapcia, J, Lubowiecka, T, Kabaciński, M.M. 2005. Main Problems of Exploitation of Sewage Systems in cities of the Southern Poland. *Gaz, Woda i Technika Sanitarna*: 39–41. (in Polish)

Wieczysty A. & Kapcia J. 2001. Analysis of the Reliability of Water-supply Systems with the Use of the Tree Failure Model. Monograph: Methods of Reliability Evaluation and Enhancement in Communal Water-supply Systems: 417–435. Published by The Committee on Environmental Engineering: Krakow (in Polish).

Environmental Engineering III – Pawłowski, Dudzińska & Pawłowski (eds)
© *2010 Taylor & Francis Group, London, ISBN 978-0-415-54882-3*

The potential of metal – complex dyes removal from wastewater the sorption method onto organic – matter rich substances

J. Kyzioł-Komosińska, C. Rosik-Dulewska & M. Pająk
Institute of Environmental Engineering, Polish Academy of Science, Zabrze, Poland

ABSTRACT: The objective of this study was to remove metal-complex dyes (Acid Blue 193 and Acid Black 194) from wastewater using organic-matter-rich substances (low-moor peats, brown coal and compost) in a batch adsorption experiment. The parameters in Langmuir and Freundlich isotherms were analysed using regression and non-regression models. The results demonstrated the potential of the investigated cost-effective sorbents for metal-complex dye removal over a wide range of initial concentrations. There was no influence of nitro groups on the effectiveness of the removal of dyes by organic sorbents. Langmuir isotherms were the best- fit of adsorption isotherm models for experimental data using non-linear regression.

Keywords: Metal-complex dyes, organic matter, adsorption isotherms, wastewater.

1 INTRODUCTION

Surface and groundwater are subject to European Union (EU) legislation mainly due to their role as sources of high-quality drinking water. Restrictive regulations concerning contamination of the aquatic environment in the EU Directives as well as an increasing trend to demand high quality water require more efficient and environmentally sound methods of water and wastewater treatment.

Dyes and pigments are an important group of organic pollutants in the aquatic environment. They are emitted into wastewater by different sectors, mainly from the dying, textile, cosmetic and paper industries. Even at very low concentrations, dyes can colour large water bodies. Due to their complex molecular structure and high substitution capacity, dyes are resistant to physical, chemical or biological decomposition. Moreover, the decomposition products may include small amounts of toxic or carcinogenic substances.

Dyes can be effectively removed from water and wastewater by adsorption on porous synthetic sorbents (e.g. activated carbon and ion-exchange resins). However, high production costs and problems related to recovery of the carbon used have resulted in a search for efficient and cost-effective natural and waste mineral and organogenic substances with a high capacity for dye adsorption from wastewater (Poots et al. 1976, McKay 1980, Namasivayam et al. 1996, Ho & McKay 1998, Robinson et al. 2002, Allen et al. 2004, Arami et al. 2005, Ozcan et al. 2006).

The principal method for describing the sorption phenomenon is an equation representing the relation between the concentration in water and a corresponding concentration in solid phase at equilibrium conditions, i.e. the sorption isotherm. Presently, the sorption parameters can be determined using mathematical models, laboratory experiments or *in situ*. Data from experiments are interpreted using a non-linear isotherm. The selection of the specific isotherm is determined by fitting a theoretical curve to empirical data; the coefficient of determination (R^2) is a measure of the fit (Ho et al. 2005, Vasanth & Sivanesan 2005).

The present analysis is aimed at studying the adsorption capacity of available cost-effective sorbents, rich in organic matter, for removal of metal-complex dyes (Acid Blue 193 and Acid Black 194) from synthetic wastewater using a batch method. The adsorbents investigated were brown coal, compost and two low-moor peats from the overburden of brown coal deposits. Linear and non-linear regressions were used to estimate the Freundlich and Langmuir coefficients and the best-fit of the theoretical isotherms to experimental data.

2 MATERIALS AND METHODS

The following organic-matter-rich substances were used as adsorbents in the experiment:

– brown coal originating from the Belchatow Brown Coal Mine (Central Poland),
– compost produced from shredded branches, leaves and grass in the Zabrze composting plant (southern Poland); ash content 69.53%,
– two low-moor peats from the overburden of brown coal deposits (central Poland): from Belchatow Brown Coal Mine, an alder forest peat with

Table 1. Physicochemical and chemical properties of sorbents.

Physicochemical properties

| Samples | Porosity n_0/n_e | Specific surface area (m²/g) | | pH (H₂O) | C(%$_{daf}$) | Fe$_{ox}$ (mg/kg) |
		External	Total			
Brown coal	0.42/0.41	9.98	267.85	5.75	66.79	1731
Alder peat	0.52/0.50	11.38	218.98	5.62	58.29	9713
Sedge peat	0.59/0.58	14.10	215.26	6.33	60.96	4913
Compost	0.49/0.48	14.54	119.13	8.04	45.68	830

Cation exchange capacity

| | CEC$_0$ cmol$_{(+)}$/kg | CEC$_t$ | Base saturation V (%) | Exchangeable ions (cmol(+)/kg) [% of total content] | | | | |
				Ca^{2+}	Mg^{2+}	Na$^+$	K$^+$	H$^+$
Brown coal	107	136	78.71	100.00 [56.22]	6.35 [79.38]	0.28 [73.23]	0.10 [29.38]	29.0
Alder peat	103	125	82.48	95.18 [58.31]	7.14 [43.26]	0.20 [9.26]	0.09 [1.92]	21.8
Sedge peat	114	117	97.36	104.20 [75.00]	8.91 [80.98]	0.80 [57.68]	0.23 [14.06]	3.04
Compost	35.9	34.5	96.10	10.3	4.22	7.12	12.86	2.56

decomposition rate 70% and ash content 20.88%; and from Konin Brown Coal Mine, a sedge peat of decomposition rate 30% and ash content 11.43%.

The physicochemical properties of the investigated adsorbents are given in Table 1. The material for experiment was dried at room temperature (23 ± 2°C), homogenized and sieved to particles smaller than 1 mm.

Major physical and physicochemical properties of the samples were determined. Total (n_o) and effective (n_e) porosity was determined by mercury porosimeter (Carlo Erba model 2000). BET(N_2) isotherm, was used to measure the external specific surface area and BET(H_2O) isotherm – to measure the total specific surface area (Fisons, Sorptomatic 1990). Sample pH was determined in deionised water (suspension ratio 1:10). Exchangeable Ca^{2+}, Mg^{2+}, K^+ and Na^+ were extracted with 1 M NH₄Ac at pH 7. The Ca^{2+} and Mg^{2+} ions were determined by atomic absorption spectrometry; and K^+ and Na^+ by flame photometry. The cation exchange capacity (CEC$_0$) was calculated from the content of exchangeable cations. Total cation exchange capacity (CEC$_t$) was determined by Ba^{2+} retention after percolation with a solution of 0.2 N BaCl₂–triethanolamine at pH 8.1 (Carpena et al. 1972). Total organic carbon was assessed in a CHN analyser (Carlo Erba) after pretreatment with HCl to eliminate carbonates (Navarro et al. 1991) and combustion at 1020°C. Free noncrystalline forms of Fe oxides (Fe$_{ox}$) were determined using acid ammonium oxalate method (Ross & Wang 1993).

Two dyes were used for research: Acid Blue 193 and Acid Black 194 produced by the Boruta-Kolor Plant in Zgierz (Poland), which is the leading domestic (and an internationally recognised) producer of synthetic organic dyes. The investigated dyes are metal-complex dyes 1:2; each containing two SO_2O groups and one atom of chromium. Additionally, Acid Black 194 contains two NO_2 groups which give higher molecular weight than Acid Blue 193. The metal-complex dye group is one of the most important groups of dyes in the textile dyeing industries. They are recommended for dyeing and printing cotton, polyamide fibres, natural silk and leather. Despite common use of these dyes for textile dyeing, there have been few studies on their removal from wastewater (Ozacar & Sengil 2005).

The key features of the dyes used are given in Table 2. For ease of identification, in addition to the common and systematic name, the CAS registry number of the dye from the ChemIDplus Lite database is provided. The database helps identify chemical compounds like dyes which may have different common names.

Sorption tests were carried out in static conditions using a batch method. The technique employed dye solutions of initial concentrations (C_0) in the range 1–1000 mg/dm³ for brown coal and compost; and 1–5000 mg/dm³ for peats. The ratio of solid phase (m):solution (V) was 1:20, and shaking time was 12 h. The c_0 and equilibrium metal concentrations (C_{eq}) in solutions were determined by UV/VIS spectrometry in a Varian Cary 50 Scan spectrometer, using a wavelength of 577 nm for Acid Blue 193 and 570 nm for Acid Black 194. The adsorbed amount of the dyes (S) was calculated from:

$$S = (C_0 - C_{eq}) \cdot V/m \text{ (mg/kg)} \qquad (1)$$

Additionally, pH was determined in equilibrium solutions.

Table 2. Characteristics of investigated dyes.

Name of dye	CAS	Systematic name	Molecular formula	Molecular weight	Structure	pH in water for 1 g/dm^3
ACID BLUE 193	12392-64-2	Chromate(3-), bis[3-hydroxy-4-((2-hydroxy-1-naphthalenyl)azo)-1-naphthalenesulfonato(3-)]-sodium hydrogen (1:2:1)	C40-H22-Cr-N4-O10-S2.H.2Na	881.8		6.70
ACID BLACK 194	57693-14-8	Chromate(3-), bis[3-(hydroxy6O)-4-[[2-(hydroxy6O)-1-naphthalenyl]azo-6N1]-7-nitro-1-naphthalenesulfonato(3-)]-trisodium	C40-H20-Cr-N6-O14-S2-2Na	970.7		5.11

The equilibrium sorption of Acid Blue 193 and Acid Black 194 onto organic matter samples was shown as S versus C_{eq} (Figure 1). Sorption parameters were estimated based on the linear forms of the Freundlich and Langmuir equations (Tor & Cengeloglu 2006, Vasanth & Sivanesan 2005). The R^2 for each system indicated the fit of the theoretical curve to experimental data. STATISTICA ver. 6.0 software was used in a nonlinear regression method to estimate the constants in the Freundlich and Langmuir equations.

3 RESULTS AND DISCUSSION

3.1 Properties of the investigated sorbents

The tested sorbents had high total porosity (n_o) in the range 0.42–0.59 and a very large total specific surface area, in the range 119.13–267.85 m^2/g. A small difference between total (n_o) and effective (n_e) porosity (d > 0.2 μm) indicated that practically all pores were actively involved in conveying water. Carbon content was 45.68–66.79% and decreased in the following order: brown coal > sedge peat ≥ alder peat > compost. Both carbon content and specific surface were highly correlated with CEC$_0$ (R^2 = 0.9088 and 0.9411, respectively).

The dominant cations in the biolithes' sorption complex were Ca^{2+} and Mg^{2+}; and K$^+$ and Ca^{2+} for compost (Table 1).

The investigated biolithe samples had acid or slightly acid reactions in water (pH 5.62–6.33) while compost was slightly basic (pH 8.04).

3.2 Sorption of dyes onto organic - matter – rich substances

The sorption isotherms of Acid Blue 193 and Acid Black 194 onto organic-matter-rich substances are presented in Figure 1.

Analysis of the course of the sorption process revealed that peats could effectively uptake even high amounts of both dyes. At the initial dye concentrations (C_0) of 1–1000 mg/dm^3, their sorption onto peats reached 98–99%. Based on these promising results, the C_0 range was extended to 5000 mg/dm^3. At the maximum initial concentration of dyes in solution, alder peat adsorbed 51 660 mg/kg of Acid Blue 193 and 59 180 mg/kg of Acid Black 194, i.e. 51.22 and 59.04%, respectively. Sedge peat showed higher sorption capacity than alder peat; at 5000 mg/dm^3, sedge peat adsorbed 59 240 mg/kg (59.04%) of Acid Blue 193 and 78 620 mg/kg (77.75%) of Acid Black 194.

Brown coal and compost had low sorption capacities for both dyes. At dye concentrations of 1000 mg/dm^3, brown coal adsorbed 11 100 mg/kg (54.53%) of Acid Blue 193 and 13 900 mg/kg (70.67%) of Acid Black 194. Compost showed the lowest sorption capacity of the tested sorbents, with uptake of Acid Blue 193 of 8 830 mg/kg (43.38%) and Acid Black 194 of 12 398 mg/kg (63.03%).

All organic-matter-rich substances had a higher sorption capacity for Acid Black 194 than for Acid Blue 193, from 12.9% adsorbed on sedge peat to 40.4% on compost. Recalculating the amounts of adsorbed dyes from mass units (mg/kg) to the amount of substances (mmol/kg) showed that the difference in adsorbed dyes was about 2 mmol/kg. Thus NO$_2$ groups in Acid Black 194 had no significant influence on the efficiency of dye removal from solution.

The pH of equilibrium solutions are presented in Figure 2, using a semi-logarithmic scale to make them more understandable. The pH of equilibrium solutions showed that the sorption of dyes by individual sorbents was practically at constant pH, due to the pH of both the dye and the sorbent suspension.

The main sorption centres of acid reaction in the tested biolithes are carboxyl and hydroxyl groups of phenols of fulvic and humic acids. They are responsible for the negative charge of the surface at the solution pH > pH$_{ZPC}$ = 3.5. Therefore they demonstrate high efficiency of cationic dye removal and a rather low sorption capacity of anionic dyes, which increases with

199

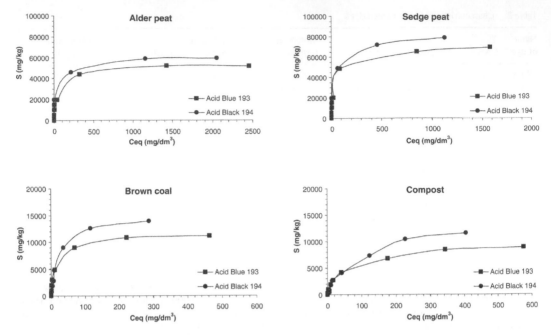

Figure 1. Sorption isotherms of Acid Blue 193 and Acid Black 194 onto organic-matter-rich substances (low-moor peats, brown coal and compost).

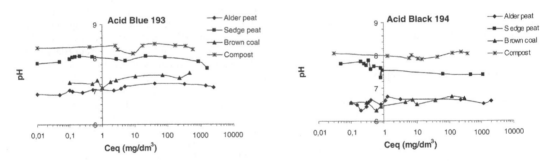

Figure 2. Changes of pH as a function of dyes concentration in equilibrium solution.

a decrease in pH of the dye solution (Ramakrishna & Viraraghavan 1997). Organogenic sorbents (particularly peats) contain amorphous iron oxides (Table 1), which are characterised by a positively charged surface at pH < pH_{ZPC} = 8.5.

Unsaturated surface groups (OH_2^+ or OH) play the role of active sorption centres. The uptake of the dyes is a result of the electrostatic force between positive charge of surface and negative charge of dye or of specific sorption which consists in the substitution of the OH surface group by an anion of dye (L^-). Thus the anionic dyes can be adsorbed in the reactions, respectively:

$FeOOH_2^+ + L <=> FeOOH_2L$, or
$FeOOH + L <=> FeOL + OH$, $(FeOOH)_2 + L^2 <=>$
$(FeO)_2L + 2 OH$.

The high removal efficiency by organogenic sorbents (especially peats) of the metal-complex dyes of anionic character, at pH in the range 6.33–8.05 (Figure 2), suggests that the hydrated iron oxides have an active role in sorption of these dyes.

3.3 Linear regression

The simplest method to estimate the sorption parameters is transformation of the isotherm equation to a linear form and then application of linear regression. In this paper the following equations were used to calculate the sorption parameters:

– Freundlich isotherm

$$S = K_F C_{eq}^n \qquad (2)$$

where C_{eq} (mg/dm^3) and S (mg/kg) = the liquid phase and solid phase concentrations of dye at equilibrium, respectively; K_F = the Freundlich constant (mg/dm^3) and n is the heterogeneity factor.

Table 3. Isotherm parameters obtained from the linear regression.

Freundlich isotherm

	ACID BLUE 193			ACID BLACK 194		
Sorbents	n	K_F	R^2	n	K_F	R^2
Alder peat	0.5833	2337	0.8315	0.5885	1804	0.6564
Sedge peat	0.8551	4121	0.8543	0.8593	23442	0.8805
Brown coal	0.6548	504.5	0.7931	0.7770	431.3	0.8605
Compost	0.5761	358.9	0.8728	0.7600	186.2	0.8718

Langmuir isotherm

	S_{max} (mg/kg)	Q (mg/kg)	K_L (dm^3/mg)	R^2	S_{max} (mg/kg)	Q (mg/kg)	K_L (dm^3/mg)	R^2
Alder peat	51660	53606	0.0399	0.9579	59180	60232	0.0144	0.9651
Sedge peat	69630	71428	0.0495	0.8987	78620	80909	0.0175	0.8968
Brown coal	11100	11628	0.0461	0.9682	13900	16667	0.0153	0.8415
Compost	8830	9259	0.0284	0.9525	12398	15873	0.0086	0.8926

– Langmuir isotherm

$$S = \frac{QK_LC_{eq}}{1+K_LC_{eq}} \tag{3}$$

where K_L (dm^3/mg) = the sorption equilibrium constant and Q (mg/kg) = the theoretical mono-layer saturation capacity.

These equations are well known and commonly used to describe sorption processes (Allen et al. 2004, Ho et al. 2005).

The constants n and K_F in the Freundlich equation were determined using the linear form of the following formula:

$$logS = logK_F + nlogC_{eq} \tag{4}$$

This provides a straight line with slope of n and an intercept of $logK_F$ when logS is plotted against $logC_{eq}$.

The constants Q and K_L in the Langmuir equation were determined based on the following equation:

$$\frac{C_{eq}}{S} = \frac{1}{Q}C_{eq} + \frac{1}{K_LQ} \tag{5}$$

A plot of C_{eq}/S versus C_{eq} should give a straight line of slope 1/Q and an intercept of $1/K_LQ$.

The estimated isotherm parameters of the dyes sorption together with the corresponding R^2 values are given in Table 3. The sorption parameters estimated according to equation 4 were the best-fit to the experimental data (i.e. highest R^2).

As determined in equation 4, the constant Q (reflecting monolayer capacity) was close to the maximum experimental sorption (S_{max}) of both dyes.

3.4 Non – linear regression

Application of linear regression is burdened with an error since, depending on the selected isotherm, the determined regression line does not minimise the total of:

$$\sum_i [S_i - K_FC_{eq}^n]^2 \text{ or } \sum_i [S_i - \frac{QK_LC_{eq}}{1+K_LC_{eq}}]^2 \tag{6}$$

The isotherm parameters best-fitting the data were acquired using non-linear regression analysis based on the classical least-squares method. The sorption parameters together with R^2 values are given in Table 4.

Higher R^2 values indicate that application of non-linear regression provided a better fit of curves to the experimental data. The Langmuir and Freundlich isotherms satisfactorily fitted the experimental data (Table 4). However, the Langmuir isotherms better fitted the adsorption data for both dyes than the Freundlich isotherms. Ozacar & Sengil (2005) had similar findings in studies of sorption of metal-complex dyes onto pine sawdust.

4 CONCLUSIONS

The present investigation confirmed the potential of low-moor peats, brown coal and compost to remove metal-complex dyes (Acid Blue 193 and Acid Black 194) from wastewater over a wide range of initial concentrations. They are efficient and cost-effective sorbents for dye removal from water. There was no influence of the NO_2 group on the amount of dye adsorbed. However, there was an active role of iron oxides in the dye sorption process. The best fitting of the isotherms to experimental data was by non-linear

Table 4. Isotherm parameters obtained by non-linear regression.

Freundlich isotherm

Sorbents	ACID BLUE 193			ACID BLACK 194		
	n	K_F	R^2	n	K_F	R^2
Alder peat	0.2834	7534	0.9629	0.2618	7180	0.9374
Sedge peat	0.2798	9366	0.9503	0.3166	10339	0.9578
Brown coal	0.3074	1092	0.9309	0.3872	1722	0.9732
Compost	0.3718	904.6	0.9741	0.5620	468.3	0.9891

Langmuir isotherm

	S_{max} (mg/kg)	Q (mg/kg)	K_L (dm³/mg)	R^2	S_{max} (mg/kg)	Q (mg/kg)	K_L (dm³/mg)	R^2
Alder peat	51660	54999	0.0162	0.9960	59180	60479	0.2014	0.9868
Sedge peat	69630	72112	0.0532	0.9923	78620	81765	0.1387	0.9780
Brown coal	11100	12081	0.0873	0.9896	13900	15319	0.0377	0.9954
Compost	8830	9278	0.0218	0.9932	12398	16617	0.0073	0.9995

regression. The Langmuir isotherm best described the adsorption of metal-complex dyes onto organic-matter-rich substances such as low-moor peats, brown coal and compost.

ACKNOWLEDGEMENTS

This study was supported by project No. N523 3509 33 from the Ministry of Science and Higher Education (Poland).

REFERENCES

Allen, S.J., Mckay, G. & Porter, J.F. 2004. Adsorption isotherm models for basic dye adsorption by peat in single and binary component system. *Journal of Colloid and Interface Science* 280: 322–333.

Arami, M., Limaee, N.Y., Mahmoodi, N.M. & Tabrizi, N.S. 2005. Removal of dyes from colored textile wastewater by orange peel adsorbent: equilibrium and kinetic studies. *Journal of Colloid and Interface Science* 288: 371–376.

Carpena, O., Lax, A. & Vahtras, K. 1972. Determination of exchangeable cations in calcareous soils. *Soil Science* 113: 194–199.

ChemIDplus Lite database. http://chem.sis.nlm.nih.gov/chemidplus/chemidlite.jsp.

Ho, Y.S., Chiu, W.T. & Wang, C.C. 2005. Regression analysis for the sorption isotherms of basic dyes on sugarcane dust. *Bioresource Technology* 96: 1285–1291.

Ho, Y.S. & Mckay, G. 1998. Sorption of dye from aqueous solution by peat. *Chemical Engineering Journal* 70: 115–124.

Mckay, G. 1980. Color removal by adsorption. *American Dyestuff Reporter* 69(3): 44–66.

Namasivayam, C., Muniasamy, N., Gayatri, K., Rani, M. & Ranganathan, K. 1996. Removal of dyes from aqueous solutions by cellulosic waste orange peel. *Bioresource Technology* 57: 37–43.

Navarro, A.F., Cegarra, J., Roig, A. & Bernal, M.P. 1991. An automatic microanalysis method for the determination of organic carbon in wastes. *Soil Science and Plant Analysis* 22: 2137–2144.

Ozcan, A., Oncu, E.M. & Ozcan, A.S. 2006. Kinetics, isotherm and thermodynamic studies of adsorption of Acid Blue 193 from aqueous solutions onto natural sepiolite. *Colloids and Surfaces A* 277: 90–97.

Ozacar, M. & Sengil, I.A. 2005. Adsorption of metal complex dyes from aqueous solutions by pine sawdust. *Bioresource Technology* 96: 791–795.

Poots, V.J.P., McKay, G. & Healy, J.J. 1976. The removal of acid dye from effluent using natural adsorbents – I Peat. *Water Research* 10: 1061–1066.

Ramakrishna, K. R. & Viraraghavan, T. 1997. Dye removal using low cost adsorbents. *Water Science and Technology* 36(2-3): 189–196.

Robinson, T., Chandran, B. & Nigam, P. 2002. Removal of dyes from an artificial textile dye effluent by two agricultural waste residues, corncob and barley husk. *Environmental International* 28: 29–33.

Ross, G.J. & Wang, C. 1993. Extractable Al, Fe, Mn and Si. In: Carter M.R. (ed.), *Soil sampling and methods of analysis. Canadian Society of Soil Science:* 239–246. Levis Publishers. Boca Raton. FL.

Tor A. & Cengeloglu Y. 2006. Removal of congo red from aqueous solution by adsorption onto acid activated red mud. *Journal of Hazardous Materials B* 138: 409–415.

Vasanth K. K. & Sivanesan S. 2005. Prediction of optimum sorption isotherm: Comparison of linear and non-linear method. *Journal of Hazardous Materials B* 126: 198–201.

Environmental Engineering III – Pawłowski, Dudzińska & Pawłowski (eds)
© 2010 Taylor & Francis Group, London, ISBN 978-0-415-54882-3

Listeria monocytogenes and chemical pollutants migration with landfill leachates

A. Kulig, A. Grabińska-Łoniewska, E. Pajor & M. Szyłak-Szydłowski
Faculty of Environmental Engineering, Warsaw University of Technology, Warsaw, Poland

ABSTRACT: Field and laboratory research identified the migration of pathogenic and potentially pathogenic microorganisms and chemical contaminants, contained in leachates from municipal solid waste stored at a waste dump without subsoil insulation, into the soil and water environment. The research determined physicochemical, chemical and microbiological parameters, including *Listeria monocytogenes* titre, in surface water and ground-water collected at various distances from the waste dump. The most common microbiological contaminant in groundwater samples was *Listeria monocytogenes*. Of the remaining bacteria in the water samples, a group of heterotrophic psychrophilic bacteria, commonly recognised as indicators of water pollution with organic compounds, was most dominant.

Keywords: Bacteria, microscopic fungi, heavy metals, landfills, leachates, *Listeria monocytogenes*, organic matter, pH, salinity, solid waste.

1 INTRODUCTION

Municipal solid waste storage facilities are a considerable potential source of chemical and microbiological contaminant emissions into the atmosphere, soil and groundwater. Research into the influence of municipal solid waste dumps on the degree of chemical contamination of the air was conducted in Germany by Jager & Kuchta 1992, Pőhle et al. 1993, and Fricke & Műller 1994. In Poland, comprehensive research into the impact of municipal solid waste dumps on the environmental medium (the air, soil and groundwater) was started in the second half of the 1970s. In the 1980s and 1990s the research was continued, taking into consideration both the need to improve research methods, and advances in waste storage technologies. The main goal of the research was to find indicators to describe the impact of the waste dumps on the air, identify the levels of atmospheric contamination in the vicinity of the waste dumps, and determine the range of the waste dump influence. All three aforementioned goals have become even more important since the introduction of the environmental impact assessment system in Poland (Kulig 1995).

In the past two decades, the problem of limiting the unfavourable impact of municipal solid waste dumps on the quality of groundwater was analysed mainly in terms of determining the amount of leachates produced (Bengtsson et al. 1994, Freund 1992), their physicochemical profile (Murray & Beck 1990, Christensen et al. 1998, Paxéus 2000), and developing leachate treatment methods (Imai et al. 1993, Chiang et al. 1995, Zaloum & Abbott 1997, Kim et al. 1997,

Amokrane et al. 1997, Welander et al. 1997, 1998, Bae et al. 1999, Lin & Chang 2000, Zamora et al. 2000).

Research conducted by Grabińska-Łoniewska et al. (2007) showed that the main sanitary and epidemiological hazard was posed by *Listeria monocytogenes* bacteria present in refluxes (titre within the range of 10^{-1} to 10^{-5}). The bacteria are classified as relative anaerobes; they are not acid-resisting and do not create capsules and endospores. The optimum pH for the growth of *Listeria monocytogenes* is neutral or slightly alkaline and the optimum temperature range is 20 to 37°C. However, the bacteria can also grow and reproduce at 2 to 4°C in conditions where other bacteria are inactivated (acidic environment, high salinity) (Ghandi & Chikindas 2007). In homeothermic animals and humans, the bacteria cause listeriosis, which is an epizootic disease because wild rodents and birds, and domestic animals are natural reservoirs of the bacteria – the infection dose is within the range of 10^2 to 10^9 colony-forming units (cfu) per 1 g of food (Grabińska-Łoniewska & Siński 2009).

At present, *Listeria monocytogenes* bacteria are considered widespread in the environment (surface water, soil, plants) and wastewater. Land irrigation with wastewater containing *L. monocytogenes* causes the microorganism to spread, leading to its presence in plants and, consequently, in the food products of animal origin (e.g. cheeses, milk or meat). The common occurrence and long survival period of *L. monocytogenes* in plants and soil point to the fact that the germ is not only spread by infected specimens but is also a saprophyte capable of growing in natural biotopes (Holt 2004). There are data saying that it can survive

295 days in the soil or even two years in dry faeces (Libudzisz et al. 2007).

Estimation of the scale of pathogenic organism migration from waste dumps to groundwater, and development of a strategy to eliminate this threat, are open issues of extreme importance from the environmental protection point of view. The morphological contents of the solid waste stored at waste dumps in Poland (the content of the organic waste, including waste food is 35%, and the share of the biodegradable fraction is 57%) make it possible to assume that the waste forms an adequate substrate for the reproduction of saprophytic microorganisms and transfer of pathogenic microorganisms brought in with waste and droppings of animals existing on or visiting the waste dump area (annelida, insects, birds and rodents).

2 OBJECTIVE AND SCOPE OF THE RESEARCH

The objective of the field and laboratory research was to identify the migration of pathogenic and potentially pathogenic microorganisms and chemical pollutants, contained in the leachates from municipal solid waste stored at a waste dump without subsoil insulation, into the soil and water environment.

The scope of the physicochemical and chemical examination included the determination of the following parameters in surface water and groundwater collected at various distances from the waste dump: pH, electrolytic conductivity, total organic carbon (TOC), five-day biological oxygen demand (BOD_5), chemical oxygen demand (COD), metal contents (Cu, Zn, Pb, Cd, Cr, Hg, Na, P, Li) and sulphate and chloride contents. The water samples were characterised in microbiological terms, taking into consideration: *Listeria monocytogenes* titre, the overall numbers of heterotrophic psychrophilic bacteria (present at 20°C), mesophilic bacteria (present at 37°C), spore-forming bacteria (present at 26°C) and microscopic fungi, the most probable number of thermotolerant coliform bacteria (present at 44°C), and *C. perfringens* titre.

The scope of the physicochemical, chemical and microbiological examination of water in the environment was determined on the basis of the analysed leachate examination results (Grabińska-Łoniewska et al. 2007).

3 MATERIALS AND METHODS

3.1 Location of sampling points and schedule of collecting the samples for examiantion

The research focused on a municipal solid waste dump without subsoil insulation located in Lipiny Stare near Wołomin. Since 1974, the waste dump (its area is about 4.5 ha) has received about 5.500.000 cubic m (about 1.300.000 Mg) of solid waste from: Wołomin, Zielonka, Marki, Ząbki, Kobyłka, Radzymin and Warsaw. This is municipal waste (i.e. that other than neutral or hazardous waste) with a high biodegradable organic matter content. In the 1980s and 1990s, the waste dump was also used for storing sewage sludge. The solid waste is stored in beds in layers 1.5 to 2 m thick and covered with insulating layers 0.15 m thick.

In the vicinity of waste dump no. 8, research locations were selected (Figure 1); their profile is shown in Table 1.

Altogether in the period of April to November 2005, eight series of field and laboratory research were conducted. The field research schedule can be found in Table 2. In assessing the impact of the examined waste dump on the soil and water environment, the results of the physicochemical, chemical and biological examination of leachates collected from the band around the waste dump during the preliminary examination were also taken into consideration (Grabińska-Łoniewska et al. 2005). Samples of the leachate seeping from the waste dump without subsoil insulation located in Lipiny Stare near Wołomin were collected from the band ditch going along the northern slope of the waste fill. In rainless periods, the ditch collects leaks seeping from under the waste dump base (w.d.b.).

For physicochemical and chemical determinations, samples of natural leachates, surface water and groundwater were collected. Physicochemical, chemical and biological examination of the water was carried out in line with the methodologies applied to the examination of leachates (Grabińska-Łoniewska et al. 2007).

3.2 Physicochemical, chemical and biological examiantion

Reaction (pH value) was determined according to the Polish standard PN-90/C-04540/01; the electrolytic conductivity in µS/cm was determined according to PN-77/C-04542; the total organic carbon (TOC) in mg C/l was determined according to PN-EN 1484; the copper content in mg Cu/l was determined according to PN-92/C-04570/01; the zinc, lead and cadmium content in mg/l and the (total) chromium content in mg Cr/l were determined according to PN-87/C-04570/08; the mercury content in µg Hg/l was determined according to PN-82/C-04570/03; the sodium and potassium content in mg/l was determined according to PN-88/C-04953; the lithium content in mg Li/l was determined according to PN-84/C-04575; BOD_5 in mg O_2/l was determined using Sapromat; COD in mg O_2/l was determined according to PN-74/C-0457803; the sulphate content in mg SO_4^{2-}/l was determined according to PN-79/C-04566 (HACH methodology); and the chloride content in mg Cl^-/l was determined according to PN-ISO 9297. Detailed characteristics of the physicochemical and chemical methods employed in the research were described in a paper by Grabińska-Łoniewska et al. (2007).

The preliminary research showed that the leachates from municipal solid waste dumps were not

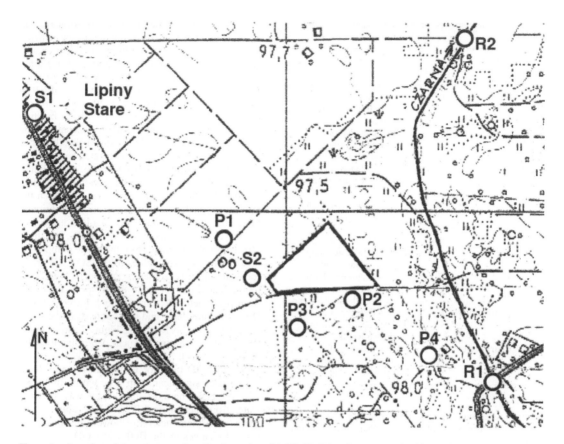

Figure 1. Location of the water sample collection points: P1, P2, P3, P4 – piezometers, S1, S2 – wells, R1, R2 – points on Czarna River.

contaminated by pathogenic bacteria *Campylobacter* sp., *Yersinia* sp. or *Ascaris lumbricoides* helminths.

A methodology recommended by Merck, described in detail by Grabińska-Łoniewska et al. (2005), was employed to determine the *Listeria* genus bacteria. The bacteria were determined in natural samples and the samples concentrated from a volume of 10 and 100 ml using the membrane filtration method. At the preliminary stage, a selective cultivation of the bacteria was conducted by means of a two-phase D.G.Al. method on a twofold diluted Fraser bouillon medium with incubation at 30°C for the period of 18 to 24 hours, followed by a cultivation on an undiluted Fraser bouillon medium with incubation at 37°C for the period of another 18 to 24 hours. The cultures showing some growth and blackening of the medium were inoculated on a selective Palcam agar medium and incubated at 37°C for the period of 24 hours. The grey and green colonies grown on the medium, surrounded by the medium blackening zone, were subject to a confirmation examination taking into consideration the cell morphology (Gram-positive rods), the ability to produce catalase and oxidase, and other biochemical qualities by means of the API-*Listeria* test.

Other microbiological determinations were conducted using natural samples in accordance with the methodology given by Grabińska-Łoniewska et al. (2005) and in the handbook by Grabińska-Łoniewska et al. (1999).

The quantities of psychrophilic, mesophilic and spore-forming bacteria, and the microscopic fungi were shown converted into cfu/ml of leachate or water; the quantities of thermotolerant coliform bacteria were shown as the most probable number (MPN) per 100 ml of leachate or water; and the quantities of *Listeria monocytogenes* and *C. perfringens* were shown as titre. The taxonomic affiliation of the microscopic fungi was determined in accordance with the methodology described by Grabińska-Łoniewska et al. (2007).

4 RESULTS AND DISCUSSION

4.1 *Physicochemical and chemical examiantion*

The results of the physicochemical and chemical examination are summarised in Table 3 and are shown in Figures 2 to 5.

The research showed that substantial variations are typical for both inorganic and organic water pollution indicators. Local underlying conditions may be the reason behind those variations but, first and foremost, the leachates seeping from under the solid waste

Table 1. Characteristics of the water sample collection points.

Item	Type of location	Marking	Location characteristics
1.	Piezometer	P_1	At a distance of about 25 m from the waste dump base
2.	Piezometer	P_2	In the ditch around the waste dump (the so-called 'band')
3.	Piezometer	P_3	At a distance of 20 m from the waste dump base
4.	Piezometer	P_4	At a distance of 30 m from the waste dump base
5.	Well	S_1	At a distance of 150 m from the waste dump base, private property
6.	Well	S_2	At a distance of 10 m from the waste dump base, by the waste dump
7.	Czarna river	R_1	At a distance of about 70 m upstream from the waste dump base
8.	Czarna river	R_2	At a distance of about 100 m downstream from the waste dump base

Table 2. Schedule of the field research around the municipal solid waste dump in Lipiny Stare near Wołomin in 2005.

Item Series no.	Water sample collection date	Location of the water sample collection points							
		P_1	P_2	P_3	P_4	S_1	S_2	R_1	R_2
1.	04/04		+	+	+				
2.	26/04	+	+	+					
3.	16/05	+	+	+					
4.	14/06	+	+	+					
5.	11/07		+	+		+			
6.	29/08		+	+			+		
7.	24/10				+			+	+
8.	21/11				+			+	+

dump have a decisive influence on the level of some variations. The variability of the examined parameters observed in the period of 4 April to 21 November 2005 showed the following dynamics:

- pH of all the examined water samples varied within the range of 6.3 to 7.7 pH, showing the highest stability at piezometers P_1 and P_4 (Figure 2a). In the remaining waters, apart from a fluctuation in springtime (26 April 2005), pH showed stable values too. However it can be pointed out that the water from piezometer P_2, located closest to the waste dump, shows the highest reaction value when compared with all other examined locations; it may indicate that the leachates from under the waste dump tend to

raise the reaction of the groundwater. An increased water reaction level at piezometer P_2 is justified by the results of the preliminary chemical examination of the leachate samples collected in Lipiny Stare in the period of March to May and August 2004; the samples showed the pH values of 6.9 to 7.8 (Grabińska-Łoniewska et al. 2005);

- Electrolytic conductivity is characterised by high variability caused by the impact of leachate on the groundwater. Figure 2b clearly shows that the water from piezometers P_2 and P_3, located on the outflow of groundwater from under the waste dump, showed substantially increased conductivity when compared with other water samples for which the indicator reflected natural background conditions;
- Sulphates and chlorides show a clear impact of the waste dump at locations P_2 and P_3, i.e. those affected by the leachate from the waste dump (Figures 2c and 2d). The values of the SO_4^{2-}/Cl^- ratio for water from piezometers P_2 and P_3 were very highly stable (Table 3). In this respect, the water from piezometer P_1 showed much lower stability. However, both SO_4^{2-} and Cl^- contents in the water were incomparably lower, see Figure 2c and 2d (and the relative errors in chemical determinations are probably bigger);
- BOD_5 to COD ratio for individual research locations varied significantly in time within the range of 0 to 0.833. The most stable value was shown in water from piezometers P_2 and P_3 (Table 3). The water from piezometer P_1 showed the most substantial changes in the BOD_5 to COD ratio but it was caused probably by low absolute values of both BOD_5 and COD and, consequently, bigger relative errors in individual determinations (Figures 3a and 3b). Such a justification is confirmed by the results of the total organic carbon examination (TOC), see Table 3;
- In general, metals also appeared at piezometers P_2 and P_3 in larger amounts than in other groundwater (Figures 4 and 5). Although the gain in the heavy metal concentrations (Cu, Zn, Pb, Cd, Cr_{tot} and Hg) (Figure 4) was not as distinct as that in chlorides, sulphates, conductivity, BOD_5 or COD, nevertheless it was significant enough to indicate that the heavy metals present in the stored solid waste are transmitted by leachate into groundwater. In the case of alkali metals, the influence of leachates was recorded mainly at location P_1, but the influence was very clear (Figure 5).

It is clear that at almost all the reception points, the biggest changes in the discussed parameters occurred in springtime: from April to May (sometimes to June).

4.2 Biological examination

The results of the biological examination can be found in Tables 4 and 5.

The smallest changes in the *Listeria monocytogenes* bacteria titre, indicating high contamination with the bacteria, were found in the groundwater from

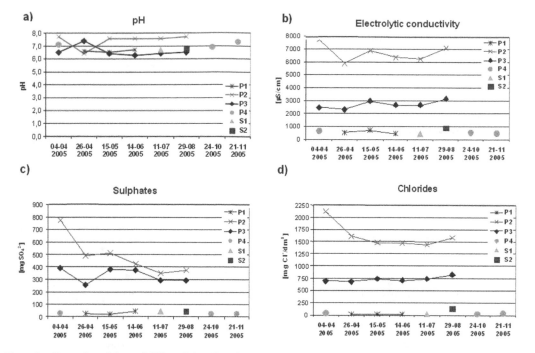

Figure 2. Dynamics of the variability of the pH and electrolytic conductivity, and sulphate and chloride concentrations in groundwater collected from piezometers and wells located around the municipal solid waste dump in Lipiny Stare near Wołomin (the period of 04/04/05 to 21/11/05) P1, P2, P3, P4 – piezometers, S1, S2 – wells, R1, R2 – points on Czarna River.

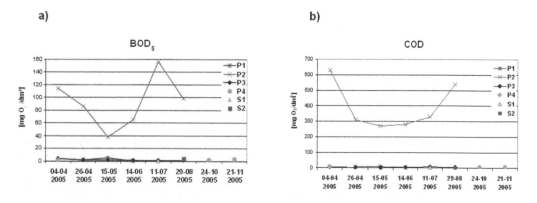

Figure 3. Dynamics of the variability of BOD₅ and COD in groundwater collected from piezometers and wells located around the municipal solid waste dump in Lipiny Stare near Wołomin (the period of 04/04/05 to 21/11/05). P1, P2, P3, P4 – piezometers, S1, S2 – wells, R1, R2 – points on Czarna River.

piezometers P_1, P_2 and P_3. Moreover, substantial quantities of the bacteria were discovered in the water from a well situated on a private property at a distance of 150 m from the waste dump base (S_1) and the water from Czarna River upstream of the waste dump (R_1).

The microbiological examination of the remaining bacteria showed that they were represented in greatest numbers in the analysed water by a group of heterotrophic psychrophilic bacteria, commonly recognised as indicators of water polluted with organic compounds. There are many species, mainly pigmented

ones, represented in this bacteria group, showing opportunistic pathogen characteristics. The range of the groundwater contaminated with the these bacteria was delimited by the locations of piezometers: P_1 situated about 25 m from the waste dump base (w.d.b.), P_2 (the so-called 'band' around the waste dump), P_3 at a distance of 20 m from w.d.b., P_4 at a distance of 30 m from w.d.b.; and the water collected from a bored well (S_2) situated by the waste dump. The Czarna River was moderately contaminated with the bacteria; similar contamination levels were found in samples

207

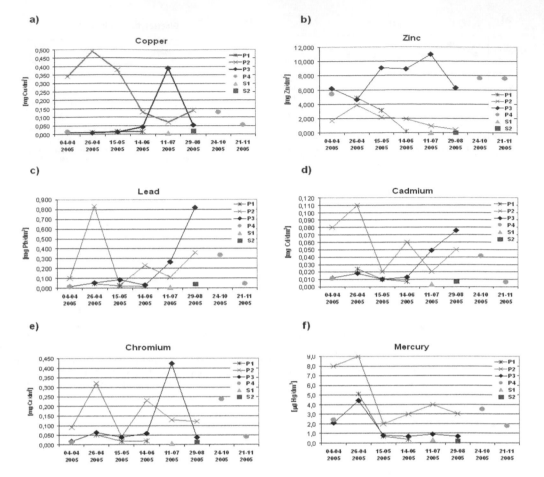

a) Copper

b) Zinc

c) Lead

d) Cadmium

e) Chromium

f) Mercury

Figure 4. Dynamics of the variability of metal concentrations (copper, zinc, lead, cadmium, total chromium and mercury) in groundwater collected from piezometers and wells located around the municipal solid waste dump in Lipiny Stare near Wołomin (the period of 04/04/05 to 21/11/05). P1, P2, P3, P4 – piezometers, S1, S2 – wells, R1, R2 – points on Czarna River.

collected at a distance of 50 to 70 m upstream of the waste dump (R_1) and about 100 m downstream of the waste dump (R_2).

Moreover, the water from piezometer P_2 was characterised by the largest number of mesophilic bacteria which may indicate that the contaminants originating from faeces penetrate the groundwater. Much lower bacterial contamination levels were found in the water from piezometers P_3 and P_4, and the bored well (S_2) situated in the vicinity of the waste dump. Small quantities of the bacteria representing this group were typical for the water collected from piezometer P_1 and the Czarna River at both research locations.

The range of the groundwater contaminated with spore-forming bacteria around the examined waste dump was delimited by the locations of piezometers: P_1, P_2 and P_3. Small quantities of the bacteria, recognised as indicators of the degree of organic compound decomposition, were found in the water collected from piezometer P_4, in the water from the wells and in Czarna River.

The mycological examination shows that the migration of the microscopic fungi into the groundwater took place only in relation to the aquiferous layers which were in direct contact with the waste dump (piezometers P_1 and P_2). Quite substantial quantities of the microorganisms found in the groundwater collected from piezometer P_4 can be considered the natural mycoflora of the soil because the fungi species found there differed from those discovered in the leachates. The foregoing conclusion is further supported by the fact that similar quantities of the microscopic fungi were found in the Czarna River and there were no pathogenic or potentially pathogenic species (usually found in leachates) in those biotopes (Table 5).

Moreover, in the examined area around the waste dump, high levels of contamination of the groundwater collected by piezometer P_2 with thermotolerant coliform bacteria (TCFB) and quite substantial contamination with *C. perfringens* was found. The quantity of the bacteria at other groundwater sampling locations and in the water collected from the wells S_1

a)

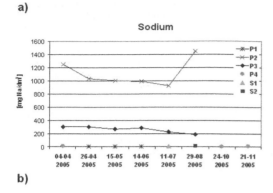

b)

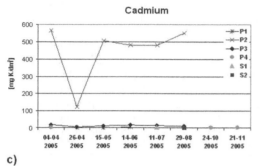

c)

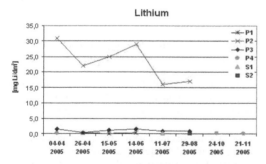

Figure 5. Dynamics of the variability of alkali metal concentrations (sodium, potassium and lithium) in groundwater collected from piezometers and wells located around the municipal solid waste dump in Lipiny Stare near Wołomin (the period of 04/04/05 to 21/11/05) P1, P2, P3, P4 – piezometers, S1, S2 – wells, R1, R2 – points on Czarna River.

and S_2 was small. At both research locations (R_1 and R_2) the Czarna River showed low contamination with *C. perfringens* and quite high contamination with thermotolerant coliform bacteria. Due to the small quantity of TCFB in the groundwater collected by piezometers P_1, P_3 and P_4 located closest to the waste dump base it can be presumed that the bacteria originated from farm facilities situated in the drainage basin of the Czarna River rather than from the waste dump.

The research results obtained, due to their specificity, are not correlated directly with the data found in related publications. The research conducted by Karwaczyńska et al. (2005) showed that the analysed leachates coming from an unsealed municipal solid

waste dump were in complete contact with groundwater in the Turonian level (the water-table 1.5 to 2.5 m below the ground level) as demonstrated by significant statistical differences between the water on the inflow and underground runoff in terms of the chemical oxygen demand, conductivity, ammonium ions, chlorides, and the quantity of psychrophilic and mesophilic bacteria; and partial contact with groundwater in the Cenomanian level (15 to 17 m below the ground level) in terms of conductivity, ammonium ions, chlorides and bacteriological indicators. The authors determined that the contamination of groundwater occurred up to around 270 m from the waste dump downstream for the Turonian level and to around 800 m for the Cenomanian level.

5 CONCLUSIONS

From the analysis of the results concerning metal concentrations in the groundwater around the municipal solid waste dump in Lipiny Stare near Wołomin, the following conclusions may be reached:

- The water in the wells S_1 and S_2 is characterised by metal concentrations considerably (about tenfold) lower than the levels permissible in C areas. In this respect, the wells can be used as reference points (defining the geochemical background).
- In piezometer P_1, levels exceeded permissible values for zinc (samples collected on 26/04/05 and 16/05/05), cadmium (samples collected on 26/04/05) and mercury (26/04/05). The examination carried out on 14/06/05 found no levels exceeding permissible values.
- In piezometer P_2, all six samples are characterised by metal concentration levels exceeding the permissible values (especially in relation to cadmium and mercury). In the examination carried out on 26/04/05 the foregoing conclusion applies to all six metals (Cu, Zn, Pb, Cd, Cr and Hg); in the examination of 14/06/05 it applies to five metals; in the examinations of 04/04/05 and 16/05/05 it applies to four metals and in the examinations of 11/07/05 and 29/08/05 it applies to three metals.
- All the six samples collected from piezometer P_3 are contaminated with zinc above the level permissible under the relevant standard (7.7 mg Zn/dm^3 on average) and the sample of 11/07/05 was contaminated to a particularly large extent (as compared with the five metals).
- Similarly, the water from piezometer P_4 is contaminated with zinc (6.9 mg Zn/dm^3 on average); especially high contamination was found in the sample of 24/10/05.
- The samples collected on 04/04/05 show an increased level of zinc and mercury concentrations in water from piezometers P_2, P_3 and P_4.

From the analysis of the numerical force of specific microorganism groups (used as indicators of environmental pollution by pathogenic or opportunistic

Table 3. Summarised results of the physicochemical examination of groundwater collected from piezometers, and the water from the wells and the Czarna River situated around the municipal solid waste dump in Lipiny Stare near Wołomin.

Item	Location of the water sample collection points				Characteristics		
					SO_4^{2-}/Cl^- ratio (−)	BOD_5/COD ratio (−)	Total organic carbon (mg C/dm^3)
1.	S_1 Well about 150 m from the w.d.b.				5.000	0	Not determined
2.	P_1 about 25 m from the w.d.b.		range	1.105 − 2.047	0 − 0.694	9.0 − 13.6	
			average	1.489	0.324	11.3	
3.	S_2 Well about 10 m from the w.d.b.				0.355	0.714	4.2
4.	P_2 Ditch around the waste dump (the so-called 'band')		range	0.237 − 0.365	0.141 − 0.472	329 − 744	
			average	0.297	0.247	465	
5.	P_3 about 20 m from the w.d.b.		range	0.352 − 0.565	0.135 − 0.833	7.4 − 19.0	
			average	0.457	0.394	11.6	
6.	P_4 about 30 m from the w.d.b.		range	0.923 − 1.438	0.278 − 0.417	13.9 - 20.0	
			average	1.148	0.330	17.3	
7.	Czarna River	R_1 about 70 m from the w.d.b.	range	0.880 − 1.143	0.357 − 0.405	12.0 − 15.7	
			average	1.012	0.381	13.8	
		R_2 about 100 m from the w.d.b.	range	0.696 − 1.231	0.328 − 0.400	11.5 − 15.3	
			average	0.964	0.728	13.4	

Abbreviations used: w.d.b. − waste dump base

Table 4. Comparison of the taxonomy of mycoflora in leachates, in groundwater collected from piezometer P_4, and in the Czarna River water.

Sample type	Taxonomy of mycoflora
Leachates	*Aspergillus fumigatus* Fres. *Penicillium expansum* Link et Gray *Penicillium janthinellum* Biourge *Penicillium verrucosum* Dierckx *Trichoderma viride* Pers et Gray *Diplodia mutila* (Fr.) Mont.
Groundwater from piezometer P_4	*Paeciliomyces lilacinus* *Scopulariopsis candida* (Gueguen) Vuillemin *Cladosporium resinae* (Lindau) de Vries
Czarna River, location: R_1	*Cladosporium resinae* (Lindau) de Vries *Penicillium claviforme* Bain *Penicillium frequetans* *Rhodotorula glutinis* (Fres.) Cohn
R_2	*Cladosporium cladosporoides* (Fres.) de Vries *Cladosporium resinae* (Lindau) de Vries *Phialophora malorum* (Kiddet Beaum.) Mc Collock *Phoma herbarum* Westend *Phodotorula glutinis* (Fres.) Cohn

pathogenic organisms) found in groundwater samples collected from piezometers located at various distances from the waste dump base, it can be concluded that, out of all the examined microorganism groups, *Listeria monocytogenes* bacteria was the most dominant contaminant of the area around the waste dump.

The fact that bacteria were present in the water collected from a well located on a private property and in the Czarna River water may indicate that both the waste dump and the farm facilities situated in the vicinity are the source of the bacteria.

The results of the microbiological examination may indicate that the municipal solid waste dump in Lipiny Stare near Wołomin has the biggest negative impact on groundwater through increasing the numerical force of heterotrophic psychrophilic bacteria in that water. The range of the bacteria spread covers the entire waste dump area, delimited by the locations of piezometers: P_1, P_2, P_3 and P_4. The range of the impact was considerably smaller in the case of mesophilic bacteria (piezometers P_1, P_3 and P_4), spore-forming bacteria (piezometers P_1, P_2 and P_3) and microscopic fungi (piezometers P_1 and P_2). It is important to point out the significant contamination of the well situated at a distance of 10 m from the waste dump base (S_2) by the aforementioned bacteria. The water in the well situated 150 m away from the waste dump base (S_1) was contaminated to a lesser extent but the values of relevant indicators excluded the use of the water for household purposes.

Table 5. Summarised breakdown of the results produced by the biological examination of groundwater collected from piezometers, and the water from the wells and the Czarna River situated around the municipal solid waste dump in Lipiny Stare near Wołomin.

Determination / Research location	S1 well 150 m from w.d.b. average	P1 about 25 m from w.d.b. range	P1 average	S2 well 10 m from w.d.b. average	P3 forest 20 m from w.d.b. range	P3 average	P2 ditch around the waste dump (the so-called 'band') range	P2 average	P4 heap 30 m from w.d.b. range	P4 average	Czarna River R1 70 m from w.d.b. range	R1 average	R2 100 m from w.d.b. range	R2 average
Listeria monocytogenes Titre	10^{-1}	$>1000-10^{-2}$	–	100	$>1000-10^{-2}$	–	$10-10^{-4}$	–	$>1000-100$	–	$1000-10^{-2}$	–	$10-1$	–
Overall number of bacteria [cfu/ml]: Psychrophilic bacteria temp. 20°C/72 hours	338	$219*10^2-475*10^5$	$158739*10^2$	44600	$9*10^2-526*10^2$	17432	$7*10^2-140*10^5$	$36297*10^2$	$1100*10^2-4800*10^2$	321000	$25*10^2-34*10^2$	2900	$27*10^2-32*10^2$	2900
Mesophilic bacteria temp. 37°C/24 hours	1800	$76-269$	166	10500	$0-757*10^2$	14784	$546-98*10^4$	$1986*10^2$	$2000-178*10^2$	8300	$225-340$	282	$105-12000$	6053
Spore-forming bacteria par temp. 28°C/7 days	273	$2-36*10^2$	1367	443	$0-486*10^2$	8402	$400-22600$	6746	$2-217$	90	$70-1800$	935	$20-1440$	730
Microscopic fungi [cfu/ml]	0	$7-150$	59	80	$0-30$	7	$0-300$	70	$30-1165$	415	$75-145$	110	$105-305$	205
Thermotolerant coliform bacteria [the most probable number per 100 ml]	<5	<5	<5	<5	$>5-23$	–	$<5->2400$	~958	<5	<5	$23-240$	131	$62-240$	151
Clostridium perfringens titre	>1	>1	>1	>1	$>1-10^{-3}$	–	$>1-10^{-2}$	–	>1	>1	>1	>1	>1	>1

Abbreviations used: w.d.b. – waste dump base, cfu – colony forming units

211

ACKNOWLEDGEMENTS

The study was financed by the Ministry of Science and Informatics funds for the years 2003–2007.

REFERENCES

Amokrane, A., Comel, C. & Veron, J. 1997. Landfill leachates pretreatment by coagulation-floculation. *Water Research* 11(31): 2775–2782.

Bae, B.U., Jung, E.S., Kim, Y.R. & Shin, H. S. 1999. Treatment of landfill leachate using activated sludge process and electron-beam radiation. *Water Research* 11(33): 2669–2673.

Bengtsson, L., Bendz, D., Hogland, W., Rosqvist, H. & Åkresson, M. 1994. Water balance for landfills of different age. *Journal of Hydrology* 158: 203–217.

Chiang, L.Ch., Chang, J.E. & Wen, T.Ch. 1995. Indirect oxidation effect in electrochemical oxidation treatment of landfill leachate. *Water Research* 2(29): 671–678.

Christensen, J.B., Jensen, D.L., Gron, Ch., Filip, Z. & Christensen, T.H. 1998. Characterization of the dissolved organic carbon in landfill leachate-polluted groundwater. *Water Research* 1(32): 125–135.

Freund, E. 1992. Anforderungen an die Sickerwasserbehandlung und einleitung. In: K. Wiemer & M. Kern (eds), *Abfall-Wirtschaft 9*: 123–134. Uniwersytet Kassel. M.J.C. Baeza-Verlag Witzenhausen.

Fricke, K. & Mûller, W. 1994. Anaerobe und aerobe Behandlung von Restmûll. In: K. Wiemer & M. Kern (eds), *Abfall-Wirtschaft 9*: 571–719. Uniwersytet Kassel. M.J.C. Baeza-Verlag Witzenhausen.

Gandhi, M. & Chikindas, M. 2007. Listeria: A foodborne pathogen that knows how to survive. *International Journal of Food Microbiology* 113: 1–15.

Grabińska-Łoniewska, A., Łebkowska, M., Słomczyńska, B., Słomczyński, T. & Sztompka, E. 1999. *General microbiology exercise book* (in Polish). Publishing House of Warsaw University of Technology: Warsaw.

Grabińska-Łoniewska, A., Kulig, A., Pajor, E., Skalmowski, A., Rzemek, W. & Szyłak-Szydłowski, M. 2005. Physicochemical and microbiological characteristics of leachates from various waste dumps receiving waste other than neutral or hazardous (in Polish): 433–442. *Monographs of Environmental Engineering Committee, Polish Academy of Science. Materials from 2nd Environmental Engineering Congress*: Lublin.

Grabińska-Łoniewska, A., Kulig, A., Pajor, E., Skalmowski, A., Rzemek, W. & Szyłak-Szydłowski, M. 2007. Physicochemical and microbiological characteristics of leachates from Polish municipal landfills. In: L. Pawłowski, M. Dudzińska & A. Pawłowski (eds), *Environmental Engineering: 327–337*. Taylor & Francis Group: London, New York, Singapore.

Grabińska-Łoniewska, A., Korniłłowicz-Kowalska, T., Wardzyńska, G. & Boryń, K. 2007. Occurrence of fungi in water distribution system. *Polish Journal of Environmental Studies* 16: 539–547.

Grabińska-Łoniewska, A. & Siński, E. 2009. *Transmission of the pathogenic and potentially pathogenic microorganisms through water distribution system. Problems and solutions*. Wydawnictwo Seidel-Przywecki: Warsaw.

Holt, J.G. 2004. *Bergey's manual of systematic bacteriology*. Williams and Wilkins: Baltimore, Hong Kong, London, Sidney.

Imai, A., Iwani, N., Matsushige, K., Imamori, Y. & Sudo, R. 1993. Removal of refractory organic and nitrogen from landfill leachate by the microorganism- attached activated carbon fluidized bed process. *Water Research* 1(27): 143–145.

Jager, J. & Kuchta, K. 1992. Geruchsemissionen von Kompostwerken. In: K. Wiemer & M. Kern (eds), *Abfall-Wirtschaft 9*: 99–121. Uniwersytet Kassel. M.J.C. Baeza-Verlag Witzenhausen.

Karwaczyńska, U., Rosik-Dulewska, Cz. & Ciesielczuk, T. 2005. Impact of leachates from an unsealed municipal solid waste dump on the quality of surface and groundwater (in Polish). Monographs of Environmental Engineering Committee. Polish Academy of Science. *Materials from 2nd Environmental Engineering Congress* 2(33): 509–515.

Kim, S.H., Geissen, S.U. & Vogelpohl, A. 1997. Landfill leachate treatment by a photoassisted Fenton reaction. *Water Science and Technology* 4(35): 239–248.

Kulig, A. 1995. Environmental impact assessment of municipal utilities. Proceedings of the Polish-British Conference. In: P. Manczarski & M. Nawalny (eds), *Environmental Engineering – British and Polish experience in linking education and research with industry*: 253–268. Warsaw, 16–18 October 1995.

Libudzisz, Z., Kowal, K. & Żakowska, Z. 2007. *Technical microbiology* (in Polish). Wydawnictwo Naukowe PWN: Warsaw.

Lin, S.H. & Chang, Ch.C. 2000. Treatment of landfill leachate by combined electro-Fenton oxidation and sequencing batch reactor method. *Water Research* 17(34): 4243–4249.

Murray, H.E. & Beck, J.N. 1990. Concentrations of synthetic organic chemicals in leachate from a municipal landfill. *Environmental Pollution* 67: 195–203.

Paxéus, N. 2000. Organic compounds in municipal landfill leachates. *Water Science and Technology* 7–8(42): 323–333.

Pöhle, H., Mietke, H. & Kliche, R. 1993. Zusammenhang zwischen mikrobieller Besiedlung und Geruchsemissionen bei der Bioabfallkompostierung. BFMT-Statusseminar. *Neue Techniken zur Kompostierung*: Hamburg 22-23.11.1993.

Welander, U., Henrysson, T. & Welander, T. 1997. Nitrification of landfill leachate using suspended- carrier biofilm technology. *Water Research* 9(3): 2351–2355.

Welander, U., Henrysson, T. & Welander, T. 1998. Biological nitrogen removal from municipal landfill leachate in a pilot scale suspended carrier biofilm process. *Water Research* 5(32): 1564–1570.

Zaloum, R. & Abbott, M. 1997. Anaerobic pretreatment improves single sequencing batch reactor treatment of landfill leachates. *Water Science and Technology* 1(35): 207–214.

Zamora, R.M.R., Moreno, A.D., Orta de Velasquez, M.T. & Ramirez, I.M. 2000. Treatment of landfill leachates by comparing advanced oxidation and coagulation-flocculation processes coupled with activated carbon absorption. *Water Science and Technology* 1(41): 231–235.

Environmental Engineering III – Pawłowski, Dudzińska & Pawłowski (eds)
© 2010 Taylor & Francis Group, London, ISBN 978-0-415-54882-3

Process kinetics and equilibrium in Cu^{2+} sorption in hydrogel chitosan granules

Z. Modrzejewska, A. Skwarczyńska & R. Zarzycki
Faculty of Process and Environmental Engineering, Technical University of Lodz, Lodz, Poland

ABSTRACT: We examined the ability of hydrogel chitosan beads to adsorb Cu^{2+}, with adsorption in the presence of SO_4^{2-} co-ions. Studies were carried out with initial pH 3.5 and 5. The process rate was defined and on this basis isotherms were determined. The process rates were described by pseudo first- or second-order reaction equations; and equilibrium by Langmuir, Langmuir–Freundlich, Redlich–Peterson, Dubinin–Radushkevich, modified Dubinin–Radushkevich and Toth equations. The Langmuir–Freundlich, Redlich–Peterson and Dubinin–Radushkevich isotherms gave better fits to experimental data.

Keywords: Chitosan, hydrogel, adsorption, copper.

1 INTRODUCTION

Chitosan, poly(1-4)-D-glucosamine, is a product of chitin deacetylation. One of the important properties of chitosan is its high absorption of metal ions, cholesterol, proteins and tumour cells. These properties result mainly from the presence of reactive NH_2 and OH groups in molecules. Researches described in literature show complex-forming abilities of chitosan, mainly for transition metal ions (Onsen & Skaugrud 1990, Kamiński & Modrzejewska 1997, 1999, Roberts 1997, Minamisawa et al. 1999, Ngah & Liang 1999, Becker et al. 2000, Bhatia & Ravi 2000, Krajewska 2001, Ly et al. 2003, Ngah et al. 2005, Chang et al. 2006, Kurita 2006, Vieira 2006, Baran et al. 2007, Campos et al. 2007, Septhum et al. 2007, Baroni et al. 2008, Chen et al. 2008, Rangel-Mendez et al. 2009).

The mechanism of chelate formation is not fully explained (Guibal et al. 2004, Debbaudt et al. 2004; Ramalho & Mercé 2008, Trimukhe & Varma 2008, 2009). Attempts to explain this are limited to the chitosan–Cu^{2+} system (Micera et al. 1985, Findon et al. 1993, Oyrton et al. 1999, Tomoyo et al. 2002, Rhazi et al. 2002 a,b, Zarzycki et al. 2003, Ben-Shalom et al. 2005, Terreux et al. 2006, Zhao et al. 2007, Vasconcelos et al. 2008, Ngah et al. 2008 We examined). Inoue et al. (1993) explained the formation of chelate compounds by the reactive action of -NH_2 groups. Non-protonated -NH_2 groups are ligands because of the presence of an unpaired pair of electrons which can form coordinate bonds with transition metal ions. Thermogravimetric studies confirmed the main role of amine groups (Oytron et al. 1999), but indicated a possible reaction of metal ions with hydroxyl groups at the third carbon atom, while Okuyama et al. (1997) showed that when chelate compounds are formed, two -OH groups and one -NH_2 group, and a water molecule

or -OH group at the third carbon atom are bonded. Interesting studies by Terreux et al. (2006) referred to the interaction between Cu^{2+} ions and one or several glucosamine monomer units. The interactions of Cu^{2+} with different possible coordination sites of glucosamine and N-acetyl glucosamine were studied. For every considered complex, the Cu^{2+} site was completed with H_2O and/or OH^- ligands to have a global neutral charge. Assuming a chemical character of the sorption, first and second-order equations are most often used to describe kinetics (Yiacoumi & Tien 1995, Guibal et al. 1998, Lee et al. 2001, Schmuhl et al. 2001, Evans et al. 2002, Sag & Aktay 2002, Nagh et al. 2004, Cestari et al. 2005, Ding et al. 2006, Shafaei et al. 2007, Qin et al. 2007, Zarzycki et al. 2007, 2008). In our research, studies on the adsorption of Cu^{2+} ions on chitosan were continued, with adsorption in the presence of SO_4^2 ions and using chitosan hydrogel. The process rate was defined and on this basis isotherms were determined. The process rates were described by equations of pseudo first- or second-order reactions, and equilibrium by Langmuir, Langmuir–Freundlich, Redlich–Peterson, Dubinin–Radushkevich, modified Dubinin–Radushkevich and Toth equations.

2 MATERIALS AND METHODS

2.1 Characteristics of the adsorbent

Adsorption of Cu^{2+} ions was carried out on a stable bed. The adsorbent was a porous hydrogel bed of chitosan beads (diameter 3.5×10^{-3} m) with 95% water in their structure.

The hydrogel beads were formed by the phase inversion method from 4% solution of chitosan with mean molecular weight 2×10^5 Da and deacetylation degree

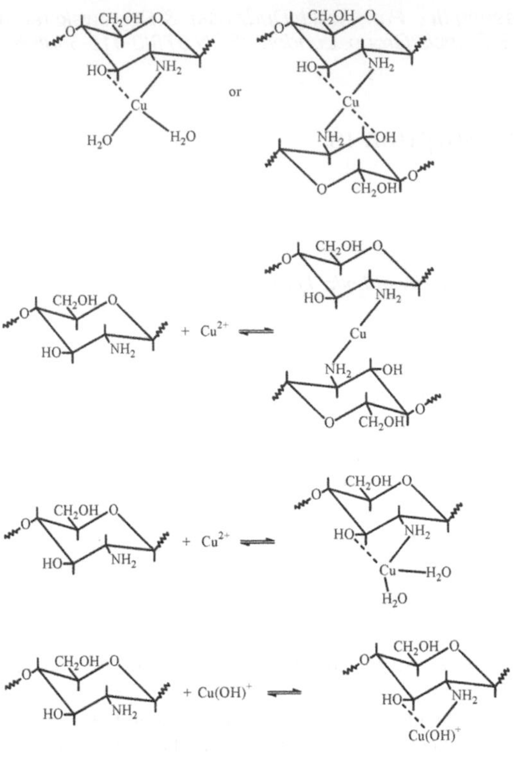

Figure 1. Scheme of the formation of chitosan-Cu^{2+} chelate compounds

of 78%, produced from Antarctic krill in the Sea Fisheries Institute – Poland. The chitosan solvent used was 2% acetic acid. The beads were formed in 10% sodium hydroxide and left there for 24 h. After that period they were washed with distilled water until reaching the neutral pH of the water in which they were then left.

2.2 Characteristics of the tested solution

The adsorption of copper was investigated in water solutions of copper sulphate ($CuSO_4$).

2.3 Analytical methods

The concentration of ions was determined by emission mass spectrometry (ICP PERKIN-ELMER – Plasma 400).

3 EXPERIMENTAL RESULTS

The research covered:

- determination of the rate of adsorption of Cu^{2+} ions on these beads.
- determination of adsorption isotherms of these ions on a hydrogel bed.

Adsorption was carried out on immobilised beads, in the system combined with a mixer (shaker, at amplitude oscillation of 8 mm), at temperature 293 K

and initial pH 5 or 3.5. The adsorbent was 20 g of chitosan beads, composed of about 1 g of pure chitosan. Tests were carried out on 0.25×10^{-3} m^3 samples, i.e. at the ratio mass of the adsorbent (m) to volume samples (V) $\frac{m}{V} = 4$. Respective adsorptions over time for $CuSO_4$ at pH 5 and 3.5 are shown in Figures 2, 3.

3.1 Determination of the adsorption rate of Cu^{2+} ions on beads.

Assuming chemisorption, the rates were described by: The pseudo first-order equation

$$q_t = q_m \left(1 - e^{-k_1 t}\right) \qquad (1)$$

and the pseudo second-order equation

$$q_t = q_e \frac{q_m k_2 t}{1 + q_m k_2 t}. \qquad (2)$$

The equation parameters are given in Table 1, and the fitting to experimental data is marked in the graphs 2 and 3 (first-order with red colour and second-order with black). It follows from the simulations that Cu^{2+} ion sorption can be described both by first and second-order reaction equations. The first-order equations (red lines) approximate the experimental data with significant probability in both cases (pH 3.5 and pH 5) at lower concentration; and the second-order equations at higher concentra- tion. However, it should be noted that reaction constants in second-order equations depend on the initial adsorptive concentration.

The process of adsorption can be divided into several stages: rapid adsorption in the first hour, then the process slowed and this lasted for about 5 hours. In subsequent hours, the adsorption was very slow and an equilibrium was established for up to several days. In the first hour, Cu^{2+} ions were combined with the reactive groups of the chitosan molecule ($-NH_2$, $-OH$) near the outer bead surface. They blocked pores on the outer surface so that the next Cu^{2+} ions had a longer diffusion path to the active groups in the bead structure. Hence the process slowed over consecutive hours.

Description of the process by a model of intramolecular diffusion also seems insufficient. The coefficient of inner diffusion is the function of initial concentration and well described the adsorption process only in specified time intervals (0–1 h or 1–5 h). To interpret the results, we assumed a multi-staged process, which can arouse doubts referring to the physical sense (Ngah et al. 2004).

Taking the above into account, a model was proposed which assumed a surface process described by the first-order reaction equation (at pH 5), slowed by diffusion in pores. Model assumptions and description are given in the study Zarzycki et al. (2007). The process was described by only one stable parameter independent of the concentration – a surface process constant. Mean value of the surface process kinetic constant was 1.44×10^{-3} (units s^{-1}). Verification of the model assumptions in the process of Cu^{2+} ion

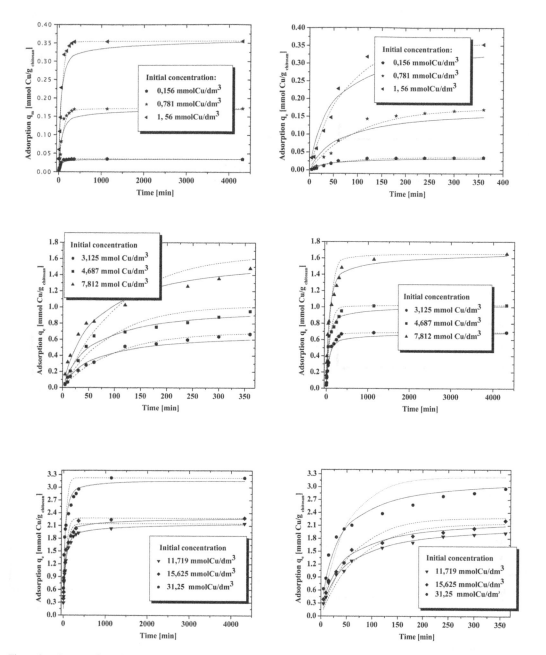

Figure 2. Copper adsorption in time Cu SO$_4$ pH = 3.5.

sorption allowed us to propose the mechanism of sorption. The results confirmed the assumption that in the case of Cu^{2+} ion sorption in hydrogel chitosan granules, we have a first-order chemical reaction slowed by diffusion in hydrogel pores.

3.2 Determination of adsorption isotherms of these ions on a hydrogel bed.

Based on kinetics it was assumed that equilibrium was established after 96 h. Taking this into account, the equilibria at initial pH 3.5 and 5 were determined.

The equilibria were specified when pH was increasing during the sorption process (Figs. 4 and 5).

Equilibrium in the water solution of metal-ions–chitosan hydrogel granules system was described with the most often used isotherm equations:

Langmuir (L) equation:

$$q_e = \frac{q_m K_L C_e}{1 + K_L C_e} \tag{3}$$

Freundlich (F) equation:

$$q_e = K_F C_e^{\frac{1}{n}} \tag{4}$$

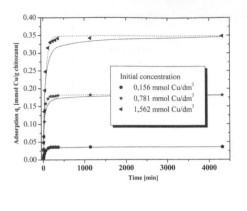
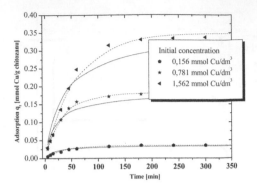
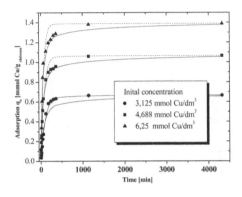
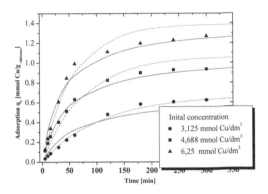
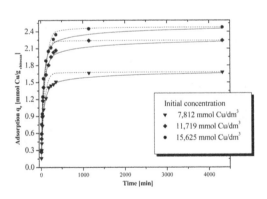
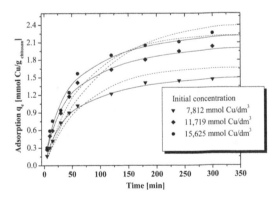

Figure 3. Copper adsorption in time Cu SO$_4$ pH = 5.

Langmuir–Freundlich (L-F) equation:

$$q_e = \frac{q_m \left(K_{LF} C_e\right)^{n_{LF}}}{1 + \left(K_{LF} C_e\right)^{n_{LF}}} \qquad (5)$$

Redlich–Peterson (R-P) equation:

$$q_e = \frac{q_m K_{RP} C_e}{1 + \left(K_{RP} C_e\right)^{n_{RP}}} \qquad (6)$$

Dubinin–Radushkevich (D-R) equation:

$$\lg q_e = \lg q_m - n_{DR} \, \lg^2 \left(K_{DR} C_e\right) \qquad (7)$$

Modified Dubinin–Radushkevich (mD-R) equation:

$$\lg q_e = \lg q_m - n_{mDR} \, \lg^2 \left(\frac{K_{mDR} C_e}{1 + K_{mDR} C_e}\right) \text{ for } C_e \geq C^*$$

$$\lg C^* = -\frac{1}{2} n_{mDR} - \lg K_{mDR} \qquad (8)$$

Toth (T) equation:

$$q_e = \frac{q_m K_T C_e}{\left(1 + \left(K_T C_e\right)^{n_T}\right)^{1/n_T}} \qquad (9)$$

216

Table 1. Parameters of equations describing the process in time.

	$q_t = q_m(1 - e^{-k_1 t})$			$q_t = q_e \dfrac{q_m k_2 t}{1 + q_m k_2 t}$		
C_o (mmol/dm^3)	q_e (mmol/g)	k_1 (min^{-1})	F	q_e (mmol/g)	k_2 (mmol/g/min)	F
CuSO$_4$. pH $=$ 3.5						
0.156	0.034	0.0163	60.38	0.0349	0.763	14.43
0.781	0.172	0.0103	48.75	0.1722	0.0973	12.17
1.562	0.355	0.0145	96.99	0.355	0.0687	15.06
3.125	0.689	0.0105	116.10	0.689	0.0265	40.36
4.688	1.021	0.0112	15.27	1.021	0.0192	53.6
6.25	1.356	0.0106	6.73	1.356	0.0143	46.64
7.812	1.649	0.0094	8.41	1.649	0.0101	56.38
11.719	2.151	0.0145	9.49	2.151	0.0117	85.9
15.625	2.279	0.0173	10.21	2.279	0.0131	78.82
31.25	3.225	0.0220	4.57	3.225	0.0114	28.7
CuSO$_4$. pH $=$ 5						
0.156	0.035	0.035	26.41	0.0355	1.795	9.16
0.781	0.181	0.0312	391.6	0.181	0.297	20.75
1.562	0.349	0.0179	167.2	0.349	0.085	20.02
3.125	0.666	0.0098	270.7	0.666	0.0248	19.29
4.688	1.065	0.0138	28.4	1.065	0.022	170.2
6.25	1.389	0.0183	28.18	1.389	0.0213	67.2
7.812	1.678	0.0146	17.63	1.679	0.0147	382.6
11.719	2.466	0.0146	17.09	2.242	0.0112	212
15.625	2.475	0.0105	17.94	2.475	0.0105	111.2

To describe the equilibrium, we estimated an equilibrium constant (K), sorption capacity (q_m) and a coefficient which determines the heterogeneity of sorbent surface (n) in the equations where this parameter occurs. Unknown values of equilibrium constants were specified by fitting experimental points to model isotherm equations by nonlinear regression based on the Marquardt–Lavenberg algorithm. To eliminate erroneous solutions, calculations were repeated several times for different starting values using two objective functions alternately; the error square sum (ERRSQ) and composite error function.

$$\text{(HYBRD):ERRSQ} = \sum_{i=1}^{NC}(q_e - e_{e,obl})^2 \qquad (10)$$

$$\text{HYBRD} = \sum_{I=1}^{NC}\left(\frac{q_e - q_{e,obl}}{q_e}\right)^2 \qquad (11)$$

The assumed criterion of model evaluation was the ratio of sorption capacity (q_m) to real value q_{max}, and analysis of curves and fitting of isotherms to experimental points on the basis of Fisher test (F) and standard deviation (σ).

$$F = \frac{(N-1)\sum\limits_{i=1}^{N}\left(q_e(i) - \dfrac{1}{N}\sum\limits_{i=1}^{N}q_e(i)\right)^2}{(N-1)\sum\limits_{i=1}^{N}(k_e(i) - k_{e,obl}(i))^2} \qquad (12)$$

Table 2. Parameters of the isotherms at pH 3.5.

CuSO$_4$pH$_0$ 3.5

	F	L	L-F	R-P	T	D-R
K_A	2.51	103.1	58.43	276.3	2.24×10^4	2.8×10^{-2}
q_{max}	–	3.19	3.58	2.21	3.745	3.38
n	0.09	–	0.36	0.957	0.2638	0.0257
F	0.3953	17.53	100.9	70.55	115.8	116.3
σ	9.239	0.287	0.1196	0.1431	0.1117	0.1114

$$\sigma = \frac{1}{N}\sum_{i=1}^{n}(q_e - q_{e,obl})^2 \quad i = 1,...,N \qquad (13)$$

Numerical values of K, q_m and n determined by a computer estimation along with standard deviation and values of the Fisher test for pH 3.5 and pH 5 are given in Tables 2 and 3, respectively. The estimated and experimental results obtained for models are shown in Figures 4 and 5.

The best fits were for equilibrium described by the Toth, Langmuir–Freundlich, Redlich–Peterson and Dubinin–Radushkevich isotherms. The results indicated that adsorption can be interpreted by the theory of volumetric filling of micropores and that the adsorption surface had a heterogeneous character. A good description of the equilibrium using the Toth and

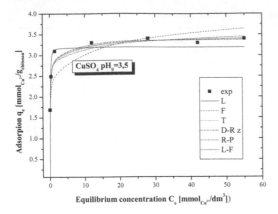

Figure 4. Equilibrium of Cu^{2+} sorption on chitosan hydrogel beads at pH 3.5.

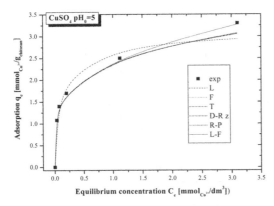

Figure 5. Equilibrium of Cu^{2+} sorption on chitosan hydrogel beads at pH 5.

Langmuir–Freundlich isotherms showed that adsorption energy distribution had a rather asymmetric quasi-Gaussian character extended towards low energies, or symmetric quasi-Gaussian nature.

The equilibrium in Cu^{2+} ion adsorption described by the Langmuir and Freundlich equation, when fitted to experimental data, had the largest error.

4 CONCLUSIONS

A good approximate description of Cu^{2+} ion sorption was given by kinetic models of a chemical reaction. The first second-order equations approximated the experimental data at lower initial concentrations ($0.15 > 3$ mmol/dm^3), and second-order equations at higher concentrations (> 4.7 mmol/dm^3). In the case of Cu^{2+} ion sorption in chitosan hydrogel, the experimental data were well described by isotherms which assumed heterogeneity of surfaces. Sorption capacity of Cu (II) ions was in the range 2.4–3.2 mmol/g chitosan. The equilibrium was best described by the Toth, Langmuir–Freundlich and

Dubinin–Radushkevich isotherms, hence the adsorption process can be interpreted by the theory of volumetric filling of micropores, and the adsorption surface had a heterogeneous character. It is most probable that there was an asymmetric adsorption energy distribution extended towards low adsorption energies or symmetric quasi-Gaussian nature.

NOMENCLATURE

q_e – ion equilibrium concentration in solid phase,
C_e – ion equilibrium concentration in liquid phase,
q_m – sorption capacity,
K – equilibrium constant,
n – constant to determine sorbent surface heterogeneity,
N – number of experimental points,
l – number of estimated parameters,
q_e – experimental values,
$q_{e,obl}$ – theoretical values

REFERENCES

Baran, A. et al. 2007. Comparative studies on the adsorption of Cr(VI) ions on to various sorbents. *Bioresource Technology* 98: 661–665.

Baroni, P. et al. 2008. Evaluation of batch adsorption of chromium ions on natural and crosslinked chitosan membranes. *Journal of Hazardous Materials* 152: 155–1163.

Becker, T. et al. 2000. Adsorption of nickel(II), zinc(II) and cadmium(II) by new chitosan derivatives. *Reactive and Functional Polymers* 44: 289–298.

Ben-shalom, N. et al. 2005. Copper-binding efficacy of water-soluble chitosans: characterization by aqueous binding isotherms. *Chemosphere* 59: 1309–1315.

Bhatia, S. et al. 2000. A magnetic study of a Fe-chitosan complex and its relevance to other biomolecules. *Biomacromolecules* 1: 413–417.

Campos, K. et al. 2007. Mercury Sorption on Chitosan, *Advanced Materials Research* 20–21: 635–638.

Cestari, A.R. et al. 2005. Determination of kinetic parameters of Cu(II) interaction with chemically modified thin chitosan membranes. *Journal of Colloid and Interface Science* 285: 288–295.

Chang, Y.C. et al. 2006. Magnetic chitosan nanoparticles: Studies on chitosan binding and adsorption of Co(II) ions. *Reactive and Functional Polymers* 66: 335–341.

Chen, A.H. et al. 2008. Comparative adsorption of Cu(II), Zn(II), and Pb(II) ions in aqueous solution on the crosslinked chitosan with epichlorohydrin. *Journal of Hazardous Materials* 154: 184–191.

Debbaudt, A.L. et al. 2004. Theoretical and experimental study of M^{2+} adsorption on biopolymers III Comparative kinetic pattern of Pb, Hg and Cd. *Carbohydrate Polymers* 56: 321–332.

Ding, P. et al. 2006. Kinetics of adsorption of Zn(II) ion on chitosan derivatives. *International Journal of Biological Macromolecules* 39(4–5):.222–227.

Evans, J.R. et al. 2002. Kinetics of cadmium uptake by chitosan-based crab shells. *Water Research* 36: 3219–3226.

Findon, A. et al. 1993. Transport studies for the sorption of copper ions by chitosan. *Journal Environmental Health* 28(1): 173–185.

Guibal, E. 2004. Interactions of metal ions with chitosan-based sorbents: a review. *Separation & Purification Technology* 38: 43–74.

Guibal, E. et al. 1998. Metal-anion sorption by chitosan beads: equilibrium and kinetic studies. *Industrial & Engineering Chemistry Research* 37(4): 1454–1463.

Kamiński, W. & Modrzejewska, Z. 1997. Application of chitosan membranes in separation of heavy metal ions. *Separation Science and Technology* 32(16): 2659.

Krajewska, B. 2001. Diffusion of metal ions through gel chitosan membranes. *Reactive and Functional Polymers* 47(1): 37–47.

Kurita, K., 2006. Chitin and Chitosan: Functional Biopolymers from Marine Crustaceans. *Marine Biotechnology* 8(3): 203–226.

Lee, S.T. et al. 2001. Equilibrium and kinetic studies of copper(II) ion uptake by chitosan-tripolyphosphate chelating resin. *Polymer* 42: 1879–1892.

Ly, A. et al. 2003. Gold sorption on chitosan derivatives. *Hydrometallurgy* 71(1–2): 191–200.

Micera, G. et al. 1985. Copper II complexation by D-glucosamine. Spectroscopic and potentiometric studies. *Inorganica Chimica Acta* 107: 45–48.

Minamisawa, H. et al. 1999. Adsorption behavior of cobalt (II) on chitosan and its determination by tungsten metal furnace atomic absorption spectrometry. *Analytica Chimica Acta* 378(1–3): 278–285.

Modrzejewska, Z.& Kamiński, W. 1999. Separation of Cr(VI) on chitosan membranes. *Industrial Engineering Chemistry Research* 38: 4946–4950.

Ngah, W.& Liang, K. 1999. Adsorption of gold (III) ions onto chitosan and N carboxymethyl chitosan: Equilibrium studies. *Ind. Eng. Chem. Res.* 38: 1411–1414.

Ngah, W.S. et al. 2004; Equilibrium and kinetics studies of adsorption of copper (II) on chitosan and chitosan/PVA beads. *International Journal of Biological Macromolecules* 34(3): 155–161.

Ngah, W.S. et al. 2005. Adsorption behaviour of Fe(II) and Fe(III) ions in aqueous solution on chitosan and cross-linked chitosan beads. *Bioresource Technology* 96(4): 443–450.

Ngah, W.S. & Fatinathan, S. 2008. Adsorption of Cu(II) ions in aqueous solution using chitosan beads, chitosan–GLA beads and chitosan–alginate beads. *Chemical Engineering Journal* 143(1–3): 62–72.

Inoue, K. et al. 1993. Adsorption of metal ions on chitosan and crosslinked copper(II)-complexed chitosan. *Bulletin of the Chemical Society of Japan* 66(5): 2915–292.

Okuijama, K. et al. 1997. Molecular and Crystal Structure of Hydrated Chitosan. *Macromolecules* 30(19): 5849–5855.

Onsoyen, E. & Skaugrud, O. 1990. Metal recovery using chitosan. *Journal of Chemical Technology and Biotechnology* 49: 395–404.

Oyrto,n A.C. et al. 1999. Some thermodynamic data on copper-chitin and copper-chitosan biopolymer interactions. *Journal of Colloid and Interface Science* 212: 212–219.

Qin, Y. et al. 2007. Absorption and release of zinc and copper ions by chitosan fibers. *Journal of Applied Polymer Science* 105: 527–532.

Ramalho Mercê, A.L. 2008. Coordination study of chitosan and Fe^{3+}. *Journal of Molecular Structure* 877(1–3): 89–99.

Rangel-Mendez, J.R. et al. 2009. Chitosan selectivity for removing cadium (II), copper (II), and lead (II) from aqueous phase: pH and organic matter effect. *Journal of Hazardous Materials* 162: 503–511.

Rhaz, M. et al. 2002a. Influence of the nature of metal ions on the complexation with chitosan. Application to the treatment of liquid waste. *European Polymer Journal* 38: 1523–1530.

Rhazi, M. et al. 2002b. Contribution to the Study of the Complexation of Copper by Chitosan and Oligomers. *Polymer* 43: 1267–127.

Roberts, G.A.F. 1997. Chitosan production routes and their role in determining the structure and properties of the product. In A. Domard & G.A.F Roberts (ed.), *Advances in Chitin Science*: 2. Lyon.

Sag, Y.I. & Aktay, Y. 2002. Kinetic studies on sorption of Cr(VI) and Cu(II) ions by chitin, chitosan and Rhizopus arrhizus. *Biochemical. Engineering Journal* 1.2: 143–153.

Septhum, C. et al. 2007. An adsorption study of Al(III) ions onto chitosan. *Journal of Hazardous Materials* 148(1–2): 185–191.

Shafaei, A. et al. 2007. Equilibrium studies of the sorption of Hg(II) ions onto chitosan. *Chemical Engineering Journal* 133(1–3): 311–316.

Schmuhl, R. et al. 2001. Adsorption of Cu(II) and Cr(VI) ions by chitosan: Kinetics and equilibrium. *Water* 27(1): 1–7.

Terreux, R. et al. 2006. Interactions study between the copper II ion and constitutive elements of chitosan structure by DFT calculation. *Biomacromolecules* 7(1): 31–37.

Tomoyo, M.T. et al. 2002. Effect of copper adsorption on the mechanical properties of chiton beads. *Journal of Applied Polymer Science* 88(13): 2988–2991.

Trimukhe, K.D.& Varma, A.J. 2008. A morphological study of heavy metal complexes of chitosan and crosslinked chitosans by SEM and WAXRD. *Carbohydrate Polymers* 71: 698–702.

Trimukhe, K.D.& Varma, A.J. 2009 Metal complexes of crosslinked chitosans. Correlations between metal ion complexatiation values and thermal properties. *Carbohydrate Polymers* 75: 63–70.

Vasconcelos, L.H. et al. 2008. Chitosan crosslinked with a metal complexing agent: Synthesis, characterization and copper(II) ions adsorption. *Reactive & Functional Polymers* 68: 572–579.

Vieira, R.S.& Beppu, M.M. 2006a. Interaction of natural and crosslinked chitosan membranes with Hg(II) ions. *Colloids Surf. A* 279: 196–207.

Vieira, R.S.& Beppu, M.M. 2006b. Dynamic and static adsorption and desorption of Hg(II) ions on chitosan membranes and spheres. *Water Res.* 40: 1726–1734.

Zarzycki, R. et al. 2003. The effect of chitosan form on copper adsorption. In L. Paw³owski, M Dudzińska & A. Paw³owski (eds), *Environmental Engineering Studies in Poland "Environmental Science Research"*: 199–206 Kluwer.

Zarzycki, R. et al. 2007. Equilibrium in Cu(II) and Ag(I) ion adsorption on chitosan microgranules formed in supercritical drying conditions. *Chemical and Process Engineering* 28: 735–745.

Zarzycki, R. et al. 2008. Model of sorption kinetics in hydrogel chitosan granules. *Chemical and Process Engineering* 29: 801–811.

Zhao, F. et al. 2007. Preparation of porous chitosan gel beads for copper(II) ion adsorption, *Journal of Hazardous Materials* 47(1–2): 67–73.

Yiacoumi, S. & Tien, C. 1995. Modeling adsorption of metal ions from aqueous solutions, I. Reaction-controlled cases. *Journal of Colloid and Interface Science* 175: 333–346.

Environmental Engineering III – Pawłowski, Dudzińska & Pawłowski (eds)
© *2010 Taylor & Francis Group, London, ISBN 978-0-415-54882-3*

Characterization of surface active properties of *Bacillus* strains growing in brewery effluent

G.A. Płaza & K. Gawior
Institute for Ecology of Industrial Areas, Katowice, Poland

K. Jangid
Department of Microbiology, University of Georgia, Athens, USA

K.A. Wilk
Department of Chemistry, Wroclaw University of Technology, Wroclaw, Poland

ABSTRACT: The research investigated the potential of utilising industrial waste to replace synthetic media for biosurfactant production. Three bacterial strains were identified by 16S rRNA gene sequencing: *Bacillus subtilis* (I'-1a), *Bacillus* sp. (T-1), *Bacillus* sp. (T'-1). The isolates were able to grow and produce biosurfactant in brewery effluent medium under aerobic and thermophilic conditions during the stationary growth phase. Biosurfactant production was indirectly evaluated by surface active properties, i.e. reduction of both surface and interfacial tension of cell-free supernatants, critical micelle concentration, emulsification of a variety of hydrocarbons, and foamability. The three isolates are good candidates for petroleum industry applications.

Keywords: *Bacillus*, biosurfactant, brewery effluent, organic waste, bioconversion.

1 INTRODUCTION

Biosurfactants are surface-active compounds produced by different bacteria, fungi and yeasts (Bodour & Maier 2002). In recent years, biosurfactants have gained attention because of their advantages such as high biodegradability, low toxicity, variable structures, ecological acceptability, high foaming, high selectivity, specific activity at extreme temperature, pH, salinity and the ability to be synthesised from renewable feed stocks (Desai & Banat 1997 Makkar & Cameotra 1997, 1999, 2002). The unique properties of biosurfactants allow their use as possible replacements for chemically synthesised surfactants in many industrial applications such as bioremediation, enhanced oil recovery, food additives, pharmaceutical and therapeutic agents, agricultural biocontrol agents, and in health and beauty products (Kosaric 2000, Singh et al. 2007).

Although biosurfactants have numerous advantages, their higher production cost compared to synthetic surfactants is a major drawback. Biosurfactants could potentially replace synthetic surfactants if raw material and process costs are lowered substantially. Achieving this goal requires finding alternative inexpensive substrates and highly efficient microorganisms for biosurfactant production. Several renewable substrates, especially from agro-industrial wastes have been intensively studied for microorganism cultivation and biosurfactant production at a laboratory scale. Agro-industrial wastes with high contents of carbohydrates or lipids meet the requirements for substrates for biosurfactant production (Mercade & Manresa, 1994, Deleu & Paquot, 2004; Maneerat, 2005; Das & Mukherjee, 2007). Some examples are olive oil mill effluent (Mercade & Manresa 1994), waste frying oil (Haba et al. 2000), oil refinery wastes (Adamczak & Bednarski 2000), soapstock (Benincasa et al. 2004, Benincasa & Accorsini 2007), molasses (Makkar & Cameotra 1997, Joshi et al. 2008), whey (Dubey & Juwarkar 2001, Nitschke et al. 2004, Joshi et al. 2008), starch wastes (Fox & Bala 2000, Thompson et al. 2000, Das & Mukherjee 2007), cassava-flour processing effluent (Nitchke & Pastore 2006) and distillery waste (Dubey & Juwarkar 2001).

The present research investigated the potential of industrial waste to replace synthetic media for biosurfactant production. We tested the biosurfactant production of previously isolated *Bacillus* strains using brewery effluent as a raw material. In this study, we report an indirect way to identify biosurfactant production by determining the surface active properties of *Bacillus* strains growing on brewery wastewater under thermophilic and aerobic conditions.

2 MATERIALS AND METHODS

2.1 Isolation, identification and characterisation of bacterial isolates

The bacterial strains (T-1, T'-1 and I'-1a) used in this study were isolated from sludge of a 100-year-old oil refinery in Czechowice-Dziedzice (Poland) as described by Berry et al. (2006) and P³aza et al. (2006). The aged sludge was acidic (pH 2) and highly contaminated with polycyclic aromatic hydrocarbons. Bacterial isolates were identified based on the 16S rRNA gene sequence analysis. A direct-colony, PCR (Polymerase Chain Reaction) was set up to amplify the 16S rRNA gene in a 30-cycle PCR using universal primers 27F and 1492R. The PCR conditions used were: initial denaturation at 95°C for 8 min, 30 cycles of denaturation at 94°C for 1 min, annealing at 55°C for 1 min and elongation at 72°C for 1 min, followed by elongation at 72°C for 10 min. The amplified PCR products were purified using the Qiagen-PCR purification kit as per the manufacturer's instructions. The purified PCR products were sequenced from both ends at the DNA Sequencing Core facility of the University of Michigan at Ann Arbor. The 16S rRNA gene sequences were analysed at the Ribosomal Database Project (RDP) II (http//:rdp.cme.msu.edu). The top 10 most homologous sequences were aligned using the CLUSTALW program v1.83 at the European Bioinformatics site (www.ebi.ac.uk/clustalw). The similarity matrix was prepared using the DNAdist program in the PHYLIP package (Felsenstein 1989) with Jukes–Cantor corrections. Isolates were identified as the genus/species to which they showed highest 16S rRNA gene sequence similarity in the RDP database.

The bacterial isolates were characterised using traditional microbiological methods (Gerhardt 2006). The biochemical characterisation was based on the API ZYM test (bioMerieux S.A.). Isolates were maintained on agar slants (SMA – Standard Methods Agar) containing 8 g peptone, 2.5 g yeast extract and 1 g glucose per L, bioMerieux) at 4°C.

Antibiotic susceptibilities of the isolates were tested against 20 different antibiotics by the disc diffusion method (Bauer et al. 1966). The strains were grown in SMA broth at 45°C for 24 h and a 100 µL aliquot was spread-plated onto Mueller–Hinton agar plates (Oxoid). The antibiotic discs were placed on these freshly prepared lawns and incubated at 37°C for 24 h. The inhibition-zone diameter was measured, and the isolates classified as resistant (R), intermediate (I) and susceptible (S) following the standard antibiotic disc sensitivity testing method (DIFCO 1984).

2.2 Characterisation of brewery effluent

Brewery raw effluent of pH 6.1 (Table 1) from manufacturing of beer was obtained and stored in the laboratory at −18°C until used.

Table 1. Composition of brewery effluent utilised in this work.

Parameters	Unit	Value
pH		9.55
Conductivity	μS/cm	2000
COD	mgO_2/dm^3	4280
BOD_5	mgO_2/dm^3	2850
Organic matter	mg/dm^3	467
SO_4^{2-}	mg/dm^3	73.2
S^{2-}	mg/dm^3	0.054
PO_4^{3-}	mg P/dm^3	13.6
P organic	mg P/dm^3	26.3
NH_4^+	mg N/dm^3	8.89
NO_2^-	mg N/dm^3	13.79
NO_3^-	mg N dm^3	1.03
Total N	mg N/dm^3	76.5
Ca	mg/dm^3	44.3
Cl^-	mg/dm^3	106.4
TOC	mg C/dm^3	1540

2.3 Growth of isolates in brewery effluent

Into 100 cm^3 of sterilised effluent, contained in a 300- cm^3 Erlenmeyer flask, 1 cm^3 of 24-h cultures of bacterial strains with a cell count of $\sim 10^4 - 10^5$ cfu/cm^3 were inoculated and incubated at 45°C for 7 d under aerobic and static batch conditions. Growth curves were obtained by monitoring the optical density at 600 nm on an UV/VIS spectrophotometer (Varian). There were three independent experiments for each bacterial strain.

2.4 Determination of surface active properties

Grown cultures of bacterial strains were centrifuged at 10 000 g for 20 min. and the supernatant used for surface activity measurements. Surface (ST) and interfacial (IT) tensions were determined with a Kruss Processor Tensiometer (model K12 Kruss, Germany) using the plate method. IT was performed against hexadecane. Critical micelle dilution (CMD^{-1} and CMD^{-2}) were measured on 10- and 100-fold diluted supernatant in distilled water, while the CMC (critical micelle concentration) was determined on serially diluted samples as described by Sheppard and Mulligan (1987).

2.5 Emulsification activity

The emulsification activity was determined as described by Cooper and Goldenberg (1987). To 4 cm^3 of supernatant in a screw-cap tube, 1 cm^3 of different, pure hydrocarbons (AR grade; Sigma Aldrich Co. and Polish Chemical Reagents S.A. Gliwice, Poland) were added, and vortexed at high speed for 2 min. The emulsion stability was determined after 24 h. The emulsification index (E24) was calculated as the percentage height of the emulsion layer to the total height of the liquid column.

2.6 Foam activity

Foamability was measured as foam volume (FV) (Das et al. 1998). Air at a constant flow rate ($50\,cm^3/min$) was passed through a measuring glass cylinder containing $20\,cm^3$ of the culture filtrate. FV was estimated as the difference between the volume occupied by the liquid-plus-foam and the volume of the liquid at rest, and was reproducible within ±5–10%. All experiments had three independent replicates.

3 RESULTS AND DISCUSSION

The three bacterial isolates are halothermotolerant Gram-positive spore-forming species affiliated to the genus *Bacillus*. The isolates were screened and selected for further studies as described by Płaza et al. (2006). The bacteria were isolated from sludge samples obtained from 100-y-old oil refinery in Czechowice-Dziedzice (Poland). On the basis of previous results, three bacteria (T-1, T'-1 and I'-1a) were chosen for identification. The 16S rRNA gene sequences showed that the isolates were *Bacillus* spp.; I'-1a was identified as *B. subtilis*, but T-1 and T'-1 were identified as *Bacillus* sp. The 16S rRNA gene sequencing could not clearly assign isolates T-1 and T'-1 to any species of *Bacillus* as both isolates showed >99% similarity to two distinct species (*B. subtilis* and *B. licheniformis* for T-1 and *B. subtilis* and *B. amyloliquefaciens* for T'-1). The morphological and biochemical characteristics of the three isolates are listed in Table 2. All three isolates were resistant to amoxycillin and nalidic acid (Table 3).

Brewery effluent was a good substrate for growth and proliferation of the isolates. The optical density increased from an initial value of 0.05 to 0.7, 0.9 and 0.6 for T-1, T'-1 and I'-1a, respectively, within 3 d of the incubation (Figure 1).

The surface-active properties of cell-free supernatants were determined as indicators of biosurfactant production by the bacterial isolates. ST of the supernatants was reduced to 27, 30 and 38 mN/m for T-1, T'-1 and I'-1a, respectively (Table 4). IT against hexadecane were reduced to 0.36, 0.68 and 0.92 mN/m for T-1, T'-1 and I'-1a, respectively.

The CMC values, determined according to the method of Sheppard and Mulligan (1987), were 0.166, 0.127 and 0.102 g/dm^3 for T-1, T'-1 and I'-1a, respectively. The determined CMC values were lower than those for synthetic surfactants such as SDS (Das & Mukherjee 2007).

The isolates emulsified a range of hydrophobic substances with average E24 values of 61.30, 35.0 and 22.63% for T-1, T'-1 and I'-1a, respectively. The emulsification activities were almost 100% for frying oil and with range 70–81% for diesel oil for all isolates (Table 5). In contrast, E24 values were the lowest for hexadecane and pristine for all isolates. These results indicate substrate specificity for emulsification by the strains. It would be advantageous for bacteria isolated from a hydrocarbon-contaminated site to possess

the ability to emulsify contaminants. The described properties make the organisms potential candidates for enhanced oil recovery and in bioremediation processes.

Among the potential biosurfactant-producing microbes, *Bacillus* species are known to produce cyclic lipopeptides including surfactins, iturins, fengycins and lichenysins as the major classes of biosurfactants (Cooper & Goldenberg 1987, Das & Mukherjee 2007). Surfactin, one of the most effective surfactants produced by *Bacillus* strains showed

Table 2. Morphological and biochemical characteristics of the tested strains.

TESTS	*Bacillus* species		
	T-1	T'-1	I'-1a
Gram-staining	+	+	+
Growth temperature:			
37°C	+	+	+
45°C	+	+	+
65°C	+	+	+
70°C	+	+	+
Sporulation	+	+	+
Salinity (% NaCl):			
2	+	+	+
4	+	+	+
8	+	+	+
12	+	−	+
Biochemical characterisation:			
Alkaline phosphatase	+	+	+
Esterase (C4)	+	+	+
Esterase lipase (C8)	+	+	+
Lipase (C14)	−	−	+
Acid phosphatase	+	+	+
Catalase test	+	+	+
Amylase	+	+	+
Indol production test	+	+	+
Citrate utilization test	+	−	−
Glucose (acidification)	+	+	+
Arginine (arginine dihydrolase)	−	−	−
Urea (urease)	−	−	−
Esculin (hydrolysis b-glucosidase)	+	+	+
Gelatine (hydrolysis)	+	+	+
Glucose (assimilation)	+	+	+
Arabinose (assimilation)	+	+	+
Mannose (assimilation)	+	+	+
Mannitol (assimilation)	+	+	+
N-acetyl-glucosamine (assimilation)	+	+	+
Maltose (assimilation)	+	+	+
Gluconate (assimilation)	+	+	+
Caprate (assimilation)	−	−	−
Adipate (assimilation)	−	+	−
Malate (assimilation)	+	+	+
Citrate (assimilation)	+	+	+
Utilisation of pectin	+	+	+
Utilisation of cellulose	−	−	−
Utilisation of sodium acetate	+	+	+
Tween 80	+	+	+
Tween 20	+	+	+

+ indicates positive reaction; − indicates negative reaction.

Table 3. Antibiotic sensitivity profile of the isolated *Bacillus* strains.

Antibiotics	Symbol (dose)	*Bacillus* strains		
		T-1	T'-1	I'-1a
		Diameter of inhibition zone (mm)		
Penicillins				
Amoxycillin	AML (25 µg)	16 (R)	NI (R)	14 (R)
Ampicillin	AMP (25 µg)	7 (R)	24 (I)	18 (I)
Cephalosporines:				
Ceftazidime	CAZ (30 µg)	19 (I)	17 (I)	21 (I)
Cephalothin	KF (30 µg)	36 (S)	60 (S)	56 (S)
Cefuroxime	CXM (30 µg)	21 (I)	19 (I)	23 (I)
Quinolones:				
Nalidic acid	NA (30 µg)	16 (R)	1 (R)	13 (R)
Aminoglycosides:				
Amikacin	AK (30 µg)	25 (S)	28 (S)	22 (I)
Doxycylin	DO (30 µg)	20 (I)	25 (S)	18 (I)
Erythromycin	E (30 µg)	34 (S)	28 (S)	34 (S)
Gentamicin	CN (30 µg)	27 (S)	30 (S)	28 (S)
Kanamycin	K (30 µg)	14 (R)	25 (S)	25 (S)
Neomycin	N (30 µg)	21 (I)	22 (I)	22 (I)
Streptomycin	S (25 µg)	17 (I)	28 (S)	21 (I)
Tobramycin	TOB (10 µg)	25 (S)	25 (S)	25 (S)
Tetracyclines:				
Tetracycline	TE (30 µg)	18 (I)	26 (S)	15 (R)
Sulfonamides:				
Trimethoprim	W (5 µg)	23 (I)	20 (I)	24 (I)
Rifampicins:				
Rifampicin	RD (30 µg)	18 (I)	21 (I)	1 (R)
Other:				
Chloramphenicol	C (30 µg)	20 (I)	21 (I)	19 (I)
Nitrofurantoin	F (200 µg)	13 (R)	11 (R)	17 (I)
Novobiocin	NV (30 µg)	19 (I)	13 (R)	20 (I)

NI-no inhibition; Letters in parenthesis indicate sensitivity: R-resistant; I-intermediate; S-sensitive.

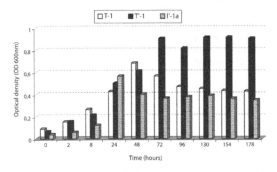

Figure 1. Growth of three *Bacillus* strains in the brewery effluent

a ST of 25 mN/m, an IT < 1.0 mN/m and CMC of 0.025 g/dm^3 (Cooper & Goldenberg 1987). The obtained values for ST, IT and CMC are slightly different from those of Das and Mukherjee (2007).

Medium composition is critical in determining biosurfactant properties (Sheppard & Mulligan 1987). Makkar and Cameotra (1997) cultivated *B. subtilis*

Table 4. Surface active properties of *Bacillus* strains growing on brewery effluent.

Properties	T-1	T'-1	I'-1a
ST (mN/m)	27.31 ± 0.12	30.25 ± 0.22	38.84 ± 0.87
IT (mN/m)	0.36 ± 0.04	0.68 ± 0.11	0.92 ± 0.09
CMD^{-1} (mN/m)	30 ± 1.67	42 ± 0.68	49 ± 0.96
CMD^{-2} (mN/m)	40 ± 2.01	58 ± 1.98	65 ± 0.87
CMC (g/dm^3)	0.166 ± 0.03	0.120 ± 0.1	0.102 ± 0.08
Foamability as FV (%)	70.56 ± 2.5	76.5 ± 1.78	57.9 ± 1.07

Data are means ± SD of three independent experiments; ST of water: 71.79 ± 0.3 mN/m;
ST of brewery effluent: 57.23 ± 0.9 mN/m.

Table 5. Emulsification activity measured by E24 (%) of the supernatants against hydrocarbons and their mixtures.

Hydrocarbons	Bacterial strains		
	T-1	T'-1	I'-1a
Hexadecane (C-16)	28 ± 0.7	9 ± 0.8	13 ± 0.66
Mineral oil	37 ± 0.65	100 ± 1.2	19 ± 1.54
Isooctane	81 ± 1.03	10 ± 0.4	9 ± 0.67
Toluene	79 ± 0.98	10 ± 0.77	13.5 ± 0.56
m-p-xylene	78 ± 0.43	12.5 ± 0.58	11.2 ± 0.78
Petrol	79.5 ± 0.5	9.6 ± 0.45	13 ± 1.01
Diesel oil	70 ± 0.77	81 ± 1.09	79 ± 1.94
Distillates:			
A-1	54.3 ± 1.01	4.5 ± 0.6	11.2 ± 0.91
A-2	86.8 ± 0.98	13.4 ± 0.78	13.4 ± 0.68
A-3	63.4 ± 0.76	6.5 ± 0.92	10 ± 0.88
A-4	83.8 ± 1.6	8.2 ± 0.89	6.4 ± 0.89
A-5	80 ± 2.32	82 ± 1.09	8.5 ± 0.24
P-1	72 ± 1.2	17 ± 0.45	25 ± 0.55
P-2	92 ± 2.04	78 ± 1.76	21.3 ± 1.63
Pristane	20 ± 0.56	19 ± 0.9	8.5 ± 0.88
Frying oil	100 ± 1.8	98 ± 1.56	100 ± 2.1
Average	61.30 ± 0.97	34.92 ± 0.89	22.63 ± 0.99

Fractions: A1 and A2 – components of petrol; A3–P2 are components of diesel oil and light fuel oil; P-2 components of high fuel oil; A – distillation under atmospheric conditions; P – distillation under vacuum pressure; Data are means ± SD of three independent experiments.

MTCC 2423 and 1427 under thermophilic conditions using molasses as a carbon source; the biosurfactant production lowered the ST of the medium to 29 and 31 mN/m for MTCC 2423 and 1427, respectively. Potato substrates were evaluated as a carbon source for surfactant production by *B. subtilis* ATCC 21332 (Fox & Bala 2000, Thompson et al. 2000); ST dropped from 71.3 to 28.3 mN/m, and CMC of 0.10 g/dm^3 was obtained. In addition, Nitschke and Pastore (2004) used a cassava-flour processing effluent as a substrate for surfactant production by *B. subtilis* LB5a and *B. subtilis* ATCC 21332, which reduced ST of the medium to 25.9 and 26.6 mN/m, respectively. Joshi et al. (2008) studied biosurfactant production using molasses and cheese whey under thermophilic conditions by four *Bacillus* strains. ST was reduced 34–37 mN/m for isolates grown under both static and shaken modes. In addition, the yield of biosurfactant produced by isolates in the present study had CMD values 30–50-fold higher compared to those reported for *B. subtilis* (Makkar & Cameotra 1997)

The three *Bacillus* strains were capable of producing biosurfactant in brewery effluent medium. Strains T-1 and T'-1 were better foam inducers (FV of 70 and 76%, respectively) than I'-1a (FV = 58%). All strains were good emulsifiers, foam inducers and ST reducers. The majority of known biosurfactants are synthesised from water-immiscible hydrocarbons. However, *B. subtilis* strains can produce surfactants from water-soluble substances that can be cheap carbon sources. All the qualities suggest that these isolates have tremendous potential in environmental protection.

4 CONCLUSIONS

Our preliminary investigation confirms that the three *Bacillus* strains, growing on brewery waste effluent as organic medium, can produce biosurfactant. Replacing traditional microbiological media with agro-industrial wastes as a substrate for biosurfactant production holds great potential. Moreover, this will reduce many management problems of processing industrial waste. The isolation and characterisation of biosurfactant-producing *Bacillus* strains growing on brewery waste should have considerable application in this regard.

REFERENCES

Adamczak, M. & Bednarski, W. 2000. Influence of medium composition and aeration on the synthesis of biosurfactants produced by *Candida Antarctica*. *Biotechnology Letters* 22(4): 313–316.

Bauer, A.W., Kirby, W.M.M., Sherris, J.C. & Turck, M. 1966. Antibiotic susceptibility testing by a standardized single disc method. *Annual Journal of Clinical Pathology* 45(2): 493–496.

Benincasa, M., Abalos, A., Oliveira, I. & Manresa, A. 2004. Chemical structure, surface properties and biological activities of the biosurfactant produced by *Pseudomonas aeruginosa* LB1 from soapstock. *Antonie van Leeuwenhoek* 85(1): 1–8.

Benincasa, M. & Accorsini, F.R. 2007. *Pseudomonas aeruginosa* LBI production as an integrated process using wastes from sunflower-oil refining as a substrate. *Bioresources Technology* 67(1): 56–63.

Berry, C.J., Story, S., Altman, D.J., Upchurch, R., Whitman, W., Singleton, D., Płaza, G. & Brigmon, R.L. 2006. Biological treatment of petroleum in radiologically contaminated soil. In J. Clayton & A. Lindner Stephenson (eds), *Remediation of hazardous waste in the subsurface. Bridging flask and field*. American Chemical Society: Washington DC.

Bodour, A.A. & Maier, R.M. 2002. Biosurfactants: types, screening methods and application. In *Encyclopedia of Environmental Microbiology* 2. Wiley: New York.

Cooper, D.G. & Goldenberg, B.G. 1987. Surface-active agents from two *Bacillus* species. *Applied and Environmental Microbiology* 53(2): 224–229.

Das, M., Das, S.K. & Mukherjee, R.K. 1998. Surface active properties of the culture filtrates of a *Micrococcus* species grown on n-alkanes and sugars. *Bioresources Technology* 63(1): 231–235.

Das, S.K. & Mukherjee, A.K. 2007. Comparison of lipopeptide biosurfactants production by *Bacillus subtilis* strains in submerged and solid state fermentation systems using a cheap carbon source: Some industrial applications of biosurfactants. *Process Biochemistry* 42(4): 1191–1199.

Deleu, M. & Paquot, M. 2004. From renewable vegetables resources to microorganisms: new trends in surfactants. *Critical Review: Chimie* 7(4): 641–646.

Desai, J.D. & Banat, I.M. 1997. Microbial production of surfactants and their commercial potential. *Microbiology of Molecular Biology Reviews* 61(1): 47–64.

Difco Laboratories Inc. 1984. *DIFCO Manual*, 10th Edition. Detroit, MI.

Dubey, K. & Juwarkar, A. 2001. Distillery and curd whey wastes as viable alternative sources for biosurfactant production. *World Journal of Microbiology and Biotechnology* 17(1): 61–69.

Felsenstein, J. 1989. PHYLIP- phylogeny inference package (version 3.2). *Cladistics* 5: 387–395.

Fox, S.L. & Bala, G.A. 2000. Production of surfactant from *Bacillus subtilis* ATCC 21332 using potato substrates. *Bioresources Technology*. 75(2): 235–240

Gerhardt P. 2006. *Manual of Methods for General Bacteriology*. American Society for Microbiology. Washington DC.

Haba, E., Espuny, M.J., Busquets, M. & Manresa, A. 2000. Screening and production of rhamnolipides by *Pseudomonas aeruginosa* 47T2 NCIB 40044 from waste frying oils, *Journal of Applied Microbiology* 88(2): 379–387.

Joshi, S., Bharucha, C., Jha, S., Yadav, S., Nerurkar, A. & Desai, A.J. 2008. Biosurfactant production using molasses and whey under thermophilic conditions. *Bioresources Technology* 99(1): 195–199.

Kosaric, N. 2000. *Biosurfactants. Production. Properties. Applications*. Marcel Dekker Inc.: New York, USA.

Makkar, R. & Cameotra, S.S. 1997. Utilization of molasses for biosurfactant production by two *Bacillus* strains at thermophilic conditions. *Journal of the American Oil Chemist's Society* 74(6): 887–889.

Makkar, R.S. & Cameotra, S.S. 1999. Biosurfactant production by microorganisms on unconventional carbon sources. *Journal of Surface Detergents* 2(1): 237–241.

Makkar, R.S. & Cameotra, S.S. 2002. An update on the use of unconventional substrates for biosurfactant production and their new applications. *Applied Microbiology and Biotechnology* 58(2): 428–434.

Maneerat, S. 2005. Production of biosurfactants using substrates from renewable-resources. *Journal of Science Technology* 27(3): 675–683.

Mercade, M.E. & Manresa, M.A. 1994. The use of agroindustrial by-products for biosurfactant production. *Journal of the American Oil Chemist's Society* 71(1): 61–64.

Nitschke, M., Ferraz, C. & Pastore, G.M. 2004. Selection of microorganisms for biosurfactant production using agroindustrial wastes. *Brazilian Journal of Microbiology* 35(1-2): 81–85.

Nitschke, M. & Pastore, G.M. 2006. Production and properties of a surfactant obtained from *Bacillus subtilis* grown on cassava wastewater. *Bioresources Technology* 97(2): 336–341.

Płaza, G., Zjawiony, I. & Banat, I.M. 2006. Use of different methods for detection of thermophilic biosurfactant-producing bacteria from hydrocarbon-contaminated and bioremediated soils. *Journal of Petroleum Science and Engineering* 50(1): 71–77.

Sheppard, J.D. & Mulligan, C.N. 1987. The production of surfactin by *Bacillus subtilis* grown on peat hydrolysate. *Applied Microbiology and Biotechnology* 27(1): 110–116.

Singh, A. & Hamme, J.D., Ward, O.P. 2007. Surfactants in microbiology and biotechnology: Part 2. Application aspects. *Biotechnology Advances* 25(1): 99–121.

Thompson, D.N., Fox, S.L. & Bala, G.A. 2000. Biosurfactants from potato process effluents. *Applied Biochemistry and Biotechnology* 84-86(5): 917–930.

Environmental Engineering III – Pawłowski, Dudzińska & Pawłowski (eds)
© *2010 Taylor & Francis Group, London, ISBN 978-0-415-54882-3*

Nitrogen and phosphorus removal paths in a sequencing batch reactor – dependence on a dissolved oxygen profile in aerobic phases

J. Podedworna & M. Żubrowska-Sudoł

Faculty of Environmental Engineering, Warsaw University of Technology, Warsaw, Poland

ABSTRACT: This paper presents results of wastewater treatment in a sequencing batch reactor operating at a dissolved oxygen (DO) setpoint of 1 mg O_2 l^{-1} in the aerobic phase. It has been demonstrated that depending on the DO concentration profile (in time) in aerobic phases, paths of N and P removal in the reactor are different. Simultaneous denitrification depends on the amount of time when DO concentration is below 1 mg O_2 l^{-1}. A relatively large uptake of easily biodegradable organic substrates in the first anoxic/anaerobic phase (I A/A) results in the reduction of nitrates in aerobic phases with internally stored organics as a carbon source. This may promote the growth of denitrifying phosphorus-accumulating organisms in the process. In the second A/A phase, synergic N and P removal was observed. The process was limited by nitrite and nitrate concentrations.

Keywords: Dissolved oxygen concentration, sequencing batch reactor, simultaneous nitrification/denitrification.

1 INTRODUCTION

Oxygen concentration is one of the most important abiotic factors influencing biochemical processes in activated sludge and biofilm systems. Its concentration affects in particular such processes as nitrification and phosphorus (ortho-P) uptake. A lack of dissolved oxygen (DO), on the other hand, promotes denitrification and release of orthophosphates (PO_4^{3-}). In the past, when designing biological treatment, bioreactors for aerobic and anaerobic/anoxic processes were clearly divided. In aerobic reactors, a decrease of DO concentrations to below 2 mg O_2 l^{-1} was regarded as inhibitory to ammonia oxidation. New scientific discoveries, however, led to the development of knowledge about cell biochemistry and the identification of different bacterial strains responsible for certain biochemical processes. Strenuous efforts to minimize the operational costs of wastewater treatment plants resulted in significant changes in their design and operation.

Nowadays, it is commonly acknowledged that highly efficient nitrification is likely to occur at DO concentration of 1 mg O_2/dm^3 and that denitrification can take place not only in separate anoxic reactors or anoxic phases, but also simultaneously with nitrification (inside the activated sludge flocs) (Munch et al. 1996; Pochana & Keller 1999, Beun et al. 2001, Satoh et al. 2003). Simultaneous nitrification/denitrification (SND) was observed in aerobic chambers of continuous flow systems and during aerobic phases in sequencing batch reactor (SBR) systems (Mosquera-Corral et al. 2005, Third et al. 2005). The discovery of denitrifying phosphorus accumulating organisms (DPAO) capable of taking up ortho-phosphates

under anoxic conditions was yet another step in the advancement of scientific awareness concerning the conditions required for N and P removal and synergy of these reactions. This synergy can lead to a possible reduction in the amount of organic carbon required for both processes and smaller surplus sludge production (Kuba et al. 1996, Broughton et al. 2008, Zhou et al. 2008). An additional advantage of denitrifying phosphorus removal is the conservation of energy, resulting from a lower oxygen demand for ortho-P accumulation in comparison with the fully aerobic ortho-P accumulation process (Kuba et al. 1996, Merzouki et al. 2001).

The results presented in this article form part of a research project on nutrient removal from wastewaters using a process of denitrifying P removal. One of the project objectives was to test the possibility of a synergic removal of N and P in an SBR. Initially, the SBR operating conditions were set up with the assumption that it was absolutely necessary to remove all easily biodegradable substrates in the anaerobic phase and to create anoxic conditions after the anaerobic phase. It was difficult to achieve this particular phase sequence in an SBR due to the lack of an inner recirculation stream such as that present in continuous flow systems. Therefore, it was assumed that if DO concentrations in aerobic phases were low, it would be possible to obtain anoxic conditions in the inner parts of the flocs (Pochana and Keller, 1999) and thus to achieve simultaneous reduction of nitrates and oxidation of ammonia. Data from the literature and results of previous research indicate that the optimum DO concentration for simultaneous nitrification/denitrification is 1 mg O_2/dm^3 (Third et al. 2003, Żubrowska-Sudoł & Podedworna 2004). Also, if the

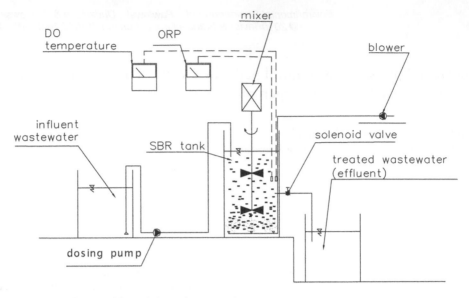

Figure 1. Schematic diagram of the experimental setup.

source of carbon is internally stored inside of a certain group of phosphorus-accumulating microorganisms, then denitrification may occur simultaneously with the accumulation of ortho-phosphates. The role of polyhydroksybutyrate (PHB) as an electron donor in simultaneous nitrification/denitrification has already been explained by Beun et al. (2001) and Third et al. (2003).

The main objective of the research was to assess nitrogen and phosphorus removal paths in a sequencing batch reactor operating at a DO setpoint of $1\,mg\,O_2\ l^{-1}$. During the experiment, it was noted that, despite providing automatic control of DO and comparable operational parameters of the SBR, the DO concentration profile (over time) varied. It was assumed that this could have an impact on nitrogen and phosphorus removal paths. Therefore, the problem was analyzed in this study.

2 MATERIALS AND METHODS

The experiments were carried out in a laboratory-scale SBR reactor with an operating volume of 28 litres (Figure 1) equipped with a mechanical paddle mixer operating continuously (with the exception of sedimentation, decant and idle phases) at a rotational speed of 150 rpm. Compressed air during aerobic phases (Ox) was supplied by an air blower and aquarium air diffusers located at the bottom of the reactor. Operation of the air blower was controlled to give a constant DO setpoint of $1\,mg\,O_2/dm^3$. DO concentration in the tank was measured with an on-line DO probe. The oxidation reduction potential (ORP) was also measured by the on-line ORP probe placed in the bioreactor. Wastewater was fed into the bioreactor by a peristaltic pump. A system of timer switches triggering the feed pump on and off, the mixer, the blower and the discharge valve

Table 1. Sequence and duration of phases/stages in a cycle.

Phase	Time [min]
I anoxic/anaerobic (I A/A):	90
fill of wastewater:	2/3 Q*
	(first of 30 minutes)
I aerobic (I Ox), in this:	105
First stage	60
Second stage	45
II anoxic/anaerobic (II A/A):	30
fill of wastewater:	1/3 Q*
	(first of 15 minutes)
II aerobic (II Ox), in this:	135
First stage	60
Second stage	75
Sedimentation (S)	90
Decant (D)	15
Idle (I)	15

* the total feed volume.

were set to control the unit operations of the SBR. The reactor was operated in three 8-hour cycles per day. The sequence and duration of phases in a cycle is shown in Table 1.

The reactor was fed with synthetic wastewater composed of peptone, ammonium acetate, glucose, starch, glycerol, acetic acid and phosphorus salts. The reactor was fed twice in a cycle at the beginning of each consecutive anoxic-anaerobic (A/A) phase. During the first feeding, two thirds of the total feed volume was added.

Chemical analyses were conducted to assess the characteristics of raw and treated wastewater (unfiltered samples) and to monitor the reactor's operation – various indicator pollutants in the bioreactor were measured after each phase of a cycle and during both aerobic phases (samples filtered through

Table 2.   Characteristic values of operational parameters.

	Activated sludge concentration [g MLSS/dm³]	Sludge loading rate (F:M) [g COD g⁻¹ MLSS d⁻¹]	Sludge retention time (SRT) [d]
Minimum	2.53	0.157	8.23
Maximum	3.94	0.244	9.70
Average	3.24	0.205	8.95

Table 3.   Raw sewage and treated effluent characteristics.

		COD [mg O₂/dm³]	N-NH₄⁺ [mg N-NH₄⁺ l⁻¹]	N-NOₓ [mg N-NOₓ/dm³]	TN [mg N/dm³]	TP [mg P/dm³]
Raw sewage	Min.	555	33.0	0.036	54.2	9.35
	Max.	737	54.0	0.372	81.1	17.7
	Mean	654	43.2	0.166	66.7	13.36
Treated effluent	Min.	17.5	0	1.72	3.95	0.60
	Max.	47.1	0.33	7.95	9.37	1.0
	Mean	29.5	0.19	4.81	7.12	0.76

Table 4.   Effectiveness of treatment processes based on influent and effluent characteristics.

		Organic C removal	Nitrification	Denitrification	Phosphorus removal
Effectiveness [%]	Min.	93.1	99.2	87.1	90.8
	Max.	97.5	100	97.3	96.2
	Mean.	95.5	99.7	92.5	94.0

0.45 μm filter papers). All chemical analyses were carried out in accordance with Standard Methods (Eaton et al. 1995). Additionally, one Phosphorus Uptake Rate (PUR) batch test was carried out in order to determine the fraction of denitrifying phosphorus-accumulating organisms (DPAO) in the total PAO biomass (Wachtmaister et al. 1997).

3   RESULTS AND DISCUSSION

3.1   Removal efficiency

The experiment was carried out at average mixed liquor suspended solids (MLSS) concentrations (at full operational volume) of 3.24 mg MLSS/dm³, a sludge retention time (SRT) of 9 days and an average sludge loading rate of 0.205 g COD g⁻¹ MLSS d⁻¹ (Table 2).

During the whole period of the experiment, very high COD removal rates with effluent COD concentrations of 17.5–47.1 mg O₂/dm³ (Table 3) and an average COD removal of 95.5% (Table 4) were observed. As expected, the DO setpoint of 1 mg O₂/dm³ was sufficient to guarantee highly efficient nitrification (Table 4) with an ammo-niacal-N concentration in the effluent less than 0.33 mg N-NH₄⁺ dm³ (Table 3). These results are in agreement with the observations of Hidaka et al. (2002), who managed to produce effluent ammoniacal-N concentrations below 1 mg N-NH₄/dm³ in the continuous flow reactor with

alternating aeration at an SRT of 10 days and DO concentrations below 1 mg O₂/dm³.

In our study, the efficiency of denitrification in the SBR has always been within the range of 87.1–97.3% (Table 4). The removal efficiency corresponded to effluent N-NOₓ concentrations of 1.72–7.95 mg/dm³ (Table 3). Total nitrogen (TN) removal ranged from 86.3 to 94.1% with effluent TN concentrations of 3.95–9.37 mg/dm³ (Table 3). The effluent total phosphorus (TP) remained ≤1 mg/dm³ (average 0.76 mg/dm³) which indicates high efficiency of biological P removal (90.8–96.2%) – Tables 3 and 4 respectively.

3.2   Monitoring of nitrogen and phosphorus removal paths – dependence on DO profile in aerobic phases in a cycle

Changes in DO concentrations in three consecutive monitoring experiments (M1, M2, M3) are shown in Figure 2. The air blower switched on when the DO concentration in the bioreactor (on-line measurement) fell below 1 mg O₂/dm³, and switched off after exceeding this level. However, due to the relatively high output rate of the air blower, DO concentrations tended to overshoot the DO setpoint to concentrations reaching 1.75 mg O₂/dm³ even at device switch-ons lasting several seconds. In the experiment M1, the DO concentration in the reactor reached 1 mg O₂/dm³ after 20 minutes of the first aerobic phase and remained equal to or greater than this value for the most part of this

Monitoring test no. 1

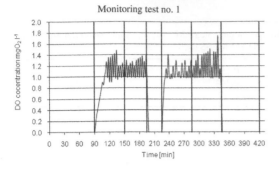

Monitoring test no. 2

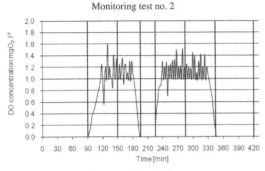

Monitoring test no. 3

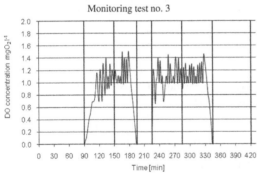

Figure 2. Changes in dissolved oxygen (DO) concentrations in consecutive phases of a cycle.

phase. In the second aerobic phase, the DO concentration rose to 1 mg O_2/dm^3 during the first 5 minutes and then remained within the range of 0.97–1.75 mg O_2/dm^3 until the end of the phase. In the monitoring experiment M1, due to the phase setup, DO at the beginning of the second aerobic phase was observed as originating from the end of the previous phase. Therefore, before the monitoring experiments M2 and M3 were carried out, the aeration system control was modified. After this modification, the air blower was switched off 15 minutes before the start of the second fill. This allowed the second anoxic/aerobic (A/A) phase to start with no initial concentration of DO. In addition, in the monitoring experiments M2 and M3, it was observed that upon the start of both aerobic phases, DO remained below 1 mg O_2/dm^3 for longer periods than those measured in M1.

Observed differences in the changes of DO concentrations over time within a cycle, as analyzed

in monitoring experiments, had an effect on N and P removal in the system. Tables 5, 6 and 7 show the following information respectively for monitoring experiments M1, M2 and M3:– phase sludge loading with COD, N-NH$_4$ and N-NO$_x$ (L_{COD}, L_{N-NH4} and L_{N-NOx}). The value of phase sludge loading was determined as a total loading of pollutant remaining in the reactor after the previous phase and loading introduced to the reactor with raw sewage (in the sewage feeding phases). Also, in the case of L_{N-NH4} and L_{N-NOx}, loading coming from ammonification and nitrification processes were included in the calculation.

– unit phase efficiencies of C removal, nitrification, denitrification and release or uptake of P-PO$_4^{3-}$ processes per gram of MLSS (E_{COD}, $E_{nitr.}$, $E_{den.}$, E_r^P, E_u^P)
– values of wastewater constituents at the end of the monitored stage or phase.

The analysis of data presented in the tables above shows that denitrification was occurring both in the A/A and Ox phases, but that the reduction of nitrates at different stages of a cycle were different for all three monitoring investigations. In the monitoring experiment M1, the process of simultaneous nitrification/denitrification could be observed only in the first stage of the first aerobic phase. However, its efficiency was very low (0.32 mg N-NO$_x$ (g MLSS·phase)$^{-1}$) – only 23.4% of overall nitrification efficiency at this stage (Tables 5, 8). The analysis of DO concentration changes over time proved that during the analyzed stage, DO concentrations were, for part of the time, below 1 mg O_2 l^{-1} (around 20 minutes each). According to other published data, those low DO concentrations are favourable for simultaneous nitrification/denitrification (Munch et al. 1996, Beun et al. 2001, Third et al. 2003, Third et al. 2005). It is known that an increase of DO concentrations in the bulk liquid leads to higher O_2 diffusion into activated sludge flocs, which is inhibitory to denitrification. It is probable that this kind of situation took place in the second stage of the first aerobic phase and during the second aerobic phase in monitoring experiment no. 1.

As the denitrification process observed in the first stage of the first aerobic phase was followed by only a slight decrease in COD concentrations (from 46.4 to 34.9 mg O_2/dm^3, Table 5), it could be assumed that microorganisms used an internal organic carbon source for denitrification (stored during the first A/A phase). It was very unlikely that soluble organic substrates present in the first stage of the first aerobic phase were not aerobically oxidized prior to diffusing into the deeper layers of the activated sludge flocs.

In monitoring experiment no. 1 (Table 5), the highest denitrification rate was observed during the second A/A phase (1.94 mg N-NO$_x$ (g MLSS·phase)$^{-1}$). Almost complete reduction of the N-NO$_x$ remaining after the first aerobic phase was noted at that time. As denitrification was associated with the uptake of orthophosphates (0.26 mg P-PO$_4^{3-}$ (g MLSS·phase)$^{-1}$), this proved the occurrence of the denitrifying P removal process. Concentration of

Table 5. Results of monitoring test no. 1.

Process parameter	Unit	I A/A	I Ox		II A/A	II Ox		S
			1st stage	2nd stage		1st stage	2nd stage	
L_{COD}	mg COD (g MLSS·phase)$^{-1}$	47.6	10.6	8.7	29.2	10.0	8.10	7.85
E_{COD}	mg COD (g MLSS·phase)$^{-1}$	37.0	2.64	–	19.1	1.95	1.82	–
COD[a]	mg O_2/dm^3	46.4	34.9	38.1	38.8	31.3	24.2	30.8
$L_{N\text{-}NH4}$	mg N-NH$_4^+$ (g MLSS·phase)$^{-1}$	6.60	3.02	1.86	2.56	1.90	0.53	0.42
$E_{nitr.}$	mg N-NH$_4^+$ (g MLSS·phase)$^{-1}$	–	1.37	1.05	–	1.65	0.19	–
N-NH$_4^+$[a]	mg N-NH$_4^+$/dm^3	13.3	7.25	3.55	5.00	0.95	1.31	0.75
NTK[a]	mg NTK/dm^3	15.2	9.20	5.00	8.20	2.80	2.40	2.60
$L_{N\text{-}NOx}$	mg N-NO$_x$ (g MLSS·phase)$^{-1}$	0.59	1.38	2.11	2.14	1.85	2.04	2.04
$E_{den.}$	mg N-NO$_x$ (g MLSS·phase)$^{-1}$	0.57	0.32	–	1.94	–	–	0.39
N-NO$_x$[a]	mg N-NO$_x$/dm^3	0.063	4.66	9.24	0.75	6.0	6.40	4.9
E_r^P	mg P-PO$_4^{3-}$ (g MLSS·phase)$^{-1}$	4.88	–	–	–	–	–	–
E_u^P	mg P-PO$_4^{3-}$ (g MLSS·phase)$^{-1}$	–	5.60	0.06	0.26	0.01	–	–
P-PO$_4^{3-}$[a]	mg P-PO$_4^{3-}$/dm^3	24.8	0.27	0.02	0.06	0.01	0.01	0.02

[a] Concentration at the end of the stage/phase of the cycle.
– not observed.

Table 6. Results of monitoring test no. 2.

Process parameter	Unit	I A/A	I Ox		II A/A	II Ox		S
			1st stage	2nd stage		1st stage	2nd stage	
L_{COD}	mg COD (g MLSS·phase)$^{-1}$	44.8	10.2	10.2	31.1	15.6	14.2	12.5
E_{COD}	mg COD (g MLSS·phase)$^{-1}$	39.3	–	0.21	15.5	1.34	1.39	7.38
COD[a]	mg O_2/dm^3	20.3	38.0	37.2	50.8	46.4	41.9	17.8
$L_{N\text{-}NH4}$	mg N-NH$_4^+$ (g MLSS·phase)$^{-1}$	6.70	3.45	1.91	3.13	2.33	0.77	0.48
$E_{nitr.}$	mg N-NH$_4^+$ (g MLSS·phase)$^{-1}$	–	1.77	1.07	–	1.84	0.28	–
N-NH$_4^+$[a]	mg N-NH$_4^+$/dm^3	9.75	6.25	3.10	7.15	1.59	1.60	0.87
NTK[a]	mg NTK/dm^3	14.8	8.20	4.20	9.40	3.40	3.40	2.20
$L_{N\text{-}NOx}$	mg N-NO$_x$ (g MLSS·phase)$^{-1}$	0.44	1.79	2.22	1.87	1.90	1.42	1.42
$E_{den.}$	mg N-NO$_x$ (g MLSS·phase)$^{-1}$	0.42	0.65	0.36	1.81	0.75	–	0.15
N-NO$_x$[a]	mg N-NO$_x$/dm^3	0.068	4.25	6.93	0.20	3.74	4.64	4.16
E_r^P	mg P-PO$_4^{3-}$ (g MLSS·phase)$^{-1}$	4.08	–	–	–	–	–	–
E_u^P	mg P-PO$_4^{3-}$ (g MLSS·phase)$^{-1}$	–	5.21	0.03	0.39	0.02	–	–
P-PO$_4^{3-}$[a]	mg P-PO$_4^{3-}$/dm^3	19.5	0.13	0.02	0.08	0.02	0.02	0.01

[a] Concentration at the end of the stage/phase of the cycle.
– not observed.

ortho-P after the second A/A phase was only 0.06 mg P/dm^3. This meant that almost the total load of orthophosphates entering the reactor during the second fill was removed under anoxic conditions.

Comparison of ortho-P accumulation rates in different stages/phases of a cycle showed that the highest rate of this process was observed in the first stage of the first aerobic phase (5.6 mg P-PO$_4^{3-}$ (g MLSS·phase)$^{-1}$). The measured effluent ortho-P concentration after this stage was only 0.27 mg P-PO$_4^{3-}$/dm^3. As it was deemed that simultaneous denitrification at this stage was occurring with use of internally stored substrates, it cannot be disregarded that a portion of N and P was then removed in the process of denitrifying P removal. The possibility of using PHB as an electron donor for simultaneous denitrification has already been reported by Beun et al. (2001) and Third et al. (2003).

DO profiles in a cycle during monitoring experiment no. 2 suggest more favourable conditions for simultaneous denitrification processes due to longer periods with DO concentrations under 1 mg O_2/dm^3 (around 60 minutes each; Figure 2). The SND process was observed in both stages of the first aerobic phase

Table 7. Results of monitoring test no. 3.

Process parameter	Unit	I A/A	I Ox 1st stage	I Ox 2nd stage	II A/A	II Ox 1st stage	II Ox 2nd stage	S
L_{COD}	mg COD (g MLSS·phase)$^{-1}$	60.5	12.3	12.9	41.0	9.91	11.6	11.9
E_{COD}	mg COD (g MLSS·phase)$^{-1}$	48.2	0.62	–	31.5	–	–	–
COD[a]	mg O_2/dm^3	35.0	33.2	36.6	23.8	24.7	29.1	30.2
L_{N-NH4}	mg N-NH$_4^+$ (g MLSS·phase)$^{-1}$	6.52	4.21	2.43	3.32	3.14	1.88	0.75
$E_{nitr.}$	mg N-NH$_4^+$ (g MLSS·phase)$^{-1}$	–	2.32	0.99	–	1.84	1.11	–
N-NH$_4^{+}$[a]	mg N-NH$_4^+$/dm^3	7.00	5.38	4.10	7.80	3.24	1.91	0.88
NTK[a]	mg NTK/dm^3	15.6	9.00	6.20	9.40	4.80	4.00	1.60
L_{N-NOx}	mg N-NO$_x$ (g MLSS·phase)$^{-1}$	0.47	2.40	1.46	1.20	1.89	2.07	2.07
$E_{den.}$	mg N-NO$_x$ (g MLSS·phase)$^{-1}$	0.39	1.92	0.28	1.16	0.93	–	1.57
N-NO$_x$[a]	mg N-NO$_x$/dm^3	0.20	1.36	3.38	0.11	2.37	5.16	1.32
E_{-r}^{P}	mg P-PO$_4^{3-}$ (g MLSS·phase)$^{-1}$	11.7	–	–	0.27	–	–	–
E_{u}^{P}	mg P-PO$_4^{3-}$ (g MLSS·phase)$^{-1}$	–	13.2	0.01	–	0.73	0.004	–
P-PO$_4^{3-}$[a]	mg P-PO$_4^{3-}$/dm^3	37.5	0.11	0.07	1.85	0.03	0.02	0.02

[a] Concentration at the end of the stage/phase of the cycle.
– not observed.

Table 8. Efficiency of simultaneous denitrification in the aerobic phases expressed as a percentage of nitrification efficiency.

Monitoring test no.:	Unit	I Ox 1st stage	I Ox 2nd stage	II Ox 1st stage	II Ox 2nd stage
1	%	23.4	–	–	–
2	%	36.7	33.6	40.8	–
3	%	82.6	28.3	50.5	–

– not observed.

and in the first stage of the second aerobic stage. Calculated SND rates were 0.65, 0.36 and 0.75 mg N-NO$_x$ (g MLSS·phase)$^{-1}$ respectively (Table 6) which constituted 36.7%, 33.6% and 40.8% of nitrification rates in these stages (Table 8). A higher denitrification rate in the first stage of the second aerobic phase, despite less favourable aerobic conditions, may be caused by a higher available load of N-NO$_x$ (Table 6). Similarly to that in the monitoring experiment M1, denitrification in the monitoring experiment M2 also took place with internally stored substrates as a C source. However, only in the first stage of the first aerobic phase could denitrification occur simultaneously with P removal, as only then could both nitrate and ortho-P removal be observed at the same time. It is important to emphasize that 92.2% of orthophosphates removed in the complete cycle occurred in this very stage and only 6.9% of orthophosphate load was accumulated in the biomass in the second A/A phase (Table 6).

During the monitoring experiment M3, very high simultaneous denitrification efficiency exceeding 82% of the efficiency of nitrification was achieved in the first stage of the first aerobic phase (Tables 7

and 8). At the same time, this stage of the experiment achieved the highest denitrification rate out of all monitoring studies (1.92 mg N-NO$_x$ (g MLSS·phase)$^{-1}$, Table 7). In this treatment cycle, a significantly high rate of simultaneous denitrification was recorded for the first stage of the second aerobic phase (0.93 mg N-NO$_x$ g^{-1} MLSS, Table 7).

Similarly to that in the previous monitoring experiment, the highest orthophosphate uptake rate (13.2 mg P-PO$_4^{3-}$ (g MLSS·phase)$^{-1}$) was observed during the first stage of the first aerobic phase, where a significant reduction of ortho-P concentra-tion from 37.5 mg P-PO$_4^{3-}$/dm^3 to 0.11 mg P-PO$_4^{3-}$/dm^3 was noted. It was also the highest observed P-PO$_4^{3-}$ uptake rate of all the monitoring studies (Table 7). It could be assumed that this high uptake rate was due to the fact that in this phase, the highest sludge loading rate (L$_{COD}$) was observed (Table 7). This may have led to higher internal organic substrate storage in the PAO mass and higher orthophosphate release rates (Table 7).

Different from previous monitoring experiments was that no P-PO$_4^{3-}$ accumulation was observed in the second A/A phase of the monitoring study no. 3. The load of ortho-P entering the reactor during the second feed was removed in the first stage of the second aerobic phase. It is suspected that the lack of P-PO$_4^{3-}$ accumulation in the second A/A phase was due to a very low concentration of N-NO$_x$ in the bulk liquid (3.38 mg N-NO$_x$/dm^3, Table 7). For comparison, in the monitoring experiments M1 and M2, nitrite and nitrate concentrations at the beginning of a corresponding stage were 9.24 mg N-NO$_x$ l^{-1} and 6.93 mg N-NO$_x$/dm^3 respectively (Tables 5 and 6). Significant dependency of anoxic P-PO$_4^{3-}$ uptake on nitrate concentration was indicated for example by Peng et al. (2006).

The results obtained show that simultaneous denitrification occurred in all monitoring experiments. The efficiencies of this process are compared to

Table 9. Percentage share of denitrification in consecutive phases in the total denitrification outcome in a cycle.

Monitoring no.:	Unit	I A/A	I Ox 1 stage	II A/A 2 stage	II Ox	Sedimentation 1 stage	2 stage	
1	%	17.7	9.94	–	60.25	–	–	12.11
2	%	10.14	15.7	8.70	43.72	18.12	–	3.62
3	%	6.24	30.72	4.48	18.56	14.88	–	25.12

– not observed.

nitrification efficiencies in Table 8, and in Table 9 are presented as a percentage share of the total denitrification process efficiency during the whole cycle. As indicated by measured data, in the monitoring experiment no. 1, simultaneous denitrifycation present only in the first stage of the first aerobic phase made up only 9.94% of total denitrification in the complete cycle (the period for which the DO concentration was lower than 1 mg O_2/dm^3 was about 20 minutes, which is 8% of the total aerobic period in the reactor per cycle). In the monitoring study M2, nitrate reduction occurred simultaneously with ammonia oxidation in both stages of the first aerobic phase with efficiencies of 15.7% and 8.7% of the total process efficiency in a complete cycle. In the second aerobic phase, simultaneous denitrification made up over 18% of total nitrate reduction in a cycle. In both aerobic phases, the share of simultaneous denitrification in the total denitrification process in a cycle was over 42.5% (the period for which the DO concentration was lower than 1 mg O_2/dm^3 was about 60 minutes, which represents 25% of the total aerobic period in the reactor per cycle). In the monitoring exercise no. 3, simultaneous denitrification occurred in the same treatment stages as for the monitoring experiment M2. Its efficiency in the first and second stage of the first aerobic phase and in the first stage of the second aerobic phase was 30.72%, 4.48% and 14.88% of total nitrate removal in a cycle respectively. This constitutes over 50% of total denitrifycation (the period for which the DO concentration was lower than 1 mg O_2/dm^3 was about 60 minutes, similar to M2). A similar contribution of SND (52%) in the total denitrification in an SBR was reported by Third et al. (2005). The experimental results presented in this article indicate that part of simultaneous denitrification in the total denitrification within a cycle increases along with the duration of periods when DO concentration falls under 1 mg O_2/dm^3 in aerobic phases of the SBR. In the experiments performed, such periods occurred mainly at the initial stages of aerobic phases when supplied oxygen was used immediately for biomass respiration. Very similar behaviour was also observed by Arnz et al. (2001) during their research on nutrient removal in a sequencing batch biofilm reactor (SBBR) reactor.

Despite the lack of anoxic conditions directly after anaerobic phases, denitrifying P removal could occur in the parts of the aerobic phases when DO

Table 10. Results of the Phosphorus Uptake Rate (PUR) test.

Name	Unit	Value
P-PO_4^{3-} accumulation under aerobic conditions	mg P-PO_4^{3-} g^{-1} MLSS	25.7
P-PO_4^{3-} accumulation under anoxic conditions	mg P-PO_4^{3-} g^{-1} MLSS	13.1
DPAO/PAO	%	51.0

concentration was around 1 mg O_2/dm^3. This assumption was confirmed by an almost unnoticeable reduction in organic substrate concentrations and the denitrification process with simultaneous phosphorrus uptake observed during these stages of the cycle. Synergic removal of N and P in aerobic phases had to take place inside the activated sludge flocs where low oxygen levels (limited by diffusion) and the presence of nitrates (the product of nitrification occurring at the surface of the flocs), led to anoxic conditions favouring the denitrification process. A Phosphorus Uptake Rate (PUR) test carried out close to the end of the experiment provided evidence that denitrifying phosphorus-accumulating organisms (DPAO) were present in the biomass and that their share of the total PAO mass stayed at a level of 51% (Table 10). The removal of nitrates in the process of simultaneous nitrification/denitrification together with phosphorus removal has also been observed in SBR reactors by Garzón-Zúñiga and González-Martínez (1996), Helness and Ødegaard (2001) and Gieseke et al. (2002).

Although conditions favourable for SND have been successfully created in the Ox phases, only at the first stage of the first aerobic phase was it possible to achieve N and P removal in a denitrifying P removal process (due to the short periods when orthophosphates were present in the liquor). Synergic removal of both nutrients was also observed in the second A/A phase. However, the process was limited by N-NO_x concentrations.

4 CONCLUSIONS

1. Unit treatment processes in the sequencing batch reactor were influenced not only by a DO setpoint

in aerobic phases but also by a DO profile over time being a function of the oxygen uptake rate in the reactor. The experiments showed that, at the DO setpoint in aerobic phases of 1 mg O_2 l^{-1}, simultaneous denitrification occurring inside activated sludge flocs depended on the length of periods in which, due to DO concentration fluctuations, this parameter fell below a setpoint value. In the case when the DO concentration was below 1 mg O_2/dm^3 in aerobic conditions for 25% of the total duration of aerobic phases, the total efficiency of simultaneous denitrification exceeded 40% of the total efficiency of nitrogen reduction.

2. A relatively large uptake of easily biodegradable organic substrates in the first anoxic/anaerobic phase (organic substrate limitation in the first Ox phase) resulted in a situation where reduction of nitrates in aerobic phases occurred with internally stored organics as a carbon source. This may potentially promote growth of DPAO in this process.

3. In the second A/A phase, synergic N and P removal was observed. Denitrifying P removal was then limited by $N-NO_x$ concentrations and was not observed at $N-NO_x$ concentrations slightly below 3.5 mg/dm^3.

4. It has been demonstrated that, depending on the evolution of DO concentrations over time in aerobic phases, the paths of N and P removal in the reactor are different. Despite this fact, similar treatment efficiencies have been attained throughout the entire research exercise:

- COD removal efficiency between 93.1% and 97.5% (95.5% on average) and effluent COD concentrations below 50 mg O_2/dm^3;
- Nitrification efficiency between 99.2% and 100% (99.7% on average) with effluent ammoniacal N concentrations below 0.5 mg $N-NH_4^+/dm^3$;
- Denitrification efficiency from 87.1% to 97.3% (92.5% on average) with effluent total nitrogen concentrations of less than 10 mg N/dm^3;
- TP removal efficiency between 90.8% and 96.2% (94% on average) and effluent TP concentrations less than or equal to 1.0 mg P/dm^3.

The results of the experiments presented here were crucial for implementing changes in the reactor operation. In the following research, during aerobic phases of the cycle, the air blower was turned off for part of the time. As a result, the period when DO concentration was below 1 mg O_2/dm^3 lengthened to 50% of the total time of aerobic phases in the cycle. Under these operating conditions, the total efficiency of simultaneous denitrification in aerobic phases exceeded 80% of the total efficiency of oxidized nitrogen reduction in the cycle (Podedworna & Żubrowska-Sudoł 2009).

ACKNOWLEDGMENTS

This research was carried out as a part of a research project entitled "Investigations of nutrient removal from wastewater in a process of denitrifying phosphorus removal" (research project number N207 018 31/1002), financed by the Ministry of Science and Higher Education (Poland).

REFERENCES

Arnz, P., Arnold, E. & Wilderer, P.A. 2001. Enhanced biological phosphorus removal in a semi full-scale SBBR, *Water Science and Technology* 43(3): 167–174.

Beun, J.J., van Loosdrecht, M.C.M. & Heijnen, J.J. 2001. N removal in granular sludge sequencing batch airlift reactor. *Biotechnology and Bioengineering* 75(1): 82–92.

Broughton, A. Pratt, S. & Schilton, A. 2008. Enhanced biological phosphorus removal for high-strength wastewater with a low rbCOD:P ratio. *Bioresource Technology* 99: 1236–1241.

Eaton, A.D., Franson, M.A.H. & Clesceri, L.S. 1995. Standards methods for the examination of water and wastewater, 19th ed., American Public Health Association/American Water Works Association/Water Environment Federation, Washington, DC.

Garzón-Zúñiga, M. A. & González-Martínez, S. 1996. Biological phosphate and nitrogen removal in a biofilm sequencing batch reactor. *Water Science and Technology* 34(1–2): 293–301.

Gieske, A., Arnz, P., Amann, R. & Schramm, A. 2002. Simultaneous P and N removal in a sequencing batch biofilm reactor: insights from reactor- and microscale investigations. *Water Research* 36: 501–509.

Helness, H. & Ødegaard H. 2001. Biological phosphorus and nitrogen removal in a sequencing batch moving bed biofilm reactor. *Water Science and Technology* 43(1): 233–240.

Hidaka, T., Yamada, H., Kawamura, M. & Tsuno, H. 2002. Effect of dissolved oxygen conditions on nitrogen removal in continuously fed intermittent-aeration process with two tanks. *Water Science and Technology* 45(12): 181–188.

Kuba, T., van Loosdrecht, M. & Heijnen, J. 1996. Phosphorus and nitrogen removal with minimal COD requirement by integration of denitrifying dephosphatation and nitrification in a two-stage system. *Water Research* 30(7): 1702–1710.

Merzouki, M., Bernet, N., Delgenes, J., Moletta, R. & Benlenmlih, M. 2001. Biological denitrifying phosphorus removal in SBR: effect of added nitrate concentration and sludge retention time. *Water Science and Technology* 43(3): 191–194.

Mosquera-Corrala, A., de Kreukb, M.K., Heijnenb, J.J. & van Loosdrecht, M.C.M. 2005. Effects of oxygen concentration on N-removal in an aerobic granular sludge reactor. *Water Research* 39: 2676–2686.

Munch, E.V., Lant, P. & Keller, J. 1996. Simultaneous nitrification and denitrification in bench-scale sequencing batch reactors. *Water Research* 30: 277–284.

Peng, Y., Wang, X. & Li, B. 2006. Anoxic biological phosphorus uptake and the effect of excessive aeration on biological phosphorus removal in the A_2O process. *Desalination* 189: 155–164.

Pochana, K. & Keller, J. 1999. Study of factors affecting simultaneous nitrification and denitrification (SND). *Water Science and Technology* 39(6): 61–68.

Podedworna, J. & Żubrowska-Sudoł, M. 2009. Próba uzyskania synergicznego usuwania N i P w SBR poprzez wprowadzanie naprzemiennego napowietrzania w fazach tlenowych, (An attempt of nitrogen and phosphorus synergic removal obtaining in a sequencing batch reactor (SBR) through the application of intermittent aeration in the aerobic phases of the cycle), *Gaz Woda i Technika Sanitarna*, no. 7–8, pp. 47–50.

Satoh, H., Nakamura, Y., Ono, H. & Okabe, S. 2003. Effect of oxygen concentration on nitrification and denitrification in single activated sludge flocs. *Biotechnology and Bioengineering* 83(5): 604–607.

Third, K.A., Burnett, N. & Cord-Ruwish, C. 2003. Simultaneous nitrification and denitrification using stored substrate (PHB) as electron donor in an SBR, *Biotechnology Bioengineering* 44(5): 595–608.

Third, K.A., Gibbs, B., Newland, M. & Cord-Ruwisch, C. 2005. Long-term aeration management for improved N-removal via SND in a sequencing batch reacto. *Water Research* 39: 3523–3530.

Wachtmeister, A., Kuba, T., van Loosdrecht, M. & Heijnen 1997. A sludge characterization assay for aerobic and denitrifying phosphorus removing sludge. *Water Research* 31(3): 471–478.

Zhou, Y., Pijuan, M. & Yuan, Z. 2008. Development of a 2-sludge, 3-stage system for nitrogen and phosphorus removal from nutrient-rich wastewater using granular sludge and biofilms. *Water Research* 41: 3207–3217.

Żubrowska-Sudoł, M. & Podedworna, J. 2004. Stężenie tlenu rozpuszczonego a efektywność eliminacji związków węgla, azotu i fosforu w sekwencyjnym reaktorze porcjowym ze złożem ruchomym (MBSBBR) (Impact of dissolved oxygen on effectiveness of carbon, nitrogen and phosphorus compounds removal in a moving bed sequencing batch biofilm reactor (MBSBBR). *Gaz Woda i Technika Sanitarna* 2: 62–68

Environmental Engineering III – Pawłowski, Dudzińska & Pawłowski (eds)
© 2010 Taylor & Francis Group, London, ISBN 978-0-415-54882-3

The possibilities of using waste compost to remove aromatic hydrocarbons from solution

C. Rosik-Dulewska

Institute of Environmental Engineering of the Polish Academy of Sciences, Zabrze, Poland

T. Ciesielczuk

Department of Land Protection, Opole University, Opole, Poland

ABSTRACT: Sorption of petroleum by-products is a fundamental process which counteracts environmental pollution. Compost derived from waste can be employed as a bio-filter to eliminate organic compounds, including undesirable odours, from air flow. This paper examines the possible use of municipal waste compost as a sorbent for mono-aromatic hydrocarbons in aqueous solution. Sorption of benzene, toluene, ethylbenzene and o-xylene was observed using the batch method. There was high sorption of pollutants from aqueous solution. The most effective sorption properties were for mixed-waste compost (19.8%), followed by green waste compost (18.0%), kitchen and garden waste (10.8%) and then green waste compost (10.5%).

Keywords: Sorption, compost, aromatic hydrocarbons.

1 INTRODUCTION

Special care with transport containers is essential to limit the detrimental effects of leakage of diesel oil, fuel oil or petrol on the natural environment. Nevertheless, these leakages still occur as a result of land or sea transport accidents or rough transit, causing water degradation and preventing production on some arable land. Sorbents are commonly used to absorb these petroleum by-products. However, in spite of their multiple merits, mineral and organic sorbents are expensive to produce due to the costs of obtaining natural rock sources and organic materials (Koh & Dixon 2001, Wefer-Roehl et al. 2001). Moreover, this process has frequently been associated with the exploitation of non-renewable resources, natural landscape devastation and disturbance of natural water processes. Legally binding regulations state that sorbents should be disposed of at hazardous landfill sites, or in the case of organic sorbents, thermally transformed.

The use of prepared sorbents is costly; specialised rescue services use a sand–sawdust mixture, which until recently was cost-effective and readily available. However, sawdust is currently being contracted many years in advance, mainly by the power industry as a renewable source of energy; this frequently results in insufficient amounts of the mixture being used. Using compost derived from waste as a sorbent has numerous advantages: compost is a product that is almost spontaneously created from mixed municipal waste, or can be separated from the biodegradable elements of municipal waste. Compost is biologically active, and acts to support the natural processes of decomposition (van Gestel et al. 2003, Tsui & Roy 2008). Its chemical state and composition also allows for long-term storage and expedient transportation to any area (Allen-King 2002, Tognetti et al. 2007). As it consists of both organic and mineral particles, compost is equivalent to a sand–sawdust mixture with the additional advantage of being a by-product of recycled waste that would otherwise be sent to a landfill site. Compost as a sorbent created, for example, in the process of dynamic composting with the MUT-Dano Method or other industrial methods is already employed to absorb undesirable odours (Otten et al. 2004, Mathur et al. 2007). With the composting mass at a suitable humidity, the sorption and decomposition of pollutants in soils contaminated with gaseous and stable hydrocarbons has been achieved (Gibert et al. 2005, McNevin & Bardford 2000, Williamson et al. 2009). Using compost to minimise the negative effects of petroleum by-product leakages can therefore be an excellent alternative to materials currently in use. The aim of this work was to investigate the sorption properties of aromatic hydrocarbons by municipal waste compost.

2 MATERIAL AND METHODS

Mono-aromatic sorption of hydrocarbons was examined in batch experiments at room temperature ($20 \pm 2°C$). Four organic compounds were examined: benzene, toluene, ethylbenzene and o-xylene (Table 1). Four types of compost were tested as potential absorbents. Each compost was produced

Table 1. Physico-chemical properties of investigated compounds.

Compound	Chemical formula	Molecular weight	Aqueous solubility (%)	Density (g/cm^3)
Benzene	C_6H_6	78.12	0.188	0.8786
Toluene	$C_6H_5CH_3$	92.14	0.067	0.873
Ethylbenzene	$C_6H_5CH_2CH_3$	106.17	0.015	0.8669
o-Xylene	$C_6H_5(CH_3)_2$	106.16	0.017	0.8969

Table 2. Basic parameters of four composts (mean of three replicates per compost).

Parameter	Ka	Za	Zy	Gru
Organic matter (%)	41.1 ± 3.0	35.3 ± 4.1	34.8 ± 1.0	28.6 ± 1.6
TOC (% d.m.)	23.6 ± 2.7	16.1 ± 2.4	17.5 ± 3.8	9.7 ± 4.6
pH in H_2O	7.86 ± 0.04	7.76 ± 0.01	7.22 ± 0.05	7.60 ± 0.03
EC (mS/cm)	1.43 ± 0.21	1.32 ± 0.14	0.922 ± 0.19	0.881 ± 0.04
Nitrogen (%N_{Kjeld})	1.18 ± 0.36	1.12 ± 0.4	1.06 ± 0.4	1.01 ± 0.20
Phosphorus (%P_2O_5)	1.4 ± 0.18	1.0 ± 0.22	1.3 ± 0.31	1.3 ± 0.13
Potassium (%K_2O)	0.68 ± 0.26	0.72 ± 0.25	0.62 ± 0.11	0.55 ± 0.19
Calcium (%)	9.01 ± 1.1	4.37 ± 0.6	5.23 ± 1.1	3.74 ± 0.5
Glass (%)	<0.1[a]	<0.1	<0.1	<0.1
Gravity (g/dm^3)	240 ± 7.8	567 ± 14.6	392 ± 20	464 ± 16.1

[a] below the detection limit
d.w. = dry weight
N_{Kjeld} = Kjeldahl nitrogen = ammonium nitrogen + organic nitrogen

differently: MUT-Dano (Ka) was derived from mixed municipal waste; MUT-Herhof (Za) was derived from a separated fraction of waste from woods and parks; Zy compost was derived from a separated fraction of organic waste from woods and parks, stabilised by calcium carbonate; and Gru was the only compost produced in individual households from kitchen and garden waste. All of the aqueous solutions of the studied compounds were prepared with distilled water. The concentration of soluble hydrocarbons in actual conditions may be considerably higher, due to the co-existence of compounds with detergent properties. No additional means were used in the experiment to increase the solubility of the examined hydrocarbons. All reagents were chromatographic-grade.

2.1 Sample preparations

Compost samples were taken directly from heaps in compost manufacturing plants. Gru compost samples were taken from compost heaps in individual households. To collect each sample a hole approximately 30 cm deep was bored and samples taken with a steel shovel. Five samples were taken from each source, and then dried at room temperature. To allow checking of the sorption properties of the composts the samples were not ground or sieved. These samples were then dried at 105°C to reach dry mass and to remove traces of benzene, toluene, ethylbenzene and xylenes (BTEX). Samples that had not been treated with fungicides were used because there were no statistically significant differences between dried samples

and samples treated with 2% mercuric chloride in pre-tests. The research was carried out using ripe compost; the samples were stored on heaps for 5–7 months before sampling. Sample standardisation and grinding was used only in order to determine basic parameters characteristic of the types of compost (Table 2). The study was carried out according to Polish analytical norms. The presence of organic carbon was determined by Alten's method, i.e. titration after mineralisation with sulphuric acid and potassium dichromate.

2.2 Sorption experiments

The study was conducted in 50 cm^3 glass containers, with a ground-glass stopper. Of the compost, 5 g was weighed and placed directly into each container. A 'BTEX stock-mixture' was first prepared from benzene, toluene, ethylbenzene and o-xylene in equal ratios (v/v). The aqueous solutions were then prepared by dissolving 50, 125, 250, 500 and 1000 μL of BTEX stock in 1000 cm^3 of distilled water. The solutions were stored at 20°C in darkness. Losses calculated for blank tests were ≤4%, and the results were not corrected in this regard. Fifty cm^3 of each solution were used; the ratio of compost to solution during the experiment was 1:10. A preliminary experiment showed that sorption reached apparent equilibrium within 70 min. Sorption time took 80 minutes, and containers were checked once per minute. When the equilibrium was established, the samples were centrifuged for 15 min, after which the supernatant was decanted and then extracted with dichloromethane within 1 h by shaking in a 4-cm^3

Table 3. Granulometric composition of four composts.

Grain size (mm)	Ka	Za	Zy	Gru
>4	34.8	19.4	14.5	15
>2	12.5	17.8	11.9	16.4
>1	20.0	16	18.5	15.8
>0.5	13.9	16.2	25.7	16.8
>0.1	16.9	27.1	26.6	33.9
<0.1	1.8	3.4	2.8	2.1

Table 4. Sorption of benzene on four composts (mmol/kg dw).

Stock (mmol/dm^3)	Ka	Za	Zy	Gru
0.217	0.447	0.029	0.245	0.286
0.310	0.740	0.562	0.299	0.270
0.502	1.068	1.026	0.395	0.562
0.852	2.472	2.119	1.265	0.893
1.38	4.710	3.954	2.215	1.107

Table 5. Sorption of benzene on four composts (mmol/kg dw).

Stock (mmol/dm^3)	Ka	Za	Zy	Gru
0.061	0.276	0.028	0.090	0.125
0.139	0.740	0.621	0.390	0.343
0.274	1.347	1.106	0.697	0.925
0.520	2.841	2.546	1.679	1.666
0.846	4.585	4.123	2.395	2.350

Table 6. Sorption of ethylbenzene on four composts (mmol/kg dw).

Stock (mmol/dm^3)	Ka	Za	Zy	Gru
0.063	0.368	0.017	0.210	0.242
0.147	1.004	0.818	0.632	0.602
0.293	1.988	1.624	1.329	1.645
0.556	4.074	3.842	2.929	3.016
0.720	4.971	4.650	2.967	3.208

Table 7. Sorption of o-xylene on four composts (mmol/kg dw).

Stock (mmol/dm^3)	Ka	Za	Zy	Gru
0.073	0.479	0.063	0.265	0.330
0.179	1.280	1.062	0.802	0.815
0.375	2.652	2.203	1.732	2.234
0.699	5.181	4.894	3.682	3.941
0.899	6.264	4.783	3.732	4.093

dark-glass screw-cap vial with a Teflon-lined septa. The background solution was decanted and the eluate dried using Na$_2$SO$_4$, which had been previously dried at 130°C. Aromatic hydrocarbons were determined in unpurified extracts by gas chromatography with flame ionization detector (GC-FID) using a ZB-5 capillary column with constant helium flow through the column at 1 cm^3/min. The injector was kept at 250°C, and the detector at 280°C. The temperature of the furnace started at 30°C, was maintained for 2 min and was subsequently heated at 4°C/min up to 80°C, and then to 170°C at a rate of 15°C/min. The final temperature of 170°C was maintained for 5 min. The detection limit was 1.0–1.5 ng for a single compound. Equipment manufacturers' instructions were adhered to in order to calibrate the curve with an initial concentration of 200 μg/cm^3 of each compound. The difference in the initial and final examined compound concentrations in the aqueous phase at equilibrium were used to calculate the corresponding concentrations in the solid phase. Isotherms were produced as the amount sorbed (mmol/kg) against the initial concentration of the solution.

3 RESULTS AND DISCUSSION

There were considerable differences in chemical composition of the resulting composts. The content of organic matter was notable in the examined materials. The highest amount of organic matter was in Ka compost (derived from mixed municipal waste); while unexpectedly, the lowest quantity was in Gru compost. Important in this respect was the coarse-grained content in Ka compost, which consisted mainly of organic substance traces; it contained >17% more organic substances than Za and Zy compost, and up to 47% more than Gru compost. The results obtained were similar to data of Tognetti et al. (2007) and Weber

et al. (2007). In spite of the calcium carbonate addition to Zy compost, the highest concentration of calcium carbonate was in Ka and the lowest in Gru compost (Table 2). Ka compost was also characterised by a different granulometric composition (>34% of its content was coarse-grained i.e. >4 mm, and only 16.9% was <1 mm), while in Gru the fraction <1 mm was predominant, at 33.9% of the content (Table 3). Due to substantial variations in weight, the coarse-grained portions of dried Ka compost tended to move to the top of the container.

In preliminary tests, composts showed substantial abilities to absorb petroleum by-products. The weight of Gru, Za and Ka samples increased by sorption of light (petrol) products by 30.3, 47.4 and 50.3%, respectively. Thus, research was focussed on hydrocarbon sorption from aqueous solution. Sorption curves of particular mono-aromatic hydrocarbons were linear, with the exception of Gru material.

Ka and Za composts had the most effective sorption properties; respectively, 4.71 and 3.95 mmol benzene/kg. In the case of Za, the exception was low sorption efficiency for the lowest concentration used. Zy and Gru composts had lower sorption capacities of 2.21 and 1.11 mmol benzene/kg, respectively. There

Table 8. Sorption parameters of BTEX on four composts.

Compost type	Benzene		Toluene		Ethylbenzene		o-Xylene	
	K_d	R^2	K_d	R^2	K_d	R^2	K_d^a	R^2
Ka	3.44	0.99	5.51	0.99	7.14	0.99	7.13	0.99
Za	3.06	0.99	5.16	0.99	7.14	0.99	7.03	0.99
Zy	1.66	0.98	2.99	0.99	$-^b$	–	–	–
Gru	0.70	0.93	2.88	0.98	–	–	–	–

[a] K_d is the linear mode distribution coefficient.
[b] – indicates that isotherms do not fit a linear model.

were differences in sorption of the various composts; all had the highest sorption for o-xylene, followed in descending order by ethylbenzene, toluene and benzene. Elimination of o-xylene and ethylbenzene from a given solution was the greatest for Ka compost, respectively, 69.7 and 69.0%; for this material the highest elimination of toluene (54.2%) and benzene (32.2%) was also noted. Nevertheless, the curve slope indicates complete saturation of the adsorbent, similar to Za and Zy compost. The lowest sorption capacity was for Gru compost, with the lowest organic matter content; the maximum amount of sorbed o-xylene and ethylbenzene was 24.2% lower than for Ka compost (which stabilised at 434 mg/kg), while ethylbenzene stabilised at 341 mg/kg. The sorption curves matched the linear model for benzene over the concentration range considered, and reached the following values: $K_d = 3.44\,dm^3/kg$, $R^2 = 0.992$; for toluene $K_d = 5.51\,dm^3/kg$, $R^2 = 0.999$; while for ethylbenzene and o-xylene $K_d = 7.13\,dm^3/kg$, $R^2 = 0.997$ (Table 8). Diversified behaviour of benzene (linear sorption according to the Langmuir equation) has been found for two bentonites (Oyanedel-Crever et al. 2007).

However, the overall amount of benzene sorbed by bentonites was 2–8 g/kg and was higher than in the composts examined (0.30–0.39 g/kg) at the equilibrium concentration of 100 mg/dm³. However, it should be emphasised that in the present study the composts simultaneously sorbed the three other examined compounds. Overall sorption of all four examined aromatic hydrocarbons was 1.98 and 1.80 g/kg for Ka and Za composts, respectively; in the case of Zy and Gru, the sorbed amounts were 1.05 and 1.08 mg/kg, respectively. At the lowest examined concentration, the sorption of examined compounds was highest for Ka compost (152 mg/kg); Zy and Gru materials had similar levels of 77.9 and 94.7 mg/kg, respectively. Surprisingly, the lowest outcome was for Za compost (13.4 mg/kg), which in contrast at higher concentrations had equal sorption values to Ka compost. Comparable values were found for compost sorption of pesticides, reaching 40–200 mg/kg (de Wilde et al., 2008), similar to the total sorption of the present study's composts. In experiments on materials with little organic matter (0.48–0.65% of organic carbon), the addition of soluble organic matter (DOM) can lower the sorption of pesticides and hydrocarbons. Due to its

properties, similar to DOM detergents, it increases the solubility of hydrophobic compounds (Li et al. 2005). In the case of composts where the amount of DOM is substantial, this phenomenon can be significant in the sorption process.

4 CONCLUSIONS

There were diverse sorption properties of composts derived from waste. There were two categories of compost distinguished. Ka and Za composts were in the first category, characterised by a higher sorption capacity. Zy and Gru were in the second category, with a clearly lower sorption capacity. The examined aromatic hydrocarbons were sorbed with different levels of efficiency. The amount of sorbed substances decreased in the following order: o-xylene > ethylbenzene > toluene > benzene. The maximum amount of total sorbed substances was 1.98 g/kg for Ka compost. The studied composts caused a substantial decrease of water-soluble monoaromatic hydrocarbons, inferring that they can be employed as an absorbent material to limit the contamination of the natural environment by petroleum by-products. Other advantages of using these materials as absorbent materials are their low price, and limiting the biodegradable elements of municipal waste that are needlessly sent to landfill sites.

REFERENCES

Allen-King, R.M., Grathwohh, P. & Ball, W.P. 2002. New modeling paradigms for the sorption of hydrophobic organic chemicals to heterogeneous carbonaceous matter in soils, sediments, and rocks. *Advances in Water Resources* 25: 985–1016.

Gestel Van, K., Mergaert, J., Swings, J., Coosemans, J. & Ryckeboer, J. 2003. Bioremediation of diesel-oil contaminated soil by composting with biowaste. *Environmental Pollution* 125: 361–368.

Gibert, O., Pablo, De J., Cortina, J.L. & Ayora, C. 2005. Municipal compost-based mixture for acid mine drainage biore mediation: Metal retention mechanisms. *Applied Geochemistry* 20: 1648–1657.

Koh, S.M. & Dixon, J.B. 2001. Preparation and application of organo-minerals as sorbent of phenol, benzene and toluene. *Applied Clay Science* 18: 111–122.

Li, K., Xing, B. & Torello, W.A. 2005. Effect of organic fertile izers derived dissolved organic matter on pesticide sorption and leaching. *Environmental Pollution* 134: 187–194.

Mathur, A.K., Majumder, C.B. & Chaterjee, S. 2007. Com bined removal of BTEX in air stream by using mixture of sugar cane bagasse, compost and GAC as biofilter media. *Journal of Hazardous Materials* 148: 64–74.

Mcnevin, D. & Barford, J. 2000. Biofiltration as an odour abatement strategy. *Biochemical Engineering Journal* 5: 231–242.

Otten, L., Afzal, M.T. & Mainville, D.M. 2004. Biofiltration of odours: laboratory studies using butyric acid. *Advances in Environmental Research* 8: 397–409.

Oyanedel-Crever, V.A., Fuller, M. & Smith, J.A. 2007. Simul taneous sorption of benzene and heavy metals onto two or ganoclays. *Journal of Colloid and Interface Science*. 309: 485–492.

Soumare, M., Demeyer, A., Tack, F.M.G. & Verlo, M.G. 2002. Chemical characteristics of Malian and Belgian solid waste composts. *Bioresource Technology* 81: 97–101.

Tognetti, C., Mazzarino, M.J. & Laos, F. 2007. Improving the quality of municipal organic waste compost. *Bioresource Technology* 98: 1067–1076.

Tsui, L. & Roy, W.R. 2008. The potential application of using compost charts for removing the hydrophobic herbicide atrazine from solution. *Bioresource Technology*. 99: 5673–5678.

Weber, J., Karczewska, A., Drozd, J., Licznar, M., Licznar, S., Jamroz, E. & Kocowicz, A. 2007. Agricultural and ecological aspects of a sandy soil as affected by the application of municipal solid waste composts. *Soil Biology & Biochemistry* 39: 1294–1302.

Wefer-Roehl, A., Graber, E.R., Borisover, M.D., Adar, E., Nativ, R. & Ronen, Z. 2001. Sorption of organic contaminants in a fractured chalk formation. *Chemosphere* 44: 1121–1130.

Wilde, De T., Mertens, J., Spanoghe, P., Ryckeboer, J., Jaeken, P. & Springael, D. 2008. Sorption kinetics and its effect on retention and leaching. *Chemosphere* 72: 509–516.

Williamson, J.C., Akinola, M., Nason, M.A., Tandy, S., Healey, J.R. & Jones, D.L. 2009. Contaminated land cleanup using composted wastes and impacts of VOCs on land. *Waste Management* 29: 1772–1778.

Environmental Engineering III – Pawłowski, Dudzińska & Pawłowski (eds)
© 2010 Taylor & Francis Group, London, ISBN 978-0-415-54882-3

Membrane bioreactor co-treatment of municipal landfill leachates and synthetic sewage

A. Świerczyńska, E. Puszczało & J. Bohdziewicz
Institute of water and Wastewater Engineering, Silesian University of Technology, Gliwice, Poland

ABSTRACT: The aim of this study was to determine the effectiveness of the co-treatment of municipal landfill leachates and synthetic sewage in a membrane bioreactor. The study determined the dependence of the degree of decontamination on the activated sludge load and the associated aeration chambers load. The share amount of leachate in the treated mixture was 5% (by volume). The concentration of the activated sludge in the membrane bioreactor varied from 3.5 to 4.0 g/dm^3; the oxygen concentration was kept constant at 4.0 g O$_2$/dm^3. The activated sludge load was varied, in a range from 0.05 to 0.8 g COD/g$_{DM}$d. The system worked as a sequential bioreactor in 12-hour cycle. The effectiveness of the process was estimated from the change in values of these parameters: COD, BOD$_5$, TOC, N$_{tot}$, N-NO$_3^-$, N-NH$_4^+$ and P$_{tot}$. The treated effluent from the membrane bioreactor was additionally cleaned by a reverse osmosis process.

Keywords: Membrane bioreactor, landfill leachate, activated sludge, reverse osmosis.

1 INTRODUCTION

The deposition of waste in landfill sites is the oldest, most widespread and universal method of their disposal. The collection of waste in landfills, even at properly designed and operated sites, can cause many environmental problems. One problem is the formation of leachates, as a result of water percolation through the dumped waste (Rosik-Dulewska 2002, Surmacz-Górska 2001). These can have a serious impact on the ground and groundwater, both in the immediate neighborhood of the landfill site and also further afield (Rosik-Dulewska 2002).

The leachates are a serious environmental hazard, which varies according to their significant load, the concentration of toxic substances, and their changeable composition and volume. The treatment of leachates from landfill sites is much more complicated than treating municipal wastewater, and it often requires the integration of physical, chemical and biological treatment methods (Surmacz-Górska 2001, Szyc 2003). A proper solution for treating leachates involves the application of membrane bioreactors combining an activated sludge method with pressure-driven membrane techniques. The presence of membrane modules in the system eliminates the necessity for using secondary settling tanks and ensures a longer retention time of virtually non-degradable high-molecular substances in the bioreactor. It also makes it possible to apply high concentrations of activated sludge, which results in a lower substantial load (Laitinen et al. 2006, Moeslang 2006, Bodzek et al. 1997).

In Europe there are over 100 wastewater treatment installations using membrane bioreactors, and in USA there are more than 200 of these facilities (Szewczyk 2007, Won-Young et al. 2002).

This study focused on the determination of the effectiveness of treatming municipal landfill leachates in a membrane bioreactor with an internal capillary membrane module. The proposed solution may become a serious competitor to the classical activated sludge method in SBR reactors, as it combines the advantages of the biological process and pressure-driven membrane techniques.

2 MATERIALS AND METHODS

2.1 The substrate

The substrate in this study were the leachates collected in the municipal landfill in Tychy (Silesia, Poland). The landfill site has two sections: an old site and a new extension. Annually 60 000–70 000 tons of waste, collected from an area of 320 km^2 inhabited by 200 000 residents, is deposited at the site. Together with the new extension built in 2004, the landfill covers an area of 7.5 ha, and has a capacity of 1 425 000 m^3.

An analysis of the leachates revealed that their composition depends on the season and atmospheric conditions. It was observed that in the summer the impurities load is significantly lower than in the winter.

The characteristics of the leachates and synthetic wastewater is shown in Table 1.

The leachates used in the process were taken from both sections of the landfill site, the old site and

Table 1. The characteristics of the leachates, synthetic sewage and permissible values of particular parameters.

Parameter	Unit	Leachates	Synthetic sewage	Permissible values *
COD	gO_2/m^3	3000–4500	680	125
BOD$_5$	gO_2/m^3	200–260	240	25
TOC	gC/m^3	300–600	118.5	30
Ammonium nitrogen	gNH_4^+/m^3	950–1550	16.5	10
Nitrate nitrogen	gNO_3^-/m^3	0–6	7	30
Total nitrogen	gN/m^3	~400	120	30
Total phosphorus	gP/m^3	0–20	15	2
pH	–	8.1–8.5	6.8–7.5	6.5–9.0

Table 2. The working parameters of the membrane bioreactor.

Parameter	Unit	Value
Biomass concentration	g/dm^3	3.5–4
Oxygen concentration (fine-bubble aeration)	$g O_2/m^3$	~4
Oxygen concentration (coarse-bubble aeration)	$g O_2/m^3$	~0.5
Volumetric flow rate	dm^3/d	3.3–14
Activated sludge load	$g COD/g_{sm}d$	0.05–0,8
Aeration chambers load	$g COD/m^3d$	200–2800
Hydraulic retention time	d	4.6–1.0
Activated sludge age	d	20–7

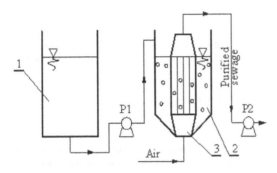

Figure 1. Schematic of aerobic membrane bioreactor. (1) raw sewage tank, (2) aeration chamber, (3) membrane module, P1, P2–pumps.

stopping its deposition on the surface of the membrane capillaries.

This protection of the membrane module is crucial for the treatment process. It allows high concentrations of the activated sludge to be used in the bioreactor chamber, and it prevents the fouling of the membrane.

The performance of the membrane bioreactor is based on the assumption that the activated sludge adsorbs and oxidises the impurities present in the treated wastewater, while the membrane acts as a filter separating biomass and refractory high molecular compounds. In this study, the biologically purified wastewaters were passed through the membrane capillaries (from the outside to the inside) and the obtained effluent was collected in the effluent tank.

the new extension, and mixed in equal volumetric ratio 1:1.

2.2 Apparatus

The apparatus set comprised of a membrane bioreactor (15 dm^3) with an internal microfiltration capillary module (by Zenon), an equalisation tank providing the required activated sludge load and an effluent tank. The capillary MF membranes were made from polyvinylidene fluoride and possessed pores of 0.04 μm diameter. The filtration area was 0.45 m^2. The construction of the membrane module enabled the backflushing of the capillaries with the permeate. The reaction chamber and the equalisation tank were equipped with sensors that measured liquid level, oxygen concentration and temperature. The scheme of the apparatus set is demonstrated in Figure 1.

Two types of aeration were used during the process: fine-bubble and coarse-bubble. The fine-bubble aeration ensured the required concentration of oxygen and intermixing of the bioreactor contents during the nitrification process. The coarse-bubble aeration was used during the denitrification process. It eliminated the need for mechanical mixing. It also prevented the sedimentation of the activated sludge as well as

2.3 The methodology

The biological treatment of the leachates was undertaken in laboratory conditions using activated sludge taken from The Municipal Wastewater Treatment Plant in Zabrze.

The membrane bioreactor acted as the SBR with a 12-hour cycle. The reactor filling took 0.5 hours, the denitrification process lasted 3.5 hours and the nitrification process went on 7 hours. The sedimentation and collection of clarified sewage took 1 hour. The working parameters of the membrane bioreactor are shown in Table 2.

The preliminary tests were designed to determine the optimum percentage share of leachates in the co-treated mixture, which also contained synthetic sewage. These tests were made with leachate shares varying from 3% to 40% (by volume). The treatment processes were carried out under a constant activated sludge load equal to 0.1 g COD/g$_{DM}$d.

The second stage of the study was focused on determining the influence of the activated sludge load on the effectiveness of impurities removal. The applied loads were 0.05, 0.1, 0.2, 0.3, 0.4, 0.6 and 0.8 g COD/g$_{DM}$d.

The effluent received during the treatment process carried out in the membrane bioreactor did not meet the conditions for treated wastewaters that could be

Table 3. The characteristics of the osmotic SS10 membrane (Osmonics catalog).

Membrane material	Cellulose acetate
Membrane type	SS10
Retention coefficient (R%)	98
pH	2–8
Transmembrane pressure, MPa	2.76–6.90
Temperature, °C	50

R – retention coefficient determined for 1 wt% NaCl solution.

directly disposed to a natural collector. It exceeded permissible concentrations of phosphorus, total nitrogen and nitrate nitrogen. An additional cleaning of the permeate was made by reverse osmosis. This process was carried out in a dead-end system with the use of GH-100-400 apparatus by Osmonics. The transmembrane pressure and rotational speed of the stirrer were kept constant (at 1.5 MPa and 200 rpm, respectively). Flat cellulose acetate membranes SS-10 (by Osmonics), with an effective area of 36.3 cm^2, were used during the process. The characteristics of the membranes are displayed in Table 3.

2.4 Analytical method

The main criterion for the estimation of the effectiveness of the treatment process was the change in the content of impurities expressed as pH, COD, BOD$_5$, TOC, IC, TC, total phosphorus, total nitrogen, nitrate nitrogen and ammonium nitrogen. The oxygen concentration was measured with an oxygen analyser CO-411, and the phosphate and nitrate concentrations were obtained using ionic chromatography (DIONEX DX-120 chromatograph by AGA Analytical was used for this purpose). The evaluation of different carbon forms content was made with the use of an carbon analyser Multi N/C by Jena Analytik. The total nitrogen and ammonium nitrogen concentrations as well as COD were using Merck analytical methods, while BOD$_5$ was analysed with the OXI Top WTW measurement system.

3 RESULTS AND DISCUSSION

In the first stage of the study, the optimum leachates share in the treated mixture containing also the synthetic sewage was determined. It was found that with the increase of the leachates share in the mixture the increase of organic compounds content could have been observed in the membrane bioreactor effluent. The obtained results are shown in Figure 2.

It was revealed that the highest decrease of COD and BOD$_5$ contents was obtained for the mixtures, in which the leachates share did not exceed 10 vol%. When the leachates share varied from 3% to 10% (by volume), the degree of removal of both COD and BOD$_5$ was constant, at 91% and 99%, respectively. These correspond to oxygen demands of 68 mgO$_2$/dm^3 for COD and 2 mgO$_2$/dm^3 for BOD$_5$.

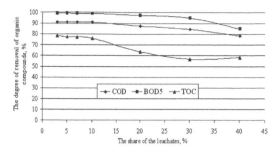

Figure 2. The dependence of the degree of removal of organic compounds on the leachates share in the mixture.

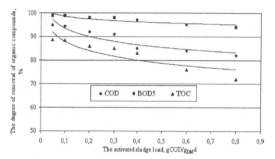

Figure 3. The relationship between the decrease of the content of COD, BOD$_5$ and TOC and the activated sludge load.

An increase in the share of leachates in the treated mixture resulted in an increased organic carbon content in the effluent, which exceeded the permissible value.

It was concluded that the treatment of mixtures containing leachate shares in the range of 3% and 5% (by volume) resulted in a sufficient decrease of organic compounds content that the membrane bioreactor effluent was suitable to be disposed directly to the natural collector.

The second stage of the study was related to the determination of the optimum activated sludge load. In these tests, the share of the leachates in the treated mixture was kept constant at 5% (by volume).

The relationship between the decrease of the content of COD, BOD$_5$ and TOC and the activated sludge load is shown in Figure 3.

It was shown that the activated sludge load had an influence on the degree of removal of organic compounds. The highest decrease of COD, which observed for a load of 0.05 g COD/g$_{DM}$d, was 94.9%, and this corresponds to a chemical oxygen demand of 43 gO$_2$/m^3. The lowest decrease of the impurities content was 80%, obtained for an activated sludge load of 0.8 g COD/g$_{DM}$d, and this corresponds to a chemical oxygen demand of 120 g O$_2$/m^3.

The influence of the activated sludge load on the BOD$_5$ parameter was negligible (Figure 3). The maximum decrease of the BOD$_5$ value was obtained for the activated sludge load in the range from 0.05 to 0.2 g COD/g$_{DM}$d. This corresponds to a biological oxygen demand decrease from 170 g O$_2$/m^3 in the

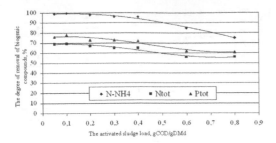

Figure 4. The relationship between the degree of removal of biogenic compounds and the activated sludge load.

Table 4. The effectiveness of the treatment process in the membrane bioreactor using the most favourable sludge load (0.1 g COD/g_{DM}d).

| Parameter | Membrane bioreactor feed, g/m^3 | Membrane bioreactor effluent | |
		Concentration g/m^3	Retention, %
COD	850	48	94.4
BOD$_5$	170	2	98.8
TOC	160	18	88.8
Total nitrogen	137	42	69.6
Ammonium nitrogen	115	0.5	99.6
Nitrate nitrogen	7	70	–
Total phosphorus	16	3.5	78.1

membrane bioreactor feed to 2 g O$_2$/m^3 in the effluent. The lowest decrease of BOD$_5$ was observed for an activated sludge load of 0.8 g COD/g_{DM}d. At 94%, this corresponds to a BOD$_5$ value in the effluent equal to 28 g O$_2$/m^3.

The maximum degree of removal of total organic carbon was 88.8% and this was obtained for the activated sludge load in the range from 0.05 to 0.1 g COD/g_{DM}d.

It corresponds to a decrease of total organic carbon concentration from 160 gC/m^3 in the crude mixture to 18 g C/m^3 in the treated mixture. The lowest removal degree of TOC of 72% was observed for the highest activated sludge load of 0.8 g COD/g_{DM}d.

The results also show that for activated sludge loads in the range from 0.05 to 0.4 g COD/g_{DM}d, the decrease in the concentration of organic compounds is sufficient to allow the effluent to be directly disposed to the natural collector. Increasing the activated sludge load above 0.6 g COD/g_{DM}d results in effluent with organic impurities that exceeds permissible values. Additionally, the higher activated sludge load produced odours and a rigid white foam appearance on the bioreactor surface. This confirmed the overload of the activated sludge.

The next stage of the study focused on the determination of the relationship between the degree of removal of biogenic compounds and the activated sludge load (Figure 4).

It was observed that the activated sludge load had a significant influence on the biogenic compounds content in the membrane bioreactor effluent. The highest total nitrogen concentration decrease was observed for an activated sludge load in the range of 0.1 to 0.2 g COD/g_{DM}d. The total nitrogen concentration of the crude mixture was 137 g N$_{tot}$/m^3, while it was 42 g N$_{tot}$/m^3 in the treated mixture. The lowest degree of removal of the total nitrogen was 56%, observed for an activated sludge load of 0.8 g COD/g_{DM}d.

The activated sludge load also limited the degree of removal of ammonium nitrogen. Its highest value of 99.6% was obtained for a load in the range from 0.05 to 0.1 g COD/g_{DM}d, while the lowest value of 75% was observed for a load of 0.8 g COD/g_{DM}d (however, in all cases the ammonium nitrogen content did not exceed the permissible value).

The other parameter used to assess the effectiveness of the treatment process was phosphorus content.

The results reveal that the highest degree of phosphorus removal was observed for an activated sludge load of 0.1 g COD/g_{DM}d. This corresponds to a decrease in phosphorus concentration from 16 gP/m^3 in the crude mixture to 3.5 gP/m^3 in the treated mixture (78.1% was removed). The highest activated sludge load of 0.8 g COD/g_{DM}d resulted in the smallest degree of removal of phosphorus at 61%.

The relationship between the nitrate nitrogen content in the effluent and the activated sludge load differed to that observed for the other biogenic compounds. It changed from 80 mg N-NO$_3^-$/dm^3 for a load of 0.1 g COD/g_{DM}d to 30 mg N-NO$_3^-$/dm^3 for a load of 0.8 g COD/g_{DM}d. For loads of 0.6 g COD/g_{DM}d and 0.8 g COD/g_{DM}d the concentration of nitrate nitrogen in the effluent decreased and nitrite nitrogen appeared. Its concentration was in the range from 20 to 35 mgN-NO$_2^-$/dm^3. This indicates that the nitrification process was slowed down. This was probably caused by the intensification of the activated sludge growth.

The results show that the effectiveness of the treatment process was not equally dependent on the activated sludge load. For a load of 0.1 g COD/g_{DM}d the lowest concentrations of biogenic compounds in the effluent were obtained, while the degree of removal of organic compounds did not change significantly with the load. Table 4 shows the characteristics of the membrane bioreactor effluent for the assumed optimum activated sludge load of 0.1 g COD/g_{DM}d.

The treatment process with an activated sludge load of 0.1 g COD/g_{DM}d resulted in obtaining effluent that was suitable for disposal to the natural collector in terms of its organic compounds content. However, this could not have been done as the permissible biogenic compounds concentrations were exceeded. It was therefore decided that the membrane bioreactor effluent should be additionally cleaned by the reverse osmosis process.

The obtained permeate fulfilled all the conditions for treated wastewaters, allowing it to be disposed

Table 5. The effectiveness of the reverse osmosis process.

Parameter	Membrane bioreactor effluent, g/m^3	Permeate after RO process	
		Concentration, g/m^3	Retention %
COD	48	5.5	88.5
BOD$_5$	2	0.0	100
TOC	18	3.0	83.3
Total nitrogen	42	3.5	91.7
Ammonium nitrogen	0.5	0.0	100
Nitrate nitrogen	70	7.5	89.3
Total phosphorus	3.5	0.05	98.6

to the natural collector. The final parameters of the treated sewage are shown in Table 5.

4 CONCLUSIONS

The results obtained in this study allow us to conclude that an increase of the leachate share in the mixture negatively influences the biological treatment process. A leachate share in the mixture in the range of 3% to 5% (by volume) enabled a decrease in the organic substances content in the membrane bioreactor effluent to levels permissible for the disposal of the treated wastewater to the natural collector.

The study revealed that the optimum activated sludge load was 0.1 g COD/g$_{DM}$d. However, the content of biogenic compounds (expressed as phosphorus, nitrate nitrogen and total nitrogen concentrations) in the effluent still exceeded the permissible values.

An activated sludge load above 0.6 g COD/g$_{DM}$d resulted in a significant decrease in the effectiveness of the treatment process as well as leading to odour production, the appearance of a rigid white foam on the bioreactor surface and nitrite nitrogen occurrence in the effluent.

The additional cleaning of the bioreactor effluent by the reverse osmosis significantly improved the quality of the treated wastewater, so that it fulfilled the conditions for disposal to the natural collector. The degree of removal of phosphorus was 98.6%, and for total nitrogen and nitrate nitrogen it reached 90%.

REFERENCES

Bodzek, M., Bohdziewicz, J. & Konieczny, K. 1997. *Membrane techniques in environmental protection.* Silesian University of Technology Publisher: Gliwice.

Flat sheet membrane chart – brochures by OSMONICS, 1996.

Laintinen, N., Luonsi, A. & Vilen, J. 2006. Landfill leachate treatment with sequencing batch reactor and membrane bioreactor. *Desalination* 191.

Linde, K., Jonsson, A., Wimmerstedt, R. 1995. Treatment of three types of landfill leachate witch reserve osmosis. *Desalination* 101: 21–30.

Moeslang, H. 2006. Membrane bioreactors (MBR) – for muncipal and industrial wastewater. *Monographie of Environmental Engineering Comitee of Polish Academy of Science* 36: 671–679.

Rosik-Dulewska. Cz. 2002. *Basics of waste management.* Scientific Publisher PWN: Warsaw. (in polish)

Surmacz-Górska, J., Miksch, K. & Kita, T. 2000. Possibilities of biological treatment of leachate from municipal landfill. *Environmental Protection Archive* 26: 42–54.

Surmacz-Górska J. 2001. Degradation of organic compounds in landfill leachate. *Monographie of Environmental Engineering Comitee of Polish Academy of Science* 5.

Szewczyk, K.W. 2007. Membrane bioreactors In environmental protection. *IX Membrane School, Membranes and membrane techniques in environmental protection".*

Won-Young, A., Moon-Sun, K., Seong-Keun, Y. & Kwang-Ho, Ch. 2002. Advanced landfill leachate treatment using an integrated membrane process. *Desalination* 149(1–3): 109–114.

Environmental Engineering III – Pawłowski, Dudzińska & Pawłowski (eds)
© *2010 Taylor & Francis Group, London, ISBN 978-0-415-54882-3*

Quality of surface run-off from municipal landfill area

I.A. Tałałaj

Bialystok Technical University, Bialystok, Poland

ABSTRACT: This paper describes the quality of landfill surface run–off. As the object of investigation, a landfill site was chosen where municipal waste is disposed from a town with a 300,000 population. The scale and character of the surface run-off pollution was observed in two points – from the circumferential ditch system at the foot of the waste hill (point P1) and from the surrounding ditch (point P2) beyond the fence of the landfill area. The conducted analyses of the surface run-off indicated strong pollution. An analysis of the variance has shown that the values of particular pollution indicators can undergo crucial changes according to the time (season of the year) of sampling. A lower pollution concentration in the surrounding ditch outside the area limited by the fence of the landfill area was observed. The greatest degree of pollution reduction was observed for organic polluters.

Keywords: Pollution, surface run-off, landfill, leachate.

1 INTRODUCTION

The main ways of pollution migration from landfills are leachates, surface run-off, and the emission of gases and dusts. The physico-chemical properties of landfill leachate as well as the quality of biogas that is emitted from landfills have been examined and described at many sites. However, there is lack of more complete information concerning the qualitative characteristics of landfill run-off from the waste dump surface.

One can find information relating to the characteristics of landfill surface run-off in the works of Zafar & Alappat (2004a,b), Cartazar & Monzan (2007) and Mangimbulude et al. (2009). According these authors, the landfill operation phase has the additional potential to affect surface run-off volume and quality by (Zafar & Alappat 2004b):

– the release of uncontrolled discharges of surface water from the site
– the break-out of leachate from the site
– the provision of an engineered drainage system including diversion ditches and cut-off drains.

The main components of landfill surface run-off are:

– major ions – calcium, magnesium, iron, ammonium, sulphate, chloride
– trace metals such as manganese, zinc, copper, chromium, nickel and cadmium
– a wide variety of organic compounds, which are usually measured as Total Organic Carbon (TOC), Chemical Oxygen Demand (COD) or Biochemical Oxygen Demand (BOD); individual compounds that are hazardous at very low concentrations may also be of concern, such us pesticides, benzene and phenol
– microbiological components.

The major ions and organic compounds are common components of a waste stream, traditionally analysed to provide an overview and characterisation of the waste stream. They are typically present in elevated concentration in landfill surface run-off and can thus often indicate the presence of leachate in unsaturated or saturated groundwaters (Jones-Lee & Lee 2003). Some organics can serve as co-substrates for microorganisms that can facilitate the conversion of hazardous chemicals to even more hazardous forms. It is estimated that about 90% of the organic material in municipal landfill surface run-off are of unknown composition. These chemicals have not been identified, and obviously their potential impacts on environment are unknown (Jones-Lee & Lee 2003).

The study is aimed at analysing the quality of the landfill surface run-off from a selected municipal landfill.

2 MATERIALS AND METHODS

A landfill that receives municipal waste from a 300,000 population town was adopted as the research object. The landfill area consists of three dumping fields: A, B and C (Figure 1)

Exploitation of the B and C dumping fields is conducted in compliance with all binding environmental protection norms. The landfill bottom under the B and C dumping fields has been sealed with 2 mm PEHD foil on which a drainage system was placed that captures the produced leachates. The leachates are then

Figure 1. Landfill area and run-off's sampling points.

routed to an intermediate pumping station from which they flow to two retention reservoirs, and later they are transported by a gully emptier fleet to the municipal sewage treatment plant.

The oldest of the dumping fields (commissioned in 1982) – the A dumping field – is not fitted with a leachate drainage. Its bottom is proofed with a 50 cm layer of clay. The dumping field scarps have been formed from a layer of construction debris covered with a layer of cultivable soil of 1 m in thickness, seeded with a mixture of grass. Because the B and C dumping fields had been full, municipal waste was temporarily placed in the A field again. While the A dumping field was functioning, liquid outflows (leachates) from the scarps, sometimes forming small streams, were observed. In order to limit the leachate migration, circumferential drainage around the scarp foot of the A field was built. This drainage collects:

– leachates flowing out from the scarp of the A field landfill
– precipitation water or meltwater flowing down the waste hill surface
– surface run-off.

The collected surface run-off is directed to the retention reservoirs where it is mixed with the leachates from the remaining two dumping fields (B and C). In the landfill under investigation there is no monitoring of the quantity of the surface run-off collected by the circumferential ditch, only the aggregate quantity of leachates routed to the retention reservoirs, comprising both the polluted waters from the circumferential drainage and the leachates from the B and C fields bottom drainage system, is calculated.

There is a network of surrounding ditches beyond the fence of the landfill area, which are within the reach

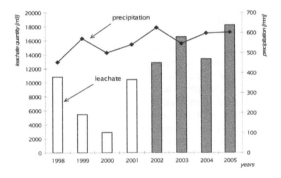

Figure 2. Total amount of atmospheric precipitation and leachate quantity in particular years (research period 2002–2005).

of subsurface and surface run-offs from the landfill area.

The analysed area is covered by a sand formation, which is underlain by a complex of boulderclay. The free groundwatertable lays 0.95 m to 5.4 m below the land surface. The landfill is underwashed on the west side by groundwater that flows down in north-easterly, south-easterly, and easterly directions.

The mean atmospheric precipitation level in the area under research amounts to 560 mm a year. Figure 2 presents the annual precipitation levels as well as the total quantity of the cumulated leachates for particular years, including the research period of 2002–2005.

The samples for analysis were taken from the leachate drainage system at the foot of the waste hill (point P1) and from the surrounding ditch (point P2) beyond the fence of the landfill area. During localisation of the sampling points, topographic features

Table 1. Quality comparison of surface run-offs collected in circumferential ditches (P1) and surface waters ditches beyond landfill area (P2).

Parameter	Unit	n	Permissible values*	Surface run-offs collected in circumferential ditches (P1 point) mean	standard deviation	Waters from surrounding ditches beyond the landfills area (P2 point) mean	standard deviation	t	p
Reaction	pH	15	–	7.88	0.91	7.71	0.84	0.438	0.665
Temperature	°C	15	–	15.33	6.47	12.86	7.33	0.909	0.372
Conductivity	μS/cm	15	–	22391.7	11,082.1	6103.6	4256.9	4.631	0.000
Nitrite nitrogen	mg/dm³	15	10	0.38	0.59	0.05	0.06	1.834	0.079
Nitrate nitrogen	mg/dm³	15	–	34.25	36.99	9.82	10.64	2.117	0.045
Ammonia nitrogen	mg/dm³	15	100–200	1533.97	1383.74	0.68	0.94	3.655	0.001
Dissolved oxygen	mg/dm³	15	–	8.06	5.33	8.19	3.11	−0.069	0.945
COD$_{Cr}$	mg/dm³	15	–	7410.87	6419.1	146.54	153.26	3.732	0.001
Phosphates	mg/dm³	15	–	19.98	21.78	1.09	1.74	2.853	0.009
Sulfates (VI)	mg/dm³	15	500	53.57	59.99	15.86	16.62	2.019	0.055
Iron	mg/dm³	15	–	7.97	3.19	0.87	0.77	7.203	0.000
Free cyanides	mg/dm³	15	0.5	0.04	0.05	0.03	0.04	0.708	0.486
Hardness, general	mg/dm³	15	–	734.93	910.02	571.27	641.37	0.510	0.615
Chlorides	mg/dm³	15	100	1788.33	876.11	1285.2	1176.82	1.252	0.223
Boron	mg/dm³	15	10	7.61	5.06	0.71	0.89	3.783	0.001
Suspended matter	mg/dm³	15	–	638.73	441.11	626.36	1134.54	0.039	0.969
Cadmium	mg/dm³	7	0.2	0.06	0.07	0.06	0.07	−0.202	0.844
Copper	mg/dm³	7	1	0.12	0.07	0.05	0.04	1.984	0.075
Zinc	mg/dm³	7	5	1.01	0.76	0.48	0.43	1.394	0.193
Nickel	mg/dm³	7	1	0.46	0.42	0.08	0.09	1.974	0.077

* Permissible values for sewage introduced to sewerage system (Jour. of Laws 2006.136.964).

as well as the direction of the surface and subsurface flows of polluted water were considered.

In order to obtain as reliable results as possible, sampling under some specific atmospheric/climatic conditions, such us after periods of intensive precipitation or following long-lasting periods without precipitation (droughts), was avoided. The sampling was done four times a year, in 3–4 month intervals. On the whole, thirty samples (fifteen from each sampling point) were taken during investigation. The physico-chemical analyses included qualitative determinations of sixteen pollution indicators (Table 1). In order to perform a more accurate analysis of the sampled waters, a concentration determination of chosen heavy metals (cadmium, copper, zinc and nickel) was also conducted. Furthermore, the data concerning the precipitation level and the total amounts of leachates produced within the research period were analysed.

All the determinations were made in accordance with Polish Norms using of a pH-meter, DR2000 spectrometer, conductometer, and an oxygen probe.

The variability of the analysed indices in time was evaluated by applying single-factor variance analysis. The relationships between the parameters were assessed by correlation analysis. The quality differentiation between the landfill surface run-off and the surfacewater from the surrounding ditches was estimated with the Student's t-test. These methods are usually used for monitoring seasonal variation of

environmental contamination (Gibbons & Coleman 2001, Uzoukwu et al. 2004).

3 RESULTS AND DISCUSSION

Throughout this article, the term surface run-off – collected by the circumferential ditches at the foot of the waste hill – should be understood as:

- the leachates breaking out from the dump hill
- the precipitation water flowing down from the waste hill surface (mixed with strongly polluted leachates)
- the run-off of polluted water from the landfill area.

The quality of the surface run-off collected in the circumferential ditches (P1) is presented in Table 1.

The presented data indicate that the collected surface run-off is characterised by a high conductivity value as well as by high content of ammonium nitrogen and chlorides. The concentration of pollution indicators does not usually exceed the permissible values for sewage introduced to sewage devices (Jour. of Laws 2006.136.964).

The results of a statistical analysis of the surface run-off showed a strong relationship among some variables. Correlations for which $r \geq |\pm 0.7|$ at $p < 0.5$ were considered as statistically significant. From the correlation analysis, one can draw conclusions about

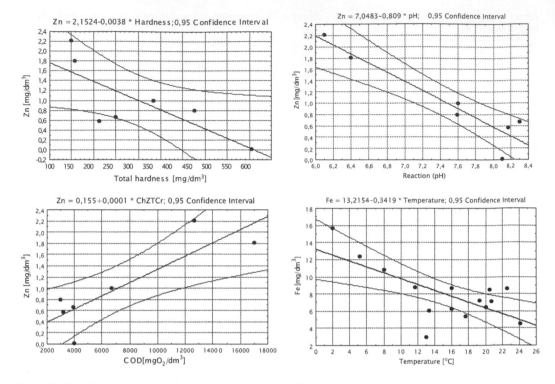

Figure 3. Correlation between chosen indicators in surface run-off water.

propitious circumstances as well as about the form of the occurrence of the chosen indicators.

A negative correlation was noted between the concentration of zinc and general hardness (Figure 3).

One of the reasons influencing the zinc concentration decrease is probably the presence of calcium carbonate, which causes the deposition of sparingly soluble compounds of zinc.

A decrease in zinc concentration is also observed together with an increase in the run-off water reaction. A strong positive correlation between zinc and the chemical oxygen demand (COD) can be indicative of the possibility of the formation of complex organic ions by zinc.

An interesting negative correlation was noted between the iron concentration and the surface run-off temperature (Figure 3). A decrease in iron concentration occurs with a rise in temperature, which can be related to the oxide-reductive conditions changing at that time.

Furthermore, a number of softer $|\pm 0.7| > r \geq |\pm 0.5|$ positive correlations were obtained, including between cyanides and the electrolytic conductivity ($r = 0.52$), cyanides and boron ($r = 0.59$), and the temperature and COD ($r = 0.61$); negative correlations were also observed between the conductivity and reaction ($r = -0.54$), and nitrogen nitrite (III) and suspension ($r = -0.61$).

For a detailed analysis of the results, the seasonal influence on the surface run-off water quality was examined. The outcomes of this variance analysis are given in Table 2 and Figure 5. From the presented data it appears that the values of the majority of the pollution indicators show similar concentrations throughout the whole year. The exceptions are temperature, COD and iron, whose pollution concentrations are changeable within the year. These dependencies were also indicated by the correlation analysis.

Because of the lack of monitoring of the amount of surface run-off in the landfill, it was not possible to calculate pollution loads. The quality of the surface run-off in the respective seasons of the year was therefore assessed against the data concerning the atmospheric precipitation level as well as the quantity of all leachates in the respective periods of the year.

Figure 4 illustrates the variability of the atmospheric precipitation level and of the leachate quantity in particular months.

Figure 4 shows that seasonal variability is more evident for precipitation than for the quantity of landfill leachates. The weight of the deposited waste and its consistence, which may have an averaging influence through its absorption properties (water absorption), exerts modifying effects on the quantity of leachates.

Considering the above, the search for the reasons for the seasonal changes in values of the indicators, as shown in the variance analysis, was focused on atmospheric changes factors rather than changes in the leachate quantity.

Figure 5 indicates that the iron concentration reaches its maximum values in the winter period. Analysing the precipitation level (and the leachate

Table 2. Variance analyses of surface run off to assess seasonal dependence.

Parameter	degrees of freedom effect	mean square effect	degrees of freedom error	mean sum square error	F	p
Reaction	3	0.45	11	1	0.48	0.70
Temperature	3	133	11	17	7.86	0.004
Conductivity	3	$88*10^6$	11	$130*10^6$	0.67	0.59
Nitrite nitrogen	3	1	11	0.3	1.77	0.21
Nitrate nitrogen	3	865	11	1505	0.57	0.64
Ammonia nitrogen	3	$4.2*10^6$	11	$1.3*10^6$	3.31	0.06
Dissolved oxygen	3	35	11	27	1.30	0.32
COD_{Cr}	3	$103*10^6$	11	$24*10^6$	4.25	0.03
Phosphates	3	613	11	436	1.41	0.29
Sulfates (VI)	3	5513	11	3076	1.79	0.21
Iron	3	29	11	5	5.88	0.01
Free cyanides	3	0	11	0.003	0.27	0.84
Hardness, general	3	$1.2*10^6$	11	$0.7*10^6$	1.77	0.21
Chlorides	3	$1.5*10^6$	11	$0.5*10^6$	2.76	0.09
Boron	3	16	11	29	0.53	0.67
Suspended matter	3	32,680	11	$0.2*10^6$	0.14	0.94
Cadmium	3	$1.0*10^{-3}$	4	0.008	0.13	0.88
Copper	3	$0.9*10^{-3}$	4	0.002	5.01	0.08
Zinc	3	0.24	4	0.73	0.33	0.74
Nickel	3	0.18	4	0.17	1.10	0.42

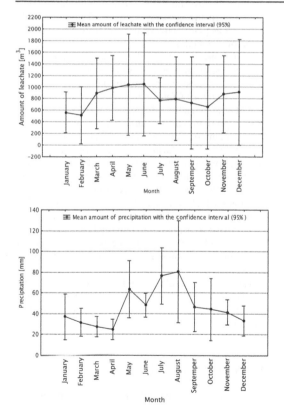

Figure 4. Variability of leachate quantity and atmospheric precipitation level in particular months.

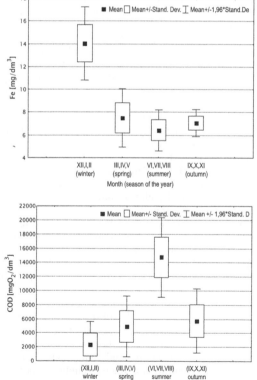

Figure 5. Seasonal variability analysis of surface run-off water quality.

quantity) distributions it might be supposed that one reason for this finding might be the pollution condensation due to a lower level of precipitation in this season of the year. This would also be supported by lower iron concentrations in the summer season, when there are maximum precipitation levels. The reasons for the iron concentration variability may also be looked for in the changing redox circumstances. In winter periods, when the amount of oxygen dissolved in the circumferential drainage water decreases (because of snow cover and the ice layer), iron is reduced to the Fe^{2+} form, which is easily soluble in water, whereas in oxidative circumstances, when for example during the circumferential drainage water is in contact with the oxygen in the air, bivalent iron is oxidised to Fe^{3+} and deposited, thus converting into sparingly soluble forms.

The increased concentrations of iron compounds are facilitated both by the presence in water of organic compounds, with which iron can form numerous complex forms, and by the presence of aggressive carbon dioxide resulting in iron hydrogen carbonates being leached into the water.

A rise in air temperature (and – as a consequence – a rise in the temperature of the surface run-off) influences the intensification of the processes taking place in the leachates. The result can be more intensive organic substance decomposition, which leads to an increase of the COD concentration limit. An increase in the amount of organic matter in the drainage water in summer is further supported by the observed liquid outflows from the landfill scarp, which are actually strongly polluted leachate. It might be the case that the greater atmospheric precipitation occurring in the summer periods results in precipitation infiltration through the waste hill and also increases the possibility of washing out pollutants, which partially reveal themselves in the form of the liquid outflows described above (it has to be remembered that the bottom under the waste hill at dumping field A does not have any leachate drainage system and is sealed with a 50 cm layer of clay).

In Table 1, the quality of the surface run-off collected in the circumferential ditches (point P1) is compared to the quality of the surface water from the surrounding ditches beyond the fence of the landfill area (point P2). The results show that in the surface waters beyond the landfill (P2), the values of the majority of pollution indicators are decidedly lower than those from the water in the circumferential ditches. Crucial statistical differences are found for these indices: electrolytic conductivity, ammonium nitrogen, COD, nitrate nitrogen, orthophosphates, iron and boron. The extent of the percentage decrease in values varies from 71% $(N - NO_3)$ to 98% (COD) and 99.9 % $(N - NH_4)$.

From this data, it appears that during pollution migration – both through subsurface and surface routes – a process of self-cleaning takes place. It suggests that some organic substances can undergo degradation to simpler mineral compounds. Some of the final decomposition products, including some metals, result from sorption during pollution infiltration into the soil-water environment.

4 CONCLUSIONS

These analyses of landfill surface run-off indicated strong pollution, particularly with nitrogen

compounds and chlorides. High conductivity values were also observed. The presence of a wide variety of pollutants in the surface run-off affects the strength and the very varied character of the correlations obtained between the particular variables. A strong correlation $r > |\pm 0.7|$ was observed between the zinc concentration and reaction, and between COD and the general hardness, as well as between the iron concentration and the temperature, and cadmium and the electrolytic conductivity.

The analysis of variance showed that the values of particular pollution indicators can undergo crucial changes according to the time of sampling (season of the year). These results suggest it is the season of the year (that is, air temperature *sensu stricto*) that has influence on the COD value and the iron concentration in the analysed surface run-off. The reasons of this variability may be related to both the changeable oxidation-reduction conditions and to the different climate conditions (air temperature, atmospheric precipitation level), which may exert influence on the intensification level of the processes taking place in the waters being researched.

During subsurface and surface pollution flow, a reduction in the concentration of all pollution indicators takes place. This results in lower concentration levels of the analysed pollution indicators in the surrounding ditches beyond the landfill compared with those in the circumferential ditches around the waste hill. The greatest degree of the pollution reduction was observed for COD and ammonium nitrogen. One should note that a high degree of concentration reduction is predominantly found for organic pollution. It is evident of further organic matter decomposition taking place during surface run-off as well as of pollution infiltration into the soil-water environment. The presence of gramineous vegetation in the pollution flow as well as vegetation with short root systems – which can be a filter/cleaner for the surface and subsurface flow – can influence (and reduce) pollution levels in the surrounding ditches. At this point, roots of higher plants can take over and transform pollution (mainly biogenic compounds) from shallow groundwater.

The obtained results are indicative of the advisability of conducting further, more detailed investigations and analyses of the quality of the landfill surface run-off.

ACKNOWLEDGMENT

This work was supported by the university internal grant W/IIŚ/38/06.

REFERENCES

Cartazar, A.G. & Monzan, I.T. 2007. Application of simulation models to the diagnosis of MSW landfills. *Waste Management. Elsevier Ltd* 27(5): 691–703.

Gibbons, D. & Coleman, D. 2001. *Statistical Methods for Detection and Quantification of Environmental Contamination*. John Wiley and Sons, Ltd.: New York.

Jones-Lee, A. & Lee G. F. 2003. Groundwater Pollution by Municipal Landfills: *Leachate Composition, Detection and Water Quality Significance. Proc. Sardinia International Landfill Symposium, Sardinia*: 1093–1103 Italy.

Mangimbulude, J.C. Breukelen, B.M. Krave, A.S. Straalen, N.M. & Röling, W.F. 2009. Seasonal dynamics in leachate hydrochemistry and natural attenuation in surface run-off water from a tropical landfill. *Waste Management*: 29(2): 829–837.

Uzoukwu, B.A. Ngoka, C. Nneji, N. 2004. Monitoring Of Seasonal Variation In The Water Quality Of Ubu River. *Environmental Management (ISSN 0364-152) Jun 33(6):* 886–898.

Zafar, M. & Alappat, B.J. 2004a. Environmental mapping of water quality of the river Yamuno in Delhi with landfill location. *Management of Environmental Quality (ISSN 1477-7835)* 15(6): 608–621.

Zafar, M. & Alappat, B.J. 2004b. Landfill surface runoff and its effect on water quality on river Yamuno. Journal of Environmental Science and Health. Part A: Toxic/Hazardous Substances & Environmental Engineering (ISSN 1093-4529) *39(2/222):* 375–384.

Environmental Engineering III – Pawłowski, Dudzińska & Pawłowski (eds)
© *2010 Taylor & Francis Group, London, ISBN 978-0-415-54882-3*

Experimental feasibility study on application of a mechanical cavitation inducer for disintegration of wastewater sludges

M. Żubrowska-Sudoł, J. Podedworna & Z. Heidrich
Faculty of Environmental Engineering, Warsaw University of Technology, Warsaw, Poland

P. Krawczyk
Institute of Heat Engineering, Warsaw University of Technology, Warsaw, Poland

J. Szczygieł
Power and Environmental Protection Research Centre, Warsaw University of Technology, Warsaw, Poland

ABSTRACT: This article discusses the initial results of research on the effectiveness of sludge disintegration in a mechanical cavitation inducer. The results presented show that using specific energy inputs of 5090–5590 kJ kg^{-1} total solids (TS), the tested mechanical cavitation inducer achieved disintegration degrees of 24.6–30.2% and an increase in soluble COD and volatile fatty acids (VFAs) in sludge water to 6490–9050 mg l^{-1} and 366–529 mg l^{-1}, respectively. The highest rates of increase in soluble organic substance concentrations in sludge water were observed for specific energy inputs of 2500–4500 kJ kg^{-1} TS.

Keywords: Degree of disintegration, disintegration, mechanical cavitation inducer, specific energy input.

1 INTRODUCTION

The construction of new wastewater treatment plants and the application of new, complex and efficient treatment processes lead to increased production of wastewater sludges. This sludge, which is itself a waste by-product, should be recycled, processed and reused according to the Polish Act on Waste (2001). Increased public awareness of environmental protection, combined with the necessity for European Union countries to adhere to progressively more stringent legislation, turns our attention towards the need to minimize wastewater sludge production and recycle sludge in an environmentally friendly manner. One of the methods enabling us to achieve both of the above-mentioned priorities is the use of wastewater sludge disintegration.

In this process, organic and mineral compounds (e.g. orthophosphates, ammoniacal nitrogen, and Ca^{2+} and Mg^{2+} ions) located inside and on the surface of the activated sludge flocs are released to the liquid. Recent literature (Chiu et al. 1997, Wang et al. 1999, Wang et al. 2006, Zhang et al. 2007) shows that sludge disintegration leads to a significant increase in soluble organic carbon concentrations, including volatile fatty acids (VFAs) in sludge water. Disintegration has been successfully applied to treat surplus activated sludge (SAS) in order to intensify its anaerobic digestion. The benefits of disintegration in this application are supported by the results of experimental lab measurements and the data obtained from full-scale

plants (Baier & Schmidheiny 1997, Chiu et al. 1997, Wang et al. 1999, Tiehm et al. 2001, Zielewicz-Madej 2003, Nowak 2006, Suschka et al. 2007). Through disintegration of SAS prior to anaerobic digestion, benefits such as smaller digested sludge mass, reduced digestion time, increased biogas yields and increased reduction of volatile solids mass can be achieved. Disintegration of wastewater sludges can also be used as an alternative method for the production of easily biodegradable organic substrates in order to improve the efficiency of biological nutrient removal processes (Müller 2000, Schmitt 2006, Dytczak et al. 2007).

Many different types of wastewater sludge disintegration have been developed. These methods differ in terms of the agent used to disrupt the floc structure (thereby causing cell lysis). The most widely used disintegration methods include ultrasonic disintegration (Wang et al. 1999, Tiehm et al. 2001, Zielewicz-Madej 2003), ball mills (Müller 2000, Nah & Kang 2000), homogenizing mixers (Fukas-Płonka & Janik 2006), pressure homogenization, thermal hydrolysis (Camacho et al. 2002) and ozonation (Dytczak et al. 2007). In practice, the choice of a disintegration method is dictated, on the one hand, by the required disintegration efficiency and, on the other hand, by the capital and operational costs of the system. As efficiency indicators, the following parameters can be used: increase of soluble COD (SCOD) in the sludge water, degree of disintegration (DD) or solubilization efficiency (α). For a comparison of energy inputs into a disintegration process, a parameter called specific energy (E$_S$)

was introduced. E_s is represented in kJ per kilogram of suspended total solids (kJ kg^{-1} TS). Müller et al. (2000) indicate that in order to attain a 25% degree of disintegration in a stirred ball mill and an ultrasonic homogenizer, the required specific energy input was 3000 and 10400 kJ kg^{-1}TS, respectively. The same disintegration degree was also attained using a deflaker with 4000 kJ kg^{-1} TS energy input (Kampas et al. 2007). El-Hajd et al. (2007), in their experiments conducted on ultrasound disintegration, achieved a DD of 27.5% at 11000 kJ kg^{-1}TS specific energy input. These results indicate that mechanical sludge disintegration methods allow us to attain similar disintegration efficiencies to ultrasound methods at a much lower energy input. This justifies the need to seek new construction designs of mechanical disintegration units, which would allow us to further reduce current energy requirements and, at the same time, maintain or even improve their disintegration efficiencies.

At the Power and Environmental Protection Research Centre of the Warsaw University of Technology, a prototype unit of a sludge disintegration apparatus has been constructed. Operation of this apparatus is based on the idea of the intentional creation of cavitation inside sludge liquor (patent application). The phenomenon of cavitation is accompanied by pressure gradients, which results in local temperature increases. Quoting Chu at al. (2001) "…both bubble explosion and the induced bulk solution temperature rise are equally important in sludge floc disintegration and cell lysis". This paper presents the results of the first stage of the innovation project, which aims to assess the applicability of this apparatus for disintegration of wastewater sludges. At this stage of research, the efficiency of disintegration as a function of specific energy input has been measured.

2 MATERIALS AND METHODS

2.1 Surplus activated sludge

Surplus activated sludge (SAS) used in this study was collected immediately after the belt thickener, from a local full-scale wastewater treatment plant that used the anaerobic-anoxic-aerobic process (A$_2$O). TS concentrations for different samples of the thickened SAS were in the range of 3.57–4.75%.

2.2 Disintegration setup

The main component of the mechanical cavitation inducer is a specially formed impeller propelled by an 11 kW electrical motor of a rotational speed of 3000 rpm. During the operation of the disintegration apparatus in certain, liquid-filled areas of the impeller (so-called active spaces), cavitation is intentionally induced. This cavitation causes disruption of the sludge flocs. The sludge is fed to the apparatus by an external feed pump at pressures of about 0.2 bar. The disintegration process is controlled by

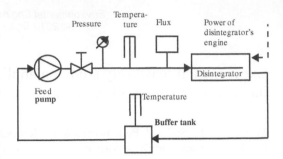

Figure 1. Schematic diagram of the laboratory set-up for carrying out batch sludge disintegration tests.

manipulating the hydraulic retention time in the active spaces (flow through the unit). During the improvement works on the disintegration unit, both the shape and the number of active spaces in the rotor have been changed (patent application). These modifications have been made in order to increase the energy efficiency of the disintegrator unit.

Each construction modification was accompanied by batch test experiments used to check the physico-chemical properties of the sludge water at different disintegration times (0–1280 seconds), which are proportionally related to energy inputs per sludge volume – E_v (kJ l^{-1}). All tests started within 1 h after sampling to prevent subsequent sludge changes. The schematic diagram of the experimental set-up is shown in Figure 1.

The batch tests were carried out by feeding the cavitation inducer 1–9 times with a 65-litre sample of SAS. The device was operating at a sludge throughput of about 2.0 m^3 h^{-1}. The process was carried out in a closed loop. After each preset disintegration time, a 0.5 l sample of disintegrated sludge was taken from a continuously stirred buffer tank. Immediately after this, the clear liquid in each sample was separated from the sludge by centrifugation and filtration (see Analytical procedures), and then SCOD and VFAs were measured in the sludge water. In an individual test, the sludge flow and instantaneous motor power were continuously controlled. On this basis, the true energy consumption was calculated for each sample of sludge. The sludge temperature was also monitored. In a disintegration process, an increase in energy input was observed to lead to a stepwise increase in sludge temperature from 10°C in the influent to 60°C in the effluent. A similar range of temperature increase in the sludge being disintegrated has been noted by Chu et al. (2001) and Grönroos et al. (2005) in their studies on ultrasonic sludge disintegration. These authors indicate that temperature increases, together with other physical and chemical processes accompanying cavitation, have a significant impact on the process efficiency. Wang et al. (2006) showed that subjecting the sludge only to thermal disintegration at temperatures up to 65°C results in a small effect on disintegration.

Table 1. Operational parameters of the mechanical cavitation inducer in each batch test.

Test no.	Solids content in the raw sludge (inlet) [%]	Flow rate through the apparatus [m³ h⁻¹]	Changes in sludge temperature during disintegration [°C]	Remarks
1	3.57	2.0	10–53	Before modernization
2	4.19	2.1	10–54	After modernization
3	4.36	2.0	10–52	
4	4.75	2.1	10–60	

The operational parameters of the mechanical cavitation inducer for wastewater sludge disintegration in each of the batch test series is shown in Table 1.

2.3 Calculation of the degree of disintegration

The sludge disintegration degree was calculated as a ratio of COD increase by mechanical disintegration in the sludge supernatant to the COD increase by chemical hydrolyzation (Müller 2000):

$$DD_{COD} = [(COD_d - COD_0)/(COD_{ch} - COD_0)] * 100\% \tag{1}$$

where COD_d and COD_0 depict soluble chemical oxygen demands for the disintegrated and the untreated samples, respectively, and COD_{ch} is the SCOD value of a sample hydrolyzed chemically in a 0.5 M NaOH solution at 20°C for 22 h (Nickel & Neis 2007).

2.4 Analytical procedures

The analytical procedures included the measurement of dry solids content of the sludge intake and the measurements of SCOD and VFA concentrations in the sludge water before and after disintegration. Sludge water was separated from the sludge mass by a 30-minute centrifugation at a g-force of 19621 g and subsequent filtration through a 0.45 μm membrane filter. All chemical tests were conducted accordingly to the Standard Methods (1995).

3 RESULTS AND DISCUSSION

In the initially designed cavitation inducer, a disintegration test was carried out for three different levels of energy input per unit volume of a 3.57% w/w sludge (SAS) (Table 2). As expected, as the energy input increased, the amount of organic substrates released from sludge flocs into the sludge water also increased. At the maximum tested E_v (211 kJ l⁻¹), a 45-fold increase in SCOD concentration in the sludge water up to a value of 4130 mg l⁻¹ was observed. The VFA

Table 2. Results of a sludge disintegration batch test before the modernization of the apparatus (total solids – 3.57%).

Name	Unit	Raw sludge	Disintegrated sludge		
Number of successive sludge throughputs through the apparatus	–	–	1	5	7
Amount of energy used for disintegration	kJ l⁻¹	–	30.2	151	211
	kJ (kg⁻¹TS)	–	846	4230	5922
Sludge water characteristics:					
Soluble COD	mg l⁻¹	91	480	2550	4130
Volatile fatty acids	mg l⁻¹	0	2	303	365
Disintegration degree	%	–	1.68	10.66	17.51

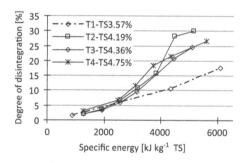

Figure 2. Degree of disintegration as a function of specific energy input (TS = total solids).

concentration was, at that point, equal to 365 mg l⁻¹. Despite a relatively large increase in SCOD and VFA concentrations in the sludge water, only a 17.5% DD was observed. The results of the following test support the results shown in the Table 2.

A comparable degree of disintegration was obtained in a deflaker by Kampas et al. (2007) at a similar E_v level of about 180 kJ l⁻¹. As one of the objectives of the project was to create a sludge disintegrator unit that was capable of producing a high DD at a relatively low energy consumption, it was decided that some construction changes should be introduced into the apparatus in order to reduce the energy input requirements for the sludge disintegration process.

The experimental results carried out after the disintegrator unit modifications had taken effect (Figure 2) showed that the modification works led to the expected operational improvements. At the specific unit energy inputs of 5090–5590 kJ kg⁻¹ TS, disintegration degrees of 24.6–30.2% were attained. This was accompanied by an increase in SCOD and VFAs in the sludge water to the levels of 6490–9050 mg l⁻¹ and 366–529 mg l⁻¹, respectively (Figure 3, Figure 4).

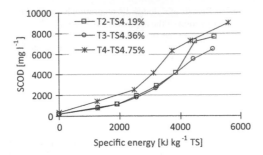

Figure 3. Changes in soluble COD (SCOD) in sludge water as a function of specific energy input.

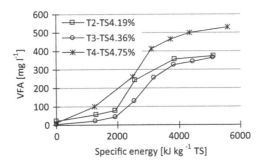

Figure 4. Changes in volatile fatty acids (VFAs) in sludge water as a function of specific energy input.

Figure 3 displays the changes in COD of sludge water as a function of specific energy input to the process. Based on this information, it was found that the most effective release of organic compounds from activated sludge flocs occurred while E_s was maintained in the range of 2500–4500 kJ kg^{-1} TS. After exposure of the sludge to the first 2500 kJ kg^{-1} TS, SCOD in the sludge water increased from 235–390 mg l^{-1} to 1790–2570 mg l^{-1}. The amount of "produced" organic substances per 1 kJ of energy input (W_{SCOD}) for this experiment was 13.8–18.5 mg SCOD kJ^{-1}. After an additional exposure to 2000 kJ kg^{-1} TS, a significant increase in the rate of organic substrates release was observed. This observation corresponded to SCOD concentrations in the sludge water of between 5510 and 7350 mg l^{-1}. W_{SCOD} for E_s of between 2500 and 4500 kJ kg^{-1} TS reached a level of 44.7–66.1 mg SCOD kJ^{-1}. A further increase in specific input energy caused a further release of organics into the sludge water, but the process occurred at lower intensities than observed at an E_s of 2500–4500 kJ kg^{-1} TS. W_{SCOD} recorded for $E_s >$ 4500 kJ TS^{-1} was 18.5–35.1 mg SCOD kJ^{-1}.

Similarly to SCOD, an increase of energy input into the sludge led to an increase in VFAs in the sludge water (Figure 4). The highest rate of VFA release was observed for specific energy inputs between about 1900 and 3800 kJ kg^{-1} TS. After the upper bound of this range was exceeded, the intensity of VFA release decreased substantially. As SCOD was still increasing after the rate of VFA release slowed down,

this indicated that the organic compounds released for $E_s >$ 3800 kJ kg^{-1} TS were slowly biodegradable. Based on this observation, it is hypothesized that this increase in slowly biodegradable SCOD might be due to lysis of bacterial cells, in the course of which the compounds used for building cell structures are released to the environment.

After the analysis of disintegration efficiencies obtained from a mechanical cavitation inducer and a deflaker (Kampas et al. 2007) at comparable energy densities, it was found that the device developed by the authors allows for a significant reduction of energy expenditures. For example, at E_v of 180 kJ l^{-1}, the values of SCOD in the sludge water from the sludge disintegrated in a mechanical cavitation inducer increased up to about 7200 mg l^{-1} (TS = 4.19%, SCOD in the sludge water before the disintegration – 235 mg l^{-1}), whereas in the case of sludge disintegration in a deflaker unit the SCOD concentration reached only 2000 mg l^{-1} (TS = 6.1% and TS = 7.2%, average and maximum SCOD in the sludge water before disintegration of 176 mg l^{-1} and 342 mg l^{-1}).

A higher release of organic substances for the mechanical cavitation inducer indicate that this device allows us to create higher disintegrating forces inside the sludge for similar energy inputs than a deflaker unit. These shear forces lead to a porous and microporous breakdown of the floc structure. It is assumed that a proportion of organic compounds released at disintegration may originate from cell lysis. In order to explain the mechanisms of sludge disintegration in the cavitation inducer, further experiments are being carried out. These experiments have been extended to include the analyses of DNA in sludge water, microscopic investigations of the sludge floc sizes and the determination of oxygen uptake rates.

4 CONCLUSIONS

1. The results obtained show that the mechanical cavitation inducer allowed us to achieve high disintegration effects of wastewater sludge. The main factor influencing the effect of the process is the specific energy input.
   At the specific energy inputs of 5090–5590 kJ kg^{-1} TS, 24.6–30.2% disintegration ratios were achieved. This correlated with an increase in SCOD and VFA concentrations in the sludge water to 6490–9050 mg l^{-1} and 366–529 mg l^{-1}, respectively. This work further supports the evidence that mechanical disintegration methods allow the achievement of disintegration efficiencies comparable to ultrasonic methods at substantially lower energy inputs.

2. The highest rate of SCOD increase in the sludge water was observed when the specific energy input was maintained in the range of 2500–4500 kJ kg^{-1} TS.

3. Results of this research allow us to formulate a hypothesis that an application of such a device for the disintegration of a portion of SAS going

into anaerobic digestion in a full-scale plant might lead to improvements in digestion efficiency, such as reduction in volatile solids mass and increased biogas yield. This hypothesis is currently being verified in further experimental studies at several full-scale wastewater treatment plants.

ACKNOWLEDGEMENTS

The experiments were carried out as part of the research project entitled "Wastewater sludge disintegration technology using mechanical cavitation inducers for application at Polish wastewater treatment plants", No. KB/73/12966/IT1-B/U/08, and financed by the "Programme of the Minister of Science and Higher Education – Technology Initiative I" (Poland).

REFERENCES

Baier, U. & Schmidheiny, P. 1997. Enhanced anaerobic degradation of mechanically disintegrated sludge. *Water Science and Technology* 36(11): 137–143.

Camacho, P., Deleris, S., Geaugey, V., Ginestet, P. & Paul, E. 2002. A comparative study between mechanical, thermal and oxidative disintegration techniques of waste activated sludge. *Water Science and Technology* 46(10): 79–87.

Chiu, Y.-C., Chang, C.-N., Lin, J.-G. & Huang, S.-J. 1997. Alkaline and ultrasonic pre-treatment of sludge before anaerobic digestion. *Water Science Technology* 36(11): 155–162.

Chu, C.P., Chang, B.-V., Liao, G.S., Jean, D.S. & Lee, D.J. 2001. Observation on changes in ultrasonically treated waste-activated sludge. *Water Research* 35(4): 1038–1046.

Dytczak, M., Londry, K., Siegrist, H. & Oleszkiewicz, J. 2007. Ozonation reduced sludge production and improves denitrification. *Water Research* 41: 543–550.

Benabdallah El-Hajd, T., Dosta, J., Marquez-Serrano, R. & Mata-Alvarez. 2007. Effect of ultrasound pretreatment in mesophilic and thermophilic anaerobic digestion with emphasis on naphthalene and pyrene removal. *Water Research* 41: 87–93.

Fukas-Płonka, Ł. & Janik, M. 2006. Homogenizacja osadu nadmiernego (Surplus activated sludge homogenization). *Forum Eksploatatora (Operater Forum)* 3: 14–16.

Grönroos, A., Kyllönen, H., Korpijärvi, K., Pirkonen, P., Paavola, T., Jokera, J. & Rintala, J. 2005. Ultrasound assisted method to increase soluble chemical oxygen demand (SCOD) of sewage sludge for digestion. *Ultrasonics Sonochemistry* 12: 115–120.

Kampas, P., Parsons, S.A., Pearce, P., Ledoux, S., Vale, P., Churchley, J. & Cartmell, E. 2007. Mechanical sludge disintegration for the production of carbon source for biological nutrient removal. *Water Research* 41: 1734–742.

Müller, J. 2000. Disintegration as a key-step in sewage sludge treatment. *Water Science and Technology* 41(8): 123–130.

Nah, W. & Kang, Y.W. 2000. Mechanical pretreatment of waste activated sludge for anaerobic digestion process. *Water Research* 34: 2362–2368.

Nickle, K. & Neis, U. 2007. Ultrasonic disintegration of biosolids for improved biodegradation. *Ultrasonics Sonochemistry* 14: 450–455.

Nowak, A. 2006. Instalacja dezintegracji osadu nadmiernego na oczyszczalni ścieków w Rzeszowie (Surplus activated sludge disintegration system at wastewater treatment plant in Rzeszow). *Mat. Konf. Gdańska Fundacja Wody(Proc. Conf. Gdansk Water Foundation).*

Schmitt, W. 2006. Raport końcowy. Zastosowanie systemu CROWN służącego do dezintegracji osadu recyrkulowanego na Centralnej oczyszczalni ścieków w Wiesbaden – stolicy Hesji (The final raport. Application of CROWN system for wastewater sludge disintegration at Central wastewater treatment plant in Wiesbaden – capital of Hesja) *Mat. Konf. Gdańska Fundacja Wody (Proc. Conf. Gdansk Water Foundation).*

Standard Methods for the Examination of Water and Wastewater; 1995, 19th ed., American Public Health Association/American Water Works Association/Water Environment Federation. Washington DC: USA.

Suschka, J., Grübel, K. & Machnicka, A. 2007. Możliwości intensyfikacji procesu fermentacji beztlenowej osadów ściekowych poprzez dezintegrację osadu czynnego w procesie kawitacji mechanicznej (Possibility of intensification of waste sludges anaerobic digestion by disintegration of surplus activated sludge in mechanical cavitation process). *GWiTS* 3: 26–28.

Tiehm, A. Nickel, K. Zellhorn, M. & Neis, U. 2001. Ultrasonic waste activated sludge disintegration for improving anaerobic stabilization. *Water Research* 35(8): 2003–2009.

Wang, Q., Kuninobu, M., Kamikoto, K., Ogawa, H.-I. & Kato, Y. 1999. Upgrading of anaerobic digestion of waste activated sludge by ultrasonic pre-treatment. *Bioresource Technology* 68: 309–313.

Wang, F., Shan L, S. & JI, M. 2006. Components of released liquid from ultrasonic waste activated sludge disintegration. *Ultrasonics Sonochemistry* 13: 334–338.

Zhang, P., Zhang, G. & Wang, W. 2007. Ultrasonic treatment of biological sludge: Floc disintegration, cell lysis and inactivation. *Bioresource Technology* 98: 207–210.

Zielewicz-Madej, E. 2003. The influence of parameters of ultrasonic disintegration on the intensification anaerobic biodegradation of organic compounds from sewage sludge. *Inżynieria i Ochrona Środowiska (Engineering & Protection of Environment)* 6(3–4): 455–467.

Neutralization of solid wastes and sludges

Environmental Engineering III – Pawłowski, Dudzińska & Pawłowski (eds)
© 2010 Taylor & Francis Group, London, ISBN 978-0-415-54882-3

Disintegration of fermented sludge – possibilities and potential gas

M. Cimochowicz-Rybicka & S.M. Rybicki
Institute of Water Supply and Environmental Protection, Cracov University of Technology,
Cracov, Poland

B. Fryzlewicz-Kozak
Institute of Chemical and Process Engineering, Cracov University of Technology,
Cracov, Poland

ABSTRACT: This paper summarises the first stage of laboratory tests on the disintegration of fermented (digested) sludge. It addresses more efficient methane recovery and better stabilisation. Respirometric tests on digested sludge from full-scale plants showed a possibility of increasing the total recovery of fermentation gas. The paper also describes the behaviour of the microorganism population and its ability to recover. The methodology presented has been tested on sludge from a large (500,000 population equivalent) municipal wastewater treatment plant. A disintegration method using ultrasound was chosen as the reference, but any disintegration technology can be tested using the proposed procedure.

Keywords: Biogas, disintegration, energy recovery, sludge digestion, sludge processing, ultrasound.

1 INTRODUCTION

The general need for sustainability of modern urban systems includes a requirement for improvement in energy and resource recovery at wastewater treatment plants (WWTPs). This has led to a greater emphasis being placed on enhancing the efficiency of anaerobic digestion producing methane-rich gas ('biogas'). A proposal for sludge disintegration has been incorporated recently into the process scheme of a WWTP to improve the hydrolysis phase of sludge digestion. Disintegration of waste activated sludge (WAS) prior to digestion is a common practice. The operational conditions and expected gains have been described by numerous authors (Müller et al. 1998, Müller 2000, Ødegaard 2004, Bougrier et al. 2005, Boehler & Siegrist 2006, Zabranska et al. 2006, Nickel & Neis 2007). Topics addressed by these authors have included:

- Increased net biogas production and thus better energy characteristics of the sludge handling system;
- Possible use of the organic matter obtained from a disintegrated WAS as a source of easily biodegradable carbon for denitrification processes;
- Minimisation of the sludge bulking effect in WWTPs.

In this research, we have examined the disintegration of digested sludge to check whether an application of this process would change gas production. The purpose of this research was to validate and discuss the possible control of gas production during the storage and handling of digested sludge. This has the potential to be important knowledge from both a technical and an ecological point of view. In addition, tests were conducted on the ability of the sludge to recover, in order to determine if such sludge could be used to seed a fermentation process.

An ultrasound method was used to disintegrate sludge prior to its digestion. This approach was selected based on our previous technical and research experience. This is one of best methods applicable for sludge disintegration and it is used in Poland more frequently than other methods such as cavitation, ball mills, thermal disintegration or ozonization.

1.1 Technical parameters of full-scale disintegration

Ultrasound disintegration of sludge flocculants can transform a significant portion of the insoluble organics into soluble forms (Bień & Szparkowska 2004, Wang et al. 2005). The mechanism of sludge particle disruption during the application of ultrasound is still being examined, which is why the correlation between technical parameters (cavitation bubbles size, ultrasound frequency and density) and sludge disruption has not been fully recognised. Changes in the structure of the flocculants are mainly the effect of the mechanical interactions associated with ultrasound cavitation as well as the chemical reactions and thermal changes. For both low and medium ultrasound intensities – up to $30\,kW/m^2$, – reversible changes in the medium and the cells take place, while at high ultrasound intensities ($>30\,kW/m^2$) irreversible damage to the cells occurs. The following parameters have a significant effect on sludge disintegration: the pH of the disintegrated sludge and its concentration, the intensity of the ultrasound, ultrasonic density and irradiation time. Moreover, the effects of applying ultrasound depend largely upon the container geometry and the probe position (Zielewicz-Madej & Sorys 2007, Fryźlewicz-Kozak & Tal-Figiel 2008).

The disruption of the sludge particles can enhance the subsequent acidogenesis, acetogenesis and methanogenesis reactions, increasing methane production and reducing sludge volume. Most of the tests on the disintegration of stabilised sludge completed in recent years have focused on improving de-watering and minimising the wet mass of sludge rather than ecological safety (stability) and potential biogas recovery (Tiehm et al. 2001, Oneyche et al. 2002, Scholz 2005, Tomczak-Wandzel et al. 2008). This paper, however, puts an emphasis on increasing the ecological safety of sludge and biogas production.

2 MATERIALS AND METHODS

2.1 Basic concepts related to measurements

An experimental procedure was developed to observe sludge activity in conditions as close to the routine operations of sludge digesters (at real-scale WWTPs) as possible. We found the respirometric tests the most applicable as they reflect the sequences of all processes related to biogas generation. Digested sludge samples from an industrial scale wastewater treatment plant (500,000 population equivalent) at Kraków Kujawy were delivered to the Cracow University laboratory for disintegration and testing. The measurements for each series of tests were taken over a time period equivalent to the sludge retention time (SRT) in real digestion chambers (i.e. 21 days). Microscopic analyses of sludge before and after disintegration were performed in parallel with the gas measurement.

2.2 Experimental protocol

The sludge was transported from the wastewater treatment plants in 20-litre thermostatic containers and delivered to the laboratory tests stand within 75 minutes of being collected. Sludge samples of $125\,cm^3$ were disintegrated using sonic equipment UD 11, nominal frequency 22.5 kHz. The ultrasound intensity was $24*10^3\,W/m^2$. This value was chosen as the optimal one for this type of sludge in previous experiments by the authors. The positioning of the sonotrode was also decided on the basis of prior experience (Lettinga et al. 1991, Cimochowicz-Rybicka & Rybicki 1999, Cimochowicz-Rybicka 2004, Cimochowicz-Rybicka et al. 2008). Tests were conducted on the sludge after 5, 7, 9 minutes of treatment with ultrasound. The degree of sludge disintegration (DD) was calculated based on the chemical oxygen demand (COD) in the sludge supernatant (equation 1).

$$DD = \frac{COD_d - COD_i}{COD_a - COD_i} *100\% \qquad (1)$$

where: $COD_d = COD$ of the centrate of the disintegrated sludge sample; $COD_i =$ initial COD of the centrate of the sludge sample (before disintegration); $COD_a =$ maximum value of COD, which can be obtained in the supernatant after alkaline hydrolysis of the sludge (chemical disintegration with NaOH).

The COD value in all the supernatant samples (COD_d, COD_i, COD_a) was determined after centrifugation and filtration (paper filter $0.45\,\mu m$). Various procedures for determining the maximum COD value after chemical disintegration can be found in the literature (Nickel & Neis, 2007, Gonze et al. 2003). Both 0.5 M and 1.0 M solutions of sodium hydroxide (NaOH) have been used by different teams; however, we found the application of a 1.0 M NaOH solution to the digested sludge supernatant (reaction time 22 hours at 20°C) yielded more accurate results.

Samples of disintegrated sludge (of known DD value) were stored in eight $500\,cm^3$ gas-tight vessels of a Challenger AER-208 (Challenge Environmental Systems, Fayetteville, AR) respirometer. As the disintegration stand allowed for the treatment of $125\,cm^3$ of sludge at the same time, this process was repeated four times under the same conditions. A $500\,cm^3$ portion was then homogenized and poured into a measuring vessel. The vessel was placed in a water bath with magnetic stirrers (approximate energy input $5\,W/m^3$), to ensure complete and thorough mixing, closely replicating the conditions of a full scale WWTP. The incubation temperature was 35°C (±0.5°C). Methanogenic bacteria are sensitive to pH; this parameter was carefully controlled in the samples and ranged from 7 (±0.1) to 7.7 (±0.1). The gas generated in each vessel was measured at two hourly intervals, the data being stored in the computer integrated into the test stand. The methane content of the fermentation gas was checked three times during each series of tests. A set of parallel tests was conducted with eight cells, i.e. each sludge sample after ultrasound treatment was poured into two vessels and exactly the same procedure was followed. These parallel tests allowed a comparison of results and acted as a check on the accuracy of the measurements.

Specific tests on the possible recovery of the microorganisms were performed simultaneously with the tests on gas production. Samples of digested sludge, both treated and untreated, were fed with substrate (reflecting the digestion chamber feed) and gas production was measured. Sample feeding was adopted from the methanogenic activity tests. That is, the samples were fed with a volatile fatty acid (VFA) mixture (based on butyric and propionic acids); the substrate concentration in the samples was approximately 5.0 g volatile suspended solids (VSS) per litre. This method was recommended by Lettinga

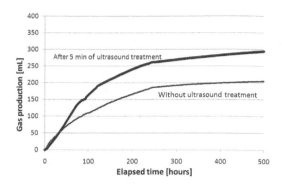

Figure 1. Gas production cumulative curve – time of sonication 5 minutes. (Samples without disintegration, initial parameters: pH = 7,05, COD = 295 mg/dm³; SS = 14050 mg/dm³; samples after 5 minute sonication, initial parameters pH = 7,55, COD = 506 mg/dm³; SS = 12780 mg/dm³).

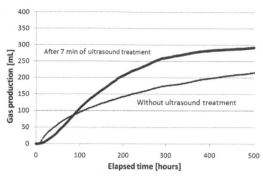

Figure 2. Gas production cumulative curve – time of sonication 7 minutes (Samples without disintegration, initial parameters: pH = 7,53, COD = 240 mg/dm³; SS = 9700 mg/dm³; samples after 7 minute sonication, initial parameters pH = 7,70, COD = 537 mg/dm³; SS = 9640 mg/dm³).

et al. (1991) and previously tested by Cimochowicz-Rybicka (2004).

3 RESULTS AND DISCUSSION

3.1 *Part I – test on potential gas production*

Experiments were conducted on digested sludge collected from a full-scale WWTP operating with an SRT of between 19 and 21 days. Tests were performed during the period Winter 2008 to Spring 2009; the plant, however, does not reflect seasonal changes in sludge quality. The degree of disintegration was calculated for various ultrasound treatment times. The DD value increased with increases in the length of the ultrasound treatment. The values of DD were 15% for an ultrasound treatment time of 5 minutes, 28% following 7 minutes of treatment and 29% after 9 minutes of treatment. Gas production was measured and then expressed as cumulative gas production curves (see Figures 1, 2 and 3). Gas production curves reflect the average value of the two identical samples as previously described. The 'untreated' terms in Figures 1, 2 and 3 refer to sludge after digestion, but without disintegration (the term 'raw sludge' is usually applied to sludge that has not undergone any processing and is inappropriate in this instance).

Ultrasonic disintegration led to additional production of fermentation gas. The specific increase for the different periods of ultrasound treatment varied between 60% and 70% more than from the 'untreated' sludge. The methane content was measured in samples taken after 240, 360, 480 hours. The results obtained were similar to those of a digestion chamber in a WWTP – an average of 68% (±9%) methane in the fermentation gas. Some differences were found for the methane content in the gas produced between 240 and 360 hours of the experiment. We will focus on this anomaly in further investigations to check whether this result can be generalised. The effects of disintegration

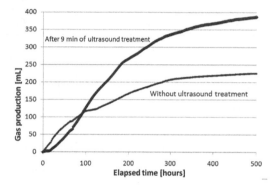

Figure 3. Gas production cumulative curve – time of sonication 9 minutes (Samples without disintegration, initial parameters: pH = 7,60, COD = 196 mg/dm³; SS = 9730 mg/dm³; samples after 9 minute sonication, initial parameters pH = 7,74, COD = 467 mg/dm³; SS = 9540 mg/dm³).

and sludge composition on gas production are summarised in Tables 1, 2 and 3. Gas production measurements were taken for up to 500 hours. It must be emphasised that unit gas production without disintegration (expressed in gas volume per mass of VSS introduced) was significantly lower than usually measured for disintegrated WAS. This can be attributed to the consumption of most of the degradable matter during anaerobic digestion. Ultrasonic disintegration increased this unit gas production as follows: 43% increase after 5 minutes of treatment, 44% after 7 minutes of treatment and 64% after 9 minutes of treatment.

The increase in the production of fermentation gas in all three series of tests was, as expected, bound to the decrease in COD concentration, rather than any changes in the concentration of suspended solids (SS) and/or decrease in the concentration of VSS.

3.2 *Part II – tests on microorganism recovery*

The main mechanism of disintegration is the destruction by the ultrasound of cell walls of the

Table 1. Effect of 5 minutes of ultrasonic disintegration on sludge parameters and biogas production.

Parameters	Units	Sludge without disintegration			Disintegration time 5 minutes		
		Co	Ce	Change [%]	Co	Ce	Change [%]
COD	mgO$_2$/dm^3	295	276	7	506	308	64
SS	mg/dm^3	14.050	13.440	5	12.780	12.000	7
VSS	mg/dm^3	7.500	7.450	1	7.350	6.410	15
Alkalinity	CaCO$_3$	1.750	1.400	25	1,150	1.750	(+34)
pH		7.05	7.68	(+8)	7.55	6.97	8
Unit gas production	dm^3/kg VSS$_{int}$	49.6			71.3		

Where: Co = value of the parameter in the samples before the respirometric tests; Ce = value of the parameter in samples after the respirometric tests; VSS$_{int}$ = VSS introduced into the reactor.
Change [%] = specific change in parameters' values, referred to Co value; values in parentheses preceded by plus (+) symbol indicate and increase in value, otherwise 'change' means decrease.

Table 2. Effect of 7 minutes of ultrasonic disintegration on sludge parameters and biogas production.

Parameters	Units	Sludge without disintegration			Disintegration time 7 minutes		
		Co	Ce	Change [%]	Co	Ce	Change [%]
COD	mg O$_2$/dm^3	240	161	33	537	270	50
SS	mg/dm^3	9.700	9.160	6	9.640	8.770	9
VSS	mg/dm^3	5.000	4.670	7	5.000	4.470	11
Alkalinity	CaCO$_3$	1.100	1.500	(+27)	1.050	1.650	(+36)
pH		7.53	6.93	8	7.70	6.96	10
Unit gas production	dm^3/kg VSS$_{int}$	83.2			120.0		

Table 3. Effect of 9 minutes of ultrasonic disintegration on sludge parameters and biogas production.

Parameters	Units	Sludge without disintegration			Disintegration time 9 minutes		
		Co	Ce	Change [%]	Co	Ce	Change [%]
COD	mg O$_2$/dm^3	196	186	5	467	186	60
SS	mg/dm^3	9.730	9.100	6	9.540	8.600	10
VSS	mg/dm^3	5.263	4.000	2	5.000	4.500	10
Alkalinity	CaCO$_3$	1.125	1.300	(+16)	1,075	1.680	(+56)
pH		7.60	6.90	9	7.74	6.96	10
Unit gas production	dm^3/kg VSS$_{int}$	93.5			153.2		

Table 4. Effect of 7 minutes of ultrasonic disintegration on sludge parameters and biogas production with volatile fatty acid (VFA) feed – 'recovery' tests.

Parameters	Units	Sludge without disintegration			Disintegration time 7 minutes		
		Co	Ce	Change [%]	Co	Ce	Change [%]
COD	mg O$_2$/dm^3	4.148	302	93	4,440	694	84
SS	mg/dm^3	9.670	12.100	(+20)	9.600	10.930	(+12)
VSS	mg/dm^3	5.000	5.280	(+5)	5.000	5.250	(+5)
Alkalinity	CaCO$_3$	3.500	7.013	(+50)	3.250	4.250	(+24)
pH		7.42	7.60	(+2)	7.40	7.56	(+2)
Unit gas production	dm^3/kg VSS$_{int}$	455			497		

microorganisms. The question we asked was whether it was possible to recover these presumably destroyed organisms as active 'producers' of a methane-rich gas. If so, these microorganisms could be used to seed sludge fermentation. This question is also ecologically important. The possible 'recovery' of these microorganisms may lead to gas production following a time of sludge storage. The testing stand was the same as

Table 5. Effect of 9 minutes of ultrasonic disintegration on sludge parameters and biogas production with volatile fatty acid (VFA) feed – 'recovery' tests.

Parameters	Units	Sludge without disintegration			Disintegration time 9 minutes		
		Co	Ce	Change [%]	Co	Ce	Change [%]
COD	mg O_2/dm^3	4.704	325	93	4.992	780	84
SS	mg/dm^3	9.730	12.370	(+27)	9.540	12.170	(+27)
VSS	mg/dm^3	5.260	5.420	(+3)	5.000	5.240	(+5)
Alkalinity	CaCO$_3$	3.500	4.700	(+25)	3.300	4.500	(+27)
pH		7.40	7.65	(+3)	7.40	7.71	(+4)
Unit gas production	dm^3/kg VSS$_{int}$	590			618		

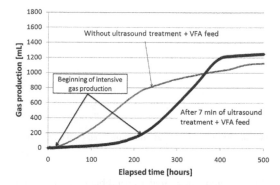

Figure 4. Gas production cumulative curve – time of sonication 7 minutes; sample with a VFA feed; (Samples without disintegration, initial parameters: pH = 7.42, COD = 4148 mg/dm^3; SS = 9670 mg/dm^3; samples after 7 minute sonication, initial parameters pH = 7.40, COD = 4440 mg/dm^3; SS = 9600 mg/dm^3).

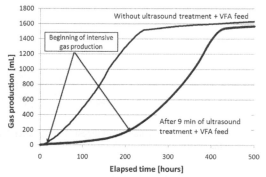

Figure 5. Gas production cumulative curve – time of sonication 9 minutes; sample with a VFA feed; (Samples without disintegration, initial parameters: pH = 7,40, COD = 4704 mg/dm^3; SS = 9730 mg/dm^3; samples after 9 minute sonication, initial parameters pH = 7.40, COD = 4992 mg/dm^3; SS = 9540 mg/dm^3).

described above. Two series of tests were completed, one for 7 minutes and one for 9 minutes of ultrasound treatment. The following procedure was adopted. VFA feed was added to both the 'untreated' and the disintegrated samples. The gas measurement procedure described above was followed simultaneously with the gas production and basic sludge parameters being measured. The results are summarised in Tables 4 and 5. The cumulative gas production curves can be seen in Figures 4, 5.

Initial parameters without treatment: pH = 7.42; COD = 4148 mg/dm^3; SS = 9670 mg/dm^3. Initial parameters after 7 minutes of ultrasound treatment: pH = 7.40; COD = 4.440 mg/dm^3; SS = 9,600 mg/dm^3.

Initial parameters without treatment: pH = 7.40; COD = 4704 mg/dm^3; SS = 9730 mg/dm^3. Initial parameters after 9 minutes of ultrasound treatment: pH = 7.40, COD = 4.992 mg/dm^3; SS = 9.540 mg/dm^3).

Both figures show that gas production after VFA substrate addition results in different dynamics for this process between the treated and untreated samples. However, both groups of samples produced similar amounts of gas, whether they had been treated or not.

There is a significant difference between the activities of microorganisms in each pair of samples. Arrows on Figures 4 and 5 show the starting point for the intensive gas production phase.

It can be seen that in the non-disintegrated samples this production starts rapidly (in the eighth hour of the experiment) while for the disintegrated samples it requires as long as 8 to 9 days to initiate this phase, i.e. approximately 200 hours. Until this time, the sludge is relatively very stable. Tables 4 and 5 summarise the 'recovery' tests. Also, in this case, COD consumption was a driving force for the process. A similar increase in gas production caused by disintegration was observed in the first part of the experiment. For example, after 7 minutes of ultrasound disintegration, the net increase in gas production was 184 ml per litre of sludge for the sample without the added VFA and 210 ml per litre for the sample with the VFA feed. This confirms that the VFA feed composition did not interfere with this process. The increase in dry mass and volatile dry mass may be credited to cell growth or to measurement discrepancies. The increase in alkalinity shown in Tables 4 and 5 is typical for the process. Figure 6 shows images of the biological preparations of the sludge particles (non-dyed on the

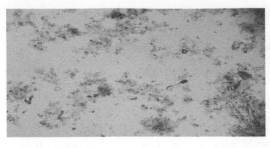

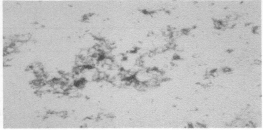

Figure 6.   Biological preparations of digested sludge before disintegration (non-dyed on the left, Gram dyed on the right).

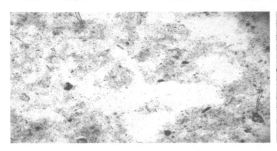

Figure 7.   Biological preparations of digested sludge after 7 minutes of ultrasound disintegration (non-dyed on the left, Gram dyed on the right).

left and Gram-dyed on the right). The original structure of the sludge flocculants is clearly visible. Images of sludge flocculants after 7 minutes of ultrasound disintegration are shown in Figure 7. The significant changes in the structure of the flocculent structure are observable – the flocculants have a specific, fuzzy appearance. These structural changes are more easily observed in the dyed samples.

4   RESULTS AND DISCUSSION

In both samples (disintegrated and non-disintegrated), the methane-rich gas production from digested sludge was powered by the chemical oxygen demand rather than by VSS decay. As was expected, it also confirmed the well known role of increased solubility in the whole process. It was observed that production of methane-rich biogas increased with an increase in the disintegration time and, also, the activity (the dynamics of methane production) was more pronounced in samples after longer treatment times.

The first phase of testing showed the possibility of an additional yield of 'biogas' from the disintegration of digested sludge followed by additional (second step) digestion or recirculation to the digestion chambers. This observation can be applied in WWTP practice to improve overall energy efficiency. Tests on the recovery of microorganisms showed that it takes about 200 hours for the cells to regenerate their gas production ability after disintegration. This demonstrates that the disintegration of digested sludge not only decreases its volume, but also makes it more stable. Long-term storage of digested and

disintegrated sludge under anaerobic conditions may lead to unwanted gas release, which is potentially harmful to the environment.

As was expected, the results showed that an increase in gas production resulted from the longer periods of ultrasound treatment. While this result is not particularly innovative, there is other information that may be important for operators and designers of WWTPs:

- Disintegration of digested sludge allows an increase in gas production by using the soluble COD released during the breakdown of the cells;
- Sludge digestion followed by disintegration makes it more stable;
- It is possible to rearrange the typical industrial-scale disintegration unit at a WWTP to perform a dual role – the disintegration of WAS prior to digestion and the disintegration of digested sludge prior to dewatering. In this latter case the rejected waters from the digestion process are COD-rich and adding these to the digestion chamber results in an increase in overall energy efficiency.

4.1   Proosals for further investigation

Further investigations will focus on the make up of the soluble phase after disintegration. Does the soluble phase contain mainly VFAs, which should be recycled through the process, or do long-chain fatty acids dominate, which favours the digestion chamber as a the receiving facility for the reject water? These investigations could lead to the development of a protocol to be routinely applied in the design of waste water treatment plants.

The differences observed between the methane content measured during the 240th and 360th hours of the experiments were presented above. Further experiments are planned to check the nature of this phenomenon.

5 CONCLUSIONS

In this paper the effect of ultrasound disintegration of digested sludge on its characteristics has been discussed. The main conclusions obtained are as follows:

- Disintegration of sludge may significantly increase recovery of a methane-containing fermentation gas generated in the anaerobic digestion chamber. Thus, it may increase the overall sustainability of wastewater treatment plants.
- Ultrasound disintegration was found to be quite feasible. However, an increase in the ultrasound treatment time beyond 7 minutes did not result in significant increases in gas yields.
- Sludge digestion followed by disintegration led to a better stabilisation of the sludge than did conventional digestion. This is a result of the more effective conversion of VSS. However, disintegrated sludge can recover its gas-producing abilities after approximately 7 to 8 days of exposure to mesophilic conditions in the presence of VFAs.

The above findings are the first steps in developing control strategies for the specific processing of sludge.

ACKNOWLEDGEMENTS

Experiments described in this paper were financed under a grant, number 1T09D02830 – 1258/H03/2006/30, from the Ministry of Informatics and Science, Poland.

REFERENCES

Bień, J.B. & Szpakowska, I. 2004. Impact of ultrasound disintegration of wastewater sludges on anaerobic stabilization performancej (in Polish). *Inżynieria i Ochrona Środowiska* 7(3–4): 341–352.

Boehler, M. & Siegrist, H. 2006. Potential of activated sludge disintegration. *Water Science and Technology* 53(12): 207–216.

Bougrier, C., Carrere, H. & Delgenes, J.P. 2005. Solubilisation of waste-activated sludge by ultrasonic treatment. *Chemical Engineering Journal* 106: 163–169.

Cimochowicz-Rybicka, M. & Rybicki, S.M. 1999. Application of sludge methanogenic activity to predict its ability to anaerobic fermentation processes: Polish experience. Proceedings of a Polish-Swedish seminar. In E. Plaza,

E. Levlin, E. & B. Hultman (eds), *Advanced Wastewater Treatment, Joint Swedish-Polish Reports*, Report (5). KTH Publishing, Report 3063: 99–109

Cimochowicz-Rybicka. M., Rybicki, S.M., Tomczak-Wandzel, R. & Mikosz, J. 2008. Application of methanogenic activity as a design tool for improvement of sludge disintegration. *Proceedings of 10th IWA World Water Congress*, Vienna, September 2008, file code 666675.

Fryżlewicz-Kozak, B. & Tal-Figiel, B. 2008. Theoretical and experimental analysis of flocs structure of activated sludge under sonication. *Chemical and Process Engineering* 29: 87–98.

Gonze, E., Pillot, S., Valette, E., Gonthier, Y. & Bernis, A. 2003. Ultrasonic treatment of an aerobic activated sludge in batch reactor. *Chemical Engineering and Processing* 42: 963–975.

Lettinga, G., Hulshoff, P. & Pol, L.W. 1991. Anaerobic reactor technology. *Proceedings of Advanced Waste Water Treatment International Course*, UNESCO-IHE Institute for Water Education., Wageningen Agricultural University, Delft, the Netherlands.

Müller, J. 2000. Disintegration as a key-step in sewage sludge treatment. *Water Science and Technology* 41(8): 123–130.

Müller, J., Lehne, G., Schwedes, J., Battenberg, S., Näveke, R., Kopp, J. & Dichtl, N. 1998. Disintegration of sewage sludge and influence on anaerobic digestion. *Water Science and Technology* 38(8–9): 425–433.

Nickel, K. & Neis, U. 2007. Ultrasonic disintegration of biosolids for improved biodegradation. *Ultrasonic Sonochemistry* 14: 450–455.

Ødegaard, H. 2004. Sludge minimization technologies – an overview. *Water Science and Technology* 49(10): 31–40.

Oneyche, T.I., Schlaefer, O. & Sievers, M. 2002. Improved energy recovery from waste sludge, Paper presented at ENVIRO 2002, *Waste conference*, 15–16 November 2002, Melbourne, Australia.

Scholz, M. 2005. Review of recent trends in capillary suction time (CST) dewaterability testing research. *Industrial and Engineering Chemistry Research* 44: 8157–8163.

Tiehm, A., Nickel, K., Zellhorn, M. & Neis, U. 2001. Ultrasonic waste activated sludge disintegration for improving anaerobic stabilization. *Water Research* 35(8): 2003–2009.

Tomczak-Wandzel, R., Mędrzycka, K. & Cimochowicz-Rybicka, M. 2008. Sewerage sludges' disintegration – a promising way to minimize negative environmental impact. In M. Pawłowska & L. Pawlowski (eds), *Management of pollutant emission from landfills and sludge*: 169–174. Taylor and Francis, London.

Wang F., Wang Y. & Ji M. 2005. Mechanisms and kinetics models for ultrasonic waste activated sludge disintegration. *Journal of Hazardous Materials* 123: 145–150.

Zabranska, J., Dohanyos, M., Jenicek, P. & Kutil, J. 2006. Disintegration of excess activated sludge – evaluation and experience of full-scale applications. *Water Science and Technology* 53(12): 229–236.

Zielewicz-Madej E. & Sorys P. 2007. Ultrasound disintegration of wasted activated sludge (in Polish), *Forum Eksploatatora* (in Polish) 2: 45–52.

Environmental Engineering III – Pawłowski, Dudzińska & Pawłowski (eds)
© 2010 Taylor & Francis Group, London, ISBN 978-0-415-54882-3

Mathematical modeling of wet oxidation of excess sludge in counter – current bubble columns

A. Chacuk & M. Imbierowicz

Faculty of Process and Environmental Engineering, Technical University of Lodz, Lodz, Poland

ABSTRACT: A mathematical model enabling a quantitative description of wet oxidation of excess sludge in continuous bubble columns is proposed. The model consists of mass and heat transfer kinetic equations, and material and heat-balance equations for gas and liquid phases flowing through the absorber. The equations refer to parallel, counter-current flows of the gas and liquid phases and take into account a complex chemical reaction in the liquid phase core. The model was used in a numerical simulation of wet oxidation in a bubble absorber for different process conditions: flow rate and composition of the gas and liquid phase, temperature and pressure, and different column heights and diameters.

Keywords: Mathematical modelling, wet oxidation, excess sludge, bubble column.

1 INTRODUCTION

The current annual production of municipal sewage sludge in Poland is more than 0.5 million tonnes dry solids (TDS) and has been increasing at an estimated 6% per year. Almost 90% of this sludge is discharged into landfill or used in agriculture; however, but these applications are being increasingly questioned because of strict regulations and the environmental impact of disposing of sewage sludge in landfill. For breaking down sewage sludge, anaerobic digestion is the most popular method, but only 20% of Polish wastewater treatment plants (WWTPs) carry out the process in airproof fermentation tanks and produce biogas, which can be used as a valuable source of 'green' energy. It is therefore necessary to increase the number of WWTPs producing biogas through anaerobic digestion of sewage sludge and to improve methane production efficiency in this process.

The yield of methane produced during sewage sludge fermentation can be improved by thermal pre-treatment. The use of wet air oxidation (WAO) as a pre-treatment step in the context of an integrated chemical–biological process of sewage sludge disposal has been investigated since the 1970s. Haug (1978) studied the wet oxidation of activated sludge at 448 K, and showed that sludge filterability and the yield of methane production during anaerobic digestion of thermally treated sludge increased significantly. Higher temperatures, high oxygen partial pressure and longer retention times increase cell rupture/breakdown and the release of soluble proteins, carbohydrates, and long chain acids into the liquid phase. Partial or complete solubilization of non-microbial organic matter also occurs during the process (Khan et al. 1999). Lissens et al. (2004) studied

the wet air oxidation of biosolids obtained during anaerobic digestion of activated sludge. They showed that thermal wet oxidation improves the anaerobic biodegradability of raw and digested sludge. For example, the yield of methane production increased from 120 dm^3 CH$_4$/kg DS to 180 dm^3 CH$_4$/kg when activated sludge was thermally treated at 458 K. It can be concluded that wet oxidation, as a process of thermal pre-treatment of activated sludge, appears to be a suitable step in integrated chemical/biological processes where anaerobic digestion is used for biogas production.

Mathematical models of the wet air oxidation process are a useful tool in designing and simulating the functioning of reactors. Many lumped kinetic models have been proposed to predict optimum operating conditions (Li 1991, Zhang & Chuang 1999, Belkacemi et al. 2000, Verenich & Kallas 2002). However, they do not include changes in insoluble biosolids in the sludge processed using WAO. Modelling biosolids concentrations during the wet oxidation process is important for predicting subsequent anaerobic digestion of the liquid and/or suspension obtained, because the lysis of microorganism cells is a controlling step in the methane fermentation process.

On an industrial scale, the wet oxidation of activated sludge is carried out in various types of bubble absorbers (Ploos van Amstel & Rietema 1973, Grean-Heedfeld et al. 1995, Daun & Birr 1996, Debellefontaine et al. 1999, Debellefontaine & Foussard 2000).

Because of the vital role of wet oxidation in an integrated thermal/biological sewage sludge disposal system, mathematical models that provide a relatively precise quantitative description of the process are also

Table 1. Kinetic parameters of wet oxidation.

Reaction	$10 \cdot k_{0i}$ $(m^3/mol)^{a+1} \cdot s^{-1}$	$10^{-4} \cdot E_i$ kJ/mol	$10 \cdot a$ —	ΔH_r MJ/mol
1	8.07	4.61	1.13	0.0
2	2040.0	7.15	1.13	−1.469
3	8.05	6.13	1.13	−0.595
4	13.9	6.10	1.13	−1.469

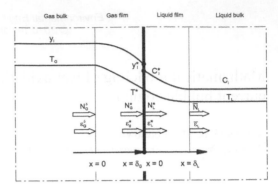

Figure 1. Schematic diagram of mass and heat transfer.

important. In this study, a mathematical model is proposed to describe quantitatively the process of wet oxidation of excess activated sludge in counter-current bubble columns.

2 CHEMICAL REACTION

The wet oxidation of organic compounds is a complex reaction whose mechanism has not been fully recognized (Takamatsu et al. 1970, Debellefontaine et al. 1999, Tettamanti et al. 2001). In this study, a sequence of reactions which take place in the system was approximated by four non-elementary chemical reactions (Imbierowicz & Chacuk 2006):

$$S + O_2 \xrightarrow{R_1} T \qquad (1)$$

$$S + O_2 \xrightarrow{R_2} P \qquad (2)$$

$$T + O_2 \xrightarrow{R_3} K \qquad (3)$$

$$T + O_2 \xrightarrow{R_4} P \qquad (4)$$

where: S = activated sludge; T = readily oxidizable organic compounds; K = sparingly oxidizable carboxylic acids; P = products of oxidation (carbon dioxide, water, etc.).

The partial reaction rates are described by the following relations (see final 'Nomenclature' section for parameter descriptions):

$$R_i = k_i C_{1L}^a C_{jL}^2, \quad i = 1, 2 \Rightarrow j = 5; \ i = 3, 4 \Rightarrow j = 6 \quad (5)$$

$$k_i = k_{0i} \exp(-E_i / RT), \quad i = 1, ..., 4 \quad (6)$$

The subscripts used with the concentration symbols C are defined as follows:

$1 \equiv O_2$, $2 \equiv P(CO_2)$, $3 \equiv H_2O$, $4 \equiv N_2$, $5 \equiv S$, $6 \equiv T$, $7 \equiv K$

Concentrations of compounds S, T and K are expressed as the mole of total organic carbon (TOC) per cubic metre.

Values of kinetic parameters and heats of reactions (1)–(4) are given in Table 1 (Imbierowicz & Chacuk 2006).

3 KINETICS OF MASS AND HEAT TRANSFER

The process of wet oxidation of organic compounds in sewage proceeds in a heterogeneous gas–liquid system under non-isothermal conditions. A mathematical description of mass and heat transfer (Zarzycki & Chacuk 1993) covers (Figure 1):

– mass transfer equations in the gas and liquid phase

$$N_G = C_G k_G \, \Xi_G^* (y - y^*) + y \sum_{i=1}^{3} N_{iG} \qquad (7)$$

$$N_L = k_L \Xi_L \left(C_L^* - C_L \right) + C_L / C_L \sum_{i=1}^{3} N_{iL} \qquad (8)$$

heat transfer equations in the gas and liquid phase

$$\varepsilon_G = \alpha_G \Xi_{HG} \left(T - T^* \right) + \sum_{i=1}^{n} H_{iG} N_{iG} \qquad (9)$$

$$\varepsilon_L = \alpha_L \left(T^* - T \right) + \sum_{i=1}^{3} \widetilde{H}_{iL}^* N_{iL} \qquad (10)$$

– conditions of mass and heat flux continuity on the interface

$$N_{iG} = N_{iG}^* = N_{iL}^*, \quad i = 1, 2, 3 \qquad (11)$$

$$\varepsilon_G = \varepsilon_G^\delta = \varepsilon_G^* = \varepsilon_L^* \qquad (12)$$

– thermodynamic equilibrium equations

$$f_i \left(y^*, C^*, T^* \right) = 0, \quad i = 1, 2, 3 \qquad (13)$$

– determinancy conditions

$$N_{4G} = 0 \qquad (14)$$

The concrete forms of equation (13) for the process of wet oxidation of excess sludge were provided in a former study (Zarzycki et al. 2000).

For the assumed state of gas and liquid phase cores (composition and temperature) in a given cross section of the column, relations (7)–(14) are used to determine conditions at the interface, mass streams of particular components and heat fluxes that are necessary for bubble column balances.

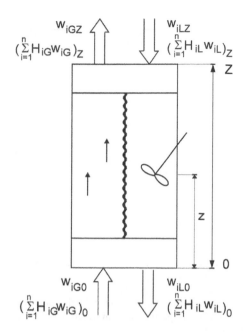

$$\left(\sum_{i=1}^{n} H_{iG} w_{iG}\right)_Z \qquad \left(\sum_{i=1}^{n} H_{iL} w_{iL}\right)_Z$$

$$\left(\sum_{i=1}^{n} H_{iG} w_{iG}\right)_0 \qquad \left(\sum_{i=1}^{n} H_{iL} w_{iL}\right)_0$$

Figure 2. Schematic diagram of bubble columns.

4 KINETICS OF MASS AND HEAT TRANSFER

A mathematical description was formulated under the following assumptions (Figure 2):

- the process is steady-state and adiabatic,
- the flow of the gas phase corresponds to the model of plug flow,
- the flow of the liquid phase corresponds to the ideal mixing model in both directions: vertical and horizontal,
- the chemical reaction takes place in the liquid phase core only,
- the temperatures of the liquid core and interfacial areas are the same,
- the column diameter and the total height, as well as all parameters of streams flowing to this column, are known.

For the assumed kinetic model of the reaction and simplifying assumptions, mass balances for particular reagents and heat balances have the following forms in different phases:

- gas phase

$$\frac{d\,W_{iG}}{dz} = -a_m N_{iG}\,, \quad W_{iG}(0) = W_{iG0}\,, \quad i = 1,2,3 \quad (15)$$

$$\frac{d\left(\sum_{i=1}^{4} \widetilde{H}_{iG} W_{iG}\right)}{dz} = -a_m \varepsilon_G\,, \quad \left(\sum_{i=1}^{4} \widetilde{H}_{iG} W_{iG}\right)(0) = \left(\sum_{i=1}^{4} \widetilde{H}_{iG} W_{iG}\right)_0 \quad (16)$$

$$y_i = W_{iG}/W_G\,, \quad i = 1,2,3\,, \quad W_G = \sum_{i=1}^{4} W_{iG} \quad (17)$$

- liquid phase

$$\left(W_{iL}\right)_Z - W_{iL} + a_m Z \overline{N_{iL}} + \beta Z \sum_{j=1}^{4} \nu_{i,j} R_j = 0\,, \quad i = 1,2 \quad (18)$$

$$\left(W_{3L}\right)_Z - W_{3L} + a_m Z \overline{N_{3L}} = 0 \quad (19)$$

$$\overline{N_{iL}} = \frac{1}{Z} \int_0^Z N_{iL}^* dz\,, \quad i = 1,2,3 \quad (20)$$

$$\left(W_{3L}\right)_Z - W_{3L} + a_m Z \overline{N_{3L}} = 0 \quad (21)$$

$$\left(W_{i+2L}\right)_Z - W_{i+2L} + \beta Z \sum_{j=1}^{4} \nu_{i-2,j} R_j = 0\,, \quad i = 3,4,5 \quad (22)$$

$$\left(\sum_{i=1,i\neq4}^{7} \widetilde{H}_{iL} W_{iL}\right)_0 - \sum_{i=1,i\neq4}^{7} \widetilde{H}_{iL} W_{iL} + a_m Z \overline{\varepsilon_L^*} = 0 \quad (23)$$

$$\overline{\varepsilon_L} = \frac{1}{Z} \int_0^Z \varepsilon_L^* dz \quad (24)$$

Equations (15)–(24) along with the auxiliary dependencies (7)–(14) constitute a mathematical description of the wet oxidation of excess sludge in a counter-current bubble column. These equations form a system of non-linear differential-algebraic equations. In the study, the system of these equations is solved using the associated method of Marquardt and Merson (Zarzycki & Chacuk 1993).

5 NUMERICAL SIMULATION

In order to test the proposed mathematical model, a numerical simulation of wet oxidation of excess sludge in a counter-current bubble column for different process parameters was carried out. Results of calculations presented in this paper refer to the process carried out under pressure $P = 6$ MPa in the counter-current bubble column of diameter $d_k = 1$ m and height $Z = 5$ m for the following parameters of gas and liquid phase streams:

- gas phase: temperature $T_{G0} = 473.15$ K, volumetric flow rate $v_{G0} = 0.04$ m^3/s, molar fraction of oxygen and nitrogen: $y_{10} = 0.5$ and $y_{40} = 0.5$, respectively;
- liquid phase: temperature $T_{LZ} = 473.15$ K, volumetric flow rate $v_{LZ} = 0.005$ m^3/s, composition: $C_{5LZ} = 0.36$ kmol/m^3, $C_{6LZ} = C_{7LZ} = 0$ kmol/m^3.

The coefficients of mass transfer in the gas and liquid phases and an appropriate interfacial area were derived from the formulae presented by Zarzycki & Chacuk (1993).

Results of numerical calculations are given in Figures 3 to 6.

The oxidation of excess sludge is a strongly exothermic process (Figure 3). Thus, the temperature of the liquid stream leaving the column is much higher than the temperature of the stream at the inlet to the column (at this stage by approximately 80 K). The gas phase temperature increases very rapidly, attaining the value

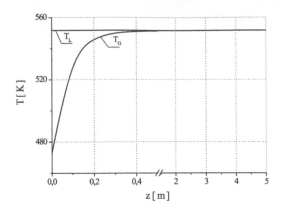

Figure 3. Temperature profiles in the bubble column G – gas phase, L – liquid phase.

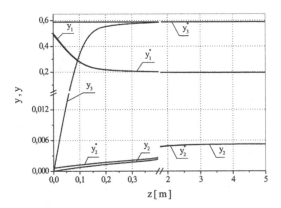

Figure 4. Composition profiles in the bubble column $1 \equiv O_2, 2 \equiv P\ (CO_2), 3 \equiv H_2O$.

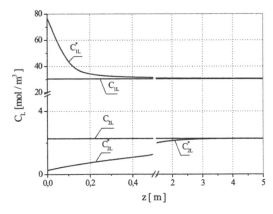

Figure 5. Composition profiles of O_2 and CO_2 in the bubble column $1 \equiv O_2, 2 \equiv P\ (CO_2)$.

of the liquid phase temperature on a very short section of the column (here on a section smaller than 0.5 m). This is caused by very intensive water evaporation into the gas stream.

Mass flow resistance of oxygen and carbon dioxide in the gas phase has no significant effect on the process

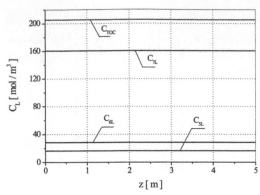

Figure 6. Composition profiles of S, T, K and global TOC in the bubble column $5 \equiv S, 6 \equiv T, 7 \equiv K$.

rate, while the effect of mass transfer resistance of water vapour is revealed on a very short segment of the column only (Figure 4).

The mass transfer resistance in the liquid phase is significant, particularly in reference to oxygen (Figure 5).

6 CONCLUSIONS

Irrespective of the process conditions of wet oxidation, and the direction of phase flow or column size, the numerical calculations show that the resistance of mass transfer of oxygen and carbon dioxide in the gas phase has no significant influence on the process rate. The effect of mass transfer resistance of water vapour in this phase is insignificant and is revealed on a very short segment of the column only. However, the mass transfer resistance is vital in the liquid phase, particularly with reference to oxygen. Heat transfer resistance in the gas phase is negligibly small. The direction of phase flow has no practical influence on the degree of excess activated sludge conversion and total organic carbon concentration in the stream of liquid leaving the column.

A comparison of the calculation results with the literature data indicates that the accepted mathematical model defines quite well the process of wet oxidation of excess sludge in the counter-current bubble column.

NOMENCLATURE

a – index
a_m – interfacial area, m^2/m^3
d_k – column diameter, m
C – molar concentration, $kmol/m^3$
E – activation energy, kJ/mol
k – chemical reaction rate constant, $m^{3a}/(kmol^a \cdot s)$
k – matrix of multicomponent mass transfer coefficient, m/s
H – molar enthalpy, kJ/mol
N – molar flux, $kmol/(m^2 \cdot s)$

N – vector of molar flux, $kmol/(m^2 \cdot s)$
R – universal gas constant, $kJ/(kmol \cdot K)$
R – molar rate of chemical reaction, $kmol/(m^3 \cdot s)$
T – temperature, K
v – volumetric flow rate, m^3/s
W – superficial molar fluid velocity, $kmol/(m^2 \, s)$
y – mole fraction of gas phase
y – vector of mole fraction in gas phase
α – heat transfer coefficient, $kW/(m^2 \, s)$
β – liquid hold-up, m^3/m^3
ε – energy flux, kW/m^2
Ξ – matrix of flux correction factors

Subscripts
i – refers to reagent
j – refers to chemical reaction
G – refers to gas phase
L – refers to liquid phase
0 – gas inlet
Z – gas outlet or liquid inlet

Superscripts
i – in the bulk of fluid
$*$ – gas–liquid interface
$\sim$ – partial value
– mean value

REFERENCES

Belkacemi, K., Larachi, F. & Sayari, A. 2000. Lumped kinetics for solid-catalyzed wet oxidation: a versatile model. *J. Catal.* 193: 224–237.

Daun, M. & Birr R. 1996. Treatment of wastewater sludge by wet oxidation with the VerTech-Deep Well Method. *Wasser Boden* 48(5): 34.

Debellefontaine, H., Crispel, S., Reilhac, P., Périé, F. & Foussard J.N. 1999. Wet air oxidation (WAO) for the treatment of industrial wastewater and domestic sludge. Design of bubble column reactors. *Chem. Eng. Sci.* 54(21): 4953.

Debellefontaine, H. & Foussard J.N. 2000. Wet air oxidation for the treatment of industrial wastes. Chemical aspects, reactor design and industrial applications in Europe. *Waste Management* 20: 15.

Grean-Heedfeld, J., Schluter, S. & Daun, M. 1995. Modeling and simulation of deep well reactor for the wet air oxidation of sewage sludge. *Chem. Eng. Proc.* 34(2): 121.

Haug, R. T. 1978. Effect of thermal pre-treatment on digestability and dewaterability of organic sludges. *J. Water Pollut. Control Fed.* 50: 73.

Imbierowicz, M. & Chacuk, A. 2006. The advanced kinetic model of the excess activated sludge wet oxidation. *Polish Journal of Chemical Technology* 8(2): 16–19.

Khan, Y., Anderson, G.K. & Elliot, D.J. 1999. Wet oxidation of activated sludge. *Water Res.* 33(7): 1681–1687.

Li, L., Chen, P. & Gloyna, E. F. 1991. Generalized kinetic model for wet oxidation of organic compounds. *AIChE J.* 37(11): 1687–1697.

Lissens, G., Thomsen, A.B., De Baere, L., Verstraete, W. & Ahring B.K. 2004. Thermal wet oxidation improves anaerobic biodegradability of raw and digested biowaste. *Environ. Sci. Technol.* 38(12): 3418–3424.

Ploos Van Amstel, J.J. & Rietema, K. 1973. Wet air oxidation of sewage sludge. Part II. The oxidation of real sludges. *Chem. Ing. Tech.* 45(20): 1205.

Takamatsu, T., Hashimoto, I. & Sioya, S. 1970. Model identification of wet air oxidation process thermal decomposition. *Water Res.* 4(1): 33.

Tettamani, M., Lasagni, M., Collina, E., Sancassani, M., Pitea, D., Fermo, P. & Fariati, F. 2001. Thermal oxidation kinetics and mechanism of sludge from a wastewater treatment plant. *Environ. Sci. Technol.* 35: 3981.

Verenich, S., & Kallas, J. 2002. Wet oxidation lumped kinetic model for wastewater organic burden biodegradability prediction. *Environ. Sci. Technol.* 36: 3335–3339.

Zarzycki, R. & Chacuk A. 1993. *Absorption: fundamentals and applications*. Pergamon Press: Oxford.

Zarzycki, R., Imbierowicz, M. & Chacuk, A. 2000. Sprawozdanie dot. projektu badawczego 3TOC05914 pt. "Badanie mechanizmu i modelowanie procesu mokrego utleniania nadmiarowego osadu czynnego". Politechnika Łódzka.

Zhang Q. & Chuang K.T. 1999. Lumped kinetic model for catalytic wet oxidation of organic compounds in industrial wastewater. *AIChE J.* 45(1): 145.

Environmental Engineering III – Pawłowski, Dudzińska & Pawłowski (eds)
© 2010 Taylor & Francis Group, London, ISBN 978-0-415-54882-3

The effect of disintegration of sewage sludge by hydrodynamic cavitation on organic and inorganic matter relase

K. Grübel, A. Machnicka & J. Suschka

Institute of Environmental Protection and Engineering, University of Bielsko-Biala, Bielsko-Biala, Poland

ABSTRACT: Large amounts of sludge are produced in biological wastewater treatment plants. As the sludge is highly contaminated, it has to undergo proper stabilization before it is disposed of or utilized in an environmentally safe way.

On the whole, the aim of bacterial-cell disintegration is the release of cell contents in the form of an aqueous extract. Mechanical disintegration activates biological hydrolysis and, therefore, it can significantly increase the stabilization rate of the secondary sludge. It has been shown that when the activated sludge was subjected to 30 min of mechanical disintegration, the COD concentration increased from 77 mg/dm^3 to more than 251 mg/dm^3 and from 57 mg/dm^3 to more than 566 mg/dm^3 in sludge supernatant and foam phase, respectively.

Keywords: Disintegration, hydrodynamic cavitation, foam, activated sludge, proteins, carbohydrates.

1 INTRODUCTION

The aim of wastewater treatment is mineralization of organic matter and nutrient removal. Activated sludge systems that are designed for enhanced nutrient removal are based on the principle of altered anaerobic and aerobic conditions for growth of appropriate microorganisms.

The application of disintegration technology into the sludge treatment process leads to reduced sludge quantities and markedly improves sludge quality. The disintegration process is realized by the application of physical or chemical methods to break down cell walls. Thus, cell walls are fragmented and intracellular compounds are released. The product can be utilized both as a substrate in aerobic as well as anaerobic biological processes.

Several disintegration processes are developed: mechanical: hydrodynamic cavitation, ultrasound, homogenizer, stirred ball mills; thermal hydrolysis (autoclave or steam heating), wet oxidation; chemical: use of enzymes, alkaline/acid hydrolysis; biological: thermophilic aerobic/anaerobic pretreatment. These disintegration methods are common for activated sludge stabilization, resulting in solubilization of sludge volatile matter and the production of biogas (Appels et al., 2008). Positive effects were shown for thermal pretreatment (Camacho et al. 2005, Kepp et al. 2000, Phothilangka et al. 2008), addition of enzymes (Barjenbruch & Kopplow 2003, Roman et al. 2006), ozonation (Carballa et al. 2007, Song et al. 2003, Weemaes et al. 2000), chemical solubilization by acidification (Woodard & Wukasch 1994) or alkaline hydrolysis (Mukherjee & Levine 1992, Vlyssides & Karlis 2004), and mechanical and ultrasonic sludge

disintegration (Antoniadis et al. 2007, Kampas et al. 2007, Kennedy et al. 2007, Müller et al. 1998, Müller 2000, Wang et al. 2006, Zhang G. et al. 2007, Zhang P. et al. 2007).

Although the methods are different in character, they all aim to achieve partial or complete lysis of bacteria cells – that is, the destruction and release of organic substances that are present inside cells to the liquid phase of the sludge. It has been shown (Neyens et al. 2004) that these methods mainly influence and degrade extracellular polymeric substances.

Disintegration by hydrodynamic cavitation has a positive effect on the degree and rate of sludge anaerobic digestion. Cavitation results in the formation of cavities (bubbles) filled with a vapour/gas mixture inside the flowing liquid, or at the boundary of constriction devices due to a drastic drop in local pressure. Subsequently, the pressure recovers down the constriction (valve or nozzle) and causes cavities to collapse. The collapse of cavitation bubbles is defined as implosion and the forces associated with this results in mechanical and physico-chemical effects. The physical effects include the production of shear forces and shock waves, generating local high temperatures and pressures, whereas the chemical effects result in the generation of radicals – such as the formation of reactive hydrogen atoms and hydroxyl radicals, which recombine to form hydrogen peroxide (Dewil et al. 2006, Vichare et al. 2000, Senthilkumar et al. 2000).

The new concept of surplus activated sludge hydrodynamic disintegration described in this paper is based on the constructed cavitation nozzle. The main aim of this article was to describe the effects of hydrodynamic cavitation on organic and inorganic matter release.

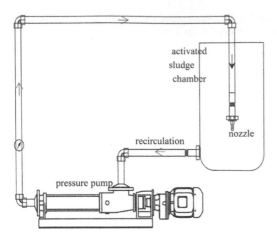

Figure 1.   Scheme of the experimental installation.

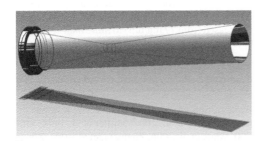

Figure 2.   Cavitation nozzle.

2   MATERIALS AND METHODS

2.1   Method of sludge disintegration

Activated sludge samples were taken from an Enhanced Biological Nutrient Removal (EBNR) full-scale municipal sewage treatment plant. Hydrodynamic disintegration was executed with the application of a pressure pump (12 bar), which recirculated sludge from a 25-litre container, through a constructed cavitation nozzle. To force 25 litres of sludge through the nozzle took 3 minutes. The process was carried out for 15, 30, 45, 60, 75 and 90 minutes, which corresponded to 3, 6, 9, 12, 15 and 18 multiplicity flow by cavitation nozzle. The scheme of the experimental installation and cavitation nozzle is shown in Figures 1 and 2, respectively. In our research, we decided to use a constructed cavitation nozzle with a diameter ratio of $\beta = d_0/d_1 = 0.30$ (d_0 – diameter narrowing; d_1 – diameter of inflow), which allows us to obtain a cavitation number of $\sigma = 0.245$, in selected flow conditions. Accordingly, the numerical results of the design of this device are relatively efficient – the calculated pressure loss is $\Delta p = 74.8$ kPa, whereas the net pressure drop ($p_{min}/\Delta p$) is almost five times greater.

2.2   Analytical methods

Chemical analyses were performed for samples before and after each point of disintegration. All chemical and physical parameters were determined according to the procedures given in the Standard Methods for Examination of Water and Wastewater (19th ed.). For colorimetric determinations, a spectrophotometer HACH DR 4000 was applied. The concentrations of potassium, magnesium and calcium were determined using an atomic absorption analysis instrument – A Analyst 100 Perkin Elmer.

The procedure provided by Lowry was used for protein determination, whereas the Anthrone method has a high specificity for carbohydrates. Both methods were performed according to Gerhardt et al. (2005).

2.3   Anaerobic process

The anaerobic digestion experiments were performed in six glass fermenters (2.5 litres), operated in parallel at a temperature of $35 \pm 2°C$ with a holding time of 22 days. The production of biogas was measured each day in two fermenters. The first reactor was fed with raw surplus activated sludge (volatile solids of 5.64 g/dm^3; percentage of the feed, 62.48%); the second reactor was fed with surplus activated sludge after hydrodynamic disintegration (volatile solids of 4.38 g/dm^3; percentage of the feed, 59.19%).

2.4   Precipitation of magnesium-ammonium phosphate

The struvite precipitation process in liquid of activated sludge after disintegration was realized on a laboratory scale. The addition of magnesium oxide was necessary to obtain an appropriate ratio of magnesium and phosphates, and ammonia nitrogen to fulfill the formula of struvite – $Mg:NH_4:PO_4 \cdot 6H_2O$.

3   RESULTS AND DISCUSSION

3.1   Organic matter release

Release of organic matter expressed as an increase in soluble COD value is considered as a tool for the measurement of bacteria cell destruction effects.

According to the methodology used, the process of hydrodynamic disintegration was carried out for 15, 30, 45, 60, 75 and 90 minutes. Thirty minutes of hydrodynamic activated sludge flocs disintegration results in a COD increase in the liquid of 218 mg O_2/dm^3 (from 77 to 295 mg O_2/dm^3). This represents an almost four-fold increase of COD. What is more, a further increase in disintegration time causes a further increase in COD value (Figure 3).

For a quantitative measurement of the effects of disintegration, a coefficient defined as Degree of Disintegration (DD) was introduced. In this case, the degree of sludge disintegration was determined according to that given by Müller (2000) and Müller et al. (1998) reading as follows:

$$DD_M = \frac{[COD_1 - COD_2]}{[COD_3 - COD_2]} \cdot 100\% \qquad (1)$$

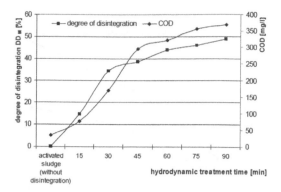

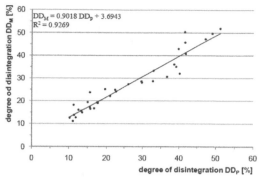

Figure 3. The effects of disintegration of activated sludge on organic matter released (expressed as COD) and degree of activated sludge disintegration.

Figure 5. Degree of activated sludge disintegration.

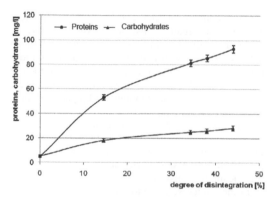

Figure 4. Proteins and carbohydrates released as an effect of surplus activated sludge disintegration.

where: DD_M is the degree of disintegration, COD_1 is the COD of the liquid phase of the disintegrated sample, COD_2 is the COD of the original sample, and COD_3 is the value after chemical lysis using a 1 N NaOH solution.

In accordance to equation (1), an increase of the DD was determined. The results are presented in Figure 3.

Within the range of examined time, between 15 min and 90 min, the degree of disintegration increased most rapidly in the first 30 min. The achieved degree of sludge disintegration was about 34%. Although the efficiency of sludge disintegration increased further (Figure 3), the rate of DD_M was definitely lower.

Moreover, destruction of activated sludge flocs in the process of hydrodynamic disintegration resulted in protein and carbohydrate release into the aqueous phase (Figure 4).

Within the first 30 min of disintegration, the release of proteins was relatively fast, as shown in Fig. 4 (DD_M was 34.3%; DD_p was 34.8%). The concentration of proteins increased to 81 mg/l. With prolongation of the disintegration time, the release of proteins was distinctively slower. It was observed that the carbohydrate concentration also increased with the time of disintegration. Similarly, as with proteins, the concentration of carbohydrates increased most rapidly in the first 30

min of disintegration. In comparison to the amount of released proteins, the release of carbohydrates was less.

On the basis of the obtained results, it was confirmed that the amount of proteins released in the process of disintegration could be adopted as a suitable parameter for assessing the rate of disintegration.

Similar to the procedure based on COD determination, the DD was based on the protein concentration in the liquid phase of the sludge before and after disintegration. The protein concentration after chemical disintegration (1 N NaOH) was the reference value. The degree of activated sludge disintegration was calculated as follows (2) and was shown in Figure 5:

$$DD_P = \frac{[P_1 - P_2]}{[P_3 - P_2]} \cdot 100\% \qquad (2)$$

where: DD_P is the degree of disintegration, P_1 is the concentration of protein in the liquid phase of the disintegrated sample, P_2 is the concentration of protein in the original sample, and P_3 is the value after chemical disintegration. Chemical disintegration of the sample was carried out according to the above methodology

The achieved DD, calculated in accordance with equation 2, correlates with the DD calculated by means of equation 1 (Müller methods) (Figure 5). The assessment of the DD based on protein concentration is simpler, quicker and less laborious. The suggested procedure can therefore be recommended. The protein test is a quick photometric determination, which can be carried out within minutes, whereas the time required for COD determination is considerably longer. Moreover, the cost of chemicals needed for COD determination is high.

3.2 Release of inorganic matter

Hydrodynamic disintegration of activated sludge resulted in an increase of the concentration of phosphates in the solution – from 4 mg PO_4/dm^3 to 28 mg PO_4/dm^3 (Figure 6). Most of the phosphates were released just within the first 30 minutes of disintegration. The difference in phosphate concentration

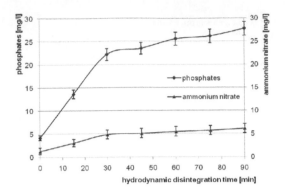

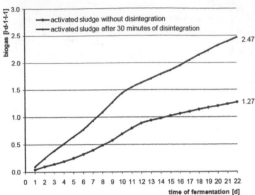

Figure 6. The effects of activated sludge disintegration on phosphate and ammonium release.

Figure 8. Production of biogas during fermentation.

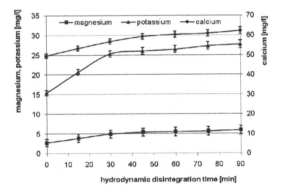

Figure 7. Release of potassium, magnesium and calcium during hydrodynamic treatment time.

between the sample after 30 and 90 minutes of disintegration was only about 6 mg PO_4/dm^3. Similarly, the concentration of ammonia nitrogen increased to 5 and 6 mg $N-NH_4/dm^3$ after 30 minutes and 60 minutes, respectively.

The difference in phosphate concentration between the sample after 30 and 90 minutes of disintegration was only about 6 mg PO_4/dm^3. Similarly, the concentration of ammonia nitrogen increased to 5 and 6 mg $N-NH_4/dm^3$ after 30 minutes and 60 minutes, respectively.

The disintegration process caused the release of cations. The concentration of potassium in the solution increased from 15 mg/dm^3 to 28 mg/dm^3, the concentration of magnesium from 3 mg/dm^3 to 6 mg/dm^3, and the concentration of calcium from 50 mg/dm^3 to 62 mg/dm^3 (Figure 7).

Disintegration for more than 60 minutes had a negligible effect on additional phosphates, nitrogen compounds and the release of metal cations. Disintegration for a longer period of time means that a higher power consumption is not compensated by the ability to achieve phosphorous recovery effects.

3.3 Practical implementation

Hydrodynamic disintegration accelerates the biological degradation of sludge. The released cell liquid

contains components that can be easily assimilated. The released organic substances (expressed here as COD or as protein and carbohydrate concentration) have the effect of activated sludge flocs disintegration, leading to a substantial increase of biogas production in the process of anaerobic sludge digestion (Figure 8).

Significantly higher amounts of biogas were produced in the fermenters that were fed with disintegrated activated sludge. The production of biogas increased by about 95% in samples after 30 min hydrodynamic disintegration, compared with samples of activated sludge without disintegration. The organic matter transferred by hydrodynamic treatment from the sludge solids into the liquid phase is readily biodegradable. The break-up of cells walls of the bacteria limits the degradation process. By applying hydrodynamic disruption, the lysis of cells occurs in minutes rather than days. The intracellular and extracellular components are set free and are immediately available for biological degradation, which leads to an acceleration of the anaerobic process.

Release of phosphates, nitrogen compounds and metal cations during the disintegration processes can be used to achieve phosphorous recovery effects – that is, struvite precipitation. Taking into account the practical possibilities of nutrient (P, N) removal and recovery, it is important to have an appropriate ratio of magnesium or potassium and phosphates, and ammonia nitrogen to fulfill the formula of struvite – $Mg:NH_4:PO_4 \cdot 6H_2O$ – of which the molar ratio of $Mg:NH_4:PO_4$ is 1:1:1. The mass ratio of $Mg:NH_4:PO_4$ is 0.25:0.19:1. Although the concentration of magnesium in the liquid phase after disintegration is not sufficient to fulfill the struvite formula and cover the magnesium deficit, addition of magnesium oxide can be used (source of magnesium ions). In addition, a magnesium oxide is a poor solubility reagent that leads to molar overdosing and thus to pH increases. Because magnesium oxide is a cheap by-product (prices comparable to calcium hydroxide = lime), its addition to cause struvite precipitation could be a viable solution for phosphorus and ammonium removal/recovery from sludge liquors. In our investigations, we added

Figure 9. Struvite crystals.

magnesium oxide (150% of stechiometric dose to formula of struvite) to the probe after 30 min of disintegrated activated sludge. Figure 9 presents an example of the precipitated struvite, which later can easily be separated from the disintegrated activated sludge.

4 CONCLUSIONS

Hydrodynamic disintegration is a suitable method for destroying the microorganisms of activated sludge. In this study, hydrodynamic cavitation was examined, with the aim of releasing organic and inorganic matter. The most important conclusions of this study are:

1. Hydrodynamic cavitation causes the transfer of organic matter from the sludge solids into the liquid phase (expressed as COD). Moreover, the disruption of the structure of microorganism cells leads to an increased release of polymers: protein and carbohydrates. The concentration of proteins increased from 5 mg/l to 99 mg/l, whereas the concentration of carbohydrates increased from 5 mg/l to 32 mg/l.
2. The hydrodynamic disintegration of activated sludge leads to a higher degree of degradation. In the first 30 min of disintegration, the achieved degree of sludge disintegration amounted to about 34%.
3. Hydrodynamic disintegration of activated sludge permits achievement of intensification of biogas production. After 22 days of the anaerobic process, the biogas production increased by about 95%, in comparison to the un-disintegrated activated sludge.
4. Disintegration also allowed phosphates, ammonia nitrogen, as well as potassium, magnesium and calcium cations, to be transferred from the activated sludge solids into the liquid phase, resulting in an enhanced biological phosphorus recovery and removal from wastewater in the form of struvite.

REFERENCES

Antoniadis, A., Poulios, I., Nikolakaki, E. & Mantzavinos D. 2007. Sonochemical disinfection of municipal wastewater. *Journal of Hazardous Materials* 146: 492–495.

Appels, L., Baeyen,S J., Degrève, J. & Dewil, R. 2008. Principles and potential of the anaerobic digestion of waste-activated sludge. *Progress In Energy and Combustion Science* 34: 755–781.

Barjenbruch, M. & Kopplow, O. 2003. Enzymatic, mechanical and thermal pre-treatment of surplus sludge. *Advances in Environmental Research* 7: 715–720.

Camacho, P., Ginestet, P. & Audic, J. M. 2005. Understanding the mechanisms of thermal disintegrating treatment in the reduction of sludge production. *Water Science and Technology* 10–11: 235–245.

Carballa, M., Manterola, G., Larrea, L., Ternes, T., Omil, F. & Lema, J. M. 2007. Influence of ozone pre-treatment on sludge anaerobic digestion: Removal of pharmaceutical and personal care products. *Chemosphere* 67: 1444–1452.

Dewil, R., Baeyens, J. & Goutvrind, R. 2006. Ultrasonic treatment of waste activated sludge. *Environmental Progress* 25: 121–128.

Gerhardt, P., Murray, R. G. E., Wood, W. A. & Krieg, N. R. 2005. *Methods for General and Molecular Bacteriology.* ASM. Washington DC.

Kampas, P., Parsons, S. A., Pearce, P., Ledoux, S., Vale, P., Churchley, J. & Cartmell, E. 2007. Mechanical sludge disintegration for the production of carbon source for biological nutrient removal. *Water Research* 41(8): 1734–1742. Kennedy, K. J., Thibault, G. & Droste, R. L. 2007. Micro wave enhanced digestion of aerobic SBR sludge. *Water SA* 33: 261–270.

Kepp, U., Machenbach, I., Weisz, N., Solheim, O. E. 2000. Enhanced stabilisation of sewage sludge through thermal hydrolysis – three years of experience with full scale plant. *Water Science and Technology* 42: 89–96.

Mukherjee, S. R. & Levine, A. D. 1992. Chemical solubilization of particulate organics as a pretreatment approach. *Water Science and Technology* 26: 2289–2292.

Müller, J. 2000. Disintegration as a key-step in sewage sludge treatment. *Water Science and Technology* 41: 123–130.

Müller, J., Lehne, G., Schwedes, J., Battenberg, S., Näveke, R., Kopp, J. & Dichtl, N. 1998. Disintegration of sewage sludge and influence on anaerobic digestion. *Water Science and Technology* 38: 425–433.

Neyens, E., Baeyens, J., Dewil, R. & De Heyder, B. 2004. Advanced sludge treatment affects extracellular polymeric substances to improve activated sludge dewatering. *Journal of Hazardous Materials* 106B: 83–92.

Phothilangka, P., Schoen, M. A., Hube, M., Luchetta, P., Winkler, T. & Wett, B. 2008. Prediction of thermal hydrolysis pretreatment on anaerobic digestion of waste activated sludge, *Water Science and Technolog* 58(7): 1467–1473.

Roman, H. J., Burgess, J. E. & Pletschke, B. I., 2006. Enzyme treatment to decrease solids and improve digestion of primary sewage sludge. *African Journal of Biotechnology* 5: 963–967.

Senthilkumar, P., Sivakumar, M. & Pandit, A. B. 2000. Experimental quantification of chemical effects of hydrodynamic cavitation. *Chemical Engineering and Science* 55: 1633–1639.

Song, K.-G., Choung, Y.-K., Ahtf, K.-H., Cho, J. & Yun, H. 2003. Performance of membrane bioreactor system with sludge ozonation process for minimization of excess sludge production. *Desalination* 157: 353–359.

Vichare, N. P., Gogate, P. R. & Pandit, A. B. 2000. Optimization of hydrodynamic cavitation using model reaction. *Chemical Engineering and Technology* 23: 683–690.

Vlyssides, A. G. & Karlis, P. K. 2004. Thermal-alkaline solubilization of waste activated sludge as a pre-treatment stage for anaerobic digestion. *Bioresource Technology* 91: 201–206.

Wang, F., Lu, S. & Ji, M. 2006. Components of released liquid sfrom ultrasonic waste activated sludge disintegration. *Ultrasonics Sonochemistry* 13: 334–338.

Weemaes, M., Grootaerd, H., SimoenS, F. & Verstraete, W. 2000. Anaerobic digestion of ozonized biosolids. *Water Research* 34: 2330–2336.

Woodard, S.E. & Wukasch, R.F. 1994. A hydrolysis/thickening/filtration process for the treatment of waste activated sludge. *Water Science and Technology* 30: 29–38.

Zhang, G., Zhang, P., Yang, J., Chena, Y. 2007. Ultrasonic reduction of excess sludge from the activated sludge system. *Journal of Hazardous Materials* 145: 515–519.

Zhang P., Zhang G., Wang W. 2007. Ultrasonic treatment of biological sludge: Floc disintegration, cell lysis and inactivation. *Bioresource Technology* 98: 207–210.

Environmental Engineering III – Pawłowski, Dudzińska & Pawłowski (eds)
© 2010 Taylor & Francis Group, London, ISBN 978-0-415-54882-3

Speciation of heavy metals in municipal sewage sludge from different capacity sewage treatment plants

J. Gawdzik & J. Latosińska

Faculty of Civil and Environmental Engineering, Kielce University of Technology, Kielce, Poland

ABSTRACT: This article presents the results of heavy metal mobility research on sludge from five municipal sewage treatment plants of different capacities, and methods of sewage sludge stabilisation. Heavy metals were present mainly in the immobile sludge fractions. The capacities of the plants did not demonstrate any explicit influence on the forms of the heavy metals present. It was hypothesised that copper forms in sewage sludge are a feature characteristic of this chemical element. The per cent contributions of zinc, lead, cadmium, nickel and chromium in the BCR fractions did not depend significantly on the method of sludge stabilisation.

Keywords: Heavy metal, metal speciation, sequential extraction, sewage sludge.

1 INTRODUCTION

In Poland, as in other European Union countries, the amount of municipal sewage sludge produced is increasing (Fytili & Zabaniotou 2008, Eurostat). One of the reasons for this situation is the intensification of the quality requirements for sewage sludge drained to lakes, rivers and other locations. (1991/271/EEC).

The content of organic substances, nutrients (N, P, K) and microelements predisposes the sewage sludge to its environmental use (Rogers 1996, Wang 1997). Except for its desirable constituents, essential for agricultural use, sewage sludge consists of toxic substances, including heavy metals.

The presence of heavy metals in sewage sludge, especially in high concentration, arises from the con-tributions of industrial sewage (e.g. tannery, enamel and metallurgical wastes) in the overall mass of urban sewage (Weiner & Matthews 2003). Additionally, heavy metals originate from domestic sewage, surface flushing and the corrosion of sewer pipes (Werther & Ogada 1999). Heavy metals are present in sewage sludge in the following states; dissolved, precipitated, co-precipitated with metal oxides, adsorbed on or asso-ciated with particles of biological remains. They can appear in the forms of oxides, hydroxides, sulphides, sulphates, phosphates, silicates, organic combinations, in the form of humic groups, and compounds with polysaccharides (De la Guardia & Morales-Rubio 1996, Werther & Ogada 1999).

The highest contribution of heavy metals is present in fermented, dehydrated sludge, which is connected with the concentration of the raw sludge solid mass in the process of fermentation (Alvarez et al. 2002). The heavy metal content of sludge generally accounts for between 0.5 and 2.0% of dry mass, but it can be as high as 4% of dry mass (Ryu et al. 2003).

The known chemical methods for heavy metals removal from sewage sludge were inapplicable due to high costs, operational difficulties and low efficiency (Ryu et al. 2003). Microbiological methods, based on the transition of heavy metals from sewage sludge to reflux, are in the testing phase (Xiang et al. 2000, Ryu et al. 2003).

While it is true to say that trace amounts of heavy metals are essential for plants and animals to live, not only are they toxic and carcinogenic in high con-centrations, they bio-accumulate in living organisms (Krogmann et al. 1999).

The limits for heavy metals for the environmental use of sewage sludge are regulated by the Ordinance of the Minister of Environment (Table 1), in accor-dance with the Council Directive 86/278/EEC. The current regulations, similar to the planned changes for 2015 and 2025 (ENV/E.3/LM) (Table 1), apply to the total content of lead, cadmium, mercury, nickel, zinc, copper and chromium. This generalisation does not provide useful information about the potential danger of heavy metal emissions to the soil–water environ-ment, because the bioavailability and toxicity of heavy metals depend on their forms of occurrence (Table 2).

The analytical techniques applied make it possible to determine the concentration of the mobile forms of the heavy metals. The speciation of heavy metals can be carried out in accordance with various procedures depending on the type of extraction matrices (Hris-tensen 1998, Pitt et al. 1999, Dahlin et al. 2002). In terms of sewage sludge testing, the procedure of the Community Bureau of Reference (BCR) is commonly used (Pitt et al. 1999, Alvarez et al. 2002):

- Step I: extraction with acetic acid (CH_3COOH) – in order to identify and measure the content of assimilable and carbonate-bound metals (fraction FI);

Table 1. Admissible fractions for heavy metals in municipal sewage sludge designed for environmental use in accordance with current standards and planned changes.

	Admissible fractions of heavy metals in sewage sludge designed for use [mg/kg d.m.]					
	In agriculture			The Ordinance of the Minister of Environment, Journal of Laws No. 134, point 1140, 2002		
Metal	1986/278/ EEC–valid	ENV/E.3/ LM – suggested changes year 2015	year 2025	In agriculture and in land reclamation for agriculture	In land reclamation for other purposes	In adaptation of lands for particular needs*
Pb	750–1200	500	200	500	1000	1500
Cd	20–40	5	2	10	25	50
Hg	16–25	5	2	5	10	25
Ni	300–400	200	100	100	200	500
Zn	2500–4000	2000	1500	2500	3500	5000
Cu	1000–1750	800	600	800	1200	2000
Cr	–	800	600	500	1000	2500

*Based on waste disposal plans, land development plans or decisions on the conditions and the land development, for cultivation of compost plants, for cultivation of plants which are not designed for consumption and fodder production.

Table 2. The relation between heavy metal fractions and eco-toxicity and bioavailability (Chen et al. 2008).

Fraction of heavy metals	Eco-toxicity	Bioavailability
Acid soluble/ exchangeable fraction I; reducible fraction II	Direct toxicity	Direct effect fraction
Oxidisable fraction III	Potential toxicity	Potential effect fraction
Residual fraction IV	No toxicity	Stable fraction

- Step II: extraction with hydroxylamine hydrochloride (NH_2OH HCl) – in order to identify and measure the content of assimilable metals bound with amorphous iron and manganese oxides (fraction FII);
- Step III: extraction with hydrogen peroxide/ammonium acetate H_2O_2/CH_3COONH_4 – in order to identify and measure the content of the organo-metallic and sulphide fraction (fraction FIII);
- Step IV: mineralisation of the residual fraction with a mixture of concentrated acids (HCl, HF, HNO_3) – in order to identify and measure the content of silicate-bound metals (fraction FIV).

The objective of the tests was the estimation of heavy metal mobility in sewage sludge supplied by chosen sewage treatment plants of different capacities and methods of sewage sludge stabilisation.

2 MATERIALS AND METHODS

2.1 Sample collection and pre-treatment

The tests were conducted on municipal sewage sludge collected (in accordance with PN-EN ISO 5667-13:2004) from five municipal sewage treatment plants located in central Poland (Table 3).

2.2 The sequential extraction

The tests were conducted in accordance with the four-step BCR sequential extraction procedure (Pitt et al. 1999, Alvarez et al. 2002), introducing a change in the method of residual fraction mineralisation, i.e. aqua regia was used in the process of mineralisation (EN ISO 15587:2002).

Step one: acid soluble/exchangeable fraction (FI)

A 2 g sample of sewage sludge was placed in a $100\,cm^3$ test-tube for centrifuging. Then, $40\,cm^3$ of 0.11-molar acetic acid solution was added. The sample was shaken for 16 hours at room temperature. The extract was separated from the sewage sludge by centrifuge (4000 rpm). The content of the soluble metals in the water was marked in the liquid.

Step two: reducible fraction (FII)

Sewage sludge was washed in $20\,cm^3$ of distilled water (shaken and centrifuged). Subsequently, $40\,cm^3$ of 0.1-molar hydroxylamine hydrochloride solution, of pH = 2, was added to the sewage sludge. Nitric acid was used for the correction of the pH value. The procedure was the same as in step one, the mixture was shaken and centrifuged. Fraction II metals were marked in the liquid.

Step three: oxidation fraction (FIII)

The sewage sludge was carried over quantitatively to a quartz evaporating dish and $10\,cm^3$ of 30% hydrogen peroxide was added. The contents of evaporating dish were heated in a water bath at 85°C for one hour. The process was repeated with the addition of $10\,cm^3$ of 8.8-molar hydrogen peroxide solution to the sewage sludge. After drying, the sewage sludge sample was transferred to test-tubes to be centrifuged and then $50\,cm^3$ of ammonium acetate solution ($1mol/dm^3$, pH = 2; nitric acid was used to correct the pH value) was added. The sample was shaken for 16 hours and afterwards the sewage sludge was separated from the extract. Fraction III metals were marked in the solution.

Table 3. Municipal sludge collected from municipal wastewater treatment plants.

Test marking	Name of the town	Sewage treatment plant type	Equivalent Population	Sludge stabilisation method	Sludge utilisation method
S1	Kostomłoty-Laskowa	M-B S	3 000	O.S.	Land reclamation
S2	Daleszyce	M-B S	5 000	O.S.	Soilless land reclamation
S3	Busko-Siesławice	MB	30 500	O.S.	Land reclamation
S4	Skarżysko-Kamienna	MB	50 000	A.S.	Isolating layers on the disposal ground
S5	Sitkówka-Nowiny	MB	275 000	A.S.	Land reclamation

M-B S – mechanical – biological SBR;
M-B - mechanical-biological; O.S. – Oxygen stabilisation;
A. S. - Anaerobic stabilisation

Table 4. Total (FI + FII + FIII + FIV) concentration of heavy metals in sewage sludge.

	Heavy metal [mg/kg d.m.]					
Sample	Cu	Ni	Cr	Pb	Cd	Zn
S1	9.3	5.72	28.35	275.17	7.32	596.02
S2	21.03	13.87	105.67	427.06	9.48	2769.77
S3	78.83	1.31	35.24	322.40	16.71	840.65
S4	21.8	28.5	2759.81	31.26	12.13	5351.14
S5	83.48	51.85	238.5	67.74	5.6	1315.37

Step four: residual fraction (FIV)

The sludge was washed and dried to a solid state. The mineralisation of the residual fraction was conducted with aqua regia; 30 cm³ of concentrated hydrochloric acid and 10 cm³ of concentrated nitric acid were added carefully to a 300 cm³ conical flask together with 0.5 g of sludge. The conical flask was heated for 30 min and subsequently evaporated to dryness. After cooling, 25 cm³ of 5% hydrochloric acid were added. The sewage sludge was dissolved, carried over to a metal measuring flask and topped up with 50 cm³ of distilled water. Then the sample was mixed and strained to a dry dish. In the filtrate, the metal forms, Fraction IV, were marked.

The heavy metals in the extracts obtained were determined in accordance with ISO 9001:2000 using a Perkin-Elmer 3100 FAAS-BG atomic absorption spectrophotometer (impact bead). Each determination was repeated four times.

3 RESULTS AND DISCUSSION

3.1 *Total concentration of heavy metals in sewage sludge*

The total concentrations of the heavy metals in the tested sewage sludge are presented in Table 4.

The levels of heavy metals in S1 and S5 sewage sludge did not exceed the current admissible limits in Poland for sludge designed for environmental use, including the agricultural use (Table 1, Table 4). S1, S2 and S3 sludge cannot be used in agriculture because the quantities of cadmium (S3, S4) and zinc (S2, S4) present exceed the admissible levels. Because of the amount of zinc, S2 sewage sludge can be used only for the reclamation of land for other than agricultural purposes. This use is not permitted for S4 sludge.

The introduction of the suggested changes in heavy metal limits (Table 1) will exclude S1, S2, S3, S4, S5 sewage sludge from agricultural use. The sewage sludge, in this case, contains above-average amounts of cadmium (S1, S2, S3, S4, S5) and zinc (S2, S3).

3.2 *The speciation of heavy metals in sewage sludge*

The sequential analyses showed that different forms of heavy metals are present in sewage sludge. The results of heavy metal speciation in sewage sludge are presented in Figures 1 to 5 and in Table 5.

The agronomic parameters of municipal sludge are present in Table 6.

Much of the sewage sludge tested had very small amounts of the mobile copper fractions (FI, FII) in comparison with the immobile ones (FIII, FIV), i.e. Below 5% (figures 1 to 5). From 40% to 72% of the copper was bound with the organic matter (FIII). As far as FIIIi is concerned, copper is temporarily immobile because its behaviour can vary according to the level of mineralisation in the ground. The immobile fraction (FIV) of copper ranged from 22% to 55% of the overall amount of copper.

Similar contributions from the individual copper fractions in sewage sludge were obtained by (Chen et al. 2008). An obvious conclusion is that the copper forms in sewage sludge are the characteristic feature of this chemical element.

The analysed samples had different concentration of nickel (from 1.31 to 51.85 mg/kg d.m.). The

Table 5. Statistical results* for each fraction of heavy metals in samples of sewage sludge.

Heavy metal [mg/kg sd.m.]	Fraction I	Speciation Fraction II	Fraction III	Fraction IV
Kostomłoty-Laskowa (S1)				
Cu	0.33 ± 0.03	0.00 ± 0.01	6.72 ± 0.55	2.25 ± 0.12
Cr	19.98 ± 0.98	3.22 ± 0.15	1.48 ± 0.09	3.67 ± 0.44
Cd	4.67 ± 0.37	0.89 ± 0.04	1.49 ± 0.09	0.27 ± 0.03
Ni	1.08 ± 0.11	0.33 ± 0.05	1.41 ± 0.18	2.9 ± 0.19
Pb	0.00 ± 0.01	0.00 ± 0.01	0.00 ± 0.01	275.15 ± 9.45
Zn	111.47 ± 9.31	109.72 ± 9.12	143.36 ± 9.59	231.47 ± 7.14
Daleszyce (S2)				
Cu	0.00 ± 0.01	0.00 ± 0.014	14.57 ± 0.89	6.46 ± 0.39
Cr	13.02 ± 0.95	4.17 ± 0.22	29.35 ± 1.55	59.14 ± 2.34
Cd	1.22 ± 0.09	1.52 ± 0.09	3.69 ± 0.11	3.04 ± 0.12
Ni	1.97 ± 0.23	0.9 ± 0.11	6.14 ± 0.48	7.68 ± 0.58
Pb	2.50 ± 0.32	0.00 ± 0.01	16.15 ± 0.27	408.42 ± 9.08
Zn	509.91 ± 8.97	447.33 ± 9.54	1119.36 ± 14.50	693.18 ± 8.36
Busko-Siesławice (S3)				
Cu	0.7 ± 0.29	1.0 ± 0.07	57.4 ± 1.05	20.73 ± 0.73
Cr	0.63 ± 0.25	0.69 ± 0.04	7.28 ± 0.73	26.64 ± 0.72
Cd	0.38 ± 0.09	0.64 ± 0.08	0.94 ± 0.11	14.75 ± 0.82
Ni	0.00 ± 0.01	0.0 ± 0.01	0.00 ± 0.01	1.31 ± 0.28
Pb	0.00 ± 0.01	0.35 ± 0.23	0.00 ± 0.01	322.05 ± 9.43
Zn	152.60 ± 0.67	75.79 ± 2.55	151.47 ± 1.52	460.79 ± 1.32
Skarżysko-Kamienna (S4)				
Cu	0.00 ± 0.01	0.00 ± 0.01	9.49 ± 2.66	12.31 ± 1.90
Cr	4.99 ± 0.42	2.32 ± 0.11	1283.75 ± 11.92	1468.75 ± 46.82
Cd	0.3 ± 0.08	0.7 ± 0.18	1.14 ± 0.25	9.99 ± 1.72
Ni	4.50 ± 0.11	0.99 ± 0.08	14.08 ± 0.27	8.93 ± 0.08
Pb	0.70 ± 0.22	0.96 ± 0.23	0.00 ± 0.01	29.6 ± 7.83
Zn	152.87 ± 0.73	144.52 ± 0.22	537.50 ± 23.58	516.25 ± 90.92
Sitkówka-Nowiny (S5)				
Cu	0.79 ± 0.11	0.00 ± 0.01	60.93 ± 1.45	21.74 ± 0.62
Cr	5.32 ± 0.61	2.83 ± 0.38	93.84 ± 1.73	136.51 ± 9.50
Cd	0.26 ± 0.07	0.54 ± 0.07	2.53 ± 0.05	2.27 ± 0.15
Ni	1.31 ± 0.17	0.00 ± 0.01	1.35 ± 0.32	49.19 ± 5.11
Pb	3.55 ± 0.12	4.00 ± 0.36	3.40 ± 0.18	56.79 ± 3.67
Zn	144.05 ± 15.97	98.29 ± 10.33	832.56 ± 24.55	240.47 ± 24.36

*Results are expressed in the form mean ± standard deviation.

Table 6. Agronomic parameters of municipal sludge.

Sample	pH	Total N [%]	Total P [%]	Total K [%]	Organic C [%]
S1	6.5	1.1	3.6	0.4	39.7
S2	6.5	1.5	2.1	0.3	57.6
S3	6.7	1.7	5.7	0.4	32.7
S4	6.9	1.8	1.7	0.1	30.6
S5	6.9	1.9	2.4	0.4	39.9

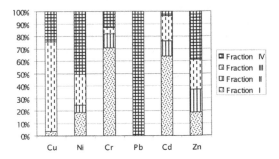

Figure 1. Speciation of heavy metals in sewage sludge from the wastewater treatment plant in Kostomłoty-Laskowa (S1).

preponderant fractions of nickel were the immobile fraction (FIV) that varied from 95% to 100% and the temporarily immobile fraction, FIII, at 49%.

The concentration of chromium in the tested sludge was very varied, ranging from 28.35 to 2759.81 mg/kg d.m. The FIV fraction was predominant for the S2, S3, S4, S5 sludge, and the FI fraction (i.e. mobile chromium) for S1 sludge. For the S4 sludge, the FI and FII fractions were unstated.

For S1 and S3 sludge, lead was present exclusively in an immobile fraction FIV. For the S2, S4 and S5

sludge, the residual fraction (FIV) of chromium also had a high contribution (from 83.8 to 95.6%). According to Chen et al. (2008) a high immobility of lead in sewage sludge results from the presence of lead (in sludge) in the form of indissoluble salts. Furthermore,

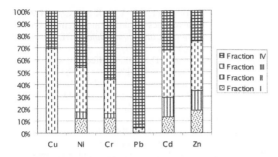

Figure 2. Speciation of heavy metals in sewage sludge from the wastewater treatment plant in Daleszyce (S2).

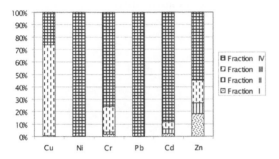

Figure 3. Speciation of heavy metals in sewage sludge from the wastewater treatment plant in Busko-Siesławice (S3).

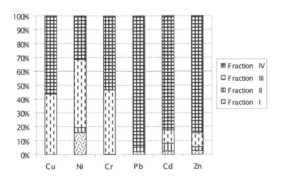

Figure 4. Speciation of heavy metals in sewage sludge from the wastewater treatment plant in Skarżysko-Kamienna (S4).

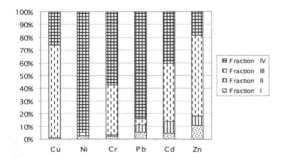

Figure 5. Speciation of heavy metals in sewage sludge from the wastewater treatment plant in Sitkówka-Nowiny (S5).

the soil's organic substance is a limiting factor in lead bioactivity in sewage sludge (Strawn & Sparks 2000).

The total concentrations of cadmium in all types of sewage sludge were at similar levels. The immobile fraction (FIV) was predominant for S3 and S4 sludge (88%). The S5 sludge contained the highest number of temporarily immobile cadmium fractions (FIII). However, S1 sludge was dominated by assimilable cadmium fractions and fractions bounded with carbonates (FI, 63.8%). A different combination of cadmium fractions was present in S2 sludge, in which the immobile fractions FII and FIV were predominant.

The zinc contents of the tested sewage sludge were the most varied (from 596.02 to 5351.14 mg/kg d.m.). In the S1, S3 and S4 sludge, the immobile fractions FIV (at 38.8; 54; 84%) had the highest contributions. The temporarily immobile fraction (FIII) had the highest per cent contribution in sludge S5 (63%) and S2 (40.4%).

4 CONCLUSIONS

The total content of a chemical element in sludge is not equivalent to the possibility of its release to the soil–water environment. The form in which it occurs is significant. The application of heavy metal speciation analysis allowed for the per cent determination of the content of water-soluble forms with reference to the total content. This information is helpful in estimating the ability of a heavy metal to migrate from sewage sludge to the soil–water environment.

For the metals tested, the proportions of the individual heavy metal fractions in the sewage sludge clearly did not depend on the method of sludge stabilisation.

The results permit the observation that the contributions of the mobile heavy metal fractions (fractions I and II) in sewage sludge are not considerable.

The authors suggest that the projected changes to the heavy metal limits in municipal sewage sludge designed for environmental usage should include heavy metal forms.

The inclusion of heavy metal fractions in the regulations controlling the environmental use of sewage sludge would permit some of the tested sewage sludge types (from Busko – Siesławice, Skarżysko-Kamienna, Sitkówka – Nowiny Sewage treatment plants) to be used in agriculture. This would capitalise on the fertilising properties of the sewage sludge.

REFERENCES

Alvarez, E.A., Mochón, M.C., Jiménez, Sánchez, J.C. & Rodríguez, M.T. 2002. Heavy metal extractable forms in sludge from wastewater treatment plants. *Chemosphere* 47: 765–775.

Chen, M., Li X., Yang, Q., Zeng, G., Zhang, Y., Liao, D., Liu, J., Hu, J. & Guo, L. 2008. Total concentration and speciation of heavy metals in sewage sludge from Changasha, Zhuzhou and Xiangtan in middle–south region of China. *Journal of Hazardous Materials* 160: 324–329.

Dahlin, C.L., Williamson, C.A., Collins, W.K., Dahlin, D.C. 2002. Sequential extraction versus comprehensive characterization of heavy metal species in brownfield soils. *Environmental Forensics* 3: 191–201.

De La Guardia, M. & Morales-Rubio, A. 1996. Modern strategies for the rapid determination of metals in sewage sludge. *Trends in Analytical Chemistry* 15(8): 311–318.

Fytili, D. & Zabaniotou, A. 2008. Utilization of sewage sludge in UE application of old and new methods – A review. *Renewable and Sustainable Energy Reviews* 12: 116–140.

Hristensen E.R. 1998. Metals, acid-volatile sulfides organics, and particle distributions of contaminated sediments. *Water Science and Technology* 37(6–7): 149–156.

Krogmann, U., Boyles, L.S., Bamka, W.J., Chaiprapat, S. & Martel, C. J. 1999. Biosolids and sludge management. *Water Environ. Res.* 71(5): 692–714.

Pitt, R., Clark, S. & Field, R. 1999. Groundwater contamination potential from stormwater infiltration practices. *Urban Water* 1: 217–236.

Rogers, H.R. 1996. Sources, behaviour and fate of organic contaminants during sewage treatment and in sewage sludge. *The Science of the Total Environment* 185: 3–26.

Ryu, H.W., Moon, H.S., Lee, E.Y., Cho, K.S. & Choi, H. 2003. Leaching characteristics of heavy metals from sewage sludge by *Acidithiobacillus thiooxidans* MET. *Journal of Environmental Quality* 32:751–759.

Strawn, D.G. & Sparks, D.L. 2000. Effects of soil organic matter on the kinetics and mechanisms of Pb (II) sorption and desorption in soil. *Soil Sci. Am. J.* 64: 144–156.

Wang, M.J. 1997. Land application of sewage sludge in China. *The Science of the Total Environment* 197: 149–160.

Weiner, R. F., Matthews, R.A. 2003. *Environmental Engineering*. Elsevier Science, Burlington.

Werther, J. & Ogada T. 1999. Sewage sludge combustion. *Progress in Energy and Combustion Science* 25: 55–116.

Xiang L., Chan L.C., Wong J.W.C. 2000. Removal of heavy metals from anaerobically digested sewage sludge by isolated indigenous iron-oxidizing bacteria. *Chemosphere* 41: 283–287.

Environmental Engineering III – Pawłowski, Dudzińska & Pawłowski (eds)
© 2010 Taylor & Francis Group, London, ISBN 978-0-415-54882-3

Copper and zinc bioleaching from galvanic sludge in mixed microbial cultures

E. Karwowska
Warsaw University of Technology, Faculty of Environmental Engineering, Department of Biology, Warsaw, Poland

ABSTRACT: Biohydrometallurgy is a promising technique for the removal of heavy metals from industrial wastes. In this research copper and zinc from galvanic sludge were bioleached using microorganisms enriched from activated sludge in various culture media. The best results were obtained in bioleaching cultures ensuring optimum growth conditions for various groups of microorganisms. The pre-adaptation of active microorganisms in bioleaching media resulted in an significant improvement of the process. The maximum metal removal was 67% for copper and 82% for zinc. The high enzymatic activity of microorganisms corresponded with increasing bioleaching effectiveness. In bioleaching media the occurrence of *Acidithiobacilus thiooxidans, Thiobacillus denitrificans, Acetobacter aceti, Acidomonas methanolica* and *Acetobacter pasteurianus* was observed.

Keywords: Galvanic sludge, heavy metals, bioleaching.

1 INTRODUCTION

The increasing production of waste containing heavy metals poses a serious of environmental pollution problem. The electronic industry is a source of 1200000 t/year of waste containing mainly As, Cr, Hg, Se, Ni and Cu. The mining and metallurgy industry produces 390000 t/year of wastes contaminated with Hg, Cr, Cu, As, Zn and Pb. The oil and coal industry introduces about 1200000 t/year of waste into the environment loaded with As, Pb, V, Cd, Ni and Zn (Veglio et al 2003).

Metal plating is an environmental risky industrial sector due to the huge amounts of metals that are introduced into the environment through galvanic wastewater and sludges (Silva et al. 2005b). Approximately 150000 tons/year of these sludges are generated in EU countries. They are classified as hazardous, basically due to the high content of leachable heavy metals (Magalhães et al. 2005).

The conventional technique for toxic metal removal from galvanic wastewaters is through precipitation, mainly as metal hydroxides. The galvanic sludge is created from the precipitate by dewatering. The composition of the sludge strongly depends on the electroplating technology used, and the type of plating and it may vary over time.

Galvanic sludge must be stabilised prior to landfilling in order to prevent the release of pollutants into the environment (Bednarik et al. 2005). One proposed method is to stabilize the solid wastes from galvanic treatment (which contain Cd, Cr and Ni) in a cementitious matrix based on calcium silicate and sulphoaluminate (Cioffi et al. 2002). However, the current level of galvanic sludge management is not satisfactory. It is mainly disposed in landfill sites as a dangerous waste (Jandova et al. 2002).

Treatment processes that reduce the environmental impact of various types of waste are of great interest today. Unfortunately, many technologies are too expensive or simply not available. Regulations regarding solid waste management, including EU directives give high priority to technologies that involve recycling precious raw materials or energy from waste (Cioffi et al. 2002).

Some research work concentrate on the possibility of recovering heavy metals from industrial waste. For gold recovery from the waste products of the electronic and jewellery industries, a combined method of thermal degradation, two-stage leaching with nitric acid and aqua regia, solvent extraction and reduction of gold from organic phase has been proposed. Nickel is removed from the ashes after heavy oil burning using acidic leaching, filtration, precipitation of ammonium-nickelous sulphate, filtration and drying. Electroplating sludges containing chromium and copper are treated in a process involving hydroxide dissolution in chromic acid and metal reuse by ion exchange (Chmielewski et al. 1997).

The most determinant step in any hydrometallurgical process is generally the first one – the removal or dissolution of the metal-containing material. In many chemical leaching processes, sulphuric acid is applied as the cheapest and most effective leachant (Veglio et al. 2003).

Neutralisation sludges produced during galvanic wastewater treatment contain heavy metals in amounts comparable to, or even exceeding, their concentration in metallurgical raw materials, so they may serve as a source for heavy metals recovery and reuse.

Heavy metal content in neutralisation sludge reaches 5–10% of sludge weight (Jandova et al., 2002). There have been some attempts to apply hydrometalurgical technologies to remove heavy metals from galvanic sludges, but the high investment and operational costs of chemical treatment are limiting factors.

A solution to the problem may be the application of biohydrometalurgical methods. They are environmentally friendly, low cost and easy to operate (Jandova et al. 2002). A variety of microorganisms are capable of mobilising and leaching metals from solid materials by producing organic and inorganic acids, redox reactions or excretion of complexing agents (Aung & Ting 2005, Łebkowska & Karwowska 2003). The most frequently used microorganisms in biohydrometallurgical processes are iron and sulphur oxidising bacteria and organic acids producing bacteria and fungi.

The aim of this research was to elaborate the optimum conditions for copper and zinc bioleaching from galvanic sludge in mixed microbial cultures.

2 MATERIALS AND METHODS

2.1 Materials

In this study, mixed, non-homogenous galvanic sludge generated during the physico-chemical treatment of wastewater produced in different plating processes was used. It was composed mainly of metal hydroxides and oxides and showed reasonably high chemical variability, depending on production conditions and stocking age. The average concentration of copper in the sludge was 2413 mg Cu/kg and of zinc 1087 mg/kg. Water content in sludge was about 20%.

Samples of surplus activated sludge were obtained from municipal wastewater treatment plant. The solids concentration in sludge was about 3–4 g of dry weight/l.

2.2 Bioleaching experiments

A fixed amount of galvanic sludge (10 g) was placed in flasks with 150 cm^3 of bioleaching medium. For the experiments, the medium in each flask was inoculated with 10% (v/v) of the enriched sludge culture, containing active strains of bioleaching sulphur-oxidising bacteria and (in some variants) supplied with 1% powdered sulphur. BSPC medium was additionally inoculated with biosurfactant-producing bacteria (1% v/v).

The flasks were put into a shaker (120 rpm) to keep the content in homogenous form at room temperature. Cultures were carried out in non-sterile conditions for 21 days. For each variant of the experiment five bioleaching series were provided. The control flasks contained the galvanic sludge in 150 ml of distilled water without inoculation.

In order to obtain an inoculum culture, indigenous sulphur-oxidising bacteria were enriched from the sludge with the addition of 1% sulphur and grown up for 2–3 weeks to lower the culture reaction to pH

Table 1. The composition of bioleaching media.

Type of medium	Medium composition
1% sulphur medium	Surplus activated sludge supplied with 1% powdered sulphur
Beer medium	Surplus activated sludge supplied with beer, as a source of organic compounds and vitamins, 2:1 ratio
Sulphur-beer medium	Surplus activated sludge supplied with beer, 2:1 ratio, with addition of 1% of powdered sulphur
BSPC medium	Surplus activated sludge supplied with addition of biosurfactants producing bacteria

2–3. After three consecutive transfers, the enriched sludge was retained as the inoculum.

In different variants of the experiment, four various bioleaching media were applied. Their compositions are summarised in Table 1.

The influence of the microorganisms' pre-adaptation on their activity in metal bioleaching was estimated. Two cultures, 1% sulphur medium and the sulphur-beer medium, were tested. In order to obtain pre-adapted culture, microorganisms from activated sludge were grown in culture media without sludge addition for two weeks. The proper amount of galvanic sludge was then added to start the bioleaching process. Non-adapted bioleaching cultures were applied as the control samples.

2.3 Control analyses

For the evaluation of metal concentration in solution during the bioleaching process, 10 ml samples were periodically drawn (after 7 and 21 days from the start of the experiments). The pH was determined according to PN-90/C-04540/01. The metal concentrations were determined in supernatant, after sample filtration through 0.45 μm filters, with flame atomic absorption (AAS Thermo-Jarrel Ash SH-1150 spectrophotometer). The metal removal efficiency was calculated as the ratio between the amount of metal in solution and initial total metal content in a sludge.

Enzymatic activity tests and adenosine-triphosphate (ATP) content were applied to characteise the activity of bioleaching microorganisms. Experiments were carried out both in presence and without galvanic sludge addition. Enzymatic tests were conducted using the dehydrogenase activity test with TTC, according to PN-82/C-046/08 (modified) and the hydrolytic activity test with fluoresceine diacetate (FDA), according to Schnürer and Rosswal (1982). ATP content was determined by the routine luminometric method, using HY LiTE® Luminometer (Merck).

2.4 Identification of microorganisms

The detection and differentiation of sulphur-oxidising bacteria and acetic acid producing bacteria were assessed by PCR-based technique. The design of the PCR primers were based on 16S rRNA genes sequences, applying AlignX program and in accordance with literature data. The most suitable sequences were identified using the AmplifiX progamme. The analysis was done in cooperation with Faculty of Biology, Warsaw University of Technology. DNA starters were synthesised at the Institute of Biochemistry and Biophysics, Polish Academy of Sciences.

3 RESULTS

During the experiments, galvanic sludge particles were colonised by biofilm-forming microorganisms (Figure 1). The sludge stepwise solubilisation was then observed (Figure 2). Metals were released into the bioleaching medium.

The first series of bioleaching experiments concerned metals removal in cultures without previous adaptation. The effectiveness of the process depended on the type of the culture (Figures 3, 4). The copper and zinc extraction yields after 21 days of the experiment were 13–37% and 31–63%, respectively (Table 2).

The greatest amounts of both copper and zinc were released from sludge in the beer medium and the sulphur-beer medium. These were significantly higher

A.
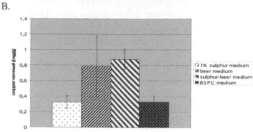

B.
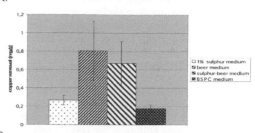

Figure 3. Average amounts of copper removed from galvanic sludge in different bioleaching cultures (A- after 7 days, B- after 21 days).

A.
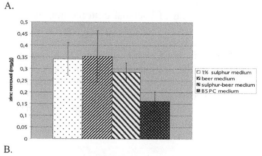

B.
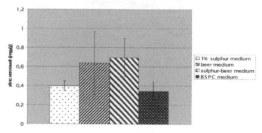

Figure 4. Average amounts of zinc removed from galvanic sludge in different bioleaching cultures (A- after 7 days, B- after 21 days).

Figure 1. The colonisation of the galvanic sludge by microorganisms.

Figure 2. Sludge solubilisation during the bioleaching process.

Table 2. Average copper and zinc removal (%) from the galvanic sludge after 21 days of bioleaching.

Bioleaching medium	Copper removal from the sludge (%)	Zinc removal from the sludge (%)
1% sulphur medium	13 ± 3	36 ± 5
Beer medium	37 ± 16	63 ± 30
Sulphur-beer medium	34 ± 5	58 ± 18
BSPC medium	14 ± 3	31 ± 9

than those obtained using a culture based mainly on sulphur-oxidising bacteria (in 1% sulphur medium).

Copper was bioleached faster than zinc. The significant bioleaching yields for copper were obtained after 7 days of the process (0.81 mg/dm³ in the beer medium and 0.67 mg/dm³ in the sulphur-beer medium). The maximum values for zinc removal were reached after 21 days of experiment (0.64 mg/dm³ and 0.69 mg/dm³ respectively).

No positive effect from the biosufactant presence in the bioleaching culture in the BSPC medium was observed. The average effectiveness after 21 days of the experiment was lower than for other bioleaching culture (0.33 mg/dm³ for copper and 0.34 mg/dm³ for zinc).

Periodical pH measurements revealed that metals were bioleached in a slightly acidic environment. It might be due to the alkalising influence of the sludge that diminished the effect of medium acidification by sulphur-oxidising bacteria (Figure 5).

The next series of bioleaching experiments was carried out simultaneously in pre-adapted and non-adapted microbial cultures. Two bioleaching media, 1% sulphur medium and the sulphur-beer medium, were applied.

Pre-adaptation of bacterial cultures improved the effectiveness of the process compared with non-adapted ones. The bioleaching effectiveness increased 2–3 fold with both tested cultures, the 1% sulphur medium and the sulphur-beer medium (Table 3). The maximum copper removal level in pre-adapted cultures reached 67% (for sulphur-beer medium). For zinc, the metal elimination level exceeded 80% in both tested bioleaching cultures.

The improved results from bacterial leaching using pre-adapted cultures were probably connected with intensified acidification (Figure 6). The pH of the pre-adapted cultures was below 4 compared with 6-7 for non-adapted cultures, as recorded during the previous experiments.

The parameters analysed to characterise the bioleaching process were the dehydrogenase and hydrolytic activity of the bioleaching cultures as well as ATP content, measured after 7 and 21 days of the process. The results are presented on Figures 7, 8 and 9.

A significant increase in the dehydrogenase and hydrolase activity after 21 days of the experiment (comparing with the results obtained after 7 days of bioleaching) was observed. The highest activity and ATP content were noticed in the cultures that were most effective in removing heavy metals from the sludge, i.e.

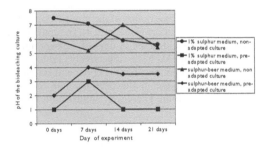

Figure 6. pH changes during copper and zinc bioleaching in pre-adapted and non-adapted cultures.

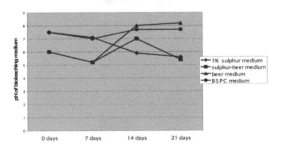

Figure 5. Changes of pH in various cultures during bioleaching process.

Table 3. Copper and zinc removal (%) in non-adapted and pre-adapted bioleaching cultures after 21 days of the experiment.

| Metal | Amount of removed metal (%) | | | |
| | 1% sulphur medium | | Sulphur-beer medium | |
	non-adapted	pre-adapted	non-adapted	pre-adapted
Copper	9	42	33	67
Zinc	32	82	25	81

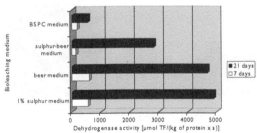

Figure 7. Dehydrogenase activity of various bioleaching cultures.

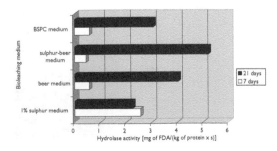

Figure 8. Hydrolytic activity of various bioleaching cultures.

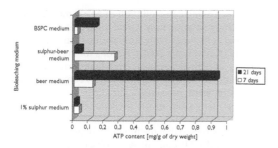

Figure 9. ATP content in various bioleaching cultures.

Table 4. Dehydrogenase activity in bioleaching cultures carried out in presence of galvanic sludge, comparing with control cultures (without sludge addition), after 21 days of experiment.

Bioleaching medium	Dehydrogenase activity (μmol TF/(kg of protein·s))	
	With galvanic sludge	Without galvanic sludge
1% sulphur medium	4883.1	134.9
Beer medium	4659.4	4145.4
Sulphur-beer medium	2808.4	5166.3
BSPC medium	542.7	76.9

Table 5. Hydrolytic activity in bioleaching cultures carried out in presence of galvanic sludge, comparing with control cultures (without sludge addition), after 21 days of experiment.

Bioleaching medium	Hydrolytic activity (mg of fluoresceine/(kg of protein·s))	
	With galvanic sludge	Without galvanic sludge
1% sulphur medium	2.248	0.826
Beer medium	4.014	6.835
Sulphur-beer medium	5.196	4.616
BSPC medium	3.056	0.528

cultures in the beer medium and in the sulphur-beer medium (metal removal effectiveness reaching 37% for copper and 63% for zinc in non-adapted cultures). The lowest dehydrogenase activity was observed in the BSPC medium, but in case of hydrolytic activity this medium was similar to the others. The ATP content value was the highest for the beer medium after 21 days of the bioleaching process.

Enzymatic activity and ATP content in bioleaching cultures were comparable in variants with and without presence of galvanic sludge (Tables 4, 5 and 6).

A PCR-mediated method applied to detect sulphur-oxidising bacteria and acetic acid producing bacteria allowed for the identification of some bacterial strains occurring in bioleaching cultures. In the culture carried out in the 1% sulphur medium, *Thiobacillus*

Table 6. ATP content in bioleaching cultures carried out in presence of galvanic sludge, comparing with control cultures (without sludge addition), after 21 days of experiment.

Bioleaching medium	ATP content (mg/g of dry weight)	
	With galvanic sludge	Without galvanic sludge
1% sulphur medium	0.020	0.016
Beer medium	0.925	0.520
Sulphur-beer medium	0.045	0.198
BSPC medium	0.146	0.002

denitrificans was detected. In the culture in the beer medium, three strains of acetic acid producing bacteria were present: *Acetobacter aceti, Acidomonas methanolica* and *Acetobacter pasteurianus*. Strains of *Acidomonas methanolica, Acetobacter pasteurianus* and *Acidithiobacillus thiooxidans* predominated in the sulphur-beer medium.

It was found that in culture media with addition of both sulphur and beer, sulphur-oxidising acidophilic bacteria coexisted with acetic acid producing heterotrophic microorganisms.

4 DISCUSSION

Some literature data prove the applicability of biohydrometalurgical methods to heavy metals reclamation from various types of industrial waste. Copper and zinc are the most frequently recycled heavy metals. They have been removed from ashes (Paul et al. 2004, Krebs et al. 2001, Ishigaki et al. 2005, Rejinders, 2005), industrial metal wastes (Breed & Hansford 1999) and also river and storm sediments (Mercier et al. 1996, Anderson et al. 1998). There is some data concerning copper bioleaching from electronic waste materials using hetrotrophic bacteria and fungi (Brandl et al 2001). According to the literature data, the effectiveness of metal removal from waste may reach 60–90% for copper and 78–98% for zinc.

Relatively few papers present data on heavy metals removal from galvanic waste. They cover mainly chemical techniques of metal leaching (Veglio et al. 2003, Jandova et al. 2002). Silva et al. (2005a) proposed the solvent extraction of metals from galvanic sludge using di-(2-ethylhexyl)-phosphoric acid (D2EHPA) and bis-(2,4,4-trimethylpentyl)-posphinic acid, both diluted in kerosene, obtaining promising results for zinc recovery. Chemical leaching of galvanic sludge with 10% sulphuric acid achieved maximal conversions of 88.6% Cu, 98% Ni and 99.2% Zn; an ammoniacal medium allowed the extraction of 70% of Cu and 50% Ni, but not Cr (Silva et al. 2005b).

Data obtained in this study shows that galvanic sludge may serve as a source of recycled copper and zinc, due to a removal effectiveness reaching about 67% and 82%, respectively, in optimum bioleaching conditions. According to the literature, from an economic point of view, nickel, zinc and copper are the most interesting metals that could be recycled after the galvanic processes (Silva et al. 2005a).

Traditional biohydrometalurgical techniques are based mainly on the application of iron and sulphur-oxidising autotrophic bacteria, mainly from genus *Acidithiobacillus*. However, it was shown that mixed cultures of autotrophic and heterotrophic microorganisms may be more effective in the removal of heavy metals from waste, due to higher biomass production and a variety of excreted methabolites (including organic acids). In this research, bioleaching of metals from galvanic sludge was carried out in four different media capable of stimulating the growth of various groups of microorganisms.

It has been previously reported that activated sludge may serve as a useful source of active strains of bioleaching bacteria (Blais et al. 1993, Karwowska 2000). Krebs et al. (2001) revealed that an addition of sewage sludge may stimulate the growth of some strains of sulphur-oxidising bacteria and fasten the acidification of bioleaching medium. In this research an activated sludge from municipal wastewater treatment sludge was applied both as a main component of bioleaching media and as a source of microorganisms.

It was shown that the pre-adaptation of bioleaching bacteria to the growth conditions present in the bioleaching medium is of great importance for the effectiveness of the process. This finding is in agreement with previous suggestions by Elzeky & Attia (1995), Rezza et al. (1997) and Shanableh & Ginige (2000).

It was proved that the best results may be obtained by ensuring optimum growth conditions for both sulphur-oxidising (autotrophic and heterotrophic) bacteria and organic acid producing heterotrophic microorganisms. Additional improvement may be achieved by way of pre-adaptation of active microorganisms in the bioleaching media. The maximum metal removal from galvanic sludge was 67% (1.6 mg/g) for copper and 82% (0.9 mg/g) for zinc and it was achieved in sulphur-beer pre-adapted bioleaching culture.

Although there is some evidence concerning the positive role of biosurfactants in the heavy metals bioleaching processes (Banat et al. 2003, Christofi & Ivshina 2002), in this research no influence of biosurfactant-producing bacteria on process effectiveness was observed.

Molecular techniques of microbial identification are very useful in cases where microorganisms in bioleaching environments, especially some autotrophic strains, are difficult to culture (Romero et al. 2003, Escobar & Godoy 2001). The PCR method based on analysis of 16S rRNA genes allowed the detection of two strains of sulphur-oxidising bacteria (*Acidithiobacilus thiooxidans, Thiobacillus denitrificans*) and three strains of acetic acid producing bacteria (*Acetobacter aceti, Acidomonas methanolica* and *Acetobacter pasteurianus*) in the bioleaching cultures.

An increasing concentration of heavy metals in the bioleaching solution may reveal toxic effects to living bacterial cells, negatively influencing their enzymatic activity (Karwowska 2000). Results for the enzymatic activity of dehydrogenases and hydrolases, as well as the adenosine-tri-phosphate (ATP) content showed that there were no significant differences between cultures carried out in the presence and without the addition of the 10% galvanic sludge. Moreover, it should be stressed that the highest values of enzymatic activity and ATP content were observed in cultures with the highest bioleaching activity. It may suggest that enzymatic tests may serve as a useful monitoring tool of the bioleaching process.

5 CONCLUSIONS

The analysis of the results allows the following conclusions to be made.

- Galvanic sludge may serve as a substrate in bioleaching processes and in consequence as a source of metals recycling.
- The enrichment of the bioleaching medium with sulphur, organic compounds and vitamins (in the sulphur-beer medium) caused a 2–3 fold increase in the bioleaching effectiveness of copper and zinc from galvanic sludge compared with traditional bioleaching techniques applying sulphur-oxidising bacteria.
- Preliminary adaptation of microorganisms to bioleaching medium conditions allowed a significant increase in the effectiveness of metals removal from galvanic sludge.
- The average effectiveness of metal removal from galvanic sludge in the sulphur-beer pre-adapted medium reached 67% for copper and 82% for zinc.
- In applied bioleaching cultures both sulphur-oxidising bacteria and acetic acid producing bacteria were detected.
- Enzymatic activity of microorganisms and ATP content in bioleaching cultures corresponded with the bioleaching effectiveness of the cultures, and the values for these parameters were greatest in case of the sulphur-beer medium.

REFERENCES

Anderson, B.C., Brown, A.T.F., Watt, W.E. & Marsalek, J. 1998. Biological leaching of trace metals from stormwater sediments: influential variables and continuous reactor operation. *Water Science and Technology* 38(10): 73–81.

Aung, M.M. & Ting Y.-P. 2005. Bioleaching of spent fluid catalytic cracking catalyst using Aspergillus niger. *Journal of Biotechnology* 116: 159–170.

Banat, I.M., Makkar, R.S. & Cameotra, S.S. 2003. Potential commercial applications of microbial surfactants. *Applied Microbiology and Biotechnology* 53: 495–508.

Bednarik, V., Vondruska, M. & Koutny, M. 2005. Stabilization/solidification of galvanic sludges by asphalt emulsions. *Journal of Hazardous Materials* B122: 139–145.

Blais, J.F., Tyagi, R.D. & Auclair, J.C. 1993. Bioleaching of metals from sewage sludge: microorganisms and growth kinetics. *Water Research* 27(1): 101–110.

Brandl, H., Bosshard, R. & Wegmann, M. 2001. Computer-munching microbes: metal leaching from electronic scrap by bacteria and fungi. *Hydrometallurgy* 59: 319–326.

Breed, A.W. & Hansford, G.S. 1999. Studies on the mechanism and kinetics of bioleaching. *Minerals Engineering* 12(4): 383–392.

Chmielewski, A.G., Urbański, T.S., Migdał, W. 1997. Separation technologies for metal recovery from industrial wastes. *Hydrometallurgy* 45(3): 333–344.

Christofi, N. & Ivshina, I.B. 2002. Microbial surfactants and their use in field studies of soil remediation. *Journal of Applied Microbiology* 93(6): 915–929.

Cioffi, R., Lavorgna, M. & Santoro, L. 2002. Environmental and technological effectiveness of a process for the stabilization of a galvanic sludge. *Journal of Hazardous Materials* B89: 165–175.

Elzeky, M. & Attia, Y.A. 1995. Effect of bacterial adaptation on kinetics and mechanisms of bioleaching ferrous sulfides. *The Chemical Engineering Journal* 56: 115–124.

Escobar, B.M. & Godoy, I.R. 2002. Enumeration of Acidithiobacillus ferrooxidans adhered to agglomerated ores in bioleaching processes. *World Journal of Microbiology and Biotechnology* 18: 875–879.

Ishigaki, T., Nakanishi, A., Tateda, M., Ike, M. & Fujita, M. 2005. Bioleaching of metal from municipal waste incineration fly ash using a mixed culture of sulfur-oxidizing and iron-oxidizing bacteria. *Chemosphere* 60: 1087–1094.

Jandova, J., Maixner, J. & Grygar, T. 2002. Reprocessing of zinc galvanic waste sludge by selective precipitation. *Ceramics –Silikáty* 46(2): 52–55.

Karwowska E. 2000. *Removal of chosen heavy metals from wastewater using activated sludge.* Doctoral dissertation. Warsaw University of Technology. Warsaw (in Polish).

Krebs, W., Bachofen, R. & Brandl, H. 2001. Growth stimulation of sulfur oxidizing bacteria for optimization of metal leaching efficiency of fly ash from municipal solid waste incineration. *Hydrometallurgy* 59(2–3): 283–290.

Łebkowska M. & Karwowska E. 2003. *Heavy metals removal from industrial wastewater and sewage sludges* (in Polish). Seria Wodociągi i Kanalizacja, 10. PZIiTS, Warszawa.

Malgalhães, J.M., Silva, J.E., Castro, F.P. & Labrincha, J.A. 2005. Physical and chemical characterisation of metal finishing industrial wastes. *Journal of Environmental Management* 75: 157–166.

Mercier, G., Chartier, M. & Couillard, D. 1996. Strategies to maximize the microbial leaching of lead from metal-contaminated aquatic sediments. *Water Research* 30(10): 2452–2464.

Paul, M., Sandström, Å. & Paul, J. 2004. Prospects for cleaning ash in the acidic effluent from bioleaching of sulfidic concentrates. *Journal of Hazardous Materials* 106b: 39–54.

Polish Standard PN-82/C-04616/08.

Polish Standard PN-90/C-04540/01.

Rezza, I., Salinas, E., Calvente, V., Benuzzi, D. & Sanz de Tosetti, M.I. 1997. Extraction of lithium from spodumene by bioleaching. *Letters in Applied Microbiology* 25(3): 172–176.

Rejinders, L. 2005. Disposal, uses and treatments of combustion ashes: a review. *Resources, Conservation and Recycling* 43: 313–336.

Romero, J., Yañez, C., Vásqeuz, M., Moore, E.R.B. & Espejo, R.T. 2003. Characterization and identification of an iron-oxidizing, Leptospirillum-like bacterium, present in the high sulfate leaching solution of a commercial bioleaching plant. *Research in Microbiology* 154: 353–359.

Schnurer, J. & Rosswal, T. 1982. Fluorescein diacetate hydrolysis as a heasure of total microbial activity in soil and litter. *Applied and Environmental Microbiology* 43(6): 1256–1261.

Shanableh, A. & Ginige P. 2000. Acidic bioleaching of heavy metals from sewage sludge. *Journal of Material Cycles and the Waste Management* 2: 43–50.

Silva, J.E., Paiva, A.P., Soares, D., Labrincha, A. & Castro, F. 2005a. Solvent extraction applied to the recovery of heavy metals from galvanic sludge. *Journal of Hazardous Materials* B120: 113–118.

Silva, J.E., Soares, D., Paiva, A.P., Labrincha, A. & Castro, F. 2005b. Leaching behaviour of galvanic sludge in sulphuric acid and ammoniacal media. *Journal of Hazardous Materials* B121: 195202.

Veglio, F., Quaresima, R., Fornali, P. & Ubaldini, S. 2003. Recovery of valuable metals from electronic and galvanic wastes by leaching and electrowinning. *Waste Management* 23: 245–252.

Environmental Engineering III – Pawłowski, Dudzińska & Pawłowski (eds)
© 2010 Taylor & Francis Group, London, ISBN 978-0-415-54882-3

Excess sludge treatment using electro-hydraulic cavitation

T.A. Marcinkowski & P.J. Aulich
Institute of Environmental Protection Engineering, Wroclaw University of Technology, Wroclaw, Poland

ABSTRACT: This paper presents a brief characterization of the method of electrochemical conditioning of sludge, using electro-hydraulic cavitation. Excess sludge from municipal wastewater treatment plants in Głogów, Bolesławiec, and Gorzów was analyzed. The settlement characteristics, hydration, dry mass content, and organic and mineral substances of each sample of sludge were measured. Additionally, the specific filtration resistance, effectiveness of filtration, the iron content of the ash, the susceptibility to thicken and the energy consumption of the process were determined.

Keywords: Conditioning, dewatering, filtration, power consumption.

1 INTRODUCTION

The research process described below is a component of a programme investigating the electrochemical conditioning of municipal sewage sludge. Mixed sludge is a combination of primary and excess sludge, in various proportions. Given the different natures of the primary and excess sludge, it was decided to divide the experimental process. The results for the primary sludge have been reported in previous articles (Aulich & Marcinkowski 2007, Marcinkowski & Aulich 2009).

Sewage sludge management is a problem that is still under investigation. New technologies are currently being tested, based on anoxic-steam gasification. Cheap thermal and electrical energy produced in the thermal processes of waste treatment can be utilized for sewage sludge conditioning in order to transform such material into fuel adapted for further treatment in anoxic-steam gasification processes.

In 2001, the amount of sewage sludge generated in Polish municipal wastewater treatment plants was 397 200 tonne (dry mass) (Bień 2002, KPOŚK 2003). The municipal sewage sludge generated in Poland in 2004 was managed as follows: for agricultural use 17%, for land reclamation use 28%, for the cultivation of compostable plants approximately 8%, for deposit over 41%, for storage approximately 6%, and for thermal treatment 0.3% (Polish Gazette 2007). According to the Central Statistical Office, the amount of municipal sewage sludge generated in Poland in 2007 was 533.4 thousand tonne d.m. (Statistical Yearbook 2007). And KPOŚK (the National Program for Municipal Sludge Treatment), anticipates the amount of dry mass of stabilized sewage sludge that will be produced by municipal wastewater treatment plants in the year 2015 will amount to 642.4 thousand tonne. Some authors claim that the process of treatment and disposal of sewage sludge consumes as much as half – and in

some cases even as much as 60% – of the total cost of wastewater treatment (Egemen et al. 2001).

Many different techniques for conditioning, dewatering, and pretreating sludge are currently in use.

Research work currently being carried out in many centres worldwide is directed at reducing the amount of sludge generated and increasing the profitability of treatment, as well as reducing the difficulties of thermal and chemical treatment techniques.

Many authors have tried to implement individual solutions, such as:

- mechanical disintegration in high-pressure mixers or ball mixers (Müller 1996);
- ultrasonic treatment (Forster et al. 1999, Bień et al. 2001);
- chemical treatment using dihydrogen dioxide, Fenton reaction (Barbusinski & Filipek 2000), ozone (Neyens et al. 2003), acids or bases (Marcinkowski 2004, Czechowski & Marcinkowski 2006);
- thermal hydrolysis (Pinnekamp 1989);
- combined thermal and chemical methods of treatment, such as Protox, Syntox, Krepro (Neyens & Baeyens 2003); and
- processes utilizing the effects of ultrasonic and electrical fields (Clark and Nujjoo, 1998), and a microwave field (Wojciechowska & Kowalik 2003).

The following are the more important obstacles in the full-scale implementation of these processes:

- the high specific consumption of electrical energy and the relatively low effectiveness, in the case of mechanical disintegration;
- problems in enlarging the scale of the process; including significant consumption of electrical energy, low durability, and the high cost of the equipment for ultrasonic treatment;
- considerable capital costs for equipment resistant to the highly corrosive environment, as well as

the emissions of noxious gases in the chemical treatment processes;
- the high environmental load and the need to recirculate a considerable part of the chemical load from thermal hydrolysis to wastewater plants; and
- the significant capital costs and high recurrent costs associated with the chemical reagents used and in heating the sludge, as well as the marketing problems for the products generated in the combined thermo-chemical processes.

2 MATERIAL AND METHODS

2.1 Fluidized bed disintegrator used in research

In order to examine the process of combined treatments, such as heat and mass exchange processes, decomposition in plasma using advanced oxygenation techniques, and high-power sound energy, a modified reactor for heat and mass exchange in barbotage conditions was proposed. Such a configuration consists of an insulating case, two feed electrodes, a fluidized bed, and a system for delivering a reactive gas (in this instance, air).

An electrical discharge is used in the reactor, which is simultaneously utilizing a number of other approaches. These include mechanical, chemical, acoustic, electro-hydraulic cavitation, sonoluminescence, plasma, and UV radiation, as well as microwave and electromagnetic radiation.

Figure 1 (Schumacher 2004) illustrates three separate phases of an electrical discharge phenomenon in a liquid as used in practice and in our reactor. These are (i) the preparation phase for ignition, (ii) the discharge phase, and (iii) the interval phase between discharges.

The preparation phase for ignition (Figures 1a–c). When the generator voltage is switched on the electric field reaches its highest strength in the region where the electrode surfaces are closest and where particles exist in the gap that can create an electric bridge. Ignition will not take place in any former discharge channels, as the bubbles remaining from the previous discharge continue to exist for a much longer period of time than that of the power on-time. The region of new discharges is defined by thin bridges of particles between the grains of the fluidized bed (Schumacher 1965, Müller 1965). These evaporate, creating high-temperature plasma.

The discharge phase (Figures 1d–f). The plasma channel that is developed has a very high pressure inside it. This causes the distribution of the shock wave in the sludge. The current passing across the gap creates a high temperature causing material evaporation on the grains of the fluidized bed. Current density and temperature quickly decrease with the continuous growth of the plasma channel. The plasma channel diameter stabilizes when an equilibrium is reached between the energy supply from the generator and the heat flow to the fluidized bed, the evaporation heat of the liquid, and the heat transfer to the sludge. The liquid evaporation continues to enlarge a gas bubble around

the plasma channel. The enlarged discharge channel is still under high pressure as the vapour generated during evaporation increases its volume by a factor of from 20 to 40. The material of the fluidized bed is melted at various points in the plasma effect. During the oscillations of the plasma, metal may be partially ejected at those moments when the pressure lowers. Singermann (1956) reports such phenomenon as 'flares'.

The interval phase between discharges (Figures 1g–i). The discharge ends when the generator is switched off. The plasma channel is de-ionised. However, the gas bubble stays in position for quite a long time. As can be seen in the ultra fast photographs, (Schumacher 1965, Hockenberry 1967), this period may be up to 25 times longer than the on-time of the generator. Along with the de-ionisation, the pressure and temperature in the plasma channel also decrease. The molten material of the fluidized bed, overheated under the discharge pressure, now starts boiling instantaneously and this is accompanied by the ejection of liquid globules. When these enter the relatively cool dielectric (the liquid sludge), they are shock hardened from the outside. After solidification, they may show hollow inner sections. This occurs in place of further metal shrinkage.

As a result of events described above, conditions for the synthesis of new chemical compounds arise. The shock wave induces an oscillating movement of particles in the fluidized bed. During this mutual friction, particles of sludge mechanically disintegrate.

2.2 Research on excess sludge conditioning

Excess sludge generated in the municipal wastewater treatment plants (MWWTP) in Bolesławiec, Głogów and Gorzów was analyzed.

The intent of the research was to compare the effects of electrochemical conditioning of excess sludge generated in municipal wastewater treatment plants in various towns.

The following indicators and parameters were measured: dry mass, ash and organic substance content, capillary suction time (CST), specific resistance of filtration, effectiveness of filtration, moisture content in the sludge cake, thickening after 24 hours, the iron content of the ash, and the energy consumption per volume unit.

The following samples of excess sludge were analyzed:

- Sludge from the municipal wastewater treatment plant in Głogów:
  Sample labelling nomenclature:
  NOG   Excess sludge (N), reference (O), WWTP in Głogów (G).
  NEK30G   Excess sludge (N), electrochemically conditioned (EK) for 30 minutes (30), WWTP in Głogów (G).
- Sludge from the municipal wastewater treatment plant in Bolesławiec:
  Sample labelling nomenclature:
  NOB   Excess sludge (N), reference (O), WWTP in Bolesławiec (B).

Preparation phase for ignition

Discharge phase

Interval phase between discharges

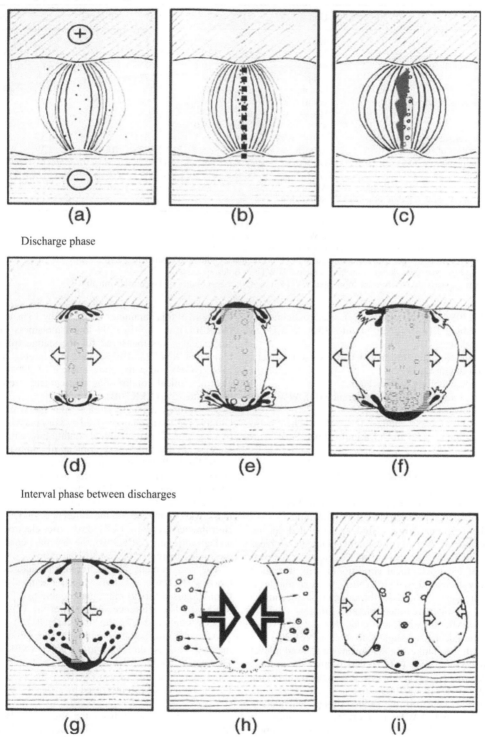

Figure 1.   The phenomenon of an electrical discharge in a liquid (according to Schumacher 2004).

Table 1. Results of research on sludge from WWTPs in Gorzów, Głogów and Bolesławiec.

Parameter	Unit	NO	NEK30	NOG	NEK30G	NOB	NEK30B
Hydration	%	99.2	99.0	99.5	99.2	98.5	98.4
Moisture content in sludge cake	%	89.4	77.2	88	79	79	75
Dry mass in sludge	g/dm^3	7.9	10.1	4.7	7.6	14.8	15.9
Ash content	g/dm^3	2.7	5.7	1.72	4.92	5.06	7.57
	% d.m.	34	56	37	65	34	48
Organic substance content	g/dm^3	5.2	4.4	2.98	2.65	9.71	8.32
	% d.m.	66	44	63	35	66	52
CST	s	8.0	9.0	6	7	17	43
Specific resistance of filtration	$\times 10^{10}$ m/kg	3.6	5.5	2.5	3	14.5	22.7
Effectiveness of filtration	$kg/m^2 h$	23.3	5.5	35.32	19.48	2.39	1.1
Thickening after 24 hours	cm^3/dm^3	550	450	184	150	656	672
Fe content in ash	g/kg	20.7	218	49	269	17	149
Energy m^3 of sludge	MJ/m^3	–	76	–	61	–	54
consumption per kg d.m. of sludge	MJ/kg d.m.	–	7.2	–	8.09	–	3.40

NO – reference excess sludge from the municipal WWTP in Gorzów,
NEK30 – excess sludge from the municipal WWTP in Gorzów, conditioned electrochemically,
NOG – reference excess sludge from the municipal WWTP in Głogów,
NEK30G – excess sludge from the municipal WWTP in Głogów, conditioned electrochemically,
NOB – reference excess sludge from the municipal WWTP in Bolesławiec,
NEK30B – excess sludge from the municipal WWTP in Bolesławiec, conditioned electrochemically

NEK30B Excess sludge (N), electrochemically conditioned (EK) for 30 minutes (30), WWTP in Bolesławiec (B).

- Sludge from the municipal wastewater treatment plant in Gorzów:
Sample labelling nomenclature:
NO Excess sludge (N), reference (O), WWTP in Gorzów (-).
NEK30 Excess sludge (N), electrochemically conditioned (EK) for 30 minutes (30), WWTP in Gorzów (-).

3 RESULT AND DISCCUSIONS

Results of analysis of the physicochemical composition and sludge properties are shown in Table 1.

The reference excess sludge, generated in the municipal wastewater treatment plant in Gorzów (NO), had a hydration level of 99.2%. After the process of electrochemical conditioning, its hydration slightly decreased, to 99.0% (NEK30). A similar level of hydration was shown by the sludge from the municipal wastewater treatment plant in Głogów – 99.5% (NOG). After the process of conditioning, the level of hydration decreased slightly, to 99.2% (NEK30G). In contrast, the sludge from Bolesławiec showed a level of hydration 98.5% (NOB), which, after the process of electrochemical conditioning, remained almost the same – 98.4% (NEK30B).

Sludge cake from the excess sludge generated in the municipal WWTP in Gorzów had a moisture content of 89.4% (NO). After 30 minutes of conditioning, its moisture content decreased by 14%, to 77.2% (NEK30). Comparable results were achieved for the excess sludge from the municipal WWTP in Głogów. The moistness of a cake made from the reference sludge (NOG) was 88.0%. After conditioning,

the value of this parameter decreased by 13%, to 79% (NEK30G). A significantly lower moistness of the sludge cake was measured for the sludge from the municipal WWTP in Bolesławiec. A reference sludge cake (NOB) had a moisture content of 79%. After electrochemical conditioning this content decreased by 4%, to 75% (NEK30B).

The reference excess sludge from the WWTP in Gorzów (NO) was characterized by a dry mass content of 7.9 g/dm^3. Electrochemical conditioning caused an increase in the dry mass content of almost 30%, to 10.1 g/dm^3 (NEK30). The reference sludge from Głogów (NOG) contained a dry mass of 4.7 g/dm^3. After conditioning (NEK30G), the dry mass increased by more than 60%, to 7.6 g/dm^3. Much more dry mass was contained in the reference sludge from Bolesławiec (NOB) – 14.8 g/dm^3. After electrochemical conditioning (NEK30B), the dry mass content in the sludge increased by 7%, to 15.9 g/dm^3.

The ash content in the reference sludge from Gorzów (NO) was 2.7 g/dm^3. Electrochemical conditioning caused this to more than double to 5.7 g/dm^3 (NEK30). The reference sludge from the WWTP in Głogów contained 1.7 g/dm^3 of ash (NOG).

After the process of conditioning, the ash content increased almost threefold, to 4.9 g/dm^3 (NEK30G). The excess sludge from the WWTP in Bolesławiec contained 5.1 g/dm^3 of ash (NOB). After conditioning the amount of ash increased by 50%, to 7.6 g/dm^3 (NK30B).

The excess sludge from the municipal WWTP in Gorzów contained 5.2 g/dm^3 organic substance (NO). Conditioning decreased this by 0.8 g/dm^3, to 4.4 g/dm^3 (NEK30). The excess sludge from the WWTP in Głogów contained 3.0 g/dm^3 organic substance (NOG). After electrochemical conditioning the amount of organic substance decreased by

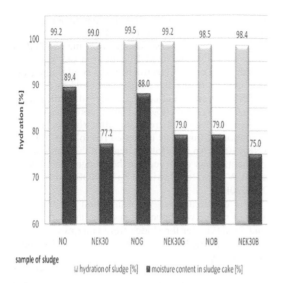

Figure 2. The hydration and moistness of sludge cakes made from the excess sludge before and after the conditioning process.

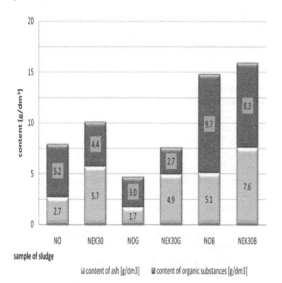

Figure 3. The ash and organic substance contents of the excess sludge before and after the conditioning process.

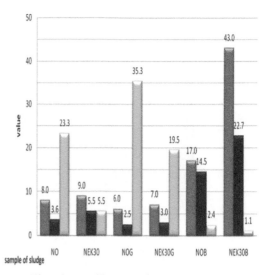

Figure 4. The CST, specific resistance of filtration, and effectiveness of filtration of excess sludge before and after the conditioning process.

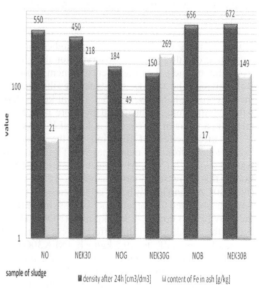

Figure 5. The susceptibility to thicken of the sludge samples before and after treatment after 24 hours of sedimentation and the iron content of the excess sludge before and after the conditioning process.

0.3 g/dm^3, to 2.7 g/dm^3 (NEK30G). The greatest reduction in the amount of organic substance, i.e. 1.4 g/dm^3, was noted for the sludge from the municipal WWTP in Bolesławiec. The reference sludge contained 9.7 g/dm^3 organic substance (NOB), and after conditioning this decreased to 8.4 g/dm^3 (NEK30B).

The CST for the reference excess sludge from Gorzów was 8 s (NO). Electrochemical conditioning caused an increase in CST to 9 s (NEK30). A slightly lower value for the CST was observed for the reference sludge from Głogów at 6 s (NOG). After conditioning, the CST increased to 7 s (NEK30G). The CST parameter for the reference sludge from the WWTP

in Bolesławiec was 17 s (NOB). After conditioning, this time increased to 43 s – an increase of 150% (NEK30B).

The specific resistance of filtration for the reference excess sludge from the municipal treatment plant in Gorzów (NO) was 3.6 × 10^{10} m/kg. Electrochemical conditioning caused an increase in specific resistance of 50%, to 5.5 × 10^{10} m/kg (NEK30). The specific resistance of filtration for the reference sludge from

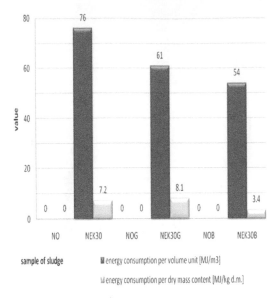

Figure 6. Energy consumption per unit volume of sludge and per dry mass content in the excess sludge.

the municipal WWTP in Głogów was 2.5×10^{10} m/kg (NOG). The process of electrochemical conditioning caused an increase in specific resistance of 20%, to 3.0×10^{10} m/kg (NEK30G). A much higher specific resistance of filtration was observed for the sludge from the municipal wastewater treatment plant in Bolesławiec. The specific resistance of filtration for the reference sludge was 14.5×10^{10} m/kg (NOB). After conditioning, it increased by more than 50%, attaining a value of 22.7×10^{10} m/kg (NEK30B).

The effectiveness of filtration for the reference excess sludge from the municipal WWTP in Gorzów was 23.3 kg/m²h (NO). After electrochemical conditioning this decreased by more than 75%, to 5.5 kg/m²h (NEK30). A significantly better effectiveness of filtration was observed for the reference excess sludge from the WWTP in Głogów – 35.3 kg/m²h (NOG). However, after conditioning this effectiveness decreased by nearly one-half, to 19.5 kg/m²h (NEK30G). A low effectiveness of filtration was noted for the reference sludge from the WWTP in Bolesławiec – 2.4 kg/m²h (NOB), but after electrochemical conditioning it decreased by more than one-half, to 1.1 kg/m²h (NEK30B).

The sludge from the municipal WWTP in Gorzów had a thickness of 550 cm³/dm³ (NO) after 24 hours of sedimentation. Electrochemical conditioning for 30 minutes caused the thickness to decrease to 450 cm³/dm³ (NEK30). Sludge from the WWTP in Głogów had a thickness of 184 cm³/dm³ (NOG) after 24 hours of sedimentation. As a result of the electrochemical conditioning, a slight decrease in its thickness was observed – 150 cm³/dm³ (NEK30G). For the reference sludge from the WWTP in Bolesławiec, its thickness after 24 hours of sedimentation was 656 cm³/dm³ (NOB). Electrochemical conditioning

did not significantly change this parameter; it was measured as 672 cm³/dm³ (NEK30B).

The reference excess sludge from the WWTP in Gorzów (NO) contained 21 g of iron per kilogram of ash (g/kg). After electrochemical conditioning the iron content of the ash increased tenfold to 218 g/kg (NEK30). A similar effect was observed for the sludge from the WWTP in Głogów. The excess sludge (NOG) contained 49 g/kg of iron, and after electrochemical conditioning this value increased fivefold, to 269 g/kg (NEK30G). An eightfold increase in the iron content of the ash was observed for the sludge generated in the WWTP in Bolesławiec; 17 g/kg NOB and 149 g/kg NEK30B.

The electrical energy consumption per unit volume of sludge was balanced and amounted to 76, 61, and 54 MJ/m³ for the sludge from Gorzów (NEK30), Głogów (NEK30G) and Bolesławiec (NEK30B), respectively. It should be noted that the electrical energy consumptions per unit of dry mass of the sludge from the wastewater treatment plants in Gorzów and Głogów were similar and amounted to 7.2 MJ/kg d.m. (NEK30) and 8.1 MJ/kg d.m. (NEK30G). However, for the sludge from Bolesławiec, this energy consumption was significantly lower – 3.4 MJ/kg d.m. (NEK30B).

The conditioning of the excess sludge from the municipal wastewater treatment plant in Gorzów achieved the following effects:

- a slight decrease in hydration, from 99.2% to 99.0%;
- a decrease in the moisture content of the sludge cake by 12.2%, from 89.4% to 77.2%;
- an increase in the dry mass content by almost 30%, from 7.9 to 10.1 g/dm³;
- a more than twofold increase in the ash content, from 2.7 to 5.7 g/dm³;
- a decrease in the organic substance content by 15%;
- an increase in the capillary suction time by 12%;
- an increase in the specific resistance of filtration by more than 50%;
- a more than fourfold decrease in the effectiveness of filtration;
- a slight decrease in the thickness after 24 hours of sedimentation;
- a tenfold increase in the iron content of the ash, from 21 to 218 g/kg of ash;
- an electrical energy consumption for conditioning of 7.2 MJ/kg d.m.

The conditioning of the excess sludge from the municipal wastewater treatment plant in Głogów resulted in the following effects:

- a slight decrease in hydration, from 99.5% to 99.2%;
- a decrease in the moisture content of the sludge cake by 9%, from 88% to 79%;
- an increase in the dry mass content by 60%, from 4.7 to 7.6 g/dm³;
- an almost threefold increase in the ash content;
- a decrease in the organic substance content by 10%;
- an increase in the capillary suction time by 16%;

- an increase in the specific resistance of filtration by 20%;
- a decrease in the effectiveness of filtration by nearly 50%;
- a slight decrease in the thickness after 24 hours of sedimentation;
- a more than fivefold increase in the iron content of the ash;
- an electrical energy consumption for conditioning of 8.1 MJ/kg d.m.

The conditioning of the excess sludge from the municipal wastewater treatment plant in Bolesławiec achieved the following effects:

- a slight decrease in hydration, from 98.5% to 98.4%;
- a decrease in the moisture content of the sludge cake by 4%, from 79% to 75%;
- an increase in the dry mass content by 7%, from 14.8 to 15.9 g/dm^3;
- an increase in the ash content by 50%;
- a decrease in the organic substance content by 15%;
- a more than twofold increase in the capillary suction time;
- an increase in the specific resistance of filtration by 60%;
- a more than twofold decrease in the effectiveness of filtration;
- no change in the thickness after 24 hours of sedimentation;
- a more than eightfold increase in the iron content of the ash;
- an electrical energy consumption for conditioning of 3.4 MJ/kg d.m.

In all the cases considered, electrochemical conditioning did not significantly affect the hydration of sludge. The slight decrease in hydration that did occur was a result of an increase in the dry mass content in sludge. It was due to an increase in the mineral fraction content.

Electrochemical conditioning caused a quite significant decrease in the moisture content of the sludge cakes. Decreases in the moistness of the sludge cakes in the range of from 5% to 14% were observed. This is connected with the change in the proportion of ash and organic substance contents. For each sludge, a decrease in the amount of organic substances, of from 10 to 15%, was observed. This was accompanied by many-fold increases in the mineral substance contents. The less dry mass the sludge contained, the greater was the increase. In the sludge from the WWTP in Głogów, a threefold increase in ash content was observed, and the initial dry mass content was 4.7 g/dm^3. For the sludge from the WWTP in Bolesławiec, the increase in ash content was 50%, when the dry mass content was 14.9 g/dm^3. No direct relationship between the decrease in the amount of organic substance and the amount of dry mass in the sludge was observed. The sludge having the lowest dry mass content showed the least significant reduction in organic substance content, whereas the sludge from the WWTP in

Gorzów and Bolesławiec showed similar decreases in organic substance content after the electrochemical conditioning process. In all cases, an increase in the dry mass of the sludge, inversely proportional to the dry mass content of the reference sludge, was observed. Sludge from the wastewater treatment plant in Głogów was characterized by an increase in its dry mass content by as much as 60%, while that from the WWTP in Bolesławiec showed only a 7% increase in dry mass. The increase in dry mass content was, in each case, a result of an increase in the ash content of the dry mass, as well an increase in the iron content of the ash.

In all cases, a prolonged capillary suction time (CST) was observed. The perceived relationship was that the more the amount of organic substance in the sludge, the greater the increase in CST. An extremely high value for the CST was observed for the sludge from the WWTP in Bolesławiec – a more than twofold increase, as compared with an approximate 15% increase in the CST for the sludge from the other water treatment plants. A similar disadvantageous influence of the electrochemical conditioning was observed on the specific resistance of filtration and the effectiveness of filtration. For the sludge from the wastewater treatment plants in Gorzów and Bolesławiec, the specific resistance of filtration increased by more than 50%, while the effectiveness of filtration decreased – by 75% in the sludge from Gorzów and by 50% in the sludge from Bolesławiec. Deterioration in the parameters characterizing the dewaterability of sludge is associated with the destruction of the flocculent structure of sludge.

For the sludge from the WWTPs in Głogów and Gorzów, an approximate 18% improvement in the susceptibility to thicken was observed. Only the sludge from Bolesławiec, containing high amounts of dry mass, deteriorated its susceptibility to thicken; a 2% deterioration. The changes in the thickness are connected with the release of some of the water which is present mainly in the macroflocule particles.

In all the sludge samples a multiple increase of the iron content of the ash was observed. For the sludge from Gorzów and Głogów, the increases in the iron content of the ash amounted to 200 g/kg. For the sludge from the WWTP in Bolesławiec, an increase in the iron content of the ash of 130 g/kg was observed. This can be attributed to the more than doubling of the ash content of the reference sludge (NOB), and hence there was only a marginal influence from the increase in iron originating from the electrochemical conditioning process.

Electrical energy consumption per unit volume of sludge was similar for all sludge samples, lying in the range 54 and 76 MJ/m^3. Such a relationship was not observed in reference to the electrical energy consumption per unit volume of dry mass. The sludge from Bolesławiec was conditioned using less than half the amount of energy required for the other two samples. This was due to the high amount of dry mass per volume unit. At the same time, it resulted in a reduction in the effects of electrochemical conditioning.

4 CONCLUSIONS

Electrochemical conditioning of excess sludge gives rise to the following outcomes:

a. a decrease in the moisture content of the sludge cakes;
b. an increase in the amount of dry mass of the sludge;
c. an increase in the amount of ash from the dry mass;
d. a decrease in the organic substance content;
e. an increase in sludge mineralization;
f. a prolongation of the capillary suction time;
g. an increase in the specific resistance of filtration;
h. a decrease in the effectiveness of filtration; and
i. a manifold increase in the iron content of the ash.

It is supposed that the deterioration in the parameters that characterize the dewaterability of the sludge may be connected with the destruction of the flocculent structure of the sludge despite the increase in mineralization. Furthermore, it was found that electrochemical conditioning does not significantly affect the susceptibility to thicken or the hydration of excess sludge.

REFERENCES

Aulich, P.J. & Marcinkowski, T.A. 2007. Disintegration of synthetic sludge and excess sludge. Comparison of results. *Sowremennyj Naucnyj Vestnik. Biologija Chimija* 15: 20–31.

Barbusinski, K. & Filipek, K. 2000. Aerobic sludge digestion in the presence of chemical oxidizing agents. Part II. Fenton's reagent. *Polish Journal of Environmental Studies* 9(3): 145–149.

Bień, J.B. 2002. *Osady ściekowe. Teoria i praktyka*. Politechnika Częstochowska.

Bień, J., Wolny, L. & Jabłońska, A. 2001. Sewage sludge preparation for dewatering with ultrasonic field application. *Inżynieria i Ochrona Środowiska* 4(1): 9–16.

Clark, P.B., Nujjoo, I. 1998. Ultrasonic sludge pretreatment for enhanced sludge digestion. *Innovation 2000 Conference*. Cambridge, UK.

Czechowski, F. & Marcinkowski, T. 2006. Sewage sludge stabilisation with calcium hydroxide: Effect on physicochemical properties and molecular composition. *Water Research* 40(9): 1895–1905.

Egemen, E., Corpening, J. & Nirmalakhandan, N. 2001. Evaluation of an ozonation system for reduced waste sludge generation. *Water Science and Technology* 44(2–3): 445–52.

Forster C.F., Chacin E., Fernandez N., 1999, The use of ultrasound to enhance the thermophilic digestion of waste activated sludge, *Environmental Technology*, vol. 21, pp. 357–362.

Hockenberry T.O., 1967, Geometrical formation of the discharge channel in narrow gaps, *SME Paper*.

Marcinkowski, T. 2004. *Alkaliczna stabilizacja osadów ściekowych*, Prace Naukowe Instytutu Inżynierii Ochrony Środowiska Politechniki Wrocławskiej 76. Seria: Monografie 43. Wrocław.

Marcinkowski, T.A. & Aulich, P.J. 2009. Porównanie osadów wstępnych przetwarzanych z wykorzystaniem oddziaływania kawitacji elektrohydraulicznej in: *Efektywne zarządzanie gospodarką odpadami, VIII Międzynarodowe Forum Gospodarki Odpadami. Poznań 2009*: 423–435.

MINISTERSTWO ŚRODOWISKA. *Krajowy Program Oczyszczania Ścieków Komunalnych (KPOŚK)*. Warszawa 2003.

MONITOR POLSKI NR 90, Poz. 946, 29. grudnia 2006 r. Krajowy Plan Gospodarki Odpadami 2010.

Müller, J. 1996. *Mechanischer Klärschlammaufschluß*, PhD thesis, Institut für Mechanische Verfahrenstechnik. Technical University of Braunschweig.

Müller, H. 1965. Contribution to spark-erosion phenomena, *Elektrowärm* 23(3).

Neyens, E. & Baeyens, J. 2003. A review of thermal sludge pre treatment processes to improve dewaterability. *Journal of Hazardous Materials* B98: 51–67.

Neyens, E., Baeyens, J., Weemaes, M. & De Heyder, B. 2003. Pilot-scale peroxidation (H_2O_2) of sewage sludge. *Journal of Hazardous Materials* B98: 91–106.

Pinnekamp, J. 1989. Effects of thermal pre-treatment of sewage sludge on anaerobic digestion. *Water Science and Technology* 21: 97–108.

ROCZNIK STATYSTYCZNY, 2007 rok, Główny Urząd Statystyczny Warszawa 2009.

Schumacher, B.M. 1965. *Removal behavior and wear when spark eroding steel with condenser- and semiconductor-pulse generators*. Thesis, RWTH, Aachen.

Schumacher, B.M. 2004. After 60 years of EDM the discharge process remains still disputed. *Journal of Materials Processing Technology* 149: 376–381.

Singermann, A.S. 1956. About the development of the discharge channel in electroerosive metal machining. *Journal of Technical Physics* 26(5): 107–112.

Wojciechowska, E., Kowalik, P. 2003. Zastosowanie promieniowania mikrofalowego do kondycjonowania osadów ściekowych. *Inżynieria i Ochrona Środowiska* 6(2): 167–178.

Environmental Engineering III – Pawłowski, Dudzińska & Pawłowski (eds)
© 2010 Taylor & Francis Group, London, ISBN 978-0-415-54882-3

Bone sludge as a raw material in the production of hydroxyapatite for biological applications

A. Sobczak, E. Błyszczak, Z. Kowalski & Z. Wzorek

Institute of Inorganic Chemistry and Technology, Krakow University of Technology, Krakow, Poland

ABSTRACT: The paper presents the results of the calcining process of deproteinised and defatted bone pulp (bone sludge). A first stage was conducted at a temperature of 600°C in a rotary kiln, while the second took place within the temperature range 600°C to 950°C. The products of the process were analysed using x-rays and a spectrophotometer. Calcium content was determined by titration, while phosphorus content was determined by spectrophotometer. In the products obtained, hydroxyapatite was the only crystalline phase identifiable by x-rays. Calcium and phosphorus contents remained at 39% and 17.5%, which corresponds to the Ca/P ratio of non-stoichiometric hydroxyapatite.

Keywords: Bone sludge, calcining, hydroxyapatite, physicochemical properties.

1 INTRODUCTION

Changes resulting from industrial development exert significant influences upon the conditions of the environment and its chemical composition. A continuous increase in population worldwide is also a reason for environmental changes on both local and global scales. For many years, all activities related to industry and production did not include a rational exploitation of natural resources or balanced waste management. Only the dramatic effects of the destruction of the natural environment led to the emergence of new activities geared to environmental protection. The reduction of all types of generated waste, as well as the processing of waste before emission, did not solve the problem altogether. They simply decreased the rate at which the surrounding environment was degraded. The new idea of 'cleaner production' then appeared, which assumed that waste should be considered as a source of resources for the production of other quality products (Pezacki 1991, Kowalski et al. 2003, Kowalski & Kulczycka 2004, Deydier et al. 2005).

Meat industry waste, and bone waste in particular, came to be perceived as a potential source of phosphorus. Hydroxyapatite is a natural component of both human and animal bones. Its spatial structure and porosity, as well as the molar ratio of calcium to phosphorus, are also similar in human and animal bones. These similarities make it potentially possible for hydroxyapatite obtained from meat industry waste to be used as a biomaterial for both surgical and stomatological implants. Hydroxyapatite obtained from meat industry waste can also be used in otolaryngology, as a material from which components of medical equipment are built and as a tissue-bearing area in tissue engineering (Kowalski &

Krupa-Żuczek 2007, Kowalski et al. 2007, Krupa-Żuczek et al. 2008).

Hydroxyapatite is considered a biomaterial due to its particular qualities, such as an appropriate chemical and phase composition, microstructure, a high level of biocompatibility as well as a lack of cytotoxic and carcinogenic effects. Hydroxyapatite demonstrates a considerable level of bioactivity, which results in an ability to join directly with bone tissue (Ślósarczyk 1997, Orlovskii et al. 2002). The mineral component of bone is similar to hydroxyapatite, but contains carbonate, fluoride, magnesium, sodium and other ions (Orlovskii et al. 2002, Wopenka & Pasterias 2005, Rey et al. 2007). It has the chemical formula $Ca_{10-x}(PO_4)_{6-x}$ $(HPO_4^{2-}$ or $CO_3^{2-})_x(OH)_{2-x}$ (Rey et al. 2007).

Using bone waste to obtain hydroxyapatite for the production of implants is very beneficial for many reasons. First, it is a solution to the problem of a huge mass of dangerous waste, as this waste is treated as a raw material for another production. Second, because of structural similarities, using natural hydroxyapatite has the potential to reduce the defence reactions of the body after the application of an implant, similar to a case of implanting an autogenic bone. Bioactivity and osteoinductance contribute to a faster and more stable process of bonding with the patient's bone. This guarantees a faster recovery, more durability and greater comfort. Economic reasons also play an important role. The costs of synthesis using wet methods are very high, mainly because of the application of very diluted substrates (Knychalska-Karwan & Ślósarczyk 1994).

This paper is an attempt to analyze the influence of the calcining temperature upon the physicochemical properties of bone ash in respect of their application as biomaterials.

2 MATERIALS AND METHODS

The meat industry bone waste analysed here is bone sludge – deproteinised and defatted bones – obtained from the Zakład Mięsny DUDA BIS company in Sosnowiec, Poland.

Following mineralisation in a mixture of concentrated hydrochloric and nitric acids, the phosphorus content was determined using the differential photometric method in accordance with the standard PN-88/C-87015, using a UV-VIS Marcel Media spectrophotometer. Calcium content was determined by titration according to the standard PN-64803:1997 with respect to the mixed indicator calcein and thymolphthalein. The phosphorus content soluble in 2% citric acid was determined by the differential photometric method.

The phase composition was determined by the x-ray method with a Philips X'Pert diffractometer equipped with a graphite monochromator, using Cu Kα 1.54 Å, Ni filter (40 kV, 30 mA). The Scimitar Series FTS 2000, a Fourier Transform Infrared (FT-IR), spectrophotometer produced by Digilab, was used in the medium infrared range, 400–4000 cm^{-1}, for the research. The sample amount weighed 0.0007 g, and this was pressed with KBr.

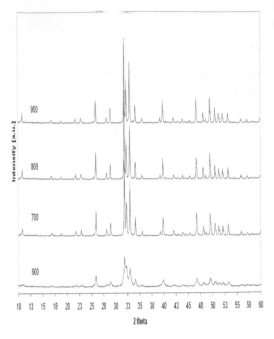

Figure 1. Diffractograms of the products of calcining bone sludge.

3 RESULTS AND DISCUSSION

3.1 Obtaining hydroxyapatite

A method of obtaining hydroxyapatite from deproteinised and defatted pork bones (often referred to as bone sludge) (Kowalski et al. 2007) was developed in the Institute of Inorganic Chemistry and Technology at the Cracow University of Technology. The material was exposed to two stages of a calcining process. The first stage was conducted at a temperature of 600°C in an air atmosphere in a rotary kiln equipped with a gaseous heater. In the second stage, the homogeneous material with a sieve fraction of less than 0.15 mm was exposed to calcining within the temperature range 600°C to 950°C ($\Delta T = \pm 50$°C). The material was exposed to the maximum temperature for 2 hours in an electrically heated stationary kiln in an air atmosphere. Two series of measurement were taken out for all the experiments.

3.2 The influence of the calcining temperature upon the psychochemical properties hydroxyapatite

The colour of the samples depended on the calcining temperature. The darkest material (dark grey) was obtained when a temperature of 600°C was applied; a light grey colour resulted from a temperature of 700°C; and a white product was obtained at the highest temperatures of 800°C and 900°C. The materials lightened in colour with increasing temperature. The samples obtained at lower temperatures are darker coloured because of traces of organic matter, which are not present in the whiter coloured samples obtained at higher temperatures.

In both measurement series performed, the increase in the calcining temperature is accompanied by a greater loss of weight. This ranged from 6.40% to 10.67%. The loss of weight is related to both the combustion of organic traces and to the decomposition of carbonate groups present in natural bone.

The analysis of the phase composition of the products resulting from bone sludge calcining at 600 to 950°C was carried out following a material exposure time of 2 hours. In all the materials obtained, only hydroxyapatite (HAp) is present in a crystalline phase, as indicated by the x-ray method. X-ray diagrams shows reflections corresponding to hydroxyapatite only (standard JCPDS 9-432), however, at higher temperatures, thermal decomposition of HAp could proceed and other phases might appear (calcium oxide (CaO), tricalcium phosphate (TCP), tubing tricalcium phosphate (TTCP)) in small amounts below the detection limits of the diffractometer. Moreover, the dehydroxylation process could also proceed, resulting in a losing of the OH$^-$ groups and the appearance of oxyapatite, which is indistinguishable from HAp using the x-ray method (Ooi et al. 2007). The degree of crystallinity of the hydroxyapatite is connected with the calcining temperature and increases with increasing temperature. This maybe related to the recrystallisation and stabilisation of the structure at a higher temperature. Figure 1 depicts *diffractograms* of the products of the calcining process.

Table 2 presents the results of analyses of the calcium and phosphorus contents as well as the molar

Table 1. Loss of weight during the calcining process in a stationary kiln.

Temperature [°C]	600	650	700	750	800	850	900	950
Loss of weight Series I [%weight]	6.81	8.00	8.99	9.42	10.20	10.28	10.64	10.38
Loss of weight Series II [%weight]	6.40	7.22	9.15	10.11	10.41	10.28	10.52	10.67

Table 2. Calcium and phosphorus content and Ca/P molar ratio in hydroxyapatite obtained from calcining bone sludge.

Calcining temperature [°C]	Phosphorus content [%weight]		Calcium content [%weight]		Molar ratio Ca/P	
	Series I	Series II	Series I	Series II	Series I	Series II
600	17.70	17.59	39.42	39.55	1.70	1.74
650	17.56	17.63	39.67	39.30	1.75	1.72
700	17.68	17.57	39.42	39.56	1.72	1.74
750	17.67	17.56	39.66	39.56	1.73	1.74
800	17.56	17.65	39.61	39.76	1.74	1.74
850	17.96	17.91	39.46	39.39	1.70	1.70
900	17.81	17.79	39.62	39.65	1.72	1.72
950	17.90	18.05	39.96	39.70	1.72	1.70

ratio of calcium to phosphorus (Ca/P). The phosphorus and calcium contents rise together slightly with increasing calcining temperatures, ranging between 17 and 18% for phosphorus and around 39% for calcium. A lower phosphorus content in relation to the stoichiometric HAp may be caused by the substitution of carbonate anions for phosphate groups in the anion sub-lattice. The Ca/P molar ratios correspond to non-stoichiometric hydroxyapatite. The essential difference between synthetic and animal origin materials was that the natural hydroxyapatite showed a higher Ca/P ratio than the synthetic material (Haberko et al. 2006, Ooi et al. 2007). In general, the Ca/P molar ratio of all samples was higher than that of stoichiometric hydroxyapatite. In the mineral phase of natural bone, in addition to phosphorus, calcium and oxygen, minor amounts of sodium, magnesium, potassium and carbon are present. These influence the Ca/P molar ratio. The increased molar ratio may also be caused by a partial decomposition of the natural hydroxyapatite with preservation of the hydroxyapatite structure (Haberko et al. 2006).

The phosphorus content soluble in 2% citric acid decreases with increasing calcining temperatures. At higher temperatures a sintering process of the material grains takes place, which makes it more difficult for the solvent particles to penetrate the HAp micropores. The decreased solubility may also be connected with an increased degree of crystallinity caused by an increase in the process temperature. The results are presented in Figure 2.

Bands characteristic of hydroxyapatite are present in all the samples obtained on the FT-IR spectrum. Bands within the wave number range 1200 to 1000 cm^{-1} correspond to the vibrations of the PO_4^{3-} group (asymmetric, stretching), which demonstrate

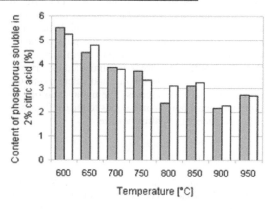

Figure 2. Phosphorus content soluble in 2% citric acid.

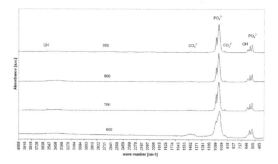

Figure 3. FT-IR spectra of selected products of the calcining process.

the highest levels of intensity. Bands of low intensity, within the wave number range 570 to 560 cm^{-1}, correspond to the vibrations of the PO_4^{3-} group (asymmetric). A small band within the range of large wave numbers, 3670 to 3570 cm^{-1}, corresponds to the stretching vibrations of the OH$^-$ group. Additional bands corresponding to the vibrations of the CO_3^{2-} groups are visible on the spectra of materials obtained at lower temperatures. Bands within the range 1457 to 1420 cm^{-1} and the band at 875 cm^{-1} correspond to the B type hydroxyapatite, where carbonate groups are incorporated in place of the PO_4^{3-} anions. The weak band at 1550 cm^{-1} is evidence of the CO_3^{2-} anions incorporated in place of the hydroxyl groups – the A type hydroxyapatite. Polymorphic types of calcium carbonate were not detected. There are no bands characteristic of aragonite (713 and 700 cm^{-1}), calcite (712 cm^{-1}) or vaterite (745 cm^{-1}).

4 CONCLUSIONS

The experiments performed on calcining bone sludge waste within the temperature range of 600 to 950°C demonstrated that it is possible to obtain hydroxyapatite as a result of the thermal utilisation of deproteinised bone waste. The product of the calcining process contains hydroxyapatite as the only crystalline phase identifiable by x-ray. From research in infrared spectrophotometry, it may be assumed that carbonate groups are present in the anion sub-lattice. The calcining temperature influences the physicochemical properties of the hydroxyapatite. An increase in the calcining temperature is accompanied by a rise in the total content of phosphorus and calcium, but the solubility in acids decreases. The molar ratio of calcium and phosphorus corresponds to the Ca/P ratio in non-stoichiometric hydroxyapatite.

REFERENCES

Deydier, E., Guilet, R., Sarda, S. & Sharrock, P. 2005. Physical and chemical characterization of crude meat and bone meal combustion residue: "waste or raw material?". *Journal of Hazardous Materials* 121: 141–148.

Haberko, K., Bucko, M., Brzezinska-Miecznik, J., Haberko, M., Mozgawa, W., Panz, T., Pyda, A. & Zarebski, J. 2006. Natural hydroxyapatite-its behaviour during heat treatment. *Journal of the European Ceramic Society* 26: 537–542.

Knychalska-Karwan, Z. Ślósarczyk, A. 1994. *Hydroksyapatyt w stomatologii*. Krakmedia: Kraków.

Kowalski, Z. & Krupa-Żuczek, K. 2007. A model of meat waste management. *Polish Journal of Chemical Technology* 9(4): 91–97.

Kowalski, Z. & Kulczycka, J. 2004. Cleaner production as a basic element for a sustainable development strategy. *Polish Journal of Chemical Technology* 6(4): 35–40.

Kowalski, Z., Wzorek, Z., Krupa-Żuczek, K. & Sobczak, A. 2007. Możliwości otrzymywania hydroksyapatytu poprzez kalcynację półproduktów kostnych z przemysłu mięsnego. *Inżynieria Stomatologiczna Biomateriały* 4(1): 7–11.

Kowalski, Z., Wzorek, Z. & Kulczycka, J. 2003. Environmentally safe production system on the example of chromium compound. *Environmental Engineering Studies. Polish Research on the Way to the EU*. Kluver Academic/Plenum Publishers: New York.

Krupa-Żuczek, K., Kowalski, Z. & Wzorek, Z. 2008. Manufacturing of phosphoric acid from hydroxyapatite contained in the ashes of incinerated meat-bone wastes. *Polish Journal of Chemical Technology* 10(3): 13–20.

Ooi, C.Y., Hamdi, M. & Ramesh, S. 2007. Properties of hydroxyapatite produced by annealing bovine bone. *Ceramics International* 33: 1171–1177.

Orlovskii, V.P., Komlev, V.S. & Barinov, S.M. 2002. Hydroxyapatite and Hydroxyapatite-Based Ceramics. *Inorganic Materials*, 38(10): 973–984.

Pezacki, W. 1991. *Przetwarzanie surowców rzeźnych. Wpływ na środowisko przyrodnicze*. PWN: Warszawa.

PN-88/C-87015 Polish Standard. 1988. Chemical fertilizer. Laboratory test of phosphate content.

PN-P-64803:1997. Polish Standard. 1997. Feedstuffs. Forage-feed phosphates.

Rey, C., Combes, C., Drouet, C., Sfihi, H. & Barroug, A. 2007. Physico-chemical properties of nanocrystalline apatites: Implications for biominerals and biomaterials. *Materials Science and Engineering* C 27: 198–205.

Ślósarczyk, A. 1997. *Bioceramika hydroksyapatytowa* Biuletyn Ceramiczny nr 13 Ceramika 51, Polskie Towarzystwo Ceramiczne, Kraków.

Wopenka, B. & Pasteris, J.D. 2005. A mineralogical perspective on the apatite in bone. *Materials Science and Engineering* C 25: 131–143.

Environmental Engineering III – Pawłowski, Dudzińska & Pawłowski (eds)
© 2010 Taylor & Francis Group, London, ISBN 978-0-415-54882-3

Reuse of coal mining wastes: environmental benefits and hazards

S. Stefaniak & I. Twardowska
Institute of Environmental Engineering, Polish Academy of Sciences, Zabrze, Poland

ABSTRACT: Mechanized mining of hard coal results in generating waste estimated at 30–50% of coal output. Although not considered a hazardous waste, it is not environmentally neutral, mostly due to the content of reactive sulfides, and in freshly generated waste also due to chloride salinity. The major factors determining the extent of the adverse environmental impact of coal mining waste include content and kind of soluble compounds, sulfate content and reactivity, neutralization potential and water balance of the dump, air penetration conditions and granulation of the material. Re-mining of waste dumps for residual coal extraction and utilization of waste as a common fill causes disturbance of the primary layers in the dump and extensive exposure of waste to air and water. This paper presents the environmental impact, benefits and hazards related to coal mining waste reuse exemplified in re-mining the Bukow dump in the Upper Silesia coal basin in Poland.

Keywords: Hard coal; coal mining waste; coal re-extraction; disposal and reuse of mining waste; environmental impact.

1 INTRODUCTION

Among mineral commodities of the world, hard coal is extracted in the highest amount (USGS, 2005). Coal is the most important global energy source, providing 26% of the world's primary energy and 41% of electricity. The top and continuously growing hard coal producer is China with 2549 Mt in 2007, while Poland with declining annual output of 90 Mt holds the eighth position among coal producers, being the biggest hard coal producer in European Union (WCI 2008). The amount of wastes generated and their characteristics depend both on the geologic conditions in the mining area, and on the method of coal extraction. Mechanized mining with shearers results in generation of 30–50% waste with respect to coal output. In Poland, coal mining waste is the largest type bulk waste. In 2007, 39.3 Mt was generated, while 587.5 Mt was stored at dumps (GUS 2008). High extent of coal mining waste re-use, mostly as common fill in engineering construction and for residual coal extraction by physical methods is associated with the re-mining of existing waste dumps and re-deposition of rock material in different conditions that significantly alters pollutants generation, mobilization and migration processes in waste, and adversely affects the aquatic environment, in particular groundwater. Long-term observations in the local ground- and surface water monitoring networks and porewater analysis along the profiles of the re-mined waste dumps, among them the one of the Anna coal mine in Bukow, Poland, showed that coal re-extraction process induces significant alteration in the hydrogeochemical profiles of the dump. A considerable intensification of sulfide decomposition

and sulfate and acidity generation along the profile of a re-disposed layer was observed, at the simultaneous increase of carbonate mineral buffering effect due to increase of their exposure extent, up to the neutralization of the previously acidic mineral. This process was called "waste activation" that resulted from the increase of exposure surface of both acidifying and buffering compounds. It should be anticipated that the observed restoration of the material buffering is only temporary (Szczepańska & Twardowska 2004). Re-disposed material, besides alteration of distribution and concentration of dissolved constituents in pore solution along the vertical profile of a waste layer, shows also a considerable increase of infiltration water flow rate in the first year of re-disposal and a general increase of contaminant loads, mostly of sulfates, and of iron and manganese migration to the groundwater.

On the other hand, re-mining of waste dumps is a form of contribution to the program of the disturbed post-mining areas' revitalization, which considers performing land management in accordance with the novel approach to the land planning. The program is aimed at conversion of often unaesthetic and harmful dumps into green or urban areas. Simultaneously, reduction of the amount of the disposed material is about 10% whereas recycling waste disposed at the dumps is almost 100%. This recycled waste yields two raw materials: coal for energy production and stone aggregate used for engineering constructions as a common fill.

The benefits from bulk waste reuse seem to be obvious, although its prerequisite is the proven environmental safety of any reuse application. For these reasons, long term studies on the environmental impact of coal mining waste have been conducted, with a focus on the identification of its extent and mechanism, in order to prevent, intercept and efficiently control possible adverse effects of waste management. The studies are aimed to enhancing benefits of bulk waste reuse in environmentally safe applications.

The presented study is a part of a larger research project carried out at different coal mining waste dumping and reuse sites, among them sites with intensive and diverse waste management practices. The impact of coal mining waste reuse on the environment, possible benefits and hazards were exemplified in three methods of waste utilization: (1) coal extraction from waste; (2) reuse of coal mining discards for construction of embankments; (3) ground leveling with waste rock disposal below the water table.

2 OBJECTS AND METHODS

2.1 Object characteristics

All these reuse methods were studied at one selected object that was a coal mining waste dump of the Anna coal mine in Bukow (Figure 1). The Bukow coal mining waste dump of a total area 44.8 hectares is located in the Western part of the Upper Silesia coal mining basin (USCB, Poland) in the area of a worked out gravel quarry in the Odra River valley.

From 1976 up to 2001, discards from coal preparation processes were disposed at the dump. The waste rock originated from the coal seams of 600 and 700 group in the carboniferous strata of the Namurian A series mined by the Anna coal mine. The disposed discards primarily comprised only coarser waste with particle size from 20 to 200 mm from heavy media suspension separation (56.6%) and finer discards with particle size ranging from 1 to 20 mm from jigs (40%).

Since 1995 besides discards, small amounts of flotation slurry with mineral particles <1 mm (4.7%) and other fine waste (1.6%, mostly fly ash and slag from the mine power plant) were also disposed at this dump. The disposed wastes are rather stable with respect to particle size and the petrologic and mineralogical composition. In the grain size distribution, a coarse fraction prevails (93.7% wt., in this fraction >50 mm comprises 43.8% wt.). Wastes are resistant to weathering decomposition to clay fractions, show high hydraulic conductivity and are permeable to air. Therefore, at this dump, free infiltration of precipitation water, and easy access of air to the internal parts of the dump occurs.

Average sulfur content accounted for $S_t = 0.82\%$ wt., while sulfide-S responsible for acid generation was as high as 0.68% wt. The material has a relatively high Neutralization Potential Ratio $NPR = NP/AP = 2.37$, which means that the buffering capacity (Neutralization Potential NP) of this waste over 2–fold exceeds Acid Potential AP determined by sulfide-S content. Thus the waste is not susceptible to acidification and ARD generation, and can be considered "relatively safe" in this respect (Price & Errington 1998).

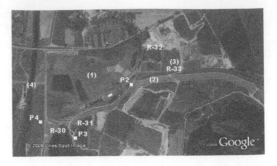

Figure 1. Objects location
(1) – coal mining waste dump in Bukow; (2) – embankments of the Buków polder; (3) – reservoir filled by mining waste (4) – the Odra river; P – sampling points for groundwater; R – water reservoir.

The residual coal extraction started in 1998 at the over 20 years' old dump and lasted for 11 years, gradually covering all the dump area, with a planned closure in 2009. Currently, the extraction operations have been completed at the almost all dump area (Twardowska & Szczepańska 1995).

Residual coal extraction comprises re-mining of the disposed waste rock, its replacement, transport, repeated screening with use of different sieve mesh, washing, elutriation in a settling cone, heavy media (magnetite) suspension enrichment process, and finally re-disposal of the processed rock material. As a result of technological processes, the disturbed waste become exposed to the atmospheric conditions (precipitation and air).

2.2 Sampling

2.2.1 Waste rock sampling
Sampling procedure was accomplished according to CEN/TR 15310-1,2,3,4 standards.

In order to define the changes occurring in the dump, chemical composition of pore solutions along the vertical profile of the dump and its alteration as a result of weathering and technological processes of re-mining, five drillings 10–17 m deep were performed at the flat top of the dump (three in older, undisturbed part of the dump, and two in the re-mined part) (Figure 1). Samples of rock material of about 25 kg each were taken at the dump from the top layer 0–0.25 m, and along the vertical profiles directly from the drill at 0.5 m, 1.0 m, 1.5 m, 2.0 m and then at every 1.0 m. After macroscopic analysis the samples were packed into double HDPE bags, sealed to prevent moisture loss and transported to the laboratory.

2.2.2 Ground water and surface water sampling
Groundwater and surface water were sampled four times during each hydrologic year since 1997 from the monitoring wells and profiles at the local streams and the Odra River within the Local Ground- and Surface Water Monitoring Network with use of a mobile laboratory van Peugeot J-51400, along with water table measurements and pH and electrical conductivity

312

analysis (EN 27888). The sampling locations are presented in Figure 1.

2.2.3 Analytical method

At the laboratory, samples were analyzed for moisture content W_n. Next, the pore solutions were extracted by pressure method, filtered through Sartorius 0.45 μm membrane filters and analyzed for pH, conductivity γ_{25}, 9 major anions and 29 cations including trace metals with use of ICP-MS (Perkin-Elmer model Elan DRC-e). Analyses were performed in accordance with EN-12506:2003.

Carboniferous rock samples from the dump profile were analyzed for hydraulic conductivity coefficient k.

Groundwater and surface water analysis in the field comprised measurements of temperature, pH and conductivity γ_{25}, filtration through Sartorius 0.45 μm membrane filters and preservation of samples for further laboratory analysis. Water samples were next transported to the laboratory in accordance with the sampling protocol and EN-5667-1, EN-5667-11, and EN-5667-6. In the laboratory, water samples were analyzed for the basic chemical composition, including pH (PN-90/C-04540/02), electric conductivity (EN ISO 27888), alkalinity (EN-ISO 9963), sulfates (ISO 9280:2002) chlorides (ISO 9297:1994) and 14 elements with use of FAAS technique (AA-Scan 1 Thermo Jarrell Ash), following ISO 7980, ISO 9964-1 and ISO-8288 standards. All analyses were subject to a routine QA/QC procedure, with an acceptable error ±5%.

3 RESULTS AND DISCUSION

3.1 Residual coal extraction from waste

The results confirmed high hydraulic conductivity of waste carboniferous rock. Mean moisture content W_n was 5.4% in the undisturbed profile (Figure 2) and 6.9% in the profiles of re-disposed waste (Figure 3).

The pattern of hydrogeochemical profiles in the undisturbed and disturbed dump layers considerably differed, while character of these differences showed high similarity to other objects of the same kind, e.g. to the Smolnica dump of the Szczyglowice mine (Szczepańska & Twardowska 1999). This confirms a regular character of the observed patterns and allows generalization of conclusions. In particular, the hydrogeochemical profiles in undisturbed waste layer have a characteristic pattern from the vertical re-distribution of dissolved constituent loads, which migrate in the anthropogenic vadose zone along with subsequent portions of infiltrating precipitation water that transports contaminant loads into the deeper parts of the layer (Figure 2). The highest concentrations and loads of constituents occurred at the bottom part of the profile, close to the dump base, if the re-distribution process comprised the whole layer. A characteristic feature of these profiles was the domination of sulfates, which are produced by sulfide oxidation, and almost complete lack or low chloride concentrations in the ionic composition.

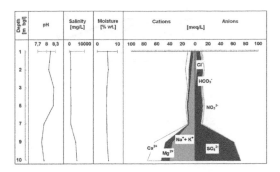

Figure 2. Hydrogeochemical profile of pore solutions in undisturbed layer of coal mining waste.

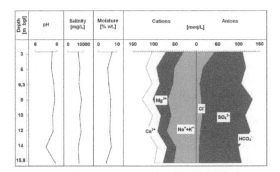

Figure 3. Hydrogeochemical profile of pore solutions in re-disposed layer of coal mining waste.

This indicates that the material is old and has been washed out several times. Cationic composition of pore solutions in general showed domination of Ca^{2+} and Mg^{2+} ions that confirms buffering with carbonate minerals (calcite, dolomite).

Inversely, the pattern of hydrogeochemical profiles in the re-disposed waste layers was uniform in the whole profile due to waste mixing and saturation with water in the technological process, which resulted in the lack of distinct phases and of a gradual saturation of water retention capacity in the layer and of the vertical redistribution of constituent loads associated with this process (Figure 3).

Pore solutions were alkaline (also in the weakly buffered material susceptible to acidification, as it occurred at the Smolnica dump (Szczepańska & Twardowska 1999), and showed higher chloride concentrations than in undisturbed profiles, and what was particularly characteristic was the higher concentrations of sulfates.

3.2 Reuse of coal mining waste for construction of embankments

Leaching of contaminants from engineering constructions made of coal mining waste was exemplificd in the construction of embankments for the antiflood Bukow polder located in the vicinity of the dump, within a distance of several tenths of meters from its eastern edge (Figure 1). An internal part

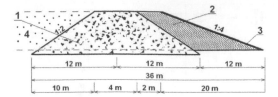

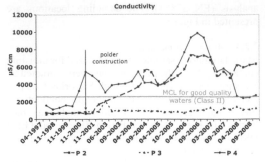

Figure 4. Scheme of the embankment for the Bukow polder. 1 – polder embankment; 2 – clay layer; 3 – fertile soil layer 10 cm thick sowed with grass mix; 4 – reclamation area

Figure 5. Conductivity distribution in groundwater in the vicinity of coal mining dump; (P – sampling points for groundwater).

of the embankment was constructed of carboniferous waste rock from the process of residual coal extraction. The total amount of re-mined waste re-used for the embankment construction was $120,000\,m^3$. The embankment is 1,400 m long and about 5 m high (Figure 4),

Regular assessment of water quality conducted since 1997 within the Local Monitoring Network for Surface- and Groundwater allowed for observations of groundwater quality of Quaternary aquifer in the area of polder embankments. Monitoring wells P-2 and P-3 (Figure 1) were primarily designed for observation of background chemical status of groundwater flowing from E direction up-gradient of the dump. Although, at the end of 2001, when construction of polder embankments started, the P-2 monitoring well became an observation point of their direct impact on the groundwater quality. The surface water quality within the possible impact of the embankments was monitored at the reservoirs R-30 and R-31 that are hydraulically connected with the same Quaternary aquifer (Figure 1). From the time of the polder construction start in 2001 up to the end of 2006, the Odra River in this reach did not exceed the alarm water level. In June 2007, the Odra River flooded the polder Bukow area that caused a contact of both polder and reservoir R-31 embankments with flood water (reservoir R-31 is a source of water for the technological process of residual coal extraction from the dump).

The embankment construction caused practically instantaneous increase of dissolved solids content TDS (indicated by electrical conductivity values) and visible decrease of pH resulting from sulfide oxidation and leaching of sulfates to the groundwater (Figure 5, 6). Before 2001, the groundwater chemical status in this area was good and typical for the Quaternary aquifer, of SO_4-Ca type, at somewhat elevated TDS (up to $680\,mg/dm^3$), mostly due to sulfate salinity. Since 2001, almost 10-fold increase of sulfate salinity occurred, reaching maximum in 2006 when electric conductivity ranged from 2740 to $7310\,\mu S/cm$. High concentrations mostly of sulfates, but also of chlorides, sodium, iron and manganese were also observed. Other trace elements did not exceed MCL, while pH values were within a permissible limit of pH 6.0–7.5.

Surface water reservoirs in the area represent in fact uncovered exposed Quaternary aquifer hydraulically connected with the Bukow polder. Flooding of the Bukow polder with the Odra River waters in July 2007,

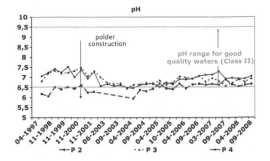

Figure 6. ph distribution in groundwater in the vicinity of coal mining dump; (P – sampling points for groundwater).

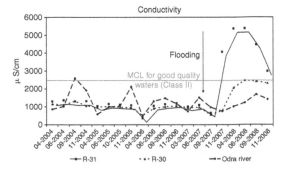

Figure 7. Conductivity distribution in surface water in the vicinity of coal mining dump; (R – water reservoir).

caused thus instant adverse qualitative alterations both in surface- and groundwater waters in the area. Such indicators of salinity as electrical conductivity and sulfate concentrations increased to levels that never had been observed from the beginning of the regular monitoring, in particular, in reservoir R-30 which stretches along the River Odra bed in S direction from the dump.

The conductivity increased to values characteristic to poor quality waters, up to $2470\,\mu S/cm$, sulfate concentrations to $635\,mg/dm^3$ and chlorides up to $390\,mg/dm^3$. Metal concentrations did not exceed MCL values. Even greater adverse water quality alterations occurred in nearby reservoir R-31, where at the end of 2007 dramatic increase of the major salinity parameters was observed.

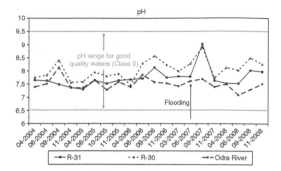

Figure 8. pH distribution in surface water in the vicinity of coal mining dump; (R – water reservoir).

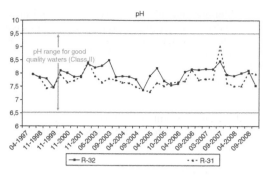

Figure 10. pH distribution in surface water in the vicinity of coal mining dump; (R – water reservoir).

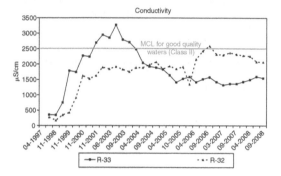

Figure 9. Conductivity distribution in surface water in the vicinity of coal mining dump; (R – water reservoir).

Conductivity values increased there up to $5400\,\mu S/cm$, sulfate concentrations up to $1390\,mg/dm^3$ and chlorides up to $880\,mg/dm^3$. Metal contents, except Mn and Fe, were within a range of good quality. The surface water deterioration resulted from the contact of flood waters with the non-insulated embankment which was constructed of coal mining wastes and had not yet been covered with a layer of impermeable material.

3.3 Ground leveling with disposal of waste below the groundwater table

Sulfidic waste re-use for ground leveling below the water table is one of the environmentally safe recycling (or disposal) options for such wastes, recognized and recommended by BREF (2004). It effectively eliminates waste contact with atmospheric air by cutting off the oxygen source required for sulfide oxidation. This situation is exemplified by filling water reservoir R-33 at the NE side of the Bukow dump with buffered coal mining waste after residual coal extraction process (Figure 1).

Figures 9 and 10 illustrate impact of such way of waste re-use on the water quality in this and in another hydraulically connected reservoir R-32 situated in the direction of groundwater flow from reservoir R-33.

Due to the high soluble salts content (mostly actively generated sulfates and residual chloride loads), waste disposal into water reservoirs, despite terminating generation of new sulfate loads, caused deterioration of water quality in the affected area, resulting from the release and migration in groundwater of salt loads that these wastes already contained. Water quality reduction is associated with the stage of filling the R-33 reservoir with coal mining waste and release of its predominant sulfate and lesser chloride loads, their diffusion to the water of the filled reservoir, and further migration in groundwater in the direction of R-32 reservoir receiving these waters. The stage of a gradual improvement of groundwater quality since 2003 resulted from a thorough exchange of water in the reservoir and was related to the rate of contaminants release from the carboniferous rock material through diffusion and convection to the groundwater stream. In the presented case, the period of water quality deterioration due to sulfate leaching lasted for four years (up to the second half of 2003). Currently, the process of a gradual decrease of sulfate salinity in the R-33 reservoir (at a lower extent due to dilution), and adequate time-lagged alterations of water quality were observed in the R-32 reservoir situated downgradient of the R-33 reservoir in the direction of groundwater flow. The presented example shows that in the case of sub-surface disposal (below the water table), the stage of release of contaminant loads occurring in a material at the moment of disposal should be also considered.

4 CONCLUSIONS

Residual coal extraction from coal mining waste disposed at the dams, and reuse of these wastes in civil engineering as common fill is associated with re-mining of existing dumps and re-disposal of waste rock material in different, altered conditions. This might exert a considerable impact on the processes of generation, release and migration of contaminants from waste and on the receiving aquatic environment. Objects constructed of re-disposed material showed strong adverse impact on the ground- and surface water mostly due to the high sulfate salinity. The presented mode of a negative impact basically of two major compounds, i.e. chlorides and sulfates, as well as iron and manganese that are products of sulfide decomposition, is a characteristic feature of coal extraction waste in the Upper Silesia coal basin (USCB), including also buffered material not susceptible to acidification. The extent of water quality deterioration caused by intensification of sulfide decomposition appeared to be very high. Long-term character of these processes is particularly environmentally problematic and should be taken into consideration at the extraction waste reuse.

REFERENCES

Appelo, C.A.J. & Postma, D. 2007. *Geochemistry, groundwater and pollution*. 2nd Edition, A.A. Balkema Publishers. Leiden: The Netherlands.

BREF 2004. Reference Document on Best Available Techniques for Management of Tailings and Waste-Rock in Mining Activities, July 2004. European Commission, DG JRC, ST/EIPPCB/MTWR_BREF_FINAL, Edificio EXPO, Sevilla, Spain, p. 563.

Directive 2006/21/EC of the European Parliament and of the Council of 15 March 2006 on the management of waste from extractive industries and amending Directive 2004/35/EC, OJ L 102, 11.4.2006, 15–34.

GUS 2008. Ochrona Środowiska. Informacje i Opracowania Statystyczne. Główny Urząd Statystyczny. Warszawa

Price, W.A. & Errington, J.C. 1998. *Guidelines for metal leaching and Acid Rock Drainage at mine sites in British Columbia*. Ministry of Energy and Mines, British Columbia, Canada.

Szczepańska, J. & Twardowska, I. 1999. Distribution and environmental impact of coal-mining wastes in Upper Silesia. *Environ. Geol. Poland*. 38(3): 249–258.

Szczepańska, J. & Twardowska, I. 2004. Mining waste. In I. Twarowska, H.E. Allen, A.A.F. Kettrup & W.J. Lacy (eds.), *Solid Waste: Assessment, Monitoring and Remediation: 319–386*. Elsevier: Amsterdam.

Twardowska, I. & Szczepańska, J. 1995. Carboniferous waste rock dump as a long-term source of groundwater pollution: monitoring studies. *Współczesne Problemy Hydrogeologii* VII(1), Kraków – Krynica. Poland: 475–483 (in Polish).

USGS 2005. International Mineral Statistics and Information (April 25, 2005).

Web site: http://minerals.usgs.gov/minerals/pubs/country/

WCI 2008. Coal Facts 2008. http://www.worldcoal.org (accessed February 5, 2009).

Environmental Engineering III – Pawłowski, Dudzińska & Pawłowski (eds)
© *2010 Taylor & Francis Group, London, ISBN 978-0-415-54882-3*

Feedstock recycling of plastic wastes and scrap rubber via thermal cracking

M. Stelmachowski & K. Słowiński

Department of Environmental Engineering, Technical University of Lodz, Lodz, Poland

ABSTRACT: The experimental results of an investigation of the thermal degradation of waste polyolefins and scrap rubber are presented in this paper. Thermal decomposition of wastes was performed in a new type of a tubular reactor with molten metal. Three products, i.e., one gaseous (below 14 wt %), one liquid (over 41 wt %) and one solid residue, were obtained during the degradation of waste rubber, and two products were formed from the conversion of polyolefins: one gaseous (8–16 wt % of the input) and one liquid (84 ÷ 92 wt %) stream. The light, "gasoline" fraction of the liquid hydrocarbon mixture (C_4–C_{10}) comprised over 50% of the liquid product for polyolefins and over 90 wt % for rubber degradation.

Keywords: Thermal degradation, pyrolysis, cracking, waste rubber, waste plastics, fuel from wastes, molten metal.

1 INTRODUCTION

The energy crisis and the environmental degradation by polymer wastes together have made it imperative to find and propose technologies for the recovery of raw materials and energy from non-conventional sources, e.g., organic wastes, plastic wastes, and scrap tires. Methods for the utilization of waste plastics and scrap rubber (mainly tires) have much in common as well as differences. A variety of methods and processes connected with global or national policies have been proposed worldwide (Stelmachowski 2003, Scheirs & Kaminski 2006, Aguado et al. 2008). The strategy of sustainable development determines the hierarchy of waste management methods best suited to decrease the environmental impact. However, prevention and waste minimization at the source, the most favorable methods, are options with limited applicability because plastics now constitute many articles of common use, having replaced natural materials such as paper, leather, glass, and even metals due to their various properties and this cannot be reversed. Similarly, the increase in the generation of used tires is a consequence of the global increase of the number of motor vehicles (cars, trucks, and tractors) and the shorter life of tires because of traffic safety requirements. Therefore, prevention is almost impossible in this case as well.

1.1 Waste plastics

Waste plastics contribute to great environmental and social problems due to the loss of natural resources, environmental pollution, and depletion of landfill space. Global production of plastics amounted to about 230 million metric tons in 2005 (about 45–50 million tons in the EU alone; Scheirs & Kaminski 2006, Aguado et al. 2008). In Japan, the consumption of plastics amounts to more than 10 million tons/per year (Nishino et al. 2008) in 2004, and in China it rose from 23.0 (in 2000) to 31.2 million tons in 2003 (an average annual growth rate of 11.8% (Xiao et al. 2007)). In Poland, plastics consumption (excluding synthetic fibers) exceeds 1.67 million tons, and in 2006 over 75% of these were polyolefins (Statistical Yearbook of the Republic of Poland, 2007). In 2005, the consumption of post-consumer plastics was estimated at 22–25 million tons in the EU, and the amount has been increasing 6–7% annually and is expected to further increase due to the current relatively low consumption of plastics in developing countries (Scheirs & Kaminski 2006, Williams & Slaney 2007, Achilias et al. 2007). The main part (over 70% by mass) of the household waste plastics stream consists of polyolefins: polyethylene, (LDPE, HDPE, LLDPE, polypropylene (PP)), and polystyrene (PS). Non-returnable (single-trip) packaging containers (STPC) are the largest fraction of post-consumer plastic wastes.

The involvement of global communities and organizations as well as national governments in enacting new legislation has driven technological development for new ways of plastic waste utilization, energy recovery and feedstock recycling (e.g., Directive 2005/20/EC on Packaging and Packaging Wastes, Japan Containers and Packaging Recycling Law).

Unfortunately, land-filling is still the basic treatment method for used plastics, although its share is slowly decreasing, at a rate of about 2% annually. In Poland, only about 25% of plastics are recycled

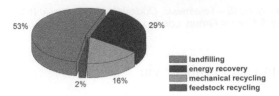

Figure 1. The distribution of treatment methods in the EU in 2005 (Aguado et al. 2008).

(Statistical Yearbook of the Republic of Poland 2007). Figure 1 presents the distribution of different treatment methods for plastic wastes in the EU in 2005 (Aguado et al. 2008). There are some technological and economic constrains that limit the full and efficient recycling of plastic wastes into useful products, e.g., contaminated waste plastics can only be partly recycled into new products.

Feedstock recycling of plastic wastes may be a profitable treatment method to eliminate negative impact on the environment and obtain useful products. During the past decade, this option has undergone an important evolution from a promising scientific idea to industrial applications. It includes chemical and thermal processing. Chemical processing by depolymerization may be applied, for selected polymers only, to generate monomers (e.g., PET into terephthalic acid (TPA) and/or dimethyltherephtalate (DMT) and ethylene glycol (EG), Pa-6 into caprolactam, Yoshioka & Grause 2006). Thermal reprocessing of plastic wastes may be performed in two fundamental ways: liquefaction via pyrolysis (thermal/catalytic cracking) or gasification. Thermal liquefaction of post-consumer plastics to hydrocarbon mixtures with the production of small amounts of gas and solid (residue) products is preferred for polyolefins (PE, PP), and sometimes PS, even when they are mixed in any ratio. The process has been gaining growing interest for 10–15 years. Numerous scientific papers have presented different problems that have been investigated (Stelmachowski 2003, Scheirs & Kaminski 2006, Aguado at al. 2008, Siddique et al. 2008). They concern the yield, selectivity, productivity, and kinetics of the degradation processes, selection of the catalyst, methods of degradation, and type of reactor. The most valuable product is a liquid mixture of hydrocarbons that may be used in the refinery industry for fuel production or electricity generation. The yields of liquid, gas, and solid products obtained via pyrolysis/cracking of plastic wastes depend on many parameters such as the composition of the waste mixture, temperature, type of catalyst, residence time in the reactor, heating rate, type of reactor, and process used. The catalyst may decrease the temperature of the conversion process, change the composition of the liquid product, or give more gas product. However, recovery and regeneration of the catalyst may be difficult or even impossible because it is often transported into the solid residue (mixed with coke and impurities from waste plastics)

during the process. Moreover, catalysts deactivate very quickly due to coking processes, which may increase the costs of the process. Fluidized-bed reactors have many advantages because difficulties with the mixing of wastes, removal of the coke and heat transfer resistance may be reduced. However, fluidized-bed reactors may be profitable probably only in large industrial-scale plants due to the investment costs (Scheirs & Kaminski 2006).

Over 30 commercial technologies have been proposed for thermal degradation of post-consumer plastics via pyrolysis or catalytic cracking to a fuel-like liquid mixture of hydrocarbons as the basic product (Scheirs & Kaminski 2006, Stelmachowski 2003; Aguado et al. 2008). They are usually carried out at temperatures of 350–430°C. However, the industrial plants are rare or have been running for a very short time, indicating that the proposed solutions are imperfect and their profitability is weak. The unfavorable situation for feedstock recycling is mostly based on the high investment costs of recycling treatments, the necessity of frequent cleaning of the reactor, costs of catalysts, and other economic circumstances, e.g., taxes. In Poland, during the period 2004–2006 over 10 small industrial plants were built and waste plastics were liquefied with yields of 70–80%. Most of them were closed due to decreased profitability during 2007. Searching for new technologies and reactors is strongly recommended. The new technologies (and plants) should have the following features:

– Low operating costs and investments costs are needed because the products have to be inexpensive; the conversion process must be profitable for investors.
– The process should be carried out without catalysts due to the difficulties and cost of their recovery.
– The yield of liquid product should be high as it is usually more valuable than gaseous product.
– The reactor cleaning frequency should be low.
– The heat-transfer resistance between waste particles and the heating medium should be minimized.
– The coking process should be minimized or even eliminated.
– The plant should have modular construction at the industrial scale. This allows for greater flexibility and enables construction of small or large plants with almost the same profitability. In some local and economic conditions, small plants may be more profitable and in others larger industrial plants will be more efficient (e.g., if they are constructed in the area of the oil refinery plant).

There are also pyrolysis technologies carried out in molten metals or molten inorganic salts. Waste plastics are decomposed to monomers and mixtures of hydrocarbons or gasified to syn-gas, hydrogen, or to simple inorganic compounds:

– Hydromax® Technology proposed by the Alchemix Corporation; the method is based on the conversion process of organic wastes to hydrogen on the surface of molten iron with the addition of tin at approximately 1,300°C (Alchemix, 2003, Stelmachowski, 2003).

- Molten salt oxidation (MSO); this is performed by injection of wastes beneath the surface of a bed of molten carbonate salts at 900–950°C. Only simple inorganic compounds are obtained by this method (Hsu et al., 2000).

- The method of recycling organic wastes, particularly waste polymers, based on thermal degradation in molten metal or on its surface (Newborough 2002, Stelmachowski 2003). The process, sometimes called the *"Clementi Process"*, is performed below 600°C (often between 350 and 550°C). The thermal degradation of wastes in molten metal is a known process and several reactors were patented (e.g., US patents: 1601777, 1709370 2459550, Domingo & Cabanero 1949, Mausre et al. 1989, Stelmachowski & Tokarz 2003). Until now, the reactors have been constructed as basin reactors with a low height of the molten metal layer. Among these technologies, this process likely has the greatest number of the desired features mentioned above.

1.2 Waste rubber. Scrap tires

At the beginning of the 20th century, 50% of post-consumer rubber was reclaimed due to the fact that rubber was almost as expensive as silver. Thus, rubber recycling processes are as old as rubber industry itself. Decreasing costs of rubber production and rising quality requirements for new tires have since made simple rubber reclaiming technologies unprofitable. Therefore, in the 1960s and 1970s only 20% of scrap tires were recycled, and at the end of the 20th century 40–70% of them were dumped in most countries.

World production at the beginning of the 21st century was ~34 million tons and 20% of tires have to be recycled each year. The integrated and estimated statistical data on the production and utilization of tires is presented in Table 1 (Sharma et al. 2000, Rodriguez et al. 2001, Conesa et al. 2004; Reschner 2007, Xiao et al. 2008, Olazar et al. 2008).

Utilization of tires is not easy due to their composition, which depends on the type of the tire and the manufacturing process and may be generally described as follows (Orr et al. 1996, Galvagno 2002, Rodriguez 2001):

natural or synthetic rubber	40–69 mass %,
carbon black	23–45 mass %,
hydrocarbon oils	2–4%,
moisture	<1.2 mass %,
zinc oxide	~2 mass %
sulfur	1–2 mass %
steel cord	<10%
others	<5%

As mentioned previously, a huge amount of scrap tires have been dumped in massive stockpiles, which may cause a fire hazard (e.g., in California in 1999, and in Poland in 2003) and other environmental or even health problems. This method of their utilization is the worst one; it is unacceptable from an environmental point of view and in the near future will be banned by law.

Tire remolding by vulcanization is now rare due to the demands of high quality and low price for new tires. A low cost of investment, pure usable product, and very low gas emission are the advantages of mechanical and cryogenic methods of their recycling. However, high consumption of energy and a limited market for the products makes these processes unprofitable; therefore, further research and development is still needed for these methods. Energy recovery by incineration (co-combustion) is the second option for post-consumer rubber utilization that takes advantage of their high energy content (~39 MJ/kg) and reduces the volume of wastes. Similarly, as for waste plastics, this process sometimes arouses public resistance to their combustion in municipal incineration plants. Incineration of scrap tires in rotary cement kilns is often employed for their utilization, with an acceptable impact on the environment. In Poland the majority of scrap tires are utilized by this method.

The most promising processes seemed to be gasification and/or liquefaction by pyrolysis (cracking) of scrap rubber (Conesa et al. 2004, Xiao et al. 2008). They enable the recovery of energy as well raw products (hydrocarbons, synthesis gas, and hydrogen).

Table 1. Production and recycling of tires in Europe in 2006 (integrated data from different articles and statistical sources).

Country	Tires production [thousands of tons/year]	% of recycled tires
Estonia + Lithuania + Latvia	29	6.90%
Ireland	40	7.50%
Hungary	46	73.91%
Romania	50	40.00%
Czech Republic + Slovakia	100	7.00%
Balkans + Greece + Cyprus + Malta	122	12.75%
Belgium + Holland + Luxemburg	129	75.19%
Poland	146	45.21%
Scandinavia	227	78.85%
Italy	380	60.79%
Portugal + Spain	397	38.79%
France	398	66.08%
Great Britain	475	59.79%
Austria + Switzerland + Germany	694	73.49%
Total in above European countries	3.233	57.70%
USA[1]	2,444	75.60%
Japan &China[1]	2.200–2.600	Not available

[1] in 2004

Gasification is considered to be a very attractive method to recover energy and raw material (e.g., carbon black) efficiently and economically. Gasification at lower temperatures seems to be more efficient (Leung & Wang 2003).

However, recycling (via gasification or pyrolysis) may only be profitable if all components of the scrap tires are reclaimed and reused – not only the mixture of hydrocarbons (liquid or/and gas) – as a product of synthetic/natural rubber degradation. The carbon black may be recycled for tire production or for production of active carbon (Ko et al. 2004). Zinc oxide and steel may also be recovered and reused. Sulfur (during the process) is transferred to all products and during their desulfurization it has to be removed and may also be recovered.

The scientific research and development for tire reclamation has been focused on pyrolysis (cracking) for at least two decades. The problem of bad conditions for heat transfer from a heating medium to the particles of disintegrated scrap tire may also be solved by a new technology based on the thermal decomposition of waste in the molten-metal bed reactor.

2 EXPERIMENTAL

2.1 *Experimental set – up*

The degradation of wastes was carried out with molten metal in a new type of the tubular reactor called a "tube-in-tube" design. The construction of the reactor differs from the known basin reactors that have been patented to date (Domingo and Cabanero 1949, Mausre et al. 1989, Stelmachowski & Tokarz 2003). It was constructed from two tubes of different diameters (Stelmachowski, 2008). The inner (input) tube was placed coaxially in the external (outflow) tube. A mobile piston was located inside the inner pipe for transporting wastes into the molten metal bed. The residence time of waste rubber in the interior volume of the liquid metal bed depends on the piston speed. A diagram of the reactor, a photograph, and the scheme of the experimental set-up are shown in Figure 2.

2.2 *Materials*

The granular commercial plastic material (HDPE or PP pellets ~3–5 mm in diameter) or disintegrated waste bicycle tires and flat rubber boards (rubber particles ~5–15 mm in size) were the feedstock for thermal decomposition in the laboratory reactor. Thirteen kilograms of an alloy of tin and lead was used to create the molten metal bed.

2.3 *Measurements and analytical process*

The liquid and gas product samples were analyzed by gas chromatography. The internal normalization method was applied for calculating the concentrations of all components (olefin and paraffin). The error for gas sample analysis was estimated at 0.5–2% depending on the component; the error for liquid sample

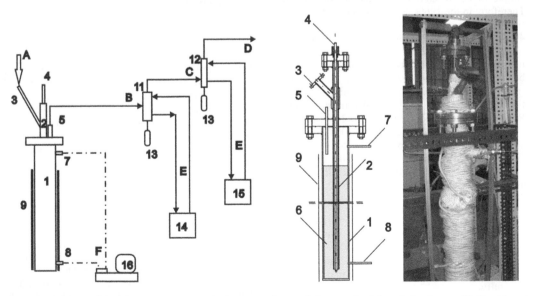

Figure 2. Scheme of the laboratory set-up, vertical cross-section, and photograph of the tubular reactor. (1) the external (outflow) pipe, (2) the inner (input) pipe, (3) the loading port, (4) the device for wastes transport, (5) the vapor outlet (total product), (6) the bed of molten metal, (7) the thermocouple (gas-phase temperature), (8) the thermocouple (molten metal), (9) electrical heating, (11) the primary cooler for the liquid product, (12) the final cooler (13) receivers of liquid products, (14,15) thermostats, (1) computer system for data acquisition, (10) thermocouples for gas-phase and molten metal bed, (11) electrical heating, A – polymers, B, C – vapor products, D – non-condensable gases to bubble flow meter and to GC, E – cooling water, F – temperature signal to the acquisition system.

analysis for hydrocarbons C_5–C_{10} was below 0.5%, for C_{11}–C_{16} 1.5%, and for others below 3.0%. The error for the gas flow measurement was estimated at 1.5% and for liquid product mass was 0.05–0.1 mass %. The heavy metal contents in the liquid products were analyzed by AAS. The concentration of Pb and Sn in the mixture of hydrocarbons was below 25 mg/kg in all liquid samples and the relative error of the analysis was below 0.1%.

2.4 Process description

The process was carried out in a semi-batch reactor. Wastes (reduced to small particles) were put through the loading port into the inner tube and then slowly transported by the piston into the liquid molten alloy at the end of the inner tube, during which they were melted and decomposed. Degradation products and un-decomposed components flowed out to the external tube and, next, to the surface of the bed. Further degradation was carried out in the molten metal bed in the external tube and on the surface of the molten alloy to give the final products – the mixture of hydrocarbons. The vapors flowed out from the reactor and were condensed in coolers. Liquid products were collected in the small receivers that allowed measurement of the liquid stream during the time of the experiments. Gaseous products, the mixture of un-condensable hydrocarbons, flowed out from the reactor to the bubble flow meter and, next, through the sampling port to the ventilation system. The temperatures of the molten metal bed, the gas phase, and the liquid product in the coolers were measured and recorded by the data acquisition system. The residues (solid products obtained only for rubber degradation) were removed from the reactor after the experiments.

3 RESULTS AND DISCUSSION

Several experiments were performed in the laboratory-scale reactor in manner described above for the degradation of rubber or plastics. The general profiles of the runs are presented in Table 2. The distributions of temperature of the molten metal bed, volume flow rates of gaseous product, and mass flow rates of liquid product are presented in Figure 3 for two representative experiments.

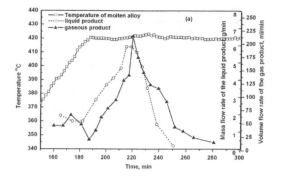

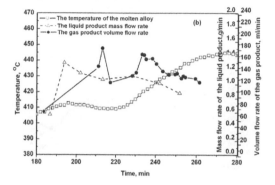

Figure 3. The distributions of product flows and temperature of molten metal bed for (a) waste plastics and (b) scrap rubber in the representative experiments.

The yield of gas product for the PE (HDPE) degradation process was over twice that for the PP degradation process. No solid residue was obtained and no coke was observed on the walls of the reactor or on the surface of molten metal, indicating that coking processes had not proceeded with the polyolefins. The solid product obtained in rubber degradation was not analyzed. The composition of all liquid and gas products was nearly stable over the time of the experimental runs, and the average products for different runs (for the same raw material) were very similar regardless of process conditions; this can be seen in Figures 4 and 5, in which the carbon-number liquid product distributions for average liquid and gas products is presented. However, the distribution profiles differed for PP and PE as well for rubber degradation, meaning that the mechanisms of the cracking process were different for different polymers.

Table 2. General profiles of experiment runs for waste plastics and rubber degradation.

Polymer		HDPE	PP	Rubber
Temperature of the process	°C	408–428	362–430	410–417
Mass of the stock	g	300.01–336.09	300.00–300.16	153.37–299.50
Yield of the gas product during full semi-batch experiment.	weight %	9.91–16.97	8.12–9.15	5.79–14.73
Yield of the liquid product during the full semi-batch experiment.		90.09–83.07	90.85–91.88	41.04–43.62

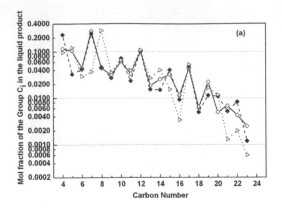

(a)

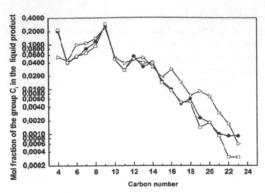

Figure 5. Carbon-number average liquid and gas product distributions in different runs for rubber degradation (liquid products).

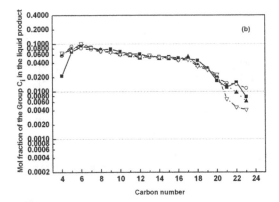

(b)

Table 3. The mean fractional composition of the products in all runs.

Polymer		PE	PP	Rubber
Fractions				
C_4–C_{10}	mol %	48.8–52.1	67.4–70.2	69.9–73.9
C_{11}–C_{16}	mol %	30.9–33.6	20.7–23.4	18.6–23.0
C_{17}–C_{24}	mol %	16.9–17.9	9.1–9.3	1.9–2.9
Fractions of paraffins and olefins				
paraffins	mol %	46.7–54.8	33.5–46.1	36.6–37.7
olefins	mol %	45.2–53.3	53.9–66.5	57.3–59.3

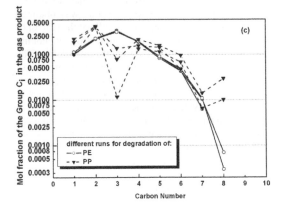

(c)

different runs for degradation of:
—○— PE
- ▼- PP

Figure 4. Carbon-number average liquid and gas product distributions in different runs for: (a) – PP degradation (liquid products), (b) – PE degradation (liquid products), and (c) – PP and PE degradation (gas products).

Three basic fractions in the liquid product are usually distinguished for polymer thermal degradation: light ("gasoline"; C_4–C_{10}), medium ("diesel" C_{11}–C_{16}); and heavy ("light waxes" C_{17}–C_{24}). The contents of these fractions in the liquid hydrocarbon mixtures are presented in Table 3. The content of light waxes was below 18 mol % for PE degradation, 10 mol % for

PP, and ~2 mol % for the rubber degradation process. This means that the product obtained is very useful as a raw material for chemical, refinery and petrochemical applications. All liquid samples recovered were fluid at ambient conditions, although the thermal cracking of polyethylene gave a denser liquid product, a "diesel", or heavy oil fraction mixed with light waxes. The product of waste rubber pyrolysis consisted mainly of light fractions and was also fluid, but a char was suspended in the hydrocarbon mixture obtained. Therefore, it would require clarification before further use.

The ratio of paraffin to olefins was almost 1:1 for PE decomposition but, for PP and rubber degradation, the fraction of olefins in the total product was much greater than the fraction of paraffin (Table 3).

The products also contained a very small amount of aromatic hydrocarbons; below 0.5 mol % for polyolefin degradation (in this case, aromatic hydrocarbons were included in the mixture of aliphatic hydrocarbons in the balances presented) and below 5 mol % for rubber degradation. The aromatic contents were estimated by IR spectrophotometry. The content of aromatic hydrocarbons in the liquid product of rubber pyrolysis is definitely lower for the process carried out in the molten metal than for processes performed in fluidized-bed reactors or other types of tube reactors. Only Williams and Brindle, 2002, reported a low content of aromatic hydrocarbons for un-catalyzed pyrolysis. However, in the presence of a catalyst, (zeolites) they also observed an increase of the amount of aromatic hydrocarbons in the liquid fraction. Other researchers (Galvagno et al. 2002, Rodriguez et al., 2001, Williams and Brindle 2002, Li et al. 2004, Olazar et al. 2008) generally noted a higher content of aromatic hydrocarbons in the pyrolysis oil, particularly if the process was performed with the presence of zeolites as a catalyst.

The solid product (residue) of rubber degradation was collected from the surface of the molten metal after the experiments, but it was not analyzed. The yield of solids was ~42–52 wt %. Recovery methods for char, zinc oxide, and steel are known, and their employment is vital for the industrial application of all methods and processes based on thermal degradation of scrap rubber (tires).

4 CONCLUSIONS

The results of the investigation of polymer pyrolysis in the new "tube-in-tube" tubular reactor design with a molten metal bed were promising in comparison with thermal and catalytic cracking in basin-vessel reactors with stirring or tubular-flow reactors. The thermal decomposition of rubber and plastics wastes in molten metal has many advantages: the process was fast and the coking process was minimized (no coke was observed on the walls of the reactor during PE and PP cracking). The main advantage of the method was the reduction in the heat transfer resistance between the molten polymer and the heating medium because of direct contact of the particle of wastes and molten metal. This is very important in polymer pyrolysis, particularly for rubber degradation due to the very low thermal conductivity of rubber. No catalyst and no stirring were needed. The liquid product contained mainly the gasoline fraction of hydrocarbons. The residence time of wastes in the reactor and the reaction time were shorter than in other tube reactors applied for rubber pyrolysis. The temperature of the gas phase in the reactor was moderate (below 230°C). Therefore, the number of recombination reactions was minimized and the fraction of aromatic hydrocarbons produced was very small, even for the rubber pyrolysis. The heavy metal contents were low, meaning that at this process temperature the surface of the molten alloy in the reactor exhibited desirable properties, such that metal particles or ions were not transferred to the liquid products. The yield of gas product was below 7.5% by weight for rubber, PP, and PE degradation, so over 90% of post-consumer polyolefins and 40–50% of rubber were converted into liquid products. The composition of the liquid product was very profitable, as it consisted of $50 \div 90\%$ of a light ("gasoline") fraction. The fraction of light waxes in the liquid product was below 18% for HDPE degradation, below 10% for PP, and ~2% for rubber degradation. The liquid product is fluid at ambient conditions, which is important for further use, pumping and transport. In the laboratory scale set-up, no solid product (residue) was derived for PP or PE decomposition. For real-world, municipal post-consumer plastics waste degradation, the solid residue will likely consist of only waste impurities. The gas product may be burned to supply the process heat for the reactor in an industrial scale set-up. Nevertheless, thermal cracking processes of rubber in a reactor with a molten metal bed, as with other thermal processes, will be cost-effective only if all of products obtained, not only the hydrocarbons, but carbon black, zinc oxide, and steel, are recycled and used.

ACKNOWLEDGEMENTS

This work was done as a part of the project "Thermocatalytic degradation of waste plastics and rubber" supported by the Ministry of Science and Higher Education: Project number 3 T09D 035 27

REFERENCES

Achilias, D.S., Roupakias, C., Megalokonomos, P., Lappas, A.A. & Antonakou, E.V. 2007. Chemical recycling of plastic waste made from polyethylene (LDPE and HDPE) and polypropylene (PP). *Journal of Hazardous Materials* 149: 536–542.

Aguado, J., Serrano, D. P., Escola J. M. 2008. Fuels from Waste Plastics by Thermal and Catalytic Processes: A Review. *Ind. Eng. Chem. Res.* 47: 7982–7992

Alchemix, Executive Summary. 2003. Private information from Alchemix Corporation and website: www.alchemix.net,

Conesa, J., Martäin—Gulloän, I., Font, R . Jauhiainen, J. 2004. Complete Study of the Pyrolysis and Gasification of Scrap Tires in a Pilot Plant Reactor Environ. *Sci. Technol.* 38: 3189–3194

Domingo, J. & Cabanero, D. 1949. Process and device for regeneration of monomer from polymethyl methacylate, Spanish patent 192909.

Galvagno, S. Casu, S., Casablanca, T., Calabrese, A. & Cornacchia, G. 2002. Pyrolysis for treatment of scrap tyres: preliminary experimental results. *Waste Management* 22 (8): 917–923.

Hsu, P.C., Foster, K.G., Ford, T.D., Wallman, P.H., Watkins, B.E., Pruneda, C.O. & Adamson, M. G. 2000. Treatment of solid wastes with molten salt oxidation. *Waste Management* 20 (5–6): 363–368.

Ko, D.C.K, Mui, E., Dward L.K, Lau, K.S.T. & McKay, G. 2004. Production of activated carbons from waste tire-process design and economical analysis. *Waste Management* 24(9): 875–88.

Lee, K.-H., Noh, N.-S., Shin, D.-H. & Seo, Y. 2002. Comparison of plastic types for catalytic degradation of waste plastics into liquid product with spent FCC catalyst. *Polymer degradation and Stability* 78: 539–544.

Leung, D.Y.C. & Wang, C.L. 2003. Fluidized-bed gasification of waste tire powders. *Fuel Process Technol.* 84(1–3): 175–96.

Mausre, W.E., Donahue, J.R., Larue, G.W., Bonney, L.H., Glanton, G.W. & Harris, W.L. 1989. Method and Apparatus for thermal conversion of organic matter. US Patent 342056.

Newborough, M., Highgate, P., Vaughan, P. 2002. Thermal depolymerisation of scrap polymers. *Applied Thermal Engineering* 22 (17): 1875–1883.

Nishino, J., Itoh, M, Fujiyoshi, H. & Uemichi, Y. 2008. Catalytic degradation of plastic waste into petrochemicals using Ga-ZSM-5. *Fuel* 87: 3681–3686.

Olazar, M., Aguado, R., Arabiourrutia, M., Lopez, G., Barona, A. & Bilbao, J. 2008. Catalyst Effect on the Composition of Tire Pyrolysis Products. *Energy & Fuels* 22: 2909–2916.

Orr, E.C, Burghard, J.A., Tuntawiron, W., Anderson, L.L. & Eyring, E.M. 1996. Coprocessing waste rubber tire material and coal. *Fuel Processing Technology* 47: 245–259.

Predel, M. & Kaminsky, W. 2000. Pyrolysis of mixed polyolefins in fluidized-bed reactor and on a pyro-GC/MS to yield aliphatic waxes. *Polymer Degradation and Stability*. 70(3): 373–385.

Reschner, K. 2007. Scrap Tire Recycling. A Summary of Prevalent Scrap Tire Recycling Methods http://www.energymanagertraining.com/tyre/pdf/ScrapTireRecycling.pdf2.

Rodriguez del Marco, I., Laresgoiti, M.F., Cabrero, M.A., Torres, A., Chomon, N.J. & Caballero, B. 2001. Pyrolysis of scrap tires. *Fuel Process Technology* 72: 9–22.

Salmiaton A. & Garforth A. 2007. Waste catalysts for waste polymer. *Waste Management* 27: 1891–1896.

Scheirs, J. Kaminski, W., (eds.), 2006. Feedstock recycling and pyrolysis of waste plastics: converting waste plastics into diesel and other fuels, Wiley Series in Polymer Sciences, John Wiley & Sons, Ltd.

Sharma, V. K., Fortuna, F., Mincarini, M., Berillo, M. & Conacchia, G. 2000. Disposal of waste tires for energy recovery and safe environment. *Applied Energy* 65: 381–394.

Siddique, R., Khatib, J. & Kaur, I. 2008. Use of recycled plastic in concrete: A review. *Waste Management* 28: 1835–1852.

Spokas, K. 2008. Plastics – still young, but having a mature impact. *Waste Management* 26: 473–474.

Statistical Yearbook of the Republic of Poland. 2007. Central Statistical Office of Poland. Warsaw.

Stelmachowski, M. 2003. Termo-catalytic degradation of polymers (in polish). Monografie, PAN Oddział w Łodzi, Komisja Ochrony Środowiska, Łódź, 2003, ISBN – 83-86492-19-8.

Stelmachowski, M. & Tokarz, Z. 2003. The method of continuous conversion of waste plastic. Patent application P-358774.

Stelmachowski, M. 2008. The reactor and method for thermal conversion of waste polymers. Patent Application P384806.

Walendziewski, J. 2005. Continuous flow cracking of waste plastics. *Fuel Processing Technology* 86 (12–13): 1265–1278.

Wiliams, P.T. & Brindle, A.J. 2002. Catalytic pyrolysis of tyres: influence of catalyst temperature. *Fuel* 81: 2425–2434.

Williams, P.T. & Slaney, E. 2007. Analysis of products from the pyrolysis and liquefaction of single plastics and waste mixtures. *Resources Conservation & Recycling* 51: 754–769.

Xiao, R., Jin, B., Zhou, H., Zhong, Z. & Zhang, M. 2007. Air gasification of polypropylene plastic waste in fluidized bed gasifier. *Energy Conversion and Management* 48: 778–786.

Xiao, G., Ni, M-J., Chi, Y. & Ke-Fa Cen, K-F. 2008. Low-temperature gasification of waste tire in a fluidized bed. *Energy Conversion and Management* 49: 2078–2082.

Yoshaki, T. & Grause, G. 2006. Feedstock Recycling of PET. In: J. Scheirs & W. Kamiński (eds), Feedstock recycling and pyrolysis of waste plastics: converting waste plastics into diesel and other fuels: 641–661 Wiley Series in Polymer Sciences. John Wiley & Sons. New York

Environmental Engineering III – Pawłowski, Dudzińska & Pawłowski (eds)
© 2010 Taylor & Francis Group, London, ISBN 978-0-415-54882-3

Application of modern research methods to determine the properties of raw minerals and waste materials

D.K. Szponder & K. Trybalski

Department of Mineral processing and Environmental Protection, Faculty of Mining and Geoengineering, AGH University of Science and Technology, Krakow, Poland

ABSTRACT: This article presents modern research methods and measuring devices used to determine physical, chemical and mineralogical properties of raw minerals and waste materials. X-ray diffraction analysis, and X-ray microanalysis were used. Characteristics of these methods, as well as their potential to measure and quantify properties of mineral waste materials (fly ashes), were presented. Preliminary results were shown as pictures, scan maps and diffraction patterns. They contain information on the samples' mineralogical and elemental composition and surface morphology. Each of the methods presents information about different physical, chemical and mineralogical properties of material samples, in this case fly ashes.

Keywords: Fly ashes, chemicophysical properties, mineralogical properties, X-ray diffraction analysis, scanning electron microscopy, X-ray microanalysis.

1 INTRODUCTION

Currently in science, there is rapid development in the use of new improved methods. Examples of these methods are electron microscopy and X-ray diffraction analysis. These two methods can be used by virtually every field of science dealing with solids. Modern methods have also been used to study the properties of raw materials and mineral wastes. The results of such tests are used to assess the feasibility of enriching raw minerals, to asses the feasibility of utilizing and treating waste, and to assess technological processes. These analyses have many applications, and a number of these are presented below.

In a study on the application of bacterial leaching in the processing of waste from an incineration plant (Fecko et al. 1996), a scanning microscope was used to obtain microscopic images of samples. These images showed the characteristics of samples before and after leaching and their morphology. This method was also used to identify grains with a high content of a particular element.

Another technique, X-ray diffraction (XRD), was used to identify sulphides in a study involving the microscopic determination of chalcocite content in mixtures with galena (Bigosiński & Drzymała 1996). After thermal modification of their surfaces, the samples were observed under the microscope. A planimetric method was used to determine the population of red particles (chalcocite after heating).

Both scanning microscopy and X-ray diffraction were used by Kêdzior et al. (2003) to examine the

products of enrichment using the floatation of copper ores. Information about the mineralogical composition of the products, their elemental composition and surface morphology was obtained in that study. Its results proved that these methods can be used to evaluate and control the processes of enrichment of copper ores and the mode of preparation of the ore; an appropriate method of enrichment can then be chosen.

2 MATERIAL AND METHODS

2.1 *Materials*

Scanning electron microscopy (SEM) is currently the most common modern method used for observing and testing microstructures. In the scanning microscope, an electron spot beam bombards the sample. The beam produces an emission of secondary electrons through a linear sweep of the sample's surface. Thus, the sample emits a variety of signals. The signals are recorded by detectors and converted sequentially into an image of the test samples or X-ray spectrum (Goldstein et al. 2003). Received images of the same sample surface have a different character depending on the type of signal that is subjected to detection. These may be electron images or X-ray images. An electron image shows inequalities and the geometry of the sample surface. In contrast, an X-ray image shows what kind of atoms are included in the sample and how they are arranged on the surface of the sample (chemical analysis) (Buršík & Brož 2009). Signals that generate information about the properties of the samples can be divided into groups:

– Secondary electron emissions – the topography and morphology of the sample, the crystalline structure of the sample, the distribution of potential and intensity of electric and magnetic fields in the sample.
– Backscattered electron emissions – the topography and morphology of the sample, the distribution of magnetic domains in the sample, composition contrast.
– Auger electron emissions – a chemical analysis of surface layers of the sample, measurement of local potentials in the sample.
– Cathodoluminescence – recombination processes in the material, identification of impurities, additives and structural heterogeneity.
– Characteristic X-ray emission – qualitative and quantitative chemical analysis of the material. Measurement techniques:
  • Qualitative analysis of the whole picture to identify the elements in a selected field
  • Element mapping – technique identifying surface elements in a micro-area
  • Linear analysis – element distribution along a line
  • Quantitative analysis at a point to determine elements present at a set point and to identify the microstructure (Goldstein et al. 2003).

The advantages of scanning microscopy include the following: the possibility of direct analysis of a fresh fracture of samples; high accuracy in mapping a surface of the test substance; the ability to observe the surface topography; a large depth of focus (about 300 times bigger than in an optical microscope); a high voltage contrast; and the possibility of observing adjacent areas of the sample, at a magnification of 100 to 500 times, and at a resolution less than 10 nm (Elssner et al. 1999). The result of a sample analysis is an image with a material surface and with marked points which can be analyzed, and a map.

In structural and phase X-ray analysis, an X-ray of known wavelength is used to test unknown crystalline phases. In X-ray phase analysis, the most frequently used technique is the powder method performed using diffractograms (X-ray apparatus). Polycrystalline samples in the form of powdered mineral or rock, and monolithic samples of fine-grained and fine-crystalline minerals (i.e. fly ash), are tested by this method. The desired size of allotriomorphic grains is 1 μm. Samples that contain grains of a larger size should be crushed before the test. This method is based on an assumption that a sample consists of grains oriented in a random way. A limited number of these grains is located relative to a monochromatic X-ray beam that falls on the sample, in such a way that the specified crystal planes (hkl) satisfy the Bragg condition for wave interference. Each crystalline phase is attributed to crystal planes that are filled in a specified way with atoms or ions forming that phase. Each substance is characterized by a corresponding deviation

from the original direction of incidence of the interference beam. The interference beam is recorded on a film (film technique) or by using a Geiger or scintillation counter (diffractometer, counter technique). The deviation of the beam is 2Θ, which allows for the identification of the test sample (Ashraf et al. 2009). The ionizing action (ionization) of the X-ray on atoms is used in X-ray diffraction analysis. In this method, reflections are tracked by a meter that constantly moves around the camera. The meter's angular velocity has to be twice as big as the angular velocity of the sample (Ahuja & Jespersen 2006). The result of the test, using the powder method, is a record of spectra in the form of a diffraction pattern. The diffraction pattern plots the diffractive extremes. It is the basis for the identification of the crystalline phases contained in the sample (Ashraf et al. 2009). The diffraction pattern is obtained in numerical and graphical forms. The mineral composition of the sample and the percentage of individual phases are identified automatically using appropriate software that includes directories of characteristic values for each phase. The result of testing is a complete qualitative and quantitative analysis of the mineral composition of tested samples.

2.2 Methods

Fly ash is an artificial puzzolana that is produced as a result of the combustion of fragmented coal. It leaves a pulverized, fuel-fired furnace with flue gases. It takes the form of fine mineral dust from light to dark grey in colour and consists primarily of oxides of silicon, aluminium and iron. In addition, it contains a variety of trace elements and a minimum trace amount of unburned carbon (Van Dyk et al. 2009).

Fly ashes from the Power Plant BOT Opole SA (sample designated as Sample O) and from the Thermal-Electric Power Station 'Cracow' SA (sample designated as Sample L) were used in the studies. Samples were collected from the fly ash retention tanks. Fly ashes on three zones of electrostatic precipitators, which are located at each of the blocks of energy plants, are collected into the tanks. Samples (approximately 1 kg) were collected with a probe (with an inside diameter of 80 mm and a length of 1,500 mm), from the hopper of a retention tank, at the time of gravitational movement of the material. The collected samples were an averaged mixture of ash produced in these plants.

For X-ray analysis, a few milligrams of ash from the original samples were collected. Then the ash was dust pressed, without the addition of a binder, in a flat container made of X-ray amorphous substance (quartz glass). Samples were placed in an X-ray diffractometer, and were tested. Also, for the scanning microscope analysis, a few milligrams of ash from the original samples were collected. The samples were dried in an electric oven. They were located on the base, which provides a sample discharge, made of a material which is a good conductor (aluminium). Samples were attached to the base by a carbon tape that conducted electrical loads. Grains were fixed to the base

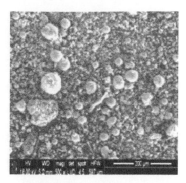

Figure 1. General view of sample.

Sample O

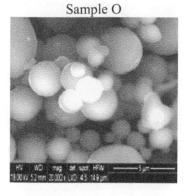

Figure 2. Smooth surface grains.

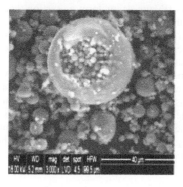

Figure 3. Plerospherical grain.

Figure 4. Agglomerate with points marked for the X-ray microanalysis.

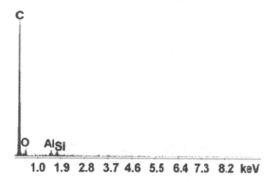

Figure 4a. X-ray microanalysis at point: No. 1 at Fig.4.

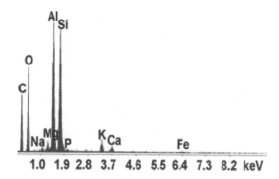

Figure 4b. X-ray microanalysis at point: No. 2 at Fig.4.

in such a way that there was no contact between them; they formed a single layer, and each grain was in contact with the tack base. Fly ashes are non-conducting materials, so they were covered by vacuum sublimation with a layer of conductive material (carbon with a thickness from 0.05 nm to 1 nm) (Szponder 2008).

3 RESULTS AND DISCUSSION

Figures 1 to 15 present the scanning microscopic images and diffraction patterns of samples of fly ashes. Due to the large number of test results, only a selection is presented in this publication. However, conclusions were reached on the basis of all results obtained. An interpretation of the results provides information on the composition of test samples. Preliminary results for the test samples are presented below.

Scanning microscopy proved that the sample O is a fine-grained ash (grain 1–100 μm), (Figure 1). Spherical grains with a smooth surface prevail in the ash (pirospherical grains – grains filled inside (Figure 2), cenospherical grains – grains empty inside, and plerospherical grains – grains filled inside with smaller grains (Figure 3), either singly (Figure 3) or combined into agglomerates (Figure 4). In Figures 4a and 4b, the results of point microanalysis performed on the

Figure 5. Smooth surface grains with crust on surface.

Figure 6. Spherical grains made of crystals with structural fabric.

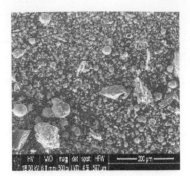

Figure 8. General view of sample.

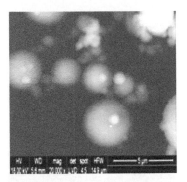

Figure 9. Smooth surface grains.

surface of this kind of grain (Numbers 1 and 2 in Figure 4) are plotted. The microanalysis revealed mainly silicon, aluminium and oxygen (mineralogical composition – silicate glass, quartz (SiO_2), and mullite ($3Al_2O_3 \cdot 2SiO_2$). Pirospherical grains are formed from a liquid alloy of these substances. Cenospherical grains are formed from a light alloy with a large portion of the gas phase. Other grains are divided into:

– Grains of irregular shape – thin plates with striations, containing calcium, magnesium, sodium, potassium, and carbon, which cement spherical grains (Figure 4), or form a crust on their surface (Figure 5).

– Spherical grains made of crystals with a structural fabric, containing iron, titanium and other metals (hematite) (Figure 6).

– Grains of irregular shape, very porous with rounded or sharp-edged borders (carbon) (Figure 4). These grains are unburned carbon residue. Pores were formed as a result of rapid degassing. As a result of the reducing atmosphere in the chamber (not enough oxygen) organic material has not been burned completely.

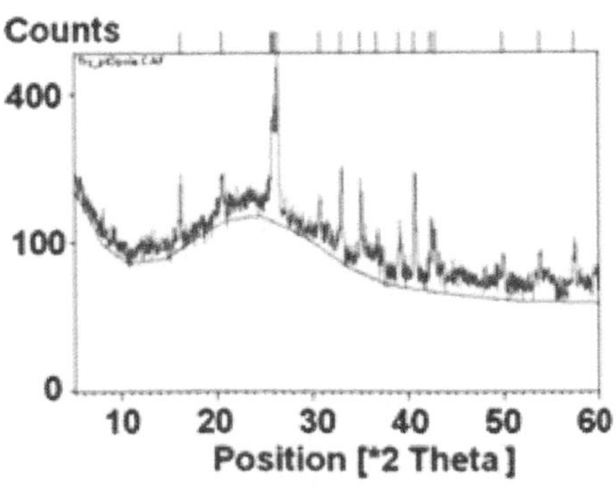

Mass [%]	25	6	62	7
Ref. Code	83-2187	88-2359	79-145	73-2444
Sc.	60	31	51	24
Compound name	Quartz	Hematite	Mullite, syn	Dolomite
Scale Fac.	0.906	0.199	0.501	0.195
Chem. Formula	SiO_2	Fe_2O_3	$Al_{4.984}Si_{1.016}O_{9.0}$	$CaMg(CO_3)_2$

Figure 7. Mineralogical composition of the ash set in X-ray analysis.

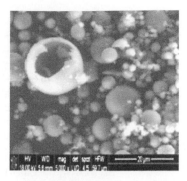

Figure 10.   Cenospherical grain.

Figure 11.   Plerospherical grain.

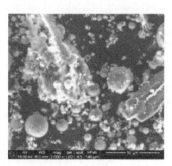

Figure 12.   Four different kinds of grain.

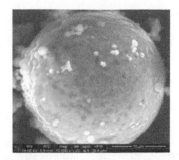

Figure 13.   Smooth surface grains with crust on surface.

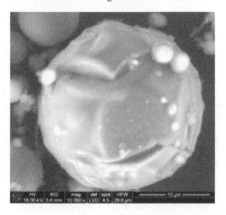

Figure 14.   Spherical grains made of crystals with structural fabric.

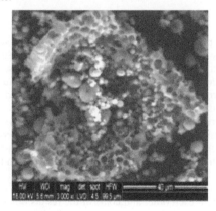

Figure 15.   Agglomerate with a high concentration of carbon.

The results of X-ray analysis are shown in Figure 7. The mineral composition of sample O comprises 62% synthetic mullite ($3Al_2O_3 \cdot 2SiO_2$), 25% quartz (SiO_2), 7% dolomite (($CaMg[CO_3]_2$)$\alpha(3Al_2O_3 \cdot 2SiO_2$ and 6% hematite (Fe_2O_3). Sample O belongs to the siliceous-aluminium ashes group, which contains small quantities of calcium in the form of dolomite. It consists of a large quantity of synthetic mullite, which was formed as a result of burning coal that contains mainly aluminosilicate, and slate and clay as the waste rock. Another component of the sample that is present in large quantities is quartz. It is a component of many types of gangue, with different characteristics and origins.

Scanning microscopy showed that sample L is a fine-grained ash (grain 1–100 μm), (Figure 8). The ash is dominated by spherical grains with a smooth surface – pirospherical (Figure 9), cenospherical (Figure 10), plerospherical (Figure 11), single (Figure 9, 10, 11) or combined into agglomerates (Figure 12). These kinds of grains consist mainly of silicon, aluminium and oxygen (glass, quartz, and mullite). Other grains are divided into:

– Grains of irregular shape – thin plates with striations, which cement spherical grains or form a crust on their surface and contain magnesium carbonate, calcium carbonate, sodium carbonate, and potassium carbonate (calcite, dolomite) (Figure 13).

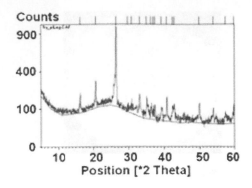

	49	47	4
Mass [%]			
Ref. Code	81-0065	82-1237	86-2336
Sc.	62	39	22
Compound Name	Quartz	Mullite, syn	Calcite Magnesium
Scale Fac.	0.896	0.202	0.069
Chem. Formula	SiO_2	$Al_{5.65} Si_{0.35} O_{9.175}$	$(Mg_{.129} Ca_{.871}) CO_3$

Figure 16. Mineralogical composition of the ash set in X-ray analysis.

– Spherical grains made of crystals with a structural fabric containing iron, titanium and other metals (hematite) (Figure 14).
– Grains of irregular shape, very porous with rounded or sharp-edged borders, containing carbon (Figure 15).

The results of X-ray analysis are shown in Figure 16. The mineral composition of sample L is 49% quartz (αSiO_2), 48% synthetic mullite ($3Al_2O_3 \cdot 2SiO_2$), and 4% calcite ($CaCO_3$), with a small dolomite impurity ($CaMg[CO_3]_2)\alpha(3Al_2O_3 \cdot 2SiO_2$). Sample L belongs to group of siliceous ashes containing a significant amount of aluminium compounds.

4 CONCLUSIONS

This paper presents the preliminary results of a study of fly ashes, which was carried out using modern measurement methods. The results – scanning microscopic images and diffraction patterns – provide valuable information on the composition of minerals and elements and the morphology of the surface samples. The results provide a characterization of the physical, chemical and mineral properties of sample materials. This knowledge allows one to choose the industries in which fly ash can be used most effectively, and the most efficient methods and technologies.

These methods, when used in parallel studies, complement each other and confirm the results obtained. The elemental composition of a sample obtained using point X-ray microanalysis (scanning microscope) corresponds with the elemental composition as determined by X-ray apparatus (elements, in the form of specific minerals).

Further studies, using the above and other advanced testing methods, will create comprehensive testing procedures for mineral raw materials and waste at a laboratory scale. Then, these procedures can be applied for *in situ* measurements of the variable properties of raw materials, products and waste on an industrial scale. The test procedure will be helpful in the continuous evaluation of the properties of raw materials, products and waste in processing plants. This will allow for improvements in enrichment

technology, to increase the yield of useful components, and to find the most economic means of disposing of waste.

REFERENCES

Ahuja, S. & Jespersen, N. 2006. *Modern instrumental analysis*. Elsevier. Amsterdam.

Ashraf, M., Naeem Khan, A., Qasair, A., Mirza, J., Goyal, A., & Anwar, A.M. 2009. Physico-chemical, morphological and thermal analysis for the combined pozzolanic activities of minerals additives. *Construction and Building Materials* 23(6): 2207–2213.

Bigosiński, J. & Drzymała, J. 1996. Microscopic determination of content of chalcocite in mixtures with galena after thermal modification of their surfaces. II Międzynarodowa Konferencja Przeróbki Kopalin. *Zeszyty naukowe Politechniki Śląskie*. no. 1349, Seria Górnictwo 231: 37–45. Ustroń.

Buršík, J. & Brož, P. 2009. Constitution of Ni–Al–Ti system studied by scanning electron microscopy. *Intermetallics* 17(8): 591–59.

Elssner, G., Hoven, H., Kiessler, G. & Wellner, P. 1999. *Ceramics and ceramic composites: Materialographic preparation*. Elsevier. Eastbourne.

Fecko, P., Kucerova, R., Stahovcova, A. & Bouchal, T. 1996. Application of bacterial leaching in the processing of waste from incineration plant from Prostejov. II Międzynarodowa Konferencja Przeróbki Kopalin, *Zeszyty naukowe Politechniki Ślłskiej*, no. 1349, Seria Górnictwo 231: 129–142. Ustroń.

Goldstein, J., Newbury, D.E., Joy, D.C., Lyman, C.E., Echlin, P., Lifshin, E., Sawyer, L.C. & Michael, J.R. 2003. *Scanning electron microscopy and X-ray microanalysis*. 3rd ed. Plenum Press. New York.

Kędzior, A., Trybalski, K. & Konieczny, A. 2003. Zastosowanie nowoczesnych metod badawczych w inżynierii mineralnej. *Inżynieria Mineralna* 3: 155–165. Kraków. in Polish.

Szponder D. 2008. *Badanie właściwości fizycznych popiołów lotnych z wykorzystaniem różnych metod i urządzeń badawczych*. Praca Magisterska. Akademia Górniczo-Hutnicza im. Stanisława Staszica w Krakowie. Wydział Górnictwa – Geoinżynierii. Katedra Przeróbki Kopalin i Ochrony Środowiska. Kraków (not published), in Polish.

Van Dyk, J.C., Benson, S.A., Laumb, M.L., & Waanders, B. 2009. Coal and coal ash characteristics to understand mineral transformations and slag formation. *Fuel:* 88(6): 1057–1063.

Environmental Engineering III – Pawłowski, Dudzińska & Pawłowski (eds)
© *2010 Taylor & Francis Group, London, ISBN 978-0-415-54882-3*

The influence of aeration rate on production of leachate and biogas in aerobic landfills

R. Ślęzak, L. Krzystek & S. Ledakowicz
Faculty of Process and Environmental Engineering,
Technical University of Lodz, Lodz, Poland

ABSTRACT: Simulation of aeration of old waste for short periods was carried out in laboratory-scale lysimeters. In parallel, the decomposition of organic matter present in old waste was performed under anaerobic conditions. The course of degradation processes after completion of aeration was also monitored. The high degree of reduction in leachate indices using an aerated lysimeter compared to an anaerobic lysimeter was observed. After completion of aeration, an increase in the values of indices for organic pollutants in the leachate, as well as the rate of biogas production and subsequently their reduction, were observed.

Keywords: Municipal solid waste, leachate, lysimeter, aerobic landfill.

1 INTRODUCTION

Landfills are treated as huge bioreactors in which long-term physical, chemical and biochemical processes occur. The products of decomposition of organic matter that are present in municipal waste constitute a serious hazard to the environment. The lack of or damaged isolating layer and the layer covering a landfill may bring about the transfer of pollutants to the environment. The acceleration of organic matter decomposition in waste may be achieved by changing the anaerobic conditions to aerobic ones.

Decomposition of organic matter that is present in waste under anaerobic conditions is a slow process and depends on many factors (Bilgili et al. 2006). Methane and carbon dioxide are generated from organic matter that is present in municipal waste under anaerobic conditions. Biogas that is generated in anaerobic landfills can be energetically utilized for about 20 years. The anaerobic processes of organic matter degradation contribute to the production of leachate with high pollutant loading. In anaerobic processes, the decomposition of organic matter present in waste may be accelerated by leachate recirculation (Pohland, 1980).

The air supplied to waste contributes to the acceleration of biological degradation of biodegradable organic components (Hantsch et al. 2003). During aerobic degradation of waste, the biodegradable mass is converted mainly into carbon dioxide and water (Mertoglu et al. 2006). The aeration of waste mass was examined both in bioreactors and in landfills. Aerobic landfill investigations were carried out by Jacobs et al. (2003) and Read et al. (2001), as well as by others. The simulation of aerobic landfill in bioreactors was conducted by Bilgili et al. (2006 and 2007), Borglin et al. (2004), Cossu et al. (2003) and Erses et al. (2008).

The influence of aeration and recirculation of leachate on organic matter decomposition rate was investigated on a laboratory scale. During waste aeration, a reduction in organic pollutants of leachate occurs, causing acceleration of waste mass settlement, lack of odor emission, an increase in temperature during aeration and reduction of methane emissions.

Investigations concerning aerobic stabilization of old landfills were carried out both in laboratories and in landfills. In Europe, the stabilization of old landfills was performed in Kuhstedt, Amberg, Milmersdorf, Modena and Legnago. The aerobic stabilization of old landfills was undertaken by Hantsch et al. (2003), Prantl et al. (2006), Ritzkowski et al. (2006) and Zieleniewska-Jastrzebska et al. (2008). Rapid waste stabilization was obtained by decomposition of organic matter remaining in waste, as shown by the investigations carried out both in landfills and in laboratories. The aeration of waste caused a reduction in odors, a decrease in methane concentration in gas and, furthermore, it accelerated landfill settlement and carbon emissions from waste through an increase in carbon dioxide (Ritzkowski & Stegmann, 2003).

To compare the processes occurring in aerobic and anaerobic landfills, tests were conducted in bioreactors. The comparison of organic matter decomposition under aerobic and anaerobic conditions was the subject of interest of research by Bilgili et al. (2007), Borglin et al. (2004) and Erses et al. (2008). From the simulations, it may be concluded that the most rapid decrease of organic load in leachate occurred in aerated bioreactors. In their examinations, Erses et al. (2008) achieved a COD reduction of 90% in the aerobic bioreactor after 70 days and in the anaerobic bioreactor after 462 days.

Prantl et al. (2006) investigated the influence of aeration and leachate recirculation on the stabilization rate of old landfills. The examinations were carried out in bioreactors that were loaded with waste originating from an old landfill. The simulation of aerobic stabilization of waste was conducted for 513 days. In one of the bioreactors, waste was aerated for 270 days and the anaerobic conditions were maintained for the consecutive 243 days. The earlier completion of aeration caused a slight increase in respiratory activity (RA_4) and ammonium concentration, indicating the existence of biologically available organic substances.

Spendlin (1991) noted that, even during a short aeration period, an organic fraction is made available for anaerobic microorganisms in the following anaerobic processes. Nevertheless, the simulation of aeration of waste of low biogas production for a short period of time and observations relating to the completion of aeration, as well as a comparison with anaerobic processes, have not yet been carried out.

The aim of this study was to carry out simulation of aeration of old waste for a short period of time in a lysimeter, and to monitor the changes on completion of aeration. In parallel, the decomposition of organic matter present in old waste was carried out under anaerobic conditions. Although the processes were carried out under both aerobic and anaerobic conditions, the composition of leachate and flue gas from lysimeters was determined. In leachate, the following indices were analyzed: biochemical oxygen demand (BOD_5), chemical oxygen demand (COD) and ammonium nitrogen concentration ($N-NH_4^+$). The concentration of oxygen, methane and carbon dioxide, as well as the amount of gas generated, were analyzed in flue gas from lysimeters.

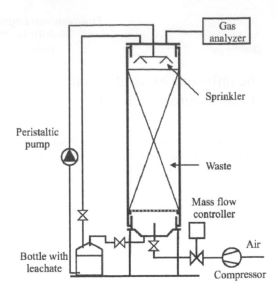

Figure 1. Experimental set-up.

2 MATERIALS AND METHODS

2.1 *Apparatus and process parameters*

The experimental simulation of a landfill was carried out in two lysimeters (S1, S2) of working volume 15 dm^3. The set-up of the experimental installation is shown in Figure 1. The lysimeters consisted of a plastic cylinder of 150 mm in inner diameter and of 1150 mm in height, which was closed on the top and bottom with stainless steel covers and equipped with pipes for leachate recirculation, which could be used to take samples for analysis, and to supply and remove gases. The moisture content in waste was maintained by a system of leachate recirculation consisting of a bottle storing leachate from the lysimeter, a peristaltic pump (type 101 U/R, WATSON MARLOW PUMPS) and a sprinkler. The rate of the air pumped into the lysimeter was controlled by a mass flow-meter (a model 5850TR, Emerson). At the outlet of gas from the lysimeter, a gas analyzer (LMS Gas Data), as well as a gas flow-meter (Ritter TG01/05 type, with EDU 32 recorder), were located.

In the lysimeters (S1, S2), the anaerobic conditions were maintained for about 6 months. During anaerobic conditions, an acidic and a methane

phase could be observed in the lysimeter. When the final methane phase occurred in the lysimeter (Zieleniewska-Jastrzebska et al. 2008), and the gas production rate was decreased ten-fold with regards to the maximum production in the methane phase, the process of aeration of waste commenced. The process of aeration of waste was carried out in lysimeter S1 for 75 days. On completion of aeration, leachate and the composition of biogas were monitored for the following 80 days. Lysimeter S2 was not aerated in order to compare the results attained.

In the methane phase, the maximum production of gas was equal to 4 dm^3/d (65% v/v CH_4, 35% v/v CO_2). The aeration started with the volumetric flow-rate of gas production equal to 0.4 dm^3/d and the composition of gas was 45% v/v methane and 55% v/v carbon dioxide. During the aeration of waste, 3 dm^3/h of the air was supplied into the lysimeter. In the course of the whole experiment, leachate was recirculated once a day for 15 minutes, with the volumetric flow rate being equal to 1 cm^3/s. The investigations in the lysimeter were repeated three times.

2.2 *Substrate*

The lysimeter was loaded with a model composition of municipal solid waste, which was defined on the basis of analysis of the morphological composition of waste for the city of Lodz (Ledakowicz & Kaczorek, 2004). The percentage composition of model waste is presented in Table 1. The acceleration of simulation processes of landfills in the lysimeter was obtained by adding compost from the Compost Facility of Green Waste in Lodz. A layer of waste shredded to the size 2–4 cm was alternately laid with a layer of compost mixture. The lysimeters were loaded with 5 kg of model mass and then 5 dm^3 of tap water were added.

Table 1. Waste composition.

Waste composition	Composition %
Organic waste	28
Paper	19
Plastics	12
Textiles	4
Compost	27
Other inorganic	10

2.3 Methodology of analysis

In leachate taken from the lysimeter, the following indices were analyzed: BOD_5 by dilution method (Greenberg et al. 1992), COD by dichromate method (Greenberg et al. 1992), and ammonium nitrogen ($N-NH_4^+$) by a distillation method in the Büchi device.

In the gas leaving the lysimeter, the concentration of oxygen (O_2), methane (CH_4) and carbon dioxide (CO_2) was measured using a gas content analyzer LMS GAS DATA, as well as gas flow rate being assessed using the Ritter flow-meter (type TG01/05, with EDU 32 recorder).

3 RESULTS AND DISCUSSION

3.1 Leachate

In the investigations, the simulation of the waste aeration process in order to define the changes in leachate composition during and after aeration was carried out. When the gas production rate decreased to the level of 0.4 dm^3/d, the waste aeration process commenced in lysimeter S1 and it was continued for 75 days. On completion of aeration, the leachate composition was analyzed in lysimeters S1 and S2 for the following 80 days.

The changes in organic substance content in leachate were characterized in an indirect way by analyzing the following indices of organic load: BOD_5 and COD.

The changes in BOD_5 in lysimeter S1 and S2 are presented in Figure 2. During aeration of lysimeter S1, the value of BOD_5 decreased from 165 to 10 mg O_2/dm^3. The greatest reduction of BOD_5 value was observed in the first 3 weeks of aeration. Borglin et al. (2004) obtained a value of BOD_5 index of 9 mg O_2/dm^3 after 370 days of aeration during aerobic landfill simulation of waste originating from the landfill in bioreactors. In lysimeter S2, which was not aerated, the BOD_5 index decreased from 121 to 60 mg O_2/dm^3. The high reduction in BOD_5 index in lysimeter S1 was evoked by the aerobic conditions in which the rapid decomposition of organic matter occurred. When aeration of lysimeter S1 was completed, the BOD_5 index was 83% lower than that of the control lysimeter S2.

After aeration of lysimeter S1, the composition of leachate in lysimeters S1 and S2 was analyzed. On

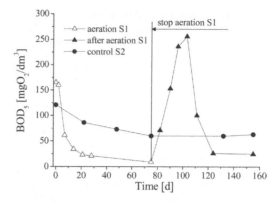

Figure 2. The changes in biochemical oxygen demand (BOD_5) in leachate from lysimeters S1 and S2.

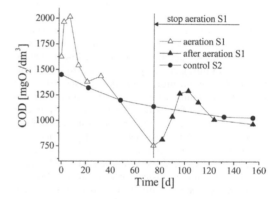

Figure 3. The changes in chemical oxygen demand (COD) in leachate from lysimeters S1 and S2.

the 28th day after aeration, the BOD_5 index value in lysimeter S1 increased to 256 mg O_2/dm^3. After 80 days of aeration of lysimeter S1, the BOD_5 value in lysimeter S1 was 25 mg O_2/dm^3. The increase in BOD_5 index on the first days after completion of aeration was brought about by the change from aerobic processes to anaerobic ones. In the control lysimeter S2, over the course of 80 days, the value of BOD_5 increased from 60 to 63 mg O_2/dm^3 after completion of aeration. On the 80th day after completion of aeration, the BOD_5 value in lysimeter S1 was lower by 60% when compared to the control lysimeter S2.

A subsequent index that defines the organic mass content in an indirect way is COD. Changes in COD value in the lysimeters are presented in Figure 3. Within the first 7 days of aeration of lysimeter S1, the increase of COD from 1630 to 2020 mg O_2/dm^3 was recorded. In lysimeter S1, the increase in COD was evoked by a rapid decomposition of organic matter present in waste and a slower process of organic carbon decomposition in leachate. On consecutive days of aeration, the value of COD in lysimeter S1 started to decrease and, after 75 days of aeration, the value of COD was 760 mg O_2/dm^3. A similar value of COD after aeration was obtained by Erses et al. (2008) in their research during the simulation of aerobic landfills in a bioreactor. By aerating waste in the bioreactor for 374 days, Erses et al. (2008) obtained a reduction of COD to 680 mg O_2/dm^3. In lysimeter S2, within 75 days the value of COD decreased from 1450 to 1170 mg O_2/dm^3. The values of COD in lysimeter S1, when aeration was completed, were lower by 34% with regards to the control lysimeter S2.

After completion of aeration of lysimeter S1, the COD value started to increase. On the 28th day after completion of aeration of lysimeter S1, the COD value increased to 1290 mg O_2/dm^3. On consecutive days, the COD value started to decrease and, after 80 days from completion of aeration, it was equal to 970 mg O_2/dm^3. In the control lysimeter S2, the COD value decreased from 1140 to 1030 mg O_2/dm^3. On the 80th day after completion of aeration, the COD value in lysimeter S1 was 6% lower than in the control lysimeter S2.

The next discussed index of leachate is ammonium nitrogen concentration, which defines the processes of amonification and nitrification. The low content of ammonium nitrogen was probably caused by the initial model composition of organic substances in the lysimeter, moisture with water and addition of compost. Within the first 7 days of aeration of lysimeter S1, the value of ammonium nitrogen increased from 12 to 15 mg N/dm^3. On consecutive days of aeration, the ammonium nitrogen concentration decreased to 4 mgN/dm^3 after 75 days of aeration. The high reduction in ammonium nitrogen in leachate was brought about by the process of nitrification. In the non-aerated lysimeter (S2), a reduction of ammonium nitrogen concentration in leachate from 15 to 13 mgN/dm^3 was observed. The value of ammonium nitrogen concentration after aeration in lysimeter S1 was 69% lower than that in lysimeter S2. The similar values of ammonium nitrogen concentration after aeration were achieved by Cossu et al. (2003), as well as by Erses et al. (2008). In their investigations, Cossu et al. (2003) aerated the

mechanical biological pretreated (MBP) refuse from the Legnago Plant (Verona, Italy) in a bioreactor for 120 days.

After completion of aeration of lysimeter S1, the ammonium nitrogen concentration in leachate started to increase. On the 49th day after completion of aeration, the ammonium nitrogen concentration increased to 13 mg N/dm^3. On the 80th day after completion of aeration of lysimeter S1, the ammonium nitrogen concentration decreased to 5 mgN/dm^3. During the investigations of aerobic waste stabilization in a bioreactor, Prantl et al. (2006) noticed that after the earlier completion of aeration, the increase in ammonium nitrogen concentration occurred and, following this, it was maintained at 40 mgN/dm^3.

In the control lysimeter S2, the ammonium nitrogen concentration in leachate decreased from 13 to 12 mg N/dm^3 after completion of aeration of lysimeter S1. On the 80th day after completion of aeration, the ammonium nitrogen concentration in lysimeter S1 was 58% lower than in the non-aerated lysimeter S2.

3.2 Gas

The composition of flue gas from the lysimeters and its quantity are important parameters defining both the impact on the environment and the processes taking place in the lysimeters. Figure 5 shows the changes in composition of flue gas from lysimeter S1, which was aerated, and the control lysimeter S2.

From the moment that the aeration of lysimeter S1 started, the concentrations of methane and carbon dioxide in lysimeters S1 and S2 were similar and were equal to 45 and 55% v/v, respectively. After the first day of aeration of lysimeter S1, the concentration of methane decreased to zero and the concentration of carbon dioxide decreased to 4.6% v/v. In the course of aeration of lysimeter S1, the concentration of carbon dioxide in flue gas was 2.8% v/v and the concentration of oxygen was 17.5% v/v. During aeration, the concentration of methane in flue gas from lysimeter

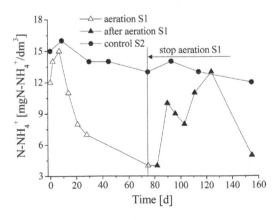

Figure 4. The changes in ammonium nitrogen content (N-NH$_4^+$) in leachate from lysimeters S1 and S2.

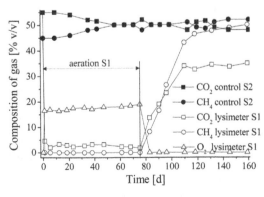

Figure 5. The changes in flue gas composition in lysimeters S1 and S2.

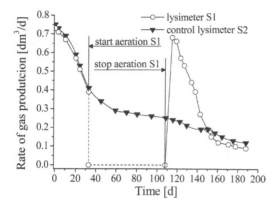

Figure 6. The changed rate of gas production in lysimeters S1 and S2.

S1 was equal to zero. In the control lysimeter S2, the concentrations of methane and carbon dioxide after 75 days were equal to 48 and 52% v/v, respectively.

After completion of aeration of lysimeter S1, the concentration of oxygen rapidly decreased to zero and the concentrations of methane and carbon dioxide started to increase. On the 35th day after completion of aeration of lysimeter S1, the concentration of methane was 43% v/v and the concentration of carbon dioxide was 34% v/v. On subsequent days, the concentration of carbon dioxide was maintained at a constant level and the concentration of methane slightly increased. After 80 days of aeration of lysimeter S1, the concentration of methane was equal to 50% v/v and the concentration of carbon dioxide was equal to 35% v/v. After 80 days of aeration of lysimeter S1, the concentration of methane was lower by 4% and carbon dioxide was lower by 27% compared with the control lysimeter S2, which was not aerated.

In addition to the composition of flue gas, an important parameter defining the influence on the environment is the quantity of gas produced. In Figure 6, the changes in gas production rate in lysimeters S1 and S2 are presented. The aeration of lysimeters commenced when the gas production rate was equal to $0.4 \, dm^3/d$. After finishing the aeration of lysimeter S1, it was noticed that, on the 7th day, the gas production rate increased to $0.7 \, dm^3/d$. On consecutive days, the gas production rate decreased and, on the 80th day from completion of aeration, it was equal to $0.09 \, dm^3/d$. After completion of aeration of lysimeter S1, the gas production rate in the control lysimeter S2 was equal to $0.25 \, dm^3/d$, and after 80 days it decreased to $0.12 \, dm^3/d$.

4 CONCLUSIONS

Simulation of landfill processes was conducted in two lysimeters filled with standard waste composition. One of the lysimeters was aerated for a short

period of time (75 days) from the moment when a final methane phase was observed and the production of biogas decreased to $0.4 \, dm^3/d$ (i.e. after around 6 months). Next, for the subsequent 80 days, the processes running under no aeration conditions were analysed. The S2 lysimeter was not aerated throughout the entire research period in order to compare aerobic and anaerobic processes running in the lysimeters. The novelty of the study involved presentation of changes in the amount and composition of the produced biogas after short-lasting, intensive waste aeration and intensive recirculation of leachate have been completed; comparison of results was performed in a non-aerated lysimeter.

During waste aeration, within 75 days in lysimeter S1, the reduction in BOD_5 value was 95%, COD was 53% and the ammonium nitrogen concentration was 67%. In the non-aerated lysimeter (S2), the reduction in BOD_5 was equal to 80%, COD was 21% and the ammonium nitrogen concentration was 13%. The high degree of reduction in leachate indices in lysimeter S1 when compared to lysimeter S2 was evoked by the rapid organic matter decomposition by aerobic microorganisms.

From the leachate analysis after 80 days from completion of waste aeration, it may be inferred that, in lysimeter S1, there could be an increase in BOD_5 value by 180%, in COD by 28% and in ammonium nitrogen concentration by 25%. In the non-aerated lysimeter (S2), there was a reduction in the leachate indices for COD by 10% and for the ammonium nitrogen concentration by 8%, whereas for BOD_5 there was an increase of 5%. Comparing the values of leachate indices, it was observed that the values of leachate indices were lower in lysimeter S1 when juxtaposed to the control lysimeter (S2): for BOD_5 by 60%, for COD by 6% and for the ammonium nitrogen concentration by 58%.

On the first days after the onset of aeration of the lysimeter, an increase of COD and $N\text{-}NH_4^+$ could be observed. On completion of aeration of lysimeter S1, all leachate indices started to increase within the first days. The increase of the indices after changing the conditions from anaerobic to aerobic ones or vice versa was brought about by changes in the organic matter decomposition rate in leachate and waste.

Lysimeter S1 started to be aerated when the gas production rate was equal to $0.4 \, dm^3/d$. In the course of aeration of lysimeter S1, the concentration of oxygen and carbon dioxide in flue gas from lysimeter S1 was equal to 17.5 and 2.8% v/v, respectively. High concentrations of oxygen and low concentrations of carbon dioxide in flue gas were evoked by the introduction of a large quantity of the air to waste. In the control lysimeter S2, while lysimeter S1 was being aerated, the concentration of methane and carbon dioxide was in the range of 45–55% v/v.

After completion of aeration of lysimeter S1, the concentration of methane and carbon dioxide commenced to increase. On the 35th day after completion of aeration of lysimeter S1, the concentration of methane was 43% v/v and the carbon dioxide concentration was 34% v/v. After 80 days since the aeration of lysimeter S1 had finished, the concentration of methane was equal to 50% v/v, whereas the concentration of carbon dioxide was equal to 35% v/v. On the other hand, in the control lysimeter S2, the concentrations of methane and carbon dioxide were in the range of 45–55% v/v.

The quantity of gas produced in lysimeter S1 was measured for 80 days after finishing the aeration. In lysimeter S1 on the 7th day after completion of aeration, the gas production rate attained the maximum value of $0.7\,dm^3/d$. On consecutive days, the gas production rate started to decrease and on the 80th day after finishing the aeration, it was equal to $0.09\,dm^3/d$. In the control lysimeter (S2), after completion of aeration of lysimeter S1, the gas production rate was $0.25\,dm^3/d$, and after 80 days it was $0.12\,dm^3/d$. The aeration of waste in lysimeter S1 contributed to the production of a much higher quantity of gas than in the control lysimeter (S2) on the first days after completion of aeration. The high gas production rate corresponded to the low concentrations of methane and carbon dioxide. A slow increase in the methane and carbon dioxide concentrations on completion of aeration was brought about by the period of adaptation of anaerobic microorganisms. The high gas production rate after completion of aeration of lysimeter S1 was evoked mainly by aerobic organisms that produced carbon dioxide for a certain period of time after finishing the aeration when oxygen was available in waste.

In the literature, there are studies that compare aerobic and anaerobic conditions in landfills with low biogas production (Prantl et al. 2006, Ritzkowski et al. 2006). In these studies, Prantl et al. (2006) conducted a process of waste aeration for 513 days and Ritzkowski et al. (2006) for 553 days. During waste aeration, there was a high reduction in organic pollutants occurring in leachate in comparison to the non-aerated lysimeters. In our study, a short period of aeration (75 days) was caused by a high aeration rate and a large amount of recirculated leachate. Despite different conditions under which the process was conducted, similar types of changes in the indices of leachate from lysimeters were observed, in which aerobic and anaerobic processes took place.

Prantl et al. (2006) presented results of the studies in which waste aeration was stopped after 270 days and, for the 234 days after the aeration had been finished, waste composition was monitored. They concluded that, after aeration had been completed, the COD value of the leachate remained on the same level, whereas the concentration of ammonia nitrogen increased.

Studies presented in this paper involve the determination of the influence of a high aeration rate, large amount of recirculated leachate and short aeration time on the processes taking place in waste after aeration has ended.

It has been concluded that, under such conditions, changes in the examined indices after completion of aeration were varied – an increase in the value of indices of organic pollutants in the leachate was observed, as well as an increase in the rate of biogas production in the first days after aeration completion and, subsequently, a decrease in both was noted.

ACKNOWLEDGEMENTS

This work was supported by grant No. PBZ MEiN 3/2/2006 founded by the Ministry of Science and Higher Education, Poland.

REFERENCES

Bilgili, M., Demir, A. & Ozkaya, B. 2006. Quality and quantity of leachate in aerobic pilot-scale landfills. *Environmental Management* 38(2): 189–196.

Bilgili, M., Demir, A. & Ozkaya, B. 2007. Influence of leachate recirculation on aerobic and anaerobic decomposition of solid wastes. *Journal of Hazardous Materials* 143(1-2): 177–183.

Borglin, S.E., Hazen, T.C., Oldenburg, C.M. & Zawislanski P.T. 2004. Comparison of aerobic and anaerobic biotreatment of municipal solid waste. *Journal of the Air & Waste Management Association* 54(7): 815–822.

Cossu, R., Raga, R. & Rossetti, D. 2003. The PAF model: an integrated approach for landfill sustainability. *Waste Management* 23(1): 37–44.

Erses, A. S., Onay, T.T. & Yenigun, O. 2008. Comparison of aerobic and anaerobic degradation of municipal solid waste in bioreactor landfills. *Bioresource Technology* 99(13): 5418–5426.

Greenberg, A. E., Clesceri, L. S. E. & Eaton, A. D. 1992. *Standard methods for the examination of water and wastewater*. 18th ed., American Public Health Association, Washington.

Hantsch, S., Michalzik, B., & Bilitewski, B. 2003. Different intensities of aeration and their effect on contaminant emission via the leachate pathway from old landfill waste – a laboratory scale study. Proceedings of Sardinia 2003, *Ninth International Waste Management and Landfill Symposium, Sardinia 2003*, Cagliari, Italy, 6–10 October 2003, Conference proceedings (CD-ROM).

Jacobs, J., Scharff, H., Van Arkel, F. & De Gier, C.W. 2003. Odour reduction by aeration of landfills: experience, operation and costs. Proceedings Sardinia 2003, *Ninth International Waste Management and Landfill Symposium, Sardinia 2003*, Cagliari, Italy, 6–10 October 2003, Conference proceedings (CD-ROM).

Ledakowicz, S. & Kaczorek, K. 2004. The effect of advanced oxidation processes on leachate biodegradation recycling lysimeters. *Waste Management & Research* 22(3): 149–157.

Mertoglu, B., Calli, B., Inanc, B. & Ozturk, I. 2006. Evaluation of in situ ammonium removal in an aerated landfill bioreactor. *Process Biochemistry* 41(12): 2359–2366.

Pohland F. G. 1980. Leachate recycle as landfill management option. *Journal of Environmental Engineering Division ASCE* 106(6): 1057–1069.

Prantl, R., Tesar, M., Huber-Humer, M. & Lechner, P. 2006. Changes in carbon and nitrogen pool during in-situ aeration of old landfills under varying conditions. *Waste Management* 26(4): 373–380.

Read A.D., Hudgins M. & Philips P. 2001. Perpetual land-filling through aeration of the waste mass; lessons from test cells Georgia (USA). *Waste Management* 21(7): 617–629.

Ritzkowski, M. & Stegmann, R. 2003. Emission behavior of aerated landfills: results of laboratory scale investigations. Proceedings of Sardinia 2003, *Ninth International Waste Management and Landfill Symposium, Sardinia 2003*, Cagliari, Italy, 6–10 October 2003, Conference proceedings (CD-ROM).

Ritzkowski, M., Heyer, K.-U. & Stegmann R. 2006. Fundamental processes and implications during in situ aeration of old landfills. *Waste Management* 26(4): 356–372.

Spendlin, H.-H. 1991. Untersuchungen zur frühzeitigen Initiierung der Methanbildung bei festen Abfallstoffen, *Hamburger Berichte*, Bd. 4, Economica Verlag, Bonn.

Zieleniewska-Jastrzebska, A., Krzystek, L. & Ledakowicz, S. 2008. Aerobic stabilization of old landfills – experimental simulation in lysimeters. In: M. Pawlowska & L. pawlowski (eds). *Management of Pollutant Emission from Landfills and Slugde*, Taylor and Francis: New York .

Environmental Engineering III – Pawłowski, Dudzińska & Pawłowski (eds)
© 2010 Taylor & Francis Group, London, ISBN 978-0-415-54882-3

Occurrence and bindings strength of metals in composted bio-waste and sewage sludge

I. Twardowska, K. Janta-Koszuta, E. Miszczak & S. Stefaniak
Institute of Environmental Engineering, Polish Academy of Sciences, Zabrze, Poland

ABSTRACT: A comparative study on Potentially Toxic Elements (PTEs): Ni, Zn, Cd, Cu, Cr and Pb content and their binding strength in selectively collected and composted bio-waste (BWC) and sewage sludge (SS) originating from the historical industrial area was carried out in order to assess their contamination potential, if applied to land. In BWC, Cd contents were found to exceed the EU median values and also mean concentrations in unpolluted soils. Concentrations of PTEs in SS appeared to be several times higher than in BWC, and about 10-fold higher than mean values in soils. Most PTEs originally occurring in BWC and SS were enriched in a reducible fraction of a moderate binding strength, but relatively high metal loads showed affinity to the labile fractions. Contents and mobility of metals in SS and in some cases also in BWC originating from historical industrial areas, suggest precautionary approach to their application to land.

Keywords: Composted bio-waste; sewage sludge; metal contents; metal mobility.

1 INTRODUCTION

Biodegradable waste (bio-waste) is the major part of municipal solid waste (MSW) stream and in the EU constitutes mostly between 30% and 40% of MSW, ranging from 18% to 60%. It comprises green waste (park and garden waste from public estates and households), food and kitchen waste from households and restaurants and waste from the food industry (ComEC 2008). Municipal waste generated in the EU 27 countries in 2007 accounted for 259.4 Mt and ranged from 309 kg to 801 kg per capita (mean value 522 kg per capita) (Eurostat 2009). The total annual generation of bio-waste in the EU is estimated at 76.5–102 Mt green and food waste of a water content 50–60% wt., and up to 37 Mt from the food industry, of a higher water content (up to 80% wt.). The overall potential for separately collected bio-waste for the EU 27 is estimated at 80 Mt that is up to 150 kg/inhabitant/year (ComEC 2008). This accounts for about 30% of current total municipal waste generation per capita. At 350–400 kg of compost production from every tone of bio-waste, the overall potential of compost production accounts for 35–40 Mt. Currently, 24 Mt of bio-waste, i.e. about 30% of the total, is collected separately and composted, which in 2005 resulted in 13.2 Mt of compost. Of this amount, 5.7 Mt (43%) was produced from green waste, 4.8 Mt from bio-waste (36%), 1.4 Mt (10.5%) from mixed waste and 1.4 Mt (10.5%) from sewage sludge. This compost is used mostly in agriculture (about 50%), for landscaping (up to 20%) and for producing blends and manufactured soil (around 20%) (ComEC 2008).

The bio-waste recycling by composting supported by separate (source) collection is generally accepted as the best management option. It has been successfully introduced in Austria, The Netherlands, Germany, Sweden, Belgium (Flanders), Spain (Catalonia) and in the northern regions of Italy (Amlinger et al. 2004, Favoino 2004). In some Member States (Denmark, France, Czech Republic), the source collection is focused on green waste only. Compost produced from mixed waste is used predominantly for land restoration and landfill cover (e.g. in Finland, Ireland, Poland). Recently released extensive state-of-the-art reviews (Amlinger et al. 2004, Smith 2009) report well documented much higher Potentially Toxic Elements (PTEs) concentrations in sewage sludge or in compost based on mechanically segregated mixed bio-wastes than in composted source-collected bio-waste.

Separate collection and composting of bio-waste, besides of being an attractive option used for soil improvement, may be regarded also as an optimum way to reduce landfilling. At present, in the EU on average 41% of MSW is landfilled, while in some new Member States (e.g. Poland, Lithuania) almost all MSW (over 90%) is disposed at landfills, thus further significant reduction of landfilled bio-waste may be achieved (ComEC 2008). It has to be noted that there is no specific directive addressing various aspects of bio-waste management, including its agricultural use, while a number of the EU legislative documents, in particular the revised Waste Framework Directive (EC 2006) is related to bio-waste to the certain extent.

On the contrary, agricultural use of another kind of biodegradable waste, sewage sludge, (SS) is regulated by the over 20-years' old Sewage Sludge Directive 86/278/EEC (EEC 1986), which encourages this means of sewage sludge use, but also sets limit values for metals (seven PTEs) in this material intended to be used in agriculture, as well as in soils to be amended with sewage sludge. While the method of collection may greatly enhance environmental safety of bio-waste, the quality of sewage sludge seems to be much less controllable due to its origin. Although enriched with nutrients (N, P, organic matter OM), sewage sludge is a sink of all possible anthropogenic pollutants, including both metals and also persistent hazardous organic compounds (Andersen-Sede 2001, Twardowska et al. 2004). Thus, application of this material to land requires using precautionary principle and should provide adequate protection against environmental risks. Due to the progressive implementation of the Urban Waste Water Treatment Directive 91/271/EEC (EEC 1991), the amount of sewage sludge generated in the EU increased from 5.5 Mt d.m. in 1992 to about 9 Mt d.m. in 2005, with prospects of further fast increase of its quantities as a result of accelerated construction of sewerage systems and Municipal Sewage Treatment Facilities (MSTF), in particular in the new Member States (EC.EUROPA, 2009). Currently, the need of the revision of the Sewage Sludge Directive is being in the course of assessment and consulting (to be completed in 2009). Sewage sludge is also a subject of several studies launched by the EC (e.g. Andersen-Sede 2001) and of many research works carried out by different authors throughout the world that are focused on the evaluating potential risk of its land application to the environment and humans, still with a large dose of uncertainty (e.g. Horn et al 2003, McBride 2003, He et al. 2007, Singh & Agrawal 2007, Bhattacharya et al. 2008, Carbonell et al. 2009, Oleszczuk 2009). According to the data from the end of 2002, reuse of SS in the EC accounted for about 40% (EC.Europa 2002), agricultural use being the preferred option in Ireland, Finland, UK, France, Luxembourg, Denmark and Spain (Andersen-Sede 2001), and in many non-EU countries of the world, e.g. in the USA, Australia and New Zealand. Both approach, experimental design, results and derived conclusions from these studies differ substantially. Several Member States have implemented more stringent standards, while other are still tolerant to higher contaminant loads; in some countries traditionally using SS as "biosolids" the level of awareness and precautionary approach is growing (Twardowska et al. 2004).

This comparative study is focused on the evaluation of metals (six PTEs – Ni, Zn, Cd, Cu, Cr, and Pb) occurrence and their binding strength in selectively collected and composted bio-waste (BWC) and sewage sludge (SS) originating from the same historical industrial area of Zabrze city in the Silesia agglomeration in Southern Poland – the area of coal mining, power plants fired by coal, steel production, coke plants and chemistry (currently partially closed or declining) and other industrial activities, which since the industrial revolution of XIX century heavily impacted the environment. The major goal of the study was to evaluate the potential of BWC and SS applied to land to contaminate the environment, mainly soil and groundwater, which depends both on the metals content and their mobility/ susceptibility to remobilization. The obtained data were analyzed on the background of values for PTEs of European national compost data and derived conclusions, presented in the final report ENV/A.2./ETU/2001/0024 for DG Environment by Working Group Compost – Consulting & Development (Amlinger et al. 2004) and in another recent review (Smith 2009). The study is considered as a contribution to the general discussion on the optimization of bio-waste and sewage sludge management in view of the environmental protection and sustainable development requirements.

The location of the research area (the Silesia industrial agglomeration in Poland) intends to draw attention to the quality of bio-waste originating from historical industrial areas in Europe, with particular regard to the new Member States of the EU. Up to now, these states including Poland have no developed municipal waste management system other than landfilling. This research is aimed to contribute also to the development of appropriate and environmentally safe system of municipal waste management in these and other countries with similar waste management status.

2 MATERIAL AND METHODS

2.1 Material sampling and preparation

2.1.1 Composted of bio – waste
The studied bio-waste was originated from Waste Segregation and Composting Facility (WSCF) in Zabrze city constructed simultaneously with MSW landfill in 1999 (Figure 1).

Bio-waste management is one of the major Facility activities. Green waste are collected from the segregation centers in the Zabrze city area and in the area of the neighboring cities of the Silesia agglomeration and transported to the WSCF. Green waste and kitchen waste from restaurants are used for production of "Eco-grunt" soil improver at a capacity of about 2200 t/year, which is utilized for degraded land restoration or in gardening and agriculture. The studied material is produced by one of the not numerous, recently established, but well managed bio-waste composting facilities situated in the area of coal extraction, coal-based power generation, coking and iron and steel metallurgy.

Figure 1. Waste Segregation and Composting Facility (WSCF) in Zabrze.

Two representative samples of freshly produced and stored (about 1 year old) compost were taken from 30 points of each compost pile, composited by thorough mixing, reduced to about 50 kg by cone-and-quartering method and transported to the storage in closed and sealed HDPE bags in accordance with CEN/TR-15310-1,2,3,4. After moisture content assessment, laboratory samples were prepared by further cone-and-quartering, air-dried and composited again.

2.1.2 Sewage sludge

Sewage sludge was sampled from the "Zabrze-Center" Sewage Treatment Facility (STF) with tertiary treatment (Figure 2). Samples were taken at the outlet of a filter-press as an average for each shift for ten consecutive days, and a representative composite sample was prepared in the same way as compost and transported to the storage, where an air-dried laboratory sample was prepared for the further analysis.

2.2 Analytical methods

The analytical methods comprised basic characterization of bio-waste and sewage sludge with respect to properties exerting a significant effect on the binding strength and susceptibility of elements to mobilization: water content, pH (H_2O), cation exchange capacity (CEC) and organic matter (OM), by methods

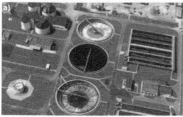

Figure 2. "Zabrze-Center" Sewage Treatment Facility (STF) with tertiary municipal sewage treatment (a) and dewatered sewage sludge outlet (b).

commonly used in soil analysis (Bednarek et al., 2004, Pansu and Gautheyrou, 2006) or provided in relevant standards for waste characterization. In particular, pH was measured electrometrically, water content according to EN 14346:2006, CEC by the ammonium acetate method at pH 7.0, and OM by digestion at 550°C.

For assessment of mobile/mobilizable forms of metals occurring in the studied bio-waste and sewage sludge, and their binding strength, sequential fractionation of metals was performed by the seven-step chemical extraction method presented in Table 1. It follows the partitioning procedure described by Kersten and Förstner (1986, 1988, 1990) being a modification of the classical method by Tessier et al. (1979). The detailed description of the procedure can be found elsewhere, also with regard to material with high amount of organic matter, where binding mechanisms may differ from the ones originally developed for a heterogeneous material with predominance of mineral phases (Twardowska & Kyziol 2003). To summarize, fractions F0, F1 and F2 represent the most labile forms of metal binding, metals in fractions F3 and F4 are moderately bound, while fractions F5 and F6 are attributed to metal bonds of the highest strength (Table 1). Besides the studied PTEs, also elements taking active part in PTE binding (Ca, Mg, Fe and Mn) were partitioned following the same sequential extraction procedure, which allowed attribution of PTE binding to certain assumed mechanisms, into which these elements were involved.

The metals in extracts were determined in triplicate by flame atomic absorption spectrometry FAAS (model TJA AA Scan 1) using standard analytical procedure in accordance with EN 12506:2003 along with QA/CC (i.e. at least once for each analytical batch with a minimum of once per 20 samples) that included blanks, spikes, duplicate samples and standard addition. Results were calculated as mean values if the deviation did not exceed 5%. The metal contents were presented in mg/kg d.m. not standardized to 30% OM d.m., as was proposed in Annex III of the 2nd draft of the DG ENV working document "Biological Treatment of Biowaste" (2001). The actual OM content

Table 1. Sequential extraction scheme for speciation of trace metals in BWC and SS with respect to binding strength and prevailing chemical forms of binding (Kersten a&Förstner 1986, 1988, 1990).

Mobility	Fractions		Extractant	Extracted forms
Labile	F0 (PS)	pore solution	Distilled water, pH 6	Water soluble
	F1 (EXC)	Exchangeable	1 M NH_4OAc, pH 7	Exchangeable ions
	F2 (CARB)	Carbonatic	0.01 M NaOAc + 1M HOAc, pH 5	Carbonate/mobile forms
Moderately bound	F3 (ERO)	easily reducible	0.01 M $NH_2OH \cdot HCl$ + 0.01 M HNO_3, pH 2	Mn/Fe–oxides
	F4 (MRO)	moderately reducible	0.2 M (HOx) + 0.2 M $(NH_4)_2$Ox, pH 2	Amorphous Fe-oxides/HS[a]
Strongly bound – Mobilizable	F5 (OM)	sulfidic/organic (oxidizable)	30% H_2O_2 + 0.02 M HNO_3, $t° = 85°C$, pH 2, next with 1 M NH_4OAc	Sulfide + organic matter
	F6 (R)	Residual	Extr. with HNO_3 conc., $t° = 130°C$ + 0.01 M HNO_3	Lithogenic crystallites (non-Si)

[a] HS – humic substances.

Table 2. Basic characteristics of studied bio-waste compost (BWC) and sewage sludge (SS).

Material	pH(H$_2$O)	Water content	CEC$_e$[a]	OM d.m.[b]
Units		% wt.	cmol(+)/kg	% wt.
BWC 1 year old	8.50	52.76	49.16	34.5
BWC fresh	8.43	58.53	63.69	30.2
SS fresh	5.48	84.95	41.42	53.1

[a] cation exchange capacity
[b] organic matter (in dry matter)

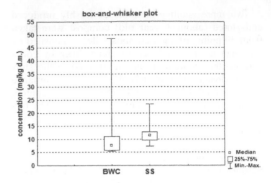

Figure 3. Box-and-whiskers plot for Cd concentrations in representative samples of bio-waste compost (BWC) and sewage sludge (SS) from the Zabrze industrial area (Silesia agglomeration, Poland) randomly sampled in 2005–2008.

in bio-waste was very close to the standard one (30.2–34.5% OM) and thus both standardized and non-standardized values might be used with satisfactory accuracy without correction.

In this paper, also the data on the source green waste and BWC survey for Cd conducted in this laboratory in 2005–2006 are discussed. The source material and BWC (grass, leaves and wood clippings) were analyzed for the total Cd content by microwave assisted digestion according to EN 13656. The obtained results were interpreted statistically with use of Statistica v.8.0 computer program.

3 RESULTS AND DISCUSSION

3.1 Metal content in BWC and SSC

Studied bio-wastes distinctly differed with regard to their basic properties (Table 2), BWC was alkaline, while SS had an acidic pH. BWC was of a somewhat higher CEC$_e$, but of lesser water and organic matter content than SS.

The mobile/mobilizable metal content (PTE and major elements) in studied bio-waste appeared to be rather stable, except Cd. This element occurred in the freshly produced BWC in the amount about 10-fold higher than in the 1-year old material, clearly showing its environmentally problematic character. In mass units, PTE contents followed the descending sequence Zn>Pb>Cd>Cu ≈ Cr>Ni in the freshly produced BWC, and Zn>Pb>Cu≈Cr>Ni>Cd in the 1 year old batch, at the concentration range from about 10 mg/kg d.m. (Ni) to around 200 mg/kg d.m. (Zn) (Table 3). Cd content in the 1-year old material was considerably above the reported mean values for the EC (Amlinger 2004, Smith 2009), while for the Cd concentration in the freshly produced material, one sample was within the outlier range (Figure 3).

Comparison of PTEs contents in both the BWC and SS samples with the corresponding value ranges in the EU for the same kind of material (Table 4) showed that the concentrations of Ni, Zn, Cu, Cr and Pb in the studied BWC were well below the weighed 90th percentile values, the contents of Ni and Cu were also below the weighed median values, while Zn, Cr and Pb contents, although exceeding the weighed median values, were relatively close to these values.

PTEs in SS represented a similar descending order as the 1-year old BWC, i.e. Zn>>Pb>Cu>Cr>Ni>Cd (Table 3). Contents of PTEs in SS were considerably higher than in BWC, although to a different extent. While Zn concentrations exceeded this metal contents in BWC at about one order of magnitude, Cu and Cd contents were 4-5 times higher, Ni, Cr and Pb occurred in SS in about 2-fold higher amounts.

Concentrations of mobile/mobilizable PTEs in SS from the same area appeared to be considerably higher than in BWC, but within the EU range of means for SS, and mostly within its lower/middle part except for Zn and Cd contents somewhat exceeded the upper limit (Table 4). It should be mentioned that the measured concentrations in this study are "pseudo-total" and did not include the silicate residual fraction that is immobile under any circumstances.

A definite problem is created by the Cd content that exceeded weighed median value from one to two orders of magnitude and even 90th percentile values for BWC derived from the EU national compost data. The survey conducted in 2005-2006 in this laboratory on Cd concentrations in source materials (grass, leaves and wood clippings) did not identify any kind of garden waste as potential source of Cd (Table 5).

All these materials contained much lower amounts of Cd than BWC sampled simultaneously. Among the source material and BWC, the lowest Cd content was in grass, and the highest was in wood clippings, which differs from the EU data that reported lower concentration range of Cd in wood clippings than in garden waste.

Higher accumulation of Cd in perennials may be attributed to the historical character of the area as a center of metallurgy and coking plants since the XIX century. Several times higher Cd content in BWC suggests another source of Cd enrichment than for major components, i.e. green waste and kitchen waste.

Other investigations exclude food residues as the most significant source of Cd in BWC, although different authors cited by Amlinger et al. (2004) report generally higher maximum Cd content in kitchen

Table 3. Speciation of trace metals (PTEs) occurring in BWC and SS with respect to binding strength.

		Ni	Zn	Cd	Cu	Cr	Pb	Ca	Mg	Fe	Mn
BWC-OLD	Total content mg/kg	13.98	193.7	5.46	26.02	22.38	81.19	26540	1272	3045	139.5
	Fractions mg/kg %										
	Labile F0+F1+F2	4.73	80.05	2.15	2.77	3.41	28.72	16820	831.9	91.65	71.91
		33.8	41.3	39.4	10.6	15.2	35.4	63.35	65.37	3.01	51.55
	Moderately bound F3+F4	4.28	94.53	0.91	19.16	8.36	16.21	5715	323.9	2449	57.65
		30.6	48.8	16.7	73.6	37.4	19.96	21.53	25.45	80.40	41.33
	Strongly bound F5+F6	4.97	19.09	2.4	4.09	10.61	36.26	4012	116.8	505.3	9.94
		35.6	9.8	43.96	15.7	47.4	44.7	15.12	9.18	16.59	7.12
BWC-FRESH	Total content mg/kg	12.3	232.5	48.53	29.03	28.39	65.40	37088	2117	3141	304.7
	Fractions mg/kg %										
	Labile F0+F1+F2	2.44	69.09	29.18	3.55	4.59	12.57	20520	1380	63.64	104.1
		19.76	29.72	60.13	12.25	16.19	19.23	55.33	65.21	2.03	34.15
	Moderately bound F3+F4	4.88	140.5	11.01	14.59	10.97	16.97	7957	438.7	2789	136.4
		39.55	60.43	22.69	50.26	38.65	25.96	21.45	20.73	88.79	44.75
	Strongly bound F5+F6	5.02	22.92	8.34	10.88	12.82	35.82	8612	297.8	288.4	64.28
		40.7	9.9	17.2	37.5	45.2	54.8	23.22	14.07	9.18	21.09
SS	Total content mg/kg	25.41	2162	23.37	122.6	53.09	132.7	28690	3218	25940	365.6
	Fractions mg/kg %										
	Labile F0+F1+F2	8.44	670.3	3.09	5.95	8.72	5.68	16860	2014	78.06	170.1
		33.23	31.01	13.21	4.85	16.42	4.28	58.77	62.59	0.30	46.51
	Moderately bound F3+F4	11.31	1360	0.77	68.03	21.65	10.30	9225	1009	24200	155.4
		44.51	62.92	3.29	55.48	40.79	7.77	32.16	31.35	93.31	42.50
	Strongly bound F5+F6	5.65	131.3	19.51	48.64	22.72	116.7	2603	194.8	1658	40.17
		22.2	6.1	83.5	39.7	42.8	87.9	9.07	6.05	6.39	10.99

waste and waste from food production (up to 1.4–2 mg/kg d.m.), and up to 3-fold enrichment in food residues compared to the edible parts (e.g. in carrots, apples and pears). In a review by Smith (2009) that comprises also data reported by Amlinger et al. (2004) and by other sources from the period of 1997–2007, the mean/median concentrations of Cd in green waste ranged from 0.28 to 1.40 mg/kg d.m., in source collected waste – from 0.32 to 4.00 mg/kg d.m., in household collected green waste and kitchen waste – from 0.60 to 8.40 mg/kg d.m. (mean 3.02 mg/kg d.m.), in green waste from household and civic amenities from 0.70 to 9.10 mg/kg d.m. (mean 3.90 mg/kg d.m.), and in mechanically segregated waste from min. 0.41 to max. 14.0 mg/kg d.m. A closer analysis of these data suggests that an influence of impurities and area character seem to be the most important factors.

According to Amlinger et al.(2004) after Gronauer et al. (1997), among the common impurities, the major extraneous component of BWC is paper (30%), plastic foils (over 7%) and metal residues (about 4% of total impurities); Cd content in paper is generally not high (0.1–0.8 mg/kg d.m.), significant in plastic foils (14–46 mg/kg), and diverse in the metal fraction. There is no data either on rate or on composition of impurities for the studied region. Data from Spain (Catalonia) indicate a wide range of impurities in source-collected food residues – from 1.8 to 18.7% (average 4.9% w.w.) at road-container system and from 0.9 to 4.0% (average 2.3% w.w.) at doorstep collection. In the analyzed area, food residues from restaurants are collected at the doorstep, thus impurities content around 4.0% w.w.

might be a rough estimation. This could cause certain Cd enrichment in BWC, nevertheless in view of actual concentrations, referred impurities are unlikely the major source of high Cd content in this material.

Most probably, the particularly high Cd enrichment in the studied BWC is attributed predominantly to the industrial character of the area. Up to now, investigations on area-originating PTE content in BWC are scarce. Amlinger et al. (2004) cite mean values of PTE concentrations in BWC for urban and rural areas in Austria and Germany, but they do not show any dramatic differences or particularly high Cd concentrations. An important observation is an increase of Cd and Zn with the traffic intensity that correlates with exceedingly high concentrations of both these metals in the freshly produced BWC (Table 3). Nevertheless, long-term industrial pollution of the area should be considered the major source of high Cd content in BWC.

Median background Cd contents in natural unpolluted soils of 11 Member States of the EU (without Poland), Norway and Romania surveyed in 2004 ranged from 0.07 to 1.48 mg/kg d.m.. In different regions of the world mean concentrations were reported to range from 0.07 to 1.05 mg/kg d.m., and in Poland from 0.07 to 0.38 mg/kg d.m. (Table 4). Therefore, background Cd concentrations in unpolluted soils are mostly low.

In polluted industrial areas of the world reported concentration ranges are broad, with maxima over 1000 mg/kg d.m. The highest indicated concentrations were found in the areas affected by metal mining

Table 4. Trace metals (PTEs) content of studied BWC and SS compared to value ranges for PTEs (mg/kg, d.m.) of the EU Member State data (after Amlinger 2004[a]) and contents in soil (after Kabata-Pendias 2001, Utermann et al. 2006).

	Ni	Zn	Cd	Cu	Cr	Pb
Zabrze industrial area (Silesia agglomeration, Poland)						
BWC 1-year old	13.98	193.7	5.46	26.02	22.38	81.19
BWC fresh	12.34	232.5	48.53	29.03	28.39	65.35
Median BWC			7.54			
SS	25.41	2162	13.37	122.6	53.10	132.7
EU						
Median BWC[b]	17.00	181.0	0.46	47.33	21.00	62.67
90th percentile BWC[b]	29.73	284.2	0.89	79.50	37.40	105.2
SS–range of means[c]	9–80	142–2000	0.4–3.8	39–641	16–275	13–221
SOIL[d]						
Poland–range	1–104	7–360	0.01–0.84	1–110	2–80	8.5–85
Poland–range of means	8–25	30–85	0.07–0.38	6–22	12–38	16–39
EU11+2[e]–range	0.4–2066	<0.1–4847	<0.01–24.3	<0.2–22360	0.2–6096	0.2–5989
EU11+2[e]–range of 90th percentile	1.5–90.0	2.2–677	0.21–5.99	7–176	13–183.5	20.2–380
EU11+2[e]–range of medians	3–48	6–130	0.07–1.48	2–32	5–68	6–79
World's range	1–450	4–770	0.04–2.7	1–323	1.4–810	1.5–280
World's range of means	4–92	27–235	0.07–1.05	6–80	7–153	8–120
Median worldwide[f]	50	90	0.35	30	70	35

[a] PTE content values are not standardized to 30% OM data;
[b] statistically weighed median and 90th percentile values for PTEs of European national compost data;
[c] ANDERSEN-CEDE on behalf of the EU DG Environment (2001);
[d] Kabata-Pendias 2001;
[e] After Utermann et al., European Commission, JRC (2006);
(Data for Austria, Estonia, Finland, France, Germany, Ireland, Italy, Lithuania, the Netherlands, Portugal, Slovak Republic, Norway and Romania – status October 2004);
[f] Bowen 1979 (after Smith 2009).

Table 5. Cadmium content (mg/kg d.m.) in different feedstock of BWC in the Zabrze industrial area (Silesia agglomeration, Poland) measured randomly in 2005-2007 compared to the relevant range for Cd of European national compost data (after Amlinger, 2004).

Garden waste		Wood	Kitchen		Median
Grass	Leaves	clippings	waste	BWC	BWC
Zabrze industrial area (Silesia agglomeration, Poland)					
0.343–0.357	0.900–0.923	1.06–1.59	–	5.46–48.52	7.54
EU-range[a]					
0.07–0.65		<0.1–0.4	0.059–1.4	0.89[b]	0.46

[a] Amlinger on behalf of the EU DG Environment (2004).
[b] 90th percentile values for PTEs of European national compost data.

and metallurgy, up to 1500 mg/kg d.m. (USA) and 1780 mg/kg d.m. (Belgium). In Poland, in the industrial Silesia agglomeration, which includes the Zabrze area, Cd concentrations in the vicinity of metallurgical plants ranged from 6 to 270 mg/kg d.m. In these areas, Zn also occurs in soils in very high concentrations: in polluted areas of the Silesia agglomeration in the vicinity of metallurgical plants it ranged from 1665 to 13800 mg/kg d.m., while in unpolluted soils, also in the same Silesia agglomeration, Zn is present

in concentrations from 7 to 150 mg/kg, at the range of means from 30 to 75 mg/kg (Kabata-Pendias 2001). This co-occurrence of these two metals in high concentrations in soils affected by long-term emission from metallurgical plants well explains also their simultaneously indicated high concentrations in BWC, e.g. in freshly produced material (Table 3).

Besides aforementioned high concentrations of metals in soils affected by the industrial pollution, a survey of background contents of trace elements in

European soils showed also exceedingly high maximum geogenic concentrations of practically all metals, which affect 90th percentile values as well. The maxima were found to occur in calcareous rocks in France and Slovak Republic (Ni, Cr), in alluvial/glacial deposits in Germany (Cd, Cu, Pb, Zn), and in crystalline rocks in Slovakia (Cu) (Utermann et.al. 2006). Nevertheless, the reported median background values in soils both in Europe and worldwide are much lower and appeared to be surprisingly similar (Table 4)

Except Cd, concentrations of other PTEs in studied BWC, comply with averaged limit values for BWC in the EU. With respect to regulations for organic farming, besides Cd median values that about an order of magnitude exceed the limits, also Zn is at the limit value or somewhat exceeds it.

Although the problem with elevated concentrations of Cd, and to the lesser extent with Zn in BWC in the EU is not so severe as in Zabrze area, the report of Amlinger et al. (2004) confirms that these two metals, in particular Cd contents, with exception of agricultural residues, in some cases create serious difficulties with compliance both with averaged limit values for BWC and with limits for organic farming in the EU.

In turn, PTE contents in the studied SS, although considerably higher than in BWC from the same area, appeared to fall within the range of means for the EU with an exception of Cd and Zn (Table 4). Mean value of Cd content in SS exceeded the EU maximum limit about 4-fold, while actual concentrations of Zn were slightly above the maximum limit. With respect to limit values set by Sewage Sludge Directive 86/278/EEC, PTE concentrations detected in SS fulfill the criteria of the upper limit for all metals, but exceed the lower limits with respect to Cd. A number of the Member States, among them Poland, introduced much more stringent PTE limits in SS intended for agricultural use (Andersen-Sede 2001, Twardowska et al. 2004). Much higher concentration range of Cd, Zn and other PTEs and their higher mobility in soils amended by SS than in natural soils summarized by Kabata-Pendias (2001), and confirmed in more recent publications (Horn et al. 2003, He et al. 2007, Singh & Agrawal 2007), thoroughly justify a precautionary approach expressed by many policy-makers and researchers, reflected in the current consultations on the need of changing the limit values and overall regulations for agricultural applications of sewage sludge.

Other metals (Ca, Mg and Mn) occurred in BWC and SS in comparable amounts, while content of iron Fe in SS was of about an order of magnitude higher than in BWC (Table 3).

3.2 Binding strength and mobility of metals in BWC and SS

The selection of Kersten and Förstner's (1986, 1988, 1990) partitioning method among a multitude of presently existing, also more recent modifications e.g. modified BCR three-step sequential extraction scheme (Rauret et al., 1999; Lamer et al., 2006), has been justified by its proven suitability to this

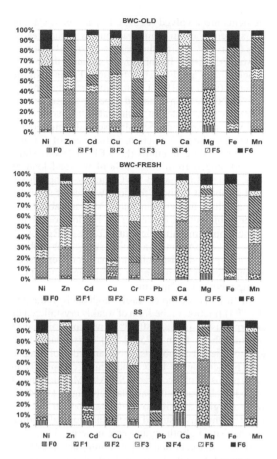

Figure 4. Speciation of metals occurring in BWC and SS with respect to binding strength and forms of binding.

kind of organogenic waste material, logical extraction sequence from the most labile to the most stable forms, and allows a direct comparison of results obtained for different matrices. The analysis of existing commonly available operationally defined methods of chemical fractionation did not prove superiority of any particular method over others and showed the need of developing a more mechanistic than operational approach (Peijnenburg et al. 2007). It should be added that forms of metal binding commonly attributed to the particular fractions, reflect prevailing binding mechanisms in predominantly inorganic materials. Metal sorption of the similar strength of binding onto organic matter may represent different mechanisms specific to its chemical structure (Twardowska & Kyziol 2003). In the studied organogenic materials, rich in both organic and inorganic matter, binding mechanisms specific for these materials are equally involved.

The distribution of metals in fractions of the different binding strength showed, besides similarities, diversity of binding mechanisms for different metals and matrices (Table 3, Figure 4). Fractionation of Ca, Mg, Fe and Mn in all studied organogenic materials showed for each metal a very similar pattern that indicated also mechanisms of binding in which these elements are involved (Fig. 4). In particular, Ca and Mg are enriched in labile fractions as exchangeable ions (F1), Ca^{2+} being the major exchangeable ion in all studied wastes. In comparable amounts these ions enriched also another labile fraction (F2) in alkaline material (such as BWC) representing mostly binding onto carbonates; Ca appeared to be the major component also of this fraction, both in alkaline matter such as BWC and in acidic SS. Mn also had a high affinity to the labile F2 fraction. Enrichment of Ca and Mg in the labile fractions (F0 + F1 + F2) was similar and ranged from 55 to 65% of the total load of these metals for studied BWC and SS, although Ca load attributed to these fractions was about an order of magnitude higher than that of Mg, and about two order of magnitude exceeding the Mn load occurring in the mobile/mobilizable forms (Table 4). Relatively weak PTEs sorption of a similar strength as carbonates may provide also bonds to small complexes such as peptides and amino acids in organic matter (Van Campenhout et al., in print).

In turn, Fe occupied easily/moderately reducible (F3+F4) fraction in the form of weakly crystallized or amorphous Fe-oxides (80–93%), with predominance of amorphous forms (75–92.5%). Particularly high load of these compounds, about an order of magnitude higher than in BWC, and almost completely in amorphous form, occurred in SS. Also Mn enrichment of reducible fractions was relatively high (41–44%), with a higher affinity to moderately bound fraction F4. It might thus assumed that the major metal binding mechanisms in both organogenic materials identified through Ca, Mg, Fe and Mn partitioning appeared to be replacing Ca^{2+} ions from exchangeable sites, bonding onto labile fraction F2 attributed to carbonates or compounds of the similar binding strength and sorption onto amorphous Fe-oxides.

Considering the organogenic character of BWC and SS, binding onto humic substances HS ("soluble organics") of different binding strength, in particular incorporation of metal ions into chelate rings of carboxylic COOH and phenolic OH groups with displacement of H^+ ions should also play a significant role in metal enrichment in reducible fractions F4. Metal enrichment in F5 fraction is being attributed in these materials mainly to formation of strong bounds onto "insoluble organics" (Twardowska & Kyziol 2003).

Binding strength analysis of metals originally occurring in BWC and SS (Table 3, Fig. 4) showed that in general, most PTEs (Ni, Zn, Cu and Cr) at the highest part were enriched in F4 fraction of a moderate binding strength, attributed mostly to sorption onto amorphous Fe-oxides abundant in all studied waste (chelating complex formation with COOH and OH functional groups of humic substances, in particular of Cu^{2+} and Cr^{3+} ions was also possible). Cd, and to a somewhat higher extent also Pb, showed a low affinity to this fraction.

The highest loads of Cd, relatively high rates of Ni and Zn and lower rates of Pb appeared to be bound in the labile F2 fractions of BWC where mostly Ca in

the form of carbonates was involved, while retention in pore solution (F0) and ion exchange (F1) fractions played a very minor role in all PTEs occurrence in compost. Cu and Cr showed low enrichment in all labile fractions (F0+F1+F2), and a higher affinity to stable bounds onto "insoluble organics" (F5) and in residual non-silicate fraction (F6). Also 35–40% of Ni was stably bound in the F6 fraction of BWC.

The comparison of 1-year old and freshly produced BWC showed higher enrichment of Ni, Zn and Pb in labile fractions (F0+F1+F2) of 1-year old BWC compared to the freshly produced material. Only Cd, occurring in the fresh BWC in high amounts, prevailed in this material in mobile forms. In fresh BWC, higher enrichment of all PTEs (except Cd) compared to the 1-year old material, showed also fractions of moderate binding strength (F3+F4) and for Ni, Cu, and Pb – also strong bounds onto organic fraction (F5).

SS showed affinity to labile forms in general comparable to BWC, except Cd and Pb that were mostly stably bound in the residual fraction F6.

The metal enrichment in the prevailing F4 fraction of moderate binding strength for Ni, Zn, Cu, and Cr represented a regular ascending sequence onto BWC: 1-year old < BWC-fresh < SS.

Opposite to Ni, Zn, Cu and Cr which show similar fractionation in the studied organogenic waste, partitioning of Cd and Pb displayed substantial differences in fractionation from other metals and between studied wastes. In BWC, Cd was mostly bound in the labile F2 fraction. Particularly high contents of Cd in the freshly generated bio-waste compost were present as labile forms (60%, that is 29.2 mg/kg d.m.) that increases the hazard resulting from the metal mobilization, while comparable concentrations of Cd in SS (83%, i.e. 19.5 mg/kg d.m.) were stably bound. Nevertheless, significantly higher contents of Cd^{2+} in pore solution (F0) and in the most labile exchangeable fraction F1 in SS compared to fresh BWC, makes this material no less problematic (Table 3, Figure 4).

Another problematic metal is Pb that occurred in SS in about two times higher concentrations than in BWC, although in SS it appears to be predominantly stably bound (88%, i.e. 117 mg/kg, d.m.), while loads associated with labile fractions in BWC were up to 6-fold higher than in SS. Still, also in SS the highest occurrence of Pb in the most mobile fractions – pore-water F0 and as exchangeable ions F1, at low contents in the labile F2 fraction clearly shows that the problem with this metal exist in both kinds of material.

In SS, also particularly high concentrations of Zn, ion of a proven phytotoxicity, requires close attention Problems were created by contents, in particular due to its substantial association with the labile fractions. In relatively higher rates than in BWC Zn enriched weakly bound exchangeable F1 fraction, while stable binding mechanisms (F5+F6) played a negligible role (Table 3, Figure 4).

Therefore, although in principle BWC is a much more environmentally acceptable potential soil fertilizer than SS, in some areas, such as the studied historical industrial region, specific pollutants occurring in the area, may exert a risk of irreversible soil contamination. This precludes the material from being considered for agricultural use in its present state. On the other hand, it is well known that even in the industrial regions like Silesia, heavy pollution occurs usually in limited areas. Identification and avoidance of such areas may lead to substantial improvement of BWC quality, whereas reduction of contamination in SS from such area is rather questionable.

The obtained results, confirm the conclusions of many compiled observations (Amlinger 2004, Smith 2009) that most metal concentrations in all kinds of waste are higher than the background concentrations in soil. On the other hand, most of metal contents in source-collected BWC, even in the historical regions of heavy industry and chemistry like Silesia agglomeration, appear to be within the mean and median range in the areas of a lesser anthropogenic pressure. Some metals like Cd in the studied area, might create a problem, and fall within the highest reported concentrations. In addition, our observations of metal fractionation with respect to binding strength show higher affinity of metal ions to more labile fractions in older BWC, than in freshly produced material.

In turn, most metals in SS except Cd and Pb, are not abundant in the residual, strongly bound F6 fraction, and show the highest affinity to the fraction F4 of a moderate binding strength; Ni, Cd, and Cr occur in relatively high rates in pore solution (F0), and enrich also labile F1 fraction, while Ni and Zn show high affinity to the mobile F2 fraction. Therefore, there is no strong basis to conclusion that the studied BWC and SS are environmentally safe.

Although separate collection and composting is an attractive option of bio-waste management, due to the relatively low production potential compared to soil improvement needs, it might be used for application onto no more than 3.2% of agricultural land in the EU (ComEC 2008). Therefore, though bio-waste composting will not solve the soil fertility upgrading problems, it may be regarded primarily as an optimum way to reduce landfilling. The amount of sewage sludge in the EU and the Member States, although fast growing, is considerably lesser than that of bio-waste, and in general much more problematic with respect to the contaminants content, among them metals.

On the other hand, a simple comparison of metal (PTEs) loads primary bound onto the studied BWC and SS, and much higher sorption capacity for these metals (partially estimated by the sequential extraction of the major elements - Ca, Mg, Fe and Mn), suggests a potential of SS and BWC from contaminated areas to be applied as abundant, cost-effective and easily available materials for contaminated lands reclamation or in protective barriers (see Naftz et al. 2002).

4 CONCLUSIONS

Studies on BWC and SS from the historic industrial area in the Silesia Agglomeration shows that both materials are environmentally problematic, BWC due to the high and unstable Cd content, 1-2 orders of magnitude higher than in uncontaminated soils, and SS due to much higher concentrations of Zn, Cd, Cu, and Pb than is usual in uncontaminated soils. In both kinds of organogenic wastes, metals predominantly enriched fractions of a moderate and weak binding strength. No shift to stronger bounds of metals in older BWC was observed: it appeared that older material showed higher metal enrichment in fractions of a weaker binding strength. While elevated metal content in BWC may be potentially controlled by avoiding highly contaminated areas for source collecting, sewage sludge is practically uncontrollable. Precautionary principle and sustainable development requirements suggest application of BWC, mostly green waste from uncontaminated areas, in agriculture as the best option, but avoiding agricultural utilization of BWC from industrial areas if content of any PTE exceeds 90th percentile, and SS as a whole. For these wastes adequately designed use for contaminated industrial lands reclamation seems to be a better solution (this was confirmed by a series of further experiments).

ACKNOWLEDGEMENTS

This work is a part of the study conducted within the statute project 1-a-83/05/06/07 of the Institute of Environmental Engineering of the Polish Academy of Sciences in Zabrze, Poland.

REFERENCES

Amlinger, F., Pollak, M. & Favoino, E. *Heavy Metals and Organic Compounds From Wastes Used As Organic Fertilizers.* ENV.A.2./ETU/2001/0024. Final Report to DG Environment, Brussels 2004.

Andersen-Sede, L. 2001. *Disposal and Recycling Routes for sewage Sludge, Part 3*, Scientific and Technical Report for EC DG Environment, Office for Official Publications of the European communities, Luxembourg.

Bednarek, R., Dziadowiec, U., Pokojska, U. & Prusinkiewicz Z. 2004. *Ecologic and Soil Analysis.* Wydawnictwo Naukowe PWN: Warszawa (in Polish).

Bhattacharyya, P., Chakrabarti, K., Chakraborty A., Tripathy S., Kim K. & Powell M.A. 2008. Cobalt and nickel uptake by rice and accumulation in soil amended with municipal solid waste compost. *Ecotoxicology and Environmental Safety* 69(3): 506–512.

Carbonell, G., Pro, J., Gómez, N., Babin, M.M., Fernández, C., Alonso, E. & Tarazona, J.V. 2009. Sewage sludge applied to agricultural soil: ecotoxicological effect on representative soil organisms. *Ecotoxicology and Environmental Safety* 72(4): 1309–1319.

CEN/TR 15310-1,2,3,4, 2006; Characterization of waste – Sampling of waste materials – Parts 1,2,3,4, *CEN – European Committee for Standardization.*

COM EC (Commission Of The European Communities). *Green Paper on the Management of bio-waste in the European Union,* COM(2008)811 final. Brussels 2008.

DG ENV (EC Directorate General Environment), *Working Document "Biological Treatment of Biowaste" 2nd draft.* DG ENV. A2/LM/biowaste/2nd draft, Brussels 2001.

EC. 2006. Directive 2006/12/EC of the European Parliament and of the Council of 5 April 2006 on waste (text with EEA relevance, *Official Journal of the European Union* 114(9): 9–21.

EC.EUROPA. 2002. Sewage Sludge *http://ec.europa.eu/environment/waste/sludge/htm*

EEC. 1986. Council Directive 86/278/EEC of 12 June 1986 on the protection of the environment, and in particular of the soil, when sewage sludge is used in agriculture. *Official Journal of the European Communities* 181(1): 6–12.

EEC. 1991. Council Directive 91/271/EEC of 21 May 1991 concerning urban waste-water treatment. *Official Journal of the European Communities* 135(30.5.1991): 40–52.

EN 12506. 2003. Characterization of waste – Analysis of eluates – Determination of pH, As, Cd, Cr(VI), Cu, Ni, Zn, Cl$^-$, NO$_2^-$, SO$_4^{2-}$, *CEN – European Committee for Standardization.*

EN 14346. 2006. Characterization of waste – Calculation if dry matter by determination of dry residue or water content. *ÖNORM – Austrian Standards Institute, Vienna.*

EN 13656. 2007. Characterization of waste – Microvawe assisted digestion with hydrofluoric (HF), nitric (HNO$_3$) and hydrochloric (HCl) acid mixture for subsequent determination of elements in waste. *CEN – European Committee for Standardization.*

EUROSTAT. 2009. Municipal waste generated. *European Community*, http://epp.eurostat.ec.europa.eu/, last update 30.01.2009.

Favoino, E. 2004. Success stories of composting in the European Union. Leading experiences and developing situations: ways to success. In: I. Twardowska, H.E.A. Kettrup, W.J. Lacy (eds), *Solid waste: assessment, monitoring and remediation.* Elsevier: Amsterdam.

He, M.-M., Tian, G.-M., Liang, X.-Q., Yu, Y.-T., Wu, J.-Y. & Zhou, G.-D. 2007. Effects of two sludge application on fractionation and phytotoxicity of zinc and copper in soil. *Journal of Environmental Sciences* 19: 1482–1490.

Horn, A.L., Düring, R.-A. & Gath, S. 2003. Comparison of decision support systems for an optimised application of compost and sewage sludge on agricultural land based on heavy metal accumulation in soil. *The Science of the Total Environment* 311(1–3): 35–48.

Kabata-Pendias, A. 2001. *Trace Elements in Soil and Plants*, 3rd edn, CRC Press. Boca Raton, FL 2001.

Kersten, M. & Förstner, U. 1986. Chemical fractionation of heavy metals in anoxic estuarine and coastal sediments. *Water Science and Technology* 18: 121–130.

Kersten, M. & Förstner, U. 1988. Assessment of metal mobility in dredged material and mine waste by pore water chemistry and solid speciation. In: M. Kersten & U. Förstner (eds) *Chemistry and biology of solid waste, dredged material and mine tailings.* Springer-Verlag: Berlin.

Kersten, M. & Förstner, U. 1990. Speciation of trace elements in sediments. In: G. E. Batley (ed.), *Trace element speciation: analytical methods and problems.* CRC Press Inc.: Boca Raton.

Larner, B.L., Seen, A.J. & Townsend ,A.T. 2006. Comparative study of optimised BCR sequential extraction scheme and acid leaching of elements in the certified reference material NIST 2711. *Analytica Chimica Acta* 556: 444–449.

Mcbride, M.B. 2003. Toxic metals in sewage sludge-amended soils: has promotion of beneficial use discounted the risks? *Advances in the Environmental Research* 8: 5–19.

Naftz, D., Morrison, S., Fuller, Ch. & Davis, J. 2002. *Handbook of Groundwater Remediation Using Permeable Reactive Barriers.* Academic Press.

Oleszczuk, P. 2009. The TENAX fraction of PAHs relates to effects in sewage sludges. *Ecotoxicology and Environmental Safety* 74(4) (in print).

Pansu, M. & Gautheyrou, J. 2006. *Handbook of Soil Analysis. Mineralogical, Organic and Inorganic Methods.* Springer: Berlin Heidelberg.

Peijnenburg, W.J.G.M., Zabolotskaja, M. & Vijver, M.G. 2007. Monitoring metals in terrestrial environments within a bioavailability framework and a focus bon soil extraction. *Ecotoxicology and Environmental Safety* 67(2): 163–179.

Rauret, G., López-Sánchez, A., Sahuquillo, A., Rubipo, R., Davidson, C., Ure, A. & Quevauviller, Ph. 1999. Improvement of the BCR three step sequential extraction procedure prior to the certification of new sediment and soil reference materials. *Journal of Environmental Monitoring* 1: 57–61.

Singh, R.P. & Agrawal, M. 2007. Effects of sewage sludge amendent on heavy metal accumulation and consequent responses of *Beta vulgaris* plants. *Chemosphere* 67: 2229–2240.

Smith, S.R. 2009. A critical review of the bioavailability and impacts of heavy metals in municipal solid waste composts compared to sewage sludge. *Environment International:* 35: 142–156.

Tessier, A., Campbell, P.G.X. & Bisson, M. 1979. Sequential extraction procedure for the speciation of particulate trace metals. *Analytical Chemistry* 51: 844–851.

Twardowska, I., & Kyziol, J. 2003. Sorption of metals onto natural organic matter as a function of complexation and adsorbent-adsorbate contact mode. *Environment International* 28: 783–791.

Twardowska, I., Shramm, K.-W. & Berg, K. 2004 Sewage sludge. In: I. Twardowska, H.E.A. Kettrup, W.J. Lacy (eds), *Solid waste: assessment, monitoring and remediation.* Elsevier: Amsterdam.

Utermann, J., Düwel, O., Nagel, I. 2006. Part II. Contents of trace elements and organic matter in European soils. In: B.M. Gawlik & G. Bidoglio (eds.), *Background values in European soils and sewage sludges, Results of a JRC-coordinated study on background values.* European Communities, JRC, Office for Official Publications of the European Communities: Luxembourg.

Van Campenhout, K., Infante, H.G., Hoff, Ph.T., Moens, L., Goemans, G., Belpaire, C., Adamu, F., Blust, R. & Bervoets, L. 2009. Cytosolic distribution of Cd, Cu and Zn, and metallothionein levels in relation to physiological changes in gibel carp from metal-impacted habitats. *Ecotoxicology and Environmental Safety* (in print).

Remediation of polluted sites

Environmental Engineering III – Pawłowski, Dudzińska & Pawłowski (eds)
© *2010 Taylor & Francis Group, London, ISBN 978-0-415-54882-3*

Methanotrophs and their role in mitigating methane emissions from landfill sites

E. Staszewska & M. Pawłowska

Faculty of Environmental Engineering, Lublin University of Technology, Lublin, Poland

ABSTRACT: For some time, environmental engineers have been utilising the capacity of methanotrophs (the bacteria abundant in nature that oxidise methane to carbon dioxide) in attempts to limit emissions of methane (CH_4) from landfills, these being the world's third-most important source of anthropogenic emissions of the gas. This study reviews the literature on the microbiological bases for methanotrophs. It characterises methanotrophic microorganisms, and presents the current system of classification of these bacteria, along with modern methods of identification based on environmental sampling. The current state of knowledge on habitat requirements is then summarised, along with qualitative characteristics of the methanotrophic bacteria found in samples collected from layers overlying landfills, and in biofilters engaged in CH_4 oxidation. This paper also offers a characterisation of microorganisms that oxidise CH_4 under anoxic conditions (i.e. where molecular oxygen is lacking).

Keywords: Methane oxidation, landfills, methanotrophs, biofilter, biocovers, anoxic oxidation, identification of microorganisms, 16S rRNA, functional genes.

1 INTRODUCTION

The need to reduce the emissions of methane (CH_4) from landfills and waste dumps is a reflection of the gas's capacity to enhance the greenhouse effect. CH_4's propensity for absorbing infrared radiation makes it 20–30 times more efficient (per molecule) than carbon dioxide (CO_2) as a greenhouse gas (Glatzel & Stahr 1998, Le Mer & Roger 2001). CH_4 is also flammable, explosively so if in a 5–15% mix by volume with air, and so can be a genuine threat to the health or lives of those who work at landfills or live in close proximity. History recalls examples of gas explosions at landfills with serious consequences. Three people died at Winston-Salem (North Carolina USA) in 1969, with a further five seriously injured, when gas accumulated in the basement of a weapons factory adjacent to a landfill (ATSDR 2001, according to USACE 1984). Then in 1983, in Cincinnati (Ohio, USA), a residential building was destroyed when gas infiltrated from a nearby waste dump (ATSDR 2001, according to EPA 1991). There are also numerous references to fires breaking out at landfills in various parts of the world, not least one of four months' duration at a dump at Ma'alaea, Maui (Hawaii, USA) in 1998; a several-week fire at Danbury (Connecticut, USA) in 1996 (U.S. Fire Administration 2001); a 2007 fire at the Fredericton Regional Landfill in Canada; and a 2006 fire at Tagarades (Greece; Vassiliadou et al. 2009).

CH_4 is mainly removed in the troposphere, by way of its oxidation by OH free radicals, which oxidise many pollutant gases in the atmosphere. Since, as Lu and Khalil (1993) and Crutzen (1994) showed, reactions with CH_4 now account for 20% of these radicals, it is likely that increased CH_4 concentrations are significant to the chemistry of the atmosphere as a whole, since less OH remains available to deal with the many other atmospheric pollutants. In the stratosphere, in turn, CH_4 reacts with the Cl_2 derived from chlorofluorocarbons thus generating hydrochloric acid molecules and CH_3 radicals (Le Mer & Roger 2001).

Currently, the concentration of CH_4 in the atmosphere is greater than 1.774 ppb, with persistent, if small differences between the northern and southern hemispheres (IPCC, 2007). Of the CH_4 removed from the atmosphere, 90% is oxidised by photochemical reactions in the troposphere; only 10% of removal is attributable to methanotroph activity (Le Mer & Roger 2001, Dalton 2005).

Atmospheric CH_4 may be of biological origin (i.e. anaerobic digestion of organic matter in natural ecosystems like swamps or bogs, and anthropogenic ones like rice fields and landfills) or non-biological origin (i.e. incomplete combustion of organic matter). An estimated 70–80% of atmospheric CH_4 is of biological origin (Wise et al. 1999, Le Mer & Roger 2001). The anthropogenic sources are dominant in the overall emissions of CH_4, being responsible for emission of 60–70% of the CH_4 in the atmosphere (Lelieveld et al. 1993, Le Mer & Roger 2001). Landfills are responsible for 6% of total CH_4 emissions (i.e. 30 Tg year^{-1}) (Le Mer & Roger 2001). However, the role played in individual countries' emissions may be very much greater. The gaseous mixture released from layers of waste comprises CH_4 ($\approx$64% of the total), CO_2 ($\approx$35%) and volatile odorous compounds accounting for <1% (Perdikea et al. 2008).

Eliminating landfill emissions of CH_4 would tangibly improve air quality, and this seems achievable. Where the CH_4 content of biogas is <30%, the biological methods for removing CH_4 might be used., i.e. to take advantage of bacterial capabilities to oxidise CH_4 to CO_2. The method entails transfer of the CH_4-containing gas to a biological deposit, most often a suitable organic or mineral–organic material (compost, natural soil, peat or bark) within which CH_4-oxidising bacteria (i.e. methanotrophs) can develop.

Precise method selection depends on the concentration and rate of generation of the CH_4 in the biogas, the anticipated period over which emissions are likely, and environmental conditions (the presence of degassing wells, and the insulation – or lack of insulation – of the layer capping the landfill). The methods applied or being introduced to limit CH_4 emissions from landfill involve the so-called bio-covers, bio-windows or biofilters, or else temporary and daily covers (Huber-Humer et al. 2008).

The bacteria used in the above technology occur widely in nature, inhabiting a variety of different environments sometimes characterised by extremes of temperature, pH or salinity. They use CH_4 as a source of both carbon and energy and are obligate aerobes, usually preferring an O_2 concentration lower than that in the atmosphere. Their capacity to oxidise CH_4 is associated with the presence of the multi-component enzyme complex known as methane monooxygenase (MMO).

Data from the IPCC (2007) showed that $\approx$5.2% of CH_4 removal from the atmosphere entails oxidation by methanotrophic bacteria in soils. It is not only removal from the atmosphere that takes place; indeed first and foremost, the bacteria prevent the initial emission of gas from the land surface. Assessments suggest that oxidation of the gas by bacteria in the upper layers of the landfill, or in the overlying layers sealing it, lower the annual emission of CH_4 from landfills by 10–25% (Nozhevnikova et al. 1993, Chanton & Liptay 2000, Stralis-Pavese et al. 2006). Estimates by Reeburgh et al. (1993) in turn suggest that the role of soils as biofilters preventing CH_4 emissions from wetland soils, rice fields and landfills is huge. Microbially-mediated CH_4 oxidation is about 200 Tg per year larger than annual rates of emission. In comparison, total annual emission from all sources is an estimated 582 Tg $CH_4 y^{-1}$ (IPCC 2007).

2 CLASSIFICATION OF THE BACTERIA OXIDISING CH_4

The bacteria oxidising CH_4 (methanotrophs or CH_4-oxidising bacteria) are obligate, Gram-negative aerobes that use CH_4 as a source of both carbon and energy (Bratina et al. 1992, Hanson & Hanson 1996). They are part of the so-called methylotroph sub-group, in that they include bacteria that utilise either one carbon compounds (other than CO_2), or multicarbon compounds lacking direct C–C bonds. The bacteria oxidising CH_4 were first identified in 1906 (Hanson & Hanson 1996).

Over time it became conventional to assign the methanotrophs to two separate taxonomic groups dubbed types I and II (Table 1). Type I belong to the γ subdivision of the Proteobacteria, while those of type II are in the α subdivision of the Proteobacteria. The type I genera are *Methylobacter*, *Methylomicrobium*, *Methylomonas*, *Methylocaldum*, *Methylosphaera*, *Methylothermus*, *Methylosarcina*, *Methylohalobius*, *Methylosoma* and *Methylococcus*, assigned to the family Methylococcaceae. Type I make use of the ribulose pathway in converting formaldehyde into multicarbon components that form part of cell biomass. The type II genera are *Methylocystis*, *Methylosinus*, *Methylocella* and *Methylocapsa*, all included among the Methylocystaceae. Type II bacteria utilise the serine pathway to convert formaldehyde into multicarbon compounds that again form part of the cell biomass.

Within type I and *Methylococcus* spp., it has proved possible to distinguish type X, resembling type I in that formaldehyde is assimilated via the ribulose pathway, but differing in that small amounts of enzyme characteristic for the serine pathway of type II are present (Chistoserdova et al. 2005, McDonald et al. 2008).

Two filamentous methane oxidizers have been described – *Crenothrix polyspora* and *Clonothrix fusca*. Both are closely related to the type I methanotrophs (McDonald 2008).

The early classification of methanotrophs was in turn based on morphological differences and types of resting stages. Today, DNA analysis (differences in numbers of G + C base pairs) offers a more scientific basis for identification of bacteria, as do references to internal structural detail (intracytoplasmic membrane organisation), as well as the nature of the pathways by which carbon is assimilated from formaldehyde. Type I methanotrophs have characteristic intracytoplasmic membrane bundles throughout the entire cell, while type II bacteria have cell membranes concentrated at the margins (Higgins et al. 1981, Putzer et al. 1991).

The most recent discoveries show that methanotrophic bacteria are more diversified taxonomically, ecologically and genetically than previously thought and this will obviously have further consequences for their classification. In essence, there are bacteria oxidising CH_4 that do not belong to phylum Proteobacteria. Such organisms were found in the geothermal waters of the Hell's Gate Reserve in New Zealand which are contaminated with high concentrations of extremely toxic compounds (Dunfield et al. 2007). The bacteria in question have been isolated successfully and appear to belong to family Verrumicrobiaceae (phylum Verrucomicrobia), rather than to phylum Proteobacteria, this affinity being suggested

Table 1. Characteristics of types I, II and X methanotrophs (according to Bowman et al. 1995, Hanson &Hanson 1996; Bowman 2006).

Characteristic	Type I		Type II
	Type I	Type X	
Cell morphology	Short rods, usually occur singly; some cocci or elipsoids	Cocci, often found as pairs	Crescent-shaped rods, rods, pear-shaped cells, sometimes occur in rosettes
Growth at 45°C	–	+	–
G + C content of DNA (mol %)	43–60	56–65	60–67
Nitrogen fixation	–		+
Resting stages formed		–	Some strains
– exospores		*Azotobacter*-like cysts	Lipid cyst
– cysts		Some strains	Some strains
sMMO* inherency	–	–	+
Pathway for methanal fixation	RuMP pathway present	RuMP pathway present, sometimes serine pathway present	Serine pathway present
The citric acid (Krebs) cycle		Incomplete cycles	Complete cycle
Benson–Calvin cycle	–	+	–
Major PLFAs	14.0, 16:1ω7c, 16:1ω5t	16:0, 16:1ω7c	18:1ω8c
Proteobacterial subdivision		γ	α
Concentration of Cu^{2+}		high	low
Membrane arrangement		Bundles of vesicular discs	Paired membranes aligned to periphery cells

*generally lacking in type II, but found in genera *Methylococcus* and *Methylomonas*.

by phylogenetic analysis of 16S rRNA genes (Islam et al. 2008).

What were first dubbed *Methylokorus infernorum* were later renamed *Methylacidiphilum infernorum*, consistent with their preference for temperature of at least 60°C, as well as extremely acid conditions (pH 2.0–2.5). These are autotrophic organisms. Analysis of their genome, which has now been sequenced in full (Hou et al. 2008) shows that the genes coding for MMO are homologous with those in methanotrophic Proteobacteria. However, the genetic module coding for the oxidation of methanol and formaldehyde was either absent or incomplete, suggesting that these bacteria utilise a metabolic pathway different to that so far identified in methanotrophs. Other bacteria assigned to the Verrucomicrobia have been found in the hot springs of Kamchatka and are known as *Methyloacida kamchatkensis* (Islam et al. 2008), and others (*Acidimethylosilex fumarolicum*, Pol et al. 2007) in the soils and muds around the southern Italian volcanoes.

2.1 Environmental conditions supporting methanotroph growth

Methanotrophs have been found in different environments that are terrestrial (soils, dried-out rice fields and landfill covers), amphibious (wetlands and early-season rice fields) or aquatic (lakes and sea). Some occur in extreme environments like hot springs (Tsubota et al. 2005) or salt lakes (Khmelenina et al. 1997). Most known methanotrophs are non-halophilic neutrophiles (sensu Hanson & Hanson 1996), developing in habitats with low salt concentrations and a neutral reaction. However, strains have been discovered that can grow in a habitat of salinity 8–9% (w/v) NaCl (Kalyuzhnaya et al. 1999, Khmelenina et al. 1999).

Methanotrophs can be grown in a wide range of pHs. The acidophilic bacteria grow best at pH 5.0–5.5 (like *Methylocella palustris*) or at a pH between 4.2–7.2 (like *Methylocapsa acidophila*), while alkaliphilic species like *Methylomicrobium* spp. at pH 7.5–10 (Trotsenko & Khmelenina 2002). Equally, the aforementioned methanotrophic bacteria assigned to the Verrucomicrobia can oxidise CH$_4$ at pH < 1 (Pol et al. 2007).

Methanotrophs develop in environments of various temperatures, species having been isolated from both hot springs and soils in the far north of the world. Optima of 42–62°C characterise thermophilic species like *Methylocaldum* spp. (type I), *Methylococcus* spp. (X), *Methylothermus* spp. (I) and *Methylokorus infernorum* (now *Methylacidiphilum infernorum*), while 25–35°C is best for mesophilic species like *Methylocystis*, *Methylomonas*, *Methylosinus* and *Methylobacter* spp. (Trotsenko & Khmelenina 2002, Halet et al. 2006, Dunfield 2007). Bacteria oxidising CH$_4$ assigned to the Verrucomicrobia (and hence both thermophilous and acidophilous) were found in the acidic waters of Kamchatka's hot springs. These bacteria do best at a temperature ≈55°C and pH 3.5 (Islam *et al.* 2008). In turn, for the psychrophilic bacteria of genera *Methylomonas* (type I), *Methylosphaera* (I) and *Methylobacter* (I), temperature 5–15°C is optimal (Trotsenko & Khmelenina 2002, Halet et al. 2006).

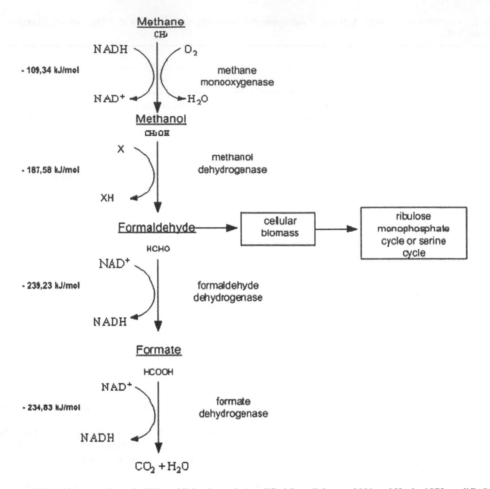

Figure 1. CH$_4$ oxidation pathway in CH$_4$-oxidising bacteria (modified from Brigmon 2001 and Hoeks 1972 modified).

Due to the presence of copper (Cu) at the active centre of MMO (as the key enzyme in the process of CH$_4$ oxidation), the correct concentration of this metal in the substrate is of major significance to the success of methanotroph culture experiments. Optimal concentrations of the element differ from one strain of methanotroph to another. However the copper concentration should not exceed 4.3 mM (Bender & Conrad 1995, Yu et al. 2003).

2.2 Biochemical transformation in the process of oxidizing CH$_4$ in aerobic conditions

The simplified formula for the reaction by which methanotrophs oxidise CH$_4$ – expressing the balance between substrates and products – is as follows (Semrau et al. 1995, Yuan et al. 1998):

$$CH_4 + 2O_2 \rightarrow CO_2 + 2H_2O$$

However, methanotrophs oxidise CH$_4$ to CO$_2$ using a series of reactions that generate such intermediates as methanol, formaldehyde and formate (Figure 1). Over-all, CH$_4$ oxidation is exoenergetic, the total amount of energy generated being the sum of the energies released at successive stages. The greatest release in fact characterises stages 3 and 4 of the overall reaction, in which formaldehyde is transformed into formate, and that into inorganic end products.

It is possible to identify two types of pathway in the oxidation of CH$_4$:

– the dissimilative pathway, whereby carbon from CH$_4$ is reincorporated into CO$_2$ released to the atmosphere and is not utilised in cell biomass.
– the assimilative pathway, whereby the carbon from CH$_4$ is built up into organic compounds building cellular biomass.

The course of the two pathways is the same up until the time of generation of formaldehyde, which may be oxidised to formate (Figure 1), or assimilated by cells via the serine or ribulose monophosphate cycles (Stępniewska et'al. 2004).

As already noted, the oxidation of CH$_4$ to methanol involves MMO methane monooxygenase, an enzyme that may be in a form associated with the cell membrane (pMMO), or else in a dissolved cytoplasmic form (sMMO). The type of monoxygenase depends

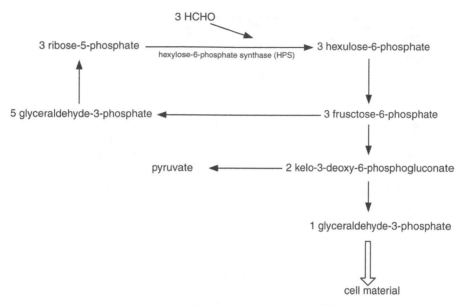

3 HCHO

3 ribose-5-phosphate ⟶ 3 hexulose-6-phosphate
hexylose-6-phosphate synthase (HPS)

5 glyceraldehyde-3-phosphate ⟵ 3 frusctose-6-phosphate

pyruvate ⟵ 2 kelo-3-deoxy-6-phosphogluconate

1 glyceraldehyde-3-phosphate

cell material

Figure 2.   RuMP pathway for formaldehyde fixation (according to Hanson & Hanson 1996).

on the Cu concentration in the substrate. sMMO is a nonhaeme Fe-containing three-component enzymatic complex comprising a hydroxylase, plus proteins B and C (Murrell et al. 2000b). pMMO is in turn built from three polypeptides of masses 46, 29 and 28 kDa (Liebermann & Rosenzweig 2004). There are CBCs (Cu-binding compounds) of masses 1.218 and 0.779 kDa in the structure of pMMO, their probable function being stabilisation and maintenance of an appropriate oxidation–reduction setup. They also could be important for sequestering copper ions (Murrell et al. 2000b). Cu is not merely an element in the enzyme, but also a regulator of enzyme expression. pMMO is also sensitive to the presence of oxygen and the availability of light (Gilbert et al. 2000).

The process by which methanol is transformed into formaldehyde involves the methanol dehydrogenase enzyme (MDH), which consists of an α-subunit of mass 60–67 kDa plus a ß-subunit of mass 8.5 kDa. The sub-units are present together in the form of an $\alpha_2\text{ß}_2$ tetramer (Hanson & Hanson 1996, McDonald & Murrell 1997b, Ward et al. 2004). Methanol present in the substrate may also derive from pectin and lignin degradation.

A further stage in the process of CH_4 oxidation, is formaldehyde conversion into CO_2, with formate as an intermediate. This process involves:

– formaldehyde dehydrogenase associated with NAD(P), which may or may not require reduced glutathione or another co-factor for its activity,
– dye-linked dehydrogenase that is measured by the reduction of such dyes as 2,6-dichlorophenol.

Formate is oxidised to CO_2 with the aid of an NAD-dependent formate dehydrogenase enzyme (Hanson & Hanson 1996).

Formaldehyde arising through the oxidation of CH_4 and methanol by methanotrophic bacteria is assimilated to form intermediates of the central metabolic cycles that are used for biosynthesis of cell material (Hanson & Hanson 1996). There are two known pathways by which formaldehyde is converted into biomass:

– the serine pathway, in which two moles of formaldehyde and one mole of CO_2 are used in the production of three-carbon compounds;
– the ribulose monophosphate (RuMP) pathway, in which three moles of formaldehyde are used in generation of the three-carbon compounds utilised in core metabolic reactions (Hanson & Hanson 1996).

After Hilger and Humer (2003), the stoichiometric equations expressing the demand for oxygen and nitrogen along the RuMP and serine pathways are as follows (where $C_4H_8O_2N$ denotes bacterial biomass):

– the RuMP pathway
$CH_4 + 1.5\,O_2 + 0.118\,NH_4 \rightarrow 0.118\,(C_4H_8O_2N) + 0.529\,CO_2 + 1.71\,H_2O + 0.118\,H^+$
– the serine pathway
$CH_4 + 1.57\,O_2 + 0.102\,NH_4 \rightarrow 0.102\,(C_4H_8O_2N) + 0.593\,CO_2 + 1.75\,H_2O + 0.102\,H^+$

Along the RuMP pathway (Figure 2), the aldol condensation reaction uses formaldehyde and ribulose monophosphate in the generation of hexulose-6-phosphate in a reaction catalysed by hexulose phosphate synthase (HPS). Hexulose phosphate isomerase (HPI) in turn converts hexulose-6-phosphate into fructose-6-phosphate. The two enzymes in question are characteristic for methanotrophs of types I and X. In the second part of the RuMP pathway, the fructose-6-phosphate is converted to

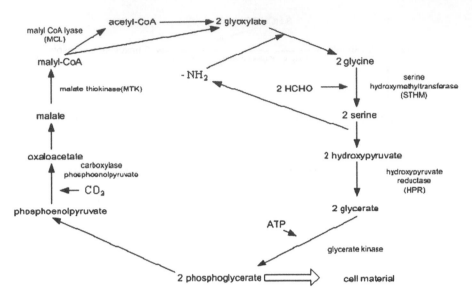

Figure 3.   Serine pathway for formaldehyde fixation (according to Hanson & Hanson 1996).

2-keto-3-deoxy-6-phosphogluconate. In further reactions, the latter is split into pyruvate and 3-phosphoglyceraldehyde. In the third part 3-phosphoglyceraldehyde and fructose-6-phosphate undergo a series of reactions which lead to the regeneration of ribulose-5-phosphate and the end of the pathway.

The enzymes of the serine pathway (Figure 3) are characteristic of type II methanotrophs. The first stage of the pathway entails a reaction, catalysed by serine hydroxylmethyltransferase (STHM), between formaldehyde and glycine, and creating serine. Serine is transaminated with glyoxylate as the acceptor of the amino group, thus generating hydroxypyruvate and glycine. Hydroxypyruvate is reduced to glycerate by hydroxypyruvate reductase (HPR). Glycerate kinase catalyses the addition of a phosphate group from ATP to produce 2-phosphoglycerate. The latter's conversion into phosphoenolpyruvate, with the binding of CO_2, is catalysed by phosphoenolypyruvate carboxylase, forming oxaloacetate, and is reduced to malate. By activity of malate thiokinase, malate is converted into malyl-CoA, which is split by malyl-CoA lyase to create acetyl coenzyme A and glyoxylate, thus bringing an end to the pathway.

3   METHODS OF IDENTIFYING METHANOTROPHS

Methanotrophs can be isolated from a large number of environmental samples, including from soil, sediment or water. Work on these bacteria is hindered by slow growth in culture, and difficulties obtaining pure cultures free of non-methanotrophs (Nishio et al. 1997).

There are two approaches to identify methanotrophs in the environment (Figure 4). The first of these (indirect) is based around the enrichment of the culture through the supply of an appropriate amount of CH_4 in culture bottles, the isolation of microorganisms, and their later characterisation. The second (direct) approach involves identifying methanotrophs directly in environmental samples, without the need for prior culture (Murrell et al. 1998, Kolb et al. 2003).

The molecular analysis of methanotrophs *inter alia* makes use of the so-called phylogenetic markers (16S rRNA), or else the functional markers (genes for MMO or for MDH), as amplified via the polymerase chain reaction (PCR) carried out on DNA isolated from pure bacterial cultures or directly from environmental sampling. Analysis of cell-wall phospholipids (PLFAs), the use of antibodies against bacterial enzymes (e.g. MMO and MDH), and assessment of overall methanotroph biomass and activity all aid the study of methanotrophs from environmental samples. Activity and growth in biomass of methanotrophs in an environmental sample is also assessable by measuring its release of $^{14}CO_2$ and incorporation of $^{14}CH_4$ (Murrell et al. 1998, Gebert et al. 2004).

3.1   *Methanotroph identification based on the analysis of housekeeping genes*

The key to identifying responsible genes in the study of methanotroph populations in the environment is the accessibility of nucleotide sequences in databases that make the design of primers possible. The 'housekeeping gene' sequence coding for a unit of ribosomal RNA (16S rRNA) is a universally-applied marker of basal metabolism in taxonomic and phylogenetic studies of bacteria, and reflects the extensive database of

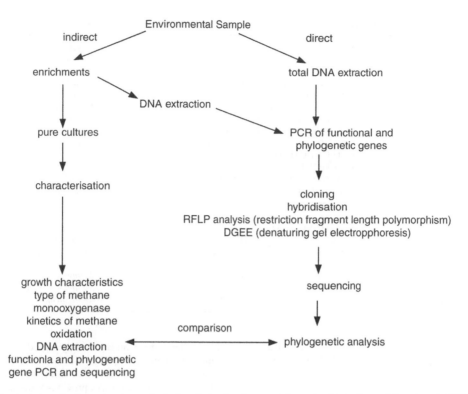

Figure 4.   A strategy for the molecular ecological analysis of environmental samples (according to Murrell et al. 1998).

sequences available for it. In the case of each newly-described organism, the 16S rRNA gene is sequenced and deposited in the GenBank base of nucleotide sequences. The gene for 16S rRNA is suitable for use in phylogenetic classification because it is present in all bacteria in the form of multigene families or operons. It contains both nucleotide sequences that are highly conserved and variable sequences. Variable sequences provide for classification at both the lowest and highest taxonomic levels. A further advantage of the gene for 16S rRNA is the limited variability in its sequences that has arisen through evolutionary time, as well as its small size (1500 bp) (Singleton 2000, Janda & Abbott 2007).

Probes specific to methanotrophs are constructed by reference to DNA sequences isolated from environmental samples, and may be used in PCR (along with universal primers for 16S rRNA). They may also be applied in hybridisation techniques like FISH (fluorescent *in situ* hybridisation; Murrell & Radajewski 2000). First probes (9α and 10γ) based on genes for 16S rRNA in investigation of methanotrophs and methylotrophs targeted genes for the serine and ribulose pathways. The drawback of these probes is that they target methylotrophs and are not methanotroph-specific. With a view to methanotrophs among the larger group of methylotrophs being identified, the specific probes 1034-Ser and 1035-RuMP were devised; however, these do not allow inter-specific differentiation to be studied (Tsien et al. 1990, Brusseau et al. 1994).

The first genus-specific primers to be designed were Mb1007, Mc1005, Mm 1007 and Ms1020; these targeted the genera *Methylobacter*, *Methylococcus*, *Methylomonas* and *Methylosinus*, respectively.

Type I and II methanotroph-specific primers (i.e. MethTdF, MethT1bR and MethT2R) were designed to examine methanotroph diversity in landfill cover soils. Further work has provided primers specific to nearly all genera of methanotrophic bacteria (Table 2).

3.2   *Methanotroph identification based on the analysis of functional genes*

The genes unique to the metabolism of methanotrophs (i.e. their functional genes) are also studied. These genes have two main advantages compared to 'housekeeping genes' relating to the main metabolic reactions. These reflect the easier procurement from environmental samples, as well as the easier identification of groups of bacteria hard to obtain as pure cultures (uncultured) (Dahllöf 2002). This all favours the application of these genes in direct analysis of environmental samples.

The gene sequences coding for the key enzymes of CH_4 metabolism (i.e. the functional genes) are available in the GenBank database, for various different methanotrophs. The identification of the conserved regions within the relevant gene sequences enabled the design of PCR primers (Table 3) that act specifically to amplify the genes for MMO or MDH in samples of DNA isolated from enriched cultures or

Table 2.   16S rRNA gene probes targeting methanotrophs (according to McDonald et al. 2008).

PROBE	Sequence (5'→3')	Target
Type I methanotroph probes		
10γ	GGT CCG AAG ATC CCC CGC TT	RuMP pathway methylotrophs
1035-RuMP	GAT TCT CTG GAT GTC AAG GG	RuMP pathway methylotrophs
MethT1dF	CCT TCG GGM GCY GAC GAG T	Type I methanotroph probes
MethT1bR	GAT TCY MTG SAT GTC AAG G	Type I methanotroph probes
Mm 1007 (r)	CAC TCC GCT ATC TCT AAC AG	*Methylomonas* spp.
Mb 1007 (r)	CAC TCT ACG ATC TCT CAC AG	*Methylobacter* spp.
Mc 1005 (r)	CCG CAT CTC TGC AGG AT	*Methylococcus* spp.
Mmb 482	GGT GCT TCT TCT ATA GGT AAT GT	*Methylomicrobium* spp.
Mcd 77	GCC ACC CAC CGG TTA CCC GGC	*Methylocaldum* spp.
Mh 996 (r)	CAC TCT ACT ATC TCT AAC GG	*Methylosphaera* spp.
Type II methanotroph probes		
9α	CCC TGA GTT ATT CCG AAC	Serine pathway methylotrophs
1034-Ser	CCA TAC CGG ACA TGT CAA AAG C	Serine pathway methylotrophs
MethT2R	CAT CTC TGR CSA YCA TAC CGG	Type II methanotroph probes
Ms1020 (r)	CCC TTG CGG AAG GAA GTC	*Methylosinus* spp.
Mcyst 1432	CGG TTG GCG AAA CGC CTT	*Methylocystis* spp.
Mcell-1026	GTT CTC GCC ACC CGA AGT	*Methylocella palustris*
Mcaps-1032	CAC CTG TGT CCC TGG CTC	*Methylocapsa acidiphila*

Table 3.   Oligonucleotide primers used for the PCR amplification of genes from DNA extracts of CH$_4$-oxidising bacteria cultures (according to Heyer et al. 2002).

Name	Target	Sequence (5'→3')
A189f	*pmoA*	GGN GAC TGG GAC TTC TGG
A682r	*pmoA*	GAA SGC NGA GAA GAA SGC
1003f	*mxaF*	GCG GCA CCA ACT GGG GCT TGG T
1561r	*mxaF*	GGG CAG CAT GAA GGG CTC CC
A166f	*mmoX*	ACC AAG GAR CAR TTC AAG
B1401r	*mmoX*	TGG CAC TCR TAR CGC TC

samples of methanotrophs in the environment. A better understanding of the numerous genes coding for the aforementioned key enzymes has allowed these phylogenetic markers to be used in identifying the CH$_4$-oxidising bacteria in different types of habitat. Currently, the following are used:

- the *pmoA* gene coding for the active sites in the molecular subunit of pMMO, which is universal to all methanotrophs except *Methylocella palustris* and *Methylocella silvestris* (*pmoA* proved not detectable from strains of these bacteria with the aid of the starters currently in use) (Dedysh et al. 2000, Dunfield et al. 2003)
- the *mmoX* gene coding for a α subunit of sMMO, which is present in a small number of methanotroph strains;
- the *mxaF* gene coding for the MDH subunit; this is universal to methylotrophic bacteria and therefore also universal, but not unique, to methanotrophic bacteria (Heyer et al. 2002).

Numerous functional genes have been used to detect methanotrophs in environmental samples. However, researchers' attention has mainly turned to the genes unique to methanotrophs in that they code for MMO, which catalyses the first stage in the transformation of CH$_4$ into methanol. As previously noted, pMMO and sMMO are the two forms of MMO. In culture, the expression of the genes for these enzymes is dependent on the Cu:biomass ratio. There are several mechanisms by which MMO is regulated in bacterial cells growing in substrates rich or poor in Cu ions (Figure 5). The genes coding for pMMO (which contains Cu) are expressed where there is a high Cu:biomass ratio in the substrate, while the genes for sMMO (containing non-haeme Fe) are expressed where Cu:biomass ratio is low (Murrell et al. 2000a, Dalton 2005). Hypothetically at a high Cu:biomass ratio, hypothetical repressor (R) and activator (A) proteins are associated with Cu, with the effect that *pMMO* genes are repressed by the R-protein, while *sMMO* genes activated by the A-protein are suppressed (Liebermann & Rosenzweig 2004). Where Cu:biomass ratio is low, the Cu–repressor–activator complex, with the result that the repressor represses the expression of *pMMO* genes, while the activator initiates expression of genes for *sMMO* (DiSpirito et al. 1998, Takeguchi et al. 1999, Murrell et al. 2000a). Research shows that a Cu concentration in solution $>1\,\mu M$ inhibits sMMO synthesis, while concentrations of $1–5\,\mu M$ stimulates pMMO synthesis (Hanson & Hanson 1996). Liebermann and Rosenzweig (2004) observed increased pMMO activity in the cells of methanotrophs in cultures in which the Cu concentration (as CuSO$_4$ or CuCl$_2$) increased in the range $0–20\,\mu M$.

The genes coding for the proteins that make up sMMO were identified and sequenced for the first time in *Methylococcus capsulatus* (Bath) (Dalton 2005). These genes form a 5.5-kb operon located on the chromosome (Figure 6). Identified among the genes coding

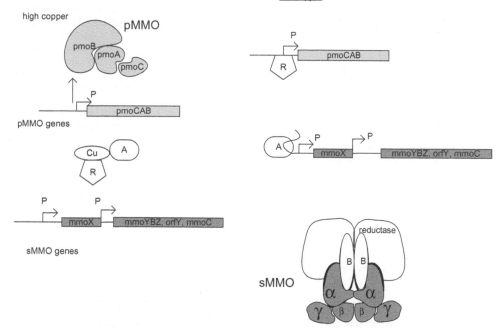

low copper

high copper

pMMO

Figure 5. Model for the regulation of MMO in *Methylosinus trichosporium* OB3b in cells grown under high and low Cu regimes (according to Murrell et al. 2000a, Dalton 2005).

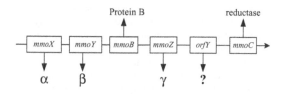

Figure 6. The soluble MMO gene (*sMMO*) clusters of CH$_4$-oxidising bacteria (according to Murrell et al. 2000a).

Table 4. Primers used for the amplification of *sMMO* genes (according to McDonald et al., 1995).

Name	Sequence 5'→3'
mmoX f882	GGC TCC AAG TTC AAG GTC GAG C
mmoX r1403	TGG CAC TCG TAG CGC TCC GGC TCG
mmoY f198	CCG ACT GGA TCG CCG GCG GCC T
mmoY r820	CGC TGG AAG AAC TCG CGG CGG
mmoZ f133	CGC CGT TCC GCA AGA GCT ACG A
mmoZ r483	TTG CGC AGC CCT TCC AGC GGC GTG
mmoB f77	AGT TCT TCG CCG AGG AGA ACC A
mmoB r369	TGC CCA GGG TGT AGG CGC GGC CGA
mmoC f542	GGT TCT GCT GTG CCG CAC C
mmoC r986	ATC CCG TGC CGC CGG CGA CG

for sMMO are *mmoX*, *mmoY* and *mmoZ*, which code respectively for the α-, β- and γ-hydroxylase subunits. The genes *mmoB* and *mmoC* code for the B protein and reductase, while gene *mmoB* is located between *mmoY* and *mmoZ* (Murrell et al. 2000a, Dalton 2005). An Open Reading Frame named *orfY*, of unknown function, separates the *mmoY* and *mmoZ* genes (Murrell et al. 2000b). The expression of these genes in *M. capsulatus* (Bath) is controlled by a Cu-regulated promoter located upstream of the *mmoX* gene (Murrell et al. 2000a, Dalton 2005).

The high degree of similarity between genes for sMMO has allowed the design of PCR primers, and thus it is now possible to amplify each of the five structural genes coding for sMMO (Table 4).

Amplification of these genes is possible among methanotrophs obtained as pure cultures, or from the overall DNA isolated from environmental samples (Murrell et al. 1998). sMMO is not an enzyme universal to all methanotrophs, it is confined mainly to *Methylosinus* and *Methylococcus* spp. (McDonald &

Murrell 1997a). For this reason, it is more effective to study the genes coding for pMMO (Murrell et al. 1998), which present in all methanotrophs except *Methylocella* spp. (McDonald et al. 2008). Genes for pMMO (*pmo*) are organised in the *pmoCAB* operon (Figure 7), in which *pmoB*, *pmoA* and *pmoC* encode polypeptides corresponding to pMMO subunits known as α (≈46 kDa), β (≈28 kDa), and γ (≈29 kDa). The α subunit has three transmembrane domains, while β has five and γ six. The *pmoCAB* operon is present in two copies (Liebermann & Rosenzweig 2004), these being identical and undergoing transcription with the aid of the type o[70] promoter that lies above the *pmoC* gene (McDonald et al. 2008). The subunit of mass 28 kDa (i.e. β) encoded by the *pmoA* gene has been shown

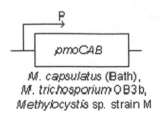

 (labels within figure:)
M. capsulatus (Bath),
M. trichosporium OB3b,
Methylocystis sp. strain M

Figure 7. The *pMMO* operon (according to Lieberman & Rosenzweig 2004).

to be highly conserved among methanotrophs, and may be used to detect these microorganisms in diversified environmental samples (Gilbert et al. 2000). Most probably it also contains active sites for MMO (Ricke et al. 2004).

Primers designed to amplify the 525-bp internal fragment of the *pMMO* gene were the A189f/A682r primers widely used in environmental studies (Murrell et al. 1998). Further work allowed for the design of PCR primers specific to different groups of methanotrophs (Table 5).

4 MICROBIOLOGICAL OXIDATION OF CH₄ IN ANOXIC CONDITIONS

There are reliable data on the oxidation of CH_4 in anoxic conditions (i.e. absence of molecular oxygen), such as in marine and lacustrine sediments, in rice fields or in the anoxic parts of refuse dumps (Stêpniewska et al. 2004, Valentine 2002, Hinrichs & Boetius 2002). This process is termed the Anaerobic Oxidation of Methane (AOM) and the evidence suggests that it involves a 'consortium' of Archaea species, as well as sulphate-reducing bacteria of the genera *Desulfosarcina/Desulfococcus*. Among the Archaea participating in the anoxic oxidation of CH_4 it is possible to distinguish two phylogenetically separate groups known as ANME-1 and ANME-2. Hitherto, certain Archaea were regarded as direct producers of CH_4, with the result that their mechanism of oxidising the gas remains poorly understood (Michaelis et al. 2002, Valentine 2002).

In theory, in anoxic conditions, CH_4 may also be oxidised by other oxygen-containing compounds, via the reactions:

$$H_4 + 8Fe^{3+} + 3H_2O \rightarrow HCO_3^- + 8Fe^{2+} + 9H^+$$

(Miura et al. 1992)

$$CH_4 + SO_4^{2-} \rightarrow HCO_3^- + HS^- + H_2O$$ (Murase & Kimura 1994; Hinrichs & Boetius 2002)

$$CH_4 + NO_3^- + H^+ \rightarrow HCO_3^- + NH_4^+$$ (Stępniewski & Pawłowska 1996)

$$CH_4 + 2NO_3^- + H^+ \rightarrow HCO_3^- + N_2O + 2H_2O$$

(Stępniewski & Pawłowska 1996)

5 POPULATION OF METHANOTROPHS IN THE LANDFILL COVERS AND BIOFILTERS

The number of methanotrophic bacteria found in landfill covers or methanotrophic biofilters tested in laboratory conditions is usually higher than in other environments. The number of methanotrophic bacteria cells in CH_4-oxidising biofilter fitted to the passive venting system of a harbour sludge landfill in Germany, were 1.3×10^8 to 7.1×10^9 cells g dw (dry weight)$^{-1}$, found after about one year of operation, while the maximum number measured in this biofilter during the experiment was 1.2×10^{11} cells g dw^{-1} (Gebert et al. 2003). The methanotroph number measured by Kallistova et al. (2007) in the cover soil of old municipal landfill in Khmetievo in Russia was in a similar range ($1.5 \pm 2 \times 10^9$ to $5.6 \pm 7 \times 10^9$ cells g dw^{-1}). The maximum number of methanotrophs measured by Ait-Benichou et al. (2009) in the uppermost layer of a passive CH_4-oxidation biocover of landfill was 1.5×10^9 cells g dw^{-1}. The number of bacteria found in a column filled by soil taken from landfill cover was 4.0×10^7 cells g^{-1}; higher in the case of earthworm cast amendments to the filling bed (4.6×10^7 to 1×10^8 cells g^{-1}; Park et al. 2008).

Methanotrophs contributed to 67.3 and 51.5% of eubacterial population in two experimental methanotrophic biofilters, respectively (Wang et al. 2008). Similar results were found by Kallistova et al. (2007), who showed that methanotrophs accounted for 50% of all bacterial population in a landfill cover soil, at depths of 40–60 cm. Wang et al. (2008) concluded that laboratory biofilters, filled with a mixture of peat and composted municipal wastewater sludge (40:60 v/v) purged by landfill gas were dominated by type I methanotrophs. In the biofilters, the average number of type I methanotrophs in the analysed layers was >5 times that of type II methanotrophs.

After purging the biofilter with landfill gas for 22 d, the maximum number of type I methanotrophs was 1.9×10^9 cells g dw^{-1} (at a depth of 25 cm), and type II methanotrophs was 2.7×10^8 cells g dw^{-1} (at a depth 15 cm). Kallistova et al. (2007) also found that type I methanotrophs dominated over type II in landfill cover soil. He et al. (2008) concluded that type I methanotrophs predominated in a clay landfill cover (ratios of type I/type II were 9.16–13.93), while in the waste soil cover the population of type II was higher (type I/type II of 0.3–1.2).

Contrary to these data, the type II methanotrophs dominated the type I in experimental lysimeters (Stralis-Pavese et al. 2004) and in two field-scale landfill gas biofilters (Gebert et al. 2004). Type I methanotrophs were found only in the top parts of lysimeters where there was relatively high oxygen and low CH_4 concentration (Stralis-Pavese et al. 2004). Wang et al. (2008) also concluded that type I methanotrophs occurred in the upper layer of a biofilter, whereas type II methanotrophs were mostly distributed in the zone with high CH_4 and low O_2 concentrations. Wang et al. (2008) observed declining numbers of type I methanotrophs with temperature increasing from 15 to 23°C, and increasing numbers with temperature lowering to 5°C. There was a slight decrease in type II methanotrophs when temperature increased from 15 to

Table 5. PCR primers used for amplification of *pmoA* genes from environmental samples (according to Jugnia et al. 2008, McDonald et al. 2008).

Primer	Sequence (5'→3')	Target group/ Product size
A189f/A682r	GGN GAC TGG GAC TTC TGG/GAASGCNGAGAAGAASGC	All genera, 525 bp
A189f/Mb661r	GGN GAC TGG GAC TTC TGG/CCG GMG CAA CGT CYT TAC C	*Methylobacter/Methylosarcina/ Methylococcus/Methylosinus/ Methylocapsa* (491 bp)
A189f/Mb601R	GGN GAC TGG GAC TTC TGG/ACR TAG TGG TAA CCT TGY AA	*Methylobacter/ Methylosarcina* (432 bp)
A189f/Mc468R	GGN GAC TGG GAC TTC TGG/GCS GTG AAC AGG TAG CTG CC	*Methylococcus* (299 bp)
II 223F/II646R	CGT CGT ATG TGG CCG AC/CGT GCC GCG CTC GAC CAT GYG	*Methylosinus* (444 bp)
A189f/Mcap630R	GGN GAC TGG GAC TTC TGG/CTC GAC GAT GCG GAG ATA TT	*Methylocapsa* (461 bp)

Table 6. Methanotrophs found in landfill biocovers and laboratory tests.

Methanotrophs	Material	Literature source
Methylobacter (type I)	experimental biocovers at landfill	Jugnia et al. (2009)
Methylocaldum (type I) and *Methylocystis* (type II)	landfill site simulating lysimeter	Stralis-Pavese et al. (2004)
Methylosarcina , *Methylococcus* (type I); and *Methylosinus*, *Methylocystis* (type II),	simulated landfill cover soil	Wang Y. et al. 2008b
Methylomonas, Methylosarcina, Methylobacter (type I); and *Methylocella, Methylocystis* (type II)	landfill cover soil	Chen et al. (2007)
Methylobacter, Methylomonas (type I); and *Methylocella, Methylocapsa, Methylocystis* (type II)	landfill cover soil	Cebron et al. (2007)

23°C, whereas cell numbers remained constant when lowered to 5°C.

Qualitative characteristics of CH_4-oxidising bacteria populations in landfill covers or in simulated landfill covers are given in Table 6.

6 CONCLUSIONS

The CH_4-oxidising bacteria are a specific Gram-negative group that grows under aerobic conditions, using CH_4 as a source of carbon and energy. The methanotrophs are divided into two types on the basis of morphological and physiological differences. Type I bacteria have been assigned to the γ-Proteobacteria and thus far include 10 identified genera. Type II bacteria include four genera and are assigned to the α-Proteobacteria.

Methanotrophs oxidise CH_4 to CO_2 by way of such intermediates as methanol, formaldehyde and formate. The differences between the types relate to the route by which formaldehyde is assimilated. In type I, this compound is transformed into poly-carbon components along the ribulose pathway, while in type II, the same processes utilises the serine pathway.

An important aspect of research on methanotrophy seeking to reduce emissions of CH_4 is to ensure the identification of the bacteria responsible for the process. While traditional techniques based on bacterial culturing are hard to use in the study of methanotroph ecology (due to difficulties with obtaining pure cultures), the problem of the free growth of CH_4-oxidising bacteria and of contamination by non-methanotrophic bacteria has proved solvable using the rapidly developing techniques of molecular biology, which provide for the identification of bacteria directly from environmental sampling. In the case of methanotrophs, these techniques are based around analysis of gene sequences used in phylogenetic analysis (*16S rRNA*), as well as genes operating in CH_4 metabolism (*pmoA, mmoX* and *mxaF*). The use of starters specific to the above genes has allowed their analysis via PCR, the determination of sequences, and in consequence the determination and study of the phylogenetic relatedness of the different bacteria.

ACKNOWLEDGEMENTS

This work was supported by Polish Ministry of Science and Higher Education, Research Project No PBZ-MEiN-3/2/2006, entitled: Inżynieria procesów ograniczania emisji oraz utylizacji gazów szkodliwych (Engineering of harmful gases emission reduction and utilisation).

REFERENCES

Ait-Benichou, S., Jugnia, L.B., Greer, C.W., & Cabral, A.R. 2009. Methanotrophs and methanotrophic activity in engineered landfill biocovers. *Waste Manage.* 29: 2509.

ATSDR (Agency for Toxic Substances and Disease Registry), Landfill Gas Primer – An Overview for Environmental Health Professionals, 2001 (http://www.atsdr.cdc.gov/hac/landfill/html/intro.html).

Bender, M. & Conrad, R. 1995. Effect of methane concentration and soil conditions on the induction of methane oxidation activity. *Soil Biol Biochem.* 27: 1517.

Bowman, J. 2006. The methanotrophs – The families *Methylococcaceae* and *Methylocystaceae.* In M. Dworkin, S. Stanley Falkow, E. Roseberg, K.-H. Schleifer, and E. Stackebrandt, Eds. *The Prokaryotes: A Handbook on the Biology of Bacteria. Volume 5: Alpha and Beta Subclasses.* New York: Springer, Inc., p. 266.

Bowman, J.P., Sly, L.I., and Stackebrandt, E. 1995. The phylogenetic position of the family *Methylococcaceae. Int. J. Syst. Bacteriol.* 45: 182.

Bratina, B.J., Brusseau, G.A., & Hanson, R.S. 1992. Use of 16S rRNA analysis to investigate phylogeny of methylotrophic bacteria. *Int. J. Syst. Bacteriol.* 42, 645.

Brigmon, R.L. 2001. Methanotrophic Bacteria: Use in Bioremediation. In G.B. Bitton, Eds., *Encyclopedia of environmental microbiology.* John Wiley & Sons: New York. Inc., p.1936.

Brusseau, G.A., Bulygina, E.S. & Hanson, R.S. 1994. Phylogenetic analysis and development of probes for differentiating methylotrophic bacteria. *Appl. Environ. Microbiol.* 60: 626.

Cebron, A., Bodrossy, L., Chen, Y., Singer, A.C., Thompson, I.P., James, I., & Murrell, J.C. 2007. Identity of active methanotrophs in landfill cover soil as revealed by DNA-stable isotope probing. *FEMS Microbiol. Ecol.* 62: 12.

Chanton, J. & Liptay, K. 2000. Seasonal variation in methane oxidation in a landfill cover soil as determined by an in situ stable isotope technique. *Global Biogeochem. Cycles* 14: 51.

Chen, Y., Dumont, M.G., Cebron, A., & Murrell, J.C. 2007. Identification of active methanotrophs in a landfill cover soil through detection of expression of 16S rRNA and functional genes. *Environ. Microbiol.* 9: 2855.

Chistoserdova, L., Vorholt, J.A. & Lidstrom, M.E. 2005. A genomic view of methane oxidation by aerobic bacteria and anaerobic archaea. *Genome Biology* 6: 208.

Crutzen, P.J. 1994. Global budgets for non-CO_2 greenhouse gases. *Environ. Monit. Assess.* 31: 1.

Dahllöf, I. 2002. Molecular community analysis of microbial diversity. *Curr. Opin. Biotechnol.* 13: 213.

Dalton, H. 2005. The Leeuwehoek lecture 2000. The natural and unnatural history of methane-oxidizing bacteria. *Phil. Trans. R. Soc.* 360: 1207.

Dedysh, S.N., Liesack, W., Khmelenina, V.N., Suzina, N.E., Trotsenko, Y.A., Semrau, J.D., Bares, A.M., Panikov, N.S. & Tiedje, J.M. 2000. *Methylocella palustris* gen. nov., sp. nov., a new methane-oxidizing acidophilic bacterium from peat bogs, representing a novel subtype of serine-pathway methanotrophs. *Int. J. Syst. Evol. Microbiol.* 50: 955.

DiSpirito, A.A., Zahn, J.A., Graham, D.W., Kim, H.J., Larive, C.K., Derrick, T.S., Cox, C.D. & Taylor, A. 1998. Copper-binding compounds from *Methylosinus trichosporium* OB3b. *J. Bacteriol.* 180: 3606.

Dunfield, P.F., Yuryev, A., Senin, S., Smirnova, A.V., Stott, M.B., Hou, S., Ly, B., Saw, J.H., Zhou, Z., Ren, Y., Wang, J., Mountain, B.W., Crowe, M.A., Weatherby, T.M., Bodelier, P.L.E., Liesack, W., Feng, L., Wang, L., & Alam, M. 2007. Methane oxidation by an extremely acidophilic bacterium of the phylum *Verrucomicrobia. Nature* 450: 879.

Dunfield, P.F., Khmelenina, V.N., Suzina, N.E., Trotsenko, Y.A. & Dedysh, S.N. 2003. *Methylocella silvestris* sp.

nov., a novel methanotroph isolated from an acidic forest cambisol. *Int. J. Syst. Evol. Microbiol.* 53: 1231.

EPA 1991, U.S. Environmental Protection Agency. Air emissions from municipal solid waste landfills: background information for proposed standards and guidelines. EPA-450/3-90/011a. March 1991.

Gebert, J., Gröngröft, A. & Miehlich, G. 2003. Kinetics of microbial landfill methane oxidation in biofilters. *Waste Manage.* 23: 609.

Gebert, J., Gröngröft, A., Schloter, M., & Gattinger, A. 2004. Community structure in a methanotroph biofilter as revealed by phospholipid fatty acid analysis. *FEMS Microbiol. Lett.* 240: 61.

Gilbert, B., McDonald, I.R., Finch, R., Stafford, G.P., Nielsen, A.K., & Murrell, J.C. 2000. Molecular analysis of the pmo (particulate methane monooxygenase) operons from two type II methanotrophs. *Appl. Environ. Microbiol.* 66: 966.

Glatzel, S. & Stahr, K. 1998. The trace gas budget of differently managed grassland using the Hohenheim chamber. In: Proceedings of the 16th World Congress of Soil Science, Montpellier/France.

Halet, D., Boon, N., & Verstraete, W. 2006. Community dynamics of methanotrophic bacteria during composting of organic matter. *J. Biosci. Bioeng.* 4: 297.

Hanson, R.S. & Hanson, T.E. 1996. Methanotrophic bacteria. *Microbiol. Rev.* 2: 439.

He, R., Ruan, A., Jiang, C. & Shen, D. 2008. Responses of oxidation rate and microbial communities to methane in simulated landfill cover soil microcosms. *Bioresour. Technol.* 99: 7192.

Heyer, J., Galchenko, F.V. & Dunfield, P.F. 2002. Molecular phylogeny of type methane- oxidizing bacteria isolated from various environments. *Microbiol.* 148: 2831.

Higgins, I.J., Best, D.J., Hammond, R.C. & Scott, D. 1981. Methane-oxidizing microorganisms. *Microbiol. Rev.* 45: 556.

Hilger, H. & Humer, M. 2003. Biotic landfill cover treatments for mitigating methane emissions. *Environ. Monit. Assess.* 84: 71.

Hinrichs, K. U. & Boetius, A. 2002. The anaerobic oxidation of methane: new insights in microbial ecology and biogeochemistry. In: G. Wefer, D. Billett, D. Hebbeln, B.B. Jørgensen, M. Schlüter, T. Van Weering (eds.). *Ocean Margin Systems.* Springer-Verlag: Berlin Heidelberg: 457–477.

Hoeks, J. 197). Effect of leaking natural gas on soil and vegetation in urban areas. Agricultural Research Reports 778, Centre for Agricultural Publishing and Documentation (Pudoc), Wageningen: Netherlands 120.

Hou, S., Makarova, K.S., Saw, J.H.W., Senin, P., Ly, B.V., Zhou, Z., Ren, Y., Wang, J., Galperin, M.Y., Omelchenko, M.V., Wolf, Y.I., Yutin, N., Koonin, E.V., Stott, M.B., Mountain, B.W., Crowe, M.A., Smirnova, A.V., Dunfield, P.F., Feng, L., Wang, L. & Alam, M. 2008. Complete genome sequence of the extremely acidophilic methanotroph isolate V4, *Methylacidiphilum infernorum*, a representative of the bacterial phylum *Verrucomicrobia. Biol. Direct* 1: 3.

Huber-Humer, M., Gebert, J. & Hilger, H. 2008. Biotic systems to mitigate landfill methane emissions. *Waste Manage. Res.* 26: 33.

IPCC 2007. Climate Change 2007: The Physical Climate Change 2007: The Physical Science Basis Contribution of Working Group I to the Fourth Assessment Report of the Intergovernmental Panel on Climate Change, S. Solomon, D. Qin, M. Manning, Z. Chen, M. Marquis, K. Averyt, M.MB. Tignor & H.L. Miller (eds.). Cambridge University Press: Cambridge, United Kingdom and New York, NY, USA. 499.

Islam, T, Jensen, S., Reigstad, L.J., Larsen, Ø. & Birkeland, N.K. 2008. Methane oxidation at 55° C and pH
2 by a thermoacidophilic bacterium belonging to the
Verrucomicrobia phylum. *PNAS* 105: 300.

Janda, J.M. & Abbott, S.L. 2007. 16S rRNA gene sequencing
for bacterial identification in the diagnostic laboratory:
pluses, perils and pitfalls. *J. Clin. Microbiol.* 45: 2761.

Jugnia, L.B., Ait-Benichou, S., Fortin, N., Cabral, A.R. &
Greer, C.W. 2009. Diversity and dynamics of methan-
otrophs within an experimental landfill cover soil. *Soil
Sci.Soc. Am.* J. 73, 1479.

Jugnia, L.B., Cabral, A.R. & Greer, C.W. 2008. Biotic
methane oxidation within an instrumented experimental
landfill cover. *Ecol. Eng.* 33: 102.

Kallistova, A.Y., Kevbrina, M.V., Nekrasova, V.K., Shnyrev,
N.A., Einola, J.K.M., Kulomaa, M.S., Rintala, J.A.
& Nozhevnikova, A.N. 2007. Enumeration of methanotrophic bacteria in the cover soil of an aged municipal
landfill. *Microb. Ecol.* 54: 637.

Kalyuzhnaya, M.G., Khmelenina, V.N., Suzina, N.E.,
Lysenko, A.M. & Trotsenko, Y.A. 1999. New methan-
otrophic isolates from soda lakes of the southeastern
Transbaikal region. *Microbiol.* 68: 677.

Khmelenina, V.N, Kalyuzhnaya, M.G., Sakharovsky, V.G.,
Suzina, N.E., Trotsenko, Y.A. & Gottschalk, G. 1999.
Osmoadaptation in halophilic and alkaliphilic methan-
otrophs. *Arch. Microbiol.* 172: 321.

Khmelenina, V.N, Kalyuzhnaya, M.G., Starostina, N.G.,
Suuzina, N.E. & Trotsenko, Y.A. 1997. Isolation and
characterization of halotolerant alkaliphilic methanotrophic bacteria from Tuva soda lakes. *Curr. Microbiol.*
35: 257.

Kolb, S., Knief, C., Stubner, S. & Conrad, R. 2003. Quantitative detection of methanotrophs in soil by novel *pmoA* –
targeted Real-Time PCR assays. *Appl. Environ. Microbiol.*
69: 2423.

Le Mer, J. & Roger, P. 2001. Production, oxidation, emission
and consumption of methane by soils: A review. *Eur. J.
Soil Biol.* 37: 25.

Lelieveld J., Crutzen P.J. & Brühl C. 1993. Climate effects of
atmospheric methane. *Chemosphere* 26: 739.

Lieberman, R.L. & Rosenzweig, A.C. 2004. Biological
methane oxidation: regulation, biochemistry, and active
site structure of particulate methane monooxygenase.
Crit. Rev. Biochem. Mol. 3: 147.

Lu, Y. & Khalil, M.A.K. 1993. Methane and carbon monoxide
in OH chemistry: the effects of feedbacks and reservoirs
generated by the reactive products. *Chemosphere* 26: 641.

McDonald, I.R., Bodrossy, L., Chen, Y. & Murrell, J.C. 2008.
Molecular ecology techniques for the study of aerobic
methanotrophs. *Appl. Environ. Microbiol.* 5: 1305.

McDonald, I.R. & Murrell, J.C. 1997a. The particulate
methane monooxygenase gene *pmoA* and its use as a functional gene probe for methanotrophs. *FEMS Microbiol.
Lett.* 156: 205.

McDonald, I.R. & Murrell, J.C. 1997b. The methanol dehydrogenase structural gene *mxaF* and its use as a functional
gene probe for methanotrophs and methylotrophs. *Appl.
Environ. Microbiol.* 63: 3218.

McDonald, I.R., Kenna, E.M. & Murrell, J.C. 1995. Detection
of methanotrophic bacteria in environmental samples with
the PCR. *Appl. Environ. Microbiol.* 61: 116.

Michaelis, W., Seifert, R., Nauhaus, K., Treude, T., Thiel, V.,
Blumenberg, M., Knittel, K., Gieseke, A., Peterknecht,
K., Pape, T., Boetius, A., Amann, R., Jørgensen, B.B.,
Widdel, F., Peckmann, J., Pimenov, N.V., & Gulin, M.B.
2002. Microbial reefs in the Black Sea fueled by anaerobic
oxidation of methane. *Science* 9: 1013.

Miura, J., Watanabe, A., Murase, M. & Kimura, M. 1992.
Methane production and its fate in paddy fields. II. Oxidation of methane and its coupled ferric in oxide reduction
subsoil. *Soil Sci. Plant Nutr. 38: 673.*

Murase, J. & Kimura, M. 1994. Methane production and its
fate in paddy fields. VII. Electron acceptors responsible
for anaerobic methane oxidation. *Soil Sci. Plant Nutr.* 40:
647.

Murrell, J.C. & Radajewski, S. 2000. Cultivation-
independent techniques for studying methanotroph ecology. *Res. Microbiol.* 151: 807.

Murrell, J.C., Gilbert, B. & McDonald, I.R. 2000a. Molecular
biology and regulation of methane monooxygenase. *Arch.
Microbiol.* 173: 325.

Murrell, J.C., McDonald, I.R. & Gilbert, B. 2000b. Regulation of expression of methane monooxygenase by copper
ions. *Trends Microbiol.* 8: 221.

Murrell, J.C., McDonald, I.R. & Bourne, D.G. 1998. Molecular methods for the study of methanotroph ecology. *FEMS
Microbiol. Ecol.* 27: 103.

Nishio, T., Yoshikura, T. & Itoh, H. 1997. Detection of *Methy-
lobacterium* species by 16S rRNA gene-targeted PCR.
Appl. Environ. Microbiol. 63: 1594.

Nozhevnikova, A.N., Lifshitz, A.B., Lebedev, V.S. &
Zavarzin, G.A. 1993. Emission of methane into the atmosphere from landfills in the former USSR. *Chemosphere*
26: 401.

Park, S., Lee, I., Cho, C. & Sung, K. 2008. Effects of earthworm cast and powdered activated carbon on methane
removal capacity of landfill cover soils. *Chemosphere* 70:
1117.

Perdikea, K., Mekrotra, A.K. & Hettiaratchi, J.P.A. 2008.
Study of thin biocovers (TBC) for oxidizing uncaptured
methane emissions in bioreactor landfills. *Waste Manage.*
28: 1364.

Pol, A., Heijmans, K., Harhangi, H.R., Tedesco, D., Jetten,
M.S. & Op den Camp H.J. 2007. Methanotrophy below
pH 1 by a new *Verrucomicrobia* species. *Nature* 450: 874.

Putzer, K.P., Buchholz, L.A., Lidstrom, M.E. & Remsen, C.C.
1991. Separation of methanotrophic bacteria by using per-
coll and its application to isolation of mixed and pure
cultures. *Appl. Environ. Microbiol.* 57: 3656.

Reeburgh, W.S., Whalen, S.C.& Alperin, M.J. 1993. The role
of methylotrophy in the global methane budget. In: J.C.
Murrell & D.P. Kelly (eds.). *Microbial Growth on C1
Compounds.* Andover: Intercept, p.1.

Ricke, P., Erkel, C., Kube, M., Reinhardt, R. & Liesack, W.
2004. Comparative analysis of the conventional and novel
pmo (particulate methane monooxygenase) operons from
Methylocystis strain SC2. *Appl. Environ. Microbiol.* 70:
3055.

Semrau, J.D., Chistoserdov, A., Lebron, J., Costello, A.,
Davagnino, J., Kenna, E., Holmes, A.J., Finch, R., Murrell, J.C. & Lidstrom, M.E. 1995. Particulate methane
monooxygenase genes in methanotrophs. *J. Bacteriol.*
177: 3071.

Singleton, P. 2000. Bacteria in Biology, Biotechnology and
Medicine. Warszawa: PWN.

Stępniewska, Z., Przywara, G. & Bennicelli, R.P. 2004.
Reakcja roœlin w warunkach anaerobowych. *Acta Agrophysica 113.*

Stępniewski, W. & Pawłowska, M. 1996. A possibility to
reduce methane emission from landfills by its oxidation in
the soil cover. In L. Pawłowski, W.J. Lacy, Ch. G. Uchrin
& M. R. Dudzińska (eds.). *Chemistry for the Protection
of the Environment.* New York: Plenum Press. 75.

Stralis-Pavese N., Bodrossy L., Reichenauer T.G., Weilharter
A. & Sessitsch A. 2006. 16S rRNA based T-RFLP analysis

of methane oxidising bacteria- Assessment, critical evaluation of methodology performance and application for landfill site cover soils. *Appl. Soil Ecol.* 31: 251.

Stralis-Pavese N., Sessitsch, A., Weilharter, A., Reichenauer, T., Riesing, J., Csontos, J., Murrel, C. & Bodrossy, L. 2004. Optimization of diagnostic microarray for application in analysing landfill methanotroph communities under different plant cover. *Environ. Microbiol.* 6: 347.

Takeguchi, M., Miyakawa, K. & Okura, I. 1999. The role of copper in particulate methane monooxygenase from *Methylosinus trichosporium* OB3b. *J. Mol. Catal. A: Chem.* 137: 161.

Trotsenko Y.A. & Khmelenina, V.N. 2002. Extremophilic and extremotolerant methanotrophic bacteria. *Arch. Microbiol.* 177: 123.

Tsien, H.C., Bratina, B.J., Tsuji, K., Hanson, R.S. 1990. Use of oligodeoxynucleotide signature probes for identification of physiological groups of methylotrophic bacteria. *Appl. Environ. Microbiol.* 56: 2858.

Tsubota, J., Eshinimaev, B.T., Khmelenina V.N., & Trotsenko, Y.A. 2005. *Methylothermus thermalis* gen. nov., sp. nov., a novel moderately thermophilic obligate methanotroph from a hot spring in Japan. *Int J Syst Evol Microbiol* 55: 1877.

U.S. Fire Administration. 2001. *Topical Fire Research Series.* 1: 18, (http://www.usfa.dhs.gov/downloads/pdf/tfrs/v1i18-508.pdf).

USACE 1984. U.S. Army Corps of Engineers. Landfill gas control at military installations, Prepared by R.A. Shafer. Publication Number CERL-TR-N-173. January 1984.

Valentine, D.L. 2002. Biochemistry and microbial ecology of methane oxidation in anoxic environments: a review. *Antonie van Leeuwenhoek* 81: 271.

Vassiliadou, I., Papadopoulos, A., Costopoulou, D., Vasiliadou, S., Christoforou, S. & Leondiadis, L. 2009. Dioxin contamination after an accidental fire in the municipal landfill of Tagarades, Thessaloniki, Greece. *Chemosphere* 74: 879.

Wang, H., Einola, J., Heinonen, M., Kulomaa, M. & Rintala, J. 2008a. Group specific quantification of methanotrophs in landfill gas-purged laboratory biofilters by tyramide signal amplification-fluorescence *in situ* hybridization. *Bioresour. Technol.* 99: 6426.

Wang, Y., Wu, W., Ding, Y., Liu, W., Perera, A., Chen, Y. & Devare, M. 2008b Methane oxidation activity and bacterial community composition in a simulated landfill cover soil is influenced by the growth of *Chenopodium album* L. *Soil Biol. Biochem.* 40: 2452.

Ward, N., Larsen, Ø., Sakwa, J., Bruseth, L. & Khouri, H.. 2004. Genomic insights into methanotrophy: the complete genome sequence of *Methylococcus capsulatus* (Bath). *PloS Biol.* 2, 1616.

Wise, M.G., McArthur, J.V. & Shimkets, L.J. 1999. Methanotroph diversity in landfill soil: isolation of novel type I and type II methanotrophs whose presence was suggested by culture-independent 16S ribosomal DNA analysis. *Appl. Environ. Microbiol.* 65: 4887.

Yu, S.S.F., Chen, K.H.C., Tseng, M.Y.H., Wang, Y.S., Tseng, C.F., Chen, Y., Huang, D. & Chan, S. 2003. Production of high-quality particulate methane monooxygenase in high yields from *Methylococcus capsulatus* (Bath) with a hollow-fiber membrane bioreactor. *J. Bacteriol.* 185: 5915.

Yuan, H., Collins, M.L.P. & Antholine, W.E. 1998. Concentration of Cu, EPR-detectable Cu, and formation of cupric-ferrocyanide in membranes with pMMO. *J. Inorg. Biochem.* 72: 179.

Environmental Engineering III – Pawłowski, Dudzińska & Pawłowski (eds)
© *2010 Taylor & Francis Group, London, ISBN 978-0-415-54882-3*

The application of spent ion-exchange resins as NO_3^- carriers in soil restoration

M. Chomczyńska & E. Wróblewska

Department of Environmental Engineering, Lublin University of Technology, Lublin, Poland

ABSTRACT: The studies aimed to compare the effect of monoionic nitrate form (prepared from a model spent ion exchanger) and conventional nitrogen fertilizer $(Ca(NO_3)_2 4H_2O)$ on plant vegetation after addition to depleted soil. A pot experiment using orchard grass as the test species was carried out. Plants were grown on the following media: untreated soil, soil with added monoionic nitrate form, soil plus $Ca(NO_3)_2 4H_2O$ and soil with Biona-312 substrate added (2% v/v). Applying the monoionic nitrate form positively affected orchard grass vegetation. Adding this form to the soil significantly increased plant height, stem wet and dry biomass. Bearing in mind that the amount of dry plant matter as source material for humus formation is crucial in soil reclamation, the effectiveness of the monoionic nitrate form was found to be similar to that of the mineral fertilizer. Biona-312 was the most efficient fertilizer used in the study, resulting in the greatest yield of orchard grass.

Keywords: Ion-exchange resins, waste ion exchangers, nitrate, soil restoration.

1 INTRODUCTION

Synthetic ion exchangers are widely used in water treatment, mainly for softening and demineralization purposes. During these processes, they lose their properties, their granules are mechanically destructed and a decrease in their ion-exchange capacity is observed due to degradation of functional groups or irreversible sorption of humic compounds. As a result, spent resins are periodically discharged by water treatment plants. The possible methods for neutralization of these materials are storing or incineration; however, the latter results in emission of SO_2 and NO_x in the case of cation exchangers and anion exchangers, respectively. The alternative, ecologically attractive method for reuse of spent resins is their application as nutrient carriers in the biological restoration of degraded grounds. After saturation with nutrient ions, they can serve as fertilizers that intensify plant development on degraded soils. During soil restoration, plant development is crucial for scarp planting and humus layer forming.

The possibility of using synthetic ion-exchange resins as nutrient carriers for plants has been confirmed in the last 70 years (Arnon & Grossenbacher 1947, Zemljanuchin et al. 1964, 1966). Since that time, interest in ion-exchange resins as nutrient carriers has been increasing, finally resulting in a preparation of ion-exchange substrates called Biona®. The substrates were prepared by Soldatov's research team from the Belarusian National Academy of Science (BNAS). They contained all of the macro- and micronutrients, and enabled plant cultivation in closed ecological

systems (such as spacecraft, arctic stations and submarines) (Soldatov et al. 1968, 1969, 1978). Parallel investigations of the effect of monoionic forms prepared from synthetic resins and containing some nutrients (Ca^{2+}, K^+, NO_3^-) on the properties of sandy soils were carried out in Bohemia (Podlesakova & Bouchal 1978, Podlesakova 1979a, b). The results of Podlesakova's studies showed that plant biomass obtained on soils fertilized with monoionic forms was higher than that obtained in the presence of mineral salts of nitrogen, potassium and calcium. The positive study results with Biona substrates initiated a new series of model experiments at Lublin University of Technology (Poland) on their applications in the restoration of degraded soil. Among a number of other findings, the study recommended the use of Biona additions (1-2% v/v) for the fertilization of barren soils (Soldatov et al. 1998).

In the works described above, only fresh, previously unused ion exchangers were studied. Taking into account the problem of spent ion-exchanger disposal, the research team from Lublin University of Technology has made the first attempts on application of monoionic forms prepared from waste resins. Chomczyńska used nitrate, phosphate and potassium forms as fertilizers introduced together into sand as a model of degraded soil. The application of the mentioned forms increased plant yield compared with that obtained on sand alone (Chomczyńska & Pawłowski 2003). In these tests, conventional fertilizers as controls were not applied. Thus, the purpose of the present study was to compare the effectiveness of fertilizing depleted soil with nitrogen bound to a spent resin model with nitrogen in the salt form – $Ca(NO_3)_2 4H_2O$.

2 MATERIALS AND METHODS

In this study, mineral soil, monoionic nitrate form, calcium nitrate and the ion-exchange substrate Biona-312 were used. The mineral soil was taken from an excavation boundary of a sand mine (in Turka near Lublin, Poland). It consisted of the following fractions: sand (1.0-0.1 mm) −78%; dust (0.1-0.02 mm) −11%; and silt and clay (<0.02 mm) −11%. According to the granulometric composition, the soil was classified as light loamy sand (Polish Society of Soil Science) or sandy loam (USDA). It contained 24.33 mg of available nitrogen per dm^3, which is considered insufficient for vegetable plants (Sady 2000). The pH of the soil in distilled H_2O was 5.39.

The monoionic nitrate form was prepared by treating the polifunctional anion-exchanger EDE-10P in OH form with HNO_3 solution. The prepared monoionic form contained 3.31 mmol N per g. This form was ground in a Retsch grinder (type S 1000) in order to produce a structure similar to that of a spent ion-exchange resin.

The ion-exchange substrate Biona-312 was prepared at the Institute of Physical Organic Chemistry of BNAS in Minsk and served as the control fertilizer in the study. It was a mixture of 56% (mass) of ion-exchange substrate Biona-111 and 44% of clinoptilolite (a zeolite of formula $(K_2, Na_2, Ca)Al_6Si_{30}O_{72}24H_2O)$). The Biona-312 substrate contained the following amounts of elements $(g\,kg^{-1})$: N − 11.206, P − 3.407, K − 17.595, Mg − 4.378, Ca − 22.244, S − 6.094, Mn − 0.220, Cu − 0.064, Zn − 0.057, Co − 0.015, Mo − 0.044, B − 0.110, Fe − 2.234, Na −1.379, Cl − 3.900.

To achieve the study purpose, a pot experiment was carried out in open air conditions. The test plant was orchard grass (*Dactylis glomerata* L.) – a species recommended as a constituent of plant recultivation mixtures. For the needs of the experiment, four series of media were prepared, including two control series: soil and soil with 2% (v/v) addition of ion-exchange substrate Biona-312; and two test series: soil with addition of the monoionic nitrate form and soil with addition of a conventional nitrogen fertilizer (Table 1). The additions of the monoionic form and $Ca(NO_3)_2 4H_2O$ contained the same amounts of nitrogen based on recommendations for fertilizing meadows on mineral soils $(100\,kg\,N\,ha^{-1})$ (Niewiadomski 1983). The addition of Biona-312 resulted from previous studies which showed that a 1-2% dose of ion-exchange substrate was sufficient for soil reclamation purposes (Soldatov et al. 1998).

The experiment started on 17 May 2006. There were five pots (of $360\,cm^3$ volume) in each series. In each pot, 0.05 g of *Dactylis glomerata* L. seeds was sown. After 14 days of growth, the number of plants in all pots was standardized. During the experiment, plants

Table 1. Characteristics of series in the vegetation experiment.

Series	Soil [cm^3 g/pot]	Mass of fertilizer [10^{-2}g/pot]	Macronutrients [10^{-2}g/pot]
Soil (S)	300 514	0.00	0.00
Soil+$Ca(NO_3)_2 4H_2O$ (S+N)	300 514	66.17	N-7.85
Soil+monoionic form (S+NI)	300 514	179.10	N-7.85
Soil+Biona-312 (S+B312)	300 514	513.84	K-9.04 N-5.76 P-1.75 Ca-11.43 Mg-2.25 S-3.13

were watered with distilled water. The temperature, as well as the relative air humidity, was monitored using a TZ-18 thermohygrograph ("Zootechnika", Kraków, Poland). The daytime air temperature varied between 13°C and 35°C. The night-time air temperature was in the range 15.5–24°C. The daytime air humidity achieved values between 19% and 95%; the night-time air humidity ranged from 62% to 86.5%. The experiment was terminated after 43 days from the time of seed sowing. The height of plants was measured and aboveground shoots were cut down. The plant roots were also separated. The wet and dry (105°C) biomass of stems and dry (105°C) biomass of roots was elicited. The results obtained were used for calculation of arithmetical mean values. Mean values for particular experimental series were compared and the significance of differences was assessed using Student's t-test or Aspin–Welch's v-test with a confidence coefficient p = 0.95 (Zgirski & Gondko 1983, Czermiński et al. 1992). The Student's t-test was used when variances for compared average values did not differ significantly. When variances differed significantly, the Aspin–Welch's v-test was applied to determine the significance of differences between mean values.

3 RESULTS AND DISCUSSION

The study results are presented in Figures 1–4. It can be seen that the addition of nitrogen as $Ca(NO_3)_2 4H_2O$ positively influenced orchard grass growth, significantly increasing most of the values of the vegetation parameters. Specifically, the plant height, wet and dry stem biomass on soil with added $Ca(NO_3)_2 4H_2O$ exceeded those obtained in the control series by 30%, 33% and 20%, respectively. The mean value for dry root biomass in the control series was 50% higher than that obtained on soil with addition of calcium nitrate. The reason for this is that there can be stronger development of root systems that penetrate nutrient-poor substrates in search of biogenic

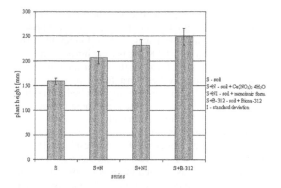

Figure 1. Mean plant height in experimental series. Note: the difference between mean values for S+NI and S+B312 was not statistically significant at p = 0.95; the differences between other mean values were statistically significant at p = 0.95.

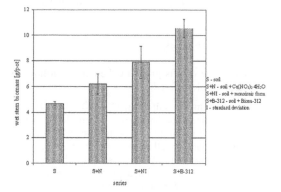

Figure 2. Mean wet stem biomass in experimental series. Note: the differences between all mean values were statistically significant at p = 0.95.

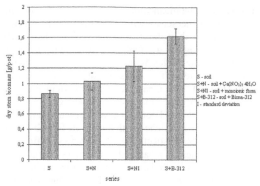

Figure 3. Mean dry stem biomass in experimental series. Note: the difference between mean values for S+N and S+NI was not statistically significant at p = 0.95; the differences between other mean values were statistically significant at p = 0.95.

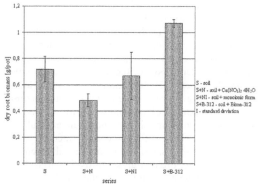

Figure 4. Mean dry root biomass in experimental series. Note: the differences between mean values for S and S+NI, as well as for S+N and S+NI, were not statistically significant at p = 0.95; the differences between other mean values were statistically significant at p = 0.95.

elements. Consequently, the root biomass of plants growing on poor soil can be higher than plant root biomass on soil enriched with nutrients. The application of the monoionic nitrate form, similarly to calcium nitrate, had a positive influence on orchard grass vegetation, significantly increasing its yield (Figures 1-3). Plant height, and wet and dry stem biomass on soil enriched with the nitrate form exceeded those on soil alone by 46%, 69% and 43%, respectively. The dry root biomass of plants growing on soil of the control series was 7% greater than that in the series with the nitrate form; however, the observed difference between the values of the parameter was not statistically significant (Figure 4).

The results analysis for the control series and the series with Biona-312 showed that the ion-exchange substrate was the best fertilizer because it increased the yield of test species the most (Figs 1–4). The mean height, mean wet and dry stem biomass, as well as mean root biomass of orchard grass on soil supplemented with Biona-312, were greater than those obtained on soil alone by 57%, 125%, 88% and 49%, respectively.

Comparing results for the series with nitrogen applied as $Ca(NO_3)_24H_2O$ and the series with nitrogen bound to the ion exchanger showed that the height and wet stem biomass of orchard grass were significantly higher on the soil supplemented with the monoionic form by 12% and 27%, respectively (Figures 1-2). The dry stem and root biomass of plants growing in series with the nitrate form exceeded those on soil fertilized with calcium nitrate by 19% and 40%, respectively; however, the differences between the mentioned parameters were not statistically significant (Figures 3–4). From a soil restoration point of view, dry matter production is important as source material for humus formation. Hence, a lack of statistically significant differences between dry biomass allows the effectiveness of both fertilizers ($Ca(NO_3)_24H_2O$ and the monoionic nitrate form) to be determined as similar.

A comparison between all the measured parameters characterizing the test species in the fertilized experimental series confirmed the greatest efficiency of

Biona-312. The mean values for almost all parameters obtained in the series with Biona-312 were significantly higher than those in series with the two other fertilizers (Figs 1–4). Only the difference in plant height for the series with Biona-312 and with the nitrate monoionic form was not statistically significant (Fig. 1). The mean wet and dry stem biomass and mean dry root biomass of orchard grass on soil supplemented with Biona-312 were greater than those obtained on soil enriched with the nitrate form by 34%, 32% and 60%, respectively. The greatest difference between values of measured parameters characterizing the series with Biona-312 and the series with $Ca(NO_3)_24H_2O$ was found for dry root biomass that was 123% higher on soil plus Biona-312 compared with soil plus calcium nitrate (Fig. 4). Other vegetation parameters – the mean plant height, and the mean wet and dry stem biomass – were higher by 21%, 70% and 57%, respectively, in the case of soil supplemented with Biona. The highest efficiency of Biona-312 as a fertilizer resulted from its high content of nitrogen and other macronutrients (Table 1).

The application of the monoionic nitrate form to soil had a positive effect on plant vegetation, which was also observed in other research team studies from Lublin University of Technology (Kloc & Szwed 1995, Wasąg et al. 2000, Chomczyńska & Pawłowski 2003). During their tests, the mentioned authors used different mixtures of monoionic forms (potassium, calcium, magnesium, nitrate, phosphate and sulphate forms), as well as different control media (garden soil, mineral soil, sand) and sometimes different test species (maize, orchard grass, birds-foot trefoil). As mentioned above, in almost all the cited works, the introduction of monoionic forms into media increased plant biomass. The lowest increase in orchard grass yield was observed in Wasąg's et al. studies in which wet stem biomass was 18% higher after application of a mixture of nitrate, phosphate and potassium forms, whereas dry stem biomass was 9% higher after introducing a mixture of potassium, calcium, sulphate and phosphate forms into mineral soil (Wasąg et al. 2000). The highest increase in plant yield was reported by Chomczyńska (Chomczyńska & Pawłowski 2003). In her tests, an addition of three monoionic forms (NPK) into sand caused four-fold increases in the wet stem biomass and dry root biomass, and an almost five-fold increase in the dry stem biomass of orchard grass.

The introduction of the nitrate form into soil did not give a significantly higher dry yield of orchard grass than that observed in the presence of nitrogen dosed in the form of calcium nitrate. Different results were obtained by Podlesakova in her studies on the effectiveness of monoionic forms (nitrate and potassium forms) compared with mineral fertilizers (NaNO$_3$ and KCl) in maize (*Zea mays* L.) cultivation on sandy soil during pot experiments (Podlesakova & Bouchal 1978, Podlesakova 1979b). In cited works, after the first vegetation period of maize, aboveground dry biomass on soil supplemented with monoionic forms was significantly greater than that observed in the presence of

mineral fertilizers. Specifically, the maize yield on soil fertilized with different K doses applied as potassium monoionic form was 7–22% higher than that obtained when parallel K doses were introduced in the potassium chloride form. In the experiment on nitrogen fertilization, Podlesakova found that the maize aboveground dry biomass on soil enriched with different N doses introduced as the monoionic nitrate form was 23–127% higher than that obtained on soil fertilized with parallel N doses applied as sodium nitrate (Podlesakova & Bouchal 1978, Podlesakova 1979b). Podlesakova also reported tests on calcium fertilization where Ca was applied in the form of CaCO$_3$ and in the form bound to a cation exchanger. She found that the aboveground dry biomass of maize on soil with the addition of the monoionic calcium form was 13% greater than that obtained on soil fertilized with CaCO$_3$. However, the observed increase in maize dry yield after the application of the monoionic form was not statistically significant (Podlesakova 1979a).

4 CONCLUSIONS

Based on the results presented herein, the following conclusions are offered:

Nitrogen in the salt form ($Ca(NO_3)_24H_2O$) positively affects orchard grass vegetation – when introduced to sandy soil, it significantly increases plant height, as well as wet and dry stem biomass. The monoionic nitrate form seems to be an effective fertilizer, increasing degraded soil productivity – mean values for almost all parameters characterizing orchard grass vegetation on soil supplemented with the nitrate form are significantly greater then those obtained on soil alone.

Biona-312, as a standard ion-exchange substrate, turns out to be the most efficient fertilizer used in the study, resulting in the highest yield of *Dactylis glomerata* L.

There is a lack of significant differences between the dry yields obtained in the series with nitrogen bound to an ion exchanger and in the series with nitrogen in salt form. This implies that the efficiency of the monoionic nitrate form is similar to that of conventional fertilizer. However, the advantage of using an anion exchanger as the nitrogen source is that the resin has the ability to gradually release NO$_3^-$ anions in exchange with plant ionic metabolites. This is important for preventing nitrate leaching from soil to groundwater or water bodies.

REFERENCES

Arnon, D.I. & Grossenbacher, K.A. 1947. Nutrient culture of crops with the use of synthetic ion-exchange materials. *Soil Science* 63: 59–182.
Chomczyńska, M. & Pawłowski, L. 2003. Utilization of spent ion exchange resins for soil reclamation. *Environmental Engineering Science* 20(4): 301–306.

Czermiński, J.B., Iwasiewicz, A., Paszek, Z. & Sikorski, A. 1992. *Statistical Methods for Chemists* (in Polish). Warszawa: PWN.

Kloc, E. & Szwed, R. 1995. *Ion exchangers as bioelements' carriers. Studies on possibilities of ion exchange applications for soil improvement* (in Polish). MSc Thesis. Lublin: Lublin University of Technology.

Niewiadomski, W. 1983. *Fundamentals of Agricultural Science* (in Polish). Warszawa: PWRiL.

Podlesakova, E. & Bouchal, P. 1978. The consequental acting of ion exchangers of czechoslovak production in sandy earth on maize production (in Czech). *Vědecké Práce Výzkumného Ústavu Meliorací v Praze* 14: 67–82.

Podlesakova, E. 1979a. An attempt at the improvement of sandy soils by in-depth application of fertilisers (in Czech). *Agrochemia* 19: 97–101.

Podlesakova, E. 1979b. Increasing the production capacity of sandy soils by ion exchangers. *Scientia Agriculturae Bohemoslovaca* 1: 1–12.

Sady, W. 2000. *Field vegetable fertilization* (in Polish). Kraków : Wydawnictwo Plantpress.

Soldatov, V.S., Terent'ev, V.M. & Periškina, N.G. 1968. Artificial soil on ion exchange materials' basis (in Russian). *Doklady Akademii Nauk BSSR* 12: 357–359.

Soldatov, V.S., Terent'ev, V.M. & Peryškina, N.G. 1969. Artificial nutrient media for plant growth on ion exchange materials' basis (in Russian). *Agrochmija* 2: 101–107.

Soldatov, V.S., Periškina, H.G. & Choroško, R.P. 1978. *Ion exchange soils* (in Russian). Minsk: Nauka i Technika.

Soldatov, V.S., Pawłowski, L., Szymańska, M., Chomczyńska, M. & Kloc, E. 1998. Ion exchange substrate Biona-111 as an efficient mean of barren grounds fertilization and soils improvement. *Zeszyty Problemowe Postępów Nauk Rolniczych* 461: 425–436.

Wasąg, H., Pawłowski, L., Soldatov, V.S., Szymańska, M., Chomczyńska, M., Kołodyńska, M., Ostrowski, J., Rut, B., Skwarek, A. & Młodawska, G. 2000. *Restoration of degraded soil by using ion exchange resins* (in Polish). Research Project KBN No 3 T09 C 105 14. Lublin: Lublin University of Technology.

Zemljanuchin, A.A., Ivanova, V.A. & Čurikova, V.V. 1964. Studies on application of ion exchange resins as nutrient carriers for plants (in Russian). In: *Teoretičeskie osnovy regulirovanija mineralnovo pitania rastenij*: Moskva: Nauka.

Zemljanuchin, A.A., Čurikova, V.V. & Ivanova, V.A. 1966. Application of synthetic ion exchangers as carriers of nutrient environment for plants (in Russian). In: *Teorija i praktika sorbcionnych processov*: Voronez: Izdatelstvo Voronezkovo Universiteta.

Zgirski, A. & Gondko, R. 1983. *Biochemical calculations* (in Polish). Warszawa: PWN.

Environmental Engineering III – Pawłowski, Dudzińska & Pawłowski (eds)
© 2010 Taylor & Francis Group, London, ISBN 978-0-415-54882-3

The increase of total nitrogen content in soilless formations as a criterion of the efficiency of reclamation measures

T. Gołda

AGH University of Science and Technology (AGH-UST), Faculty of Mining Surveying and Environmental Engineering, Department of Management and Protection of Environment, Krakow, Poland

ABSTRACT: Reclamation activities and processes are often carried out on soilless formations to produce new fully matured soil. This, however, takes centuries and only then can soil assessment indices be fully applied. We proposed using the annual increase of total nitrogen content in soilless formations as a criterion to assess soil-forming activity, to differentiate accumulation of organic matter, and to allow classification of dynamic development of soil formation. The amount of accumulated nitrogen depends on the potential properties of the soil formed, but mostly on the applied reclamation methods.

Keywords: Accumulation of total nitrogen, reclamation of mine soils, organic matter.

1 INTRODUCTION

A basic feature allowing the distinguishing between soil and soilless formations is the larger content of organic matter (OM) in soil. Soilless formations contain $<10\,\mathrm{Mg\,OM\,ha^{-1}}$, while soils have from 18 to $>200\,\mathrm{Mg\,ha^{-1}}$. One of the main purposes of the reclamation of spoil dumps, slag heaps and decantation ponds is creating conditions that allow for the introduction and survival of selected plant species, the development of which affects the dynamics of many processes leading to soil formation. The outer layers of the spoil dumps, slag heaps and decantation ponds predominantly consist of soilless formations, characterised by the lack or a little of natural nutrients necessary for the development of microorganisms and higher plants. This negative feature requires new methods of reclamation to provide adequate nutrients for the fast development of vegetation and, in turn, to accumulate and leave more nutrients/OM and systematically improve the soil. Converting soilless formations into soils with positive features is a long-term process, estimated at several centuries in the soil sciences; only after this period can soil science criteria be justifiably used (Jenny 1980). In the period of soil formation, there are no unambiguous methods for assessing the dynamic changes connected with a proper course of reclamation.

There are no scientific reports indicating the possibility of diagnostic levels in soilless formations, other than the OM level of overburden humus (litter – ecto-humic); for woodlands this is usually an initial, thin humic level and for agriculture, an arable-humic level. Thus, only the process of accumulating OM is relatively dynamics, during the initial period of changing the original morphologic profile of the soilless formation. Studying the formation of a new soil is an opportunity to assess the speed of some changes; from these the increase in OM allows an assessment of the outcome of the remediation measures.

We proposed using the annual increase of total nitrogen (TN) from a defined period, analysed by the Kjeldahl method, to assess reclamation measures. Nitrogen (N) accumulation is definitely a sensitive and comprehensive index of the course of soil-formation processes as influenced by reclamation treatments. It well describes the rate of OM accumulation, which provides a basic reservoir of nourishing substances. Only after accumulating a certain amount of these substances, do soilless formations obtain one basic feature of soil – the ability to sustain vegetation and allow it to reproduce. Soilless formations obtain the status of 'soil', when well-formed humic or ecto-humic levels can be found, and the content of OM reaches 40–$50\,\mathrm{Mg\,ha^{-1}}$. At lower contents, the conversion of the newly formed OM into specific humic compounds can be ambiguous and difficult to interpret, due to unstable properties of the formations, influencing the random predominance of humification over mineralisation or vice versa.

2 MATERIALS AND METHODS

2.1 Current methods of analysing the content of OM in reclaimed formations

Classical methods of analysing individual humus fractions are very time-consuming and have low repeatability, especially at low contents. Additionally, post-mining formations are very variable, and the content of organic substances is usually documented with indirect methods, e.g. the decrease of sample mass during annealing at temperatures from 450 to 550 °C. However, results using this quick and simple method are influenced by many factors that, especially at a lower content of OM, make unambiguous interpretation very difficult. This is mainly due to the analysis results containing additional errors, from other related changes during annealing.

Another indirect measure of OM content is the value based on the quantity of organic carbon oxidised with Tyurin's method. The obtained value, multiplied by the empirical coefficient (i.e. 1.724, corresponding to the 58% carbon content of humus compounds) gives the OM quantity. However, in many formations treated with reclamation measures, major overestimates of organic carbon contents occur (given by the sum of all oxidised compounds). This is often due to the occurrence of hard coal, lignite or xylite, and small quantities (although usually greater than in arable soils) of reduced compounds. The content of other compounds (e.g. carbonates) can also affect accuracy.

Increases in OM quantity should be a basic and objective indicator of the efficacy of reclamation, since its increase or decrease results from many properties of the formations such as microbiological activity, aeration, hydration, nutrient quantities, and the influence of cultivation procedures. The change in OM content in definite time-intervals allows conclusions on the present trend of changes (humification-mineralisation), enabling assessment of the dynamics of gaining new properties and production capabilities. In Poland, humus content is used to assess the degree of soil degradation and soil remediation; humus produced 'in situ' and that provided from outside, located on the surface of the reclaimed formations, are treated equally. According to this classification, well-reclaimed soils should contain 50–60 Mg ha^{-1} of humus, with the thickness of the humus layer being 25 cm (Siuta 1995).

2.2 Content of TN, calculated by the Kjeldahl method, in formations subjected to reclamation

The analysis mentioned above and interpretation difficulties with defining the OM content – in particular the content of specific humus compounds in reclaimed ground – indicate that another indicator could provide better information on the real increases in OM. Based on my own studies and the analysis of results for formations on reclaimed areas, it was proposed that the dynamics of new soil development be assessed by the TN content as calculated by the Kjeldahl method. This analysis is very sensitive, with detectability level 0.00X% (0.0X g kg^{-1}), good repeatability and with a much smaller influence on the 'matrix' on the results as compared to carbon quantity based method. Moreover, this well-known and widely applied method is already significantly automated. Usually, the TN content in the OM is nearly constant at 5% (ratio of TN to OM is 1:20). In soil science it is assumed that content of

TN <0.02% (<0.2 g kg^{-1}) is characteristic of soilless formations, while in upper layers of most arable soils the range in TN content is 1–3 g kg^{-1}. Generally, more OM is formed the greater the increase in the TN, and the dynamics of this process depend on the level of applied reclamation measures (such as the amounts of nitrogen, phosphate and potassium, the character of cultivation). A systematic decrease in the annual increase in N would certainly indicate the degradation of the formation caused by improper reclamation measures or agro-techniques, or a change in the conditions influencing accelerated mineralisation of already accumulated OM. A documented smaller accumulation of N would indicate that additional measures were required to halt this unfavourable trend.

The TN content in formations of different types of reclamation is presented in many publications. On the tailing heaps of East German mines, in forest reclamation the maximum annual increase in N was 33–42 kg ha^{-1} under *Robinia pseudoacacia* and Scotch pine (*Pinus sylvestris*), and the minimum of 11 kg ha^{-1} with the lime tree *Tilia* sp. (Katzur et al. 1999a, b). Katzur's results document mean annual growth of plantations over periods of tree growth of 25–70 y, and represent the longest studied periods for reclamation plots contained in the literature. A similar increase in the N content was documented in forest plantations on the heaps of the lignite mines in the Czech Republic, while the greatest increase was during 10–40 y of growth, at 30–40 kg ha^{-1} y^{-1} (Sourkova et al. 2005). Much higher values were calculated from studies of formations on the heaps of the Polish mine 'Machów'; after 23 years the mean annual growth of N reached 96 kg ha^{-1}, as the sum of N accumulated on the level of ecto-humus and endo-humus. These were plots covered with trees, with a high participation (number of trees of a certain species divided by the cumulative number of trees on the observed area) of approximately 30% of grey alder (*Alnus incana*) and black alder (*Alnus glutinosa*) (Kowalik 2004). The calculated annual growth of TN on other reclaimed objects, like the tailings heaps of opencast mines in Poland, varied within 15–70 kg ha^{-1} and depended on the granulometric composition of the soil and the cultivated species (Bykov 2003, Węgorek 2003). The highest TN accumulation was documented in formations of copper mining heaps on plots with *Alnus glutinosa*. After eight years the total amount was 905 kg TN ha^{-1} in the level of the ecto-humus overburden and the initial level of humus, which corresponds to a mean annual increase of about 115 kg TN ha^{-1} (Wójcik & Kowalik 2006). Similar annual increases in N of 110–130 kg ha^{-1} were found on the heap of the mine 'Adamów' in 25-year-old tree stands with a very high participation (70–100%) of grey or black alder (Wójcik et al. 2006).

The results presented by different authors indicate the fluctuations in the annual increases of TN of from 10 to >100 kg ha^{-1}, given as the sum of the content in ecto-humus and endo-humus. The type of agricultural use of the ground of the heaps has an influence;

the annual increase in N ranged from about 20 kg ha^{-1} for extensive use, to about 40–60 kg ha^{-1} with well adjusted doses of fertiliser (Gilewska et al. 2001, Gołda 2007). The accumulation of N, and thus OM, in this narrow range shows the natural limits of the process, with a range of between 10 and 50 kg. ha^{-1} y^{-1}, and usually depends on the properties of the formation and the cultivated species. Only by using plant species able to fix atmospheric nitrogen can this process be more dynamic, reaching a maximum of about 100–130 kg TN y^{-1}, which under Polish climatic conditions is probably the maximum natural value possible in reclamation measures.

3 RESULTS AND DISCUSSION

3.1 *Mean annual increase of TN as an indicator of the success of reclamation measures*

The studies carried out by the author, as well as literature reports, indicate the possibility of using TN content in assessing the development of newly formed initial soils. However, due to the great differences in TN content in formations at the start of reclamation, it is proposed that assessment of the changes be based on the mean annual increase calculated over study period. The calculated increase in content, but not the content itself, eliminates any error resulting from different initial contents. Moreover, the difference method eliminates the influence of samples from greater depths. It should be emphasised that there is a significant increase in N exclusively in the layers of OM accumulation that are easily recognised in the field. Beyond these layers, the N content changes little. The results of several subsequent research series have shown the dynamics of the process, the influence of factors on its course, and the direction of changes.

The proposed five-degree scale of the indicator of soil-forming activity (Table 1) allows objective assessment of the accumulation of N compounds, which depends not only on the properties of the formations, but particularly on the methods applied for improving the development of the soil. The increased TN content, and consequently organic compounds, also depend on the range of better or poorer processes or natural factors influencing for example microbiological activity, which is very important in the humification process. The created scale reduces the margin of error and limits subjectivity of assessment. An annual increase of TN of between 50 and 70 kg ha^{-1} indicates the dynamic process of OM accumulation possible with the application of special reclamation solutions (e.g. increased organic and mineral fertilisation, and greater use of various species of nitrogen-fixing plants), while the annual increase of TN of between 10 and 20 kg ha^{-1} would show low soil-forming abilities of the formation, or improper or too extensive remediation measures. Noticing this relationship allows qualification of the assessed formation, and, with correct sampling, classification into a proper class of soil-forming

Table 1. The criteria for assessing the soil-forming activities in the reclaimed formations, based on the annual increase in total nitrogen (TN) content.

Annual increase in TN content		
[mg kg^{-1}]	[kg ha^{-1}]	Soil-forming activity
>22	>70	very high
16–22	50–70	high
9.4–16	30–50	mean
3.1–9.4	10–30	low
<3.1	<10	very low

activity (the rate of the development or success of reclamation measures). The proposed assessment will be most useful when the differentiation of the profiles of initial soils does not fulfil soil-related bonitation criteria.

To assess soil-forming activity, the mean value of the annual increase of TN in the arable humic layer, or the sum of N contained in ecto-humic and endo-humic level (in kg ha^{-1}) according to the scale proposed in Table 1, is applied.

Due to the accuracy of the Kjeldahl method, studies of the rate of TN increase should include at least five years of a comparative period to minimise the influence of natural variability and accidental errors. To increase the reliability of assessment, a uniform procedure should be established that defines the time (vegetation period), method and place of sampling.

The proposed assessment of the TN increase also allows comparison of the dynamics of the development of newly formed soils using the different methods used on the reclaimed formations, by giving values as kg ha^{-1} or content as mg g^{-1}. The quantitative characteristic of the changes also allows comparative analysis with natural soils and determination of which stage of development the assessed formation has reached. In cases where the content of TN or the calculated OM content does not meet soil criteria, an approximate assessment can be made of the time necessary to achieve certain values, or the time after which the following measurement series could be achieved. After the reclaimed formations reach values comparable with soils, the application of soil science classification criteria is then justifiable. The dependence between annual increase in N and accumulated OM, depending on the time of the reclamation plantations is shown in Figure 1; in this example, TN represented 5% of OM. More objective assessment of some changes in properties of formations, taking place in the reclaimed areas from the stage of 'raw soilless formation' – through the stage of 'anthropogenic soils of the unformed profile', to at least the stage of 'very well reclaimed initial soils' and providing objective qualification criteria will facilitate making numeric maps or models of soil development. The dynamics of TN accumulation will be very important in this assessment, indicating the increase of both the contents of this basic macro-element for plant development and of OM.

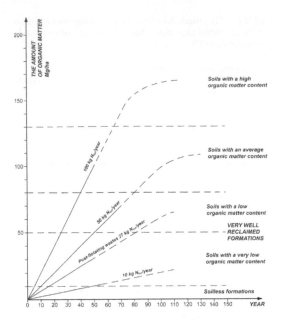

Figure 1. Examples of the relationships between the duration of the reclamation period and the amount of accumulated organic matter with different annual accumulations of total nitrogen in initially soilless formations.

and assess the effect of the applied solutions. If the increase in TN content remains low for a long period, or even becomes smaller, this can indicate degradation of the formation, which can lead to complete retreat of vegetation and so create a secondary wasteland. The calculated increase of TN content allows the determination of which of the five levels of soil-forming activity has been attained.

4 CONCLUSIONS

This paper proposed the assessment of the rate of development of newly formed soils from soilless formations, based on annual increase of TN content. This can facilitate forming judgements of the efficiency of any reclamation measures. Reports on the formation of fully mature soil from soilless formations on reclaimed areas are unknown. The long period required to make soils and the too short history of implemented reclamation measures makes the situation of when and where soils reach the climax stage uncertain. Most research papers refer to the period when there is increased accumulation of OM as the period of development of soil.

The quantity of accumulated OM depends on the properties of the initial formation, as well as on the type of reclamation measures. Analytic difficulties and low repeatability of results make it difficult to widely apply analyses allowing the full description of the humification process, and determining when soilless formations attain new and usually favourable properties. The application of a simple and widely applied method of N determination can, with a good approximation, describe the speed of the development

REFERENCES

Bykov, R. 2003. Lithological and Morphological Conditions of the Soil Processes within the Reclamated Area of "Piaseczno" Sulphur Mine inner Excavation Heap. *PhD Thesis. Manuscript. AGH University of Science and Technology in Krakow.*

Gilewska, M., BendeR, J., Drzymała, S. 2001. Organic Matter Formation in Post Mining Soils in Center Poland. ISCO 99. In D.E. Stott, R.H. Mohtar, & G.C. Steinhardt (eds.), *Sustaining the global farm, Selected papers from the 10th International Soil Conservation Organization Meeting held May 24–29,1999 at Purdue University and USDA-ARS National Soil Erosion Research Laboratory,* Purdue University and USDA-ARS National Soil Erosion, West Lafayette, IN: 623–626.

Gołda, T. 2007. Initial Soil Creating Processes Taking Place in Post-Flotation Slime in Result of Reclamation Cultivation and Long-Term Agriculture Use. *Dissertations and Monographs 164.* Krakow: AGH-UST.

Jenny, H. 1980. *The soil resource: Origin and behaviour. Ekol. Studies 37.* Springer-Verlag, New York–Heidelberg–Berlin.

Katzur, J., Böcker, L., Knoche, D. & Mertzig, C.-C. 1999A. Untersuchungen zur Optimierung der Meliorationstiefe für die forstliche Rekultivierung schwefelsaurer Kippenböden. *Beitrage für die Forstwirtschaft und Landschaftsökologie* 33:171–179.

Katzur, J., Böcker, L. & Stahr, F. 1999B. Humus- und Bodenentwicklung in Kippen- Forstokosystemen. *AFZ. Der Wald 25*: 1339–1341.

Kowalik, S. 2004. Chemical Properties of Anthrosoils of Agricultural and Forestry Management of the Waste Heap of Sulphur Mine „Machów". *Roczniki Gleboznawcze* LV(2): 239–249. Warsaw: PWN.

Siuta, J. 1995. *The Soil – the condition and hazards diagnosis.* Warszawa: Instytut Ochrony Środowiska.

Sourkova, M., Frouz, J. & Szntuckova,H. 2005. Accumulation of carbon, nitrogen, phosphorus during soil formation on alder spoil heaps after brown-coal mining, near Sokolov (Czech Republic). *Geoderma* 124: 203–214.

Węgorek, T. 2003. Changes in Certain Properties of Earth Material and Phytocenoses Development on the External Soils Bank of a Sulfur Mine as a Result of Sylvan Destination Land Reclamation. *Rozprawy Naukowe Akademii Rolniczej w Lublinie 275.* Lublin: AR.

Wójcik, J., Kowalik, S. 2006. The Formation of Selected Properties of the Initial Soil on the Waste Heap from Copper Mining under Forest-Directed Reclamation. *Inżynieria Środowiska* 11(1): 87–99. Kraków: AGH-UST.

Environmental Engineering III – Pawłowski, Dudzińska & Pawłowski (eds)
© 2010 Taylor & Francis Group, London, ISBN 978-0-415-54882-3

Classification of reclaimed soils in post industrial areas

S. Gruszczyński

Faculty of Mining Surveying and Environmental Engineering, AGH – University of Science and Technology, Krakow, Poland

ABSTRACT: This paper presents an algorithm for qualitative classification (valuation) of reclaimed areas. Methods of conventional soil-quality valuation are useless for classification of reclaimed soils, which are deprived of macroscopic qualitative attributes. An alternative algorithm, based on the results of laboratory determinations, is desirable. The valuation model presented herein uses premises resulting from the database of soil patterns contained in soil-class charts, empirical data from observations of variability in soil profiles, and assumptions concerning the impact of land morphology on soil value for use.

Keywords: Reclaimed areas, soil quality, classification.

1 INTRODUCTION

Reclaimed areas constitute special types of land that can be used for agriculture and forestry. Constituting urban soils, they have considerably different properties to naturally formed soils, such as lack of humus accumulation, nutrient scarcity and variable crop yields. Nevertheless, as areas in use they should be qualitatively classified, at least due to the need to evaluate the quality of land reclamation works and development. However, qualitative classification of this land type requires the development of a special assessment algorithm adjusted for land characteristics, which would, as far as possible, allow land to be properly positioned within the series of soil-quality valuation classes.

In Poland, a production point-of-view prevails in the field documentation of soils. Farmland (arable land, meadows and pastures) and privately-owned forests are subject to qualitative classification, using a multigrade (six- or nine-grade), discrete quality scale (Ministry of Agriculture 1963, Strzemski 1972). State forests are subject to typological classification of the type of forest but not the type of soil. The criteria and requirements specified in the soil-valuation-class chart make it possible to present relationships between soil-quality valuation scales for individual land types (Ministry of Agriculture, 1963). Comparing different soil-quality valuation-class series requires a common scale that covers a continuous range of point values. A range of 0–10 was used (Figure 1 and Table 1), assuming maximum point value for the best soils. Point values assigned to individual classes illustrate their economic value according to official conversion factors.

Relationships between classes are shown in the form of a membership function for fuzzy sets, in order to emphasise their heterogeneity and the existence of

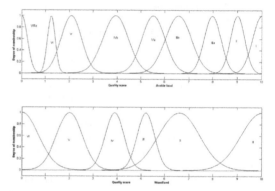

Figure 1. Fuzzy relations of Polish soil-quality valuation classes for arable lands and forest lands quality scale. Membership functions (curves marked by Roman numerals) indicate measure of similarity to typical soil valuation class patterns.

Table 1. Relationship of soil-quality valuation series for woodland areas and arable land according to the interpretation of soil-class chart (Ministry of Agriculture, 1963); extreme values in brackets.

Forestland class	Arable land class	Point scale range
I	I–II(III)	8–10
II	III–IV(II)	5–8
III	IV	4.5–5
IV	IV–V(VI)	3.5–4.5
V	V–VI	1.0–3.5
VI	VI	0.0–1.0

diversity in quality within a class. Thus, the point range covering full variability of soil quality (in the language of fuzzy sets the 'universe of discourse') is a numerical scale allowing their approximate classification.

Aim and scope of the work.

The purpose of this research was to develop a classification algorithm for reclaimed areas, allowing the evaluation of essentially raw soils and anthropogenic soils.

The soil-quality classification of natural soils serves comparative and fiscal (land tax) functions. Soil-quality classification might be used as the criterion for reclamation and development activities, and the following could be very important detailed decision-making criteria related to reclamation methods: the configuration of the area meant for reclamation (e.g. slope aspect and land morphology), the choice of material that could be used to reclaim the area, and the depth of the fertile part of the surface layer. The obvious condition on which to base the soil-quality classification algorithm are soil purity standards (Ministry of Environment 2002).

As far as possible, this algorithm should indicate where soils should be placed in the Polish soil-quality valuation series (Ministry of Agriculture 1963). The algorithm must take into account measurable characteristics for mineral material of soils, while its structure requires the selection of appropriate quality coefficients and development of a method for their integration.

In the literature (Schoenholz et al. 2000), authors specify numerous characteristics of soils that affect their quality, and it is very difficult to combine these characteristics into an integrated model (Strzemski 1972). In Poland for soil-quality classification, the system used is of verbally defined patterns, which may be determined macroscopically in field conditions (Ministry of Agriculture 1963). The soil-class chart can be assumed to be a database showing interrelationships between macroscopic soil properties and their classification. Thus, for an algorithm adequate for raw or technogenic soil formations subject to reclamation, reference to valuation criteria for natural soils is highly desirable. In this regard, the most difficult aspect is that attributes of technogenic properties exceed the scale used for natural soils (e.g. different lithology, lack of organic carbon and low fertility).

Four groups of criteria affect the assessment of soil-quality class: land morphology, soil morphology, and physical and chemical properties of soil.

Land morphology (slope, aspect, and configuration of surroundings) may reduce soil quality valuation due to increased exposure to erosion, excessive soil drying, or excessive wetting. Thus morphological factors may reduce soil quality.

Profile morphology illustrates the progress of soil-forming processes, and the intensity of soil accumulation and degradation processes. In particular, the thickness and properties of organic carbon accumulation are characteristics that determine soil quality valuation to a large extent. Organic carbon accumulations form a certain resource of nutrients (mainly nitrogen), but also affect cation exchange capacity (CEC), as well as the water-holding capacity and structure of soils.

Lithological characteristics form specific conditions for progress in soil-forming processes, and they codetermine current air and water factors. For example, grain-size distribution in the soil profile determines the drainage conditions (e.g. compact formations on light ground) and difficulties with plant growth. There is an obvious relation between grain-size distribution and many other soil properties, e.g. porosity, CEC, water-holding capacity and permeability.

Although extremely important in soil fertility formation, chemical properties generally do not find application as soil-quality criteria, at least in Polish practice. The exception here is the reaction and symptoms of reduction processes (gleying and soil concretions) indicating unfavourable air–water relationships. Natural soils in use for many years are characterised by certain levels of accumulated nutrients. In the case of reclaimed soils the situation is highly variable, depending on mineral material and reclamation method. However, there are usually considerable deficiencies of one or more macroelements (usually nitrogen and phosphorus), thus making it necessary to fertilise soils, systematically and extensively, for long periods, aiming to develop a nutrient level comparable to that of arable soils. Certainly, this lack of nutrients increases the cost of reclaiming soil, and reduces profitability and also the soil-quality valuation class. Therefore, this factor should not be omitted in a quality valuation algorithm for reclaimed soils.

2 MATERIALS, METHODS AND RESULTS

The required classification method should refer to the soil-quality valuation system used in Poland; however, this valuation system is not appropriate for classifying reclaimed soils.

There are many approaches to the soil-quality valuation problem, including the system based on patterns used in Poland. There are also many models used in other countries, with the factor distinguishing them being the method used to integrate various soil-quality coefficients into one general index. Some of the integration methods considered differ greatly in transforming measurable soil properties, e.g. regression models (additive and multiplicative) and fuzzy models. Among other things, they differ in approach to the problem of substitution, i.e. elimination (in the quality assessment model) of the impact of factors reducing soil value. In practice, this is hard to avoid in regression models, but it is appropriate for the minimum model (Strzemski 1972, Paris & Knapp 1989, Paris 1992). The construction of an appropriate model was therefore based on the following: analysis of a database consisting of patterns converted from the Polish soil-class chart (in total almost 800 patterns) and observations of selected chemical properties for soils characterised by different quality valuations, and assumptions regarding the impact of morphology on soil value. The main algorithm assumptions are as follows (Gruszczyński et al. 2008):

1. Valuation criteria are factors related to mineral material lithology (field water capacity of soil); factors indicating the sorptive capacity of soils (e.g. CEC); soil content of basic macroelements; and land surface slope.
2. Quality is measured by an integrated numerical index $W(z)$ covering the range 0–10. The index may be used to assess soil position in a soil-quality valuation series by employing a fuzzy model of classes.
3. As far as possible it is appropriate to avoid substitution of characteristics in the valuation algorithm.

In practice, fulfilling the third assumption indicates a general classification model consisting of three modules, according to the following formula:

$$W(z) = \min[L(m), L(l), L(c)] \quad (1)$$

where $W(z) = $ an integrated soil-quality coefficient; $L(m) = $ a numerical land morphology index; $L(l) = $ a numerical index depending on mineral material lithology; $L(c) = $ a numerical soil fertility index.

Considering the availability of necessary information, individual algorithm modules need to be developed according to three procedures. A land morphology index may be obtained from theoretical analysis and formalisation of recommendations concerning the significance of land-slope for qualitative classification of soils. An index depending on lithological characteristics may be derived from analysis of soil patterns within a valid soil-class chart. A fertility index requires the analysis of actual relations between concentration of macroelements and soil-quality valuation classes.

2.1 Index depending on land morphology – L(m)

Location is an extremely important element in forming soils and determining their value for use (Strzemski 1972). Slope is the easiest to interpret of various indices potentially characterising location. It determines erosion hazard, cultivation potential and acceptable uses. It seems justified to assume that absolutely flat land is characterised by quality determined only by mineral material. Increasing slope gradually lowers the value of use to the limit where it is completely excluded from certain uses. Since individual types of use are related to different degrees of slope-related difficulties, it is appropriate to use two different estimations of $L(m)$: $L(m)_R$ for arable use, and $L(m)_L$ for forest use. The decision on classification of permanent grassland areas requires explanation, despite their frequent occurrence in reclaimed areas. However, the occurrence of grassland is usually forced by circumstances (excessive slope or cultivation complications) justifying this type of use in areas not useful for other purposes. In this climatic zone, permanent grasslands are in locations favouring high levels of soil wetness, which is infrequent for reclaimed areas.

Taking into account approximate limits for potential use and assuming a linear decline in value due to slope, we can estimate $L(m)$ for arable use:

$$L(m)_R = \begin{cases} 10 - 0.66 \cdot \alpha & \text{for } \alpha \leq 15 \\ 0 & \text{for } \alpha > 15 \end{cases} \quad (2)$$

and for forest use:

$$L(m)_L = \begin{cases} 10 - 0.16 \cdot \alpha & \text{for } \alpha \leq 60 \\ 0 & \text{for } \alpha > 60 \end{cases} \quad (3)$$

In each of the formulas, α indicates land-slope in degrees. The models satisfy accepted assumptions concerning the index value. Land-slope modifies typological classification of forestlands only to a limited extent; it depends to a much larger extent on soil properties (e.g. soil as a mineral material). However, undoubtedly, slope itself (above a certain level) poses a substantial difficulty for forest growth.

2.2 Index depending on chemical characteristic – L(c)

The presence of a module dependent on chemical characteristics is determined by deficiencies in macroelements in reclaimed soils, which require higher fertilisation. This results in extra costs, and thus reduces profitability. It is difficult to model such an index; however, we may assume that land reclamation and development has a real influence on the levels of three components: nitrogen, phosphorus and potassium. Certainly, levels of these three components in arable soils are subject to strong fluctuations, depending on land use. There is only a weak relationship between soil fertility and that soil's position in a soil-quality valuation series.

We analysed 129 humus horizons in soil pits representing all soil-quality valuation classes to obtain the model. Some chemical properties of samples taken are shown in Table 2. Data on soil accumulation levels are relatively few, and thus more data is necessary for more representative chemical soil characteristics results, not just those representing southern Poland.

A statistical, model was developed, based on laboratory determinations of total nitrogen content and available forms of phosphorus and potassium. Data concerning classes were converted to a numerical scale of range 0–10. A linear (polynomial) model of the relationship between NPK content and point value did not have satisfactory parameters. The determination coefficient (R^2) = 0.25 indicated a poor model. Successive approximations, using the *MODEL* procedure (OLS algorithm) of the SAS statistical package (Paris, 1992; SAS Institute Inc., 2004), allowed development of a better model of final form:

$$L(c) = \min(9.75; 238 \cdot Z_N; 0.74 \cdot Z_K + 2.4) \quad (4)$$

where $L(c) = $ soil-quality valuation index according to chemical characteristics; $Z_N = $ total nitrogen content in percentage by weight in a 0–50-cm layer of

Table 2.   Statistics for some chemical determinations of macroelements in accumulation levels of different soil-quality valuation classes.

Statistics	S (%)	C (%)	N (%)	Mg[a] (mg/100 g)	K[a] (mg/100 g)	P[a] (mg/100 g)
Average	0.009	0.81	0.086	7.83	9.55	7.65
Std. dev.	0.007	0.53	0.055	4.66	10.14	7.81
Minimum	0.001	0.033	0.004	0.12	0.08	0.07
Maximum	0.044	3.063	0.306	22.29	54.25	31.4
Lower quartile	0.003	0.283	0.039	4.52	2.93	1.65
Upper quartile	0.014	1.193	0.120	10.60	13.35	11.02

[a] indicates the available form of an element. Number of samples = 129.

soil; Z_K = available potassium content in mg/100 g of soil in a 0–50-cm layer. All parameters of equation (4) were statistically significant ($P < 0.0001$), while mean square error of regression ($RMSE$) = 1.8, absolute average error (MSE) = 3.25, and $R^2 = 0.54$. In the step-by-step procedure for $L(c)$ model construction, the module referring to available phosphorus content was removed, since it lowered statistical significance of the equation.

2.3   Index depending on lithology – L(1)

Grain-size distribution in the soil profile is the most important index dependent on lithology. It is a substitute for many other characteristics directly or indirectly related to it, including water-holding capacity, air content, a set of properties related to soil sorption, permeability, susceptibility to erosion, and also indirectly fertility. In the case of reclaimed areas, grain size distribution may not be a good indicator of these characteristics, because its distribution most often results from mechanical waste recovery (Hearing et al. 1993, Bykov 2003, Gołda 2007). In such conditions, the properties of individual soil fractions do not correspond to characteristics of these fractions in natural soils. It was assumed that the most important properties that were derivatives of soil grain-size distribution were soil water-holding capacity and CEC. They are indicators of the cultivation requirements of soil water and nutrients.

Development of an appropriate model required transformation of soil patterns contained in the soil-class chart into a database with data expressing water capacity distribution and CEC in soil-quality valuation classes. For this purpose we used the PedoTransfer Function developed for European Union soils on the basis of the van Genuchten regression models (van Genuchten 1980, Woesten et al. 1999, McBratney et al. 2002). Independent variables for the required model, based on the database obtained from the converted soil-class chart were as follows: Rp_{050} is soil moisture (cm^3/cm^3) in the 0–50-cm layer in field retention condition (i.e. corresponding to a suction pressure of 34 kPa); Rp_{50100} is soil moisture (cm^3/cm^3) in the 50–100-cm layer in field retention condition; Rp_{100150} is soil moisture (cm^3/cm^3) in the 100–150-cm layer

in field retention condition; and CEC in cmol$^{(+)}$/kg, in the 0–50-cm layer. The division into 50-cm layers resulted from the need to define the vertical variation of reclaimed areas, and is based on the soil-valuation classification practice in Poland, which takes into account the soil lithological characteristics to a depth of 150 cm.

It should be emphasised that the model was relatively accurate for actual configuration of field retention and CEC, since the analysed samples of soils (380 pieces of data) for the models had $R^2 = 0.58$–0.74.

To obtain an adequate $L(l)$ model, we examined several forms of statistical models, and the best parameters were obtained with the model optimised using the GMDH method (Aksyonova et al. 2003), in the form:

$$L(l) = -25.88 \cdot Rp_{050}^2 - 370.5 \cdot Rp_{050} \cdot Rp_{50100}^2 + 233.4 \cdot Rp_{050} \cdot Rp_{50100} - 7.99 \cdot 10^{-4} CEC^2 \tag{5}$$

The model is relatively complex, $R^2 = 0.75$, $RMSE = 1.56$, and $MAE = 1.22$. Using the $MODEL$ module in the SAS package (SAS Institute Inc., 2004) gave a model with slightly better properties:

$$L(l) = \min\left[9.75; 67.5 \cdot Rp_{050} - 7.33; \frac{1}{\exp(-7.27 \cdot Rp_{50100} - 0.38)}; 1.31 \cdot CEC\right] \tag{6}$$

All model (6) parameters were significant ($P < 0.0001$), $RMSE = 1.51$, $MAE = 1.26$, and $R^2 = 0.78$.

However, the following values of resulting indices should be reduced if the 0–50-cm or 50–100-cm layer does not fit the standards, e.g. when a rock formation is the base (substrate). It is clear that formula (6) needs some modification.

2.4   Problems in using the algorithm

The quality-assessment algorithm for reclaimed land was intentionally divided into three modules, due to many possibilities regarding modification and observation of factors for determining classification. Land morphology depends greatly on the intended land use, and may be optimised to some extent.

Much cover heaping-up prevailed in strip mines in the 1960s and as a result the form of waste heaps depended on geotechnical properties of the heaped-up material. Scarps with high inclination degree constituted a considerable portion of reclaimed areas (Bykov 2003), and this excluded land use for agricultural purposes. The classifying algorithm as presented in this paper facilitates decisions concerning the form of reclaimed areas, because – assuming certain extrapolation of existing soil conditions – it is feasible to assess potential ground quality and to observe the rule that, if possible, land morphology should not negatively affect the classification.

In part of the 'Jeziórko' sulphur mine, the reclamation process covers post-flotation sludge as the product of mechanical processing of sulphuric ore. Some data concerning sludge properties is available in Gołda (2007), showing that at the initial state the sludge was characterised by a CEC of up to 4.1 cmol/kg and standard deviation of 0.5. This shows that sludge reached approximately 5.7 in the point-scale, corresponding to intervals III and IV of the soil-quality valuation class. After the 20 years of reclamation the nitrogen content in the reclaimed sludge increased from approximately 0.021% to > 0.042% (Gołda 2007), and the corresponding score increased from 4.99 to 9.9. At the same time, available potassium content dropped from 8.2 to approximately 5.3 mg/100 g of soil, with corresponding scoring dropping from approximately 8.4 to 6.3. To sum up this analysis, the algorithm showed that (omitting elements connected with water retention and land morphology), available potassium content was the current quality-limiting factor. The estimated quality of these formations was approximately 4.0–5.7 points, placing them in interval IV of the soil-quality valuation class.

Waste-heap soils are characterised by a mosaic character related to random storage of soil masses. Quaternary sands and Tertiary clays were the two formation types constituting the cover for sulphur mines (Bykov, 2003). These waste heaps are diverse in regard to morphology, and slopes within scarp systems reach 45°. In light waste-heap formations, Bykov (2003) found nitrogen contents changed from approximately 0.008% to 0.02% after 20 years of land reclamation; the corresponding scoring increased from 1.9 to >4.7. In the same formations, available potassium content fluctuated within 0.9–7.0 mg/100 g, with scoring range 3.1–7.6. The CEC for these formations fluctuated within 0.3–6.8 cmol/kg, giving a score of 0.4–8.8. Field water capacity for light formations range was 0.04–0.2 cm^3/cm^3, giving scores from <0 to 6.2; here the algorithm reveals a defect due to the properties of reclaimed soils going beyond the limits of characteristics for natural soils. A correction is necessary that involves maintaining the scoring within the assumed range (e.g. if point value is <0, then scoring is set to 0). We may assess that, as regards examined properties, the scoring for light formations is in the range of 0–4.7 points (soil-quality valuation classes IV–VI, or wasteland).

Tertiary clays occupy a larger area than Quaternary sands in waste-heaps of sulphur mines. As a result of over 20 years of reclamation, they accumulated nitrogen (contents increasing from 0.02% to 0.08%), and increased their scores from 4.76 to 9.75. Concentration of available potassium was very high, with range 26–80 mg/100 g. Thus, the presence of potassium did not restrict quality assessment. The sorptive capacity was characterised by very high values (28–40 cmol/kg), with a similar situation for field capacity (0.35–0.45 cm^3/cm^3). Thus, currently the only limitation given by the model was variable nitrogen content. However, the occurrence of very light formations at depths of 30–70 cm may be an important quality-reducing factor in addition to land morphology.

3 CONCLUSIONS

This paper presents a theoretical–empirical quality valuation model for reclaimed soils. It consists of three separate modules covering land morphology, lithological characteristics of soils, and indices of accumulation processes for macroelements. The individual modules may be examined separately as indicators of the adequateness of decisions made regarding land reclamation and development. It is appropriate to continue this research; however, adding other modules will considerably complicate the estimation process. Certain problems are evident, related to the evaluation of indices for reclaimed areas with extremely faulty properties. For example, extremely low water retention of light waste-heap formations may lower scoring below the minimum of 0 (due to classifying of very coarse-grained, gravel and stony formations). In these cases the model should not be treated too formalistically and the value should be altered to the limit of 0 points.

ACKNOWLEDGEMENTS

The paper was financed from the Funds for Science in 2005–2008 as a research project: 4 T12E 041 29 'The Principles of Valuation of Industry-Affected Soils on Reclaimed Areas'. The calculations were carried out in ACK CYFRONET AGH-UST on the computer HP Integrity Superdome 'Jowisz'. Calculation grant: MNiSW/HP_I_SD/AGH/052/2007.

REFERENCES

Aksyonova, T.I., Volkovich, V.V. & Tetko, I.V. 2003. Robust polynomial neural networks in quantitative-structure activity relationship studies. *System Analysis Modelling Simulation* 43(10): 1331–1339.

Bykov, R. 2003. *Litologiczne i morfologiczne uwarunkowania procesów glebowych na terenie zrekultywowanego zwałowiska zewnętrznego Kopalni Siarki, Piaseczno*. Praca doktorska. Akademia Górniczo-Hutnicza w Krakowie.

van Genuchten, M.T. 1980. A closed form equation for predicting the hydraulic conductivity of unsaturated soils. *Soil Science Society of America Journal* 44: 892–898.

Gołda, T. 2007. *Inicjalne procesy glebotwórcze zachodzące w szlamach poflotacyjnych w wyniku upraw rekultywacyjnych i wieloletniego użytkowania rolnego*. Uczelniane Wydawnictwa Naukowo-Dydaktyczne AGH. Kraków.

Gruszczyński, S., Eckes, T., Gołda, T., Trafas, M., Wojtanowicz, P. & Urbański, K. 2008. Raport końcowy z realizacji projektu: Zasady klasyfikacji gruntów poprzemysłowych obiektów rekultywowanych. AGH – University of Science and Technology. Cracow.

Hearing, K., Daniels W. & Roberts J. 1993. Changes in mine soil properties resulting from overburden weathering. *Journal of Environmental Quality* 22: 327–337.

Mcbratney A.B., Minasny B., Cattle S.R., Vervoort R.W., 2002, From pedotransfer functions to soil inference systems, *Geoderma*, vol. 109, no. 1–2, pp. 41–73.

Ministry of Agriculture. 1963. *Komentarz do tabeli klas gruntów w zakresie bonitacji gleb gruntów ornych terenów równinnych, wyżynnych i nizinnych wraz z regionalnymi instrukcjami dotyczącymi bonitacji gleb ornych terenów górzystych i komentarzami dotyczącymi bonitacji gleb użytków zielonych i gleb pod lasami dla użytku klasyfikatorów gleb i pracowników kartografii gleb*. Warszawa.

Ministry of Environment. 2002. *Rozporządzenie z dnia 9. Września 2002 w sprawie standardów jakości gleby oraz jakości ziemi*. Warszawa.

Paris, Q 1992. The von Liebig hypothesis. *American Journal of Agricultural Economics* 74(4): 1019–1028.

Paris, Q. & Knapp, K. 1989. Estimation of von Liebig response function. *American Journal of Agricultural Economics* 71(1): 178–186.

Sas Institute Inc, 2004, *SAS/ETS® 9.1 user's guide*. SAS Institute Inc. Cary. NC.

Schoenholtz, S.H., Van Miegroet, H. & Burger, J.A. 2000. A review of chemical and physical properties as indicators of forest soil quality: challenges and opportunities. *Forest Ecology and Management* 138: 335–356.

Strzemski M. 1972. *Przyrodniczo-rolnicza bonitacja gruntów ornych*. Instytut Uprawy Nawożenia i Gleboznawstwa. Puławy.

Woesten, J.H.M., Lilly, A., Nemes, A. & Le Bas, C. 1999. Development and use of a database of hydraulic properties of European soils. *Geoderma* 90(3): 169–185.

Environmental Engineering III – Pawłowski, Dudzińska & Pawłowski (eds)
© *2010 Taylor & Francis Group, London, ISBN 978-0-415-54882-3*

Improvements in industrial waste landfilling at the solid waste landfill site in Krakow-Pleszow, Poland, implemented in order to obtain an integrated permit

K. Grzesik

Department of Management and Protection of Environmental,
AGH – University of Science and Technology, Karkow, Poland

ABSTRACT: The landfill site for metallurgical slag and other solid waste in Krakow-Pleszow, Poland, is one of the biggest industrial landfill sites in the Malopolska region and the only site for solid waste generated in production processes in the ArcelorMittal Unit in Krakow. To comply with the European IPPC Directive and Polish regulations, the Pleszow site had to obtain an integrated permit. As waste landfilling at the site began in the 1970s, it was unable to meet current legal requirements. The process of applying for and issuing the integrated permit was difficult, complicated and time-consuming but was finally successful in March 2008.

Keywords: Waste management, landfilling, integrated permit, Best Available Techniques – BAT.

1 INTRODUCTION

The ArcelorMittal Unit in Krakow is one of two steel-works in Poland with a full steel production process. The full production process is carried out in an integrated steelworks, which covers sinter plants, coke oven plants, blast furnaces and basic oxygen furnaces including continuous casting. Steel is further processed by hot rolling and cold rolling to manufacture a great variety of finished products. In each of the mentioned installations, a huge amount of waste is generated. The metallurgical industry is extremely consuming of both materials and energy. About half of the raw materials used are removed as waste after processing. Part of the metallurgical waste is recovered, and reused in a sinter plant or a basic oxygen furnace. After at least six months, metallurgical slag is processed into crushed stone, which is used for road constructions, and some types of waste are transferred to external receivers to be recovered. However, the remaining solid waste, as well as slurries, needs to be landfilled.

The landfill site in Pleszow for blast-furnace and oxygen-furnace slag and other waste, commonly called 'the heap in Pleszow' is the only solid waste landfill site for the ArcelorMittal Unit in Krakow, the former Sendzimir Steelworks. Waste landfilling at this site began in the 1970s, when millions of tonnes of raw iron ore, fluxing agents, and coal were delivered to the steelworks to be processed. The effect of the high-waste processes running at the steelworks was the accumulation of 30 million tonnes of waste by the end of 2005. The vast majority of waste deposited at this landfill was blast-furnace slag and oxygen-furnace slag. In the past, there was no conception that deposited slag could be utilized. In the 1990s, 'slag recycling' was introduced into the landfill as a new enterprise, which deals with the recovery of slag accumulated in previous years as well as from current production.

At the moment, the landfill site is used for storage of the blast-furnace and oxygen-furnace slag and for landfilling other solid waste. The site covers an area of 156 hectares and stands up to 23 metres high above ground level. The facility is divided into three parts:

– part B – an area of 8 hectares dedicated to the storage and recovery of slag from basic oxygen steelmaking,
– part C – an area of 89 hectares dedicated to the storage and recovery of slag from blast furnaces,
– part D – an area of 59 hectares for other waste landfilling.

A map of the landfill site is shown in Figure 1.

On parts B and C of the landfill site, recovery of slag is being carried out. The exploitation of deposited slag to produce crushed stone and a magnetic fraction is conducted using two plants: one mobile and one stationary. The crushed stone production process consists of screening the slag for appropriate grain classes, and separating the magnetic ferrous elements from the mineral fraction.

On part D of the site, solid waste from the Arcelor-Mittal Unit in Krakow is landfilled.

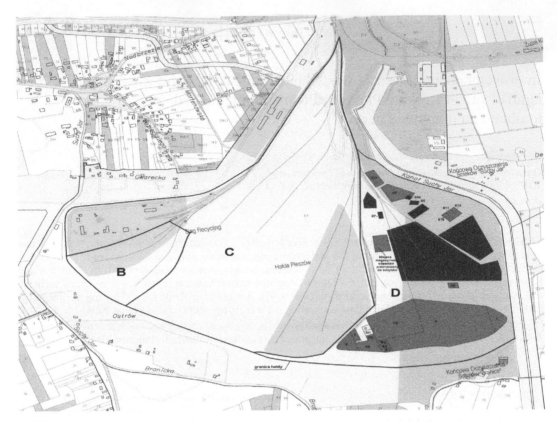

Figure 1. Map of the landfill site in Krakow-Pleszow: Part B – for the storage and recovery of slag from basic oxygen steelmaking; Part C – for the storage and recovery of slag from blast furnaces; part D – for other waste landfilling.

2 THE REQUIREMENT FOR AN INTEGRATED PERMIT FOR WASTE LANDFILLS

Council Directive 96/61/EC codified by Directive 2008/1/EC (European Commission, 2008) concerning integrated pollution prevention and control (IPPC) introduced the idea of an integrated approach to environmental protection in the industry (Schoenberger 2009). The idea assumed that the operator of an industrial installation would ensure full control of industrial processes. The objective of the control is to eliminate systematically, or reduce, threats to the environment. Two basic tools serve to achieve the directive's aims: Best Available Techniques (BAT) and integrated permits.

The objective of an integrated permit is to lay down conditions for operating an installation so as to prevent emissions from the installation, and, where this is not possible, to minimise emissions in order to achieve a high level of protection for the environment as a whole (Schoenberger 2009).

Carrying out an industrial activity in accordance with Best Available Techniques (BAT) means the application of such techniques as are available for the industry sector which allows the activity to achieve a high and effective level of environmental protection, while taking into consideration economic criteria. BAT

serves to define the limit values for emissions from an IPPC installation (Honkasalo et al. 2005).

The idea of an integrated permit was implemented within Polish law by the Act on environmental protection (Poland 2001b). The types of installation which should obtain an integrated permit were set out in an ordinance on types of installation which could create a potentially serious threat to the environment.

In waste management, an integrated permit is required for landfill sites receiving more than 10 tonnes per day or with a total capacity exceeding 25,000 tonnes, excluding landfills of inert waste. For landfill sites which had begun operations before 30th October 2000, the deadline for obtaining an integrated permit was set for 30th April 2007.

3 BEST AVAILABLE TECHNIQUES FOR WASTE LANDFILLING

Best Available Techniques reference documents (so called BREFs) have been produced. The European IPPC Bureau produced in 2006 the reference document for waste treatment installations, "Reference Document on Best Available Techniques for the Waste Treatments Industries" (European Commission, 2006). However, this document does not cover waste landfilling. Another reference document deals with waste incineration, but there is no reference document available on waste landfilling.

Requirements for the disposal of waste by landfill were laid down in the Directive on the landfill of waste 1999/31/EC (European Commission 1999). This directive was translated into Polish law by the Act on Waste (Poland, 2001a) and several ordinances on waste landfilling.

Based on the aforementioned law regulations, the identification of BAT standards was carried out for the landfill site in Krakow-Pleszow, for the following:

• location of the geological barrier, and artificial sealing liner,
• drainage system, leachate collection, prevention of surface water entering the landfilled waste,
• equipment,
• methods of operating,
• types of deposited waste,
• monitoring.

4 DISCREPANCIES BETWEEN BAT AND THE OPERATION OF THE LANDFILL IN KRAKOW-PLESZOW

The Pleszow landfill site is an old facility, which began receiving waste in the 1970s. Because of this, the site could not meet the currently set out requirements, regarding many aspects of environmental protection, including location, construction and equipment. The landfill site is located on the Vistula river terrace, on first rate soils. It has no artificial sealing liner but only a natural geological barrier, which unfortunately is not sufficient.

For many years, on part D of the landfill, waste was deposited in a non-selective way; hazardous and non-hazardous as well as municipal waste, including biodegradable waste, were deposited together. Separate sections for hazardous waste were not foreseen; neither were landfill gas control installations nor devices for washing and disinfecting vehicles' wheels.

Blast-furnace and oxygen-furnace slag have been delivered in a selective way into parts B and C of the landfill site. After six months of storage, slag is transformed into crushed stone, so the slag treatment is regarded as storage and recovery rather than landfilling.

In 2002, an ecological overview was carried out for the landfill. In a so-called 'after overview decision' issued by the regional Malopolska authority, the owner of the facility was obliged to reconstruct and adjust part D of the landfill site to meet legal requirements. The decision recommended:

• equipping the landfill site with devices for washing and disinfecting wheels of vehicles leaving the facility;
• equipping sections on which biodegradable waste is deposited with gas control installations;
• operating the site to ensure

– the area of deposited waste exposed to the weather is limited,
– the waste does not blow away
– the site is geotechnically stable;
• assigning separate sections for hazardous waste so that their areas do not exceed 2500 square metres;
• equipping the sections for hazardous waste with a separate drainage system;
• constructing the sections intended for hazardous waste in a way which makes it impossible for hazardous waste to come into contact with non-hazardous waste.

The landfill site owner, ArcelorMittal, incorporated the landfill reconstruction measures into their intended investments. In October 2005, a conception document for the Pleszow landfill site reconstruction was produced. This set out a two-stage programme of landfill modernization:

• Stage I – prior to the reconstruction of the landfill site, improving the process of waste disposal by landfilling,
• Stage II – reconstruction of the landfill site after 2009, in which new sections will be built fulfilling all the legal requirements.

5 OBTAINING THE INTEGRATED PERMIT AND IMPROVING OPERATIONS

At the time of working out the application for an integrated permit for the period 2006–2007, the landfill site, part D in particular still did not meet the legal requirements relating to environmental protection. The first application for an integrated permit for all the installations of the ArcelorMittal Unit in Krakow was submitted to the regional Malopolska authorities in July 2006. The next version of the application was prepared separately for the landfill facility in Krakow-Pleszow and this was submitted in November 2006. At this time, negotiations for obtaining the integrated permit began. In those negotiations various stakeholders participated: authors of the applications, representatives of the University of Science and Technology Department of Protection and Management of Environment, representatives of the regional Malopolska authorities and even representatives of the Ministry of Environment.

In a document issued by the Ministry of Environment, dated 24 July 2007, all doubts were finally cleared up. According to the document, obtaining the integrated permit would be possible if certain criteria were met: sections where hazardous waste was deposited should be closed down and water from precipitation should be prevented from entering into these sections by surface sealing. However, making a separate drainage system for the hazardous waste sections would be groundless and inexpedient, because one operating drainage system already exists for the whole landfill site. It was required that the landfill site should be equipped with a gas control installation, consisting of a network of out-gassing wells. It was also stated that if the landfill site operator should abandon the deposition of municipal waste, including biodegradable waste, equipping the landfill with washing and disinfection devices would not be necessary.

In negotiation with regional Malopolska authorities, some improvement works have been carried out:

1. The method of depositing waste was adapted to meet legal requirements. Appropriate sections were assigned soil or mineral embankments and dedicated to particular groups of waste that needed selective landfilling.
2. The landfill site in Pleszow received a decision permitting the closure of sections containing hazardous waste.
3. In March–April 2007, preliminary measurements of the quantity and quality of landfill gas were taken in sections where biodegradable waste had been deposited in the past. In October 2007, monitoring probes were installed.
4. Deposition of municipal waste including biodegradable waste has been stopped.

Besides modernizing works, some current action has been taken to reduce threats to the environment. On parts B and C of the landfill site, where slag is stored, two modern plants are operating: one mobile, the other stationary. These are recovering the slag transforming it into crushed stone. Sieve devices working at those plants are equipped with anti-noise covers.

Additionally, some non-exploited slag material was left on the borders of parts B and C of the landfill site, making an acoustic screen to reduce noise. Prepared crushed stone is piled up on the borders of the landfill site and acts as an additional barrier against noise. In order to minimize widespread dust emissions, water is poured onto temporary roads within the site.

On part D of the landfill site, the appropriate embankments of individual sections are used to limit the area of deposited waste. Moreover, deposited waste is instantly covered by isolating layers to prevent it blowing away. Waste is no longer deposited near the sloping edges of the site, and the inclination of side slopes has been decreased to eliminate the threat of disturbing the site's geotechnical stability.

After accomplishing all the necessary modernization works, a revised version of the application for an integrated permit was produced. The revised version included all changes introduced in landfill operations. The final application was submitted in November 2007. The integrated permit for the landfill site in Pleszow was issued in March 2008.

After completing the reconstruction of the whole landfill site, which will be carried out until 2010, the owner of the landfill site the ArcelorMittal Unit in Krakow will produce a new application for an integrated permit, for new sections to be built in the future.

6 CONCLUSIONS

The integrated permit is an important tool, the objective of which is to ensure protection of the environment as a whole. Obtaining the integrated permit for installations which have been operating for decades is a very difficult process, as older installations do not fulfil requirements laid down nowadays. Modernizing and adjusting those installations to meet legal requirements involves a huge organizational and investment effort. It is particularly difficult in the case of landfill sites, where waste was deposited on unprotected ground without sufficient barriers, where separate sections for individual types of waste did not exist, and where wastes were deposited in a non-selective and disordered way.

The process of obtaining an integrated permit for the landfill site in Pleszow was time-consuming. However, the efforts undertaken by ArcelorMittal, improvements introduced in landfill operations and also the willingness of regional Malopolska environmental authorities and the Ministry of Environment to engage in the negotiations produced a positive decision – to issue an integrated permit, nearly two years after submitting the first version of the application.

ACKNOWLEDGEMENTS

The work was completed within the scope of AGH-UST statutory research for the Polish Department of Management and Protection of Environment No. 11.11.150.008.

REFERENCES

European Commission. 1999. Directive 1999/31/EC of 26 April 1999 on the landfill of waste (OJ L 182, 16.7.1999, as amended), Brussels.
European Commission. 2006. Reference document on best Available techniques for the waste treatments industries. Integrated Pollution Prevention and Control. European Commission, August 2006.
European Commission. 2008. Directive 2008/1/EC of the European Parliament and of the Council of 15 January 2008 concerning integrated pollution prevention and control (Codified version) (OJ L 24/8 29.1.2008), Brussels.
Honkasalo, N., Rodhe, H. & Dalhammar, C. 2005. Enviromental permitting as a driver for eco-efficiency in the dairy industry: A closer look at the IPPC directive. *Journal of Cleaner Production* 13(10–11): 1049–1060.
Poland. 2001a. Act of 27 April 2001 on waste (Dz.U. 2007.39.251 as amended), Warsaw.
Poland. 2001b. Act of 27th April 2001 the Environmental Protection Law (Dz. U. 2008. 25. 150, as amended), Warsaw.
Schoenberger H. 2009. Integrated pollution prevention and control in large industrial installations on the basis of best available techniques – the Sevilla Process. *Journal of Cleaner Production* doi: 0.1016/j.jclepro.2009.06.002.

Environmental Engineering III – Pawłowski, Dudzińska & Pawłowski (eds)
© *2010 Taylor & Francis Group, London, ISBN 978-0-415-54882-3*

Application of a 2-D flow model to the analysis of forest stability in the Vistula valley

T. Kałuża

Department of Hydraulic Engineering, University of Life Sciences in Poznan, Poznan, Poland

ABSTRACT: Using a numerical model, this paper analyses conditions in which forests may loose stability. n application has been developed to calculate the critical moments for the stability of trees. The model of forest stability has been tested on riparian forests in the Vistula River Valley near Puławy. In the summer of 2001, during a fierce storm, these forests suffered heavy damage. The gale, which struck the Puławy area, coincided with a flood wave on the Vistula River. Using hydrogeological and meteorological data, simulations were carried out of conditions in which the trees had toppled.

Keywords: Critical moments for trees, forest stability.

1 INTRODUCTION

Forest damage on flood lands is often related to both strong winds and high flood water levels. Uprooted trees floating down the river threaten the hydrotechnical structures and other infrastructure, e.g bridges. Risk assessment of tree stability is, therefore, an important factor in the efficient management of vegetation on floodplains and may help to minimise potential losses during floods and storms (e.g. Nicoll et al. 2005, Achim et al. 2005).

This paper presents the results of a tree stability assessment. The results of investigations uprooting of the trees were later applied to test and validate a theoretical model of the resistance of trees in a forest to being pulled out of the ground. Equations describing the moments of the forces due to the pressure of wind or water are used to analyse forest damage in the Vistula Valley near Puławy.

In order to analyse the likelihood of a tree toppling over, one must compare the moments of the external forces acting upon the tree to the critical moment (e.g. Hartge & Bohne 1985, Dupuy et al. 2005), beyond which the tree will collapse. The tree will not be toppled as long as the following condition holds:

$$\frac{M_c}{M_W + M_Q} \rangle 1 \qquad (1)$$

where: M_c = critical moment; M_Q = toppling moment due to water; M_W = toppling moment due to wind.

When analysing the phenomenon of a tree toppling over, both water pressure and wind pressure must be taken into account (Figures 1 and 2). The toppling force depends on the size of the tree and equals the resistance to air and water offered by the vegetative structure (e.g. Achim et al. 2005, Gardiner et al. 2000). Assuming

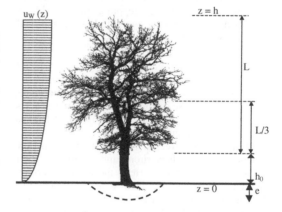

Figure 1. Influence of the dynamic pressure of wind on a tree.

that the directions of the wind and water coincide and that the water and wind velocities, as well as the water and air drag coefficients, are all constant, the toppling moments of the forces due to wind and water may be written as follows (MOC 1994):

$$M_Q = \frac{1}{2} \cdot \rho_W \cdot C_D \cdot u^2 \cdot S_Q \cdot \left(e + \frac{H}{2}\right) \qquad (2)$$

$$M_W = \frac{1}{2} \cdot \rho_P \cdot C_K \cdot u_W^2 \cdot S_W \cdot \left(e + h_o + \frac{L}{3}\right) \qquad (3)$$

where: H = water depth [m]; ρ_W = water density [kg/m³]; C_D = drag coefficient for water around a tree [−]; u = water velocity [m/s]; S_Q = tree surface influenced by water [m²]; e = depth of the turning axis of the root system [m]; ρ_P = air density [kg/m³]; C_K = drag coefficient for air around a tree [−];

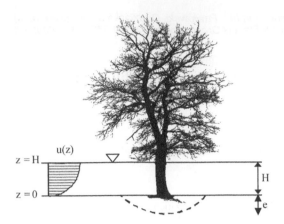

z = H   u(z)

z = 0

H

e

Figure 2. Influence of the dynamic pressure of water on a tree.

u_W = air velocity [m/s]; S_W = tree surface influenced by wind [m²]; h_0 = tree trunk height [m]; L = tree crown length [m].

Resistance to toppling depends on the plant root system, which in turn is a characteristic of the species and soil type. The critical toppling moment, M_c, is described in the literature by the following empirical formula (MOC 1994):

$$M_c = \alpha \cdot D^a \qquad (4)$$

where a and α are constants determined experimentally. Values of these constants for various tree species can be found in the literature (MOC 1994). D denotes breast height diameter [m].

The reactions of trees to wind, snow and debris flow depend largely on how strong and deformable their anchorage in the soil is. Here, the resistive turning moment M_c of the root–soil system as a function of the rotation φ at the steam base plays a major role (Gregory 2006). The Mmoment, $M(\varphi)$, describes the behaviour of the root-soils system when subject to a rotation moment, with the maximum, $M(\varphi)$, indicating the anchorage strength, M_c, of the tree the tree. Lundström et al. (2007) studied 45- to 170-year-old Norway spruce (*Picea abies* L. Karst.) that were pulled over with recordings being made simultaneously of φ and M made simultaneously. In the following studies refer to location with the high-elevation site (HE) and the low-elevation site (LE). Among the four regression models proposed by Lundström et al. (2007), $M_c = f(D^2 \cdot H)$ seems to be the most useful, as it requires only two records of tree size. In addition, $D^2 \cdot H$ is frequently referred to in the literature, which enables comparisons of the M_c values for different trees to be made. Referring to the comparison with other M_c studies of conifers in Lundström et al. (2007), it appears that the Norway spruce at the LE site exhibit among the highest M_c values, whereas the M_c values at the HE site are close to the average. However, there are only limited possibilities for comparing the M_c values obtained in this study with existing data, as most

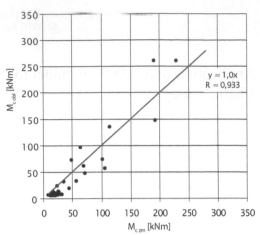

Figure 3. Calculated critical moment ($M_{c\,obl}$) vs. measured critical toppling moment ($M_{c\,zm}$) for pine.

previous studies have focused on smaller and younger trees. Moreover, the regression equations for the M_c values were not always forced through the origin, and $M^g(\varphi)$ (the stem base moment due to the overhanging weight (g) of the stem and crown) was not systematically included in M_c in the studies. It confirm that $M^g(\varphi)/M_c$ increase with decreasing tree size by an amount $(D^2 \cdot H)^{-0.5}$ and can reach as high as 70%. When modelling mechanical tree stability, it is probably essential to account for all contributions to the applied moment at the stem base when analyzing $M(\varphi)$ as well as its maximum M_c. This should also enable comparisons between M_c from experiments with different winching heights or tree characteristics.

Experiments carried out at the Poznań University of Life Sciences (Department of Hydraulic Engineering and Department of Geotechnics), which involved pulling out trees, allowed for a model to be developed which helps to estimate the critical toppling moments, taking into account both soil and tree parameters (Gardiner et al. 1997). Experiments were conducted using poplars, pines, alders and oaks. The model was tested and calibrated using the previously obtained critical moment values (Figure 3) (Kałuża & Leśny 2009). Based on the experimental results, the model was subsequently complimented with empirical relationships, providing a solution dependant on tree characteristics (tree species, weight and breast height diameter) and on basic geotechnical soil parameters (inner friction angle and compactness). Eventually, this led to the following model:

$$M_C = 12.83 \cdot D^{1.30} \cdot \frac{4.13 \cdot D^{0.71} + D}{3} \cdot \left[\frac{0.5 \cdot tg\varphi' \cdot Q \cdot b}{13.36 \cdot D^{1.41}} + c_u \right] \qquad (5)$$

where: M_C = critical moment [kNm]; D = breast height diameter [m]; Q = tree weight [N]; b = empirical coefficient dependant on tree species; φ' = effective soil inner friction angle [°]; c_u = soil compactness [kPa].

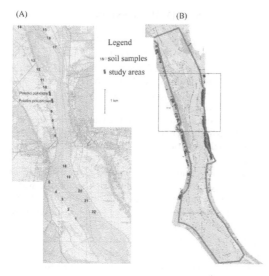

(A)   (B)

Legend

soil samples
study areas

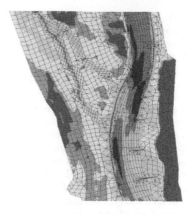

Figure 5. Digital terrain model and finite element mesh for the Vistula River at Puławy.

Figure 4. Location of the section studied and a map of research points (A). Vector map of land use and plant coverage (B).

2 MATERIALS NAD METHODS

The area of the study (Figure 4A) is located in the Puławy section of the Middle Vistula River, between the cross-sections at km 367.47 (the village of Adamówka) and at km 376.95 (the village of Łęka). The distances ca. 9.5 km are from measure along the axis of river. The predominant species in this area is white poplar, a 25 to 30 m high variety of poplar, and willow. On 4 August 2001 at 8:15 pm local time, a fierce storm with gale force winds of up to 20 m/s passed through this area. The heaviest damages occurred in the left bank forest near the villages of Adamówka and Jaroszyn. The gale on 4 August 2001 coincided with the passage of a flood wave on the Vistula River. The peak value measured on 29 July 2001 at the water-level gauge at Puławy was 121.12 m above sea level (ASL). This level corresponds to a discharge of $Q = 7650\,m^3/s$, which is close to the hundred year high water flow, $Q_{1\%} = 7520\,m^3/s$. On 4 August 2001 the water level receded to 118.36 m ASL.

On-site inspection of the Vistula River section under study was augmented with the information contained in the analysis conducted by Hydroprojekt, Warsaw . The study made use of the latest maps of the Middle Vistula section ordered by the Regional Water Management Office in Warsaw. This collection of 1 : 10,000 maps included topographical maps, land use and plant coverage maps, and colour topographical orthophotomaps. In addition, 33 hydrometric cross-sections were available. For the post at Puławy, a hydrological survey was also available as well as information on the hydrological situation on this section during the passage of the flood wave in the summer of 2001.

A survey of the geotechnical parameters of the soil included test bores and probes at 22 points, distributed as uniformly as possible on both banks of the Vistula River (Figure 4A.). All measurement points were located on the same flooded terrace. This was done so that the variability of the soil within the same morphological element could be studied. This survey was carried out to a depth of 2 m. This stratum was assumed to be decisive for tree stability. The second stage of the survey consisted of a meticulous investigation of the soils in the region, particularly in those areas where there were many fallen trees. In a smaller area, 22 research points were chosen, localised in two regions, one to the south and the other to the north (Figure 4A.).

The digital vector map of the Vistula section under study was developed using digital photogrammetry and based on a vectorization of available cartographic data and hydrometric cross-sections. This work was carried out using the ArcViewGIS 3.2 software. As an example of the vectorization process, Figure 4B shows a map of land use and plant coverage of the section of the Vistula River under study. Because of the length of this section and the map scale, further research was undertaken on a subsection of this map (defined by the rectangular frame in Figure 4). This section covered the places of particular interest for this analysis, i.e. those with highest numbers of fallen trees. The digital terrain model was generated using the Tiegris programme, taking into account the mesh for branches and old river beds as well as the model of the floodplains and the main channel.

The next stage in the research was to develop the calculation mesh. For the area studied a mesh was generated using the Tiegris programme with sides of approx. 30 m. Additionally a mesh of 80 m wide rectangular elements was generated along the axis of the main channel. In order to improve the model, the calculation mesh took into account several check structures, such as repelling spurs, groins and wing dams. The parameters of these structures were taken from the inventory maps of check structures. These data were used in the Tiegris programme to generate the finite element mesh (Figure 5.). The final results were a digital terrain model and a calculation mesh including check structures. The mesh consisted of 11.623

elements based on 29,628 nodes. It included both triangular and rectangular elements.

The assessment of the roughness of the flow area (both the main channel and the flood lands) was based on an analysis of the distribution of various vegetative clusters, maps of land use and plant coverage, aerial photos and direct site inspections. Fourteen different types of surface were identified. A numerical map of vegetation types was developed. Using this map, the relevant section of the Vistula River was analysed and the proportion of the area occupied by each surface type was determined. The overall surface of the section evaluated was approximately 8.23 km^2, with 2.90 km^2 (35%) of it being water at the medium water level. The analysis of land use types in the area revealed that the predominant form of vegetation was shrubs (approximately 20%) and fields and meadows (about 20.3%), including approximately 7% fields and meadows with shrubs or with isolated trees.

Each element of the finite element grid was assigned a certain roughness parameter (Table 1), according to the analysis of the 1997 flood wave conducted for this river section. This procedure provided a numerical terrain roughness model. The strip of water near the road bridge in Puławy is considered as a separate type due to the different hydraulic properties of the riverbed caused by the bridge pillars. Roughness is strongly influenced by the trees and shrubs on the flood lands and on several islands occurring in this area. The influence of the vegetation has been calculated using the vegetative drag coefficient, λ_V, proposed by Lindner (1982) and modified by Pasche (1984):

$$\lambda_V = \frac{4 \cdot h \cdot d_p}{a_x^2} \cdot c_{WR} \qquad (6)$$

where: h = flow depth; d_p = substitute plant diameter; a_x = substitute plant spacing; c_{WR} = non-dimensional drag coefficient for flow around plants.

Table 1. High water channel, types of roughness identified.

n [m$^{-1/3}$s]	d_p [m]	a_x [m]	Types of roughness
0.018	0.00	0.00	main channel
0.020	0.00	0.00	sands
0.022	0.02	0.30	sands with thin bushes
0.030	0.00	0.00	meadows
0.030	0.50	5.00	meadows with trees and bushes
0.035	0.05	1.00	wasteland with bushes
0.025	0.00	0.00	cultivated areas
0.030	0.50	3.00	orchards and gardens
0.010	15.00	30.00	built-up areas
0.010	0.00	0.00	roads
0.030	0.35	5.00	compact, deciduous young forests
0.035	0.50	12.00	compact, deciduous mature forests
0.035	0.50	12.00	compact, coniferous mature forests
0.040	10.00	100.00	bridge

3 RESULTS AND DISCUSSION

Hydraulic calculations were carried out using the Rismo model (Schröder 1997), developed by the Institut für Wasserbau und Wasserwirtschaft, Rheinisch – Westfälische Technische Hochschule, Aachen and currently being further improved by Valitec, Aachen. It is a two dimensional modelling system, used for transient flow simulations with a free water table. The programme's hydrodynamic module solves Navier-Stokes differential equations for flow and the continuity equation.

Boundary conditions were given as the discharge at the starting cross-section (km 367.47), and the water level at the ending cross-section (km 376.95), determined from the flow rate curve. Transient flows were analysed using the July 2001 flood wave hydrograph data. The main output data provided by the model include water levels, distributions of flow velocities in the river and on the flood lands and also the stress from the river bottom. Figure 6 shows the results of flow simulations, in terms of flow depth and velocity distribution for the 7500 m^3/s discharge.

Based on the results of the geotechnical survey and using the ArcViewGIS 3.2 programme, a numerical model was built of the variability of the strength parameters for irrigated soils. This data included soil parameters for both low water levels and long-term flooding. Tree diameters and areas covered with trees were assumed to be the same as those used for the numerical terrain roughness model. For prolonged flood water levels, the critical toppling moments for trees in the Vistula section under study were calculated from Equation 5 (Figure 7).

In order to verify the hypothesis on the significance of the dynamic pressure for forest stability, tree-toppling moments were calculated for the dynamic pressure of water and wind (from Equation 2 and 3). Depths and velocities on the flood lands were determined based on the results of flow calculations carried out using the Rismo programme. Based on these

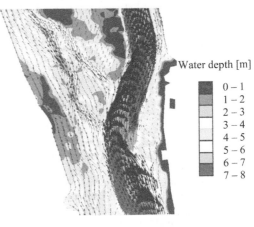

Figure 6. Water depths and velocity vectors for the 7500 m^3/s discharge.

results the toppling moments due to water pressure were then calculated (Figure 8.).

During the storm, water in the area under investigation was not particularly deep (from several centimetres to approximately 1 m at the deepest point – so it did not reach the branches). Water velocity in this area did not exceed 1 m/s. The maximum moments of the force due to the water are not very high and do

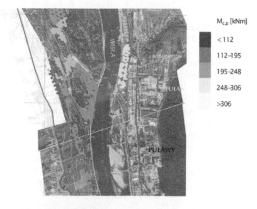

Figure 7. Distribution of critical moments in the area studied.

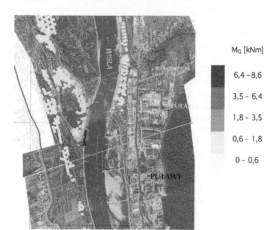

Figure 8. Distribution of critical moments due to hydrodynamic pressure of water in the area studied.

not exceed 9 kNm in the vicinity of the main channel, where the velocity and depth are the highest. On the research sites with the highest numbers of fallen trees, the moments of the force due to water varied between 0.6 kNm and 3.5 kNm, being significantly lower than the critical values.

The influence of the wind in this area was also analysed using Equation.3. The research conducted detailed analyses of the destruction on the research sites. The average trunk diameters of the fallen trees were measured, as well as tree lengths. The surfaces of tree crowns were determined using scaled photos of the trees in the neighbourhood and from field measurements of the leaf area index (LAI). This index allows the surfaces of the branch structures with leaves to be estimated. The size of the tree crown was correlated to the trunk diameter. Trunk height, from the ground to the first branches of the crown, h_0, was, on average, approximately 5 m. Holes made by toppled trees helped to determine the dimensions of the root systems; these were between 4 and 5 m in diameter and approximately. 0.4 m in depth. This shallowness of rooting was due to the high groundwater table in the area (the trees had their roots only in the surface stratum). With no information on the real peak wind velocity (the weather station in Puławy broke down during the storm) it was assumed to be $u_w = 25$ m/s, i.e. the winds were of class III (gale force) according to the Polish Institute of Meteorology and Water Management (IMGW) scale, 'Classification of peak wind velocities in Poland and possible damage'. Winds of class III may pull out trees with their roots, severely damage buildings, tear roofs off houses and break towers and electricity poles.

The drag coefficient, C_k, in Equation. 3 depends on the density of the vegetative structure. Typical values range from 0.6 to 1.2. The average for natural vegetative structures is $C_k = 0.7$ (MOC 1994).

Table 2 summarises the values of the critical moments for selected trees in the research area calculated from Equation 5. The calculations were carried out with two variants. The first determined the values of the critical moments, M_c, at low water level. In the second, ($M_{c,P}$), the change in the soil parameters due to the water was taken into account. This table also includes the values of the overall toppling

Table 2. Calculated critical moments for selected trees.

No.	D [m]	H [m]	M_c [kNm]	$M_{c,P}$ [kNm]	M_W [kNm]	M_Q [kNm]	$M_W + M_Q$ [kNm]
1 Populus	0.8	25	141.5	81.8	84.7	0.5	85.2
2 Populus	0.7	22	93.1	53.1	57.9	0.4	58.3
3 Populus	0.8	29	161.3	131.1	133.8	0.5	134.2
4 Populus	0.8	28	159.2	124.1	130.8	0.5	131.3
5 Populus	0.3	22	60.1	49.2	50.7	0.2	50.8
6 Populus	0.2	26	33.4	27.2	29.5	0.3	29.8
7 Populus	0.9	27	205.6	120	122.8	0.5	123.3
8 Populus	0.5	24	39.2	18.2	23.3	0.3	23.6
9 Salix	0.5	15	39.2	18.2	19.4	0.3	19.7
10 Salix	0.55	16	52.7	24.6	25.3	0.3	25.6

moments resulting from both water and wind pressure. In the cases under study, the overall toppling moment exceeded the critical moments for trees in the studied areas; this result is valid for long-term flooding and related soil parameter changes. Computational results confirm that the critical moments decrease due to the pressure of the water (Table 2).

4 CONCLUSIONS

Using a numerical model based on a two-dimensional flow model, the conditions under which the riparian forest in the Vistula River Valley near Puławy was destroyed, were reconstructed. An assessment of the soil resistance parameters on the flood terraces of the Vistula River was carried out. Using hydrogeological and meteorological data, simulations were conducted of the conditions in which trees had been toppled. his data was used to check whether the assumed stability model may be correlated with the actual extent of the forest devastation.

Our calculations lead us to believe that the influence of the toppling moments due to the dynamic pressure of the water were negligible; these values were relatively small compared to the moments resulting from the wind pressure.

High flood water levels had an influence on, and actually changed the soil resistance parameters. In this sense the flood wave on the Vistula River, which had passed through the area five days before the gale, did have an influence on the scale of the destruction. The analysis of the 2001 flood wave hydrograph revealed that the trees in the area studied had been submerged in water for at least nine days (from 26 July to 4 August 2001). During this period, all the soil pores may have been filled with water and the resistance parameters in the root stratum may have been significantly changed.

ACKNOWLEDGEMENTS

This work was carried out under the research grants no: N305 078 32/2740 and N N523 450236 and financed by the Polish Ministry of Science and Higher Education.

REFERENCES

Achim, A. R. J-C., Gardiner, B. A., Laflamme, G. & Meunier, S. 2005. Modelling the vulnerability of balsam fir forests to wind damage. *Forest Ecology and Management* 204: 35–50.

Dupuy, L., Fourcaud, T. & Stokes, A. 2005. A numerical investigation into the influence of soil type and root architecture on tree anchorage *Plant Soil* 278: 119–134.

Gardiner, B. A., Stagey, G. R., Belcher, R. E. & Wood, C.J. 1997. Field and wind tunnel assessments of the implications of respacing and thinning for tree stability. *Forestry* 70: 233–252.

Gardiner, B., Peltola, H. & Kellomäki, S. 2000. Comparison of two models for predicting the critical wind speeds required to damage coniferous trees. *Ecological Modelling* 129(1): 1–23.

Gregory, P. 2006. *Plant roots: growth, activity, and interaction with soils.* Blackwell. Publishing, Oxford.

Hartge, K. H. & Bohne, H. 1985. Zur gegenseitigen Beeinflussung von Baum und Bodengefüge. *Allgemeine Forstzeitschrift* 11: 235–237.

Kałuża, T. & Leśny, J. 2009. An analysis of tree stand stability relative to Institute of Meteorology and Water Management (IMGW) classification of maximum wind velocities. *Journal of Water and Land Development* 13a: 103–113.

Lindner, K. 1982. *Der Strömungswiderstand von Pflanzenbeständen. Mitteilungen des Leichtweiss – Institut für Wasserbau. Heft 75.* Technische Universität. Braunschweig.

Lundström, T., Jonsson, M.J. & Kalberer, M. 2007. The root–soil system of Norway spruce subjected to turning moment: resistance as a function of rotation. *Plant and Soil* 300(1): 35–49.

Ministry Of Construction (MOC). 1997. *Proposed Guidelines on the Clearing and Planting of Trees in Rivers.* Sankaido Tokyo.

Nicoll, B., Achim, A., Mochan, S. & Gardiner, B. 2005. Does steep terrain influence tree stability? A field investigation. *Canadian Journal of Forest Research* 35: 2360–2367.

Pasche, E. 1984. *Turbulenzmechanismen in naturnahen Fließgewässern und die Möglichkeiten ihrer mathematischen Erfassung. Mitteilungen des Instituts für Wasserbau und Wasserwirtschat. Heft 52.* Rheinisch-Westfälische Technische Hochschule. Aachen.

Schröder, P. 1997. *Zur numerischen Simulation turbulenten Freispiegelströmungen mit ausgeprägt dreidimensionaler Charakteristik, Mitteilungen des Instituts für Wasserbau und Wasserwirtschaft. Heft 108.* Rheinisch-Westfälische Technische Hochschule. Aachen.

Environmental Engineering III – Pawłowski, Dudzińska & Pawłowski (eds)
© *2010 Taylor & Francis Group, London, ISBN 978-0-415-54882-3*

The content of heavy metals in soils and *Populus nigra* leaves from the protective zone of the Głogów copper smelter

J. Kostecki

Department of Land Protection and Reclamation, Institute of Environmental Engineering,
University of Zielona Góra, Zielona Góra, Poland

ABSTRACT: The operation of industrial plants such as copper smelters can lead to different forms of environmental pollution. In the case of the Głogów copper smelter, the main pollutant, despite the modernisation of production lines, is a metal dust that accumulates in the soil and, as a result of changes occurring to the dust, can enter the food chain through accumulation in plants.
This paper presents the results of monitoring the contents of selected heavy metals in soil samples and *Populus nigra* leaves taken within the protective zone of Głogów copper smelter.

Keywords: Głogów copper smelter, heavy metals, protective zone, *Populus Nigra*, phytoextraction.

1 INTRODUCTION

Despite the use of modern, environmentally friendly technologies, the emission of pollutants from industrial facilities constantly causes the degradation of the natural environment. This degradation results from the migration of pollutants emitted into the atmosphere and their final accumulation in the environment. Soil has a special meaning in this process: on the one hand, it performs a function as the final recipient of pollutants; on the other hand, soil may be the beginning of new migration routes for xenobiotics, both in the depths of the profile as well as in the biomass of plants growing on the site.

The protective zone for the Głogów copper smelter was established in 1990 and initially covered 2,840 hectares. As a result of emissions reduction, in 2001 the land area was reduced by 6% (to 2,660 ha).

Land pollution is often caused by more then one pollutant, especially in metal-contaminated soils (with one dominating xenobiotic) (Ernst 2005, Ernst et al. 2000, Huynh et al 2008). The main pollutants from the Głogów copper smelter are SO_2, CO and metal dusts (KGHM Głogów 2008).

Plants mainly uptake all necessary life ingredients from the soil. In the ion form they are gathering elements which are essential to life, as well as those that are harmful (including heavy metal compounds). All of these elements are generally transported to the above-ground parts of plants. The ability to uptake heavy metals through roots, transport them to shoots and accumulate them in above-ground parts is a very specific process that depends on many circumstances (Castiglione et al. 2009, Adriano et al. 2004, Ernst 2005, Assunção et al. 2001 & 2003).

The aim of this paper is to present the results of an analysis of the concentration of heavy metals in samples of soil and leaves of black poplar *Populus nigra* taken from the protective zone of the Głogów copper smelter.

2 MATERIALS AND METHODS

In 2007, three points were selected as representative on the land exposed to the negative impact of the Glogów copper smelter. Main reasons were: location of the smelter facilities, distance from the smelter and location and condition of the plant cover.

In these locations in autumn 2008, soil exposures (from plant litter to 150 cm depth) were made. From each genetic horizon, soil samples were taken for further analysis.

The first point, where soil samples were taken (Żukowice I), was located about 50 m from copper smelter, and the second point (Żukowice II) was about 200 m from the industrial facilities. The third place (Bogomice) was situated about 700 m from the smelter.

Air-dry samples were examined for:

- grain composition – using the Casagrande method with Pruszyński modification
- pH – using a CP502 pH-meter in H_2O and 1M KCl (Mocek et al. 1997)
- conductivity – using a WTW Multiline P4 universal meter (Mocek et al. 1997)
- organic carbon – using Tiurin's method (Mocek et al. 1997)
- the total content and form of heavy metals (Cu, Pb, Ni) – using an atomic absorption spectrometer AAS

FL after mineralisation in aqua-regia (McGrath & Cunliffe 1985).

In the area close to the smelter (Żukowice I) there were mixed plantings with black poplar (*Populus nigra*) and European weeping birch (*Betula pendula*) as dominant species, with Siberian peashrub (*Caragana arborescens)*, black locust (*Robinia pseudoacacia*) and self-spreading black pine (*Pinus nigra*) in the undergrowth; there were no plants on the forest floor.

About 200 m from the smelter (Żukowice II) the dominant species was black poplar (*Populus nigra*). There was no undergrowth and on the forest floor garden sorrel (*Rumex acetosa*), dandelion (*Taraxacum officinale*), yarrow (*Achillea millefolium*), mugwort (*Artemisia vulgaris*), vetch (*Vicia sativa*) and yellow bedstraw (*Galium verum*) were found.

Near Bogomice village the dominant species was black poplar (*Populus nigra*). In the undergrowth European weeping birch (*Betula pendula*) and pedunculate oak (*Quercus robur*) were found. In the forest floor, male fern (*Dryopteris filix-mas*), field horsetail (*Equisetum arvense*), stinging nettle (*Urtica dioica*), wild buckwheat (*Fallopia convolvulus*), sheep's fescue (*Festuca ovina*) and garden sorrel (*Rumex acetosa*) were growing.

To examine the content of heavy metals in the plant biomass, leafs were selected. Plants samples were taken from the same locations as soil samples in 2007 and 2008.

In an effort not to distort the study, the leaves were not washed, and the whole mass of leaves (petiole and blade) were used in the analysis. Air-dry samples were mixed and used to extract representative samples, which were then burned at 550°C. The ash was then mineralised in aqua-regia (9 cm^3 HCl to 3 cm^3 HNO$_3$), burned in a sand bath and percolated. The total content and form of Cu, Pb and Ni in the resultant ooze was analysed using an atomic absorption spectrometer AAS FL.

3 RESULTS

Changes in the physical and chemical properties of tested soil samples from the Żukowice and Bogomice protective zones are compared in Tables 1 and 2. A profile of the exposure made in closest to the copper smelter (Żukowice I) contains mixed horizons of sandy loam (A1, A2, AC, C3) with loamy sand (C1, C2).

The exposure made 200 m from the smelter (Żukowice II) contains only sandy loam. This was similar to the exposure in Bogomice village (only sandy loam).

Conductivity of soil horizons in the exposure made in Żukowice I were between 401 and 450 μS. The conductivity of the top horizons was greater then the horizons located further down in the profile. The same situation was found with pH, except in the litter where the pH was similar to horizons situated in the bottom of the exposure. The pH was slightly acid (except for the AC horizon, where an acid reaction was determined).

The content of organic carbon was greater in the top horizons. The smallest content was found in the C2 horizon (only 0.12%). A similar finding occurs with heavy metals – the concentration of Cu in top horizons was over thousand times greater and the concentration of Pb over two hundred times greater. However, only a slightly higher concentration of Ni was found in the litter horizon.

The exposure made about 200 m from the smelter (Żukowice II) showed few differences from the exposure made in Żukowice I. The pH was slightly alkalic (with neutral reaction in the litter horizon), conductivity was close in horizons lower down the exposure (<20 cm). In the top horizon, conductivity was much smaller (310 μS · cm^{-1}), and in the litter horizon much higher (670 μS · cm^{-1}). The content of organic carbon was higher then in the Żukowice I profile, in the range 1,2–1,86%.

The concentrations of Cu and Pb were slightly lower then in the soil at Żukowice I, but the profile was similar in that there were higher concentration in the top horizons (up to 20 cm). The content of Ni was slightly higher in soil horizons (11–23 mg · kg^{-1} d.m.).

The exposure done in Bogomice village showed a well-made soil profile, with acid pH (and a neutral reaction in the litter horizon). Conductivity was about 400–440 μS · cm^{-1}, except at the bottom horizon (585 μS · cm^{-1}) and in litter horizon (650 μS · cm^{-1}).

The levels of organic carbon, Cu and Pb were higher in top horizons (up to 20 cm). There was no differences in the concentration of Ni in the soil horizons.

The heavy metals content in the tested leaves of *Populus nigra* from the Żukowice and Bogomice protective zones are compared in Table 3.

The concentration of heavy metals varies according to the location and the year in which the samples were taken. In 2007 the highest concentration of Cu was found in Zukowice II and the highest concentration of Pb was in Bogomice. In the other samples the concentrations of these two metals was about 50–70 mg · kg^{-1} of dry leaf mass of *Populous nigra*. The concentration of Ni was not that high at 3.3–11.6 mg · kg^{-1} of dry mass.

In the samples taken from Zukowice I and Zukowice II in 2008, the concentrations of Cu and Pb in leaves were over two times higher than in 2007. The concentrations of Cu and Ni in samples from Bogomice were similar in both years, but Pb was two times lower in 2008.

4 DISCUSSION

The pH of soils located closest to the copper smelter was acid and slightly acid. Only the samples taken from Zukowice II had an alkalic reaction. An acid reaction was found in the years just after the smelter was built (Roszyk & Szerszeń 1988), but over the whole area the pH of soils shows various rates (pH 3.6–7.9) (Sałecki 1997). The pH level has a significant bearing on heavy metals uptake by plants (Huynh et al. 2008,

Table 1. Grain-size composition of tested soils.

Area	Soil horizon	Depth cm	Grain-size composition, %					
			$2.0 \div 0.1$	$0.1 \div 0.05$	$0.05 \div 0.02$	$0.02 \div 0.005$	$0.005 \div 0.002$	< 0.002
			mm					
Żukowice I	Plant litter	$0 \div 4$	–	–	–	–	–	–
	A1	$4 \div 24$	37	23	27	10	3	0
	A2	$24 \div 29$	39	18	27	13	3	0
	AC	$29 \div 36$	35	15	35	10	5	0
	C1	$36 \div 64$	53	18	18	7	3	1
	C2	$64 \div 89$	73	4	12	6	3	2
	C3	$89 \div 150$	52	14	16	10	5	3
Żukowice II	Plant litter	$0 \div 2$	–	–	–	–	–	–
	A ($2 \div 95$)	$2 \div 20$	19	24	39	13	5	0
		$20 \div 50$	21	16	46	11	6	0
		$50 \div 95$	42	9	24	18	7	0
	DG	$95 \div 150$	14	17	48	16	5	0
Bogomice	Plant litter	$0 \div 3$	–	–	–	–	–	–
	A	$3 \div 20$	30	10	24	22	9	5
	Br1	$20 \div 50$	31	11	25	21	9	3
	Br2	$50 \div 62$	28	12	22	25	8	5
	CG	$62 \div 150$	37	9	14	20	11	9

Table 2. Changes of physical and chemical properties of tested soil samples from Żukowice and Bogomice protective zones.

Area	Soil horizon	Depth, cm	pH		Conductivity, $\mu S \cdot cm^{-1}$	Tiurin's organic carbon*, % d.m.	Content of total form of heavy metals, mg $\cdot$ kg^{-1} d.m.		
			H$_2$0	KCl			Cu	Pb	Ni
Żukowice I	Plant litter	$0 \div 4$	6.2	5.8	450	–	5249.89	1289.60	23.58
	A1	$4 \div 24$	6.6	5.9	427	0.94	1213.09	324.99	8.54
	A2	$24 \div 29$	6.8	6.3	434	0.66	1380.28	257.95	8.89
	AC	$29 \div 36$	5.1	4.4	401	0.13	3.68	7.29	6.97
	C1	$36 \div 64$	6.1	4.7	404	0.29	2.84	6.31	6.74
	C2	$64 \div 89$	6.2	5.1	415	0.12	3.57	8.12	9.14
	C3	$89 \div 150$	6.4	5.4	418	0.16	40.22	21.54	14.75
Żukowice II	Plant litter	$0 \div 2$	6.9	6.6	670	–	3320.68	960.07	23.18
	A ($2 \div 95$)	$2 \div 20$	8.0	7.4	310	1.38	997.36	202.23	12.58
		$20 \div 50$	8.0	7.6	474	1.20	5.74	22.37	11.33
		$50 \div 95$	8.1	7.5	496	1.43	5.09	19.86	19.54
	DG	$95 \div 150$	8.0	7.5	479	1.86	8.30	15.19	21.70
Bogomice	Plant litter	$0 \div 3$	7.0	6.5	650	–	3967.66	1474.11	12.09
	A	$3 \div 20$	5.1	4.1	397	1.43	1347.86	275.22	21.73
	BBr1	$20 \div 50$	5.7	4.7	408	2.86	16.99	21.58	24.28
	BBr2	$50 \div 62$	6.0	4.6	440	0.63	23.89	24.90	21.80
	CG	$62 \div 150$	5.6	4.5	585	0.98	12.41	12.01	21.64

*Tiurin's organic carbon was not determined in plant litter.

Domínguez et al. 2008, Navarro 2006, Vitkova ct al. 2009).

The conductivity of the tested soils was about 400–500 $\mu S \cdot cm^{-1}$, with the exception of the litter horizon and the bottom horizon in Bogomice.

The highest changes in organic carbon content was determined in soils from Żukowice I (ranging from 0.12 to 0.94%) and Bogomice (0.63–2.86%). Similar organic carbon content was found in horizons from the exposure made in Żukowice II (1.2–1.86%). Higher concentrations of organic carbon may result from the fact that in the past this area was used for agriculture. The variation in the concentration of organic carbon in soil horizons from the exposures may also be caused by their structure – horizons with poor organic carbon content (substratum horizon) are mixed with horizons which are rich in carbon (organic matter, subsoil, gley horizon). Other authors (such as Barajas-Aceves 1999) suggest that high concentrations of heavy metals can affect organic matter decomposition.

The highest concentration of heavy metals was found in the organic matter horizon (for Cu

Table 3. The heavy metals content in tested leaves of *Populus nigra* from Żukowice and Bogomice protective zones.

Area	Content of total form of heavy metals, mg · kg⁻¹ d.m.					
	2007			2008		
	Cu	Pb	Ni	Cu	Pb	Ni
Żukowice I	53.41	67.55	6.47	192.13	106.03	5.33
Żukowice II	121.05	54.04	3.32	217.16	168.21	8.10
Bogomice	50.22	94.08	11.58	46.40	48.23	8.60

3320.68–5249.89 mg · kg⁻¹ d.m., for Pb 960.07–1474.11 mg · kg⁻¹ d.m. and for Ni 12.09–23.58 mg · kg⁻¹ d.m.), slightly smaller concentrations were found in horizons 20–30 cm, but concentrations in these horizons were much higher then in all the other horizons. Only the Ni concentration was under the threshold value designated in Order MOŚ RP of 9 September 2002 (Dz. U. Nr 165, poz. 1359) and background values IUNG (after Siuta 1995). The concentrations of heavy metals in the bottom horizons are similar to values presented by Kabata-Pendias (Kabata-Pendias & Pendias 1992). The concentrations of Cu, Pb and Ni in the top horizons of the exposures are much higher then in the bottom, which can be associated with the concentration of organic matter.

During two years of environemntal monitoring in the protective zone of the Głogów copper smelter differences in the concentration of heavy metals in the leaves of *Populous nigra* were found. In 2008 trees from Żukowice I accumulated over four times more copper more in the leaf biomass than in 2007 (53.41–192.13 mg · kg⁻¹ d.m.), and in Żukowice II two times more (121.05–217.16 mg · kg⁻¹ d.m.). In Bogomice the copper content was slightly lower in 2008 compared to 2007 (50.22–46.40 mg · kg⁻¹ d.m.). The concentration of lead was similar to copper – in Żukowice I trees had accumulated over two times more lead in 2008 (67.55–106.03 mg · kg⁻¹ d.m.) and in Żukowice II three times more in 2008 (54.04–168.21 mg · kg⁻¹ d.m.). In Bogomice the concentration of lead was over two times smaller in 2008 compared to 2007 (48.23–94.08 mg · kg⁻¹ d.m.). The concentration of nickel was similar in all samples (in Żukowic I 5.33–6.47; Żukowic II 3.32–8.10; in Bogomice 8.60–11.58 mg · kg⁻¹ d.m.).

According to Siuta (1998), the content of copper, lead and nickel in plants can vary significantly. Other authors presents similar results (Domínguez et al. 2008, Kabata-Pendias & Pendias 1992, Ostrowska & Porębska 2002, Yanqun et al. 2004]. The fact that different metal ions can be accumulated in varying rates by different plant species can be explained by the meteorological conditions, the bioavailability of selected elements in the soil and the various means of transport in plants.

As we can see, a process called phytoextraction is going on in this area (the accumulation of heavy metals

in plant biomass). The biomass containing heavy metals is not removed from the site of the protective zone of the smelter, which limits the possibility of reducing the content of heavy metals in the top horizon of the soil (the circulation of matter). The annual increase of plant biomass confirms the thesis that heavy metals do not necessarily harm the growth and development of plants (Greinert & Greinert 1999), however there is a risk that contaminants may enter the food chain (Rozema et al. 2008).

5 CONCLUSIONS

Some conclusions may be drawn from the results of this research.

1. The main soil contaminants in the tested area were Cu and Pb.
2. The highest concentrations of copper and lead were found in the top horizons of soil profiles (up to 20–30 cm), and were much greater than concentrations in horizons on the bottom of profiles.
3. The conductivity of tested soils were mainly in the range 400–500 μS.
4. The levels of organic carbon in soil horizons were determined mainly by the soil construction.
5. Grain composition of tested profiles shows that they were mainly built by sandy loam. Only in Żukowice I was sandy loam mixed with loamy sand.

REFERENCES

Adriano, D.C., Wenzel, W.W., Vangronsveld, J. & Bolan, N.S. 2004: Role of assisted natural remediation in environmental cleanup. *Geoderma* 122: 121–142.

Barajas-Aceves, M., Grace, C., Ansorena, J., Dendooven, L. & Brookes, P.C. 1999. Soil microbial biomass and organic C in a gradient of zinc concentrations in soils around a mine spoil tip. *Soil Biology and Biochemistry* 31: 867–876.

Castiglione, S., Todeschini, V., Franchin, C., Torrigiani, P., Gastaldi, D., Cicatelli, A., Rinaudo, C., Berta, G., Biondi, S. & Lingua, G. 2009. Clonal differences in survival capacity, copper and zinc accumulation, and correlation with leaf polyamine levels in poplar: A large-scale field trial on heavily polluted soil. *Environmental Pollution* 157: 2108–2117.

Domínguez, M.T., Teodoro Marañón, T., Murillo, J.M., Schulin, R. & Robinson, B.H. 2008. Trace element accumulation in woody plants of the Guadiamar Valley, SW Spain: A large-scale phytomanagement case study. *Environmental Pollution* 152: 50–59

Dz. U. Nr 165. poz. 1359. Rozporządzenie Ministra Środowiska z dnia 9 września 2002 r. w sprawie standardów jakości gleby oraz standardów jakości ziemi

Ernst, W.H.O. 2005. Phytoextraction of mine wastes – Options and impossibilities. *Chemie der Erde* 65: 29–42

Greinert, H. & Greinert, A. 1999. *Ochrona i rekultywacja środowiska glebowego*. Wydawnictwo Politechniki Zielonogórskiej: Zielona Góra.

Huynh, T.T., Laidlawa, W.S., Singh, B., Gregory, D. & Baker, A.J.M. 2008. Effects of phytoextraction on heavy metal concentrations and pH of pore-water of biosolids determined using an in situ sampling technique. *Environmental Pollution* 156: 874–882

Kabata-Pendias, A. & Pendias, H. 1992. *Trace elements in soils and plants, 2nd edition*. CRC Press: USA.

KGHM Głogów, 2009. Protection of the air, www.kghm.pl

McGrath, S.P. & Cunliffe, C.H. 1985. A simplified method for the extraction of the metals Fe, Zn, Ni, Pb, Cr, Co, Mn from soils and sewage sludges. *J Sci Food Agric* 36: 794–8

Mocek, A., Drzymała, S. & Maszner, P. 1997. *Geneza, analiza i klasyfikacja gleb*. WAR: Poznań.

Nawarro, M.C., Perez-Sirvent, C., Martinez-Sanchez, M.J., Vidal, J. & Marimon, J. 2006. Lead, cadmium and arsenic bioavailability in the abandoned mine site of Cabezo Rajao (Murcia, SE Spain). *Chemosphere* 63: 484–489.

Ostrowska A. & Porębska G. 2002. *Skład chemiczny roślin, jego interpretacja i wykorzystanie w ochronie środowiska*. Komitet Wydawniczy Instytutu Inżynierii Środowiska: Warszawa.

Roszyk, E. & Szerszeń, D. 1988. Nagromadzenie metali ciężkich w warstwie ornej gleb strefy ochrony sanitarnej przy hutach miedzi. Część II "Głogów". *Roczniki gleboznawcze* XXXIX(4): 147–158. PWN: Warszawa.

Rozema, J., Notten, M.J.M., Aerts, R., van Gestel, C.A.M., Hobbelenb, P.H.F. & Hamers, T.H.M. 2008. Do high levels of diffuse and chronic metal pollution in sediments of Rhine and Meuse floodplains affect structure and functioning of terrestrial ecosystems? *Science of the total environment* 406: 443–448.

Sałecki T. *Ocena zmian stanu środowiska województwa legnickiego w latach 1985–1995*. 1997. Studio Komputerowe KZS Computer Solutions: Legnica.

Siuta, J. 1998. *Gleba. Diagnozowanie stanu i zagrożenia*. Komitet Wydawniczy Instytutu Inżynierii Środowiska: Warszawa.

Siuta J. 1998. *Rekultywacja gruntów – poradnik*. Komitet Wydawniczy Instytutu Inżynierii Środowiska: Warszawa.

Vítková, M., Ettler, V., Sebek, O., Mihaljevica, M., Grygar, T. & Rohovec, J. 2009. The pH-dependent leaching of inorganic contaminants from secondary lead smelter fly ash. *Journal of Hazardous Materials* 167: 427–433.

Yanqun, Z., Yuan, L., Schvartz, C., Langlade, L. & Fan, L. 2004. Accumulation of Pb, Cd, Cu and Zn in plants and hyperaccumulator choice in Lanping lead–zinc mine area. China. *Environment International* 30: 567–576.

Environmental Engineering III – Pawłowski, Dudzińska & Pawłowski (eds)
© *2010 Taylor & Francis Group, London, ISBN 978-0-415-54882-3*

Treatment of alkaline waste from aluminium waste storage site and method for reclamation of that site

Z. Kowalski, K. Gorazda & A. Sobczak

Institute of Inorganic Chemistry and Technology, Krakow University of Technology, Krakow, Poland

ABSTRACT: In the defunct 'Górka' quarry there is both a waste disposal site of solid waste generated in the process of aluminium oxide production, and a pond containing alkaline wastewater. The proposed solution involves pumping out and treating effluents and using the bottom slurry in the production of self-solidifying mixtures. In the next stage, outcropping the feed-water sources located in the northern part of the old heading is used to reconstruct the original flow system from the sources to the Ropa river. The third stage involves reclamation of the entire post-marl Górka quarry and its transformation into a landscaped park.

Keywords: Alkaline waste, infiltration, quarry, land reclamation.

1 INTRODUCTION

In the defunct 'Górka' quarry, located in Trzebinia near Cracow, Poland, there is both a waste storage site with an area of 6.7 ha containing waste generated in the processes of cement and aluminium oxide production, and a settling pond with an area of 3 ha containing alkaline wastewater (Figure 1). Total waste quantities are as follows (Rudnicki et al. 2006):

– ~600,000 m³ of solid wastes from aluminium oxide production,
– alkaline infiltrate, which has formed a pond that occupies the southern part of the heading, currently 400,000 m³,
– approx. 57,000 m³ of alkaline sediment at the bottom of the pond.

The fundamental problem posed by the wastewater-filled Górka pond is the high alkalinity of the water (pH 12–14). The most critical environmental problems are surface outflow from the pond that has resulted in surface water pollution and infiltration of wastewater from the pond into underground water (Rudnicki et al. 2006).

The objective of our investigations was to develop a complex reclamation project for the Górka waste storage site.

The main stages of this project are as follows (Żelazny et al. 2006, Kowalski et al. 2007):

– treatment of the over-sludge water pumped out; purified water will then be discharged to the surface water or to the sewage system and residual 'concentrates' of contaminants (brine, sludge) will be solidified and stored 'outside' in a safe waste dump;

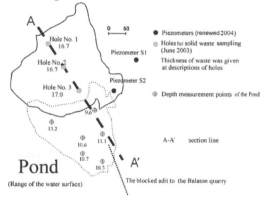

Figure 1. Map of the Górka storage site (the solid line marks the limit of the solid waste dump, the dotted line marks the limit of the free water surface of the pond).

– removal and neutralization of sludge collected at the settling pond bottom;
– exposing sources in the northern part of the former quarry excavation; the waste stored there will then be moved to the remaining part of the dump;
– restoring the original conditions of water flow, from the point where flows enter the quarry to the point where they exit
– macrolevelling of the site (into an amphitheatric system) including fertile soil layer restoration, natural lake construction and sowing with a suitably selected grass mixture;
– local underground water monitoring.

Table 1. Chemical analysis of infiltrate saturating the solid waste heap and wastewater from the Górka pond.

Parameter	Unit	Infiltrate					
		Hole No. 1, depth		Hole No. 2, depth		Hole No. 3 depth	
		2.1 m	16 m	2.3 m	16 m	2 m	16 m
pH		12.52	12.83	12.68	13.02	12.42	12.65
σ	mS/cm	11.5	27.91	16.14	48.48	10.36	25.15
Ca	mg/dm^3	108.2	21.8	5.6	0.8	146.8	55.2
Na	mg/dm^3	1264	2559	1504	3916	1116	3042
Cl$^-$	mg/dm^3	38	120	70	110	80	210
SO$_4^{2-}$	mg/dm^3	239	712	144	1184	218	1790
SiO$_2$	mg/dm^3	92	170	76	196	108	234
P total	mg/dm^3	0.56	0.78	0.38	0.24	1.25	0.76
B	mg/dm^3	1.43	3.44	1.37	5.26	1.31	8.85
Fe	mg/dm^3	52.13	9.30	2.98	0.05	67.73	12.01
Mn	mg/dm^3	0.26	0.04	0.02	<0.01	0.81	0.13
Zn	mg/dm^3	0.12	0.01	0.01	0.02	0.77	0.23
Al	mg/dm^3	70.9	28.3	45.5	28.1	153.6	45.6
Cr	mg/dm^3	0.18	0.50	0.14	1.28	0.14	0.03
As	mg/dm^3	0.20	0.93	0.28	1.46	0.41	1.00
Ni	mg/dm^3	0.10	0.15	0.09	0.13	0.34	0.32
Pb	mg/dm^3	0.05	0.01	0.01	<0.01	0.29	0.10

2 RESULTS AND DISCUSSION

2.1 Profile of the Górka quarry

The former Górka quarry is currently filled with wastewater to a level of approximately 355 m asl. The excess water is carried through an adit into the Ropka water race that has its source in the Balaton lake. The hydrological balance of the Górka quarry involves:

- a total inflow quantity of 165.2 m^3/day which includes a direct inflow of 78.8 m^3/day (rainfall on the surface of the quarry minus evaporation) and lateral inflow of 86.4 m^3/day (inflow of underground water from the Jurassic water-bearing stage, based on average measurement results before swelling);
- an overflow of 93.6 m^3/day;
- infiltration of 71.6 m^3/day.

The amount of infiltrates which have accumulated in the stockyard has been estimated to be approx. 434,000 m^3. This includes infiltrates in the pond and those saturating the solid waste heap. The profile of the infiltrate saturating the solid waste is presented in Table 1 (Kowalski et al. 2007).

All multi-elemental analyses presented in the paper were carried out using the AAS Analyst 300 Perkin Elmer flame spectrometer.

2.2 Treatment of alkaline wastewater from the Górka pond

Wastewater was treated in a system based on the reverse osmosis principle (Ahn et al. 1999, Porter 1990, Kang et al. 2000, Prats et al. 2000). Tests on a quarter-commercial scale were performed using several types of membranes and various acids for neutralization. The best result was achieved for a system including the osmotic membrane SG2540PLUSCHT at a pressure of 6 MPa, using concentrated pure hydrochloric acid as the neutralization agent. The original composition of the wastewater, and the results of wastewater treatment are presented in Table 2.

On the basis of the experiments carried out, a technological process was designed (Figure 2). The capacity of the membrane system (SG8040PLUSCHT type) should be equal to 70 m^3/h of permeate and 23.33 m^3/h of brine concentrate. Hence, the hourly pumping rate is 93.33 m^3/h and the recovery is 75%.

The 24-hour pumping rate, taking into consideration the usual breaks in operation needed for technical reasons, is 2,146.59 m^3, which will allow the required amount of wastewater to be pumped out within 180 days of continuous work. Permeate quality will correspond to the 4th class of water quality (Rudnicki et al. 2006).

Preliminary wastewater treatment will be accomplished by means of filters. After filtration, wastewater is transferred to a pH correction system (Rudnicki et al. 2006, Qina et al. 2002, Ozaki et al. 2002, Qdaisa & Moussab 2004).

The task of filtration, using porous-bed rapid filters, is to remove all contaminants generated in the pH correction process. The task of filtration using polypropylene fine filters is to clear wastewater thoroughly before it reaches the reverse osmosis (RO) modules. Two four-candle filter constructions were designed equipped with 60MCHT-1EC filter elements ensuring a filtration efficiency of 1 μm.

To protect the membranes from a build up of calcium carbonate and silica deposits, a dosage system

Table 2. Chemical composition of wastewater from the Górka pond and results of reverse osmosis (RO) treatment; feed – 1.7 m³/h at 6 MPa and 288 K.

| Parameter | Unit | Wastewater | | Reverse osmosis | | |
| | | Sampling depth | | | RO permeate recovery | |
		0 m	bottom ~11 m	RO feed	50%	75%
pH		11.69	13.34	5.4	3.86	3.88
σ	mS/cm	14.398	71.130	35.789	0.352	0.712
SO_4^{2-}	mg/dm³	802	1962	1658	0.0	0.0
COD	mg/dm³	–	1311	765	36.9	53.4
B	mg/dm³	2.73	8.40	5.54	<0.01	0.02
Fe	mg/dm³	0.25	53.04	0.38	0.025	0.216
Al	mg/dm³	25.2	206.7	0.080	0.073	0.183
Cr	mg/dm³	0.145	0.407	1.5	<0.005	<0.005
As	mg/dm³	0.90	4.72	4	<0.01	<0.01
Cd	mg/dm³	0.011	0.026	<0.001	<0.001	<0.001
Pb	mg/dm³	0.006	0.008	0.40	<0.01	0.02

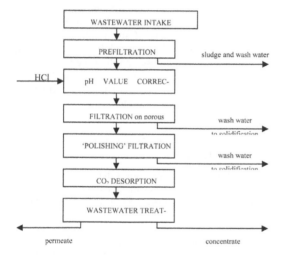

Figure 2. Treatment of wastewater from the Górka pond (RO = reverse osmosis).

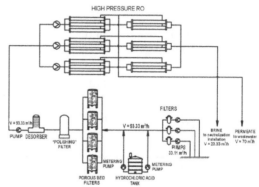

Figure 3. Flow sheet of the reverse osmosis (RO) system.

for an antiscale agent was designed. For the flow rate of 93.33 m³/h, the dosage is 373.32 g/h.

2.3 Solidification of brine solution from alkaline wastewater treatment

When the recovery of the RO process is 75%, 25% of the concentrate generated is of a concentration four times higher than that of the raw wastewater.

The concentrate contains a small amount of heavy metals, a large amount of dissolved sodium and potassium chlorides, and organics corresponding to the wastewater colour compounds (Rudnicki et al. 2006). From the three proposals analyzed (Rudnicki et al. 2006), the solidification process was accomplished by mixing brine concentrate with gypsum, burned lime, and ashes. Both burned lime and ash are ballast necessary for water binding. The time required for water

excess outlet and evaporation, mixture solidification and stabilization is 3–5 days.

The research on concentrate solidification using calcium sulphate (Table 5) indicates that water-leaching of heavy metals from the solidified material is insignificant (their concentrations in the leachate are below the standard values). The colour of the leachate is weak and NaCl and KCl concentrations are low, approximately ~1000 mg/dm³.

2.4 Solidification of waste bottom sludge from the Górka reservoir

The removal of the bottom sludge from the pond, its conversion (stabilization) into quasi-concrete material, and the application of a product for land reclamation (particularly to seal up the surface of high-aluminium waste dumps) are key processes which may allow the problem of the Górka waste storage site to be solved (Kowalski et al. 2007, Żelazny et al. 2006).

The chemical composition of the bottom sludge from the Górka pond (as dry mass) is presented in

Table 3. Dynamic extraction (with distilled water) of brine solidified with gypsum (20s%).

Parameter	Unit	Value	Standard value for introduction to water and ground
pH of leachate	–	9.2	6.5–9.0 (12.5)
Co	mg/dm^3	0.01	1.0
Cr	mg/dm^3	0.47	0.5
Cu	mg/dm^3	0.06	0.5
Ni	mg/dm^3	0.21	0.5
Pb	mg/dm^3	0.12	0.5
Zn	mg/dm^3	0.71	2.0
Hg	mg/dm^3	not found	0.06
conductivity	mS/cm	8.2	–

Table 4. Characteristics of bottom sludge from the Górka pond (as dry mass).

Component	Sludge from pond bottom [%]	Suspension in infiltrates (depth of 10.0–10.6) [%]
SiO_2	28.59	30.31
Fe_2O_3	1.47	1.71
Al_2O_3	6.26	6.76
CaO	25.86	22.6
Na_2O	1.58	1.62
SO_2	0.48	0.44
P_2O_5	0.17	0.18
Cl	0.068	0.063
	[ppm]	[ppm]
Sr	119.1	126.8
Cr total	243.6	304.7
Mn	347.0	293.5
Zn	131.7	141.5
Pb	29.3	35.0
Cd	0.5	0.6
Cr^{6+}	<0.03	<0.03
As	76.7	119.1

Table 4. The average moisture content of the sludge is 51% (a mass loss after drying for 2 hours at 105°C).

Phase composition determined by the X-ray method using the Philips X'Pert diffractometer with the graphite monochromator PW W 1752/00 confirmed that the main components of the waste are calcium carbonate, quartz and small amounts of Al_2O_3 and sodium aluminosilicates.

Wet sludge and building additives were used in the composite preparation. Cement containing: CaO – 65%, SiO_2 – 27%, Al_2O_3 – 6%, and Fe_2O_3 – 2% was used as the binding material. The material was formed into small beams, which were stored at an air humidity of 100%. After 28 days of seasoning, the samples were subjected to crushing strength tests performed on a testing machine EDB-60 (Table 5).

The effect of water on the composite sample was estimated by measuring its stability after 72 hours of watering. We can conclude that the samples with a minimum of 15% binding material addition are characterized by a crushing strength ensuring their stability and load transmission.

Another requirement for the solidified mixture is leachate alkalinity. Table 5 shows pH and conductivity values of water leachates from some composites. Higher pH values were caused by two factors: high alkalinity of the sludge (pH over 12) and 'free' CaO formation as a product of the hydrolysis reaction of alite present in the composite. However, the second factor was a transient phenomenon (Żelazny et al. 2006).

The properties of material exposed to water were also examined by studying the leaching of chemical contaminants. Water leaching tests of the composites were carried out in compliance with Polish standards (Polish standard PN-G-11011, 1998). To a sample of material, water was added at a weight ratio of 1:10; it was then agitated for 48 hours. Table 6 shows concentrations of compounds leached from a composite prepared by mixing 80% of the sludge and 20% of the binding material compared with the standard values.

One of the products of the binding process is $CaSO_4 \cdot 2H_2O$. This phase immobilizes heavy metals; therefore they do not pass into the liquid phase. The results of heavy metal leaching tests indicate that composites produced from the waste sludge and cement do not pose a threat to groundwater.

2.5 Solution to the problem of the infiltrate from the quarry

Archival geological data shows that the lateral water supply to the quarry occurs in the vicinity of the source zone located in the north-western waste-covered part of the quarry.

First, to uncover the source zone output, approximately 130,000 m^3 of the waste would have to be displaced. Water flowing from the sources would be directed through an artificial profiled channel at the bottom of the quarry to a small lake situated at the lowest point of the quarry. The waste will be transferred to the remaining part of the stockyard. Proper formation and complete reclamation of the quarry involves (Figure 4) (Kowalski et al. 2007, Żelazny et al. 2006):

- covering the remaining solid waste stockyard area with a layer of solidified slurry removed from the bottom of the pond, which will allow it to be completely sealed. The surface of the covered stockyard should be 40,000 m^2, which means (assuming a sealing layer thickness of 50 cm) consumption of about 20,000 m^3 of the solidified slurry,
- sealing the stockyard with a ground layer approximately 30–50 cm thick,
- managing this area according to the proposed programme.

The proposed locations of the channel and the water reservoir would allow management of water flowing into the quarry. Excess water from the reservoir will

Table 5. Crushing strength of composite vs. sludge and binding material content and pH and conductivity values of water leachate (Żelazny et al. 2006)

| Sample | Composite | | | | Water leachate | |
	Binding material [%]	Sludge[%]*	Water [g/100 g]	Crushing strength R_s [MPa] after 28 days	pH	Conductivity [mS]
1	10	90	in sludge	2.6	11.6	2.08
2	15	85	in sludge	3.2	11.8	2.11
3	18	82	in sludge	3.8	12.0	3.10
4	20	80	114	5.9	10.4	1.9
5	30	70	115	8.0	11.5	2.56
6	40	60	117	10.8	10.1	1.53

* dry mass.

Table 6. Analysis of water leachate (sample no. 4 in Table 5).

No.	Contaminant	Unit	Standard value	Concentration [%]
1	SO_4^{2-}	[mg/ SO_4^{2-}/dm^3]	500.0	0.054
2	As	[mg As/dm^3]	0.2	0.004
3	Cr (III)	[mg Cr/dm^3]	0.5	0.47
4	Cr (VI)	[mg Cr/dm^3]	0.2	–
5	Cd	[mg Cd/dm^3]	0.1	0.00014
6	Pb	[mg Pb/dm^3]	0.5	0.016
7	Zn	[mg Zn/dm^3]	5.0	0.036
8	Al	[mg Al/dm^3]	3.0	5.78
7	pH		6.0–12.0	12

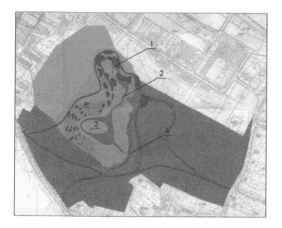

Figure 4. Final proposed form of the Górka quarry. 1 – uncovered sources, 2 – artificial rivulet, 3 – lake, 4 – water outlet into the adit.

be channelled into a drain adit using an overfall to stabilize the water level in the future lake in case of continual lateral inflow and periodic surface rainwater flow.

The creation of a park in the quarry was proposed. Slopes in the south-western part of the valley would be formed into wide (4 m) ground terraces 2 m in height, forming an amphitheatric system. The terraces would be formed on a substructure of ground masses consisting of the solid waste excavated from above the Jurassic sources zone and the channel of the water reservoir hollowed out in the bottom of the quarry.

A draining system layer, an insulation layer and a 30-cm-thick layer of rich soil would be placed on the substructure. At the bottom of the valley, in the axis of the amphitheatrically shaped slopes of the heading, an ellipsoidal glade is planned, surrounded by a lake with plants (rushes). The future lake will be a hollow in the bottom of the quarry and will act as a receiver of water coming from the eastern slope of the heading as well as from rainfall. The presence of rushes would encourage birds to inhabit the area and support the formation of a secondary biotope. The lake with a steady water flow would be connected with an adit draining off the excess water into the Ropa river.

Revitalization of this area is all the more desirable as it is situated in the city centre.

3 CONCLUSIONS

The constant inflow of infiltration waters from the waste storage site into the Górka pond is one of the fundamental problems which need to be solved before the reclamation programme for the whole quarry heading can be implemented. The proposed solution assumes firstly liquidation of the existing wastewater pond and wastewater treatment using the reverse osmosis method, and the removal and solidification of the bottom sludge from the reservoir. The next stage involves uncovering the Jurassic water sources and constructing an artificial rivulet to drain the water to the Ropa river.

The third stage involves reclamation of the entire post-marl Górka quarry and its transformation into a landscaped park. Such a project would eliminate the negative influence of the surface and underground water on the environment and also eliminate the risk of serious accidents associated with uncontrolled outflow of almost 0.5 million m^3 of alkaline wastewater from the existing reservoir. The total implementation time is estimated to be between 3 and 4 years, and the total investment cost would be PLN 52.170,000.

REFERENCES

Ahn K.H., Song K.G., Cha H.Y. & Yeom I.T. 1999. Removal of ions in nickel electroplating rinse water using low-pressure nanofiltration. *Desalination* 122(1): 77–84.

Kang M., Kawasaki M., Tamada S., Kamei T. & Magara Y. 2000. Effect of pH on the removal of arsenic and antimony using reverse osmosis membranes. *Desalination* 131(1–3): 293–298.

Kowalski Z., Strzelecki R., Wolski P., Kulczycka J., Rudnicki P. & Sobczak A. 2007. Protection of water from infiltration of alkaline waste from the Górka quarry. *Archives of Environmental Protection* 33(2): 58–70.

Porter M.C. (Editor), 1990. *Handbook of industrial membrane technology*. Park Ridge, NJ: Noyes Publication.

Prats D., Chillon-Arias M.F. & Rodriguez-Pastor M. 2000. Analysis of the influence of pH and pressure on the elimination of boron in reverse osmosis. *Desalination* 128(3): 269–273.

Qdaisa H.A. & Moussab H. 2004. Removal of heavy metals from wastewater by membrane processes: a comparative study. *Desalination* 164(2):105–110.

Qinâ J.J., Wai M.N., Oo M.H. & Wong F.S. 2002. A feasibility study on the treatment and recycling of a wastewater from metal plating. *Journal of Membrane Science* 208(1–2): 213–221.

Ozaki H., Sharmab K. & Saktaywirf W. 2002. Performance of an ultra-low-pressure reverse osmosis membrane (ULPROM) for separating heavy metal: effects of interference parameters. *Desalination* 144(1-3): 287–294.

Rudnicki P., Mioduszewski A., Kowalski Z. & Fela K. 2006. Technological conception of the treatment of alkaline wastewater collected in the "Górka" reservoir situated in Trzebinia commune area. *Polish Journal of Chemical Technology* 8(2): 26–33.

Żelazny S., Fela K., Jarosiński A. & Kowalski Z. 2006. Dump area reclamation with the use of bottom sludge suspension. *Polish Journal of Chemical Technology* 8(2): 48–53.

Environmental Engineering III – Pawłowski, Dudzińska & Pawłowski (eds)
© 2010 Taylor & Francis Group, London, ISBN 978-0-415-54882-3

Research on the mechanical durability and chemical stability of solidified hazardous waste

T.A. Marcinkowski & K.P. Banaszkiewicz

Institute of Environmental Protection Engineering, Wroclaw University of Technology, Wroclaw, Poland

ABSTRACT: This paper presents the results of studies on the solidification of galvanic sludge using Portland cement and mixtures of cement and fly ash from coal combustion. One aim was to use the maximum amount of sludge that could be effectively solidified by the binding material. As a threshold effectiveness of the process, a value of 75% of sludge in the total mass was assumed. The effectiveness of the processes was evaluated after 28 and 56 days of stabilisation, based on measurements of mechanical compression strength and chemical analysis of eluates from leaching tests performed according to the standard PN-EN 12457-4:2006.

Keywords: Galvanic sludge, neutralisation, fly ash, Portland cement.

1 INTRODUCTION

Solidification and stabilisation (S/S) processes rank among the best methods for the neutralisation of many types of hazardous mineral waste because of their low cost, the ready availability of materials and their effectiveness in binding numerous metals. Chemically stable solid substances are obtained by mixing waste with cement or cement-ash mixtures. Contaminants (mainly metals) are permanently bound as a result of sorption, precipitation and chemical incorporation of cement hydration products (Li et al. 2001). The result of the solidification/stabilisation process is a reduction in the mobility of many metal ions and improved mechanical properties (Srivastava et al. 2008, Marcinkowski & Banaszkiewicz 2008, Banaszkiewicz & Marcinkowski 2008). A definite disadvantage of solidification is the increased mass and volume of waste, because of the use of cement and fly ash as binding agents.

The most important parameter to consider when applying solidification technologies is the effectiveness of metal immobilisation in the binding mixture used. In order for such an assessment to be made, monoliths are subjected to contaminant leaching tests. One essential factor that has a significant effect on the course of metal extraction is the pH of the leaching liquid. The most popular test in the United States and in other countries is the TCLP (toxicity characteristic leaching procedure) test. This test simulates conditions of the co-disposal of 95% of municipal waste, together with 5% of industrial waste (Trezza & Ferraiuelo 2003, Omotoso et al. 1996, U.S. EPA 1990). According to the methodology of the TCLP test, extraction of toxic contaminants is performed with one of two acetate buffers (pH = 2.88 or 4.93), in the proportion liquid/solid phase = 20/1.

In Poland, and in the EU, the binding standard is a procedure described in standard PN-EN 12457-4:2006 (PN-EN, 2006). In this methodology, solidified waste material is subjected to 24 h extraction with distilled water, at the proportion liquid/solid phase = 10/1.

In the research presented here, galvanic sludge resulting from the neutralisation of wastewater from chrome, nickel and copper plating and bethanizing processes, was solidified and stabilised with mixtures of various proportions of Portland cement and fly ash. The monoliths produced were subjected to chemical and mechanical tests. Our studies aimed to characterise the changes in their physicochemical properties over time. Two identical series of samples were processed, and their physicochemical properties were examined after 28 and 56 days of stabilisation.

The effectiveness of the immobilisation of selected metals was evaluated based on an analysis of their concentration in eluates after 24 h extraction with distilled water (according to the procedure described in standard PN-EN 12457-4:2006).

2 MATERIAL AND METHODS

2.1 Materials

Materials used in the research – galvanic sludge (GS), fly ash (FA) and Portland cement (PC) – came from the region of Lower Silesia. Portland cement CEM 32.5 R was the main binder, being the basis of all solidifying mixtures. Fly ash, used as an additive, came from coal combustion in the local heat and power generating plant.

2.2 Characterisation of the compotents used

To determine the heavy metals content, samples of materials used were mineralised wet in the presence of nitric and hyperchloric acids. Leachability of toxic contaminants was determined based on tests according to the procedures described in the standard PN-EN-12457-4:2006. Quantitative analysis of metals (Cr, Cu, Zn and Ni) was performed using flame atomic adsorption with an ICP-AES Liberty 220 atomic emission spectrometer (Varian). The hydration of all materials used was determined by drying 20 g samples for 24 h at a temperature of 105°C.

2.3 Methodology of the solidification/ stabilisation process

Galvanic sludge was 'solidified' with pure Portland cement and with three cement-ash mixtures in the proportion PC:FA, 70:30, 50:50 and 40:60. The weight of sludge to binding mixture was 3.0 in all cases. All components were mixed mechanically in a Tecnotest B205/X5 cement mortar mixer. Sufficient tap water was added to achieve a good consistency. The resultant paste was placed in cylindrical steel forms with a diameter and height of 8 cm, and then thickened for 180 s on a vibrating table to remove air bubbles. These samples were stored for 28 or 56 days in a ST-6 B 60 thermostatic cabinet at 20°C. After the 28 or 56 day period of stabilisation, the monoliths were subjected to physicochemical examination, in order to evaluate their mechanical compression strength and the level of extraction of the selected metal ions.

2.4 Measurement of mechanical compression strenght

The compression strength of the samples was examined on the 28th and 56th days of stabilisation. In both cases, three samples of identical composition were crushed. The results presented in Table 2 and Figure 1 are average values.

2.5 Contaminant leaching test

The effectiveness of the binding of the selected metal ions was evaluated based on their concentration in eluates after extracting a sample of material with a granulation <10 mm (according to the procedure described in standard PN-EN 12457-4:2006). Quantitative analysis of metals (Cr, Cu, Zn and Ni) was performed using flame atomic adsorption with an ICP-AES Liberty 220 atomic emission spectrometer (Varian). All concentrations presented in the tables are average values from analyses performed on three identical samples. Results were used to calculate the effectivity of immobilisation (*EoI*) of Cr, Cu, Zn and Ni (SI), according to Equation (1) (U.S. EPA 1992a, b):

$$EoI = \left[1 - (1 + PoA) \cdot \frac{C_{solidif.}}{C_{raw}}\right] \cdot 100\% \qquad (1)$$

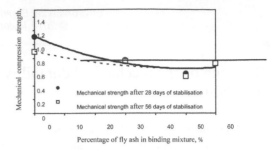

Figure 1. Mechanical compression strength of solidified galvanic sludge after 28 and 56 days of stabilisation.

$$PoA = \frac{\text{Mass of additives}}{\text{Mass of waste}}$$

where *PoA* = percentage of additives; $C_{solidif}$ = concentration in solidified waste; C_{raw} = concentration in raw waste.

3 RESULTS

3.1 Characteristics of the components used

Table 1 presents the chemical composition of each component and results of analyses of the concentration of selected metal ions in eluates from tests performed according to the standard PN-EN 12457-4:2006.

Sludges subjected to neutralisation were characterised by high levels of hydration (75.28%) and by high concentrations of Cr, Cu, Zn and Ni. Evaluation of their toxicity, performed by analysing the leachate from the PN-EN test, showed that for all the metals tested, maximum concentrations exceeded established levels for treated hazardous waste permitted for disposal at landfills assigned for waste other than inert and hazardous (LAW GAZETTE 2005; European Community 2003). Levels of none of the metals tested exceeded acceptable levels in the components of the binding mixtures. Hydration of Portland cement (CEM I 32.5 R) and of fly ash from coal combustion was 1.58 and 0.45%, respectively.

3.2 Mechanical compression strength

Mechanical tests confirmed that mechanical strength decreased as the percentage of ash in the binding mixture increased.

More water was needed in samples containing 100% or 50% of cement, due to poor workability. More water positively affects workability of the mixture, but unfortunately negatively affects the mechanical properties of cement mortars (Jasiczak & Mikołajczak 2003). This might be one reason for the deterioration in the mechanical parameters of samples solidified with cement-ash mixtures of PC:FA = 50:50, compared with samples where a more ash-rich mixture was used (40:60).

Table 1. Concentrations of selected metals in materials applied and in eluates from the tests performed according to the standard PN-EN 12457-4:2006.

| Parameter | Chemical composition of material | | | Chemical analysis of eluate from PN-EN test | | | |
	Sludge mg/kg$_{dm}$	Fly ash mg/kg$_{dm}$	Cement mg/kg$_{dm}$	Sludge mg/dm^3	Fly ash mg/dm^3	Cement mg/dm^3	Limit values mg/dm^3
Cr	78 075	103.83	39.42	2.1	0.28	0.21	1
Cu	54 712	197.68	22.82	31.8	BDL[a]	0.005	5
Zn	37 939	375.20	107.88	655	0.17	0.041	5
Ni	2 196	155.75	838.17	21.6	0.32	0.01	1
pH of eluate				5.46	11.41	13.27	>6

[a] BDL – below detection limit.

Table 2. Mechanical compression strength of solidified galvanic sludge after 28 and 56 days of stabilisation.

| Sample[a] | Index W/PC | Mechanical strength | |
		on 28th day of stabilisation MPa	on 56th day of stabilisation
GS:PC 3:1	2.76	1.11	0.892
GS:PCFA 3:1 70:30	2.68	0.765	0.750
GS:PCFA 3:1 50:50	2.76	0.567	0.534
GS:PCFA 3:1 40:60	2.68	0.708	0.722
Acceptable values		>0.05	

[a]GS   – galvanic sludge.
PC   – Portland cement.
FA   – fly ash from pit-coal combustion.
3:1   – ratio of sludge to binding mixture.
70:30   – ratio of cement to fly ash in binding mixture.

Table 3. Concentration of selected metals in eluates from tests performed after 28 days of maturation, according to the standard PN-EN 12457-4:2006.

Sample[a]	pH of eluate	Cr mg/dm^3	Cu mg/dm^3	Zn mg/dm^3	Ni mg/dm^3
GS:PC 3:1	11.1	5.08	0.73	0.009	0.03
GS:PCFA 3:1 70:30	10.71	2.00	1.53	0.009	0.03
GS:PCFA 3:1 50:50	10.52	1.35	0.29	0.009	0.05
GS:PCFA 3:1 40:60	9.51	0.32	0.34	0.009	0.03
Acceptable values	>6	1	5	5	1

[a]GS   – galvanic sludge.
PC   – Portland cement.
FA   – fly ash from pit-coal combustion.
3:1   – ratio of sludge to binding mixture.
70:30   – ratio of cement to fly ash in binding mixture.

Some deterioration in mechanical properties was observed in three samples of solidified galvanic sludge crushed after 56 days of stabilisation. In a sample that was solidified with pure cement (GS:PC 3:1) the decrease in mechanical strength was about 20%, compared with the value observed on the 28th day of stabilisation. In the case of the remaining samples, GS:PCFA 3:1 50:50 and GS:PCFA 3:1 70:30, the difference was not so significant, ranging from 2% to 6%. In one sample (GS:PCFA 3:1 40:60), strength was observed to have increased (by about 2%) on the 56th day of maturation.

One reason for the deterioration in mechanical parameters of samples examined after 56 days of stabilisation might be that too much water and too little cement was used. Excess water that was not chemically bound to the cement could have evaporated and contributed to the contraction phenomenon in samples characterised by a longer time of homogenisation. Another cause of deteriorated mechanical compression strength was the phenomenon of sulphatic corrosion. Products of sulphates are gypsum and ettringite (also called Candlot salt). Generation of gypsum causes an increase in the volume of the sample, resulting in increased strain and fracture of the solidifying mixture.

All stabilizates met the requirements concerning minimum axial compression strength, established in the Directive of the Economics and Work Ministry (LAW GAZETTE 2005) at the level of ≥0.05 MPa.

3.3 Extraction of selected metal ions

Results of 24 h extraction of solidified sludge on the 28th and on the 56th day of stabilisation are presented in Tables 3 and 4, and in Figures 2 and 3. Examination confirmed the direct effect of the presence of ash in the binding mixture on changes in the pH of the eluates. A decrease in the pH of eluates was observed as the percentage of fly ash in binding mixture increased (Li et al. 2001). Liquids extracted from samples after 28 days of maturation had a pH in the range of 9.51 to 11.1. For solidified sludge analysed on the 56th day of maturation, the pH of liquids extracted with distilled water ranged from 9.11 to 11.55.

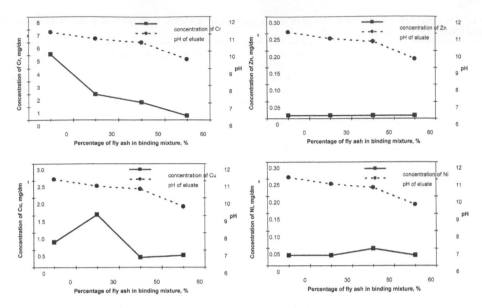

Figure 2. Effect of pH on the leachability of selected metals from solidified galvanic sludge after 28 days of stabilisation (liquid/solid phase: 10; time of contact: 24 h; temperature: 19.5°C).

Table 4. Concentration of selected metals in eluates from tests performed after 56 days of maturation, according to the standard PN-EN 12457-4:2006.

Sample[a]	pH of eluate	Cr mg/dm³	Cu mg/dm³	Zn mg/dm³	Ni mg/dm³
GS:PC 3:1	11.55	4.11	1.49	0.009	0.03
GS:PCFA 3:1 70:30	10.64	2.94	0.871	0.009	0.03
GS:PCFA 3/1 50:50	9.54	1.41	0.6	0.047	0.03
GS:PCFA 3/1 40:60	9.11	1.36	0.3	0.009	0.03
Acceptable values	>6	1	5	5	1

[a] GS   – galvanic sludge.
PC   – Portland cement.
FA   – fly ash from pit-coal combustion.
3:1   – ratio of sludge to binding mixture.
70:30 – ratio of cement to fly ash in binding mixture.

Chemical analysis of extracts showed that the cement-ash mixtures used proved to be poor immobilisers of chromium. Only in one sample, GS:PCFA 3:1 40:60 after 28 days of maturation, did the level of Cr in the eluate not exceed the maximum value (1 mg/dm³), established for treated hazardous waste permitted for disposal at landfills assigned for depositing waste other than inert and hazardous (LAW GAZETTE 2005, European Community 2003). In a sample of identical composition, examined after 56 days of stabilisation, the level of chromium in the eluate according to the PN-EN test, increased by more than 1 mg/dm³ and exceeded the value established by Polish and European regulations. The highest concentrations of Cr (4–5 mg/dm³) were observed in leaching liquids from samples solidified with pure Portland cement. Chromium hydroxide proved to be very soluble under alkaline conditions. Examination confirmed the positive effects of an increase in the percentage of fly ash in the binding mixture (the pH value decreased from 11.55 to 9.11), leading to a decrease in the solubility of chromium compounds.

The levels of copper in the eluates were relatively low. The acceptable value of 5 mg/dm³ (Law Gazette 2005, European Community 2003) was not exceeded in any of the eluates. Analysis of liquids extracted from samples stabilised for 56 days showed a slight decrease in the level of binding of copper over time. Reduction in the pH of the extraction environment caused by an increase in the percentage of fly ash in the binding mixture had a favourable effect on the immobilisation of copper. With a decrease in pH value in the range examined (from 12 to 9), there was about an 80% decrease in Cu concentration in the eluates. The effectivity of Cu immobilisation (Table 5), calculated according to Equation (1) ranged from 93% to more than 98%. This confirmed the poor solubility of copper at pH 9 to 10.5 (Marcinkowski & Banaszkiewicz 2008, Banaszkiewicz & Marcinkowski 2008).

Studies of the levels of zinc and nickel in the post-extraction fluids, showed that the binding mixtures used were very effective at the pH range studied (pH 9–12). The degree of immobilisation of Zn and Ni did not change over time. The effectiveness of the immobilisation of Zn and Ni was more than 99.8%. This confirms the poor solubility of Zn and Ni in an alkaline environment (Marcinkowski & Banaszkiewicz 2008, Banaszkiewicz & Marcinkowski 2008, Asavapisit et al. 2005, Chang et al. 2001).

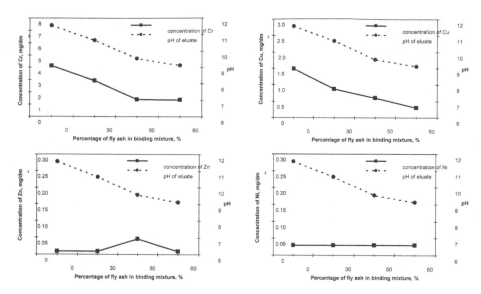

Figure 3. Effect of pH on the leachability of selected metals from solidified galvanic sludge after 56 days of stabilisation (liquid/solid phase: 10; time of contact: 24 h; temperature: 19.5°C).

Table 5. Effectiveness of immobilisation of selected metal ions in binding mixtures.

Sample[a]	pH value after 28 d	pH value after 56 d	Cr after 28 d	Cr after 56 d	Cu after 28 d	Cu after 56 d	Zn after 28 d	Zn after 56 d	Ni after 28 d	Ni after 56 d
		-		%		%		%		%
GS:PC 3:1	11.1	11.55	−222.540	−160.952	96.939	93.753	99.998	99.998	99.815	99.815
GS:PCFA 3:1 70:30	10.71	10.64	−26.984	−86.667	93.585	96.348	99.998	99.998	99.815	99.815
GS:PCFA 3:1 50:50	10.52	9.54	14.286	10.476	98.784	97.484	99.998	99.990	99.691	99.815
GS/PCFA 3:1 40:60	9.51	9.11	79.683	13.651	98.574	98.742	99.998	99.998	99.815	99.815

[a]GS   – galvanic sludge.
PC   – Portland cement.
FA   – fly ash from pit-coal combustion.
3:1   – ratio of sludge to binding mixture.
70:30 – ratio of cement to fly ash in binding mixture.

4 CONCLUSIONS

Galvanic sludges are classified as hazardous waste and in their raw form cannot be deposited in the environment. Satisfactory immobilisation of toxic metals present in such sludges can be achieved by the well-known technique of solidification using Portland cement and/or fly ash from coal combustion.

Our research showed that some heavy metals could be very effectively bound in cement-ash mixtures. In no sample was the effectiveness of immobilisation of Cu, Zn and Ni below 93%, 99.99% and 99.6%. The level of extraction of contaminants from solidified galvanic sludge depends on their concentration in the waste being neutralised and on the pH of the leaching liquid. Our experiments confirmed the poor solubility

of Cu, as well as the very poor solubility of Zn and Ni in an alkaline environment (pH from 9 to 12). It was shown the physicochemical properties of monoliths might deteriorate over time. Chemical analysis of eluates using the PN-EN test revealed an enhanced extraction of Cr and Cu from some samples examined after 56 days of stabilisation. An increased proportion of fly ash in the binding mixture caused a reduction in the pH value of the immobilisation environment that had a positive effect on immobilising metals that create hydroxides that are only slightly soluble at neutral pH.

Acid contamination of the atmosphere should be considered a serious threat to the durability of solidified waste. The influence of such contaminants on construction materials used in S/S processes is widely known. Too much fly ash has a direct effect in reducing pH, as well as long-term effects on environmental factors, e.g. acid falls, which can also significantly shorten durability of monoliths. Long-term effect of an acidic environment will also cause the faster depletion of the cement-ash matrix and will accelerate leaching of metals that were bound earlier.

Research on the mechanical strength of samples showed deteriorated values after 56 days of seasoning. The reason for such a phenomenon was the use of too much water and too little Portland cement. Excess water not bound chemically with cement, evaporated rapidly, causing mechanical contraction. The phenomenon of sulphatic corrosion caused by the presence of sulphates in galvanic sludges had an additional effect on the deterioration in the mechanical properties of solidified samples. This in turn caused the formation of gypsum and ettringite (also called Candlot salt). Gypsum also increased the volume of the samples, causing an increase in mechanical strain and resulting in the fracture (and destruction) of samples of solidified waste.

ACKNOWLEDGEMENTS

This work was supported financially by a European grant carried out at Wroclaw University of Technology: "Detectors and sensors for measuring factors hazardous to environment – modelling and monitoring of threats", No. POIG.01.03.01-02-002/08.

REFERENCES

Asavapisit, S., Naksrichum, S. & Harnwajanawong, N. 2005. Strength, leachability and microstructure characteristic of cement-based solidified plating sludge. *Cement Concrete Research* 35: 1042–1049.

Banaszkiewicz, K.P. & Marcinkowski, T.A. 2008. Evaluation of usefulness of hydrated lime in process of stabilization of sludges from electrocoating. *Environment Protection Engineering* 34(2): 115–124.

Chang, E.E., Chiang, P.C., Lu, P.H. & Ko, Y.W. 2001. Comparisons of metal leachability for various wastes by extraction and leaching methods., *Chemosphere* 45: pp. 91–99.

EUROPEAN COMMUNITY. 2003. Council Decision 2003/33/EC establishing criteria and procedures for the acceptance of waste at landfills pursuant to article 16 of and Annex II to Directive 1999/31/EC. *Official Journal of European Communities L11*.

Jasiczak, J. & Mikołajczak, P. 2003. Technology of Concrete Modified With Admixtures and Additions. Poznań University of Technology, Alma Mater.

Li, X.D., Poon, C.S., Sun, H., Sun, I.M.C. & Kirk, D.W. 2001. Heavy metal speciation and leaching behaviors in cement based solidified/stabilized waste materials. *Journal of Hazardous Materials* A82: 215–230.

LAW GAZETTE. 2005. No. 186, Item 1553; Directive of the Minister of Economics on 7 September 2005, on criteria and procedures of qualifying waste for deposition on specific types of landfills.

Marcinkowski T.A. & Banaszkiewicz K.P. 2008. Efficiencyof Chromium, Copper, Zinc and Nickel Ions Immobilization During Stabilization/Solidification of Electroplating Sludge *Ochrona Środowiska* 4: 53–56.

Omotoso, O.E., Ivey, D.G. & Mikula, R. 1996. Quantitative X-ray diffraction analysis of chromium (III) doped tricalcium silicate pastes. *Cement and Concrete Research* 26(9): 1369–1379.

PN-EN 12457-4:2006. Polish Standard. Characterization of waste. Elution. Consistency checking in respect of leaching granular waste and sludge, Part 4: One-step batch test at the proportion of liquid/solid phase 10 l/kg in case of materials having a granularity below 10 mm (with or without size reduction).

Srivastava, S., Chaudhary, R. & Khale, D. 2008. Influence of pH, curing time and environmental stress on the immobilization of hazardous waste using activated fly ash. *Journal of Hazardous Materials* 153: 1103–1109.

Trezza, M.A. & Ferraiuelo, M.F. 2003. Hydration study of limestone blended cement in the presence of hazardous wastes containing Cr(VI). *Cement and Concrete Research* 33: 1039–1045.

U.S. EPA. 1990. Toxicity Characteristic Leaching Procedure. *Federal Register* 55(61): 11798–11877.

U.S. EPA. 1992a. EPA/540/AR-92/010. Silicate Technology Corporation's. Solidification/stabilization Technology for Organic and Inorganic Contaminants in soils. In: *Applications Analysis Report*. Risk Reduction Engineering Laboratory. Office of Research and Development. U.S. Environmental Protection Agency. Cincinnati, OH 45268.

U.S. EPA. 1992b. EPA/600/K-92003A. Seminar on the Use of Treatability Guidelines in Site Remediation. Stabilization/Solidification. Office of Research and Development. Washington, DC 20460.

Environmental Engineering III – Pawłowski, Dudzińska & Pawłowski (eds)
© 2010 Taylor & Francis Group, London, ISBN 978-0-415-54882-3

Efficiency of microbiological oxidation of methane in biofilter

M. Pawłowska

Faculty of Environmental Engineering, Lublin University of Technology, Lublin, Poland

ABSTRACT: The efficiency of the microbiological oxidation of methane CH_4 is known to depend on the type of material forming the deposit in which the process takes place, since this influences conditions as regards the supply of water and air, nutrient availability and the size of the surface area available to be colonised by bacteria. The work presented here has therefore assessed efficiency of methane oxidation in three different types of medium containing material of waste origin (compost from municipal refuse and glass cullet), as well as ceramsite in the role of a mineral skeleton. The biofilter bed consisting of a 4:1 mixture of ceramsite and compost was found to sustain the greatest CH_4 oxidising capacity, equal to 527.3–$534.3\,dm^3\,m^{-2}\,d^{-1}$. In comparison, a bed comprising compost alone was only 40% as efficient, while one made up of glass cullet plus organic horticultural substrate was only 67% as efficient.

Keywords: Oxidation of methane, biofiltration, compost, ceramsite, glass cullet, landfill.

1 INTRODUCTION

When making its way through the aerated upper layers of waste or porous material representing biological filter bed, methane arising through the fermentation of the organic fraction of wastes dumped at landfill sites undergoes microbiological oxidation to carbon dioxide and water. However, depending on a landfill's size and level of organisation, as well as its age and the type of waste dumped, the form and dimensions of the biological deposit referred to may vary greatly.

While at small, old landfills, the deposit may be an overlayer (i.e. one placed directly above waste, covering the entire surface of the site), new landfills do not have this kind of deposit, instead complying with the regulations in force by virtue of their being insulated from the environment once exploitation has ceased. The insulation in question involves a seal of PEHD plastic. In such cases, the methanotrophic filter bed may assume the form of a biofilter (separated cell filled with material constituting the deposit), a bio-window (fragment of cover with removed PEHD sheeting filled with porous material that is colonised by methanotrophic bacteria), or a modified landfill cover (in which the layer of filter bed is placed above the artificial insulation, with the biogas being directed to it via an appropriate system of pipes).

The oxidation of methane achieved by methanotrophic bacteria is a process taking place under various different habitat conditions. It is to be observed in both terrestrial and aquatic ecosystems, be these natural or anthropogenic; in soils used in various ways, e.g. in cultivated fields (Willison et al. 1995, Boeckx et al. 1997), in forests (Suwanwaree & Robertson 2005, Jang et al. 2006), on meadows (Willison et al.

1995; Boeckx et al. 1997, in marshy areas (Whalen & Reeburgh 2000, Kravchenko 2002), or in the upper parts of landfills (Kightley et al. 1995, Bogner et al. 1997, Kallistova et al. 2005).

Methanotrophic bacteria are used to limit emissions of methane from such anthropogenic sources of the gas as landfill sites, wastewater treatment plants and methane-generating coalmines. Research shows that the bacteria capable of using this methane as a source of carbon and hence energy are a large group of aerobes, primarily microaerophilic species (i.e. ones displaying a preference for O_2 concentrations below those in the atmosphere – Mancinelli (1995). Such bacteria develop best where the substrate has a pH of 6–8 (King 1990, Bender & Conrad 1995, Arif et al. 1996, Min et al. 2002), and where the temperature is in the range 25–35°C (Whalen et al. 1990, Boeckx & Van Cleemput 1996, Reay et al. 2001, Min et al. 2002).

However, it is also possible to observe methane oxidation taking place in environments markedly less suitable than the above optimal conditions for methanotroph growth and development, e.g. at pH values of around 1 (Pol et al. 2007) or even 11 (Kaluzhnaya et al. 2001), and at temperatures in excess of 55 degrees (Trotsenko & Khmelenina 2002, Jaeckel et al. 2005, Islam et al. 2008). In fact, the latest discoveries are making it clear that the methanotrophs are a much more diverse group from the taxonomic, ecological and genetic points of view than had been suspected earlier. Bacteria were recently found living in the boiling geothermal waters of the Hell's Gate Reserve in New Zealand (Dunfield et al. 2007), in the hot springs of Kamchatka (species *Methyloacida kamchatkensis* – Islam et al. 2008), and in the soils and muds around the volcanoes in southern Italy (*Acidimethylosilex fumarolicum* (Pol et al. 2007), this being assigned to the *Verrucomicrobia* (rather than the *Proteobacteria*, as had been the case with all the methanotrophs described previously).

The efficiency of the process of methane oxidation is further determined by air and water conditions prevailing in the zone of oxygenation, by the availability of biogenic compounds, and by the surface area available to be colonised by bacteria. Most of the parameters influencing the rapidity of the process are linked directly with the type of matter in which the methanotrophic bacteria are developing. Thus, the most important properties of the deposit conditioning the rate of oxidation are its porosity (affecting both the diffusion of gases and the capacity to retain water, in relation to the grain size of the material), reaction, and the content of nutrients and substances toxic to bacteria.

The subject literature describes different kinds of infill of methanotrophic biofilters or materials constituting the overlying layers of landfill sites. These are most often composts produced from green waste, refuge or sewage sludge, these displaying the greatest oxidising capacity (Streese & Stegmann 2003, Wilshusen et al. 2004, Powelson et al. 2006, Abichou et al. 2009; Philopoulos et al. 2009). As material containing large amounts of organic matter, composts are characterised by a large specific surface, considerable buffering capacity, and a neutral or slightly alkaline reaction. They are also sources of the nutrients required by bacteria. However, a drawback is a very high capacity to retain water, this resulting in less favourable properties as regards hydration and aeration of the deposit in the course of its work, as well as a consequent vulnerability to biochemical degradation that changes the structure of the substrate over time. Such changes ongoing in compost ensure that the oxidising capacity of the deposit declines markedly in the course of its exploitation. For example, Streese and Stegmann (2003) noted how the rate of oxidation of methane in a compost layer was of $1314\,dm^3\,m^{-2}\,d^{-1}$ in the preliminary phase of their experiment, this declining within the space of 7 months to just $320\,dm^3\,m^{-2}\,d^{-1}$.

In the experiment whose results were described in the study it was assumed that the use of ceramsite expanded clay pellets or glass cullet would offer a mineral skeleton resistant to microbiological decay and thus capable of improving the properties of the biofilter bed as regards aeration and water content capacity. As a coarse-grained product, ceramsite provides for the continued presence of the numerous, large-diameter pores indispensable if easy permeation of gases through the biofilter bed is to be assured. Furthermore, the interiors of the granules themselves also support large numbers of small pores, this greatly increasing the specific surface of the material, and increasing its water holding capacity.

The organic materials used in the work described here were a compost derived from municipal refuse, as well as an organic horticultural substrate called *Pokon*. Both of these contain nutrients necessary for the growth of methanotrophic bacteria, and they also serve to increase water retention in the biofilter bed, as well as the specific surface area of the material.

The aim of the research was to determine which of the fillings under study would be best able to support methane removal by oxidation from a gas mixture simulating landfill gas. The results of the research were the basis for the selection of material to be used in the actual filling of a biofilter installed at a landfill, at the time of carrying out of a pilot-scale experiment.

2 METHODS

2.1 *The experimental setup*

Research on the efficiency of the microbiological oxidation of methane was carried out on the laboratory scale in two stages:

Stage I – model dynamic studies run in 6 columns, through which a 1:1 mixture of methane and carbon dioxide simulating the composition of landfill gas was passed for 5 months. The rate of flow of gas through any given column was of $20\,cm^3/min$, this translating into a methane load of $815\,dm^3\,m^{-2}\,d^{-1}$ (i.e. a value around twice as great as the mean given in relevant literature for the oxidation of CH_4, as measured in composts). Continuity and constancy of composition of gas-mixture discharge into the columns was assured through the use of 8 digital gas flow controllers from Brooks Instruments. In the course of the work, changes in the concentrations of CH_4, CO_2 and O_2 down the profiles of the columns were analysed, the efficiency of the oxidation process being assessed through the measurement of CH_4 emissions from the surface of the column fillings, in relation to differences in the temperature down the profile.

Stage II – kinetic studies performed on the basis of a static layout in glass vials, the aim being to determine the kinetic parameters underpinning the oxidation of methane in materials taken from columns at the end of Stage I.

The experimental layout comprised 6 columns made of Plexiglas (of height 1 m and diameter 15 cm), in which the studied material was placed with two replicates for each kind. A diagram showing the structure of one of the columns is presented in Fig. 1. At the base of each column, on opposite sides were valves *via* which the gas mixture was supplied to the column. At 10 cm intervals down the columns there were apertures closed off fully with rubber stoppers, through which samples of gas could be taken as necessary. The lower part of each column was filled with free-draining material (coarse gravel of 30–50 mm diameter), allowing for even access of the gas introduced to the whole of the filling, as well as free runoff of excess amounts of water from the column. Between the gravel layer and the layer of studied biofilter bed a plastic plate was placed, this having been drilled with holes 5 mm in diameter. Above this was nylon netting to ensure that material from the filter bed did not reach the free-draining layer.

During the period of research the columns were not closed off from above, so air could diffuse constantly into the layer of filling, as this was required for the development of methanotrophic bacteria as obligate aerobes. Closure of the columns (with the aid of lids with gas taking ports) only occurred at times when measurements of the capacity of the material layer to oxidise methane were being made, the space above the filling and below the hole in the lid then serving as the gas sampling site. For safety reasons, the experimental setup was located for the entire duration of the experiment beneath a mechanical extractor system.

2.2 Characteristic of the materials serving as the methanotrophic beds

The fillings for the experimental columns were:

1) compost made from refuse produced by KOM-EKO Sp. z o.o. in Lublin, sieved through a 1 cm mesh,

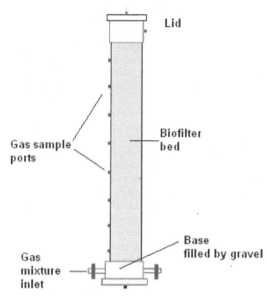

Figure 1. Structure of a single column.

2) a mixture of 8–12 mm garden ceramsite and compost (of the above origin) at a ratio of 2:1,
3) a mixture of (<2 mm) glass cullet from KOM-EKO Sp. z o.o. in Lublin, plus POKON organic horticultural substrate, at a ratio of 1:1.

The materials used in preparing the methanotrophic deposits are presented in Photo 1.

2.3 The determination of gas – concentration profiles

Analysis of the gas collected through the membranes distributed along the columns allowed for the determination of changes in CH_4, CO_2 and O_2 concentrations along the profiles in the different columns. Samples of gas of volume 100 μl were taken with the aid of a Hamilton gastight syringe, then being subjected to analysis using a Shimadzu 14B gas chromatograph with packed columns equipped with a thermal conductivity detector. Determinations for CH_4 and CO_2 were made using a 3 m glass column of inner diameter 3.2 mm, containing Porapak Q.

Oxygen concentrations were in turn determined using a 3 m column of 3.2 mm inner diameter filled with molecular sieve 5A. The parameters for the gas chromatography were: column temperature 40°C, injector 40°C, detector 60°C, and current bridge 150 mA. The carrier gas was helium at a flow rate of 40 cm³/min. Calculations of the areas of peaks involved the 1.0b version of the CHROMAX 2007 computer integration program.

2.4 Assessment of the efficiency of methane oxidation in the layer

The increase over time in methane concentrations – as measured in the space above the column fillings following their closure from above – provided the basis for the determination of methane emissions from the surface of the filling (following passage through the layer of material). During measurement – a process lasting 20 minutes – a thick needle was placed in the

Photo 1. Materials used in the research to fill the biofilters: a) ceramsite mixed with compost, b) municipal refuse compost, c) glass cullet.

Table 1. Oxidising capacity and efficiency of CH_4 removal in the studied material as measured in the 5th month of the dynamic experiment.

	Compost	Ceramsite + compost	Glass cullet + organic horticultural substrate
Oxidising capacity [$dm^3 \, m^{-2} \, d^{-1}$]	181.31–200.49	527.31–534.27	283.71–369.33
Efficiency [% of oxidised CH_4]	22.3–24.6	64.7–65.6	34.8–45.3

cap to serve in the equalisation of pressure within the measuring chamber. Samples of gas were taken every 5 minutes, and the increase in the CH_4 concentration over time analysed. The value for the increase, combined with knowledge of the volume in which the gas had accumulated plus the cross-sectional area allowed for calculation of the CH_4 emission from the surface of the material. Comparison of this value with that for the rate of inflow of CH_4 into the column served as the basis for calculating the CH_4 oxidising capacity of the given filling, as well as a percentage figure for oxidation of CH_4 in the layer in relation to the amount of the gas introduced into the column (i.e. oxidation efficiency). Measurements were made at 2-3-week intervals.

2.5 Determination of temperature profiles in the deposit layers

Changes in temperature down the column profile were analysed with the aid of a Therma CAM™ E45 thermo-imaging camera from the firm FLIR. The camera's operating parameters were: temperature of surroundings 23°, emissivity 0.9. Measurements were made through the column wall, from a distance of c. 50 cm, following removal of the aluminium foil.

2.6 Assessment of the influence of CO_2 on methanotrpohic activity

2 g samples of material were taken from 10 cm down in one of the columns filled with the cullet-compost mixture and placed in 40 ml vials. The vials were closed tightly shut with the aid of rubber stoppers, before CO_2 was injected in different amounts to ensure initial concentrations of the gas equal to 1.5, 6, 13, 15 and 19%. There were three replicates for each initial concentration of CO_2. Each vial was then injected with CH_4 until a 12% initial concentration was obtained, methanotrophic activity being calculated by reference to the rate of decline in CH_4 concentrations in the vials.

3 RESULTS

3.1 Efficiency of methane oxidation in the layer of material

Five months into the experiment, the greatest oxidising capacity was to be noted for the mixture of ceramsite with compost, the most limited for compost alone. The oxidising capacities in the different materials (expressed in terms of the dm^3 amount of CH_4 oxidised by the 80 cm layer of filling with a given surface area per unit time (day)) were as shown in Table 1. The rates measured in the ceramsite-compost filling were more than 2.5 times those in compost alone, as well as ca. 1.5 times as great as those for the mixture of glass cullet and organic horticultural substrate.

The greatest capacity to oxidise methane – expressed as the amount of CH_4 oxidised as a percentage of the amount entering a given column – was the figure of 65% noted for the mixture of ceramsite and compost. This is a very high efficiency bearing in mind the fact that the loads of CH_4 supplied into the columns were twice as large as the potential oxidising capacity for the gas given in relevant literature for the most active compost fillings. Studying the rate of methane oxidation in an artificially aerated biofilter filled with compost, Wilshusen et al. (2004) obtained a value of $590 \, dm^3 \, m^{-2} \, d^{-1}$ of CH_4, this being maintained at a relatively stable level for around 80 days, beginning some 4 weeks into their experiment. In the later phase, the rate declined to $150 \, dm^3 \, m^{-2} \, d^{-1}$. In turn, Powelson et al. (2006), whose deposit was compost from refuse mixed 1:1 by volume with polystyrene spheres, obtained a maximum rate of oxidation of CH_4 equal to $640 \, dm^3 \, m^{-2} \, d^{-1}$. In described experiment, in which the biofilter used a mixture of ceramsite with compost in the ratio 4:1, the rate of oxidation equal to ca. $530 \, dm^3 \, m^{-2} \, d^{-1}$ was even maintained 5 months into the experiment, this attesting to the very favourable properties of the proposed biofilter filling.

3.2 Gas-concentration profiles in the columns

Irrespective of the type of filling studied, concentrations of CH_4 in it were progressively higher further down the profile. Figure 2 shows the profiles for CH_4 concentrations to be observed at the end of the experiment.

A more marked increase in CH_4 concentration expressed per unit of profile depth characterised the upper part of the column – down to a depth of 30 cm. Below this, concentrations of methane were characterised by only a slight further increase, this suggesting that the zone of most intensive oxidation of the gas – in all the studied materials – was in the upper part of the columns. It was the ceramsite-compost filling that had the lowest average value for the CH_4 concentration throughout the profile, equal to 35.6%. The largest average value was the 37.4% noted for the compost filling, the intermediate value of 36.7% being noted for the mix of glass cullet and organic horticultural substrate.

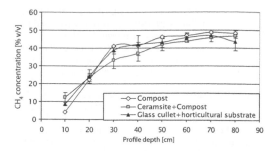

Figure 2. Comparison of CH$_4$ concentration profiles with the different fillings (after 5 months of incubation).

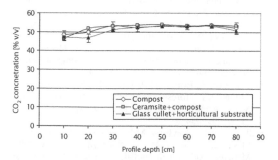

Figure 3. A comparison of the CO$_2$ concentration profiles in the studied types of filling (5 months into the experiment).

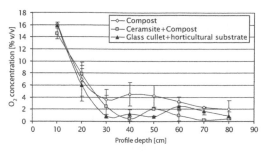

Figure 4. Comparison of profiles for oxygen concentrations in the three types of filling beds (5 months into the experiment).

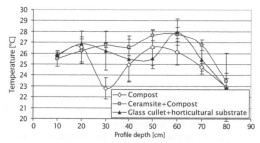

Figure 5. Temperature differences down column profiles (5 months into the experiment).

Concentrations of CO$_2$ are relatively stable in the profiles with no statistically significant differences being noted between columns at given depths (Figure 3). The mean CO$_2$ concentrations were in fact of 52.01% in the compost, 52.44% in the ceramsite-compost filling, and 51.00% in the mixture of glass cullet and gardening mulch. Nevertheless, the analysis of changes in CO$_2$ concentrations could not serve as any basis for drawing conclusions regarding methanotroph activity. While CO$_2$ is an oxidation product, it is also generated in the course of mineralisation of the kind of organic matter present in abundance in the experimental fillings, as well as being a product of respiration among microorganisms colonising the substrate.

Statistically significant differences ($\alpha = 0,05$) in the distribution of O$_2$ concentrations down the profile were to be observed in the columns containing compost. It was here that oxygen managed to diffuse deepest (Figure 4), this despite the high moisture level of the material, and its lower porosity than ceramsite-compost mixture. The cause of such a distribution for O$_2$ concentrations through the compost profile may have been the more limited use of oxygen in the methane-oxidation process. This suggests that the process in question was here limited by some other factor, probably excessive moisture content resulting from high water holding capacity of the compost.

The two remaining column fillings were characterised by more intensive utilisation of oxygen. In the case of the ceramsite-compost filling, the lower part of the profile featured trace quantities of oxygen only.

There was a strong inverse correlation between concentrations of O$_2$ and CH$_4$ in the cases of all the materials studied. Values of correlation coefficients were -0.96012 for compost; -0.9507 for the mixture of ceramsite and compost and -0.94026 for the mixture of glass cullet and horticultural substrate.

3.3 Determination of temperature profiles in the bed layer

Differences in temperature through a layer of examined material attest to ongoing exothermic processes in the column. One such process releasing heat energy is the oxidation of methane, hence the supposition that higher temperatures can indicate the zones of most intensive oxidation. In fact, 2 peaks of temperature were to be observed down profiles of each filling material, though the first peak is rather indistinct in the case of the ceramsite-compost mixture (Figure 5).

After 5 months of incubation, it was the mean temperature of the ceramsite filling that was highest, at 26.3°C. The lowest mean temperature was in turn the 25.3°C noted in compost. The temperature in the cullet-horticultural substrate mixture was intermediate between the other two, at an average of 25.9°C.

3.4 Assessment of the influence of CO$_2$ concentrations on methanotrophic activity

Carbon dioxide is always present in landfill gas, and it also arises as CH$_4$ is oxidised. As a compound easily soluble in water as an acidic anhydride, it

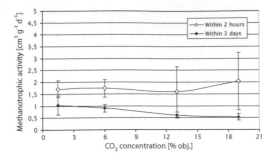

Figure 6. The influence of CO_2 concentration on methanotrophic activity in a mixture of glass cullet and an organic substrate.

acts to modify pH in the substratum, thus potentially influencing the rate of CH_4 oxidation.

Figure 6. presents the influence of CO_2 concentration on methanotrophic activity in the mixture with glass cullet.

In the course of the first two hours following closure, vials were characterised by very variable sample-to-sample activity levels (yielding very high standard deviations). In consequence, it did not prove possible to assess the influence of CO_2 concentrations on the process by which CH_4 is oxidised. An inhibitory influence of CO_2 only manifested itself after a longer period of incubation of the methanotrophs in a CO_2-rich environment. A visible decline in methanotrophic activity under the influence of higher CO_2 concentrations was to be observed following three-day incubation of samples under different carbon dioxide concentrations. This probably links up with a delayed response on the part of bacteria to changes in conditions in the substrate brought on by the dissolving of CO_2 in water contained in the sample. Successively higher levels of CO_2 across an 8-fold range (from 2.5–19%) were found to be associated with an approximate halving of methanotrophic activity.

ACKNOWLEDGEMENTS

This research was financed from a grant for the implementation of commissioned research project MEiN No. 3/2/2006 entitled: "The engineering of emission abatement processes and the utilisation of harmful and greenhouse gases".

REFERENCES

Abichou, T., Mahieu, K., Yuan, L., Chanton, J. & Hater, G. 2009. Effects of compost biocovers on gas flow and methane oxidation in a landfill cover. *Waste Management* 29(5): 1595–1601.

Arif, M.A.S., Houwen, F. & Verstraete, W. 1996. Agricultural Factors Affecting Methane Oxidation in Arable Soil., *Biology and Fertility of Soils* 21(1–2): 95–102.

Bender, M. & Conrad, R. 1995. Effect of CH_4 concentrations and soil conditions on the induction of CH_4 oxidation activity. *Soil Biol. Biochem.* 27(12): 1517–1527.

Boeckx, P., Van Cleemput, O. & Villaralvo, I. 1997. Methane Oxidation in Soils with Different Textures and Land Use. *Nutrient Cycling in Agroecosystems* 49: 91–95.

Bogner, J., Spokas, K. & Burton, E. 1997. Kinetics of methane oxidation in a landfill cover soil: temporal variations, a whole-landfill oxidation experiment, and modelling of net CH_4 emissions. *Environ. Sci. Technol.* 31(9): 2504–2514.

Dunfield, P.F. Yuryev, A, Senin, P, Smirnova, A.V., Stott, M.B., Hou, S., Ly, B., Saw, J.H., Zhou, Z., Ren, Y., Wang, J., Mountain, B.W., Crowe, M.A., Weatherby, T.M., Bodelier, P.L.E., Liesack, W., Feng, L, Wang, L. & Alam, M. 2007. Methane oxidation by an extremely acidophilic bacterium of the phylum Verrucomicrobia. *Nature* 450: 879–882.

Islam, T., Jensen, S., Reigstad, L.J., Larsen, Ø. & Birkeland, N.K. 2008. Methane oxidation at 55°C and pH 2 by a thermoacidophilic bacterium belonging to the *Verrucomicrobia* phylum. *Proceedings of the National Academy of Sciences (PNAS)* 105: 300–304.

Jaeckel, U., Thummes, K. & Kaempfer, P. 2005. Thermophilic methane production and oxidation in compost. *FEMS Microbiology Ecology* 52(2): 175–184.

Jang, I., Lee, S., Hong, J.H. & Kang, H. 2006. Methane oxidation rates in forest soils and their controlling variables: A review and a case study in Korea. *Ecological Research* 21(6): 849–854.

Kallistova, A.Yu., Kevbrina, M.V., Nekrasova, V.K., Glagolev, M.V., Serebryanaya, M.I. & Nozhevnikova, A.N. 2005. Methane oxidation in landfill cover soil. *Microbiology* 74(5): 608–614.

Kaluzhnaya, K., Khmelenina, V., Eshinimaev, B., Suzina, N., Nikitin, D, Solonin, A., Lin, J-L, McDonald, I, Murrell, C. & Trotsenko, Y. 2001. Taxonomic Characterization of New Alkaliphilic and Alkalitolerant Methanotrophs from Soda Lakes of the Southeastern Transbaikal Region and description of *Methylomicrobium buryatense* sp.nov. *Systematic and Applied Microbiology* 24(2): 166–176.

Kightley, D., Nedwell, D.B., Cooper, M. 1995. Capacity for methane oxidation in landfill cover soils measured in laboratory-scale soil microcosms. *Applied and Environmental Microbiology* 61(2): 592–601.

King, G.M. 1990. Dynamics and controls of methane oxidation in a Danish wetland sediment. *FEMS Microbiology Ecology* 74(4): 309–324.

Kravchenko, I.K. 2002. Methane oxidation in boreal peat soils treated with various nitrogen compounds. *Plant and Soil* 242(1): 157–162.

Mancinelli, R.L. 1995. The regulation of methane oxidation in soil. *Annual Reviews of Microbiology* 49: 581–605.

Min, H., Chen, Z.Y., Wu, W.X. & Chen, M.C. 2002. Microbial aerobic oxidation of methane in paddy soil, Nutrient. *Cycling in Agroecosystems* 64(1–2): 79–85.

Philopoulos, A., Ruck, J., McCartney, D. & Felske, Ch. 2009. A laboratory-scale comparison of compost and sand-compost-perlite as methane-oxidizing biofilter media. *Waste Management & Research* 27(2): 138–146.

Pol, A., Heijmans, K., Harhangi, H.R., Tedesco, D., Jetten, M.S.M. & Op den Camp, H.J.M. 2007. Methanotrophy below pH 1 by a new Verrucomicrobia species. *Nature* 450: 874–878.

Powelson, D.K., Chanton, J., Abichou, T. & Morales, J. 2006. Methane oxidation in water-spreading and compost biofilters. *Waste Management and Research* 24(6): 528–536.

Reay, D.S., Nedwell, D.B. & McNamara, N. 2001. Physical determinants of methane oxidation capacity in a temperate soil. *Water, Air, & Soil Pollution: Focus* 1(5–6): 401–414.

Streese, J. & Stegmann, R. 2003. Microbial Oxidation of Methane from Old Landfills in Biofilters. *Waste Management* 23: 573–580.

Suwanwaree, P. & Robertson, G.P. 2005. Methane Oxidation in Forest, Successional, and No-till Agricultural Ecosystems: Effect of Nitrogen and Soils Disturbance. *Soil Sci. Soc. Am. J.* 69: 1722–1729.

Trotsenko, Y.A. & Khmelenina, V.N. 2002. Biology of extremophilic and extremotolerant methanotrophs. *Arch Microbiol.* 177: 123–131.

Whalen, S.C. & Reeburgh, W.S. 2000. Methane Oxidation, Production, and Emission at Contrasting Sites in a Boreal Bog. *Geomicrobiology Journal* 17(3): 237–251.

Willison, T.W., Webster, C.P., Goulding, K.W.T. & Powlson, D.S. 1995. Methane oxidation in temperate soils: Effects of land use and the chemical form of nitrogen fertilizer. *Chemosphere* 30(3): 539–546.

Wilshusen, J.H., Hettiaratchi, J.P.A. & Stein, V.B. 2004. Long-term behavior of passively aerated compost methanotrophic biofilter columns. *Waste Management* 24(7): 643–653.

Environmental Engineering III – Pawłowski, Dudzińska & Pawłowski (eds)
© *2010 Taylor & Francis Group, London, ISBN 978-0-415-54882-3*

Leaching of soluble components from fertilizers based on sewage sludge and ashes

C. Rosik-Dulewska
Institute of Environmental Engineering of the Polish Academy of Sciences, Zabrze, Poland

U. Karwaczyńska & T. Ciesielczuk
Department of land Protection, Opole University, Opole, Poland

ABSTRACT: This research examines the susceptibility to leaching of soluble components from ash-sludge pellets obtained from municipal sewage sludge and fly ash from brown coal. Pellets are often used as fertiliser; therefore, they were enriched with potassium using potassium chloride and sulphate at 450 g of potassium per kg dry weight of pellets. Susceptibility to leaching of soluble components in water was examined using 1- and 3-step tests. Results from both leaching tests show that the most polluted load was leached from pellets with added K_2SO_4, and the least polluted from unmodified pellets. The total level of leached pollutants from pellets determined using the 3-step test was much higher than that determined using the 1-step test.

Keywords: Leaching, fertilisers, sewage sludge, fly ash.

1   INTRODUCTION

Both technical and civil development have led to increased waste production, resulting in increasing international and national legislation aimed at mitigating the adverse effects of this waste on the natural environment (Regulation 2001).

The Polish power industry relies primarily on coal and generates millions of tonnes of combustion by-products each year. Each tonne of coal burnt creates over 280 kg of solid waste, consisting of 250 kg of ash and slag and 30 kg of products resulting from the desulphurisation of flue gases. During 2004, in EU countries alone, 63 million tonnes of waste from energy fuel combustion was produced, of which 68% was fly ash.

Currently, there are numerous ways of recycling fly ash (Zhang 2002). The most common is neutralisation by an appropriate disposal method – from 2003 to 2008, 90% of fly ash was disposed of economically. Nevertheless, there is an on-going search for new solutions that can increase the number of possibilities for its reuse (Su 2003).

An additional problem presented by urban development is the escalation in quantities of sewage sludge from municipal (and industrial) sewage treatment plants (Regulation 2002). Innovative methods of neutralising and recycling municipal sewage sludge must be drawn up individually for each treatment plant, taking into account the unique properties of the sludge produced. Sewage sludge processing should aim for the maximum, economically justifiable reduction in its mass and volume to limit its negative impact on the environment. Moreover, it is necessary to exploit the fertiliser (Debosz et al. 2002) and energy potential of

sludge to recover at least part of the investment in the building and running of sewage processing equipment, which can represent up to 50% of the overall costs of a sewage treatment plant.

One ecologically and economically justifiable method is its environmental use. The main topics in sewage sludge management are complex methods of processing and systems for its environmental use and environmental monitoring (Kuchar et al. 2005, Xu 2008).

Sewage sludge is difficult to manage due to its high bulk and its chemical composition. Particular difficulties are caused by its hygiene aspects and the high levels of heavy metals it contains (Dai et al. 2007). Consequently, all methods of permanent and safe neutralisation are usually worth considering.

Nevertheless, with the aforementioned environmental concerns in mind a solution should closely heed ecological policy. Excessive amounts of waste force authorities to create new policies for management and neutralisation, which must be tempered by a concern for civil and urban development.

2   MATERIAL AND METHODS

The materials used in out study were ash-sludge (G) pellets made of municipal sewage sludge (70%) from a sewage treatment plant on the outskirts of Zabrze, and from fly ash from brown coal (30%) from the Bełchatów power plant. The pellets were made using a ring matrix pelleter. The properties of the sewage sludge and fly ash are shown in Tables 1 and 2. So that the pellets could be used as fertiliser, they were

Table 1. Characteristics of sewage sludge from the Zabrze Municipal Sewage Treatment Plant

Parameter	Unit	Value
Humidity	% w	86.00
Mineral matter	% dw	58.13
Organic matter	% dw	41.87
Cu	mg/kg dw	155.68
Zn	mg/kg dw	3569.35
Cd	mg/kg dw	3.09
Ni	mg/kg dw	20.74
Pb	mg/kg dw	187.07
Cr	mg/kg dw	36.89
Hg	mg/kg dw	1.99

Table 2. Characteristics of fly ash from brown coal (from Bełchatów power plant)

Parameter	Symbol	Unit	Value
Ash	A^a	%	98.2
Ignition loss	–	%	1.8
Silon oxide (IV)	SiO_2	%	50.36
Aluminium oxide	Al_2O_3	%	22.92
Calcium oxide	CaO	%	12.88
Iron oxide (II)	Fe_2O_3	%	4.48
Sulphur oxide (IV)	SO_3	%	4.03
Magnesium oxide	MgO	%	1.52
Titanium oxide (IV)	TiO_2	%	1.16
Phosphorus oxide (V)	P_4O_{10} (P_2O_5)	%	0.26
Potassium oxide	K_2O	%	0.21
Sodium oxide	Na_2O	%	0.11
Manganese oxide (II and III)	Mn_3O_4 ($MnO \cdot Mn_2O_3$)	%	0.03

enriched with potassium, which is easily leached during sewage processing. After calculating the ratio of potassium to pellets (450 g/kg dry weight), the dry ash was enriched with potassium salts – in one batch with KCl (G+KCl) and in the second with K_2SO_4 (G+K_2SO_4). All values were calculated using the dry mass of modified and unmodified pellets.

Sewage sludge is an alternative to natural organic fertilisers, however in comparison to traditional agricultural waste (e.g. slurry, manure) it is richer in nitrogen and phosphorus, and has a much lower potassium content.

Leaching tests were carried out to measure the potential release of pollutants from ash-sewage pellets, due to their susceptibility to leaching in water.

Two methods were used to examine the aqueous extracts:

1. The amended government directive of 21 December 1999 regulating waste disposal (Regulation 1999, No 110, pos. 1263 – currently suspended). This is a 1-step leaching test. It is carried out over 24 hours: 6 hours of intensive mixing and 18 hours of standing time.
2. The PN-Z-15009, 1997, standard for the preparation of an aqueous extract from solid waste. This is a 3-step leaching test. It is carried out over 78 hours:

24 hours of intensive shaking and 54 hours of standing time.

In both methods, the proportion of ash-sludge mixture to distilled water was 1:10. Aqueous extracts were analysed according to the legally binding Polish Standards for macro and micro indicator content. Macro elements were determined using a Philips PU 8620 spectrophotometer and metal content (Cu, Zn, Cd, Ni, Pb, and Cr) by *atomic absorption spectroscopy* using a Unicam-Philips PU 9100 X spectrometer. All reported values are an average of three, or five single measurements.

The research results were compared against the requirements laid out by the Ministry of the Environment directive of 24th July 2006 that specifies the conditions for releasing sewage into waters or soil (Regulation 2006).

3 RESULTS AND DISCUSSION

To assess potential damage to the environment from the pellets, the 1-step leaching test was conducted on the three samples of ash-sludge mixture G (ash:sludge = 3:7), G+KCl and G+K_2SO_4.

The aqueous extracts were characterised by an alkaline pH – pH 8.4 was the highest for unmodified pellets, which decreased after the addition of potassium to 8.25 in G+KCl, and to 7.04 in G+K_2SO_4 – and high conductivity corresponding to the increased potassium content, i.e. lowest for pellets without added potassium, 1970 µS/cm, increasing to 16280 µS/cm in G+KCl, and 44800 µS/cm in G+K_2SO_4.

The ammonium nitrate content was measured at 93.3 mgN_{NH4}/dm^3 in the aqueous extract from unmodified pellets, 73.2 mgN_{NH4}/dm^3 in that from G+KCl, and 136.5 mgN_{NH4}/dm^3 in that from G+K_2SO_4. In all cases it exceeded the permissible level (10 mgN_{NH4}/dm^3) laid down by the 2006 directive.

The nitrite nitrogen content of aqueous extracts from unmodified pellets was high – 246 mgN_{NO2}/dm^3, decreasing to 35 mgN_{NO2}/dm^3 in that from G+KCl, and to 8.4 mgN_{NO2}/dm^3 in that from G+K_2SO_4. In all cases, it exceeded the permissible level, which the directive gives as 1 mgN_{NO2}/dm^3.

The nitrate nitrogen content of the extracts was relatively low in comparison to the permissible level (30 mgN_{NO3}/dm^3), measuring 3.91 mgN_{NO3}/dm^3 in the aqueous extract from unmodified pellets, 0.52 mgN_{NO3}/dm^3 in that from G+KCl, and 2.13 mgN_{NO3}/dm^3 in that from G+K_2SO_4.

The content of o-phosphates is not covered by any directives. It increased from 24.9 mg$_{PO4}$/dm^3 in the aqueous extract from unmodified pellets to 43.6 mg$_{PO4}$/dm^3 and to 398 mg$_{PO4}$/dm^3 in those from G+KCl and G+K_2SO_4, respectively.

The level of chloride in the two extracts was low in comparison to the permissible level, which is 1000 mg$_{Cl}$/dm^3 – 58 mg$_{Cl}$/dm^3 for the extract from unmodified pellets and 237 mg$_{Cl}$/dm^3 in that from G+K_2SO_4, while it was very high, 9900 mg$_{Cl}$/dm^3 in that from G+KCl, which is understandable.

There was a high sulphate content in extracts from unmodified pellets – about 2.6 times higher than the magnesium content – in those from G+KCl, 1.3 times higher, and from G+K$_2$SO$_4$, 2.9 times higher. Calcium and magnesium increased from 578 mg$_{Ca}$/dm^3 and 226 mg$_{Mg}$/dm^3 in the aqueous extracts from unmodified pellets, to 1102 mg$_{Ca}$/dm^3 and 850 mg$_{Mg}$/dm^3 in those from G+KCl, and 390 mg$_{Ca}$/dm^3 and 136 mg$_{Mg}$/dm^3 in those from G+K$_2$SO$_4$. These elements are not covered by any directives.

The potassium content rose from 164 mg$_K$/dm^3 for the extract from the unmodified pellets to 30,075 mg$_K$/dm^3 in that from G+KCl and 4368 mg$_K$/dm^3 in that from G+K$_2$SO$_4$. This result was twice the permissible level (80 mg$_K$/dm^3) for the extract from the unmodified pellets, and 50× and 376× the permissible level in those from G+K$_2$SO$_4$ and G+KCl, respectively. With this in mind, pellets could be a potential source of potassium for plants.

In all extracts, contents of the microelements nickel, lead and chromium did not exceed permissible levels (0.5 mg/dm^3 for Ni and Pb, 1 mg/dm^3 for Cr). However, an increase was observed in extracts from the modified pellets, similar to that seen with cadmium (results not covered by any directives) and copper, whose content in G+K$_2$SO$_4$ extracts was 11x over the permissible level (0.1 mg$_{Cu}$/dm^3).

Zinc content was variable; it was 0.98 mg$_{Zn}$/dm^3 for the extract from unmodified pellets, 0.23 mg$_{Zn}$/dm^3 in that from G+KCl, and 14.75 mg$_{Zn}$/dm^3 in that from G+K$_2$SO$_4$, 7× over the permissible level (2 mg$_{Zn}$/dm^3). It is likely that the potassium salt used to enrich the pellets was contaminated with zinc.

The level of pollutants leached from the pellets in mg/kg dry weight (dw) was calculated by analysing the aqueous extracts (Table 3). These extracts were characterised by:

- High levels of leached ammonium nitrogen (627 – 1184 mg/kg dw) and leached nitrites – 1652 mg/kg dw from the unmodified extract, decreasing to 235 mg/kg dw in that from G+KCl and 73 mg/kg dw in that from G+K$_2$SO$_4$. A proportionately low level of nitrates: 26.3 mg/kg dw for the extract from unmodified pellets, decreasing to 3.5 mg/kg dw in that from G+KCl and 18.5 mg/kg dw in that from G+K$_2$SO$_4$.
- A very high level of leached sulphates which increased in modified extracts, from 5812 mg/kg dw for the extract from unmodified pellets, to 6121 mg/kg dw in that from G+KCl and 91027 mg/kg dw in that from G+K$_2$SO$_4$.

Levels of o-phosphates, sulphates, potassium, copper, nickel, lead and chromium increased in the following order: in the extract from unmodified pellets, in that from G+KCl and in that from G+K$_2$SO$_4$. The order was reversed for nitrate nitrogen.

In the case of ammonium nitrogen, nitrate nitrogen, chlorides, calcium, manganese, zinc and cadmium, variable levels were recorded from the unmodified

Table 3. Loads of leached pollutants in one-stage leaching test of ash-sludge pellets (G) with added KCl and K$_2$SO$_4$ (mg/kg dw).

Parameter	Leached loads [mg/kg dw]		
	G	G+KCl	G+K$_2$SO$_4$
Ammonium nitrogen	627	490	1184
Nitrite nitrogen	1652	235	73
Nitrate nitrogen	26.3	3.5	18.5
o-phosphates	168	292	3453
Chlorides	390	66330	2057
Sulphates	5812	6121	91027
Ca	3887	7380	2030
Mg	1520	5695	1310
K	1100	201503	34214
Cu	0.4	0.54	9.46
Zn	6.56	1.54	128
Cd	0.034	0.034	0.09
Ni	0.24	0.24	2.52
Pb	0.17	0.30	4.25
Cr	0.14	0.30	2.1

ash-sludge mixtures and those with added potassium salts.

The second leaching test consisted of three steps (Table 4). The aqueous extracts in all leaching steps were characterised by an alkaline pH, pH 8.09 being the highest in the extract from unmodified pellets, which decreased after adding potassium salts to pH 7.16 – 3rd step – in that from G+K$_2$SO$_4$. They also had high specific conductivity, i.e. lowest for the extract from unmodified pellets (2110 μS/cm – 1st step), but over 8× higher in that from G+KCl (18400 μS/cm) and almost twice as high (4500 μS/cm) in that from G+K$_2$SO$_4$, decreasing considerably over the leaching steps.

The ammonium nitrate content at all steps in extracts was very high and in all cases exceeded the permissible level (10 mg$_{NH4}$/dm^3), but did decrease significantly over the leaching steps. The lowest values were found in extracts from unmodified pellets (97.2 mg$_{NH4}$/dm^3 – 1st step and 41.9 mg$_{NH4}$/dm^3 – 3rd-step) and in those from G+KCl (99.7 mg$_{NH4}$/dm^3 – 1st step and 36.1 mg$_{NH4}$/dm^3 – 3rd step), while it was twice as high in those from G+K$_2$SO$_4$ (211 mg$_{NH4}$/dm^3 – 1st -step and 84 mg$_{NH4}$/dm^3 – 3rd step).

Nitrite nitrogen was not found at any of the leaching steps in extracts from unmodified pellets, but in modified pellets, contents were very high, exceeding permissible levels (1 mg$_{NO2}$/dm^3), but decreased considerably over the leaching steps (in those from G+KCl to 147.4 mg$_{NO2}$/dm^3 – 1st - step and 0.63 mg$_{NO2}$/dm^3 – 3rd step and in those from G+K$_2$SO$_4$ to 10.4 mg$_{NO2}$/dm^3 – 1st step, to 7.4 mg$_{NO2}$/dm^3 – 3rd step).

Nitrate nitrogen levels in the extracts were significantly lower than the norm (30 mg$_{NO3}$/dm^3). The maximum level found was 7.08 mg$_{NO3}$/dm^3 – 2nd step in extracts from G+KCl to a minimum

Table 4. Chemical composition of aqueous extracts from the 3-step leaching test of ash-sludge pellets (G) with added KCl and K_2SO_4.

Parameter	Unit	Value G 1°	G 2°	G 3°	G+KCl 1°	G+KCl 2°	G+KCl 3°	G+K_2SO_4 1°	G+K_2SO_4 2°	G+K_2SO_4 3°	Accepted values according to regulation
pH	–	8.09	8.06	8.09	8.30	8.04	8.02	6.50	6.87	7.16	6.5-9
EC	µS/cm	2110	846	435	18400	3280	1070	4500	1340	430	na
Ammonium nitrogen	mgNNH_4/dm³	97.2	76.8	41.9	99.7	107	36.1	211	155	84	10
Nitrite nitrogen	mgNNo_2/dm³	0	0	0	147.4	7.68	0.63	10.4	8.2	7.4	1
Nitrate nitrogen	mgNNo_3/dm³	1.32	3.54	0.75	2.75	7.08	3.13	5.3	2.9	1.2	30
o-phosphates	mgPO_4/dm³	97.2	96.1	81.5	24.3	67.0	64.4	347	307	272	–
Chlorides	mgCl/dm³	272	22	8.1	7950	1740	680	145	42	18	1000
Sulphates	mgSO_4/dm³	872	173	91	922	189	148	15700	19000	22000	500
Ca	mgCa/dm³	332	120	344	1180	123	188	389	124	162	–
Mg	mgMg/dm³	167	105	53	925	103	71	136	69.9	39.4	–
K	mgK/dm³	214	56	33	29700	1205	505	4368	1149	218	80
Cu	mgCu/dm³	0.100	0.040	0.060	0.115	0.035	0.020	0.795	0.49	0.89	0.1
Zn	mgZn/dm³	0.48	0.23	0.33	0.46	0.31	0.16	10.65	9.85	17.5	2
Cd	mgCd/dm³	<0.00	0.025	<0.00	<0.00	<0.00	<0.00	0.010	0.010	0.010	–
Ni	mgNi/dm³	0.100	0.040	0.015	0.050	0.030	0.035	0.465	0.200	0.130	0.5
Pb	mgPb/dm³	0.070	0.045	0.020	0.145	0.035	0.075	0.250	0.300	0.650	0.5
Cr	mgCr/dm³	0.010	<0.02	0	0.060	<0.02	<0.02	0.145	0.090	0.240	1

of $0.75\,mg\,N_{NO3}/dm^3$ – 3rd step in extracts from unmodified pellets.

O-phosphates content is not covered by any directives, and was highest in all aqueous extracts from $G+K_2SO_4$, decreasing at each step (from $347\,mg_{PO4}/dm^3$ – 1st step to $272\,mg_{PO4}/dm^3$ – 3rd-step).

Chloride content decreased significantly at each step and was relatively low in comparison to the permissible level $(1000\,mg_{Cl}/dm^3)$ in extracts from unmodified pellets and those from $G+K_2SO_4$ (maximum $272\,mg_{Cl}/dm^3$ – 1st step from unmodified pellets), while it was understandably very high $(7950\,mg_{Cl}/dm^3$ – 1st step) in those from $G+KCl$.

A high sulphate content was observed in all extracts at the first step, almost twice as high as the permissible level $(500\,mg_{SO4}/dm^3)$, decreasing over the steps in the extract from unmodified pellets, five times at the 3rd step, and $3.5\times$ in the extract from $G+KCl$, while it increased significantly to $15,700\,mg_{SO4}/dm^3$ – 1st step and $22,000\,mg_{SO4}/dm^3$ – 3rd step in extracts from $G+K_2SO_4$, which is also understandable.

The calcium content of extracts from unmodified pellets was from two (1st step) to $6.5\times$ (3rd step) higher than the manganese content, in those from $G+KCl$ from 1.3 (1st step) to $2.7\times$ (3rd step) higher, and those from $G+K_2SO_4$ from 2.8 (1st step) to $4\times$ (3rd step) higher. These results are not covered by any standards or directives.

The potassium content in all extracts (except the 2nd and 3rd steps of leaching from unmodified pellets) were high and exceeded the permitted level $(80\,mg_K/dm^3)$ from $2.7\times$ (1st step in extracts from unmodified pellets) to 55 times (those from $G+K_2SO_4$) and as high as $371\times$ in those from $G+KCl$. In each case, potassium content decreased significantly over the leaching steps.

Among heavy metals (Table 4), the lowest (from 0.005 to $0.025\,mg_{Cd}/dm^3$) levels in the extracts were found for cadmium (not subject to any directive), and nickel, lead and chromium (in all extracts) did not exceed permissible levels. The highest values were found for zinc and copper, but only in extracts from $G+K_2SO_4$ (at all steps), and these values exceeded permissible levels (for Zn $2\,mg/dm^3$ and for Cu $0.1\,mg/dm^3$). High zinc levels at all steps confirmed contamination of the potassium sulphate with zinc compounds.

- The levels of leached pollutants in mg/kg dw in the aqueous extracts were calculated from the analyses of the 3-step test (Table 5). The extracts were characterised by:
- A very high level of leached ammonium nitrate, which decreased proportionally with each step of leaching, for example, in the extracts from $G+K_2SO_4$ from $1705\,mg/kg\,dw$ – 1st step to $672\,mg/kg\,dw$ – 3rd step, together with a high level of leached nitrites (except in the unmodified pellets – $0\,mg/kg\,dw$), which decreased considerably with each step, for example, in the extracts from $G+KCl$, from $938\,mg/kg\,dw$ – 1st step to $4.7\,mg/kg\,dw$ – 3rd

step, and a proportionally low level of nitrates, from $42.8\,mg/kg\,dw$ – 1st step in the extract from $G+K_2SO_4$ to $5.45\,mg/kg\,dw$ – 3rd step in the extract from unmodified pellets.
- A very high level of chlorides, especially in the extracts from $G+KCl$, from $50586\,mg/kg\,dw$ – 1st leaching step to $5109\,mg/kg\,dw$ – 3rd step. The value decreased over successive leaching steps to 59 and $144\,mg/kg\,dw$ – 3rd step in the extracts from unmodified pellets and those from $G+K_2SO_4$.
- A high sulphate level, which decreased from $5886\,mg/kg\,dw$ – 1st leaching step to 659 and $1112\,mg/kg\,dw$ – 3rd step for extracts from unmodified pellets and those from $G+KCl$, while a very high value was observed in extracts from $G+K_2SO_4$, $126856\,mg/kg\,dw$ – 1st leaching step, increasing to as high as $176000\,mg/kg\,dw$ – 3rd step.

In the case of ammonium nitrate, sulphates, potassium, zinc, lead and chromium, values were lowest in the unmodified extract, higher in the extracts from $G+KCl$, and highest in the extracts from $G+K_2SO_4$.

Comparing the values obtained from the two tests – the 1-step and 3-step (as a sum of loads from the 1st, 2nd and 3rd step of leaching; see Table 6) – the data clearly show that the 3-step test produced higher values for leached pollutants.

The lowest levels of heavy metals in all extracts with both the 1- and 3-step tests were Cd (from 0.034 to $0.25\,mg/kg\,dw$), Cr (from 0.14 to $3.82\,mg/kg\,dw$), Ni (from 0.24 to $6.43\,mg/kg\,dw$) and Pb (from 0.17 to $9.64\,mg/kg\,dw$). Cu content was slightly higher (from 0.40 to $23.9\,mg/kg\,dw$), while the highest value was obtained for Zn (from 1.54 to $306\,mg/kg\,dw$). The highest values obtained for all metals were in extracts from $G+K_2SO_4$.

4 CONCLUSIONS

The lowest levels of heavy metals in all extracts with both the 1- and 3-step tests were Cd (from 0.034 to $0.25\,mg/kg\,dw$), Cr (from 0.14 to $3.82\,mg/kg\,dw$), Ni (from 0.24 to $6.43\,mg/kg\,dw$), Pb (from 0.17 to $9.64\,mg/kg\,dw$). Cu content was slightly higher (from 0.40 to $23.9\,mg/kg\,dw$), while the highest value was obtained for Zn (from 1.54 to $306\,mg/kg\,dw$). The highest values obtained for all metals were in extracts from $G+K_2SO_4$ an excessively high level (when measured against directives on the release of sewage sludge to waters and soil) of Cu and Zn was noted which was likely to be a result of contamination. From the results obtained in both leaching tests it was concluded that the highest level of pollutants were leached from $G+K_2SO_4$, while the lowest was leached from unmodified pellets. The total levels of pollutants leached from pellets determined by the 3-step test were much higher than those determined by the 1-step test. The most trace metals were noted in extracts from $G+K_2SO_4$, which is likely to have resulted from decreasing chemical reactions in comparison to unmodified extracts. Further research is being conducted to find a suitable

Table 5. Loads of parameters in 3-step leaching test for ash-sludge pellets (G) with added KCl and K_2SO_4 [mg/kg dw].

Parameter	Unit	Value								
		G			G+KCl			G+K_2SO_4		
		1°	2°	3°	1°	2°	3°	1°	2°	3°
Ammonium nitrogen	mg/kg dw	656	555	303	634	793	271	1705	1254	672
Nitrite nitrogen	mg/kg dw	0	0	0	938	57	4.7	84	66.3	59.2
Nitrate nitrogen	mg/kg dw	8.91	25.6	5.45	17.5	52.5	23.5	42.8	23.5	9.6
o-phosphates	mg/kg dw	656	694	590	155	497	484	2804	2483	2176
Chlorides	mg/kg	1836	159	59	50586	12899	5109	1172	340	144
Sulphates	mg/kg	5886	1250	659	5867	1401	1112	126856	153663	176000
Ca	mg/kg	2241	867	2490	7508	912	1412	3143	999	1296
Mg	mg/kg	1127	759	384	5886	764	533	1099	566	315
K	mg/kg	1445	405	239	188981	8933	3794	35293	9225	1744
Cu	mg/kg	0.68	0.29	0.43	0.73	0.26	0.15	6.42	3.96	7.12
Zn	mg/kg	3.24	1.66	2.39	2.93	2.30	1.20	86.1	79.7	140
Cd	mg/kg	0.034	0.18	0.036	0.032	0.037	0.038	0.081	0.081	0.080
Ni	mg/kg	0.68	0.29	0.11	0.32	0.22	0.26	3.76	1.62	1.04
Pb	mg/kg	0.47	0.33	0.15	0.92	0.26	0.56	2.02	2.42	5.20
Cr	mg/kg	0.068	0.15	0	0.38	0.15	0.15	1.17	0.73	1.92

Table 6. Comparison of parameter loads in 1- and 3-step leaching tests for ash-sludge pellets (G) with added KCl and K_2SO_4 [mg/kg dw]

Parameter	Unit	Value					
		G		G+KCl		G+K_2SO_4	
		1°	3°	1°	3°	1°	3°
Ammonium nitrogen	mg/kg dw	627	1514	490	1698	1184	3631
Nitrite nitrogen	mg/kg dw	1652	0	235	1000	73	210
Nitrate nitrogen	mg/kg dw	26.3	40	3.5	93.5	18.5	75.9
o-phosphates	mg/kg dw	168	1940	292	1136	3453	7463
Chlorides	mg/kg dw	390	2054	66330	68594	2057	1656
Sulphates	mg/kg dw	5812	7795	6121	8380	91027	456519
Ca	mg/kg dw	3887	5598	7380	9832	3373	5438
Mg	mg/kg dw	1520	2270	5695	7183	1179	1980
K	mg/kg dw	1100	2089	201503	201708	37888	46332
Cu	mg/kg dw	0.4	1.4	0.54	1.14	6.90	23.9
Zn	mg/kg dw	6.56	7.29	1.54	6.43	92.4	306
Cd	mg/kg dw	0.034	0.25	0.034	0.107	0.087	0.242
Ni	mg/kg dw	0.24	1.08	0.24	0.80	4.03	6.43
Pb	mg/kg dw	0.17	0.95	0.30	1.74	2.17	9.64
Cr	mg/kg dw	0.14	0.22	0.30	0.68	1.26	3.82

underlined values exceed values from regulation (Regulation 2006 no. 137).

mix for the pellets, that is, the appropriate proportions of sewage sludge from sewage treatment plants and ash from brown coal. Research is also being conducted into factors influencing the chemical reactions. Analyses of the test results show that the ash-sewage pellets studied could be hazardous to the environment when used as an organic-mineral fertiliser. Nevertheless, plants could use potential biogens and they could therefore be introduced to limit potential threats to the environment. Conducted in real and simulated time, lysimetric research allows an understanding of the pace at which macro and microelements are released. It would, therefore, be possible to determine the quantity of pellets that could be used to provide a continuous supply for plants in a secure ground-water environment.

REFERENCES

Dai, J., Xu, M., Chen, J., Yang, X. & Ke, Z. 2007. PCDD/F, PAH and heavy metals in the sewage sludge from six wastewater treatment plants in Beijing, China. *Chemosphere* 66: 353–361.

Debosz, K., Petersen, S., Kure, L.K. & Ambus, P. 2002. Evaluating effects of sewage sludge and household compost on soil physical, chemical and microbiological properties. *Applied Soil Ecology* 19: 237–248.

Kuchar, D., Vondruska, M., Bednarik, V., Kojima, Y. & Matsuda, H. 2005. Stabilization/solidification of sludge by means of coal fly ash as a binder. *Environment protection engineering* 2. Technical University of Wroclaw.

PN-Z-15009, 1997. *Polish National Standard. Solid aste: ater extract preparation. 1999. Leachibility test regulation of Polish Ministry Council 1999*, no. 110, pos. 1263.

Regulation 2001. *Waste directive from 27th April 2001*, Dz.U. o. 62, pos. 628.

Regulation 2002. *Polish Ministry of Environment, 1st August 2002 for municipal sewage sludge*, Dz.U. o. 134, pos. 1140.

Regulation 2006. *Polish Ministry of Environment, 24th July 2006*, Dz.U.2006. o. 137, pos. 984.

Su, D.C. & Wong, J.W.C. 2003. Chemical speciation and phytoavailability of Zn, Cu, Ni and Cd in soil amended with fly ash-stabilized sewage sludge. *Environment International* 29: 895–900.

Xu, G.R., Zou, J.L. & Li, G.B. 2008. Stabilization of heavy metals in ceramsite made with sewage sludge. *Journal of Hazardous Materials* 152: 56–61.

Zhang, F., Yamasaki, S. & Kimura, K. 2002 Waste ashes for use in agricultural production: II. Contents of minor and trace metals. *Science of the Total Environment* 286: 111–118.

Environmental Engineering III – Pawłowski, Dudzińska & Pawłowski (eds)
© *2010 Taylor & Francis Group, London, ISBN 978-0-415-54882-3*

Microbiological enhancement of CLEANSOIL method of soil remediation

A. Tabernacka, A. Muszyński, E. Zborowska & M. Łebkowska
Department of Biology, Nowowiejska, Warsaw University of Technology, Warsaw, Poland

E. Lapshina, Y. Korzhov & D. Khoroshev
Ugra State University, Khanty-Mansiysk, Russian Federation

ABSTRACT: The newly developed CLEANSOIL method was successfully applied to treat soil contaminated with hydrocarbons. The highest efficiency of the soil remediation was obtained in close vicinity of the sorbent – up to 12% for the sorbent without bacteria and 58% for the sorbent colonised with bacteria. A significant decrease in the number of microorganisms in the sorbent – up to 6 orders of magnitude – was observed during the remediation. The combination of physical and biological methods improved the overall effectiveness of the CLEANSOIL process. It is suggested that the ability of microorganisms to produce stable emulsions should be taken into consideration when choosing microorganisms for the soil treatment.

Keywords: Hydrocarbons, sorbents, soil remediation, immobilised bacteria.

1 INTRODUCTION

Various processes have been developed to remove petroleum hydrocarbons from contaminated areas. Soil treatment technologies may be classified as physical, chemical or biological. However, it should be noted that in many cases it is advantageous to use a combination of different techniques (Bhandari et al. 2007, Hester & Harrison 1997, Huang et al. 2005, Nyer 1998, Scullion 2006, Stegmann et al. 2001, Wise et al. 2000, Zhou et al. 2005).

Biological soil treatment systems are based on the activity of microorganisms or plants. Plants are used in the phytoremediation processes, which embrace phytoextraction (mainly for metals removal), phytostabilisation and phytodegradation (Huang et al. 2005, US EPA 2000).

Bioremediation involves the use of bacterial or fungal microorganisms, either by addition of nutrients and water to ensure the optimal growing conditions for indigenous microorganisms (biostimulation) or by inoculation of soil with microorganisms active in pollutants degradation (bioaugmentation). A treatment process provided by microorganisms may occur as landfarming, biopiling (in specially constructed aerated and watered piles), composting and in bioreactors (Alexander 1999, Bhandari et al. 2007, Cookson 1995, Romantschuk et al. 2000, Suthersan 2002).

The newly developed CLEANSOIL method is a simple, cost-effective and innovative technique for the *in situ* remediation of polluted soil under existing infrastructures. Horizontal boreholes are drilled in contaminated soil and filled with adsorbent. The remediation includes the transfer of pollutants from the contaminated area to the adsorbent and their subsequent adsorption. In the case of biodegradable organic substances (such as hydrocarbons) the treatment process may be enhanced by immobilising microorganisms in the sorbents. The microorganisms utilise adsorbed pollutants as a source of carbon and energy required for their life processes, which results in the 'natural reutilisation' of sorbents and recovery of adsorption capacity. After a period of time sufficient to attain the desired remediation effect, the pipes with sorbent are removed and the sorbent is regenerated for further application (Muszyński et al. 2008a, b).

Sorbents have been widely used for removal of petroleum derivatives from water and roads (Zborowska & Kurek 2008, Teas et al. 2001, Choi & Cloud 1992). Many natural organic and mineral sorbents as well as low-cost waste materials, such as straw, ashes, sawdust and peat, have been applied and patented for the removal of hydrocarbons (Gleizes 1996, Krasznai & Takats 1982, Mazet et al. 1995, Ross et al. 1992). Haussard et al. (2003) tested modified pine bark to remove synthetic emulsion of fatty acids and spent diesel motor oil from water, resulting in 97% effectiveness of the process. Research performed by Ake et al. (2001) proved a high sorption capacity of porous organoclay composite for removal of polycyclic aromatic hydrocarbons and pentachlorophenol. Teas et al. (2001) compared the effectiveness of petroleum sorption on natural and expanded perlite, polypropylene and cellulosic fibre. Polypropylene showed the highest oil sorption capacity from artificial seawater, followed by expanded perlite and cellulosic fibre.

Polypropylene sorbents have high oil sorption capacity and low water uptake, and therefore are ideal materials for oil removal (Wei et al. 2003). Johnson et al. (1973) examined the sorption capacity of cotton, synthetic and modified cellulosic fibres for removal of crude oil from water. The efficiency of oil sorption on all sorbents was similar, and in the case of polypropylene fibres was 3,9 g/g of sorbent. It should be noted that no data concerning the use of polypropylene materials for soil remediation were found in the literature.

The newly developed CLEANSOIL method is especially applicable for large areas of polluted land and causes minimum site disturbance. Therefore, the system is applicable to the remediation of soil under buildings, roads, pipelines and railroads that is subject to both local and/or diffuse contamination, and can even be used for preventive applications (Muszyński et al. 2008a, b). The CLEANSOIL avoids the destruction of the existing infrastructure, which is necessary with other methods.

The main goal of this study was to determine and compare the remediation efficiency of the combined process of adsorption and biodegradation of pollutants on the chosen sorbent colonised with microorganisms with the process of using adsorption alone. The processes were used on soil polluted with petroleum hydrocarbons. The test site was an oil product storage facility, and the existing infrastructure (oil storage tanks) had been in operation for 27 years. The CLEANSOIL method was used to clean the test site.

Due to the fact that the pollutants at the test site were easily degradable, it was assumed that the introduction of microorganisms active in pollutant biodegradation to the system might enhance the efficiency of the remediation process. However, it was not possible to introduce microorganisms in the form of bacteria suspension in the polluted soil. Therefore, it was decided to immobilise bacteria active in pollutant degradation on the sorbents used in the CLEANSOIL method. It was assumed that the microorganisms would degrade the adsorbed pollutants, thus regenerating the sorbent and prolonging its working lifetime.

2 MATERIALS AND METHODS

2.1 Test site

The test site (dimensions 200 m × 200 m) was located on a floodplain of the Irtysh river, 3 km from Khanty-Mansiysk (Siberia, Russian Federation). There were two buildings located on the test site – an oil product storage tank and a metal cage. The soil was composed of sand and loam polluted with oil products in the range of 5–12 g/kg of dry weight of soil. The test site had been owned by a petroleum storage company since 1982. The oil product storage facility operated all year round and it stored 16.000 m³ of oil products .

Khanty-Mansiysk is in the central taiga region of the West Siberian province. The climate is sharply continental with the average annual precipitation of 460–620 mm. The average annual air temperature is minus 4.5°C.

2.2 Sorbent

Irvelen is a polypropylene-based sorbent for removing oil product spills, developed by "Runo+", a company based in Tomsk (Russian Federation). It is nontoxic, inert to acids and bases, and can be regenerated up to 40 cycles.

2.3 Microorganisms

Microorganisms active in the biodegradation of hydrocarbons were isolated from the polluted soil taken from the test site by using routine microbiological procedures and the standardised API identification system (Holt & Krieg 1989). The most active strains were selected on the basis of the growth intensity at 26°C on a solidified mineral medium with 0.5% [v/v] addition of diesel oil as a sole source of carbon and energy (Łebkowska & Kańska 2000, Muszyński et al. 2008a). The same medium was used to store the identified strains in a refrigerator at 4°C.

The emulsifying capacity of bacteria strains was determined by using a method described by Bosch et al. (1988) and Willumsen & Karlson (1997). An emulsification index EI was read after 2 hours (EI_2) and 24 hours (EI_{24}). A good emulsifying capacity was defined when $EI_2 > 0.4$ after 2 hours, and the emulsion was considered stable when EI_{24} after 24 hours was at least 50% of EI_2.

2.4 Immobilisation

Microorganisms were proliferated for 48 hours at a temperature of 26°C in Erlenmeyer flasks containing a liquid mineral medium with 1% [v/v] addition of diesel oil as the sole source of carbon and energy (Grundmann & Rehm 1991, Łebkowska & Kańska 2000, Muszyński et al. 2008b). The obtained bacteria suspension was used as an inoculum in the bioreactor (Figure 1), which was filled with 2 m³ of cultivation

Figure 1. Bioreactor tank.

medium as described by Muszyński et al. (2008a). The content of the bioreactor was aerated for 48 hours.

The liquid culture obtained after 48 hours was used for the colonisation of the sorbent. Prior to the immobilisation, a flat 5 m by 5 m area was prepared and covered with a PVC geomembrane (1.5 mm thick) to avoid any outflow of the bacteria suspension into the ground. A layer of sorbent was put onto the geomembrane to a depth of 10 cm, then moistened with the bacteria suspension and left for 1 hour. During that time sorbent was colonised with bacteria.

2.5 Soil remediation

Due to the fact that the infrastructure was present on the test site, the newly developed CLEANSOIL method of soil remediation was applied. Parallel horizontal holes were drilled in the contaminated soil under the infrastructure and then filled with Irvelen (the sorbent) that had been packed into sockets. In order to determine the efficiency of soil remediation through the combined process of chemical sorption and biodegradation of pollutants, two treatment processes were studied – one using the sorbent without bacteria and the other using the sorbent colonised with bacteria. The sorbent was introduced into the soil in June and the remediation process was carried out for 6 weeks. Control chemical and microbiological analyses were performed at the start-up and after 6 weeks.

2.6 Analytical methods

Soil samples were collected from 6–8 random points and mixed to obtain one average sample. Average air-dried sieved soil samples of 1.0 kg (2 mm fraction) were used for the analyses.

The concentrations of total petroleum hydrocarbons (TPH) ($C_6 - C_{35}$) and total organic substances (TOS) (including hydrocarbons, tars and bitumens) in the soil were determined by a gravimetric method following quantitative extraction of pollutants from soil with n-hexane and chloroform, respectively, performed in accordance with the Method EPA 1005. The chemical analyses were performed at Ugra State University.

Enumeration of bacteria active in the biodegradation of hydrocarbons in the sorbents was carried out in accordance with the routine microbiological procedures (plating method), by using agar with diesel oil as the sole source of carbon and energy. Examined sorbents were shaken in a sodium pyrophosphate solution, and the obtained suspension was then properly diluted and spread onto a medium surface. Plates were incubated at 26°C for 7 days. The calculation of results was performed in accordance with EN ISO 7218:2007.

3 RESULTS AND DISCUSSION

The most active bacterial strains in the biodegradation of hydrocarbons were *Arthrobacter sp.* and *Pseudomonas fluorescens*. However, neither the culture nor the culture supernatant of *P. fluorescens*

showed emulsifying capacity (Table 1), while EI_2 measured for *Arthrobacter sp.* was 1.0 (100% emulsification in both cases). Furthermore, the obtained emulsions were stable for both *Arthrobacter sp.* culture and the culture supernatant (EI_{24} equalled 0.8 and 0.7, respectively).

Cameotra & Singh (2008) observed that the solubilisation by surfactants increases the bioavailability of petroleum derivatives to microbial cells. Among them, the most efficient accelerators for the biodegradation of hydrocarbons are bioemulsifiers (biosurfactants) produced by bacteria. Therefore, hydrocarbons degrading bacteria with high emulsifying capacity seem to be the most efficient in the soil remediation process (Wei et al. 2005). The stability of the produced emulsions also plays an important role in the performance and the effectiveness of the emulsifier (Willumsen & Karlson 1997).

Arthrobacter sp. was found to be the most active bacteria strain among other isolated strains in the biodegradation of hydrocarbons and was capable of effectively producing stable bioemulsifiers. Therefore, it was selected for the immobilisation and the soil treatment process.

During the chemical monitoring of soil remediation two parameters were measured: TPH and TOS. Nonpolar and/or low polar oil hydrocarbons extractable by n-hexane are the main component of crude oil, and their content in the soil was expressed as TPH. However, the oil pollutants present in the soil also contained oil tars and bitumens – high-molecular and high-boiling substances, which are accumulated in the soil and can be extractable by chloroform. They are less mobile and hardly biodegradable. It can be assumed that TOS contained TPH, high-molecular oil tars and bitumens as well as the intermediate metabolites such as slightly polar organic acids and ethers, extractable by chloroform. The type of soil present at the test site (sand and loam) contained hardly any humic substances. It seemed advantageous for the soil treatment process because hydrocarbons are strongly bound to humic substances and clay minerals (Richnow et al. 1995), thus reducing their mobility as well as their bioavailability.

Results of the chemical analyses of TPH and TOS content in the soil during the remediation process are presented in Figures 2 and 3, respectively.

At the beginning of the process the concentration of TPH in the soil at depths of 0–20 cm was 5.2–5.4 g/kg d.w., and this was nearly two times lower than the concentration at depths of 20–40 cm (8.8–9.3 g/kg d.w.).

Table 1. Capacity of the bacteria strains to emulsify diesel oil.

Bacteria strain	EI_2		EI_{24}	
	culture	culture supernatant	culture	culture supernatant
Arthrobacter sp.	1.0	1.0	0.8	0.7
P. fluorescens	0.0	0.0	0.0	0.0

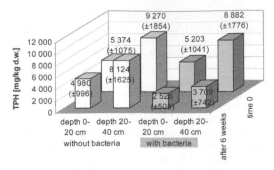

Figure 2. Changes of total petroleum hydrocarbon content in soil during the remediation.

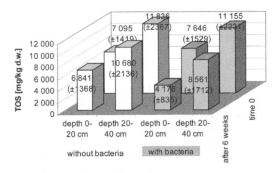

Figure 3. Changes of total organic substances content in soil during the remediation.

The concentration of TOS in the soil at a depth of 0–20 cm was 7.1–7.6 g/kg d.w., while at depths of 20–40 cm it reached 11.1–11.8 g/kg d.w. After six weeks of remediation using sorbent without bacteria, 7% of TPH and 4% of TOS was removed from the soil at depths of 0–20 cm. In the soil at depths of 20–40 cm, the elimination of TPH and TOS was a little higher – at 12% and 10%, respectively. On the testing area where the sorbent with immobilised bacteria was used, the effectiveness of the soil remediation (measured as the percentage decrease of TPH content) at depths of 20–40 cm was 58%, which was almost five times higher than for the sorbent without bacteria. The elimination of TPH content from the soil at depths of 0–20 cm was more than seven times higher when using the sorbent colonised with bacteria than when the sorbent without bacteria was applied. Overall, the TPH removal effectiveness was higher in the soil in close vicinity of the sorbent (at depths of 20–40 cm).

TOS removal efficiency for the sorbent with bacteria was 45% at depths of 0–20 cm and only 23% at depths of 20–40 cm despite the better results of TPH removal at these depths. The removed amount of TOS measured as g per kg d.w. was 3.5 at depths of 0–20 cm and 2.6 at depths of 20–40 cm.

The results for the physical removal of TPH and TOS from the test site by means of sorbents are not very high. They are, however, comparable to the results obtained by other authors who have applied physical and biological methods. Machackova et al. (2008)

remediated soil at a former Air Force base, which operated from 1940 to 1991. The complex remediation of the 28 ha area started in 1998 and was continued for 7 years. Several remediation technologies were used – soil vapour extraction, the "pump and treat" method, air sparging and bioremediation. The most effective method was biodegradation, removing 93% of pollutants, while a combination of soil vapour extraction and air sparging removed only 2% of contaminant.

Microorganisms metabolise petroleum compounds, but high molecular weight aromatics are detectable in forest soils, for example, for 5 years. Generally, about 40% to 80% of crude oil can be degraded by microbial action (Hoffman et al. 1995). By using indigenous microorganisms, Franco et al. (2004) tested in laboratory conditions the biodegradability of crude oil in a soil that had been adjusted to have a water holding capacity of 50%. In 2 months, the mean amount of oil degraded only reached 34%.

Huang et al. (2005) treated soil spiked with hydrocarbons to a concentration of 50 g/kg of soil, and achieved less than 30% TPH removal by bioremediation in 17 weeks. The final effect after 34 weeks was 40%. These results are lower than those obtained using the CLEANSOIL system with immobilised bacteria for 6 weeks (58% and 51% removal of TPH from soil at depths of 20–40 cm and 0–20 cm, respectively).

Löser et al. (1999) proved that hydrocarbons may even be strongly adsorbed on organic-free sandy soil due to its microporosity. They are no longer bioavailable for pollutant-degrading microorganisms and, as a result, the final concentrations of pollutants in soils after bioremediation are almost independent of the initial loading. The residual concentration of diesel oil after four weeks of the process in a pilot-scale percolator system was 1680 mg/kg of soil (44% elimination).

Cameotra & Singh (2008) proved that the presence of bioemulsifiers is significant for the effectiveness of bioremediation, and more important than addition of the nutrient mixture. They obtained the removal of 52.2% of hydrocarbons after 8 weeks of bioremediation by only using a bacteria inoculum. The additional introduction of nutrients and biosurfactants increased the process effectiveness to 63.4% and 73.9%, respectively. However, Urum et al. (2004) showed that the addition of surfactants has a greater impact on the effectiveness of the treatment in non-weathered soils.

In the study described in this paper, a significantly higher rate of hydrocarbon removal was observed when the sorbent colonized with *Arthrobacter sp.* was introduced into the contaminated soil compared to when the soil was treated with the sorbent without bacteria. These results seemed to difficult to justify because the main limitation of the CLEANSOIL method concerns the mobility of pollutants, which limits the transfer of contaminants from soil to the sorbent. The general principle for microbiological enhancement is the recovery of adsorption capacity caused by the utilisation of adsorbed pollutants as a source of carbon and energy for microorganisms. However, it can be assumed that the bioemulsifiers and biosurfactants

Table 2. Changes in number of bacteria active in hydrocarbons degradation in the sorbent during the remediation process.

Sampling time	Number of bacteria active in hydrocarbons degradation in sorbent	
	not subjected to immobilisation CFU/g d.w.	subjected to immobilisation CFU/g d.w.
time 0	non detectable	$1.6(\pm 0.6)*10^{10}$
after 6 weeks	$1.0(\pm 0.5)*10^{3}$	$12.0(\pm 1.5)*10^{3}$

produced extracellularly by *Arthrobacter sp.* changed the surface tension and interphase tension, allowing for a higher mobility of the pollutants in the soil and resulting in the process being more effective.

The results of microbiological analyses (presented in Table 2) revealed that the number of bacteria active in hydrocarbon degradation in the colonised sorbent was very high, which indicates that the immobilisation process was very efficient. After 6 weeks of bioremediation, however, the number of bacteria decreased by 6 orders of magnitude. After six weeks, microorganisms had also colonised the sorbent not subjected to the immobilisation, but their number was still lower than the number of bacteria in the sorbent subjected to the immobilisation.

Huang et al. (2005) and Cameotra & Singh (2008) suggest that many microorganisms introduced to soil as biocatalysts are unable to compete with the indigenous soil population and their activity may deteriorate in field conditions. As a result, it is very difficult to achieve sufficient biomass for effective bioremediation. The fast decrease in the number of bacteria immobilised on the sorbent is probably caused by the reduction of substrates (pollutants) used by microorganisms. Microbial growth was further limited by other factors, such as insufficient sources of nitrogen and phosphorus in the tested soil as well as a limited access to oxygen. Therefore, in order to improve the process, it is necessary to reintroduce microorganisms to the polluted soil. Unfortunately, this is not possible using the CLEANSOIL method.

4 CONCLUSIONS

The CLEANSOIL method can be successfully applied for the treatment of soils contaminated with hydrocarbons. It causes minimal disruption to infrastructure existing in the polluted area. It should be noted that the greatest efficiency of the soil remediation process – up to 12% for the sorbent without bacteria and 58% for the sorbent colonised with bacteria – is obtained in close vicinity of the sorbent, and this therefore needs to be considered when designing the scheme of the expensive horizontal drilling.

A combination of different techniques, such as physical and biological methods, improves the overall effectiveness of the CLEANSOIL process. However, a decrease in the number of microorganisms in the sorbent during the remediation demands a developing a method for reintroducing microorganisms into the sorbent. Moreover, when choosing microorganisms for the soil treatment, both the biochemical activity of microorganisms in hydrocarbon degradation and also their ability to produce stable emulsions should be taken into consideration.

ACKNOWLEDGEMENTS

The study was carried out as a part of European Union funded project INCO-2005-013420 "An innovative method for the on-site remediation of polluted soil under existing infrastructures (CLEANSOIL)" (6th Framework Programme).

REFERENCES

Ake C.L., Wiles M.C., Huebner H.J., McDonald T.J., Cosgriff D., Richardson M.B., Donnelly K.C.& Phillips T.D. 2003. Porous organoclay composite for the sorption of polycyclic aromatic hydrocarbons and pentachlorophenol from groundwater. *Chemosphere* 51(9): 835–844.
Alexander M. 1999. *Biodegradation and bioremediation.* San Diego, London, Boston, New York, Sydney, Tokyo, Toronto: Academic Press.
Bhandari A., Surampalli R., Chanpagne P., Ong S.K., Tyagi R.D. & Lo I.M.C. (eds), *Remediation technologies for soils and groundwater.* 2007 Reston (Virginia): ASCE.
Bosch M.P., Robert M., Mercadé M.E., Espuny M.J., Parra J.L. & Guinea J. 1988. Surface active compounds on microbial cultures. *Tenside Surfactants Detergents.* 25(4): 208–211.
Cameotra S.S. & Singh P. 2008. Bioremediation of oil sludge using crude biosurfactants. *International Biodeterioration and Biodegradation.* 62(3): 274–280.
Choi H.-M. & Cloud R.M. 1992. Natural sorbents in oil spill cleanup. *Environmental Science & Technology.* 26(4): 772–776.
Cookson J.T. 1995. *Bioremediation engineering: design and application.* New York: McGraw-Hill Inc.
Franco I., Contin M., Bragato G. & De Nobili M. 2004. Microbiological resilience of soils contaminated with crude oil. *Geoderma.* 121(1–2): 17–30.
Gleizes R. 1996. *Cendres d'origine charbonnière appliqués au traitement de divers milieux et installations de mise en œurve.* French Patent no. FR 2 734 175-A1.
Grundmann R., Rehm H.-J. 1991. Biodegradation of diesel fuel. Science and Technology 44.
Haussard M., Gaballah I., Kanari N., De Donato Ph., Barrès O. & Villieras F. 2003. Separation of hydrocarbons and lipid from water using treated bark. *Water Research.* 37(2): 362–374.
Hester R.E. & Harrison R.M. (eds). 1997. *Contaminated land and its reclamation.* Issues in Environmental Science and Technology. Cambridge: Royal Society of Chemistry.
Hoffman D.J., Rattner B.A., Burton G.A. & Cairns J. (eds). 2003. *Handbook of ecotoxicology.* Boca Raton: Lewis Publishers/CRC Press.
Holt J.G. & Krieg N.R. 1989. B*ergey's manual of systematic bacteriology,* Baltimore, Hong-Kong, London, Sydney: Wiliams and Wilkins Company.

Huang X.-D., El-Alawi Y., Gurska J., Glick B.R. & Greenberg B.M. 2005. A multi-process phytoremediation system for decontamination of persistent total petroleum hydrocarbons (tphs) from soils. *Microchemical Journal.* 81(1): 139–147.

Johnson R.F., Manjrekar T.G. & Halligan J.E. 1973. Removal of oil from water surface by sorption on unstructured fibers. *Environmental Science & Technology.* 7(5): 439–443.

Krasznai M. & Takats A. 1982. *Removal of organic contaminants from wastewaters.* Hungarian Patent no. HU 2993 A850828.

Löser C., Seidel H., Hoffmann P. & Zehnsdorf A. 1999. Bioavailability of hydrocarbons during microbial remediation of a sandy soil. *Applied Microbiology and Biotechnology.* 51(1): 105–111.

Łebkowska M. & Kańska Z. 2000. *Microbiological remediation of soils contaminated with petroleum derivatives* (*in Polish*: Sposób mikrobiologicznej remediacji gruntów z produktów naftowych). Polish Patent no. Pl 180141B1.

Machackova J., Wittlingerova Z., Alk K., Zima J. & Linka A. 2008. Comparison of two methods for assessment of in situ jet-fuel remediation efficiency. *Water, Air, & Soil Pollution.* 187(1–4): 181–194.

Mazet M., Couillault P., Castillo J.M. & Mathies G. 1995. *Procéedé pour l'élimination des graisses contenues dans les efluents aqueux.* French Patent no. FR 2 708 288-A1.

Muszyński A., Tabernacka A. & Karwowska E. 2008a. Research on optimum composition of media for cultivation of *Arthrobacter* sp. strain active in soil bioremediation. 10th *InternationalUFZ-Deltares/TNO Conference on Soil-Water Systems ConSoil 2008.* Leipzig: F&U Confirm.

Muszyński A., Tabernacka A. & Łebkowska M. 2008b. Immobilization of biocatalysts in sorbents used for soil remediation, 10th *InternationalUFZ-Deltares/TNO Conference on Soil-Water Systems ConSoil 2008.* Leipzig: F&U Confirm.

Nyer E.K. 1998. *Groundwater and soil remediation: practical methods and strategies.* Chelsea (Michigan): Ann Arbor Press.

Richnow H.H., Seifert R., Kästner M., Mahro B., Horsfield B., Tiedgen U., Böhm S. & Michaelis W. 1995. Rapid screening of PAH-residues in bioremediated soils. *Chemosphere.* 31(8): 3991-3999.

Romantschuk M., Sarand I., Petanen T., Peltola R., Jonsson-Vihanne M., Koivula T., Yrjala K. & Haahtela K. 2000. Means to improve the effect of in situ bioremediation of contaminated soil: an overview of novel approaches. *Environmental Pollution.* 107(2): 179–185.

Ross A., Shoiry J. & Narassiah S. 1992. *Sand filter containingwood ash for wastewater treatment.* American Patent no. US 5108614 A.

Scullion J. 2006. Remediating polluted soils. *Naturwissenschaften.* 93(2): 51–65.

Suthersan S.S. 2002. *Natural and enhanced remediation systems.* Boca Raton: Lewis Publishers/CRC Press.

Stegmann R., Brunner G., Calmano W. & Matz G. (eds). 2001. *Treatment of contaminated soil. Fundamentals, analysis, applications.* Berlin Heidelberg New York: Springer.

Teas Ch., Kalligeros S., Zanikos F., Stournas S., Lois E. & Anastopoulos G. 2001. Investigation of the effectiveness of absorbent materials in oil spills clean up. *Desalination.* 140(3): 259–264.

Urum K., Pekdemir T. & çopur M. 2004. Surfactants treatment of crude oil contaminated soils. *Journal of Colloid and Interface Science.* 276(2): 456–464.

U.S. EPA. 2000. *Introduction to phytoremediation.* EPA/600/R-99/107. Washington.

Wei Q.F., Mather R.R. & Fotheringham A.F. 2005. Oil removal from used sorbents using a biosurfactant. *Bioresource Technolog.* 96(3): 331–334.

Willumsen P.A. & Karlson U. 1997. Screening of bacteria, isolated from PAH-contaminated soils, for production of biosurfactants and bioemulsifiers. *Biodegradation.* 7(5): 415–423.

Wise D.L., Trantolo D.J., Eichon E.J., Inyang H.I. & Stottmeister U. (eds). 2000. *Remediation engineering of contaminated soils.* New York, Basel: Marcel Dekker Inc.

Zborowska E. & Kurek M. 2008. Use of natural sorbents and waste materials in technologies of environmental protection (*in Polish:* zastosowanie sorbentów naturalnych i materiałów odpadowych w technologiach ochrony środowiska). *Inżynieria i Ochrona Środowiska.* 11(4): 471–490.

Zhou Q., Sun F. & Liu R. 2005. Joint chemical flushing of soils contaminated with petroleum hydrocarbons. *Environment International* 31(6): 835–839.

430

Environmental Engineering III – Pawłowski, Dudzińska & Pawłowski (eds)
© 2010 Taylor & Francis Group, London, ISBN 978-0-415-54882-3

Emissions of trace compounds from selected municipal landfills in Poland

J. Czerwiński & M. Pawłowska

Faculty of Environmental Engineering, Lublin University of Technology, Lublin, Poland

ABSTRACT: Environmental impacts associated to different waste treatments (also landfilling) are of interest in the decision-making process at local, regional and international level. However, all the environmental burdens of inorganic waste biological treatment are not always considered. Real data on gaseous emissions released from full-scale landfills, composting plants and so one are difficult to obtain. These emissions are related to the composting technology and waste characteristics and therefore, an exhaustive sampling campaigns necessary to obtain representative and reliable data of a single plant.

In this article Volatile Organic Compounds (VOC) were determined in landfill gas from four landfill sites. It was shown that the composition and concentration profiles of VOC were different in each landfill. These differences were connected with age of characterized landfills, composition of landfilled material and weather conditions. In between of VOC three major group compounds were detected chlorinated compounds, aromatic hydrocarbons and organometalloid compounds.

Keywords: Landfill gas emissions, volatile trace compounds, SPME-GC-MS analysis.

1 INTRODUCTION

It is well known that municipal waste landfills emitting to the atmosphere greenhouse gases e.g. carbon dioxide and methane which has influence on greenhouse effect more than 20 times higher than CO_2. Methane content ranged from 48% to 65%, carbon dioxide from 36% to 41%. Apart these two gases content of other major compounds nitrogen and oxygen in landfill gas ranged from <1% to 17% and <1% respectively (Tchobanoglous 2002, Rasi et al. 2007). As a minor compounds in landfill gas were identified ammonia (0.1–1%), hydrogen sulfide (0–1%), hydrogen (0–0.2%) and non methane organic compounds (NMOC) at the concentrations varied from 0.01 to 0.6% (Arnold 2009, Rasi 2009, Urban et al. 2009).

According to UK EPA report (Parker et al. 2003) EPA from landfills to the atmosphere more than 530 compounds is emitted at a trace concentrations. But in the table in these report is easy to found the same compounds named systematically and traditionally, additionally some of compounds presented in these table posses vapor pressures which are not predicted to be present in a gas phase (e.g. higher chlorinated dibenzodioxins and dibenzofurans). Polychlorinaded dibenzodioxins and dibenzofurans are presented typically in condensed phase and can be co-determined as a compounds adsorbed on microparticles of dust.

Composition of volatile organic compounds emitted from landfills is connected with age and type of landfilled waste material (Schuetz et al. 2004, 2008).

While the emission of volatiles during phase I of the combined anaerobic/aerobic composting process was measured in a full-scale composting plant, the aerobic stages of both composting techniques were performed in pilot-scale composting bins. Similar groups of volatile compounds were analysed in the biogas and the aerobic composting waste gases, being alcohols, carbonyl compounds, terpenes, esters, sulphur compounds and ethers. Predominance of alcohols (38% wt/wt of the cumulative emission) was observed in the exhaust air of the aerobic composting process, while predominance of terpenes (87%) and ammonia (93%) was observed in phases I and II of the combined anaerobic/aerobic composting process, respectively. In the aerobic composting process, 2-propanol, ethanol, acetone, limonene and ethyl acetate made up about 82% of the total volatile organic compounds (VOC)-emission (Takuwa et al., 2007). Next to this, the gas analysis during the aerobic composting process revealed a strong difference in emission profile as a function of time between different groups of volatiles. The total emission of VOC, NH_3 and H_2S during the aerobic composting process was 742 g ton^{-1} biowaste, while the total emission during phases I and II of the combined anaerobic/aerobic composting process was 236 and 44 g ton^{-1} biowaste, respectively (Smet et al., 1999). Taking into consideration the 99% removal efficiency of volatiles upon combustion of the biogas of phase I in the electricity generator, the combined anaerobic/aerobic composting process can be considered as an attractive alternative for aerobic biowaste composting because of its 17 times lower overall emission of the volatiles mentioned (Ritzkowski et al. 2009, Saral et al. 2009, Shafi et al. 2006, Szymanski et al. 2007).

2 MATERIALS AND METHODS

2.1 *Characteristic of sampling sites*

The NMOCs concentrations were measured in landfill gas samples taken from 50 cm depth in the four municipal solid waste landfills located in the Lubelskie Region. A characteristic of these sites is presented in Table 1.

The samples were taken from old, closed basins in the cases of the Landfill 2 (waste deposited 10–4 years ago) and the Landfill 3 (waste deposited 13–5 years ago), and from a basin still operating after 6 years in the case of the Landfill 1. Sewage sludge from the local wastewater treatment plants were deposited on all examined landfills.

The Landfills 2, 3 and 4 were passively vented, while the Landfill 1 was actively vented and used the biogas for energy production.

The examinations were conducted in summer and autumn of 2009. In this period samples were taken four times (except for the analysis of siloxanes which was carried out only once).

Table 1. Description of examined landfill sites.

Site	Year of opening	Area covered; total number of inhabitants	Site area [ha]	Annual deposit in 2006 [Mg]	Annual deposit in 2007 [Mg]	Cumulative deposit (as of Dec. 2007) [Mg]	Waste pre-treatment system
Site 1 Rokitno	1994	city + surroundings; 396.345	11.62	107.729.96	120.041.28	1.500.806.04	1994–2002 – no pre-treatment. Limited segregation in the city since 2002 (separation of glass, plastics and metals).
Site 2 Leczna	1994	town + surroundings; 64.181	3.20	5.992.1	9.335.94	68.315.8	Segregation plant (separation of recyclable materials: plastic, aluminium, glass). The remaining (ballast) deposited in the landfill.
Site 3 Pulawy	2001	town + surroundings; 117.167	3.17	11.394.1	18.723.19	75.215.83	Segregation plant (separation of recyclable materials: plastic, metals, glass and biodegradable fraction). Biodegradable fraction is digested with sewage sludge. The remaining part of the thick fraction is compressed and deposited in the landfill.
Site 4 Jawidz	1981 Closed in 1998	city + surroundings; 396.345	12.00	–	–	6.968.8	No pretraeatement

Table 2. Operating conditions of Trace Ultra/Polaris Q GC-MS system.

The operation conditions of the chromatograph (TRACE ULTRA):

Injector:	constant temperature PTV (splitless mode) @ 270°C
Capillary column:	RTx 5 (Restek) 60 m × 0.25 mm $d_f = 0.25 \mu m$
Oven temperature programming:	45°C (2 min hold) ramp 5°/min to 270°C, 10 min hold
Carrier gas:	He (99.9996%) @ 40 cm/s

The MS operating conditions (POLARIS Q):

The ion source temperature	250°C		
The transfer line temperature	275°C		
Scanning mode I: Full Scan	42.0–450.0 amu		
Scanning mode II:	SIM: quantitation ions (Q) and qualifiers (QAL)	Q	QAL
	Benzene	78	77
	Toluene	91	92
	Ethylbenzene	106	91
	Xylenes	106	91
	Cumene	120	105
	Mesithylene	120	105
	Trichloroethylene	130	132
	Tetrachloroethylene	166	168, 164
	Dichlorobenzenes	146	148
	Trichlorobenzene	180	182

The NMOCs being studied were divided into two groups: hydrocarbons and halogenated hydrocarbons. Additionaly concentrations of CH_4 and CO_2 were described.

Table 3. Results of measurements of selected analytes in a home prepared gas standard mixture.

Compound	Measured concentration [$\mu g/m^3$]	RSD [% of measured value]
Benzene	54.2	9.1
Toluene	52.1	7.3
Ethylbenzene	51.0	5.3
m-Xylene	56.8	8.1
propylbenzene	49.2	3.2
Mesitylene	47.0	12.1
Trichloroethylene	56.9	7.1
Tetrachloroethylene	51.4	4.3
1,4-Dichlorobenzene	53.2	7.2
1,3,5-Trichlorobenzene	49.7	4.1

2.2 Methods of examination

A portable infrared landfill gas analyzer GFM 430 (OMC Envag, Poland) was used for the direct determination of methane and carbon dioxide concentrations in LFG.

Additionally gas samples were taken to the 5 dm^3 Tedlar bags with soil gas sampler (Bosh, Germany) and portable membrane pump with membrane made from Teflon. Samples were transported to the laboratory and analytes were adsorbed on SPME fiber coated with 100 μm PDMS stationary phase during 15 minutes exposition. Siloxanes were analysed with using DVB coated fibers with the same exposition and desorption time.

A GC-MS system was used to determine of the other gases. Afterwards analytes were desorbed directly in a PTV injector of the GC-MS system. The desorption time was 2 minutes. The GC-MS system (Trace Ultra – PolarisQ, USA) was operated under the conditions shown in Table 2.

Direct determination of volatile organics via isothermal GC-PID is not commonly used in landfill

Table 4. Results of determination VOC in LFG in Rokitno landfill.

Compounds	m/z	Concentration of microconsituents in samples taken from 50 cm depth [$\mu g/m^3$]					Concentration in the samples taken from surface area
					Mean	SD	
Methane [%]		4.1	9.1	4.6	6.7		
Carbon Dioxide [%]		3.5	7.4	5.2	5.9		
Benzene	78	71	24	14	36.33	30.44	Nd
Toluene	91	97	61	31	63.00	33.05	12
Xylenes	106	211	135	24	123.33	94.04	Nd
Ethylbenzene	106	134	124	102	120.00	16.37	Nd
Trimethylbenzene	120	12	7	2	7.00	5.00	0.2
n-Propylbenzene	120	58	42	31	43.67	13.58	Nd
t-Butylbenzene	134	2	0.9	Nd	1.45	0.78	Nd
o-Ethyltoluene	120	7	0.54	0.24	2.59	3.82	Nd
m-Ethylotoluene	120	11	7.2	8.4	8.87	1.94	Nd
p-Etylotoluen	120	0.4	0.2	Nd	0.30	0.14	Nd
Naphthalene	128	11	9.4	0.8	7.07	5.49	Nd
Chlorobenzene	112	2	1.4	3.2	2.20	0.92	Nd
1.2-Dichlorobenzene	146	0.4	0.2	Nd	0.30	0.14	Nd
1.3-Dichlorobenzene	146	14	31	12	19.00	10.44	Nd
1.4-Dichlorobenzene	146	34	42	54	43.33	10.07	11
1.2.3-Trichlorobenzene	178	41	51	26	39.33	12.58	Nd
1.2.4-Trichlorobenzene	178	2	9	7	6.00	3.61	Nd
Hexachlorobenzene	322	Nd	0.2	0.7	0.45	0.35	Nd
1.1-Dichloroethane	62	Nd	0.2	Nd	0.20	–	Nd
1.2-Dichloroethane	62	2	0.4	Nd	1.20	1.13	Nd
1.1.1-Trichloroethanc	132	0.4	0.8	0.8	0.67	0.23	Nd
1.1.2-Trichloroethane	132	0.7	21	12	11.23	10.17	Nd
1.1.2.2-Tetrachloroethane	168	14	24	9	15.67	7.64	Nd
Trans-1.2-dichloroethane	130	0.2	17	11	9.40	8.51	Nd
Cis-1.2-dichloroethane	130	2	0.4	Nd	1.20	1.13	Nd
Trichloroethylene	160	19	3.1	Nd	11.05	11.24	14
Tetrachloroethylene	195	7	6.2	4.4	5.87	1.33	Nd
Dichloromethane	86	3	11	7.8	7.27	4.03	Nd
Trichloromethane	127	17	4	6.2	9.07	6.96	Nd
Tetrachloromethane	152	2	3	nd	2.50	0.71	Nd
Hexamethylcyclosiloxane	207	29	34	17			
Octamethylcyclosiloxane	207	14	12	9			

Table 5. Results of determination VOC in LFG in Pulawy landfill.

Compounds	m/z	Concentration of microconsituents in samples taken from 50 cm depth [μg/m³]			Mean	SD	Concentration in the samples taken from surface area
Methane [%]		0.9	3.7	1.4	2.4		
Carbon dioxide [%]		0.7	2.4	0.9	1.8		
Benzene	78	24	13	9.3	18.50	7.78	Nd
Toluene	91	7	41	22	23.33	17.04	7.12
Xylenes	106	32	58	24	38.00	17.78	Nd
Ethylbenzene	106	22	41	ND	31.50	13.44	Nd
Trimethylbenzene	120	14	7	2	7.67	6.03	0,8
n-Propylbenzene	120	24	42	ND	33.00	12.73	Nd
t-Butylbenzene	134	ND	ND	ND			Nd
o-Ethyltoluene	120	7	0.22	2.7	7.00		Nd
m-Ethylotoluene	120	ND	ND	ND			Nd
p-Etylotoluen	120	ND	ND	ND			Nd
Naphthalene	128	9	12	3	8.00	4.58	Nd
Chlorobenzene	112	2	1.4	3.2	2.20	0.92	Nd
1.2-Dichlorobenzene	146	0.4	0.2	Nd	0.30	0.14	Nd
1.3-Dichlorobenzene	146	14	31	12	19.00	10.44	Nd
1.4-Dichlorobenzene	146	34	42	54	43.33	10.07	19
1.2.3-Trichlorobenzene	178	41	51	26	39.33	12.58	Nd
1.2.4-Trichlorobenzene	178	2	9	7	6.00	3.61	Nd
Hexachlorobenzene	322	Nd	0.3	Nd	0.30		Nd
1.1-Dichloroethane	62	Nd	0	Nd	0.00		Nd
1.2-Dichloroethane	62	Nd	31	Nd	31.00		Nd
1.1.1-Trichloroethane	132	22	21	54	32.33	18.77	Nd
1.1.2-Trichloroethane	132	13	19	31	21.00	9.17	Nd
1.1.2.2-Tetrachloroethane	168	5.2	5.7	4	4.97	0.87	Nd
Trans-1.2-dichloroethane	130	Nd	0.2	0.1	0.15	0.07	Nd
Cis-1.2-dichloroethane	130	Nd	0.2	Nd	0.20		Nd
Trichloroethylene	160	2	0.4	Nd	1.20	1.13	Nd
Tetrachloroethylene	195	0.4	0.7	0.8	0.63	0.21	32
Dichloromethane	86	11	14	7.2	10.73	3.41	Nd
Trichloromethane	127	4	11	9	8.00	3.61	Nd
Tetrachloromethane	152	6.2	12	14	10.73	4.05	Nd
Hexamethylcyclosiloxane	207	11	41	26			Nd
Octamethylcyclosiloxane	207	4	17	9			Nd

gas examination. SPME-GC-MS technique was previously applied by Davoli et al (2003), Paschke et al. (2006) and Pawlowska et al. (2008).

For the five independent analyses, static gas standard mixture contained analytes at the concentration 51 μg m^{-3} with GC-MS system the data given in Table 3 were obtained.

3 RESULTS AND DISCUSSION

Results of determination VOC in LFG are given in tables 4–7.

3.1 Volatile organic compounds

The highest total concentrations of the non methane hydrocarbons (236–540 mg/m³) were found in LFG from the largest landfill (Landfill 1).

From among the analysed hydrocarbons benzene, toluene and styrene were detected in the examined biogases. They represent a group of aromatic compounds which are hazardous for the environment. Deposition of products like: packaging, electrical and thermal insulation, fibreglass, pipes, car parts, drinking cups and other food-use items could be a source of styrene. Low levels of styrene occur naturally in a variety of foods, such as fruits, vegetables, nuts, beverages, and meats. Additionally concentration of 1,4-dichlorobenzene was relatively high in almost all determinations it suggest that this compound is stable and difficult undergo biochemical changes.

3.2 Halogenated compounds

Between 22 and 25 individual halocarbons were identified and all were relatively low molecular weight compounds of one or two carbon atoms (with the exception of dichlorobenzene). Chlorofluorocarbons are included in this group and were thought to be derived from aerosol propellants and refrigerants. In our investigations the main groups of these compounds

Table 6. Results of determination VOC in LFG in Leczna landfill.

Compounds	m/z	Concentration of microconsituents in samples taken from 50 cm depth [$\mu g/m^3$]					Concentration in the samples taken from surface area
					Mean	SD	
Methane [%]		0.2	2.1	0.4	0.87		
Carbon dioxide [%]		0.1	0.9	0.3	0.45		
Benzene	78	14	9	6	9.67	4.04	Nd
Toluene	91	38	52	12	34.00	20.30	7,12
Xylenes	106	611	14	924	516.33	462.33	Nd
Ethylbenzene	106	0.7	41	ND	20.85	28.50	Nd
Trimethylbenzene	120	ND	ND	ND			0,8
n-Propylbenzene	120	ND	ND	ND			Nd
t-Butylbenzene	134	ND	ND	ND			Nd
o-Ethyltoluene	120	ND	ND	ND			Nd
m-Ethylotoluene	120	ND	ND	ND			Nd
p-Ethyltoluene	120	ND	ND	ND			Nd
Naphthalene	128	ND	ND	ND			Nd
Chlorobenzene	112	Nd	0.3	Nd	0.30		Nd
1.2-Dichlorobenzene	146	Nd	Nd	Nd			Nd
1.3-Dichlorobenzene	146	Nd	Nd	Nd			Nd
1.4-Dichlorobenzene	146	26	31	21	26.00	5.00	19
1.2.3-Trichlorobenzene	178	Nd	Nd	Nd			Nd
1.2.4-Trichlorobenzene	178	Nd	Nd	Nd			Nd
Hexachlorobenzene	322	Nd	Nd	Nd			Nd
1.1-Dichloroethane	62	Nd	Nd	Nd			Nd
1.2-Dichloroethane	62	Nd	Nd	Nd			Nd
1.1.1-Trichloroethane	132	Nd	Nd	Nd			Nd
1.1.2-Trichloroethane	132	Nd	Nd	Nd			Nd
1.1.2.2-Tetrachloroethane	168	Nd	Nd	Nd			Nd
Trans-1.2 dichloroethane	130	Nd	Nd	Nd			Nd
Cis-1.2-dichloroethane	130	Nd	Nd	Nd			Nd
Trichloroethylene	160	1.4	2.9	2.5	2.27	0.78	Nd
Tetrachloroethylene	195	Nd	Nd	Nd			32
Dichloromethane	86	Nd	Nd	Nd			Nd
Trichloromethane	127	Nd	Nd	Nd			Nd
Tetrachloromethane	152	Nd	Nd	Nd			Nd

were chloromethanes and chloroethylenes. Changes in concentration profiles of these compounds suggest that during landfilling they undergo degradation by lousing of chlorine atom.

3.3 Organic compounds

Organometallic (metalloid) compounds are mainly anthropogenic origin. They represent several classes compounds including alkylated tin, mercury and lead compounds also organoarsenic, organoantimony and organosilica compounds (Acetola et al., 2008; Gates et al.1997, Krupp et al. 2007 Lindberg et al., 2001;2005; Merisowski 2002; Mester 2005; Kot 2000; Pinel et al., 2008). Here, among others, the compound class of organotin compounds (OTC) has come into focus, because due to their broad industrial production and use as PVC-stabilisers, fungicides, agro-chemicals, wood preservatives and anti-fouling agents, they are widely distributed throughout the terrestrial and marine biogeosphere. In the last decades, the annual production of organotin compounds rose from around 50 t in 1950, to 2000 t in 1960, 16,000 t in 1970

and 40,000 t in the mid-1980s. In 1992, the world-wide production of OTC amounted to about 50,000 t (Hoch, 2001). To date, OTC are basically used as PVC stabilisers (about 70%) encompassing mono-and di-methyl and mono- and di-butyl tin (MMT, DMT, MBT and DBT) and agro-chemicals (tri-phenyl tin, TPhT) and general biocides (e.g., tri-butyl tin, TBT, and TPhT) (about 20%). Some methylated derivatives of metal(oids) (e.g., mercury, arsenic) are basically produced by chemical transalkylation or biologically mediated methylation (mercury, antimony) in natural environments (Acetola et al. 2008, Gates et al.. Lindberg et al. 2001, 2005, Merisowski 2002, Mester 2005).

Siloxanes are a family of man-made organic compounds that contain silicon, oxygen and methyl groups. Siloxanes are used in the manufacture of personal hygiene, health care and industrial products. As a consequence of their widespread use, siloxanes are found in wastewater and in solid waste deposited in landfills. At wastewater treatment plants and landfills, low molecular weight siloxanes volatilize into digester gas and landfill gas. When this gas is combusted to

Table 7. Results of determination VOC in LFG in Jawidz landfill.

Compound	m/z	Concentration of microconsituents in samples taken from 50 cm depth [μg/m³]		Concentration of microconsituents in samples taken surface
Benzene	78	9	6	Nd
Toluene	91	52	3.1	0.2
Xylenes	120	3.4	3.8	Nd
Trichlorethylene	160	0.9	0.6	Nd
Tetrachlorethylene	195	Nd	Nd	Nd

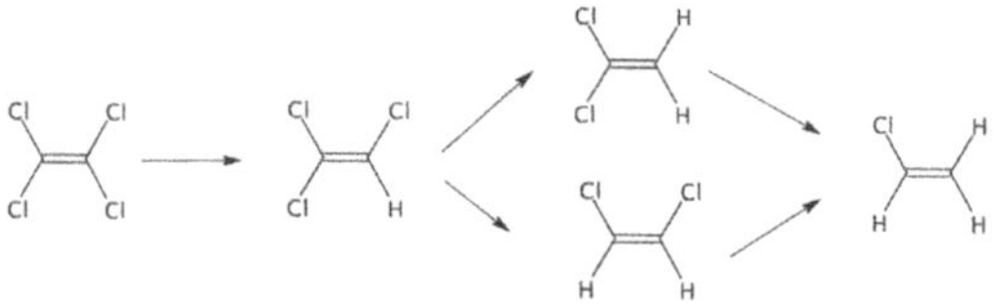

Figure 1. Anaerobic degradation of tetrachloroethylene.

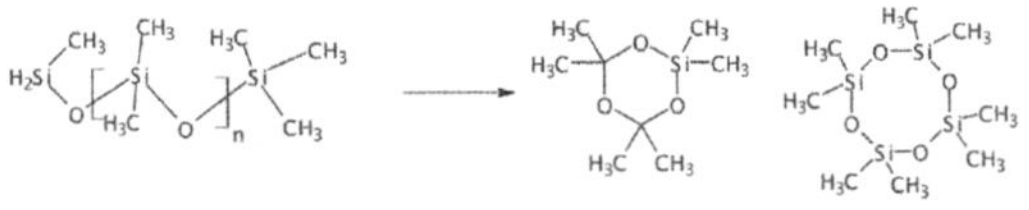

Figure 2. Formation of cyclic siloxanes in anaerobic conditions.

generate power (such as in gas turbines, boilers or internal combustion engines), siloxanes are converted to silicon dioxide (SiO_2), which can deposit in the combustion and/or exhaust stages of the equipment.

Cyclic siloxanes determined in landfill gas can be formed from their linear analogues via reaction given in Figure 2.

ACKNOWLEDGEMENTS

Authors thanks to Prof. Lucjan Pawłowski for the fruitful discussions and consultations.

This work was supported by Minister of Science and Higher Education, Research Project No PBZ-MEiN 3/2/2006 Inżynieria procesów ograniczania emisji oraz utleniania gazów szklarniowych i cieplarnianych. (Engineering of harmful gases emission reduction and utylisation).

REFERENCES

Accettola, F. Guebitz, G.M. & Schoeftner R. 2008. Siloxane removal from biogas by biofiltration: biodegradation studies. Clean Techn. Environ. Policy 10:211–218.

Arnold, M. 2009. Reduction and monitoring of biogas trace compounds, Espoo, VTT Tiedotteita – Research – Notes 2496. 1–74.

Chiriac, R. Carre, J. Perrodin, Y. Fine, L. & Letoffe, J-M. 2007. Characterisation of VOCs emitted by open cells receiving municipal solid waste. J. Haz. Mat. 149:249–263.

Chiriac, R. Carre, J. Perrodin, Y. Vaillant, H. Gasso, S. & Miele P. 2009. Study of the dispersion of VOCs emitted by a municipal solid waste landfill. Atmos. Environ. 43:1926–1931.

Davoli, E. Gangai, M.L. Morselli, L. & Tonelli, D. 2003. Characterisation of odorants emissions from landfills by SPME and GC/MS. Chemosphere 51:357–368

Dewil, R. Appels, L. & Baeyens, J., 2006. Energy use of biogas hampered by the presence of siloxanes. Energy Conversion Manage. 47:1711–1722.

Dudzinska, M. & Czerwiński, J. 2010. Persistent Organic Pollutants (POPs) in Leachates from Municipal Landfills. Int. J. Environ. Eng. (accepted for print)

Durmusoglu, E. Taspinar, F. & Karademir, A. 2009. Health risk assessment of BTEX emissions in the landfill environment. Journal Hazardous Material doi:10.1016/j.jhazmat.2009.11.117.

Finocchio, E. Montanari, T. Garuti, G. Pistarino, C. Federici, F. Cugino, M. & Busca, G. 2009. Purification of Biogases from Siloxanes by Adsorption: On the Regenerability of Activated Carbon Sorbents. Energy Fuels 23:4156–4159.

Gates, P.N. Harrop, H.A. Pridham, J.P. & Smethurst B. 1997. Can microorganisms convert antimony trioxide or potasium antimonyltartarate to methylated stilbenes? Sci. Total Environ. 205:215–221.

Glindemann, D. Morgenstern, P. Wennrich, R. Stottmeister, U. & Bergmann, A. 1996. Toxic oxide deposites from the combustion of landfill gas and biogas. Environ. Sci. Pollut. Res.3:75–77.

Hoch, M. 2001. Organotin compounds in the environment – an overview. Applied Geochem.16:719–743.

Ilgen, G. Glindemann, D. Herrmann, R. Hertel, F. & Huang, J-H. 2008. Organometals of tin, lead and mercury compounds in land?ll gases and leachates from Bavaria, Germany. Waste Manage. 28:1518–1527.

Iwakiri, R. Yoshihira, K. Futagami, T. Goto, M. & Furukawa, K. 2004. Total degradation of Pentachloroethane by an Engineered Alcaligenes Strain Expressing a modified Camphor Monooxygenase and a Hybrid Dioxygenase, Biosci. Biotechnol. Biochem. 68:1353–1356.

Kim, H-J. Yoshida, H. Matsuto, T. Tojo, Y. & Matsuo, T. 2009. Air and landfill gas movement through passive gas vents installed in closed landfills. Waste Manage. doi:10.1016/j.wasman.2009.10.005.

Kim K-H. 2006. Emissions of reduced sulfur compounds (RSC) as a landfill gas (LFG): A comparative study of young and old landfill facilities. Atmos. Environ. 40: 6567–6578.

Kim, K-H. Baek, S.O. Choi, Y-J. Sunwoo, Y. Jeon, E-C. &Hong, J.H. 2006. The Emission of Major Aromatic VOC as Landfill Gas from Urban Landfill Sites in Korea. Environ. Monitor. Asses. 118: 407–122.

Kot, A. & Namieśnik, J. 2000. The role of speciation in analytical chemistry. TRAC 19:69–79

Krachler, M. Emons, H. & Zheng, J. 2001. Speciation of antimony for the 21st century: Promises and pitfalls. TRAC 20:79–90.

Krupp, E.M. Johnson, C. Rechsteiner, C. Moir, M. Leong, D. & Feldmann, J. 2007. Investigation into the determination of trimethylarsine in natural gas and its partitioning into gas and condensate phases using (cryotrapping)/gas chromatography coupled to inductively coupled plasma mass spectrometry and liquid/solid sorption techniques. Spectrochim. Acta B 62:970–977.

Lindberg, S.E. Southworth, G. Prestbo, E.M. Wallschlager, D. Bogle, M.A. & Price, J. 2005. Gaseous methyl- and inorganic mercury in landfill gas from landfills in Florida, Minnesota, Delaware, and California. *Atmos. Environ.* 39:249–258.

Lindberg, S.E. Wallschlager, D. Prestbo, E.M. Bloom, E.M. Price, J. & Reinhart D. 2001. Methylated mercury species in municipal waste landfill gas sampled in Florida, USA. *Atmos. Environ.* 35:4011–4015.

Lombardi, L. Carnevale, E. & Corti A. 2006. Greenhouse effect reduction and energy recovery from waste landfill. *Energy* 31:3208–3219.

Maillefer, S. Lehr, C.R. & Cullen, W.R. 2003. The analysis of volatile trace compounds in landfill gases, compost heaps and forest air. *Appl. Organometal. Chem.*17:154–160.

Mersiowsky, I. 2002. Long-term fate of PVC products and their additives in land?lls. *Prog. Polym. Sci.* 27:2227–2277

Mester, Z. & Sturgeon, R. 2005. Trace element speciation using solid phase microextraction. *Spectrochim. Acta B.* 60:1243–1269.

Michalzik, B. Ilgen, G. Hertel, F. Hantsch, S. & Bilitewski, B. 2007. Emissions of organo-metal compounds via the leachate and gas pathway from two differently pre-treated municipal waste materials – A land?ll reactor study. *Waste Manage.* 27:497–509.

Muradov, N.Z. & Veziroglu, T.N. 2006. From hydrocarbon to hydrogen–carbon to hydrogen economy. *Intern. J. Hydrogen Energy* 30:225–237.

Nagamori, M. Ono, Y. Kawamura, K. Yamada, M. Ishigaki, T. & Ono Y. 2008. Changes in the Composition of Gases Emitted from a Final Landfill Site, Proc of the 17th Annual Conference of The Japan Society of Waste Management Experts, 244–254.

Nagatomi, Y. Yamamoto, H. &Yamaji, K. 2007. Effects of Regional Characteristics and Measures of Greenhouse Gas Reduction on Waste Disposal System. *J. Jpn. Inst. Energy*, 86:693–699.

Parker, T. Dottridge, J. & Kelly S. 2002. Investigation of the Composition and Emissions of Trace Components in Landfill Gas, UK EPA, R&D Technical Report P1-438/TR.

Paschke, A. Vrana, B. Popp, P. & Schüürmann, G. 2006. Comparative application of solid-phase microextraction fibre assemblies and semi-permeable membrane devices as passive air samplers for semi-volatile chlorinated organic compounds. A case study on the landfill "Grube Antonie" in Bitterfeld, Germany. *Environ. Pollut.* 144:414–422.

Pawłowska, M. Czerwiński, J. & Stępniewski, W. 2008. Variability of the non-methane volatile organic compounds (NMVOC) composition in biogas from sorted and unsorted landfill material. *Archives Environ. Prot.* 34(3): 287–298

Pinel-Raffaitin, P. Amouroux, D. LeHecho, I. Rodriguez-Gonzalez, P.& Potin-Gautier, M. 2008. Occurrence and distribution of organotin compounds in leachates and biogases from municipal landfills. *Water Res.* 42:987–996.

Popat, S.C. & Deshuses, M.A. 2008. Biological Removal of Siloxanes from Landfill and DigesterGases: Opportunitiesand Challenges. *Environ. Sci. Technol.* 42:8510–8515.

Rasi, S. 2009. Biogas composition and upgrading to biomethane, *Jyvaskyla Studies in Biol. Environ. Sci.* 202:1–135

Rasi, S. Veijanen, A. & Rintala, J. 2007. Trace compounds of biogas from different biogas production plants. *Energy* 32:1375–1380.

Read, A.D. Hudgins, M. Harper, S. Phillips, P. & Morris, J. 2001. The successful demonstration of aerobic landfilling The potential for a more sustainable solid waste management approach? *Resources, Conservation and Recycling* 32:115–146.

Ritzkowski, M. & Stegmann, R. 2007. Controlling greenhouse gas emissions through landfill in situ aeration. *Internat. J. Greenhouse Gas Control*, 1:281–288.

Saral, A. Demir, S. & Yildiz, S. 2009. Assessment of odorous VOCs released from a main MSW landfill site in Istanbul-Turkey via a modelling approach. *J. Haz. Mat.* 168:338–345

Scheutz, C. Mosbæk, H. & Kjeldsen, P. 2004. Attenuation of Methane and Volatile Organic Compounds in Landfill Soil Covers, *J. Environ. Qual.* 33:61–71.

Scheutz, C. Bogner, J. Chanton, J.P. Blake, D. Morcet, M. Aran, C. & Kjeldsen, P. 2007. Atmospheric emissions and attenuation of non-methane organic compounds in cover soils at a French landfill, *Waste Manage. (Oxford)*, doi:10.1016/j.wasman.2007.09.010.

Schöler, H.F. & Keppler, F. 2003. Abiotic Formation of Organohalogens During Early Diagenetic Processes, [in:]*The Handbook of Environmental Chemistry* Vol.3,Part P, 63–84.

Shafi, S. Sweetman, A. Hough, R.L. Smith, R. Rosevear, A. & Pollard S.J.T. 2006. Evaluating fugacity models for trace components in landfill gas. *Environ. Pollut.* 144:1013–1023.

Shin, H.C. Park, J.W. Park, K. & Song, H.C. 2002. Removal characteristics of trace compounds of landfill gas by activated carbon adsorption. *Environ. Pollut.* 119:227–236.

Smet, E. Van Langenhove, H. & De Bo, I. 1999. The emission of volatile compounds during the aerobic and the combined anaerobic/aerobic composting of biowaste. *Atmospheric Environ.* 33:1295–1303.

Song, S-K. Shon, Z-H. Kim, K-H. Kim, S.C. Kim, Y-K. & Kim J-K. 2007. Monitoring of atmospheric reduced sulfur compounds and their oxidation in two coastal landfill areas. *Atmos. Environ.* 41:974–988.

Stoddart, J. Cox, A.G. & McLeodb, C.W. 2000. Elemental analysis of landfill gas by ICP emission spectrometry – new approach for monitoring organochlorine compounds. *J. Anal. At. Spectrom.*15:1498–1500.

Szymański. K. Sidełko, R. Janowska, B. & Siebielska, I. 2007. Landfills Monitoring. 8th National Scientific Conference on Complex Problems of Environmental Engineering, Koszalin-Darłówko, Poland. 75–136.

Takuwa Y. Matsumoto T. Oshita K. Takaoka M. Takeda N., 2007. Trace constituent in landfill gas. Kankyo Eisei Kogaku Kenkyu (Environmental and Sanitary Engineering Research) 21, 155–158.

Urban, W. Lohmann, H. & Salazar Gómez, J.I. 2009. Catalytically upgraded landfill gas as a cost-effective alternative for fuel cells. *J. Power Sources* 193:359–366.

Zou, S.C. Lee, S.C. Chan, C.Y. Ho, K.F. Wang, X.M. Chan, L.Y. & Zhang, Z.X. 2003. Characterization of ambient volatile organic compounds at a landfill site in Guangzhou, South China. *Chemosphere* 51:1015–1022.

Water quality and supply

Environmental Engineering III – Pawłowski, Dudzińska & Pawłowski (eds)
© *2010 Taylor & Francis Group, London, ISBN 978-0-415-54882-3*

A combined 2D-3D seismic survey or fracturing geothermal systems in central Poland

A.P. Barbacki

Department of Mining Surveying and Environmental Engineering, University of Mining and Metallurgy, Krakow, Poland

ABSTRACT: Drilling costs often limit geothermal development, despite favourable hydrogeological conditions. Reducing such costs could increase the competitiveness of geothermal energy. This goal is achievable if geophysical surveys detect the presence of fluids in geothermal systems before drilling begins. A non-porous Triassic reservoir in central Poland was investigated using integrated 2D-3D seismic surveys. This reservoir is associated with unevenly-distributed fractures characterised by high permeability. The study's objective was to deliver a detailed picture of fault geometry in near an existing geothermal well near Łowicz city, Poland. The study identified zones with high fracture permeability for future sitings of geothermal wells.

Keywords: Geothermal reservoir, seismic survey, Central Poland.

1 INTRODUCTION

Worldwide, many exploration and exploitation wells were drilled because they looked promising in terms of high-temperature rock formations, but they turned out to lack sufficient permeability to sustain commercial water production. So, the major problem to be faced is how to detect fractures and high permeability zones.

Seismic methods are based on the propagation of elastic waves created artificially (by dynamite, vibrating machines, etc.), or by natural phenomena such as earthquakes. In geothermal exploration they are used primarily for structural definition, but they also serve to characterise the reservoir, as the elastic parameters of wave propagation (compression velocity, shear velocity, density) and wave attenuation and frequencies are related to lithology, fracturing, temperature, fluid content, pressure and saturation.

In a typical crystalline environment (volcanic, metamorphic and granite setting) seismic methods are rarely used (Honjas et al. 1997, Unruh et al. 2001, Cameli et al. 2000, Fiordelisi et al. 2005, Bertini et al. 2005, Cappetti et al. 2005). But in sedimentary basins, seismic reflection provides the best geometrical resolution among surface geophysical methods, of horizontal and weakly dipping layers or structures, and hence is invaluable in characterising reservoirs (Hersir & Bjornsson 1991).

Seismic reflection was used for oil exploration in the sedimentary basins of central Poland (Łowicz region) during the 1970s. In 2008, much work was done on reprocessing the oil-industry 2D seismic line using the new capabilities of modern software packages. The goal of this reprocessing was to provide structural information for 3D geological modelling and the initial identification of fault zones. A special 2D-3D reflection seismic survey was designed to produce a more detailed picture of reservoir geometry and fault zones, especially in the vicinity of the existing geothermal well Kompina-2, considered a promising for geothermal use (Figure 1).

The Kompina-2 well has a free outflow of brine with temperatures exceeding 100°C from Early Triassic formations and it is located close to an operational geothermal heat plant, interested in exploiting geothermal energy for electricity production.

The aim of the research at this site was to elaborate and test seismic acquisition methods for investigating faulting and fracturing geothermal systems within 2 km of the Kompina-2 well. This area was considered suitable for potential injection wells for the Kompina-2 well.

This work was performed within the EU 6th Framework Programme: 'Integrated Geophysical Exploration Technologies for deep fractured geothermal systems' (acronym I-GET).

2 MATERIALS AND METHODS

2.1 *Geological and geothermal setting*

The Łowicz area is part of the Warsaw synclinorium, a structure created during the Mesozoic period. It is located close to the boundary between two major tectonic units – the Precambrian platform (Baltic Plate) and the Palaeozoic platform (Caledonides and Variscan belts). This boundary is called the Teisseyre-Tornquist zone and was subject to tectonic movements,

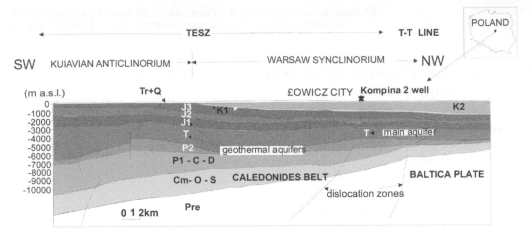

Q+Tr - Quaternary & Tertiary, **K2** - Late Cretaceous, **K1** - Early Cretaceous, **J3** - Late Jurassic, **J2** - Middle Jurassic, **J1** - Early Jurassic, **T** - Triassic, **P2** - Zechstein, **P1-C-D** - Rotliegendes+Carboniferous+ Devonian, **Cm-O-S** - Cambrian+Ordovician+Silurian, **Pre** - Precambrian, **TESZ**- Trans-European Sature Zone, **T-T line** - Teisseyre-Tornquist line

Figure 1. Geological cross-section of the Łowicz area (after Dembowska and Marek, 1986 – modified).

usually vertical, during the Variscan and Alpine orogenesis (Figure 1).

In the Łowicz area, the top of the Precambrian crystalline basement is located at a depth of about 8 km (Ryka 1978). Geothermal aquifers with high flow rates (more than 100 m³/h) occur in Late Triassic, Early Jurassic and Early Cretaceous formations. Early Jurassic sandstones are considered among the major geothermal aquifers with output exceeding 150 m³/h, but reservoir temperatures are rather low, at about 80°C. Aquifers of the Middle and Early Triassic exhibit somewhat lower output, but the reservoir temperatures there exceed 100°C. The general temperature gradient is close to 2.7°C/100 m. An average heat-flow density for the Łowicz area was estimated to be 65 mW/m² (Paczyński & Sadurski 2007).

The most interesting results from hydrogeological testing of the Triassic formations were obtained from the Kompina-2 well, where a free outflow of brine with a salinity of 337 g/l, and a temperature of 107°C was obtained at a depth of 4110 to 4115 m (Hajto 2007, Bujakowski et al. 2005).

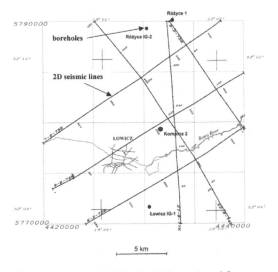

Figure 2. Location of 2D seismic lines selected for reprocessing in the Łowicz area.

2.2 Reprocessing of 2D archival seismic surveys

About 100 km of reflection seismic profiles were obtained in the Łowicz area during the mid-1970s, and four deep wells were drilled for oil-exploration (Figure 2).

Commonly gamma-ray, resistivity, calliper and, locally, temperature measurements were made in these boreholes. Acoustic measurements were available for two wells. The reliability of the temperature measurements is, however, low, because of insufficient delay for thermal stabilisation.

The studies focused on structure and tectonics near the Kompina-2 well where three geothermal aquifers have been identified. Because of the high temperatures

and outflows recorded at the Triassic aquifer (probably connected with its tectonic engagement and the presence of faults), the investigation of this aquifer was the most important. This part of the research was considered to provide a first approximation of geological structure.

Six 2D archive seismic lines (208.40 km long in total), and logs from nine wells near the Kompina-2 well were selected for reprocessing and reinterpretation.

The main procedures during reprocessing using the Omega Seismic System included the following steps: trace balance, surface consistent deconvolution, surface consistent amplitude compensation, RMS gain,

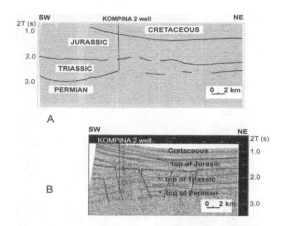

A

B

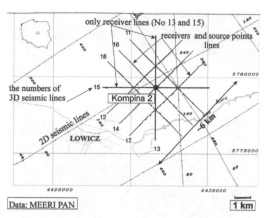

Figure 3. A. Time seismic section 9-8-76K before reprocessing (location of profile on Figure 2); B. Fragment of
section 9-8-76K after reprocessing (with migration).

Figure 5. Location of seismic sources (red dots) and
receiver points (blue dots) in the combined 2D/3D acquisition
scheme.

2.3 2D-3D seismic survey

The conceptual 3D model constitutes an efficient
tool for selecting reliable geothermal targets prior to
drilling and additionally 3D seismic ones enable better
target definition and significantly reduce exploration
risk (Cappetti et al., 2005).

In this project, the combined 2D-3D seismic method
was used mainly to visualise the fault zones and, consequently, make more reliable geothermal borehole
sites.

In designing the seismic survey for detailed recognition of geological tectonics around the Kompina-2
well (area 6 × 6 km), the seismic layout was projected
in two ways (Figure 5):

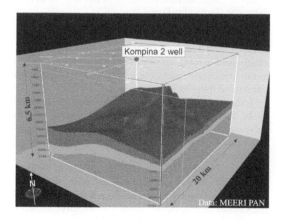

Figure 4. Spatial model of the top of Buntsandstein performed after reprocessing (seismic lines visible on the top of
cube related to Figure 2).

random noise attenuation, direct reflection statics,
reflection miser and filtration; and after stacking:
zero phase deconvolution, RAP and migration procedure. Figure 3 shows an example of a selected
seismic section before and after the reprocessing
steps.

After reprocessing, several 3D geological models
were developed to describe the geological characteristics of the area around Łowicz city, focusing on the
structural pattern of the reservoir horizons. These models are based on data from four wells, all deeper than
4000 m, and the six reprocessed 2D seismic sections
mentioned above. The general 3D model includes an
area of 400 km^2 with altitude ranging from 0 m to
−6500 m a.s.l., including strata from the Cretaceous
to Rotliegendes age. It gives an overview of the general geological setting, encompassing a fault pattern
including the dominant faults and thickness variations
of stratigraphic sub-formations, as for example within
the Triassic sediments in Figure 4.

1. as six crossing 2D seismic lines on a 1 km grid and
   nominal fold detected along the receiver lines and
   on bin lines located between receiver lines (fold on
   receiver lines: 132); and
2. as a 3D survey with homogeneous fold area of 2.44
   sq. km in the centre of the sketch (fold in central
   part of survey area: 6).

The designed maximum offset of $X_{max} = 6188$ m
allows recognition of the deepest geological horizons
in the survey area (i.e. about 7000 m). The project
included a combined 2D dimensional seismic survey
consisting of eight crossings of seismic lines, each
6760 m in length. The total length of the planned
seismic lines was 54 km. Six lines were sources and
receivers, while eight were receivers (two receiver
lines – nos. 13 and 15 were supplementary, Figure 5).
Correct distribution of the nominal fold was the main
objective during the field work. A record length of 6
seconds was planned because of long maximum offset.
Four lines intersected at the location of the Kompina-2
well. Due to the environmental requirements, 4 vibro-
seis with a sweep length of 16 s and with a frequency
range of 8–100 Hz for one seismic source were used.
Many traces (total number per record: 1020) with short
offsets (less then 1000 m) assured good recognition
of geological horizons over the shallowest target (at a

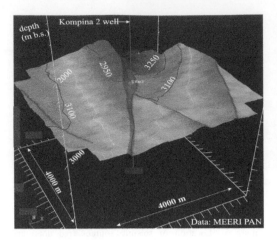

Figure 6. Spatial image of the top of Triassic formation where outflows of thermal waters are related to fracture zones near dominant faults.

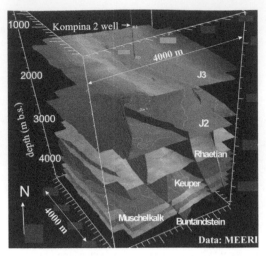

Figure 7. Selected seismic boundaries and fault surfaces in 3D image (J2, J3 as in Figure 1).

depth of 1100 m – Early Cretaceous) and some part of long offsets, longer then 4.5 km, assured projection of geology below the deepest target (~7000 m).

Processing of seismic data was performed by standard procedures similar to those used for hydrocarbon exploration with additional extensions for geothermal purposes (appropriate time interval to grasp water-bearing layers in seismic attributes analysis).

The interpretation used a preliminary geological model, stratigraphic linking of seismic interfaces with geological well-data from the Kompina-2 well, structural interpretation of recognised tectonics by migration with FK filtration, setting up of a velocity model and time-depth transformation performed by using log-data from the Kompina-2 well and construction of synthetic seismograms.

The analysis confirms the complicated geological structure of the geothermal site studied caused by vertical movements and salt tectonics related to the Zechstein formation. A structural map of the Triassic horizons (Figure 6) shows tectonic features like normal faults developed along a NW-SE axis.

3 RESULTS AND DISCUSSION

The results indicate that this seismic method is an important and useful tool for geothermal research.

This paper discusses only one small structural aspect of the potential uses of seismic methods in geothermal exploration. Additionally, maps of average and interval velocity distribution for distinguished boundaries, time slices and seismic attributes (RMS amplitude, instantaneous phase, cosine instantaneous phase, acoustic impedance) were created by the project for a better understanding of the deeper nature of the geothermal reservoir.

However, seismic methods have not been widely used in geothermal exploration because of their high cost. This special 2D-3D seismic method is much cheaper than standard 3D imaging, thus the methodology used in this project can be applied to other areas at the local scale, particularly if 2D surveys had been conducted there in the past.

Today's equipment enables thousands of channels to be recorded simultaneously, so allowing for an increase in the fold area in the bin space without extending the acquisition time (improvement in seismic resolution).

Seismic methods, especially the reflection method, have much higher resolution than most other geophysical techniques. They deliver a detailed picture of the crustal layers, depth, slope and location of faults and displacements as well as folding structures in sediments.

It must be acknowledged that no particular geophysical method is universally applicable, and methods should be chosen carefully to suit the situation (most geophysical methods display a progressive reduction in resolution at greater depths). Furthermore, geophysical exploration cannot stand alone, but must be applied along with geology, geochemistry, hydrogeology and borehole data to resolve the nature of geothermal systems.

REFERENCES

Bertini, G., Casini, M., Ciulli, B., Ciuffi, S. & Fiordelisi, A. 2005. Data revision and upgrading of the structural model of the Travale geothermal field (Italy). In: *Proceedings of the World Geothermal Congress*. Antalya, 24–29 April.

Bujakowski, W. & Barbacki, A.P. & Pająk L. 2005. Geological and technical conditions for developing geothermal energy in balneology and space heating in Gostynin city. *Technika Poszukiwań Geologicznych Geosynoptyka i Geotermia* 6: 3–11.

Cameli, M.C., Ceccarelli, A. & Dini, I. 2000. Contribution of the seismic reflection method to the location of deep fractured levels in the geothermal fields of Southern Tuscany

(Central Italy). In: *Proceedings of the World Geothermal Congress 2000.* Kyushu-Tohoku: Japan, May 28–June 10: 1025–1029.

Cappetti, G. & Fiordelisi, A. & Casini, M. & Ciuffi, S.& Mazott, A. 2005. A new deep exploration program and preliminary results of a 3D seismic survey in the Larderello-Travale geothermal field (Italy). In: *Proceedings of the World Geothermal Congress.* Antalya, 24–29 April.

Dembowska, J., Marek, S. 1986. *Profiles of the deep geological wells: Łowicz IG-1, Raducz IG-1.* Wydawnictwa Instytutu Geologicznego: Warszawa.

Fiordelisi, A., Moffatt, J., Ogliani, F., Casini, M., Ciuffi, S. & Romi, A. 2005. Revised processing and interpretation of reflection seismic data in the Travale geothermal area (Italy). In: *Proceedings of the World Geothermal Congress.* Antalya, 24–29 April.

Hajto, M., 2007. Geothermal energy resources in mesozoic and paleozoic formations of the Polish Lowlands. *Technika Poszukiwań Geologicznych, Geotermia, Zrównoważony Rozwój* 2: 57–60.

Hersir. G.P. & Bjornsson. A. 1991. *Geophysical exploration for geothermal resources.* The United Nations University: Reykjavik.

Honjas, W., Pullammanappillil, S.K. & Lettis, W.R. 1997. Predicting shallow earth structure within the Dixie Valley geothermal field, Dixie Valley, Nevada, using a nonlinear velocity optimization scheme. In: *Proceedings of the Twenty-First Workshop on Geothermal Reservoir Engineering.* Stanford University, Stanford, CA, January 27–29: 153–160.

Paczyński, B. & Sadurski, A. 2007. *The regional hydrogeology of Poland (part II).* Polish Geological Institute. Warszawa.

Ryka, W. 1978. *Lithofacial and paleogeographic atlas of Permian deposits in the platform regions of Poland.* Instytut Geologiczny: Warszawa.

Unruh, J., Pullammanappallil, S.K., Honjas, W. & Monastero, F. 2001. New seismic imaging of the Coso geothermal field, Eastern California. In: *Proceedings of the Twenty-Sixth Workshop on Geothermal Reservoir Engineering.* Stanford University, Stanford, CA, January 29–31, SGP-TR-168: 164–170.

Environmental Engineering III – Pawłowski, Dudzińska & Pawłowski (eds)
© 2010 Taylor & Francis Group, London, ISBN 978-0-415-54882-3

Effect of the van Genutchen model tortuosity parameter on hydraulic conductivity calculations

M. Iwanek, I. Krukowski, M. Widomski & W. Olszta

Faculty of Environmental Engineering, Lublin University of Technology, Lublin, Poland

ABSTRACT: The van Genuchten–Mualem model is one of the most popular indirect methods that is used to determine unsaturated hydraulic conductivity $K(h)$. One of the parameters of the model is L. Investigators usually assume $L = 0.5$, but it may vary widely.

The objective of this paper is to test the influence of the van Genuchten–Mualem L value on the conformity of the model and empirical $K(h)$ results for eight hydrogenic soil samples. The obtained results indicated that using an estimated L value other than 0.5 improved the conformity mentioned above, but this improvement was not significant in most of the considered cases.

Keywords: Hydraulic conductivity coefficient, van Genuchten–Mualem model, tortuosity parameter, peat soils.

1 INTRODUCTION

Unsaturated hydraulic conductivity – the pressure head relationship $K(h)$ – is one of the basic functions connected with water flow and pollution transport in soils. Many methods – both in the field and laboratory – have been developed and tested over time for the direct determination of this curve (e.g. Gardner 1956, Hillel et al. 1972, Stephens 1985, Stolte et al. 1994, Butters & Duchateau 2002, Fujimaki & Inoue 2003, Meadows et al. 2005), but they are expensive and time-consuming (e.g. Russo & Bresler 1981, Vereecken 1995, Chen & Payne 2001, Minasny et al. 2004, Lazarovitch et al. 2007). Hence, estimation methods for the determination of $K(h)$ based on more easily measured soil parameters has gained popularity and have been developed in parallel with the direct methods.

There are several algorithms to indirectly determine $K(h)$ (e.g. Burdine 1953, Gardner 1958, Brooks & Corey 1966, Schuh & Bauder 1986, Wösten & van Genuchten 1988, Kosugi 1996), and the van Genuchten–Mualem model (Mualem 1976, van Genuchten 1980) is one of these. The hydraulic conductivity $K(h)$ function for unsaturated conditions and the water retention curve in this model are given by equation 1 and 2, respectively:

$$K_r(h) = K_0 \cdot \left[\frac{1}{1+(\alpha \cdot h)^n} \right]^{\left(1-\frac{1}{n}\right) \cdot L} \left\{ 1 - \left[\frac{(\alpha \cdot h)^n}{1+(\alpha \cdot h)^n} \right]^{\left(1-\frac{1}{n}\right)} \right\}^2 \quad (1)$$

$$\frac{\theta_s - \theta_r}{\theta(h) - \theta_r} = \left[1 + (\alpha \cdot h)^n \right]^{\left(1-\frac{1}{n}\right)} \quad (2)$$

where: $K(h) =$ unsaturated hydraulic conductivity function, mm day^{-1}; $K_0 =$ hydraulic conductivity coefficient acting as a matching point at $h = 0$, mm day^{-1}; $K_r =$ dimensionless relative hydraulic conductivity coefficient; $h =$ pressure head, cm; α ($\alpha > 0$) = parameter related to the inverse of the air entry pressure, cm^{-1}; $L =$ dimensionless constant that accounts for pore tortuosity and pore connectivity; n ($n > 1$) = dimensionless measure of the pore-size distribution; $\theta(h) =$ volumetric water content corresponding to the pressure head h [cm], cm$^3 \cdot$ cm^{-3}; θ_r and θ_s [cm$^3 \cdot$ cm^{-3}] = residual and saturated water content.

Recently, the advanced progress in computer techniques has led to inverse procedures of $K(h)$ determination becoming attractive to scientists (Bitterlich et al. 2004). In these methods, unknown parameters are estimated by matching a numerical model to measured data (Finsterle 2004). Most inverse technique applications for unsaturated water flow use the van Genuchten–Mualem functions (formulas 1 and 2) to describe hydraulic properties (Bitterlich et al. 2004, Finsterle 2004, Ippish et al. 2006, Lazarovitch et al. 2007). Solving the inverse problem to determine an unsaturated hydraulic conductivity relationship results in the best fit, when K_0 and L (formula 1) are treated as unknown parameters (Durner 1994). On the other hand, the simultaneous estimation of large number of parameters can cause some problems (Schwartz & Evett 2002, Schwärzel et al. 2006, Lazarovitch et al. 2007), so it is possible to reduce the occurrence of such problems by expressing precise values of unknown parameters (1). To this aim, many investigators (e.g. Young et al. 2002, Zhang et al. 2003b, Ippish et al. 2006, Lazarovitch et al. 2007) assume $L = 0.5$ (as proposed by Mualem 1976) and use saturated hydraulic conductivity K_s for K_0 in expression (2) as optimal values, but the use of default values for L and K_0 can cause several limitations of this equation (1) (Schaap & van Genuchten 2006). Moreover, estimation of the two parameters from the measured curves results in different values than the abovementioned.

The published results of investigations have indicated that exponent L may be both positive (Chen & Payne 2001, Shinomiya et al. 2001, Schwartz & Evett 2003, Tuli et al. 2005) and negative (Schaap & Leij 2000, Spohrer et al. 2005, Schaap & van Genuchten 2006, Iwanek et al. 2007) and, moreover, it is related to the organic matter content (Wösten et al. 1995).

K_0 can be estimated as a fitting parameter without physical meaning (Vereecken 1995, Šimůnek & van Genuchten 1997, Schaap & Leij 2000, Schwärzel et al. 2006) and, when obtained this way, the value is smaller (often much smaller) than the measured K_s value. Some investigators (van Genuchten & Nielsen 1985, Luckner et al. 1989, Olszta & Zaradny 1991, Shinomiya et al. 2001, Spohrer et al. 2005, Schaap & van Genuchten 2006) recommend assuming a matching point at slightly unsaturated conditions.

The objective of this paper is to test the influence of the van Genuchten L value on the compatibility of the fitting $K(h)$ curve obtained using the van Genuchten–Mualem model. Investigations were conducted for the chosen hydrogenic soils sampled at Wyżyna Lubelska, Poland. To verify the validity of our findings, statistical procedures such as correlation and determination were applied.

2 MATERIAL AND METHODS

The laboratory data for this study, including water retention characteristics (so-called pF-curves and unsaturated hydraulic conductivity in relation to pressure head for chosen peat soils monoliths) were available from The Institute for Land Reclamation and Grassland Farming department in Lublin. The following soil profiles were tested: Krowie Bagno 15 (Mt I aa), Sosnowica III (Mt I ba), Krowie Bagno 16 (Mt I bb), Krowie Bagno 476 (Mt I bb), Zienki Bukaciarnia 2 (Mt I b1), Krowie Bagno 20 (Mt I b3), Hanna-Holeszów 64/80 (Mt I bc), Dobroszyce 110/9 (Mt II cc) and Bełchatów (Danielów) 89/11(Mt II b4).

Water retention curves were determined using the sand box method and unsaturated hydraulic conductivity relationships $K(h)$ were determined using the drying monolith method. A detailed description of the abovementioned methods is given by Olszta (2004).

In numerical tests, $K(h)$ relationships were evaluated using the van Genuchten–Mualem model on the basis of formula (1). Parameters α and n were estimated using nonlinear regression analysis on the basis of relationship (2). Residual water content θ_r value was assumed to be equal to 0, because previous analyses (Iwanek et al., 2004; Iwanek et al. 2007) indicate that the θ_r value does not affect $k(h)$ calculation results essentially, and its value is near to zero for hydrogenic

soil. Some investigators (Hopmans & Overmars 1986, Wessolek et al. 1994, Weiss et al. 1998) also propose that $\theta_r = 0$ in calculations.

The values of L and K_0 were estimated for all considered hydrogenic soil monoliths using formula (2) in nonlinear regression analysis. Unsaturated hydraulic conductivity $K(h)$ was also calculated for these samples with $L = 0$ and $L = 0.5$ (as proposed by Mualem 1976). K_0, in these latter two cases, was assumed to be equal to the value estimated in the first case.

Statistical calculations were conducted using the STATISTICA program, with the significance level at $p = 0.05$.

3 RESULTS AND DISCUSSION

The results of measurements and calculations, including pF-curves and hydraulic conductivity, are shown in Figures 1–3. Parameters α and n of function (2), approximating the water retention curve gained on the basis of regression analysis using the nonlinear least squares method, are presented in Table 1. The coefficient of total determination R^2 was higher than 0.98 in all considered cases. Parameters α and n appeared to be highly statistically significant.

The shape of water retention curves evaluated on the basis of approximating function (2) (Figures 1a–3a) suggests that there is a good fit in the measured values of soil moisture in all but two cases (Krowie Bagno

Table 1. Estimated parameters of function (2) approximating water retention curve.

Object	A	p for α	n	p for n	R^2
Krowie Bagno 15 Mt I aa	0.0073	0.0171	1.4210	0.0000	0.9855
Krowie Bagno 16 Mt I bb	0.0055	0.0142	1.4266	0.0000	0.9865
Krowie Bagno 476 Mt I bb	0.0219	0.0003	1.2662	0.0000	0.9957
Zienki Bukaciarnia 2 Mt I b1	0.0245	0.0178	1.2205	0.0000	0.9859
Krowie Bagno 20 Mt I b3	0.0120	0.0004	1.3479	0.0000	0.9965
Hanna Holeszów 64/80 Mt I bc	0.0188	0.0000	1.2794	0.0000	0.9983
Dobroszyce 110/9 Mt II cc	0.0380	0.0336	1.1878	0.0000	0.9827
Bełchatów (Danielów) 89/11 Mt II b4	0.0126	0.0101	1.2567	0.0000	0.9915

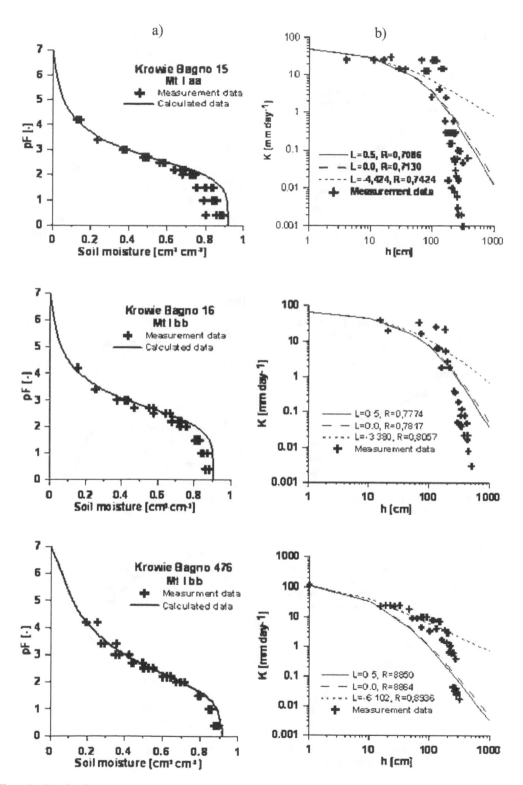

Figure 1. Results of measurements and calculations for Krowie Bagno 15, Krowie Bagno 16 and Krowie Bagno 476: a) Water retention curves, b) Diagram of unsaturated hydraulic conductivity as dependent on pressure head *h*.

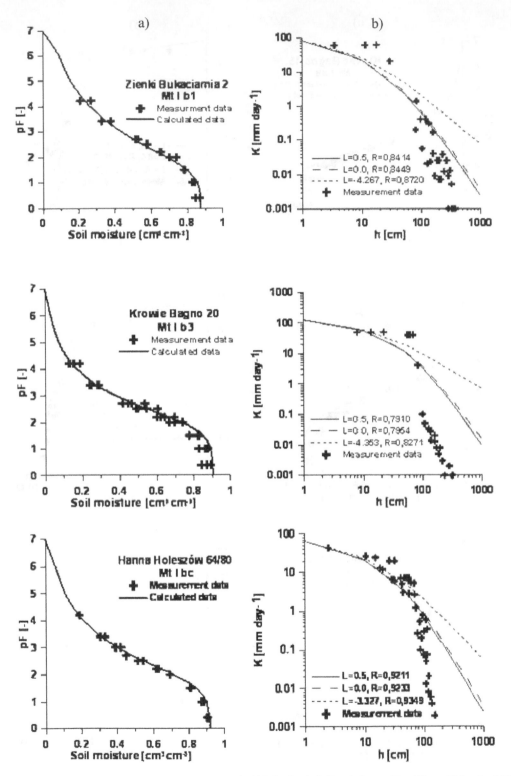

Figure 2.   Results of measurements and calculations for Zienki Bukaciarnia 2, Krowie Bagno 20 and Hanna Holeszów 64/80:
a) Water retention curves, b) Diagram of unsaturated hydraulic conductivity as dependent on pressure head h.

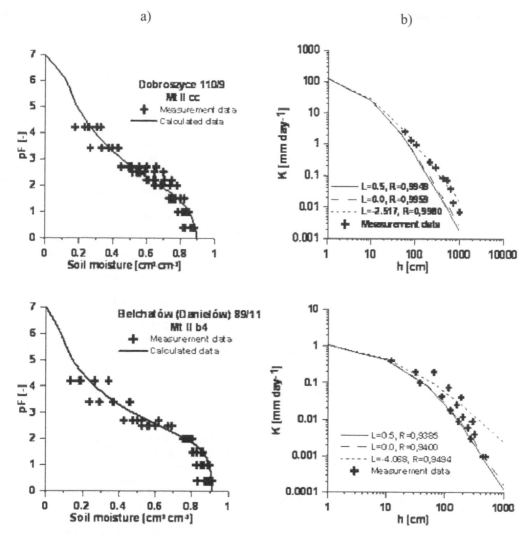

a)

b)

Figure 3. Results of measurements and calculations for Dobroszyce 110/9 and Bełchatów (Danielów) 89/11: a) Water retention curves, b) Diagram of unsaturated hydraulic conductivity as dependent on pressure head h.

15 and Krowie Bagno 16), in which some discrepancies occurred for volumetric water content that was near $0.80\ cm^3 \cdot cm^{-3}$. This was confirmed by regression analysis (Table 2). The correlation between the measured and calculated results depends on the h values of soil moisture and was significant at $p < 0.05$ for all cases; the coefficient of total determination R^2 was larger then 0.98 for all cases except for Bełchatów (Danielów) 89/11. Y-intercept b_0 values were lower then 0.02 in most cases. Slope coefficients b_1 of the regression line were a little lower then 1 in all cases; values that were most different were for Krowie Bagno 15 and Krowie Bagno 16. The best fit of water retention curves into measured values was obtained for Hanna Holeszów 64/80 (b_0 values were nearest to 0, b_1 values were nearest to 1 and $R^2 > 0.99$).

The values of exponent L and K_0 estimated on the basis of formula (1) are presented in Table 3. All values appeared to be statistically significant – the highest p-value (0.0107) was gained for the estimated L value for Krowie Bagno 16. The rest of the p-values were lower or, in most cases, much lower then 0.01. The coefficient of total determination R^2 fitted in the range 0.5511–0.9959. Negative L values were obtained in all considered cases.

As mentioned in the Introduction, some researchers found that $L < 0$ in their investigation and attempted to explain this result (Schaap & Leij 2000, Zhang et al. 2003a, Iwanek et al. 2007). Because, theoretically, L should not be less then 0, it could be treated as an empirical parameter without any physical interpretation.

451

Table 2. Regression line parameters for calculated and measured moisture.

Object	Y-intercept b_1	Slope b_0	R^2
Krowie Bagno 15 Mt I aa	0.9007	0.0173	0.9895
Krowie Bagno 16 Mt I bb	0.9260	0.0131	0.9864
Krowie Bagno 476 Mt I bb	0.9479	0.0162	0.9880
Zienki Bukaciarnia 2 Mt I b1	0.9591	0.0287	0.9863
Krowie Bagno 20 Mt I b3	0.9496	0.0116	0.9845
Hanna Holeszów 64/80 Mt I bc	0.9642	0.0057	0.9964
Dobroszyce 110/9 Mt II cc	0.9476	0.0302	0.9290
Bełchatów (Danielów) 89/11 Mt II b4	0.9559	0.0154	0.9420

Table 3. Values of exponent L and K_0 estimated on the basis of formula (1).

Object	L	p for L	K_0	p for K_0	R^2
Krowie Bagno 15 Mt I aa	−4.4249	0.0000	63.4656	0.0000	0.5511
Krowie Bagno 16 Mt I bb	−3.3803	0.0107	85.4527	0.0000	0.6471
Krowie Bagno 476 Mt I bb	−6.1019	0.0000	266.6203	0.0000	0.9753
Zienki Bukaciarnia 2 Mt I b1	−4.2666	0.0055	277.1357	0.0000	0.8062
Krowie Bagno 20 Mt I b3	−4.3529	0.0064	198.4666	0.0000	0.6819
Hanna Holeszów 64/80 Mt I bc	−3.3273	0.0011	143.1272	0.0000	0.8722
Dobroszyce 110/9 Mt II cc	−2.5174	0.0024	606.3391	0.0001	0.9959
Bełchatów (Danielów) 89/11 Mt II b4	−4.0679	0.0094	2.3675	0.0000	0.9007

The next part of the research determined the unsaturated hydraulic conductivity for estimated L, $L = 0$ and $L = 0.5$ (as proposed by Mualem 1976). A comparison of measured and calculated unsaturated hydraulic conductivity values (Figures 1b–3b, Table 4) indicated a discrepancy in the whole range of the considered

Table 4. Regression line parameters for measured and calculated $K(h)$ depend on L.

Object	L	b_1	b_0	R^2
Krowie Bagno 15 Mt I aa	0.5000	0.5462	0.3395	0.5021
	0.0000	0.5501	0.4650	0.5084
	−4.4249	0.5457	3.0798	0.5512
Krowie Bagno 16 Mt I bb	0.5000	0.6096	0.5511	0.6044
	0.0000	0.6146	0.7246	0.6111
	−3.3803	0.6260	2.9608	0.6491
Krowie Bagno 476 Mt I bb	0.5000	0.5230	0.3112	0.7833
	0.0000	0.5254	0.4047	0.7858
	−6.1019	0.5384	3.8900	0.7895
Zienki Bukaciarnia 2 Mt I b1	0.5000	0.4296	0.1139	0.7079
	0.0000	0.4340	0.1530	0.7139
	−4.2666	0.4697	0.9629	0.7603
Krowie Bagno 20 Mt I b3	0.5000	0.6259	0.6647	0.6256
	0.0000	0.6323	0.8863	0.6326
	−4.3529	0.6590	5.1818	0.6841
Hanna Holeszów 64/80 Mt I bc	0.5000	0.8272	−0.3546	0.8484
	0.0000	0.8316	−0.2119	0.8525
	−3.3273	0.8495	1.2120	0.8741
Dobroszyce 110/9 Mt II cc	0.5000	0.5293	−0.0368	0.9898
	0.0000	0.5886	−0.0361	0.9917
	−2.5174	1.0002	−0.0050	0.9959
Bełchatów (Danielów) 89/11 Mt II b4	0.5000	0.8135	−0.0077	0.8808
	0.0000	0.8229	−0.0067	0.8837
	−4.0679	0.8877	0.0096	0.9013

pressure heads and for all considered L values for all samples except for Dobroszyce 110/9. The mentioned case was the only one in which using an estimated L in formula (1) gave much better results then for $L = 0$ and $L = 0.5$, and these results indicated a very good fit for calculated and measured $K(h)$ values (b_0 values were about 0, b_1 values were about 1 and $R^2 > 0.99$). A good fit ($b_0 = 0.096$, $b_1 = 0.8877$ and $R^2 = 0.9013$) was also obtained for Bełchatów (Danielów) 89/11 for the estimated L value, but these results were comparable to values gained for $L = 0$ and $L = 0.5$ in this case.

For the rest of the soil samples (Table 4), intercept b_0 values were distinctly different from 0 and the worst results were obtained for the estimated L values – one order of magnitude larger then for $L = 0$ and $L = 0.5$. Slope coefficients b_1 were about twice less then 1 and were comparable for all L values. Nevertheless, the largest values of b_1 (the best results) were for estimated L values, with the exception of Krowie Bagno 15, where b_1 was nearest 1 for $L = 0$. For all L values, the coefficients of total determination R^2 were comparable, and the largest values were obtained for the estimated L values.

The calculations and analysis in this study indicate that exponent L does not affect the fitting conformity of the model and empirical $K(h)$ values are essentially near saturation point. This is essential for pressure heads that are greater than about 10 cm.

4 CONCLUSIONS

The van Genuchten–Mualem model is one of the most popular methods to determine the $K(h)$ relationship and has been the basis of our investigations. Function (2) approximated the pF curve of the tested profiles well, and two parameters (α and n) estimated on this basis appear to be statistically significant. Exponent L, which was evaluated using formula (1), can also be deemed to be statistically significant and was negative in value for all tested soil monoliths.

Evaluation of the L values improved the conformity of the model and the fit of the empirical $K(h)$ results, but this improvement was significant in one considered case only. Thus, a large divergence between the calculated and measured $K(h)$ results for all but one chosen peat soil profile indicates the necessity of improving the calculation method of $K(h)$ determination for peat soil, taking into consideration exponent L.

However, the presented research results reveal a certain regularity and the results were obtained for eight chosen soil samples only. Thus, further research on a wider range of testing material, including both peat and mineral soils, is necessary to make definite conclusions.

NOMENCLATURE

b_0	Y-intercept
b_1	slope of the regression line coefficient
h	pressure head (cm)
$K(h)$	unsaturated hydraulic conductivity function (mm/d)
K_0	hydraulic conductivity coefficient acting as a matching point at $h = 0$ (mm/d)
K_r	dimensionless relative hydraulic conductivity coefficient
L	dimensionless constant that accounts for pore tortuosity and pore connectivity
n	dimensionless measure of the pore-size distribution ($n > 1$)
p	probability level
R^2	coefficient of total determination
α	parameter related to the inverse of the air entry pressure ($\alpha > 0$) (cm^{-1})
$\theta(h)$	volumetric water content corresponding to the pressure head h (cm$^3 \cdot$ cm^{-3})
θ_r	residual water content, cm$^3 \cdot$ cm^{-3}
θ_s	saturated water content, cm$^3 \cdot$ cm^{-3}

REFERENCES

Bitterlich S., Durner W., Iden S. C., Knabner P., 2004, Inverse estimation of the unsaturated soil hydraulic properties from column outflow experiments using free-form parameterizations. *Vadoze Zone Journa*, 3: 971–981.

Brooks R. H., Corey A. T., 1966, Properties of porous media affecting fluid flow. *Journal of Irrigation Drainage Division.* In: *Proceedings of the American Society of Civil Engineering (IR2)* 92: 61–88.

Burdine N. T., 1953, Relative permeability calculations from size distribution data. *Trans AIME* 198: 71–78.

Butters G. L., Duchateau P., 2002, Continuous flow method for rapid measurement of soil hydraulic properties: I. Experimental Considerations. *Vadoze Zone Journal* 1: 239–251.

Chen C., Payne W. A., 2001, Measured and modeled unsaturated hydraulic conductivity of a walla walla silt loam. *Soil Science Society of America Journal* 65: 1385–1391.

Durner W., 1994, Hydraulic conductivity estimation for soils with heterogeneous pore structure. *Water Resource Research* 30(2): 211–223.

Finsterle S., 2004, Multiphase inverse modeling: Review and iTOUGH2 applications. *Vadoze Zone Journal* 3: 747–762.

Fujimaki H., Inoue M., 2003, A transient evaporation method for determining soil hydraulic properties at low pressure, *Vadoze Zone Journal* 2: 400–408.

Gardner W. R., 1956, Calculation of capillary conductivity from pressure plate outflow data. *Soil Science Society of America Journal* 20: 317–320.

Gardner W. R., 1958, Some steady state solutions of the unsaturated moisture flow equation with application to evaporation from a water table. *Soil Science* 85: 228–232.

Hillel D., Krentos U. D., Stylianou Y., 1972, Procedure and test of internal drainage method for measuring soil hydraulic characteristics in situ. *Soil Science* 114: 395–400.

Hopmans J. W., Overmars B., 1986, Presentation and application of an analytical model to describe soil hydraulic properties. *Journal of Hydrology* 87: 135–143.

Ippish O., Vogel H.-J., Bastian P., 2006, Validity limits for the van Genuchten–Mualem model and implications for parameter estimation and numerical simulation. *Advances in Water Resources* 29: 1780–1789.

Iwanek M., Kowalski D., Olszta W., 2004, Obliczanie współczynnika przewodnictwa hydraulicznego metod¹ van Genuchtena-Mualema w oparciu o parametry krzywej retencji wodnej. *Acta Agrophysica,* 3(3): 487–500 (in Polish, with English abstract).

Iwanek M., Kowalski D., Olszta W., 2007, The influence of a chosen parameter on the hydraulic conductivity calculation. In Pawłowski L., Dudzińska M. and Pawłowski A. (ed.): *Environmental engineering:* 465–469. New York: Taylor & Francis Group.

Kosugi K., 1996, Lognormal distribution model for unsaturated soil hydraulic properties, *Water Resource Research* 32: 2697–2703.

Lazarovitch, N., Ben-Gal, A., Šimůnek, J., Shani U, 2007, Uniqueness of Soil hydraulic parameters determined by a combined Wooding inverse approach. *Soil Science Society of America Journal* 71: 860–865.

Luckner L., Van Genuchten M. Th., Nielsen D. R., 1989, A consistent set of parametric models for the two-phase flow of immiscible fluids in the subsurface. *Water Resource Research* 25: 2187–2193.

Meadows D. G., Young M. H., Mcdonald E. V., 2005, A laboratory method for determining the unsaturated hydraulic properties of soil peds. *Soil Science Society of America Journal* 69: 807–815.

Minasny B., Hopmans J. W., Harter T., Eching S. O., Tuli A., Denton M. A., 2004, Neural networks prediction of soil hydraulic functions for alluvial soils using multistep outflow data. *Soil Science Society of America Journal* 68: 417–429.

Mualem Y., 1976, A new model for predicting the hydraulic conductivity of unsaturated porous media. *Water Resource Research* 12: 513–522.

Olszta W., 2004, *Podstawy inżynierii wodnej środowiska*. Lublin: Wydawnictwa Uczelniane (in Polish).

Olszta W., Zaradny H., 1991, *Pomiarowe i obliczeniowe metody określania współczynnika przewodności hydraulicznej gleb przy niepełnym nasyceniu*. Falenty: Wydawnictwo IMUZ (in Polish).

Russo D., Bresler E., 1981, Soil hydraulic properties as stochastic processes: I. An analysis of field spatial variability. *Soil Science Society of America Journal* 45: 682–687.

Schaap M. G., Van Genuchten M. Th., 2006, A modified Mualem–van Genuchten formulation for improved description of the hydraulic conductivity near saturation. *Vadoze Zone Journal:* 5: 27–34.

Schaap M. G., Leij F. J., 2000, Improved prediction of unsaturated hydraulic conductivity with the Mualem–van Genuchten model. *Soil Science Society of America Journal* 64: 843–851.

Schuh W. M., Bauder J. W., 1986, Effect of soil propertieson hydraulic conductivity-moisture relationshops. *Soil Science Society of America Journal* 50: 848–855.

Schwartz R. C., Evett S. R., 2002, Estimating hydraulic properties of a fine-textured soil using a disc infiltrometer. *Soil Science Society of America Journal* 66: 1409–1423.

Schwartz R. C., Evett S. R., 2003, Conjunctive use of tension infiltrometry and time domain reflectometry for inverse estimation of soil hydraulic properties. *Vadoze Zone J.* 2: 530–538.

Schwärzel K., Šimůnek J., Stoffregen H., Wessolek G., Van Genuchten M. Th., 2006, Estimation of the unsaturated hydraulic conductivity of peat soils: laboratory versus field data. *Vadoze Zone Journal* 5: 628–640.

Shinomiya Y., Takahashi K., Kobiyama M., Kubota J., 2001, Evaluation of the tortuosity parameter for forest soils to predict unsaturated hydraulic conductivity. *Journal of Forest Research* 6: 221–225.

Spohrer K., Herrmann L., Ingwersen J., Stahr K., 2005, Applicability of uni- and bimodal retention functions for water flow modeling in a tropical acrisol. *Vadoze Zone Journal* 5: 48–58.

Šimůnek J., Van Genuchten M. Th., 1997, Estimating unsaturated soil hydraulic properties from multiple tension disc infiltrometer data. *Soil Science* 162: 383–398.

Stephens, D. B., 1985, A field method to determine unsaturated hydraulic conductivity using flow nets. *Water Resource Research* 21: 45–50.

Stolte J., Freijer J. I., Bouten W., Dirksen C., Halbertsma J. M., Van Dam J. C., Van Den Berg J. A., Veerman G. J., Wösten J. H. M., 1994, Comparison of six methods to determine unsaturated hydraulic conductivity. *Soil Science Society of America Journal* 58: 1596–1603.

Tuli A., Hopmans J. W., Rolston D. E., Moldrup P., 2005, Comparison of air and water permeability between disturbed and undisturbed soils. *Soil Science Society of America Journal* 69: 1361–1371.

Van Genuchten M. Th., 1980, A closed-form equation for predicting the hydraulic conductivity of unsaturated soils. *Soil Science Society of America Journal* 44: 892–898.

Van Genuchten M. Th., Nielsen D. R., 1985, On describing and predicting the hydraulic properties of unsaturated soils. *Annals of Geophysics* 3: 615–628.

Vereecken H., 1995, Estimating the unsaturated hydraulic conductivity from theoretical models using simple soil properties. *Geoderma* 65: 81–92.

Wessolek G., Plagge R., Leij F. J., Van Genuchten M. Th., 1994, Analysing problems in describing field and laboratory measured soil hydraulic properties. *Geoderma* 64: 93–110.

Weiss R., Alm J., Laiho R., Laine J., 1998; Modeling moisture retention in peat soils. *Soil Science Society of America Journal* 62: 305–313.

Wösten J. H. M., Finke P. A., Jansen M. J. W., 1995. Comparison of class and continuous pedotransfer functions to generate soil hydraulic characteristics. *Geoderma* 66: 227-237.

Wösten J. H. M., Van Genuchten M. Th., 1988, Using texture and other soil properties to predict predict the unsaturated soil hydraulic functions. *Soil Science Society of America Journal* 52: 1762–1770.

Young M. H., Karagunduz A., Šimůnek J., Pennell K. D., 2002, A modified upward infiltration method for characterizing soil hydraulic properties. *Soil Science Society of America Journal* 66: 57–64.

Zhang Z. F., Ward A. L., Gee G. W., 2003a, A tensorial connectivity-tortuosity concept to describe the unsaturated hydraulic properties of anisotropic soils. *Vadoze Zone Journal* 2: 313–321.

Zhang Z. F., Ward A. L., Gee G. W., 2003b, Estimating soil hydraulic parameters of a field drainage experiment using inverse techniques. *Vadoze Zone Journal* 2: 201–211.

Environmental Engineering III – Pawłowski, Dudzińska & Pawłowski (eds)
© 2010 Taylor & Francis Group, London, ISBN 978-0-415-54882-3

Bioindicative studies of pecton in selected facilities of the Hajdów Wastewater Treatment Plant – a case study

G. Łagód, K. Jaromin, A. Kopertowska, O. Pliżga, A. Lefanowicz & P. Woś

Institute of Environmental Protection Engineering, Lublin University of Technology, Lublin, Poland

ABSTRACT: Physico–chemical and/or biological analyses may be used in the management of the activated sludge process in wastewater treatment plants. Bioindication based on cross-species morphological–functional groups is easy and quick, and provides information of a value similar to that of species identification. In bioindication, more complete information on process conditions from the application of several mutually connected indices can be obtained. The present study discusses the results of the bioindicative measurements performed in the facilities of the technological line at the Hajdów WWTP in Lublin, Poland. The biocenotic indices calculated are presented with the background of physico-chemical indicators.

Keywords: Bioindication, saprobes community, biofilm, sewage parameters, wastewater treatment plant.

1 INTRODUCTION

A pecton is a set of microorganisms in the form of a biofilm on the surface of solid bodies submerged in water or sewage. Pectons consist of bacteriophyta, microscopic fungi, algae, protozoa and metazoa (Eikelboom & van Buijsen 1983, Klimowicz 1983, Salvado et al. 1995, Buck 1999, Eikelboom 2000, Bitton 2005). Microflora (bacteria and fungi) decompose organic substances into simple compounds, while microfauna (protozoa, rotifers and nematodes), regulate the abundance of bacteria and fungi by feeding on them (Hartmann 1996, Allan 1998, Buck 1999, Evans & Furlong 2003, Bitton 2005).

Diverse groups of pecton settled in wastewater treatment plant (WWTP) facilities can use various substrates, and so eliminate these substrates from sewage, resulting in increased biofilm mass. Simultaneously pecton adapts to variable environmental conditions, with the following factors affecting development:

- temperature,
- pH,
- concentration of toxic substances,
- concentration of mineral salts,
- high load of organic substances,
- type of material on which pecton develops,
- variable level of sewage surface,
- depth of submergence under the sewage surface (stratification with a constant level).

The connections between the sets of biofilm organisms (pecton) are shaped similarly to those in an activated sludge (Eikelboom & van Buijsen 1983, Hartmann 1996, Buck 1999). However, biofilms are characterised by a greater variety of heterophyte food-chain organisms at the individual trophic levels.

Transformations of substrates in the settled biomass of pecton take place in aerobic, oxygen deficient and anaerobic conditions, and consist of the oxygenation and mineralization of organic compounds present in wastewater, mainly by the contribution of microorganisms.

Based on the available literature, an attempt was made to establish the relationships between microscopic observations of biological material and the values of pollution indicators in individual facilities of the WWTP technological line (screens, grit chambers, primary settlers, anaerobic and aerobic zones in bioreactor and final settlers), and the quality of influent and effluent wastewater. Analysis showed there were few of such studies conducted, and so there were a limited number of references to compare individual values.

The main task of the present study was to investigate the structure of the sets of saprophils and saprophages settling in the facilities of the municipal Hajdów WWTP, and to establish possible correlations between biocenotic structure and selected contamination parameters in the plant. Thus, we attempted to establish and verify possible dependencies between the technological parameters of a WWTP or contaminant indices, and the occurrence of selected morphological–functional groups.

The bioindicative analysis and physico–chemical methods do not have to be always consistent and correlated in full technical scale at a treatment plant. Generally, physico–chemical analysis reflects the current situation of when the sample was taken, while bioindicative analysis represents a certain time interval, shown by the structure of the biological material examined (Quevauviller et al. 2006, Montusiewicz et al. 2007).

2 MATERIALS AND METHODS

The bioindication method based on occurrence of particular species does not provide information quickly enough for the functioning of the individual elements of a WWTP (e.g. particular devices and interceptors). This is due to the need for many specialists to identify individual microorganisms and classify them into genera. Therefore, based on trends in worldwide research (Oliver & Beattie 1993), the use of morphological–functional groups was proposed (Łagód et al. 2007, Montusiewicz et al. 2007, Łagód & Sobczuk 2008, Chomczyńska et al. 2009). The abundance of morphological–functional groups is a parameter that alters with the environmental conditions, similar to abundance of individual species of microorganisms. The basic criterion for selection of these groups was speed and ease of microorganism classification, possibly even after only a basic preparation and training of the researcher. It should be emphasised that identification of microorganisms to species level can be difficult even for a specialist.

The samples of pecton for microscopic analysis were taken at 11 sites at the Hajdów WWTP, which purifies sewage from Lublin and Świdnik – cities of eastern Poland (daily discharge Q_d mean year value about 60 000 m^3/d). A series of samples were taken on the 12 July 2006 (at 09:00 h). The study was performed on a Wednesday, as we could obtain from the plant management the values of technological parameters, performed on that day as a standard.

Sampling sites were located in the following sites:

1. chamber inlet in front of screen,
2. chamber outlet behind screen,
3. chamber inlet in front of grit chamber,
4. chamber outlet behind grit chamber,
5. outlet from primary settler,
6. inflow to anaerobic zone in bioreactor,
7. outlet from anaerobic bioreactor chamber,
8. inlet into aerobic bioreactor chamber,
9. outlet from aerobic chamber bioreactor,
10. outlet from final settler,
11. outflow into Bystrzyca River.

Pecton samples were covered with sewage up to 33% of container volume (maximum capacity 250 mL) taken at the sampling site, and transferred to the laboratory at the Lublin University of Technology, where microscopic analysis was performed and data archived. In one preparation per sample, for each of 10 fields of vision two photographs and one 10-s segment of video footage were taken. Microorganisms were identified in the fields of vision (under a light microscope) and classified into the selected morphological–functional groups. The photographs and video sequences recorded at the same time allowed verification of results, and could be used in future analyses expanding or introducing changes in the selected morphological–functional groups.

During the microscopic studies the following groups of microorganisms were differentiated in the collected pecton samples: fungi; flagellates; attached, crawling and swimming ciliates; rhizopods (amoebas); rotifers; nematodes; green algae; and diatoms.

The slides were analysed with a conventional light microscope at 400 × magnification. In each preparation, 10 fields of vision were chosen and occurrence of representatives of adequate morphological–functional groups registered using a previously prepared protocol. Magnification of 400 × applied is the optimum to identify the morphological–functional groups. However, if identification was based on species, magnification would need to be increased, which would be associated with a change in the number of fields of vision, as well as the problem of observing larger protozoa and multicellular microorganisms in the field.

Based on the organisms counted, a number of biocenotic indicators and indices were calculated (Łagód et al. 2006, Montusiewicz et al. 2007, Chomczyńska et al. 2009). These were as follows:

Relative abundance of i-th morphological–functional group (percentage contribution of microorganisms calculated in groups):

$$\Pi_i = \frac{n_i}{n_T} \tag{1}$$

where: Π_i – fraction contribution of microorganisms calculated in groups, n_i – abundance of individual microorganisms in a morphological–functional group, n_T – sum of microorganisms in all morphological–functional groups.

Shannon's Index (H):

$$H = -\sum_{i=1}^{S^*} \Pi_i \log_2 \Pi_i \tag{2}$$

where: H – biodiversity index, calculated based on morphological–functional groups, S^* – number of morphological–functional groups.

Evenness Index (V):

$$V = \frac{H}{H_{max}} \tag{3}$$

H_{max} – value of H provided that all morphological–functional groups represented cover the same number of individuals.

H_{max} is calculated as:

$$H_{max} = \log_2 S \tag{4}$$

MacArthur's Index (E):

$$E = z^H \tag{5}$$

where: z – the base of the logarithm applied ($\log_2$ in the present study), H – Shannon's Index for the sample examined.

Table 1. Abundance of microorganisms according to morphological–functional groups in particular facilities of the Hajdów Wastewater Treatment Plant.

Pecton sampling site	Microorganisms										Σ Sum
	a	b	c	d	e	f	g	h	i	j	
1		1	3	22	5				6		37
2				31	3				1		35
3			1	46				4		13	64
4		1	1	34	6	3			1		46
5			76	51				136		174	437
6		1	6	18	45		2	1			73
7			2	39				4		141	186
8	1	6	3	12	42		2	3		105	174
9		1	6	1	42	3	3	6		52	114
10			110	1	2			7		209	329
11		2	2	12	3	4		3		38	64

Morphological–functional groups: a – attached ciliates, b – crawling ciliates, c – swimming ciliates, d – flagellates, e – rhizopods, f – nematodes, g – rotifers, h – green algae, i – fungi, j – diatoms.

Table 2. Pollution indicators and technological parameters of sewage at selected sites in the Hajdów Wastewater Treatment Plant (total amount of raw sewage 54 630 m^3/d).

Sampling site	pH	BOD_5 (g/m^3)	COD (g/m^3)	TSS (g/m^3)	ESS (g/m^3)	$N-NH_4$ (g/m^3)	$N-NO_2$ (g/m^3)	$N-NO_3$ (g/m^3)	TKN (g/m^3)	$N_{tot.}$ (g/m^3)	$P_{tot.}$ (g/m^3)	BOD_5/COD
1	7.5	470	1020	498	24	47.1	0.009	0.06	75	75.07	11.5	0.4608
5	7.6	367	699	141	0.3	55.5	0.004	0.04	75.6	75.64	14.5	0.5250
11	7.9	11	108	14	–	6	2.24	13.5	9.9	25.64	2.71	0.1018

Proportionality Index (P):

$$P = \frac{E}{S} \cdot 100 \qquad (6)$$

While calculating P the multiplication of E/S^* by 100 was omitted for more convenient presentation in graphs.

The presented biocenotic indices were used for bioindication purposes. The values of pollution indicators were scaled by $\log_2$ prior to inclusion in radar-style graphs for better comparison and connection with biocenotic indices.

3 RESULTS AND DISCUSSION

The abundance of microorganisms recognised and classified into morphological–functional groups in selected facilities of the Hajdów WWTP in Lublin were determined under the microscope (Table 1). The quantitative distribution of microorganisms in individual facilities of the technological line shows the changes in abundance of the morphological–functional groups of organisms.

The values of selected pollution indicators were measured at the inflow to the Hajdów WWTP, after the mechanical and biological sections (Table 2). Biocenotic indices were developed (Table 3), based on data from Table 1. Subsequently, parameters presented

Table 3. Values of biocenotic indices developed from abundance of microorganisms.

Sampling site	H	H_{max}	V	E	P	S^*
1	1.696	2.322	0.731	3.241	0.648	5
2	0.605	1.585	0.382	1.521	0.507	3
3	1.153	2.000	0.577	2.224	0.556	4
4	1.323	2.585	0.512	2.501	0.417	6
5	1.854	2.000	0.927	3.614	0.904	4
6	1.536	2.585	0.594	2.901	0.483	6
7	0.965	2.000	0.482	1.952	0.488	4
8	1.687	3.000	0.562	3.220	0.403	8
9	1.891	3.000	0.630	3.708	0.463	8
10	1.133	2.322	0.488	2.193	0.439	5
11	1.876	2.807	0.668	3.670	0.524	7

in Table 2 were transformed by $\log_2$ to values that enabled easy inclusion into one graph, together with biocenotic indices (Table 4).

At the beginning of the technological line (the mechanical section), where the concentrations of biochemical oxygen demand (BOD_5), chemical oxygen demand (COD), total nitrogen (N_{tot}), and total phosphorus (P_{tot}) reached high values, the richness of morphological–functional groups S^* (Table 3) was low, i.e. 4–5 groups. The flagellates were the dominant group. Along the technological line, the sewage

Table 4. Selected pollution indicators and their calculated values.

Sampling site	Indicators (g/m³)	Real value (g/m³)	Calculated value
1	BOD₅	470	8.877
	COD	1020	9.994
	N_{tot}	75.07	6.230
	P_{tot}	11.5	3.524
5	BOD₅	367	8.520
	COD	699	9.449
	N_{tot}	75.64	6.241
	P_{tot}	14.5	3.858
11	BOD₅	11	3.459
	COD	108	6.755
	N_{tot}	25.64	4.680
	P_{tot}	2.71	1.438

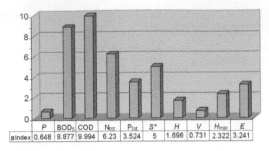

	P	BOD₅	COD	N_{tot}	P_{tot}	S*	H	V	H_{max}	E
Index	0.648	8.877	9.994	6.23	3.524	5	1.696	0.731	2.322	3.241

Figure 2. Pollution indicators and biocenotic indices measured in front of the rake screens, presented as a bar graph.

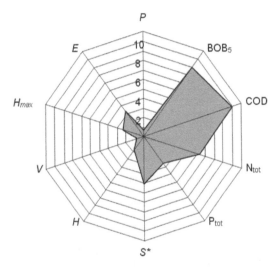

Figure 1. Correlation plot of measured and simulated.

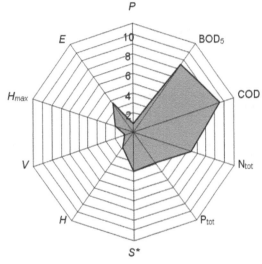

Figure 3. Pollution indicators and biocenotic indices measured behind the primary settler, presented as a radar graph.

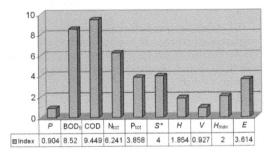

	P	BOD₅	COD	N_{tot}	P_{tot}	S*	H	V	H_{max}	E
Index	0.904	8.52	9.449	6.241	3.858	4	1.854	0.927	2	3.614

Figure 4. Pollution indicators and biocenotic indices measured behind the primary settler, presented as a bar graph.

was subjected to purification and the pollution indicators decreased (Table 2), reflected in $S*$ value which in the bioreaction chambers indicated 4–8 groups. This was due to the transformation of the organic substances inflowing with sewage in the previous facilities of the line, into forms more easily absorbed by microorganisms, as well as to decreased contaminant concentration.

Using the biocenotic indices and pollution indicators, radar graphs were made (Figures 1, 3 and 5) where the selected values were presented on 10 axes; also as respective bar graphs (Figures 2, 4 and 6).

The calculated biocenotic indices (H, H_{max}, V, E, P and $S*$) from the biological material which overgrow the facility are shown in Figure 1, compared to the selected indicators (BOD₅, COD, N_{tot} and P_{tot}) prior to subjecting the passing sewage to the purification process.

Figure 3 presents values equivalent to the parameters of pecton and sewage, which passed by the technological line via the whole mechanical section, with the sample taken at the outlet of the preliminary sedimentation tank.

The values of biocenotic indices developed based on data in Table 3, were compared to pollution indicators in Table 4 (Figure 5). The results reflect the situation

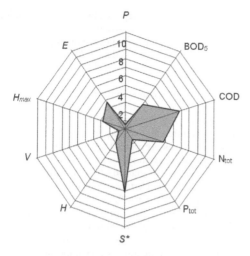

Figure 5. Pollution indicators and biocenotic indices measured at the river outflow, presented as a radar graph.

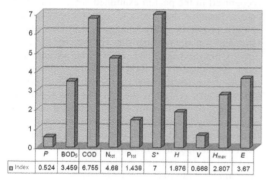

	P	BOD$_5$	COD	N$_{tot}$	P$_{tot}$	S*	H	V	H$_{max}$	E
▣ Index	0.524	3.459	6.755	4.68	1.438	7	1.876	0.668	2.807	3.67

Figure 6. Pollution indicators and biocenotic indices measured at the river outflow, presented as a bar graph.

after the biological section of the Hajdów WWTP at the outflow into the Bystrzyca River. The microorganisms creating the distinct morphological–functional groups were most diverse in the biological section, where the range in S* was 3–8, and pollution indicators were considerably reduced.

The values of the Shannon's Index (H) based on fraction contributions (Π) and number of microorganisms (S*) were from 0.605 behind the screen to 1.891 at the outlet from the aerobic chamber.

In sewage of the lowest concentration of BOD$_5$ (as well as COD, N$_{tot}$ and P$_{tot}$; i.e. at the outflow of the plant), the number of morphological–functional groups was S* = 7. With increased BOD$_5$ concentration (i.e. at the outflow of the plant), S* = 4. A decrease in the number of morphological–functional groups was associated with increased value of contaminants. At the preliminary stage of treatment, there was a decrease in the number of morphological–functional groups (Table 1), as well as the number of individual microorganisms belonging to a given group in comparison to the biological part. There was an opposite and

decreasing trend for flagellates, which were abundant at the beginning of the purification process. This was caused by richness of nutrients (i.e. high contamination of sewage), as well as by the fact that flagellates win the nutritional competition with other protozoa due to their ability to absorb nutrients in the form in which they enter the treatment plant. In the course of the purification process their abundance rapidly decreased.

The values of H (Table 3) increased with the progress of the purification process, i.e. with decreased values of pollution indicators. An exception was at the outflow of the plant, where it was high (H = 1.696), which might have been caused by a greater abundance of groups (S*), and by a relatively stable fraction contribution of microorganisms calculated in a group (Π). This may have resulted from physical and biochemical processes of sewage biodegradation taking place in gravitation conduits supplying wastewater to the WWTP. Then, H increased in subsequent facilities, as far as the inlet to the anaerobic chamber, where it decreased. H increased again in the aerobic chamber, decreased in the secondary sedimentation tank, and increased at the outflow into the river.

MacArthur's Index (E) increased (Figures 1 and 2) with the decrease in pollution indicators (BOD$_5$, COD, N$_{tot}$ and P$_{tot}$). It was difficult to establish correlation between V, P and E indices and the parameters of the sewage treated in the facility, compared to the similar relationships of H. This would require long-term and regular studies, based on which it would be possible to obtain average measurements.

For a long time methods for precise, accurate, and economical characterisation of the sewage purification process in individual facilities, as well as monitoring of wastewater quality entering WWTPs have been sought (Eikelboom & van Buijsen 1983, Eikelboom 2000, Bitton 2005, Quevauviller et al. 2006). Published analyses of the biological environment, especially water and sewage environments, suggest that the abundance of microorganisms should be determined within morphological–functional groups instead of at species level (Oliver et al. 1993, Łagód et al. 2007, 2008, Montusiewicz et al. 2007, Chomczyńska et al. 2009). This seems to be sufficient for analysis of the structure examined, and provides similar, if not the same, values as reports based on indicative species.

Knowledge of the composition of microorganisms in the studied material may allow the drawing of conclusions concerning conditions in the facility, where the material was sampled, as well as the character of the flowing sewage (Eikelboom & van Buijsen 1983, Buck 1999, Eikelboom 2000, Bitton 2005, Quevauviller et al. 2006, Montusiewicz et al. 2007, Chomczyńska et al. 2009).

The present study showed that saprobiont microorganisms settle all surfaces of WWTP technological devices that were in direct contact with flowing sewage. The composition of microorganisms forming the pecton differed in their dependence on characteristics of flowing sewage as well as the source of incoming sewage (inflowing from sewage collector or inflowing as a mixture with recirculated sludge). The composition mentioned above is also closely dependent on available solar energy. These factors influence not only the values of indicators and biocenotic indices but also the trophic characteristics of organisms creating pecton.

Photoautotrophs (chlorophyta and diatoms) appear only in devices allowing contact with solar light, especially in sewage treated mechanically – after the primary sedimentation tank (136 different chlorophyta and 174 diatoms), which simultaneously was the community with the highest flagellate population (51 individuals) and ciliates (76). Fungi in the samples were settled only in devices containing raw sewage, i.e. the entrance to the WWTP, screens and grid chamber. Rotifers were observed only in the pecton communities of the second, biological stage of wastewater treatment. Flagellates were observed in devices of mechanical sewage treatment in anaerobic conditions, and rarely in sewage treated aerobically. The Arcella rhizopods were commonly observed in the recirculated sludge stream entering the anaerobic and aerobic zones of the bioreactor. Swimming ciliates routinely appeared in the pecton sampled in sedimentation tanks. The settling tanks were the richest in representatives of all morphological–functional groups; 473 representatives in the primary and 329 in the secondary settling tank.

The lowest values for microorganism groups were before and after the screens, 37 and 35 representatives, respectively. There were higher values in the two samples taken in the grid chamber, 64 and 46 individuals; and the highest in the three samples taken in the biological part of the WWTP technological line; 186, 174 and 114 representatives. The number was relatively low at the outflow to the Bystrzyca River; 64 representatives of the studied morphological–functional groups.

In the mechanical part of WWTP, there was mean Shannon Index $(H) = 1.2337$ with standard deviation 0.4457. In the biological part of the WWTP, at the second stage of wastewater treatment process, $H = 1.5147$ with standard deviation 0.3539. The lowest values were in anaerobic volumes; $H = 0.605$ after the screens and $H = 0.965$ in the anaerobic chamber. The highest values were in pecton sampled in aerobic sewage volumes: $H = 1.891$ in the aerobic chamber of the bioreactor; $H = 1.876$ in the outflow to the wastewater receiver; $H = 1.854$ in the outflow from the primary settling tank; and characteristically of a gravitational sanitation system, in the outflow from the collector before the screens where $H = 1.696$.

The most even communities, in terms of representative shares (V), were those after the primary settling tank (0.927), at the main conduit outflow (0.731), at the outflow to the river (0.668) and at the outlet of aerobic chamber (0.630). The highest diversities in the pecton communities were at outlets of anaerobic volumes: 0.382 for the outflow from screens and 0.482 at the outlet of the anaerobic part of the bioreactor.

The values of Proportionality Index (P) for the studied communities were generally all similar, attesting the even distribution of representatives of the studied morphological–functional groups.

The diversification of oxygen and trophic conditions seemed clearly reflected in the biocenotic indices and indicator values calculated based on the studied communities' structures.

As a result of human activity an increasingly larger number of poisonous substances are produced, which cannot be quickly recognised (Quevauviller et al. 2006). Currently, the known methods of physico–chemical analyses are poorly detailed, and there is a lack of analytical methods to respond to all poisons. Moreover, the determination of the level of contamination with a given substance does not provide much information about its harmfulness to living organisms, without prior investigation of its effect (Quevauviller et al. 2006). Therefore, the application of biological methods has many advantages, and for this reason the process of bioindication is very important. Hence, physico–chemical examinations of sewage should be supplemented with biological analyses. The organisms recognised in the pecton examined are good bioindicators, because they react to substances present in sewage, providing a direct image of the effect of contaminants on microorganisms (Montusiewicz et al. 2007, Chomczyńska et al. 2009). The advantage of these bioindicators is their resistance to breaks in inflow of wastewater and decreased inflow concentration (load), as well as the fact that they reflect the occurrence of breakdowns, i.e. uncontrolled short discharge of highly concentrated or poisonous wastewater (Quevauviller et al. 2006). Pecton is not washed out from the plant, as in some critical cases happens with activated sludge, because it adheres to the surfaces of the plant facilities in contact with sewage. Hence, even in the case of serious breakdowns, when the majority of activated sludge flows out of the plant, it is possible to identify the causes of the breakdown by examining the pecton.

4 CONCLUSIONS

Based on the studies in the facilities of the technological line at the Hajdów WWTP on 12 July 2006, we made the following conclusions:

1. The pecton communities in WWTP devices settle at the surfaces contacting the raw sewage (screens), treated mechanically (grid chamber and outlet of primary settling tank), treated biologically (bioreactor chambers, secondary settling tank outlet) as well as those in the sewage after treatment process, i.e. outflow to the receiver.
2. The biocenosis of the distinct elements of the technological line of the WWTP (inflow into the plant, outlet from the primary settler, outflow from the plant to the Bystrzyca River) differs, despite having the same morphological–functional groups. The mentioned differences in pecton formations were observed in both the trophic and structural senses.

3. Biocenotic structures of the studied formations depend on the combination of settling conditions such as light, oxygen and trophic conditions, as well as pollutant load and probably many others not analysed in our studies.

4. In the material examined, there was a simultaneous occurrence of 2–9 of the 10 distinct morphological–functional groups. The largest number of morphological–functional groups was in the aerobic chambers and at the plant outflow. The smallest number was at the inflow to the plant and in the sand separator.

5. The values of indices and indicators describing the structures of pecton communities, calculated based on the morphological–functional groups seemed to greatly depend on trophic and oxygen conditions in the particular devices of the WWTP technological line.

6. The usage of only one index or technological parameter, for the characteristics of the examined structure of biocenosis settling the facilities in the plant, does not provide sufficient information about the set of microorganisms in pecton.

7. The characteristics of the changes in sewage parameters based on the bioindicative method by means of pecton are not always consistent with the characteristics of a specific physico–chemical analysis of wastewater. This is due to great variation in technological parameters and pollution indicators over time, while in the plant physico–chemical examinations are calculated on a daily average.

ACKNOWLEDGEMENT

This work was supported by the Ministry of Science and Higher Education of Poland, Grant No: 4949/B/T02/2008/34

REFERENCES

Allan, J.D. 1998. *Ekologia wód płynących*. Warszawa: PWN.

Bitton, G. 2005. *Wastewater microbiology*. Hoboken, New Jersey: John Wiley & Sons Inc.

Buck, H. 1999. *Mikroorganizmy W Osadzie Czynnym*. Szczecin: Seidel-Przywecki.

Chomczyńska, M., Montusiewicz, A., Malicki, J. & Łagód G. 2009. Application of saprobes for bioindication of wastewater quality. *Environmental Engineering Science* 26(2): 289–295.

Eikelboom, D.H. & Van Buijsen, H.J.J. 1983. *Microscopic sludge investigation manual, 2nd edition*. Delft: TNO Research Institute for Environmental Hygiene.

Eikelboom, D.H. 2000. *Process control of activated sludge plants by microscopic investigation*. London: IWA Publishing.

Evans, G.M. & Furlong, J.C. 2003. *Environmental biotechnology*, Chichester: John Wiley & Sons Ltd.

Hartmann, L. 1996. *Biologiczne oczyszczanie ścieków*. Warszawa: Wydawnictwo Instalator Polski.

Klimowicz, H. 1983. *Znaczenie mikrofauny przy oczyszczaniu ścieków osadem czynnym*. Warszawa: Zakład Wydawnictw Instytutu Kształtowania Środowiska.

Łagód, G., Chomczyńska, M., Malicki, J. & Montusiewicz, A. 2006. Quantitative methods of description estimation and comparison of microorganism communities in urban sewer system. *Ecological Chemistry and Engineering* 13(3–4): 255–263.

Łagód, G., Malicki, J., Chomczyńska, M. & Montusiewicz, A. 2007. Interpretation of the results of wastewater quality biomonitoring using saprobes. *Environmental Engineering Science* 24(7): 873–879.

Łagód, G. & Sobczuk, H. 2008. The number and size of samples required to measure the saprobe population at various pollutant concentrations in sewage. *Archives of Environmental Protection* 34(3): 281–285.

Montusiewicz, A., Malicki, J., Łagód, G. & Chomczyńska, M. 2007. Estimating the efficiency of wastewater treatment in activated sludge systems by biomonitoring. In L. Pawłowski, M.R. Dudzińska & A. Pawłowski (eds) *Environmental Engineering*. London: Taylor & Francis Group.

Oliver, I. & Beattie, A.J. 1993. A possible method for the rapid assessment of biodiversity. *Conservation Biology* 7: 562–568.

Quevauviller, P., Thomas, O. & Van Der Beken, A. 2006. *Wastewater quality monitoring and treatment* Chichester: John Wiley & Sons Ltd.

Salvado, H., Garcia, M.P. & Amigo, J.M. 1995. Capability of ciliated protozoa as indicators of effluent quality in activated sludge plants. *Water Research* 29: 1041–1050.

Environmental Engineering III – Pawłowski, Dudzińska & Pawłowski (eds)
© 2010 Taylor & Francis Group, London, ISBN 978-0-415-54882-3

Influence of valve closure characteristic on pressure increase during water hammer run

A. Kodura

Environmental Engineering Faculty, Warsaw University of Technology, Warsaw, Poland

ABSTRACT: Pressure increases caused by the complex water hammer phenomenon often occurs in pipelines. However, for many decades, these pressure increases have been calculated using theoretical equations without concurrent in-depth experimental analysis. In this paper, the pressure increases caused by the water hammer phenomenon for a ball valve and circular gate valve are measured; gate closure characteristics are varied to assess whether the rapid or slow shutting at various points in the closing process affects the timing and amount of pressure increase. The results indicate that the speed of gate closure, especially as the gate approaches full closure, can significantly impact the amount of pressure created by closure. Analysis shows that existing methods of maximum pressure that increase calculation do not accurately account for pressure during gate closure, which suggests that a new method is needed that accounts for the speed characteristics of gate closure.

Keywords: Water hammer, complex water hammer, closure valve characteristic, pressure wave, transient flow.

1 INTRODUCTION

The water hammer phenomenon occurs in systems with transient flow in pressure pipelines. As a valve or gate is closed in a pressure pipeline, the flow of fluid inside increases rapidly in speed, which in turn creates pressure that spreads in the pipeline in the form of a disturbing wave. The speed of such disturbance is influenced by a number of elements, including but not limited to fluid type, the precise bulk modulus of the fluid, friction forces, inertia forces and characteristics of the pipe wall (including its material composition, thickness and geometry) (Streeter et al. 1998, Nielacny 2002, Mitosek 2007). This phenomenon is undesirable in pressure pipe networks because it creates pressure increases of significant value, as well as significantly higher wave frequency. When pressure and frequency increase rapidly at the same time, damaged armature elements often occur, as well as disturbances in the process of fluid transportation (Ramos & de Almeida 2002).

The water hammer phenomenon presents a major problem for practical hydraulic systems. More than 80% of breakdowns in water supply systems can be attributed to the water hammer phenomenon (Ilin 1987). However, despite its prevalence, scientists and engineers currently rely on theoretical equations to calculate the pressure increases caused by gate closure. Because these theoretical models fail to account for the characteristics of gate closure and because pressure dynamics caused by this phenomenon have not been measured experimentally, scientists and engineers have lacked a specific description for how the water hammer phenomenon affects pressure. A lack

of experimental data measuring this phenomenon has limited engineers' ability to improve pipeline systems to mediate the water hammer phenomenon.

The existing literature on the water hammer phenomenon, which stretches back to the nineteenth century, describes water hammer as "slow gate closure" (Streeter et al. 1998, Thorley 2004). Rapid water hammer describes situations in which the gate closure's duration is less than one wave period. Slow water hammer/slow gate closure describes gate closing in a time frame that is longer than the wave period, hence some researchers refer to it as "complex water hammer". This study will adopt the term complex water hammer to describe slow water hammer/slow gate closure.

The pressure increases caused by complex water hammer – a problem that is very common in hydraulic systems – is still being calculated today using equations from Michaud (1878), Wood and Jones (1966), and Sharp (1967). There has been little additional work on complex water hammer since the 1960s. The maximum pressure increase of rapid water hammer is still being calculated using Joukowsky's equation (1898). However, during more than 100 years, many numerical methods have been introduced and developed (Parmakian 1955, Evett & Liu 1989, Wylie et al. 1993, Marcinkiewicz et al. 2008).

All of these equations that describe complex water hammer have limited or no ability to account for the differing pipe and system characteristics. Today, it has become common to use water supply systems that operate with several different types of materials; these systems are continuously being rebuilt or renovated as the cities above them grow and shrink. The systems probably include a combination of steel pipes and pipes made of newer materials that were unavailable before the 1960s, including plastics such as PE and PVC. These newer materials behave differently under stress than the steel pipes that are assumed to be used by many of these equations, and thus the equations may not accurately predict the water hammer phenomenon in these newer systems.

The complex water hammer phenomenon, in particular, appears often in pipeline networks, especially in rapid networks, because of the geometrical parameters of gate closure (both number of sets and times of closure) and pump closure (stopping). One way to protect a pipeline system from the potential damage caused by the water hammer phenomenon is to alter the timing of the gate closure, particularly at pumps and pump stations where the water hammer phenomenon often emerges (Thorley 2004). However, this seemingly simple solution of altering gate closure is complicated by the existence of many different kinds of gates and valves, including automatic valves and self-closing gates. Among the most common types of valves are circular gate, globe, needle, square gate, butterfly and ball valves. This study will experimentally address two common types of gate closure – ball valve and circular gate valve – in order to infer which gate characteristics may affect the water hammer phenomenon. Various speeds of gate closure (including differing speeds of closure at the beginning and end phases of gate closure) will be examined.

1.1 Evaluating Existing Equations for Calculating Pressure Increases Due to Water Hammer

Although the existing equations for calculating the water hammer phenomenon do not account for gate closure characteristics, they present an important starting point for understanding the water hammer phenomenon and its pressure effects. Joukowsky's formula describes the pressure increases in a rapid water hammer situation as (Parmakian 1955, Streeter et al. 1998):

$$\Delta p = \Delta v \cdot \rho \cdot a \tag{1}$$

where: Δv = water velocity change, ρ = liquid density, a = wave propagation celerity.

When the phenomenon consists of fluid impede, it is called positive water hammer. When it consists of acceleration, it is called negative water hammer. In positive water hammer conditions (in which fluid is impeded to the point of absolute flow retaining), the ending speed is equal to 0 $v_e = 0$ – that is, $\Delta v = v_0$. The emerging pressure increase achieves maximum value due to a total exchange of kinetic energy into potential energy. The speed of disturbance propagation a expresses the relationship between pressure wave period and pipeline length (Parmakian 1955, Streeter et al. 1998):

$$a = \frac{2L}{T} \tag{2}$$

where: T = wave period, L = pipeline length and a = wave propagation celerity.

With regards to Equation 2, however, there is another phenomenon's classification that does not carry out this condition – rapid water hammer, which is a phenomenon in which the liquid braking process (or acceleration) will end before the return of a deflected wave. This condition can be expressed as a relationship between valve closure time T_c and wave period T. For rapid water hammer, a valve at the downstream end of a pipeline is closed in a time less than the wave period:

$$T_c < T \tag{3}$$

When the valve closure is longer than the wave period during the gate closing, a deflected wave of the opposite sign is created. This wave reduces the pressure increase that resulted from the valve closure. If liquid velocity change is linear and the total pressure increase is less than 220% of pressure in steady conditions before the water hammer phenomenon takes effect, then the generated value of pressure increase can be calculated using Michaud's equation (Mitosek 2007):

$$\Delta p = \frac{2 \cdot \rho \cdot v_0 \cdot L}{T_c} \tag{4}$$

where ρ = liquid density, v_0 = velocity in steady condition before water hammer run, L = pipeline length, and T_c = time of gate closure.

In comparison to Joukowsky's equation, the value of wave propagation celerity in Michaud's equation was substituted by quotient $\frac{2 \cdot L}{T_c}$.

By keeping a linear change of liquid velocity, the pressure increase also remains linear.

Michaud's equation, however, assumes a linear velocity change, which cannot ever be achieved. It also fails to account for the effect of emerging pressure waves, which can interact with each other to create a process of energy dissipation. It is for that reason that pressure increase is lower during slow gate closure than during the rapid valve closure.

Wood and Jones' method represents an improvement on Michaud's model because it introduces the concept of dimensionless valve closure time and dimensionless maximum transient pressure change (Wood & Jones 1973, Thorley 2004). In Wood and Jones' method, the dimensionless pressure change is defined by a transformed version of Joukowsky's formula:

$$\Delta p_m = \frac{\Delta p_{max}}{\Delta v \cdot \rho \cdot a} \tag{5}$$

where Δp_m = dimensionless maximum pressure change, Δp_{max} = maximum pressure head, Δv = liquid speed change, ρ = liquid density, and a = wave propagation celerity.

The dimensionless valve closure time is then calculated from this equation (Wood & Jones 1973):

$$t_c = T_c \bigg/ \frac{2 \cdot L}{a} \tag{6}$$

where t_c = dimensionless valve closure time, T_c = valve closure time, L = pipeline length, and a = wave propagation celerity.

In a significant improvement over Michaud's equation, Wood's and Jones' method accounts for gate type by introducing a dimensionless initial condition parameter α that applies to the most commonly used types of valves (circular gate, globe, needle, square gate, butterfly and ball valves). This parameter is defined as (Wood & Jones 1973):

$$\alpha = \frac{g \cdot h_0}{\Delta v \cdot a} \tag{7}$$

The value h_0 expresses the head drop across the valve under the initial steady flow conditions, which can be observed just before the water hammer phenomenon occurs. Wood's and Jones' method accounts for a system with a downstream gate that generates pressure loss during steady flow conditions.

To employ Wood's and Jones' method, the first step is to select the most appropriate chart for the valve in question, then calculate the values of α parameter and dimensionless valve closure time T_c using equations (6) and (7). The next step is to determine dimensionless maximum transient pressure change by reading the appropriate chart. The last step is to re-arrange equation (5) and calculate the maximum transient pressure change (Thorley 2004).

By considering gate type, Wood's and Jones' method offers a significant advantage over using Michaud's equations. However, Wood's and Jones' method is limited in application; it assumes the introduction of a resistance value in initial steady conditions and applies the same resistance coefficient in transient flow (Sharp 1974). As a result, this model does not account for the characteristics of gate closure. The method also excludes friction resistance in dynamic unsteady conditions of water hammer run.

Both Michaud's and Wood's and Jones' calculation methods represent attempts to simplify the description of the complex water hammer phenomenon by assuming a linear valve spindle. Wood and Jones do provide a chart for calculating pressure change due to accelerated movement of the valve spindle during circular gate valve closure, but differences between their two charts for linear valve closure and accelerated movement are not highly significant. In addition, the work by Wood and Jones lacks information about the characteristics of that accelerated movement.

By contrast, Sharp's method accounts for the time duration of safety valve closure, for which the maximum allowed value of pressure is not exceeded (Sharp 1969). The gate characteristic parabola is expressed by Toricelli's equation:

$$v = b \cdot \sqrt{h} \tag{8}$$

where v is flow speed in pipeline, h the head lost across the valve, and b is a gate parameter. b is expressed as (Sharp 1969, 1981, Sharp & Sharp 2003):

$$b = \frac{\sqrt{2 \cdot g \cdot C_d \cdot A_v}}{A_p} \tag{9}$$

where C_d is a coefficient of discharge for an area of valve A_v, and A_p is the pipe area.

Thus, gate characteristics can be expressed by (Sharp 1969, 1981, Sharp & Sharp 2003):

$$v = \frac{\beta \cdot v_1 \cdot \sqrt{h}}{\sqrt{h_1}} \tag{10}$$

where the subscript "1" refers to the fully open conditions and β:

$$\beta = \frac{b}{b_1} = \frac{C_d \cdot A_v}{C_{d1} \cdot A_{v1}} \tag{11}$$

In conditions when the gate is fully open, the pressure losses across the gate are ignored because they are negligibly small. Thus:

$$v_n = \beta_n \cdot K \cdot \sqrt{Y_n} \tag{12}$$

where: v_n = momentary flow velocity for momentary degree of closure open and Y_n = momentary pressure head in valve's cross section. The expression $K \cdot v^2$ describes friction resistance along the pipeline and Y_n is the pressure head in the valve's cross section, whereas K is given by:

$$K = \frac{v_1}{\sqrt{h_1 - k \cdot v^2}} \tag{13}$$

At the beginning of calculation, the value of coefficient C_d has to be determined. There are two proposed ways to describe the value: the first is using the value of the discharge coefficient as seen in steady conditions, with acceptance of the unrealistic assumption for transient flow. The second possibility, as recommended by Sharp (1969, 1981), is to use the constant value $C_d = 1$. The value of β, therefore, may be taken as the ratio of the gate area to the fully open condition based on curves II and III from a chart (Sharp 1969, 1974, 1981), which includes information that determines the influence of gate characteristics on water hammer run, including current cross-section area, local losses of energy and physical gate position in relation to the pipeline flow velocity.

However, Sharp's suggested solution does not account for the problem of friction losses in unsteady flow. In addition, Sharp's solution simplifies the value of coefficient β (which expresses flow resistance through the gate) by expressing it as a ratio of current gate area to global gate area at maximum opening. In this method, there is no relationship shown between gate type and the value of coefficient β. This solution leads to the creation of a continuous process; it allows us only to sum up "snapshots" of different degrees of closure and different degrees of frictions without providing a continuous measure across the entire time span of gate closure. It thereby fails to consider inertia forces and pressure change in the gate area over time. Nonetheless, Sharp's method is a reasonable alternative to Michaud's equation, which describes gate closure characteristics only by a strictly determined value of allowed maximum pressure.

1.2 Examining Gate Speed as a Possible Contributor to Water Hammer

During the process of gate closure, there are two phenomena that interact to produce the water hammer effect. The first is the beginning of a disturbance in the form of a pressure wave that spreads in a pipeline. The second is an unsteady outflow from a pipeline, which varies depending on the actual active cross-section of a gate and the momentary pressure in the gate area at any given time. Another relevant issue is the influence of gate closure characteristics on the generated pressure wave. During the process of closure, a choked element has to be moved, which means that it must overcome the flowing liquid resistance. Each change of flow speed generates inertia forces. In the early stages of gate closure, the non-linear speed change is highly probable, just as it is in the ending phase. It is likely that the last phase of gate closure has the greatest influence on pressure increase values. A changing of closure speed can therefore be an important factor that determines the value of the pressure increase.

At the times that Joukowsky, Michaud, Wood and Jones, and Sharp were developing their equations, it was not possible to experimentally measure the effects of gate closure on water hammer phenomenon. Today, computerized measurement of pressure is available (Pires et al. 2005, Marcinkiewicz et al. 2008), but industry typically uses these computers to measure current pressure, not necessarily changes in pressure over time. This study uses this computerized measurement of pressure to record the pressure changes in a pipeline system during the complex water hammer phenomenon. Data are gathered for both ball valve and circular gate valve closure types, and those values are compared to the pressure predictions of the available theoretical equations from Michaud, Joukowsky, Wood and Jones, and Sharp. The study hypothesizes that varying the speed of gate closure may ameliorate some of the pressure effects of the water hammer phenomenon. The experiments were made for comparable values of initial steady conditions of pressure and speed. The analyzed phenomenon concerns total gate closing from full valve opening to full closing.

2 MATERIALS AND METHODS

2.1 Laboratory Experiments

2.1.1 Model

A water pipeline model was built to experimentally analyze the complex water hammer phenomenon. This model was built in the laboratory at the Department of Water Engineering and Hydraulics at Warsaw University of Technology (Poland). The laboratory station (Figure 1) consists of straight linear MDPE pipeline (1) of 35.60 m length, 50 mm external diameter and 4.6 mm wall thickness. The pipeline was fixed to the ground in order to minimize pipe movements caused by rapid pressure increases. An electronic pressure indicator (3, range from 0 to 1.2 MPa) was installed directly upstream from valve (2). The impulse from that indicator (3) was strengthened by an amplifier (4) and sent to computer (5) through an analog-digital card μDAQ.

At the same time, the analog-digital card sent data to the computer about the degree of gate closure from the downstream gate (2).

Water was supplied to the experimental system by a distribution system (8) through an air vessel (7) that stabilized pressure in the pipeline. The main task of this air vessel was to protect the distribution system against water hammer in order to make the transient phenomenon run only in the analyzed pipeline (1). During experimentation, the disconnection of the water distribution system occurred simultaneously with the completion of downstream valve closure. A flow meter with Pelton turbine (6) was installed at the defined distance needed to perform the quality measurements. By using this equipment, it was possible to achieve comparable values of the flow in initial steady flow conditions.

2.1.2 Experiments

Experiments were performed in two series for two different types of gates: a circular gate valve and a ball valve. Both of these valves were modified to provide unambiguous descriptions of closure degree over time. Electronic equipment designed specifically for this experiment sent data about the degree of closure through an analog-digital card to a computer.

A number of experiments were made for both gate types to investigate the water hammer phenomenon in comparable initial conditions (initial speed in steady flow ranged from 1.25 m/s to 1.35 m/s). Individual series of data differed in length for the closure time, which was extended for each new run while holding constant the quasi-linear characteristics of the closure. The first experiments were made for closure times that were shorter than the wave period in order to produce the rapid water hammer phenomenon; a second set of experiments produced the complex water hammer phenomenon.

3 RESULTS AND DISCUSSION

Figure 2 shows four chosen pressure characteristics for circular gate valve for different times of closure. The black continuous line represents the pressure change characteristics over time. The grey line shows the characteristics of gate closure expressed as a ratio of the valve's momentous active cross-section area to the valve's global cross-section area (At/Ag). In each

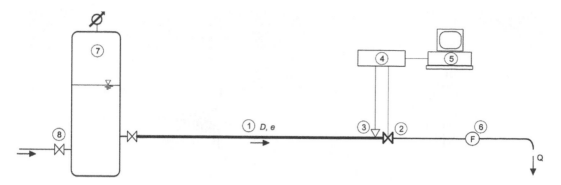

Figure 1. Draft of experimental model.

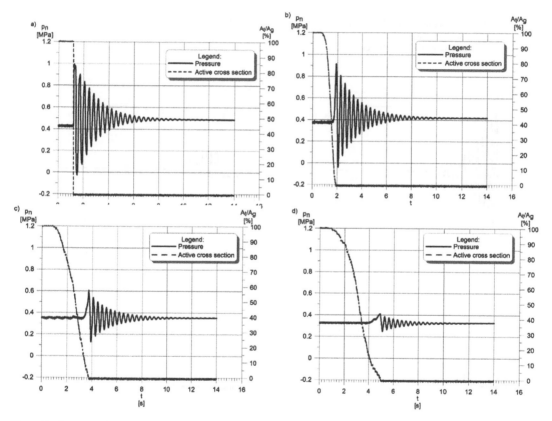

Figure 2. Pressure change characteristics during water hammer for increasing times of circular gate valve closing. a) rapid water hammer, C-1 series – time of gate closure $T_c = 0.062$ s, b) C-2 series, $T_c = 1.18$ s, c) C-3 series, $T_c = 2.85$ s, d) C-4 series, $T_c = 4.18$ s.

series, the wave characteristic of the phenomenon that characterizes decreasing amplitude during time and a steady period is clearly visible. Figure 2a shows the classic characteristics of rapid water hammer – in this case, the time of gate closure equal to 0.062 s is smaller than the wave period $T = 0.179$ s.

Figures, 2b, 2c and 2d show how the phenomenon changes when the length of gate closure is extended to 1.18 s, 2.85 s and 4.18 s, respectively. When the time of gate closure increases, the wave amplitude decreases and the total duration of the water hammer phenomenon shortens. This is the effect of additional water hammer energy dissipation, because it pits waves with opposite signs against themselves. This is clearly visible in Figure 2d as the two next peaks of waves in the first phase.

The pressure characteristics that were started by nonlinear gate closure are shown in Figure 3. Figure 3a shows pressure characteristics that were created by beginning with slow closing speed and ending with

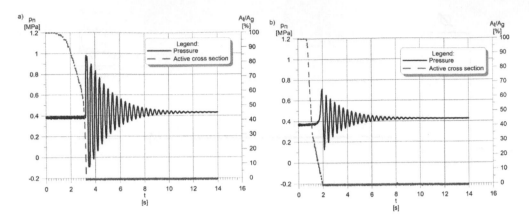

Figure 3. Pressure characteristics during complex water hammer for different methods of circular gate valve closing. a) C-5 series – slowly closing at the beginning and rapid shutting in the ending phase, $T_c = 2.46$ s, b) C-6 series – rapid shutting at the beginning and slowly closing in the ending phase, $T_c = 1.27$ s.

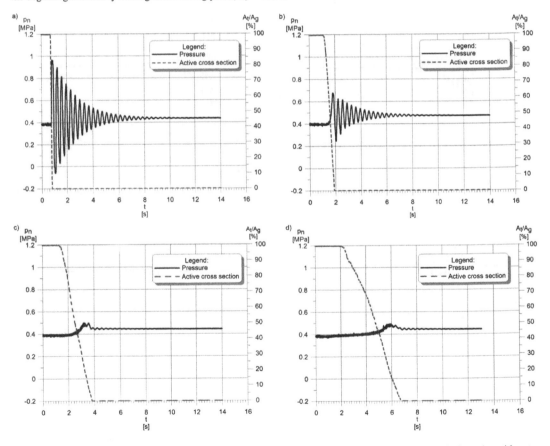

Figure 4. Pressure change characteristics during water hammer for increased time of ball valve closing. a) rapid water hammer – B-1 series – time of closing $T_c = 0.12$ s, b) B-2 series, $T_c = 0.84$ s, c) B-3 series, $T_c = 2.75$ s, d) B-4 series, $T_c = 5.03$ s.

rapid closing speed. Even though the closure time is equal to 2.46 s, the pressure wave's generated value is significantly higher than in the case of linear gate closure in a comparable time (Figure 3b).

Figure 3b shows a closing in the shortest time (1.27 s) but with inversed characteristics of the closing

process, meaning rapid shutting at the beginning and slow closing at the end. In those conditions, the pressure increase is significantly lower — approximately 40% less than seen in Figure 3a ($T_c = 2.46$ s).

Analogous experiments were made for a ball valve closure. Figure 4 shows pressure characteristics for

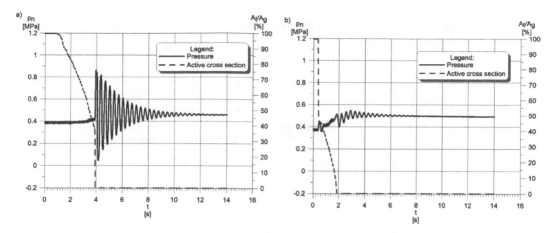

Figure 5.   Pressure characteristics during complex water hammer for different methods of ball valve closing. a) B-5 series, slowly closing at the beginning and rapid shutting at the end, $T_c = 3.00$ s, b) B-6 case, rapid shutting at the beginning with slowly closing at the end, $T_c = 1.60$ s.

Table 1.   Comparison of selected experimental results with theoretical calculations.

Series [–]	Wave period [s]	Closing time [s]	Steady flow speed [m/s]	Pressure increase			
				Experimental [MPa]	Joukowsky [MPa]	Michaud [MPa]	Wood & Jones [MPa]
Circular gate valve							
C-1	0.176	0.06	1.33	0.5052	0.5358	–	–
C-2	0.173	1.18	1.33	0.4965	–	0.0817	0.5200
C-3	0.173	2.85	1.27	0.2525	–	0.0321	0.3502
C-4	0.174	4.18	1.31	0.0827	–	0.0226	0.2305
C-5	0.170	1.27	1.32	0.2880	–	0.0751	0.4976
C-6	0.171	2.46	1.32	0.5510	–	0.0386	0.4012
Ball valve							
B-1	0.174	0.12	1.31	0.5316	0.5402	–	–
B-2	0.171	0.84	1.27	0.2069	–	0.1086	0.4230
B-3	0.171	2.75	1.28	0.0580	–	0.0335	0.1066
B-4	0.171	5.03	1.31	0.0509	–	0.0187	0.0458
B-5	0.172	3.00	1.28	0.4026	–	0.0307	0.0954
B-6	0.169	1.60	1.31	0.0573	–	0.0590	0.2263

gate closing with increasing periods of time for gate closure for 4a, 4b, 4c and 4d. The first case (Figure 4a) presents the rapid water hammer phenomenon in which the time of closing is equal to 0.12 s and the wave period $T = 0.174$ s.

The following figures illustrate the increased closing times of $T_c = 0.84$ s (Figure 4b), $T_c = 2.75$ s (Figure 4c) and $T_c = 5.03$ s (Figure 4d). In all four cases, the linear characteristic of gate closing was applied. This characteristic expresses the ratio of the momentous active cross-section area to the global cross-section area of valve At/Ag. In a similar pattern to that produced by the circular gate valve case, extending the time of closure for a ball valve creates a significant decrease of wave amplitude and a decrease in the duration of the water hammer phenomenon.

The next figure (Figure 5) shows pressure characteristics registered during nonlinear ball valve closing, including beginning with slow closing and ending with

the rapid shutting (Figure 5a) and beginning with rapid shutting and ending with slow closing (Figure 5b). For the ball valve closure series, as with the circular gate valve closure series, rapid closure at the end phase produced significantly larger pressure increases than seen in slow shutting at the end phase.

3.1   Comparing Experimental Results to Existing Theoretical Models

In order to compare the experimental results with existing theory models, a selection of experimental results were compared with corresponding theoretical calculations drawn from Joukowsky's formula for rapid water hammer phenomenon, as well as Michaud's equation and Wood's and Jones' methods for complex water hammer phenomenon. This comparison is presented in Table 1.

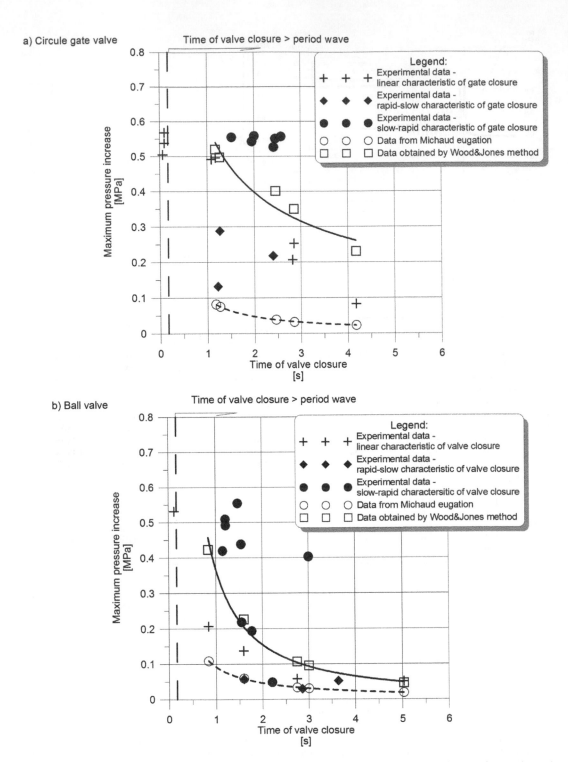

Figure 6.   Comparison of relationship between time of gate closing and maximum value of pressure increase for experimental and computational data.

Figure 6 shows calculations of the experimental data made using Michaud's and Wood's and Jones' methods compared with the experimental results.

4 CONCLUSIONS

This paper presents the effects of pressure change during the water hammer phenomenon, with simultaneous measurement of closing degrees for a ball valve and a circular gate valve. The experiments were made for comparable values of initial steady conditions of pressure and speed. The analyzed phenomenon concerns total gate closing from full valve opening to full closing.

During experimentation, the strong influence of closure speed (meaning speed of closure at the beginning, middle and end of the process) on pressure increase values was observed. Based on results, the most important phase for this pressure increase value is the end of gate closure. Accelerating the closing speed in this final phase of gate closure can generate pressure increases that reach comparable values to pressure increases during rapid water hammer (Figure 6).

A comparison of experimental data with calculations by using Michaud's equation and Wood's and Jones' methods (Table 1 and Figure 6) produced the following results:

- The results of pressure increase calculated by Joukowsky's formula are consistent with experimental effects. Differences up to 6% were observed, but the experimental data are lower than computational values.
- Regarding linear characteristics of gate closing, the measured pressure values appear in an area that is limited at the bottom by the curve produced by Michaud's equation and at the top by the curve produced with Wood's and Jones' methods. It should be emphasized that the curve determined by Wood's and Jones' methods plotted on a chart as a function of time shows that values of pressure increases are strongly related to the total time of gate closing.
- For sufficiently long durations of gate closure, calculated values of pressure increases by Michaud's equation and Woods' and Jones' method will be comparable.
- In the case of closing characteristics that begin with rapid shutting and end with slowly closing, measured pressure increases are comparable and even lower than pressure increases caused by linear closing characteristics.
- In the case of closing characteristics that end with rapid shutting, the measured pressure increases are significantly higher than calculated and comparable with the values of pressure increases caused by the rapid water hammer phenomenon.
- The calculations made using Michaud's formula lead to significantly lower values of pressure increases in comparison to experimental data – experimental data measured pressures of 300% greater than predicted by Michaud's formula.

However, in the case of circular gate valve, the difference is more pronounced. Thus, the consequence of applying Michaud's equation is a significant miscalculation of actual pressure increase values. There is a very clear relationship between total time of closing and obtained values of pressure using Michaud's equation, but the equation's inability to account for characteristics of gate closure produce the discrepancy with experimental results.

- The calculations made by using Wood's and Jones' method lead to high values of pressure increase in comparison to the linear characteristics of gate closing or slowly closing in the ending phase; similarly to Michaud's formula, the relationship between total time of closing and the obtained value of pressure increase is visible, but the relationships are more clear for the ball valve. For the circular gate valve, the differences are larger.
- Neither Wood's and Jones' nor Michaud's methods take time-changeable closing gate characteristics into consideration (Figure 6). Instead, both methods treat extended time of gate closure as a factor that lessens significant obtained pressure increases.
- The laboratory experiments demonstrate a considerable influence of gate closing method on the value of obtained pressure increase. The most important influence is acceleration of the closing process in the ending phase, whereas slowing down of this process affects obtained pressure increases to a lesser extent.
- Taking the criteria of total gate closing time as being fundamental to calculating pressure increases during water hammer is inappropriate; a new method that will take a closing characteristic of gate into consideration should be devised.

These experimental results suggest that methods of determining pressure increase, which are applied in practice, lead to results that are never consistent with reality. Applying Michaud's equation often leads to significantly smaller values of pressure. The Wood's and Jones' method allows us to obtain higher values of pressure increases on the condition that there is a linear change in the active cross-section area.

A key limitation of this experiment is that it addresses only situations of full gate closure. The problems associated with the last stage of valve closing on pressure increases should be analyzed by applying the closing gate from a partial closing stage.

Building a better understanding of the water hammer phenomenon through continued comparisons of experimental data to existing equations is crucial for future experimentation and practical application. Although this experiment measured the effects of gate closure characteristics on only ball valves and circular gate valves, it can be inferred from the results that closure characteristics for other types of valves can also be altered to reduce the water hammer phenomenon. Producing experimental results to verify this inference will provide further insights into developing better equations for calculating these pressure increases and to determine whether there are universal gate closure characteristics to reduce the water hammer phenomenon across many valve types.

REFERENCES

Evett, J. B. & Liu, C. 1989. *2500 Solved problems in fluid mechanics and hydraulics*. McGraw-Hill: New York.

Ilin, Jo. A. 1987. *Rasczet nadzieżnosti podaczi wody*. Stroizdat. Moscow (in Russian).

Marcinkiewicz, J., Adamowski, A. & Lewandowski, M. 2008. Experimental evaluation of ability of Relap5, Drako®, Flowmaster2™ and program using unsteady wall friction model to calculate water hammer loadings on pipelines. *Nuclear Engineering and Design. doi:10.1016/j.nucengdes. 2007.10.027*. NED 4960.

Mitosek, M. 2007. *Mechanika Płynów w Inżynierii i Ochronie Środowiska*. WNT: Warsaw (in Polish).

Nielacny, M. 2002. *Uderzenie hydrauliczne,*. WPP: Poznan (in Polish).

Parmakian, J. 1955. *Waterhammer analysis*. Prentice-Hall Inc.: New York.

Pires, L. F. G., Laidea, R. C. C. & Baretto, C.V. 2005. Transient flow analysis of fast valve closure in short pipelines. *Proceedings of International Pipeline Conference, October 4–8, 2004*. Calgary, Alberta: Kanada.

Ramos, H. & De Almeida, B. A. 2002. Parametric analysis of water hammer effects in small hydro schemes. *Journal of Hydraulic Engineering* 128(7): 689–696.

Sharp, B. B. 1974. Discussion of water hammer charts for various types of valves. *Journal of Hydraulic Division* 100(2): 323–326.

Sharp, B. B. 1981. *Water hammer: problems and solutions*. Edward Arnold Publisher Ltd.: London.

Sharp, B. B. 1969. Water hammer gate characteristics. *Water Power* 21: 352–354.

Sharp, B. B. & Sharp, B. D. 2003. *Water hammer – practical solutions*. Butterworth Heinemann: Oxford.

Streeter, V. L., Wylie, B. E. & Bedford, K. W. 1998. *Fluid mechanics*. WCB McGraw-Hill: New York.

Thorley, A. R. D. 2004. *Fluid transients in pipeline system: a guide to the control land suppression of fluid transients in liquids in closed conduits*. ASME Press: New York.

Wood, D. J. & Jones, S. E. 1973. Water-hammer charts for various types of valves. *Journal of Hydraulic Division* 99(1): 167–178.

Wylie, B. E., Streeter, V. L. & Suo, L. 1993. *Fluid transients in systems*. Englewood Hills New Jersey. Prentice Hall.

Environmental Engineering III – Pawłowski, Dudzińska & Pawłowski (eds)
© *2010 Taylor & Francis Group, London, ISBN 978-0-415-54882-3*

Water need of Energy Crops – one of the environmental problems of Poland

P. J. Kowalik
Department of Sanitary Engineering, Gdansk University of Technology, Gdansk, Poland

R. Scalenghe
Dipartamento di Agronomia Ambientale e Territoriale, Universita degli Studi di Palermo, Palermo IT E.U.

ABSTRACT: This paper presents updates of Polish government policy, showing legal and financial instruments for governing bioenergy developments. Within a country, the strategy for renewable energy must be coordinated with the strategy of water management. The water needs are given for the cool climate and extensive agriculture of Central and Eastern Europe. They are 2–3 times lower than for hotter conditions of Southern Europe. Despite this, the rainfall of $600\,mm\,y^{-1}$ would be insufficient for the water supply of most energy crops.

Keywords: Biomass, crop-water needs, European Union, reference yields, subsidies.

1 INTRODUCTION

Abbreviations:
CEE Central and Eastern Europe
EU European Union
ASTM American Society for Testing and Materials
EUR Euro
PLN Złoty
FAO Food and Agriculture Organization of the United Nations

Due to global warming and the production of energy crops, water is expected to become scarce in Central and Eastern Europe (CEE), a region of rain-fed agriculture. Therefore, there is a need to increase the efficiency of water use for crops in CEE. In the present paper, the growth, production and water use efficiency of energy crops in Poland was investigated. Using Poland as an example is justified by its role in CEE, and Polish experiences might be relevant for other Europeans, from the Baltic to Mediterranean Seas. Energy crops may use more water than other crops so their impact on water resources could be negative. The extensive use of water by the energy crops is an issue for our research.

When reducing emissions of carbon dioxide, the agriculture sector could contribute by different strategies: (a) swap from food to non-food (energy) crops to partially replace fossil fuels for electricity production, heating or transport; (b) use existing agricultural crops as energy sources instead of their current use; and (c) utilise agricultural wastes for energy production (Powlson et al. 2005). Potential competition for both water and soil between the energy and the food crops is one concern (Muller 2008). In fact, soils perform crucial functions which are environmentally, economically and socially in competition (Bouma 2006). Usually, various functions of soils are distinguished, namely: the production, the carrier, the filter, the resource, the habitat and the cultural function (European Commission 2006a).

This article concentrates solely on one of these functions – the production of soil (viz. agroforestry) and on the opportunities for greater production of crops grown specifically as sources of renewable energy, like perennial grasses or short rotation coppice of tree species. The purpose of this study is to determine the potential water needs of energy crops under the current conditions. Increasing the potential of energy crops, requires optimising the dry matter and energy yield of these crops per area of land and to account for water needs and water resources for crop growth.

In 2000, 6% of Europe's (EU15) primary energy came from renewable sources, of which 3.7% was from biomass (approx. 2.2 EJ) (Cannell 2003, Tuck et al. 2006). The percentage supplied by biomass is about to increase. The European Commission's White Paper on Renewables proposes a target of doubling the contribution of renewables to 12% of the EU's total primary energy needs by 2010 (European Commission 1997). This would require a capital investment of 165 billion EUR. In 2003, the European Parliament approved a Biofuel Directive (European Commission, 2003) that sets targets for the use of biofuels in the European transport fuel market at 2% by the end of 2005 and 5.75% by the end of 2010. Although these targets have not yet been met (European Commission, 2007) there is renewed impetus by European policy makers to fully implement them. In the future, more land in Europe will probably become available for the production of bioenergy crops (Tuck et al. 2006).

The European Commission (2006b) declared the goal '3 × 20% + 10%', meaning that in the year 2020 emissions of carbon dioxide will be reduced by 20% in comparison to 1990, renewable energy will be 20% of primary energy supply and the consumption of energy will be reduced by 20% in comparison to 1990, by energy-savings measures and increased energy efficiency. The participation of biofuels in transportation energy use will be 10% in year 2020. These goals were accepted in the context of climate change mitigation during several political meetings (e.g. participation of 27 member states of the EU in March 2007 in Brussels).

Poland and later Romania and Bulgaria joined the EU with abundant agricultural lands; a subsidised market for biofuels derived from the agricultural crops could help stimulate their economies. However, the broader biofuel market will increase the area of land under intensive agriculture and will provide an impetus for the capture of the key ecological resources such as water sources and soils (Lovett 2007).

2   STATE AND DECLARATIONS OF POLISH BIOMASS MARKET

In the present paper the data on yields of energy crops and their transpiration coefficients were selected. Multiplication of yield by transpiration coefficient gives the water requirements of the selected crop. We used the data on the low reference yields of energy crops in Poland, published by the Polish Ministry of Agriculture and Rural Developments for 2006 and 2007. Data on the transpiration coefficients were taken from Debski (1970) and Penning de Vries et al. (1989). The calculated water needs were compared with the data of Brouwer and Heibloem (1986) for all of Europe.

The limitation of the method used is related to the low yields established by the Polish Government for farmers producing energy crops, and used as a reference yields. If yield was higher, then water needs increase in each case. Only the general situation in the Polish market is evaluated in this paper.

Declarations on the development of renewable energy in Poland were formulated by the Polish Parliament on 23 August 2001. The Polish Government issued the document entitled 'Energy policy of Poland until year 2025' on 4 January 2005. The next document was 'Strategy for electro-energy developments' issued on 28 March 2006. According to these documents the strategy of the country is to get 7.5% of total primary energy supply from renewable energy by 2010, and 14% by 2020. For electric energy at least 7.5% should be supplied from renewable sources by 2010. The Minister of Industry prepared the decree of 1 July 2005 related to the energy policy of Poland until 2025 (Monitor Polski, M.P. 2005/42/562). According to this decree the main source of renewable energy in Poland is biomass. It indicates limitations of access to some biomass, e.g. the very limited amounts of wood residue in Poland. It notes the possible additional sources of biomass from industrial and communal residues rich in organic matter.

The Minister of Agriculture and Rural Development prepared a few decrees on energy crops produced by Polish farmers. Energy crops are subject to subsidies, so it was necessary to state which crops were to be used for energy production and which reference levels of yield were to be included in the policy on payments of subsidies. Decrees were issued on 14 March 2007 (Official Journal O.J. 2007/55/364) and 16 October 2007 (O.J. 2007/195/1410; M.P. 2007/53/607), and 14 March 2008 (O.J. 2008/44/267 and O.J. 2008/44/268). They implemented the EU directives related to land use for energy crops (O.J. L 345/1, 10/11/2004).

In the documents of the Polish Ministry of Agriculture and Rural Development there is a statement about the surface area of the energy crops in Poland. In the year 2007 it was 10 000 ha, but by 2015 is expected to be 500 000 ha (~4% of the total Polish arable land). The Polish Government produces optimistic percentages that are far from reality. The initial investment in plantations in Poland of energy crops is estimated at 8000–15 000 PLN (2000–4000 EUR) per ha.

The Minister of Industry prepared the first draft of the document entitled 'Energy policy of Poland until 2030', of 4 September 2008. According to this document there should be 15% of primary energy from renewable sources by 2020 and 20% by 2030. The participation of biofuels in the transport fuel market should be 10% by 2020. Future development should consider the equilibrium of fuel production from non-food agriculture, agriculture-for-food and sustainable forestry.

According to the abovementioned legal documents, Poland will develop production of biomass for energy. In year 2006, renewable energy sources supplied 210 513 TJ of energy, about 6.5% of Poland's total energy production (3 253 PJ), equivalent to <1 000 000 t of black coal (25 PJ). The energy from biomass was >91.3% of the total renewable energy, where hydropower was 3.5%, liquid biofuels 3.3%, biogas 1.2%, wind energy 0.4% and geothermal energy 0.3%. The solid, liquid and gas biomass supply 96% of the total renewable energy in Poland.

Production of biomass for energy by farmers is related to subsidy policies, giving 45 EUR per ha for energy crops. Subsidies are paid if some conditions are met. The surface area of the energy crops must be > 0.3 ha per farm; the biomass may be used on farm or sold to a company utilising biomass for energy, e.g. in co-firing biomass pellets with black coal. A detailed description of Polish energy crops is given in Kosciuk (2003).

3   REFERENCE YIELD OF THE ENERGY CROPS IN POLAND

The Polish Minister of Agriculture and Rural Development issued two decrees on the required yields of energy crops in Poland (Decree of 14 March 2007, concerning representative yields of energy crops, O.J. 2007/55/364 and Decree of 16 October 2007, supplementing the previous decree on representative yields of energy crops, O.J. 2007/195/1410 and O.J. 2008/44/268). The subsidies are paid only if the real yield is higher than described in these decrees.

Oil crops with reference yields in Poland (O.J. 2007/55/364) are: oil rapeseed (*Brassica napus* L. ssp. *oleifera*) with minimum yield (dry biomass) for subsidies of 2.5 t ha^{-1}; turnip (*Brassica rapa* L.) 2.5 t ha^{-1}; linseed 1 t ha^{-1}, 4 t ha^{-1} of grain; linseed dodder (false flax) 1 t ha^{-1}; and white mustard 1 t ha^{-1}.

Cereals with reference yields in Poland (O.J. 2007/55/364) are: wheat with minimum yield (dry biomass) for subsidies of 3.8 t ha^{-1}; rye 2.4 t ha^{-1}; barley 3.1 t ha^{-1}; oats 2.4 t ha^{-1}; mixed crops 2.7 t ha^{-1}; triticosecale 2.7 t ha^{-1}; maize 70 t ha^{-1} of green biomass and 5.5 t ha^{-1} of grain; millet 2.3 t ha^{-1}; buckwheat 1.2 t ha^{-1}; pea 2.0 t ha^{-1}; and amaranth 1.5 t ha^{-1} of grain.

Starch and sugar crops with reference yields in Poland (O.J. 2007/55/364) are: potato with minimum yield (dry biomass) for subsidies of 17.5 t ha^{-1}; sugar beet 40 t ha^{-1} and Jerusalem artichoke 20 t ha^{-1}.

Cellulose crops with reference yields in Poland (O.J. 2007/55/364) are: sida (*Sida* spp.) with minimum yield (dry biomass) for subsidies of 15.0 t ha^{-1}; silver grass (*Miscanthus*) 20 t ha^{-1}; prairie cordgrass (*Spartina pectina* Link) 17 t ha^{-1}; reed canary grass 8 t ha^{-1}; giant knotweed 20 t ha^{-1}; and mixed grasses 10 t ha^{-1}. One should note that the minimum yield for *Miscanthus* is 20 t ha^{-1} this is a yield rarely harvested. Reality is closer to 9 t ha^{-1}, even on good soils, non-irrigated plants are unlikely to provide a harvest of 15 t ha^{-1}.

Polish reference wood yields (O.J. 2007/55/364 and O.J. 2007/195/1410) are: willow 8 t ha^{-1} of dry wood; multiflora rose (*Rosa multiflora* L.) 12 t ha^{-1}; black locust (*Robinia pseudoacacia* L.) 8 t ha^{-1}; poplar (*Populus* spp.) 10 t ha^{-1}; alder (*Alnus* spp.) 8 t ha^{-1}; birch (*Betula* spp.) 8 t ha^{-1}; and hazelnut (*Corylus avellana* L.) 8 t ha^{-1}. For most of the traditional trees the yields are not established. They are: pine (*Pinus* spp.), oak (*Quercus* spp.), spruce (*Picea* spp.), beech (*Fagus sylvatica* L.) and fir (*Abies* spp.).

4   WATER USE CROPS

After considering the reference yields of energy crops it is necessary to calculate the water and potential irrigation needs. Vegetation continuously uses water for transpiration, which is measured as the transpiration coefficient, the amount of water (kg or l) used to produce one unit of biomass (kg) of growing plants. Transpiration coefficient (relation between growth of biomass, in kg, and transpired water, in kg or l) can be defined as the total amount of water transpired divided by the amount of harvested dry biomass produced (kg kg^{-1}) (Penning de Vries et al. 1989). There are

Table 1. The transpiration coefficient of crops and trees in Poland (data from Debski 1970).

Herbaceous crops	Transpiration coefficient (kg kg^{-1})	Arboreal crops	Transpiration coefficient (kg kg^{-1})
rye	724	larch	1165
oats	614	beech	1043
wheat	507	linden	1038
alfalfa	859	oak	616
barley	511	spruce	242
maize	358	pine	123
millet	273	fir	86

Table 2. Water needs in mm/season in Poland for energy crops.

Energy crop	Yield (kg ha^{-1})	Transpiration coefficient (kg kg^{-1})	Water need (kg ha^{-1})	Water need (mm season^{-1})
Rye	2 400	724	1 737 600	174
Wheat	3 800	507	1 926 600	193
Maize	5 500	358	1 969 000	197
Millet	2 300	273	627 900	63
Potatoes	17 500	180	3 150 000	315
Cover crops (mixed species)	10 000	600	6 000 000	600
Arboreal crops$^{\#}$	8 000	242	1 936 000	194

$^{\#}$ Spruce or willow.

considerable differences in transpiration coefficients between environments and species (Kowalik & Perttu 1989). The synthesis of knowledge in Poland (Debski 1970), including crops and trees is summarised in Table 1. The transpiration coefficient can vary by a factor >2 and is entirely empirical. It relates to estimated water use and biomass yield as a first approximation of water needs, and requires more experimentation in future.

The transpiration coefficient given here is in kg water kg^{-1} biomass. It is clear that the cereals (including maize) with transpiration coefficient <600 kg kg^{-1} are water-saving crops in comparison to alfalfa (859 kg kg^{-1}) or trees (linden, beech and larch). Beech has a transpiration coefficient of 1043 kg kg^{-1}, spruce several times less (242 kg kg^{-1}), while pine is a very water-saving tree (only 123 kg kg^{-1}). The production of biomass for biofuels should be aimed at plants with relatively low transpiration coefficients. In general, the transpiration coefficient for most crops is 150–600 kg kg^{-1} (Penning de Vries 1989). Transpiration coefficients of potato varieties in the Netherlands were 125–180 kg kg^{-1} (Bodlaender 1986).

Water needs were calculated by multiplying yield by transpiration coefficient (Table 2).

The water needs for wheat, corn or willow were very similar (193–197 mm), while for cover crops (mixed species) they were three times greater (600 mm). The yield of willow was 16 t of wood and would need 400 mm of water during the growing season. The production of biogas from fermented grass biomass is very water demanding, three times greater than traditional agricultural crops. Cover crops require irrigation as the average rainfall in Poland is about 600 mm y^{-1}. It is clear that if yields were increased by 2–3 times, the water needs would also increase 2–3 times; this method can be used to calculate irrigation needs and scheduling.

5 DISCUSSION

If globally the supply of key share of energy occurs by producing all the needed amount of the biomass, it is difficult to know in advance whether this will be sustainable for ecosystems (Muller 2008). For instance, taking into consideration the production of ethanol or biodiesel in terms of crop choice: (i) maize and soybean require one third more fossil energy than ethanol or biodiesel fuel produced, respectively; (ii) switchgrass requires 50% more fossil energy than ethanol produced; (iii) wood biomass requires 57% more fossil energy than ethanol produced; and (iv) sunflower requires 118% more fossil energy than biodiesel produced (Pimental & Patzek 2005).

In terms of water, "the crop water need or crop evapotranspiration consists of transpiration by the plant and evaporation from the soil and plant surface (of intercepted water). When the plants are fully grown the transpiration is more important than the evaporation", but "at planting and during initial stage the evaporation is more important than the transpiration" (Brouwer & Heibloem 1986). The crop water need is considered when plants are fully grown, meaning that only the transpiration as a main growth factor is calculated, without early spring weeks. In Poland, evaporation from soil is usually <10% of transpiration, and evaporation of intercepted water is <30% of intercepted rain (Kowalik & Eckersten 1984). Evapotranspiration is certainly a little higher than transpiration; this ratio is difficult to calculate, since it is fairly complicated, and so is not introduced here. Only the crop water needs that occur in the following conditions are considered: low sunshine of cloudy sky, cool temperatures, high air humidity and little wind. In such conditions transpiration is low.

An average daily water need of a standard grass during an irrigation season is about 1–2 mm d^{-1} if the climatic zone is humid and main daily temperature is <15°C (Brouwer & Heibloem 1986). Other crops are compared with such a standard grass. Thus for barley, flax, millet, oats, peas, potatoes, sugarbeet and wheat, the crop water needs are 10% higher than for standard grass. If (introducing here free hand expert approach) 2–3 mm d^{-1} is used as an indicative value of the total growing period of 150 d in Poland,

this gives 300–450 mm per growing season (Dziezyc 1989).

Some analogies can be seen between the calculations of Table 2 and data of Brouwer and Heibloem (1986), whose results over total growing periods were: barley, oats and wheat 450–650 mm; maize 500–800 mm; potato 500–700 mm; sugarbeet 550–750 mm; and alfalfa 800–1600 mm. It should be emphasised that our results are related to transpiration in cool climates and relatively extensive agriculture (in Poland). However, Brouwer and Heibloem (1986) gave values for main European conditions and good agricultural practices with the highest possible yields, 2–3 times the reference yields in the Polish legal regulations. The low yield threshold and the extensive land use in Poland will lead to low water use.

The expected final yields were higher in the calculations of Brouwer and Heibloem (1986) than in the present paper. The water needs of cereals (barley, oats, wheat and maize) were about 450–800 mm (Brouwer and Heibloem, 1986) compared to our calculations of <200 mm for transpiration for rye and wheat in the temperate climate of Poland (2–4 times lower). In hotter climates, corn requires 500–800 mm of water; however, our calculations for a cool climate were 197 mm, or 3–4 times less. In our calculations, potato needs 315 mm but Brouwer and Heibloem (1986) obtained 500–700 mm, or twice as much. Rape seed for biofuel is a water demanding crop, in Poland requiring 600–700 mm per growing season. Lower water supply can reduce final yield by about 20% (Berbec & Malicki 1989). We do not have a transpiration coefficient for rape seed.

The transpiration coefficient for willow plantations is similar to spruce (Persson 1995). The transpiration coefficient of willow was proposed as 1.9–4.9 g dry matter mm^{-1} (Kowalik & Eckersten 1989). Using 2–5 g m^{-2} for 1 l m^{-2} gives 2–5 g of biomass kg^{-1} of water. The transpiration coefficient will thus be 200–500 kg of water kg^{-1} of biomass. We calculated the transpiration coefficients for spruce and willow as 242 and 200–500 kg kg^{-1}, respectively. It is clear that water use in Poland is quite different from hotter countries.

6 CONCLUSIONS

The implications of these findings are important for the best agricultural practice of Polish farmers producing crops for energy. Growth limitations are related to the extensive land use in Poland. There are no other relevant studies dealing with water needs of energy crops. In conclusion, we emphasise some additional points:

(1) We do not distinguish between the land-use options: non-food agriculture, agriculture-for-food and sustainable forestry;
(2) We do not discuss legal instruments, market fluctuations and financial subsidies;
(3) We do not address specific comments on hydrological properties of local soils;

(4) The biomass from energy crops is relatively moist, up to 30–50% water content. Although this can be a disadvantage, the moist biomass can also be used to produce biogas;

(5) The heat value of biomass is 15–24 MJ kg^{-1} dry matter, thus 2 t of biomass may substitute for 1 t of black coal, reducing harmful emissions from coal combustion;

(6) The biomass may be used in many technological processes (i.e. for heat, electricity and biofuels) without large changes in existing land use and agricultural practice;

(7) The target of 7.5% of primary energy from biomass in the year 2010 in Poland is expected to be realised, taking into consideration the subsidies and the certificates of the origin of energy from the biomass;

(8) Water needs of biomass produced for energy in Poland should be 2–3 times lower than in Southern Europe, but 20–30% greater than rainfall during the growing season.

ACKNOWLEDGEMENTS

We used data from older published sources (Debski 1970, Brouwer & Heibloem 1986) and official documents of the EU and the Polish Government. We acknowledge them fully.

REFERENCES

Berbec, S. & Malicki, L. 1989. Potrzeby wodne roslin przemyslowych n: J. Dziezyc (ed). *Potrzeby wodne roslin uprawnych*: 85 – 118 PWN: Warsaw().

Bodlaender, K.B.A. 1986. Effects of drought on water use, photosynthesis and transpiration of potatoes. 1. Drought resistance and water use. In: *Potato Research of Tomorrow*: 36–43. Pudoc: Wageningen.

Bouma, J. 2006. Soil functions and land use. In: G. Certini & R. Scalenghe (eds), *Soils. Basic Concepts and Future Challenges*: 211–221. Cambridge University Press: Cambridge.

Brouwer, C. & Heibloem, M. 1986. *Irrigation Water Management: Irrigation Water Needs. Training Manual No. 3*. FAO Natural Resources Management and Environment Department: Rome.

Cannell, M. 2003. Carbon sequestration and biomass energy offset: theoretical, potential and achievable capacities globally, in Europe and the UK *Biomass Bioenergy* 24: 97–116.

Debski, K. 1970. *Hydrologia:* 368. Arkady: Warsaw.

Dziezyc J. 1989. *Water needs of agricultural crops:* 419. PWN: Warszawa.

European Commission. 1997. Energy for the Future: Renewable Sources of Energy. *White Paper for Community Strategy and Action Plan* (COM(97)). Office for Official Publications of the European Communities: Luxembourg.

European Commission. 2003. European Biofuels Directive 2003/30/EC. Commission of the European Communities: Brussels.

European Commission. 2006a. *Thematic Strategy for Soil Protection. 22.9.2006* EU COM(2006) 231 (final).

European Commission. 2006b. *Renewable Energy Road Map – Renewable Energies in the 21st Century – Building a more Sustainable Future.* EU COM (2006) 848 (final).

European Commission. 2007. *Biofuels Progress Report – Report on the Progress Made in the Use of Biofuels and Other Renewable Fuels in the Member States of the European Union.* Commission of the European Communities: Brussels. Kosciuk, B. 2003. *Rosliny energetyczne:* 146. Wydawnictwo Akademii Rolniczej w Lublinie: Lublin ().

Kowalik, P.J. & Eckersten, H. 1984. Water transfer from soil through plants to the atmosphere in willow energy forest *Ecol. Model.* 26: 251–284.

Kowalik, P.J. & Eckersten, H. 1989. Simulation of diurnal transpiration from willow stands. In: K.L. Perttu, P.J. Kowalik (eds), *Modelling of Energy Forestry: Growth, Water Relations and Economics:* 97–119. Pudoc: Wageningen.

Kowalik, P.J. & Perttu, K.L. 1989. Introduction to modelling of plant water conditions. In: K.L. Perttu, P.J. Kowalik (eds). *Modelling of Energy Forestry: Growth, Water Relations and Economics:* 89–96. Pudoc: Wageningen.

Lovett J.C. 2007. Biofuels and ecology *Afric. J. Ecol.* 45: 117–119.

Muller A. 2008. Sustainable agriculture and the production of biomass for energy use *Climatic Change* DOI 10.1007/s10584-008-9501-2.

Penning De Vries, F.W.T., Jansen, D.M. & Ten Berge, H.F.M. 1989. Simulation of ecophysiological processes of growth in several annual crops: 271.Pudoc: Wageningen.

Persson, G. 1995. *Water Balance of Willow Stands in Sweden.* Reports and Dissertations 20. Swedish University of Agricultural Sciences: Uppsala.

Pimental, D. & Patzek, T.W. 2005. Ethanol production using corn, switchgrass, and wood; biodiesel production using soybean and sunflower *J. Nat. Resour. Res.* 14: 65–76.

Powlson, D.S., Riche, A.B. & Shield, I.2005. Biofuels and other approaches for decreasing fossil fuel emissions from agriculture *Ann. Appl. Biol.* 146: 193–201.

Tuck, G., Glendining, M.J. & Smith, P. 2006. The potential distribution of bioenergy crops in Europe under present and future climate. *Biomass Bioenergy* 30: 183–197

Environmental Engineering III – Pawłowski, Dudzińska & Pawłowski (eds)
© *2010 Taylor & Francis Group, London, ISBN 978-0-415-54882-3*

Modified Ghmire and Barkdoll method of quantitive sensors location in a water distribution system

D. Kowalski, M. Kwietniewski, B. Kowalska & A. Musz
Department of Environmental Protection Engineering, Lublin University of Technology, Lublin, Poland

ABSTRACT: The monitoring of quantitive parameters is essential to the proper operation of a water distribution system. To obtain a suitable calibration, it is essential to identify good monitoring point locations in the water distribution system (representative location) to ensure that effective information is generated. A Ghmire and Barkdoll method of locating sensors (at a representative location) in a water distribution system is developed. The method is formulated as a two-part problem, first the pressure measuring point is located, then the flow rate measuring points are described. The analysis revealed that the new calibration method presented in this work produced close agreement between measured and simulated pressures.

Keywords: Hydraulic model calibration, heuristic method, location of monitoring points.

1 INTRODUCTION

Hydraulic simulation models for water distribution network analysis are nowadays the most valuable tools for engineers, enabling them to analyse the hydraulic performance of water delivery systems, to study fire – flow, to plan water supply systems, to solve the optimization problems regarding development/repair costs and to monitor the operating water system.

In recent years there have been a number of computer programs available commercially. These programs can differ in implemented solutions, the algorithms which are based either on steady-state or on dynamic approachesIrrespective of their mathematical representation of the real physical system, the model usually must be calibrated. Calibration of the hydraulic model requires adjustment to the measured input data which describes the physical characteristics of the system, so that the data with model-predicted performance has an acceptable accuracy. Calibration is a procedure of determining individual unknown parameters of a hydraulic model, which minimizes the differences between the measurements performed on a real water distribution system and the results of the hydraulic model. Some model parameters can be determined with reasonable accuracy if there is measurement data already available. Performing a comprehensive network calibration is a highly complex problem and a time consuming procedure and it also needs a sound knowledge of the water distribution system and extensive experience in the validation of models.

Calibration of the hydraulic model requires information about such variables as pressures and flow rates at selected locations, tank water levels, and outputs of internal and external sources. To obtain a suitable calibration, it is essential to identify good monitoring point locations in the water distribution system (representative location) to ensure that effective information is generated.

The aim of this article is to describe the modified Ghmire and Barkdoll method of quantitive sensors location (at a representative location) in water distribution system. The study involves an example of an application of this method in a real water distribution system.

2 CALIBRATION METHODS

Basic information about numerical models of water distribution systems and guidelines for network model calibration were described in the "Calibration Guidelines for Water Distribution System Modeling" (Proc. AWWA 1999 Imtech Conference). In this handbook a review of the various sources of error that will produce differences between measured and calculated network performance, was discussed. These sources can be the following: errors in input data, including the typographical errors made by typing the wrong geographical data, incorrect descriptions of the pipe internal roughness values, differences in water demands caused by grouping water use at nodes instead of placing it at its actual location, accuracy of the water system maps, elevation data differences, information on times for tank water levels, the level of system simplification (a skeletal version instead of one which contains every pipe in the network), inaccurate pump characteristic curves, incorrect settings for pressure reducing valves, poorly calibrated measuring equipment etc.

Calibration methods were developed in the early seventies according to developmentf computer techniques. Generally they can be divided into three groups:

- iteration (or trial and error) methods,
- explicit methods – hydraulic simulation models,
- implicit methods – optimization models.

The trial-and-error methods are based on adjusting the data describing the model until a suitable match is obtained (Rahal et al. 1980, Walski 1986). The procedures are as follows: skeletalization of distribution network, evaluation of the calibration parameters (typically pressures, flow rates and pipe roughness), selection of representative locations and adjusting data until a suitable match is obtained. Explicit methods of hydraulic model calibration imply direct solving of extended mass and energy conservation equations (Boulos & Wood 1991). The main set of equations describes steady-state conditions and additional equations are obtained from the measurements of pressure and flow rates. This approach requires the number of parameters to be estimated to be equal to the number of equations or measures. In this case the Newton-Raphson method is popular. The implicit methods of calibration are formulated and solved as optimization problems. The objective function allows the minimization of the differences between measurements and the results of the numerical model (Tucciarelli et al. 1999). Mathematical formulations of error function are usually non-linear. To minimize the error function different search methods were developed such as the genetic algorithm method (Savic & Walters 1997, de Schaetzen et al. 2000, Kozelj et al. 2006, Wu & Walski 2005), which has been popular in recent years.

All the aforementioned methods are used in steady state or quasi-steady (extended period simulations EPS) conditions. Although these methods are still predominantly the methods of choice, some unsteady (dynamic) methods have appeared. They are able to calibrate distribution systems and perform leak detection simultaneously. Difficulties occur because data inputs need to be very accurate, but time-consuming field studies are not required (Simpson & Vitkovsky 1997, Ferrante & Brunone 2001).

3 LOCATION OF MONITORING POINTS

It is essential in every calibrating method both to collect all necessary measurement data and to choose representative locations of monitoring points. Generally, the main tasks for planning the sensor location are:

- environmental monitoring (consistent with the applied method)
- analysis of the problem to solve
- coverage rate (measurements required to operate the system correctly)
- scientific research (according to individual feature requirements)

Designs of the systems of collecting data to calibrate the water numerical system are included in the area of scientific research (Bush & Uber 1998). The quality of information collected has a great influence on the calibration accuracy of the model. One of the first researchers to identify guidelines for collecting calibrating data was Walski (1983). He suggested the following guidelines;

- to place the sensors adjacent to the nodes with maximum water demand
- values of pressure recorded in the nodes furthest from the water sources can be used to make a skeleton network
- to use the fire flows (repeated at different nodes of the network) by opening the most possible hydrants
- to collect data from monitoring points recording both flow rates and pressures

Lee and Deiniger (1992) have developed a method enabling the tracking of the water course from the source to the demand node. With this method monitoring point locations should be set in the places in which maximum hydraulic information is possible. Yu and Powell as well as Nagar and Powell (Yu & Powell 1994, Nagar & Powell 2000), proposed designing the measuring system as an optimal continuing monitoring of the water distribution system. The first aim of this method involves the minimization of the difference between the simulated model and the field measured hydraulic parameters (pressures and flow rates). Then the objective is to minimize both the investment and operational costs. This problem was resolved with the aid of the shortest path tree (SPT). De Schaetzen et al. (2000), suggested three new sampling design approaches. The first two approaches are based on the shortest path algorithm and the rank of potential measurement locations. The third approach solves the optimization problem by building on the maximization of Shannon's entropy. There are also heuristic methods, which are still popular, that use a sensitivity-based heuristic procedure. The procedure accounts for uncertainties in measurements, their impact on model parameters, and finally, their impact on water distribution systems model predicted state variables (Ahmed et al. 1999). A similar approach is presented by Alzamora & Atala (&006) who made use of their own algorithm through topological analysis. Berry et al. (Berry et al. 2006) proposed a "p-median" algorithm related to the heuristic method. Ghimire and Barkdoll adopted a heuristic method to sensor locations at maximum demand junction nodes. The main advantage of heuristic methods is the relatively short time taken to select the representative locations of monitoring points compared to advanced computer programs. It is also possible to apply these methods to very large distribution systems (in contrast to majority of optimization methods) and for technical staff in a water company to use them.

De Schaetzen et al. (2000) proved that the optimal number of measurement points is the function of a certain level of accuracy. In view of the investment

Table 1. Values of a, b, c, d coefficients.

value	a	b	c	d
1	low settlement	residents	above 15 m H_2O	up to 25% maximum predicted by model
2	medium settlement	schools, dormitories, service centers	up to ± 15 m H_2O	up to 50% maximum predicted by model
3	high settlement	shopping centers, sports-entertainment halls, industrial yards, health centers	up to ± 10 m H_2O	up to 75% maximum predicted by model
4	town centers, industrial areas	hospitals, fire-stations	up to ± 5 m H_2O	above the maximum predicted by model

and exploitation costs they proposed that the number of measuring points should oscillate between 1% and 2% of all nodes calculated in the computer model.

As demonstrated in the literature review it is essential to identify good monitoring point locations in the water distribution system (representative location) before the calibration procedure is performed. Among lots of calibration methods, the heuristic method is still very popular, therefore, a new modified heuristic method is proposed in the next section.

4 CALIBRATION OF A REAL DISTRIBUTION SYSTEM

4.1 *The modified Ghmire and Barkdoll method*

Ghimire and Barkdoll (2006) suggested a heuristic demand-based approach in which sensors are located at the junctions with the highest demands, or the highest mass released (Ghimire & Barkdoll 2006). The total number of monitoring points is ranked according to modified water demands at the nodes.

In this paper a modified Ghmire and Barkdoll method is presented. The method is formulated as a two-part problem, firstly the location of pressure measuring points is performed, and then the flow rate measuring points are described.

4.2 *Location of pressure measuring points*

This approach involves the following tasks:

- The area is divided into the parts with the specific kind of the settlement, inhabitants, water demand quantity, ground elevations
- These parts should be a minimum 1% of all network nodes used for a calibration hydraulic model
- In each part the pressure sensors are placed at the maximum modified demand junction nodes.

The modified water demand at the node can be calculated from the following equation:

$$W1 = Q \cdot a \cdot b \cdot c \cdot d \tag{1}$$

where: Q is the water demand at the node, a, b, c, d are coefficients depend on: a kind of settlement (a), sort of water con consumers (b), minimum difference between allowable pressure in the network

and the average pressure calculated by the model (c), maximum day pressure deviation calculated by the numerical model (d). The values of the coefficients can be assigned from table 1.

Sensors are ranked according to modified water demands at the nodes estimated by means of equation 1, and finally located at places physically possible to access.

4.3 *Location of flow measurement points*

Similarly to the selection of locations for pressure measuring points, finding the locations for flow rate sensors involves some tasks of which the first two are identical to those in the aforementioned procedure. During the second phase of design, in each selected part, the monitoring points are assigned according the following relation

$$W2 = Q_{max} \cdot e \cdot f \tag{2}$$

where: Q_{max} is the day maximum water flow in the pipe predicted by computer model, e, f are coefficients depend on: percentage part of the day water demand in analyzed area (e), day flow rate deviation in the pipe calculated by the numerical model. The values of the coefficients are put in the table 2. Sensors are ranked according to modified water demands at the nodes estimated by means of equation 2, and finally located at places physically possible to access. Flow rate gauges are a part of the pipe (stationary gauge) or they can be placed on the pipe, in which case the pipe segment needs to be the length of 5 to 15 pipe diameters.

5 RESULTS AND DISCUSSION

The proposed method was applied to the middle-sized water distribution system of a Polish city with a population of around fifty thousand.

The city is situated in some hills with a difference in elevation of up to 30 m. A total system demand for an average day is 7500 dm^3/d. The length of the network is about 150 km, including 50 km of pipe connections to the houses. The network consists of 10 to 40 year old unlined cast iron pipes.

Table 2. Values of e, f coefficients.

Value/coeff.	e	f
1	up to 25% day water demand in selected area	up to 25% maximum predicted by model in selected area
2	between 25% and 50%	up to 50%
3	between 50% and 75%	up to 75%
4	between 75% and 100%	above 75%

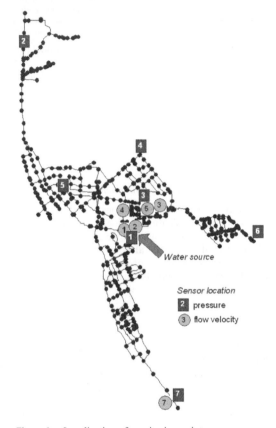

Figure 1. Localization of monitoring points.

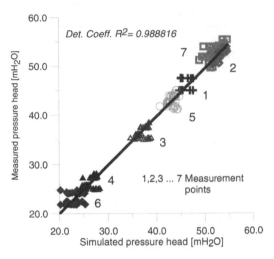

Figure 2. Correlation plot of measured and simulated.

The numerical model of the distribution network was constructed on the basis of available documentation and field measurements,. The hydraulic analysis was performed with EPANET 2 software (Rossman 2000), which models water distribution piping systems, developed by EPA's Water Supply and Water Resources Division. The model consists of 458 nodes and 601 pipes, water tank, and does not include pipe connections to the houses. EPANET tracks the flow of water in each pipe, the pressure at each node, and the height of the water in each tank. For the calibration procedure the pipe roughness values and nodal time-varying water demands have been estimated by means of the trial and error method described earlier. Pressures and flow rates as calibration parameters were assigned in a field measurements. Before calibration the water system was divided into six areas, which is 1.1% of all the analyzed network nodes.

The location of the monitoring points is shown in Figure 1. The pressure sensors type CellBox (Biatel Poland) were placed at chosen hydrants. Sensors were equipped with loggers registering the measuring data. The data was recorded every ten minutes. The Micronics Portaflow 300 and Metron ultrasound devices were used for non-invasive measurement of water flow. Selecting locations of these gauges was more difficult than for pressure sensors because of the lack of a water chamber at the selected monitoring points in the network. The instruments were attached to the outside of the available pipes using the straps. The timing of measurements was restricted because of the battery capacity of the measuring devices so data were collected for less than 58 hours. There was not sufficient time to perform an exact flow calibration. The results of the pressure calibration process are presented in Figure 2.

A correlation between the pressure head of measured and computed values at the monitoring points is good. The average roughness coefficient was equal to 1 mm and varied from 0.2 to 2.7 mm. The root mean squared error (RMS) of pressure measurements was 1.219 m H_2O and the average square root error (ASRE) was 0.863, which is less than recommended by the American Engineering Computer Applications Committee AWWA (1999). The determination coefficient between observed and simulated pressures was $R^2 = 0.989$.

6 CONCLUSIONS

According to the literature review and field research analysis it is clear that the most important factor in successful calibration is to identify good monitoring point locations (representative location) in the water distribution system. The analysis revealed that the new calibration method presented in this work has given a high level of agreement between measured and simulated pressures. Some difficulties were caused by the lack of suitable places for putting measuring devices. As some water chambers and useful trenches had to be constructed, the water company should be more cooperative the next time when the flow measurements at the selected points are taken.

Similar measurements should be carried out in a different water distribution network to ensure the soundness of the method presented here. Further investigations should be carried out to collect a correct set of pipe roughness values . Roughness values can be established as an exact match between the predicted calibration method and field measured values.

ACKNOWLEDGEMENTS

The paper has been created in NN 523 494 234 research program.

REFERENCES

Ahmed I., Lansey K. & Araujo J. 1999. Data Collection for Water Distribution Network Calibration, In D. A. Savic & G. A. Walters (eds.), *Proc. Water Industry Systems: Modelling and Optimisation Applications,* Exeter, UK, vol.1, pp. 271–278.

Alzamora F.M. & Ayala H.B. 2006. Optimal sensor location for detecting contamination events in water distribution systems using topological algorithms, *Proceedings of the 8thAnnual Water Distribution SystemAnalysis Symposium Cincinnati,* Ohio, USA.

AWWA Engineering Computer Applications Committee: Calibration Guidelines for water distribution system modeling. *Proceedings of the 1999 AWWA Information Management and Technology Conference,* New Orlean, Louisiana, April 1999.

Berry J. W., Hart W.E., Phillips C. A., & Watson J. P. 2006. A facility location approach to sensor placement optimization, *Proceedings of the 8th Annual Water Distribution System Analysis Symposium,* Cincinnati, Ohio, USA.

Boulos P.F. & Wood D.J. 1991. An Explicit Algorithm for Calculating Operating Parameters for Water Networks. *Civil Engineering Systems* 8: 115–122.

Bush C.A. & Uber J.G. 1998. Sampling Design Methods for Water Distribution Model Calibration. *Journal of Water Resources Planning and Management, ASCE* 124(6): 334–344.

De Schaetzen W., Walters G.A. & Savic D.A. 2000. Optimal Sampling Design for Model Calibration Using Shortest Path, Genetic and Entropy Algorithms. *Urban Water Journal* 2: 141–152.

Ferrante M. & Brunone B. 2001. Leak Detection in Pressurised Pipes by Means of Wavelet Analysis. In A. Lowdon (ed.), *Proc. 4th International Conference on Water Pipeline Systems,* York, UK: 243–255.

Ghimire S.R. & Barkdoll B.D. 2006. Heuristic method for the battle of the water network sensors: demand-based approach. *Proceedings of the 8th Annual Water Distribution System Analysis Symposium,* Cincinnati, Ohio, USA.

Kozelj D., Steinman F. & Banovec P. 2006. Calibration of water distribution models by genetic algorithms, *International conference on Computing and Decision Making in Civil and Building Engineering,* Montreal, Canada.

Lee B.H. & Deininger R.A. 1992. Optimal Locations of Monitoring Stations in Water Distribution Systems, *Journal of Environmental Engineering, ASCE* 118(1): 4–16.

Loaiciga H.A., Charbeneau R.J., Everett L.G., Fogg G.E., Hobbs B.F. & Rouhani S. 1992. Review of Ground-Water Quality Monitoring Network Design. *Journal of Hydraulic Engineering, ASCE* 118(1): 11–37.

Nagar A.K. & Powell R.S. 2000. Absorbability Analysis of Water Distribution Systems Under Parametric and Measurement Uncertainty. In R. H. Hotckiss & M. Glade, (eds.), *Proc. ASCE 2000 Joint Conference on Water Resources Engineering and Water Resources Planning and Management,* Minneapolis, USA, CDROM Edition.

Rahal C.M., Sterling M.J.H. & Coulbeck B. 1980. Parameter Tuning for Simulation Models of Water Distribution Networks. *Proceedings of Institution of Civil Engineers,* Part 2, 69: 751–762.

Rossman L.A. 2000. *EPANET User manual, Risk Reduction Engineering Laboratory,* US Environmental Protection Agency, Cincinati, Ohio.

Savic D. & Walters G. 1997. Genetic Algorithms for the Least-Cost Design of Water Distribution Networks. *Journal of Water Resources Planning and Management, ASCE* 123(2): 67–77.

Simpson A.R. & Vitkovsky J.P. 1997. A Review of Pipe Calibration and Leak Detection Methodologies for Water Distribution Networks. *Proc. 17th Federal Convention, Australian Water and Wastewater Association,* Australia, vol. 1: 680–687.

Tucciarelli T., Criminisi A. & Termini D. 1999. Leak Analysis in Pipeline Systems by Means of Optimal Valve Regulation. *Journal of Hydraulic Engineering, ASCE* 125(3): 277–285.

Walski T.M. 1983. Technique for Calibrating Network Models. *Journal of Water Resources and Planning Management, ASCE* 109(4): 360–372.

Walski T.M. 1986. Case Study: Pipe Network Model Calibration Issues. *Journal of Water Resources Planning and Management, ASCE* 112(2): 238–249.

Wu Z.Y. & Walski T.M. 2005. Diagnostic error prone application of optimal model calibration. *International Conference of Computing and Control in the Water Industry,* Exeter, UK.

Yu G. & Powell R.S. 1994. Optimal Design of Meter Placement in Water Distribution Systems. *International Journal of Systems Science* 25(12): 2155–2166.

Calibration Guidelines for Water Distribution System Modelling, *Proc. AWWA 1999 Imtech Conference.*

Environmental Engineering III – Pawłowski, Dudzińska & Pawłowski (eds)
© 2010 Taylor & Francis Group, London, ISBN 978-0-415-54882-3

Inhibition of the growth of *Microcrystis aureginosa* by phenolic allelochemicals from aquatic macrophytes or decomposed barley straw

B. Macioszek, D. Szczukocki & J. Dziegieć

Department of General and Inorganic Chemistry, University of Lodz, Lodz, Poland

ABSTRACT: Harmful algal blooms occur in eutrophic water bodies. It is important to remove or inhibit cyanobacterial blooms because cyanotoxins can cause poisoning in humans and animals. Natural extracts of macrophytes or barley straw are known to strongly inhibit cyanobacteria growth. Because this important area of aquatic ecology is still little understood we investigated the specific activity of four selected compounds released by aquatic macrophytes or during the decomposition of barley straw, in relation to population growth and microcystin production by *Microcystis aeruginosa*. This laboratory study shows that algal growth was significantly inhibited by the addition of ellagic acid.

Keywords: Cyanobacteria; microcystin; aquatic macrophytes; barley straw; polyphenols.

1 INTRODUCTION

Cyanobacteria blooms occur worldwide, their intensity depending on climatic and hydrochemical conditions (Codd 1995). Harmful algal blooms give the water an unpleasant smell (earthy, musty) and colour and may lead to the accumulation of various toxins (hepatotoxins, such as microcystins and nodularins, or neurotoxins, such as anatoxin-a and saxitoxins) in the water column (WHO 1999). Cyanobacterial toxins can bioaccumulate in aquatic organisms (fish, mussels, and zooplankton) and consequently, they can poison humans, livestock and pets (Kankaanpää et al. 2002, Ozawa et al. 2003, Saker et al. 2004). Effective management of algal blooms is therefore highly desirable to prevent or at least mitigate their potential health hazards. One of the oldest intervention techniques is the use of algicides (e.g. copper(II) sulphate, potassium permanganate, chlorine, sodium oxochlorate). Copper sulphate ($CuSO_4$, the most commonly applied) is considered effective, easy to apply and economical (Cooke et al. 2005) but, unfortunately, it has widespread ecological impacts (McKnight et al. 1983). Application of $CuSO_4$ causes some changes in species succession (Effler et al. 1980, Soldo et al. 2000, van Hullebusch et al. 2002), and copper, like other heavy metals, is not biodegradable and is likely to accumulate in bottom sediments (Sanchez et al. 1978, Hanson et al. 1984, Garcia-Villada et al. 2004). For these reasons, other methods are needed to inhibit or limit cyanobacterial bloom. The natural extracts of some aquatic plants (macrophytes) (Mulderij et al. 2005, Xian et al. 2006) or decomposing barley straw have algicidal effects (Barrett et al. 1992, Martin et al. 1999), and may

be an inexpensive, environmentally-friendly alternative to the use of algicides such as copper sulphate that have undesirable side effects. The growth of blue-green algae (*Anabaena* sp., *A. variabilis*, *Microcystis aeruginosa*) has been shown to be strongly inhibited by freshwater macrophytes (*Ceratophyllum demersum*, *Elodea nuttallii* and *Myriophyllum spicatum*) in a number of independent studies (Saito et al. 1989, Gross et al. 1996, 2003; Nakai et al. 1999). In particular, Erhard and Gross (2006) found that the growth of *Pseudanabaena cf. catenata* Myr 980, *Synechococcus* sp. Cha 9817 and *S. nidulans* Pot 9801 was inhibited by an average of 56–92% by extracts of *Elodea nuttallii* or *Elodea canadensis*. However, more quantitative environmental studies are needed to investigate the potential impact of macrophyte extracts or decomposed barley straw on target organisms and the level of toxins released to the environment.

Phenolic compounds are naturally occurring secondary metabolites in plants. They may be present in the leaves, seeds, bark, fruits and flowers. Phenolics that are important building blocks of plant polymers (and are found in macrophytes and barley straw) include phloroglucinol, resorcinol, ellagic acid and (+)-catechin (Nakai et al. 2000). Phloroglucinol itself is not abundant in plants but its derivatives (e.g. in flavonoids) are distributed widely in the plant world (Lee et al. 2003). High levels of these compounds can be found in the bark of *Eucalyptus globulus* and in the leaves and flowers of *Hypericum brasiliense* (Rocha et al. 1995, Mohamed et al. 2007). Resorcinol in its free form has been found in the broad bean and in extracts of tobacco leaves (Hahn et al., 2006). Ellagic acid may occur in plants in either its free form or as ellagitannins. High levels of this dilactone of hexahydroxydiphenic acid are found in fruits (raspberries, strawberries, cranberries) and nuts (walnuts, pecans) (Vattem et al. 2005, Bala et al. 2006). The tea plant, *Camellia sinensis*, green tea, apples and grapes contain high levels of the polyphenol (+)-catechin, which belongs to the flavon-3-ols family (Guyot et al. 1996).

In the present study, we incubated a culture of the cyanobacteria *Microcystis aeruginosa* PCC 7820 with four selected phenols: phloroglucinol, resorcinol, ellagic acid and (+)-catechin. Algal growth was monitored daily by cell counting in a haemacytometer and microcystin-LR production was determined by chromatography.

2 MATERILAS AND METHODS

2.1 *Phenolic compounds and cyanobacteria*

Four phenols, phloroglucinol, resorcinol, ellagic acid and (+)-catechin, were purchased from Sigma-Aldrich. Concentrations used in the study were between 2.0 and 7.0 µg/ml. Algal assays were performed using the most common cyanobacteria in Poland, *Microcystis aeruginosa* PCC 7820 strain, obtained from the microbial collection of the Department of Ecophysiology and Plant Development (University of Lodz, Poland). In order to evaluate the dose-response inhibitory effects of selected phenols we carried out algal assays following the addition of different amounts of each compound. Substances were dissolved in methanol, filtered through a membrane filter (0.22 µm, Durapore PVDF, Millipore), added to 400 ml of BG_{11} medium (Watanabe et al. 1995) and *M. aeruginosa* was immediately inoculated (the initial density of culture was about 13500 cells/µL). Blue-green algae (tested cultures and mother control) were cultivated in triplicate for 15 days at 25°C under a light intensity of 5000 lux. Culture growth was monitored by determining cell numbers in the haemacytometer.

2.2 *Microcystin-LR determination*

To determinate the concentration of microcystin-LR (MCYST-LR) in the cultures studied, we centrifuged the medium at 4000 g for 20 min (Ward et al. 1997), lyophilized the algal biomass in the homogenizer and enriched the supernatant and extract from algae cells by solid phase extraction (SPE) (Meriluoto 1997, Rapala et al. 2002). After preconcentration using the SPE technique, the samples were evaporated to dryness under reduced pressure and dissolved in acetonitrile-ammonium acetate buffer (74:26 v/v) filtered through a 0.45-mm filter and separated by RP–HPLC (HP, USA) (Meriluoto 1997).

2.3 *Statistical analyses*

Statistical analyses were performed using Statistica 6.0 software. The statistically significant differences

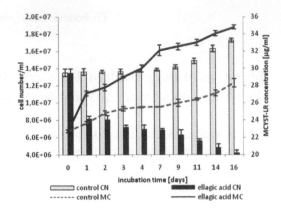

Figure 1. Changes in the concentration of MCYST-LR and in cyanobacterial cell number in *M. aeruginosa* culture in the presence of ellagic acid (c = 5.0 µ g/ml). Bars indicate a standard deviation (n = 3). The solid line shows the MCYST-LR concentration in culture with ellagic acid and the dotted line in the control. The black columns show the cell number in *M. aeruginosa* culture with the addition of ellagic acid and the grey columns in the control. (CN = cell number; MC = microcystin-LR).

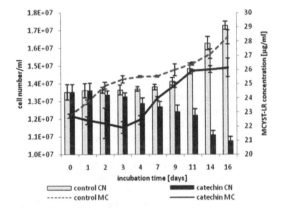

Figure 2. Changes in the concentration of MCYST-LR and in cyanobacterial cell number in *M. aeruginosa* culture in the presence of (+)-catechin (c = 5.5 µg/ml). Bars indicate a standard deviation (n = 3). The solid line shows the MCYST-LR concentration in culture with (+)-catechin and the dotted line in the control. The black columns show the cell number in *M. aeruginosa* culture with the addition of (+)-catechin and the grey columns in the control. (CN = cell number; MC = microcystin-LR).

between MCYST-LR concentrations and the cell numbers among the control and treatment over time were determined by ANOVA ($p < 0.05$).

3 RESULTS AND DISCUSSION

Ellagic acid and (+)-catechin significantly inhibited the growth of *M. aeruginosa* (Figure 1, 2).

The number of algal cells in the samples including these substances was lower after 16 days than at the beginning of incubation and also lower than in the control. It is important to note, however, that decreases in the number of *M. aeruginosa* cells following the addition of ellagic acid was accompanied by a simultaneous increase in the level of microcystin that at the end of the incubation time was higher than in the control. There are two possible explanations for this situation. The increase in MCYST-LR concentration may arise as a result of passive release from damaged cells or the remaining *M. aeruginosa* cells may increase their toxin production in some sort of stress-related response. A rise in microcystin production by *M. aeruginosa* under stress may be an undesired side effect of using plants allelochemicals to combat cyanobacteria blooms. On the other hand, passive exudation of toxins from dying blue-green algae cells may only be a transient effect, with no long-term ecological consequences. Nakai et al. (2001) showed that the concentration of several polyphenols decreased in alkaline medium containing di- or trivalent metal ions. Under such conditions, phenols are autoxidized and radicals are produced. The BG$_{11}$ medium we used for *M. aeruginosa* cultivation also contains metal ions, therefore radicals may also be responsible for the inhibitory effects seen with the phenols we studied. The results obtained suggest that ellagic acid has algicidal properties and destroys cyanobacterial cells allowing toxins to flow into the medium because the microcystin-LR concentration rose rapidly. This hypothesis does not exclude other mechanisms of action for this phenol.

How do polyphenols inhibit cyanobacterial growth? There are two possible modes of action: polyphenols can form complexes with algal exoenzymes like alkaline phosphatase and so inactivate them (Gross et al. 1996). The other possibility is inhibition of photosynthetic electron transport. Both these possible modes of action affect the photosynthetic oxygen pathway and photosystem II activity (Leu et al. 2002).

We also evaluated the inhibitory effects of phloroglucinol and resorcinol on *M. aeruginosa* growth and toxin production. Algal assays showed that these compounds decreased the cyanotoxin concentration in the medium. Furthermore, higher concentrations of the compounds also led to a reduction in the number of cells (Figure 3, 4). It should be stressed here that a decrease in the number of cyanobacterial cells resulted in a consequent decrease in microcystin levels.

The reason for the reduction in cyanotoxins is not known yet. It could be due to a reaction between phloroglucinol or resorcinol and microcystin-LR. The most likely cause of such results could also involve the phenols' ability to pass through cell membranes and inhibit phytoplanktonic enzyme activity as well as interrupting other life functions.

4 CONCLUSIONS

Allelopathic compounds produced by aquatic plants or released during the decomposition of barley straw (ellagic acid, (+)-catechin, phloroglucinol, resorcinol)

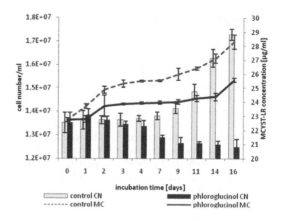

Figure 3. Changes in the concentration of MCYST-LR and in cyanobacterial cell number in *M. aeruginosa* culture in the presence of phloroglucinol (c = 7.0μg/ml). Bars indicate a standard deviation (n = 3). The solid line shows the MCYST-LR concentration in culture with phloroglucinol and the dotted line in the control. The black columns show the cell number in *M. aeruginosa* culture with the addition of phloroglucinol and the grey columns in the control. (CN = cell number; MC = microcystin-LR).

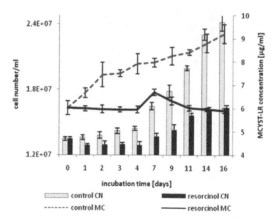

Figure 4. Changes in the concentration of MCYST-LR and in yanobacterial cell number in *M. aeruginosa* culture in the presence of resorcinol (c = 2.0 μg/ml). Bars indicate a standard deviation (n = 3). The solid line shows the MCYST-LR concentration in culture with resorcinol and the dotted line in the control. The black columns show the cells number in *M. aeruginosa* culture with the addition of resorcinol and the grey columns in the control. (CN = cell number; MC = microcystin-LR).

affect growth and toxin production in the cyanobacterium *M. aeruginosa*. In the present study we found that ellagic acid and (+)-catechin, in particular, reduced the cyanobacterial biomass but did not cause a decrease in the levels of the cyanotoxin microcystin-LR in the medium. On the other hand, the presence of phloroglucinol or resorcinol in algal cultures led to a reduction in toxins but only moderate inhibition of cyanobacterial growth. The differences observed in the antialgal activity of the polyphenols studied may also be related to the rate of autoxidation (Nakai et al. 2001).

Our results indicate the feasibility of controlling cyanobacterial biomass and toxin production using polyphenols released in two ways: by aquatic macrophytes or during the decomposition of barley straw. The study suggests the potential use of phloroglucinol and resorcinol in the management of algal blooms. Nevertheless, further work needs to be done to fully explain the inhibitory activity of phloroglucinol and resorcinol and to determine whether this activity could be extrapolated to the field, as well as to assess the potential impact of these compounds upon other aquatic organisms.

ACKNOWLEDGEMENTS

Two anonymous reviewers made valuable comments on the manuscript. This study was supported by a grant from the Ministry of Science and Higher Education, Poland (project number N N523 1024 33).

REFERENCES

Bala, I., Bhardwaj, V., Hariharan, S. & Ravi Kumar, M.N.V. 2006. Analytical methods for assay of ellagic acid and its solubility studies. *Journal of Pharmaceutical and Biomedical Analysis* 40: 206–210.

Barrett, P.R.F. & Newman, J.R. 1992. Algal growth inhibition by rotting barley straw. *British Phycology Journal* 27: 83–84.

Codd, G.A. 1995. Cyanobacterial toxins: occurrence, properties and biological significance. *Water Science Technology* 32(4): 149–156.

Cooke, G.D., Welch, E.B., Peterson, S.A. & Nichols, S.A. 2005. *Restoration and management of lakes and reservoirs*. 3rd ed. Taylor and Francis, Boca Raton.

Effler, S.W., Linen, S., Field, S.D., Tong-Ngork, T., Hale, F., Meyer M., Quirk M., 1980, Whole lake responses to low level copper sulphate treatment, *Water Research*, vol. 14, pp. 1489–1499.

Erhard, D., Gross, E. M., 2006, Allelopathic activity of *Elodea canadensis* and *Elodea nuttallii* against epiphytes and phyto plankton, *Aquatic Botany*, vol. 85, pp. 203–211.

Garcia-Villada, L., Rico, M., Altamirano, M., Sanchez-Martin, L., Lopez-Rodas, V. & Costas, E. 2004. Occurrence of copper resistant mutants in the toxic cyanobacteria *Micro cystis aeruginosa*: characterisation and future implications in the use of copper sulphate as algaecide. *Water Research* 38: 2207–2213.

Gross, E.M., Meyer, H. & Schilling, G. 1996. Release and eco logical impact of algicidal hydrolysable polyphenols in *Myriophyllum spicatum*. *Phytochemistr* 41: 133–138.

Gross, E.M., Erhard, D. & Iványi, E. 2003. Allelopathic activity of *Ceratophyllum demersum* L. and *Najas marina* ssp. *intermedia* (Wolfgang) *Casper, Hydrobiology* 506–509: 583–589.

Guyot, S., Vercauteren, J. & Cheynier, V. 1996. Structural de termination of colourless and yellow dimers resulting from (+)-catechin coupling catalysed by grape polyphenoloxi dase. *Phytochemistry* 42(5): 1279–1288.

Hahn, S., Kielhorn, J., Koppenhöfer, J., Wibbertmann, A. & Mangelsdorf, I. *Resorcinol*. Concise Internat. Chem. Assess ment Doc. 71. WHO. 2006.

Hanson, M.J. & Stefan, H.G. 1984. Side effects of 58 years of copper sulphate treatment of the Fairmont Lakes, Minnesota. *Water Research* 20: 889–900.

Kankaanpää, H., Vuorinen, P.J., Sipiä, V. & KEinänen, M. 2002. Acute effects and bioaccumulation of nodularin in seatrout (*Salmo trutta m. trutta L.*) exposed orally to *Nodu laria spumigena* under laboratory conditions. *Aquatic cology* 61: 155–168.

Lee, S.M., Na, M.K., An, R.B., Min, B.S. & Lee, H.K. 2003. Antioxidant activity of two phloroglucinol derivatives from *Dryopteris crassirhizoma*. *Biological Pharmaceutical Bul letin* 26(9): 1354–1356.

Leu, E., Krieger-Liszkay, A., Goussias, C.H. & Gross, E.M. 2002, Polyphenolic allelochemicals from the aquatic angio sperm *Myriophyllum spicatum* inhibit photosystem II. *Plant Physiology* 130: 2011–2018.

Martin, D. & Ridge, I. 1999. The relative sensitivity of algae to decomposing barley straw. *Journal of Applied Phycology* 11: 285–291.

McKnight, D.M., Chisholm, S.W. & Harleman, D.R.F. 1983. $CuSO_4$ treatment of nuisance algal blooms in drinking water reservoirs. *Environmental Management* 7: 311–320.

Meriluoto, J. 1997. Chromatography of microcystins. *Analytica Chimica Acta* 352: 277–298.

Mohamed, G.A. & Ibrahim, S.R.M. 2007. Eucalyptone G, a new phloroglucinol derivative and other constituents from *Eucalyptus globulus* Labill. *Archive for Organic Chemistry:* 281–291.

Mulderij, G., Mooij, W.M., Smolders, A.J.P. & Van Donk, E. 2005. Allelopathic inhibition of phytoplankton by exudates from *Stratiotes aloides*. *Aquatic Botany* 82: 284–296.

Nakai, S., Inoue, Y., Hosomi, M. & Murakami, A. 1999. Growth inhibition of blue-green algae by allelopathic effects of macrophytes. *Water Science Technology* 39(8): 47–53.

Nakai, S., Inoue, Y., Hosomi, M. & Murakami, A. 2000. *Myriophyllum spicatum*-released allelopathic polyphenols inhibiting growth of blue-green algae *Microcystis aerugi nosa*. *Water Research* 34(11): 3026–3032.

Nakai, S., Inoue, Y. & Hosomi, M. 2001. Algal growth inhibition effects and inducement modes by plant-producing phenols. *Water Research* 35(7): 1855–1859.

Ozawa, K., Yokoyama, A., Ishikawa, K., Kumagai, M., Watanabe, M.F. & Park, H.-D. 2003. Accumulation and depuration of microcystin produced by the cyanobacterium *Micro cystis* in a freshwater snail. *Limnology* 4: 131–138.

Rapala, J., Erkomaa, K., Kukkonen, J., Sivonen, K. & Lahti, K. 2002. Detection of microcystins with protein phosphatase inhibition assay, high-performance liquid chromatography- UV detection and enzyme-linked immunosorbent assay. Comparison of methods. *Analytica Chimica Acta* 466: 213–231.

Rocha, L., Marston, A., Potterat, O., Auxiliadora, M., Kaplan, C., Stoeckli-Evans, H. & Hostettmann, K. 1995. Antibacterial phloroglucinols and flavonoids from *Hypericum brasiliense*. *Phytochemistry* 40(5): 1447–1452.

Saito, K., Matsumoto, M., Sekine, T., Murakoshi, I., Morisaki, N. & Iwasaki, S. 1989. Inhibitory substances from *Myrio phyllum brasiliense* on growth of blue-green algae. *Journal of Natural Products* 52(6): 1221–1226.

Saker, M.L., Metcalf, J.S., Codd, G.A., & Vasconcelos, V.M. 2004. Accumulation and depuration of the cyanobacterial toxin cylindrospermopsin in the freshwater mussel *Ano donta cygnea*. *Toxicon* 43: 185–194.

Sanchez, I. & Lee, G.F. 1978. Environmental chemistry of copper in Lake Monona, Wisconsin. *Water Research* 12: 899–903.

Soldo, D. & Behra, R. 2000. Long-term effects of copper on the structure of freshwater periphyton communities and their tolerance to copper, zinc, nickel and silver. *Aquatic Toxicology* 47: 181–189.

Van Hullebusch, E., Deluchat, V., Chazal, P.M. & Baudu, M. 2002. Environmental impact of two successive chemical treatments in a small shallow eutrophied lake: Part II. Case of copper sulphate. *Environmental Pollutio* 120: 627–634.

Vattem, D.A. & Shetty, K. 2005. Biological functionality of el lagic acid: A review. *Journal of Food Biochemistry* 29: 234–266.

Ward, C.J., Beattie, K.A., Lee, E.Y.C. & Codd, G.A. 1997. Colorimetric protein phosphatase inhibition assay of laboratory strains and natural blooms of cyanobacteria: compare sons with high performance liquid chromatographic analysis for microcystins. *FEMS Microbiology Letter* 153(2): 465–473.

Watanabe, M.F., Harada, K.-I., Carmichael, W.W. & Fujiki, H. *Toxic microcystic*. CRC Press. Inc. Florida. 1996.

WHO 1999. *Toxic cyanobacteria in water*. E & FN Spon Press: London.

Xian, Q., Chen, H., Zou, H. & Yin, D. 2006. Allelopathic activity of volatile substance from submerged macrophytes on *Microcystin aeruginosa*. *Acta Ecologica Sinica* 26(11): 3549-3554.

489

Environmental Engineering III – Pawłowski, Dudzińska & Pawłowski (eds)
© 2010 Taylor & Francis Group, London, ISBN 978-0-415-54882-3

Biofilm sampling for bioindication of municipal wastewater treatment

A. Montusiewicz, M. Chomczyńska, J. Malicki & G. Łagód
Institute of Environmental Protection Engineering, Lublin University of Technology, Lublin, Poland

ABSTRACT: The paper describes methods for the sampling and analysis of pecton (biofilm), which occupies the walls of facilities in wastewater treatment plants. The minimum sample area and number of measurements are determined. The study results show that a subsample of $9\,cm^2$ is sufficient for bioindication of wastewater treatment based on the numbers and abundances of morphological–functional groups within biofilm communities. For quantitative studies, a total sample containing six subsamples and equal to an area of $54\,cm^2$ is also sufficient. If a significance level $\alpha = 0.1$ and an accuracy $d = 0.1$ is assumed, and there are minimal numbers of taxa, occurring in equal proportions, 67 test lines on one slide is sufficient. Thus, under real circumstances, 45 measurements should be sufficient. The time for determining the proportions of morphological–functional groups in biofilm samples and calculating the values of useful indices generally does not exceed 30 minutes.

Keywords: Sampling methodology, bioindication, biofilm (pecton, epilithon), wastewater treatment plant.

1 INTRODUCTION

The use of suspended-growth models in describing wastewater treatment processes requires that they are appropriately calibrated and verified using data obtained from full-scale operating systems. The suitability of such models verified by process and/or object responses can be determined by comparing two sets of values including, respectively, theoretical response (or dependent) variables and real values of the parameters measured.

Efficient monitoring of the wastewater treatment process should be quick, cheap and repeatable. Some contaminants, such as available and unavailable carbon compounds, total suspended solids, or nitrogen and phosphorus compounds, may be considered as indicators for the efficiency of wastewater treatment processes; their removal is reflected in changes in values measured physically or chemically. Measurements can be undertaken at hourly or daily intervals and require both special equipment and suitable reagents.

Processes occurring in bioreactors, including the activated sludge process, depend both on the presence and concentrations of the substrates supplying the system and on the condition of the activated sludge, which is commonly considered to be an important process factor. Therefore, bioindication strategies (suitable for particular conditions) can reflect the levels of the parameters which characterize the processes in question (Montusiewicz et al. 2007).

The quality of the wastewater treatment plant inflow may be evaluated on the basis of autochthonous saprobe formations, washed away from the biofilm occupying the sewer walls, as well as allochthonous prokaryotic and eukaryotic individuals included in the

total suspended solids in raw sewage (Huisman 2001, Hvitved-Jacobsen 2002, Łagód et al. 2004). When toxicants are present at high concentrations, no visible mobility is observed for most (or all) of these individuals, indicating their lifelessness.

Both the quality of the substrate and the "diagnostic class of the activated sludge" (Curds & Cockburn 1970a, b, Martin-Cereceda et al. 1996, Salvado et al. 1995) influence the scenario run of the wastewater treatment process. The diagnostic class of the sludge includes, among other things, the "biological quality of the sludge" identifiable using the SBI (sludge biotic index, Madoni 1994), sludge retention time (SRT) usually correlated with species richness (Błędzki 2007 a, b, Gotelli et al. 2006), biodiversity indices (Magurran 1988, Friedrich et al. 1992, Gove et al. 1994, Lydy et al. 2000, Ravera 2001, Łagód et al. 2004, 2007, Montusiewicz et al. 2007) and, to some degree, with saprobic indices measured using the Pantle and Buck method (Pantle & Buck 1955).

Wastewater treatment plants operating with two or more stages contain different devices, such as screens, grit chambers, primary and secondary clarifiers, bioreactors separated into anaerobic, anoxic and aerobic phases or zones, pump stations with recirculation sewers, as well as devices essential for the treatment and utilization of sludge (Roelevelt et al. 1997). Wastewater is characterized by its physicochemical properties, which are continuously changing in the wastewater stream flowing through the successive devices. Changes in parameter values may also be observed and measured using bioindication methods (Montusiewicz et al. 2007). Considering that the surfaces of the components in contact with wastewater are covered by biofilm (pecton), bioindication based on this ecological formation seems to be possible; the taxonomic composition of biofilm (pecton) reflects both the availability of necessary substrates and electron acceptors, and the presence of inhibitors affecting different saprobe groups (Montusiewicz et al. 2007). It should be noted that the taxonomic composition of biofilm (pecton) characterizes both the stage of a process run at a given moment and an identified point in the technological system (i.e. the type of facility).

The common presence of biofilm (pecton) in wastewater treatment devices allows, through observations and measurements, the monitoring of the efficiency of wastewater purification, resulting in the calibration and verification of models describing wastewater treatment processes.

Determining the necessary area and number of samples of biofilm, as well as the number of measurements required for any sample examined, is of essential importance for bioindication that is used in municipal wastewater treatment processes and that is based on eukaryotic saprobes representing biofilm (pecton) formations.

The aims of this paper are therefore to ascertain both the minimum required surface area (SRSA) for biofilm sampling and the lowest necessary number of coexistence measurements (SNM) for eukaryotic saprobes in biofilm.

2 MATERIALS AND METHODS

2.1 Material

The material used for the present pilot study was collected as samples of biofilm (pecton) that occupied device walls and that was in contact with wastewater undergoing purification. Representatives of the biofilm community were divided into the following morphological–functional groups: swimming ciliates (*Holotricha*), attached ciliates (*Peritricha*), crawling ciliates (*Hypotricha*), flagellates, amoebas, nematodes, rotifers, green algae, fungi, diatoms and bacteria. The specified groups represent a group of different taxa or different taxonomic categories (from order, through subclass, class, phylum-division to kingdom), being morphologically and behaviorally diversified formations that are easy to identify. According to Olivier & Beattie (1993) and Pullin (2004), such formations are called Recognisable Taxonomic Units – RTUs. Pecton samples were obtained from a full-scale, mechanical–biological wastewater treatment plant (WWTP) in Lublin (Poland). During the study, the biofilm examined was removed from the walls of the following devices: screen and grit chamber (inflow and outflow channels), primary settler, anaerobic and aerobic chamber (inflow and outflow), final clarifier and outflow channel from WWTP. At the time of sampling, each biofilm sample was located 5 cm below the level of the wastewater. The sampling strategy was based on the suggestions of Barnett (1986). It was assumed that units of the general population were present in the sample examined, which included subsamples of the biofilm (pecton) communities established on the walls of particular technical components of the wastewater treatment system. Guidelines for "simple sampling, without turns and with arbitrary probability" indicate that it is important to ensure the following conditions: each sample should be taken from the surface of one object (one habitat); the surfaces sampled should be typical for the object examined; and the area of the surfaces sampled should be representative for the object (Barnett 1986).

2.2 Determining the smallest representative area of subsamples

To inventory the morphological-functional groups the area of the biofilm samples was primarily equal to 5 cm × 10 cm. Subsequently, the smallest representative subsample area (SRSA) for biofilm was determined based on the differential quotient method (Matuszkiewicz & Wydrzycka 1972). The numbers of morphological–functional groups were determined in pecton subsamples with areas equivalent to the successively larger quadrat areas of the Greig-Smith grid (areas were the n-th powers of 2 (2^n) in cm^2) (Greig-Smith 1952, Fehmi & Bartolome 2001).

During our study, pecton was sampled using an iron blade, always following the same schedule for the successive devices at the same time of day. Each biofilm (pecton) subsample was placed in a sterile plastic container filled with sewage to two-thirds of its volume (about 6.7 cm^3). Sewage was taken from the same device locations as the biofilm. Air filled about the one-third of the container's volume, to provide oxygen for the organisms. The collected samples were immediately transported to the laboratory. Material from each container was mixed then sewage was decanted and finally pecton subsample was taken from the container and used to prepare three slides of living organisms. Numbers of morphological–functional groups at particular sampling locations were determined using an optical light microscope with a bright field of view and a magnification of 10×10 (a magnification of 10×40 was used only when there were problems identifying individuals).

The smallest representative subsample area was evaluated using the following equation (Matuszkiewicz & Wydrzycka 1972):

$$\lg_2 X_{SRSA} = \lg_2 X_0 + \frac{(\Delta y / \Delta x)_0 - 1}{(\Delta y / \Delta x)_0 - (\Delta y / \Delta x)_1} \quad (1)$$

where: Δx is the increase in area between two quadrats, Δy is the increase in the mean number of morphological–functional groups connected with an increase in area, $(\Delta y / \Delta x)_0$ is the superior limit containing value 1, $(\Delta y / \Delta x)_1$ is the low limit containing value 1, and X_0 is the area between $(\Delta y / \Delta x)_0$ and $(\Delta y / \Delta x)_1$ (cm^2).

The most favorable smallest area is the one in which no new morphological–functional groups appear.

The standard error (*SE*) for the mean number of morphological–functional groups (*y*) was determined using the following equation (Bethea et al. 1985):

$$SE = \frac{\sigma}{\sqrt{n-1}} \qquad (2)$$

where: σ is a standard deviation, *n* are degrees of freedom.

2.3 Determining the smallest number of measurements (SNM)

It is reasonable to expect that wastewater treatment plants differ with regard to numbers of morphological–functional groups and their relative abundances. Thus, to ensure higher certainty, the smallest number of measurements (SNM), or test lines was determined for cumulative samples prepared by combining six subsamples – each one corresponding to the size of the smallest representative area evaluated (SRSA). Microscopic analysis of slides prepared from cumulative samples was carried out as described above. Organisms were counted along 400 test lines with the distance of 55 μm in a slide (Pluta 1982, Hopkins 1976). Only those individuals which crossed the lines or touched them from the right-hand side were counted. In the case of counting individual colonies, each member was treated as one organism. The density of individuals in slides did not affect the relative abundances (proportions) of morphological–functional groups. The SNM was established with an assumed significance level of 0.1. Since measurements were connected with proportions of the morphological–functional groups, the smallest number of measurements was calculated based on a model used in structure analysis (Bethea et al. 1985):

$$\eta = \frac{t_\alpha^2 \cdot p \cdot q}{d^2} \qquad (3)$$

where: t_α is the value from the Student's t-distribution for assumed α and degrees of freedom; *p* is the dominant's proportion, *q* is the influent's proportion (or $q = 1 - p$), and *d* is the assumed precision of estimation ($d = 0.1$).

3 RESULTS AND DISCUSSION

The study results, which allow the smallest representative subsample area to be determined, are shown in Table 1 and in Figure 1.

Table 1 and Figure 1 include the results of observation for the most diverse pecton community which was sampled form outflow channel of the final clarifier. In this case, the superior limit $(\Delta y/\Delta x)_0$ was 1.0425, the low limit $(\Delta y/\Delta x)_1$ was 0.0425 and the area between these two values (X_0) was 8 cm^2 ($8 \cdot 10^{-4}$ m). Hence:

$$lg_2 X_{SRSA} = 3 + \frac{1.0425 - 1}{1.0425 - 0.0425} = 3.0425$$

Table 1. Mean number of morphological–functional (m–f) groups as a function of biofilm sample area.

2^n	*x*	*a*	Δx	$\bar{y}$	*SE y*	Δy	$\Delta y/\Delta x$
	–	–	1.00	–	–	2.00	2.0000
2^0	1.00	1.00		2.00	0.26	–	–
	–	–	1.00	–	–	1.50	1.5000
2^1	2.00	1.41	–	3.50	0.36	–	–
	–	–	2.00	–	–	2.16	1.0800
2^2	4.00	2.00	–	5.66	0.33	–	–
	–	–	4.00	–	–	4.17	1.0425
2^3	8.00	2.83	–	9.83	0.31	–	–
	–	–	8.00	–	–	0.34	0.0425
2^4	16.00	4.00	–	10.17	0.40	–	–
			16		–	0.33	0.0206
2^5	32	5.66	–	10.50	0.22	–	–

a – side of quadrat (cm), *x* – area (cm^2), Δx– area increase, $\bar{y}$ – mean number of m–f groups, *SE y* – standard error of *y*, Δy– increase in mean number of m–f groups.

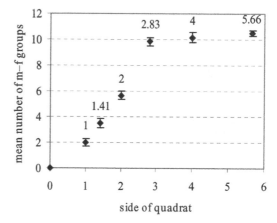

Figure 1. Accumulation curve with an increase in number of morphological–functional (m–f) groups.

and $X_{SRSA} = 8.239$ cm^2. Thus, a 9 cm^2 ($9 \cdot 10^{-4}$m^2) subsample of biofilm was taken. At the calculated value of X_{SRSA}, the area of the cumulative sample was 54 cm^2 ($5.4 \cdot 10^{-3}$ m^2 – six subsample areas combined into one sample).

For particular examined devices, the smallest number of morphological–functional groups found in the cumulative samples was 3 and the highest was 11 (without green algae, diatoms, fungi and bacteria these values were 3 and 7, respectively). In the least favorable case (equal proportions) at taken values $\alpha = 0.1$, $t_\alpha = 1.645$ so $t_\alpha^2 = 2.706$. For three morphological–functional groups (each proportion equal to $1/3 = 0.33$), the value of SNM (η) is:

$$\eta = \frac{1.645^2 \cdot 0.33 \cdot 0.33}{0.1^2} = \frac{2.706 \cdot 0.1089}{0.01} = 29.47 = 30$$

or

$$\eta \frac{2.706 \cdot 0.33 \cdot 0.66}{0.01} = 58.94 = 59$$

It is worth noticing that the SNM values achieved using the above equations differed as a result of the various values of the parameter q assumed for calculations made according to equation 3 (see the explanation in Experimental protocol).

For seven morphological–functional groups (each proportion equal to $1/7 = 0.1429$) the smallest number of measurements in a sample is:

$$\eta = \frac{2.706 \cdot 0.1429 \cdot 0.1429}{0.1^2} = \frac{2.706 \cdot 0.0204}{0.01} = 5.52 = 6$$

or

$$\eta = \frac{2.706 \cdot 0.1429 \cdot 0.8571}{0.01} = 33.14 = 34$$

During the pilot studies, in the case of a device with three morphological–functional groups, the proportion for the most dominant group was 0.903 and that for the influent was 0.032. Therefore,

$$\eta = \frac{2.706 \cdot 0.029}{0.01} = 7.85 = 8$$

or

$$\eta = \frac{2.706 \cdot 0.903 \cdot 0.097}{0.01} = 23.70 = 24$$

In the case of a device with seven morphological–functional groups, the proportion for the most dominant group was 0.553 and that for the influent was 0.0026. Therefore,

$$\eta = \frac{2.706 \cdot 0.015}{0.01} = 4.06 = 5$$

or

$$\eta = \frac{2.706 \cdot 0.553 \cdot 0.447}{0.01} = 66.89 = 67$$

The average of these last two results (8 and 5, or 24 and 67) show that individuals on 7 or 46 test lines should be counted on a slide from a cumulative sample of total area equal to $54 \, cm^2$. For bioindication purposes in the devices studied, 45 test lines (with the distance of $490 \, \mu m$) were taken in one slide for one sample, because analysis of the results obtained from the pilot studies and theoretical considerations indicated that the highest number of measurements was 67, while the smallest was 5. The number of test lines (the smallest number of measurements) taken was substantially lower than the number of test lines along which individuals were counted during the pilot studies. For the number of measurements taken ($\eta = 45$), the time period needed to determine the proportions of morphological–functional groups in biofilm samples generally does not exceed 20 minutes. Using appropriate software, the values of all indices useful in bioindication can be calculated within 30 minutes of a sample arriving in the laboratory.

It should be noted that, despite its complexity, the problem of calculating the area of a representative sample in ecological studies has been investigated for decades. Several solutions have been found. The simplest one is to assume a constant size of plant patches (depending on the formation described) under analysis in phytosociological studies. The areas sampled in this regard range from $0.1–1 \, m^2$ for lichen groups to $500–2500 \, m^2$ and $100–200 \, m^2$ for rich mixed forests, in the tree and ground cover layers, respectively (Fukarek 1967). For the purposes of soil zoology, when studying mesofauna the necessary sample area is $10–35 \, cm^2$, while for macrofauna and megafauna the recommended values are $0.5–1 \, m^2$ (Górny 1975). Studies of biofilm from trickling filters require the use of ten fragments of packing materials of $2–3 \, cm$ in diameter. Therefore, the average sample of pecton and periphiton has a surface area ranging from $31.42 \, cm^2$ to $70.68 \, cm^2$, which usually amounts to about $50 \, cm^2$. This surface area could also be taken as a surface corresponding to the area of several stones, leaves or microscopic slides submerged in water (Fisher & Dunbar 2007, Obolewski 2005, Villbaste 2001).

Determining the representative sample size is slightly more complicated based on Beklemiszew's criterion. The criterion determines the relation between the area of the sample and the number of identified species, and is described by the following equation (Górny & Gruma 1981):

$$N = ay^c \tag{4}$$

where: N is the area of the sample, y is the number of taxa, and a and c are constants dependent on the area and the object studied.

Since the plot of this equation is a curve approaching an asymptote, an inflection can be appointed using Calley's method modified by Matuszkiewicz & Wydrzycka (1972) and used to determine the smallest representative sample area (SRSA).

Determining the smallest representative sample area is also possible using the relation between species number and individuals number, or between species number and sample number, which enables rarefaction curves to be plotted (Gotelli & Cowell 2001).

It is generally assumed that the smallest number of measurements in samples should be counted as the average values of three measurements. A sample consisting of 30 measurements is sometimes treated as a large one. The proposed method allows the number of measurements to be calculated based on variances from an initial analysis (set of pilot samples), at an assumed significance level α (reliance coefficient $1 - \alpha$) and also at an assumed level of measurement precision (half-range of the reliance or 0.1 of the arithmetic average value).

4 CONCLUSIONS

Based on the results presented in this paper, the following conclusions are offered:

1. A subsample area of $9 \, cm^2$ is sufficient for the bioindication of wastewater treatment based on

the numbers and abundances of morphological–functional groups that belong to biofilm (pecton) occupying the walls of particular devices in wastewater treatment plants.

2. For quantitative studies of biofilm (pecton), a total sample containing 6 subsamples and equal to $54\,cm^2$ ($5.4 \cdot 10^{-3}\,m^2$) is sufficient.

3. Sixty-seven test lines are sufficient in the following conditions: significance level $\alpha = 0.1$, accuracy $d = 0.1$, minimal numbers of taxa, and their occurrence in equal proportions. Thus, an SNM equal to 45 is sufficient in real circumstances.

4. The time period needed to determine the proportions of morphological–functional groups in biofilm samples, and to calculate the values of all useful indices, generally does not exceed 30 minutes.

ACKNOWLEDGEMENT

This work was supported by the Ministry of Science and Higher Education of Poland, Grant No: 4949/B/T02/2008/34

REFERENCES

Barnett, V. 1986. *Elements of Sampling Theory*. London, Sydney, Auckland, Toronto: Hodder & Stoughton.

Bethea, R.M., Duran, B.S. & Boullion, T.L. 1985. *Statistical Methods for Engineers and Scientists*. New York, Basel: Marcel Dekker Inc.

Błędzki, L.A. 2007a. Method for comparing species richness and species diversity. Part I (in Polish). *Bioskop* 01: 18–22.

Błędzki, L.A. 2007b. Method for comparing species richness and species diversity. Part II (in Polish). *Bioskop* 02: 20–23.

Curds, C.R. & Cockburn, A. 1970a. Protozoa in biological sewage-treatment processes – I. A survey of the protozoan fauna of British percolating filters and activated-sludge process. *Water Research* 4: 225–236.

Curds, C.R. & Cockburn, A. 1970b. Protozoa in biological sewage-treatment processes – II. Protozoa as indicators in the activated sludge processes. *Water Research* 4: 237–249.

Fehmi, J.S. & Bartolome, J.W. 2001. A grid-based method for sampling and analysing spatially ambiguous plants. *Journal of Vegetation Science*. 12(4): 467–472.

Fisher, J. & Dunbar, M.J. 2007. Towards representative periphitic diatom sample. *Hydrol. Earth. Syst. Sci.* 11(1): 399–407.

Friedrich, G., Chapman, D. & Beim A. 1992. The use of biological material in water quality assessments. In: *Water quality assessment - a guide to use of biota, sediments and water in environmental monitoring*. UNESCO/WHO/UNEP.

Fukarek, F. 1967. *Phytosociology* (in Polish). Warszawa: PWRiL.

Gotelli, N.J. & Colwell, R.K. 2001. Quantifying biodiversity: procedures and pitfalls in the measurement and comparison of species richness. *Ecology Letters*. 4: 379–391.

Gotelli, N.J. & Entsminger, G.L. 2006. *EcoSim: Null models software for ecology. Version 7*. Jericho: Ackuiret Intelligence Inc. & Kesey-Bear. VT 05456.

Gove, I.H., Patil, G.P., Swindel, B.F. & Taillie C. 1994. Ecological diversity and forest management. In G.P. Patil, & C.R. Rao (eds), *Handbook of Statistics 12*. Amsterdam: Elsevier Science B.V.

Górny, M. 1975. *Zooecology of forest soils* (in Polish). Warszawa: PWRiL.

Górny, M. & Gruma, L. 1981. *Methods applied in soil zoology* (in Polish). Warszawa: PWN.

Greig-Smith P. 1952. The use of random and contiguous quadrats in the study of the structure of plant communities, *Annals of Botany* 16: 293–316.

Hopkins, B.M. 1976. A quantitative image analysis system. *Optical Engineering*. 15: 236-240.

Huisman, J.L. 2001. Transport and transformation process in combined sewers. *IHW Shriftenfreihe*. 10: 1–180.

Hvitved-Jacobsen T. 2002. *Sewer Processes. Microbial and Chemical Process Engineering of Sewer Networks*. London: CRC Press.

Lydy, M. J., Crawford, C.G. & Frey J. W. 2000. A comparison of selected diversity, similarity, and biotic indices for detecting changes in benthic-invertebrate community structure and stream quality. *Archives of Environmental Contamination and Toxicology* 39: 469–479.

Łagód, G., Malicki, J., Montusiewicz A. & Chomczyńska M. 2004. Application of saprobionts for bioindication of wastewater quality in the sewage systems (in Polish), *Archives of Environmental Protection* 30(3): 3–12.

Łagód, G., Malicki, J., Chomczyńska, M. & Montusiewicz A. 2007. Interpretation of the results of wastewater quality biomonitoring using saprobes. *Environmental Engineering Science* 24 (7): 873–879.

Madoni, P. 1994. A sludge biotic index (SBI) for the evaluation of the biological performance of activate sludge plants based on microfauna analysis. *Water Research* 28: 67–75.

Magurran, A. 1988. *Ecological diversity and its measurement*. London: Croom Helm.

Martin–Cereceda, M., Serrano, S. & Guinea, A. 1996. A comparative study of ciliated protozoa communities in activated-sludge plants. *FEMS Microbiology Ecology* 21: 267–276.

Matuszkiewicz, W. & Wydrzycka, U. 1972. The number of species as a function of surface area and the problem of the representative area of a phytocoenosis. *Phytocoenosis* 1(2): 95–120.

Montusiewicz, A., Malicki, J., Łagód G. & Chomczyńska M. 2007. Estimating the efficiency of wastewater treatment in activated sludge systems by biomonitoring. In L. Pawłowski, M.R. Dudzińska & A. Pawłowski (eds) *Environmental Engineering* London: Taylor & Francis Group.

Obolewski, K. 2005. Epiphitic macrofauna on water soldiers (*Stratiotes aloides* L.) in Słupia river oxbows. *Oceanological and Hydrobiological Studies* 34(2): 37–54.

Oliver, I. & Beattie, A.J. 1993. A possible method for the rapid assessment of biodiversity. *Conservation Biology* 7: 562–568.

Pantle, R. & Buck, H. 1955. Die Biologische Uberwachung der Gewaser und die Darstellung der Ergebinsse. *Gas und Wasserfach*. 96: 604.

Pluta, M. 1982. *Optical microscopy* (in Polish). Warszawa: PWN.

Pullin, A.S. 2002. *Conservation biology*. Cambridge: Cambridge University Press.

Ravera, O. 2001. A comparison between diversity, similarity and biotic indices applied to the macroinvertebrate community of a small stream: the Ravella river (Como Province, Northern Italy). *Aquatic Ecology*. 35: 97–107.

Roelevelt, P. J., Klap Wijk, A., Eggels, P.G., Rurkens, W. H. & Starkenburg, W. 1997. Sustainability of municipal wastewater treatment, *Water Science and Technology* 35(10): 221–228.

Salvado, H., Garcia, M.P. & Amigo, J.M. 1995. Capability of ciliated protozoa as indicators of effluent quality in activated sludge plants. *Water Research*. 29: 1041–1050.

Thompson, H.R. 1958. The statistical study of plant distribution patterns using a grid of quadrats. *Australian Journal of Botany* 6(4): 322–342.

Vilbaste, S. 2001. Benthic diatom communities in Estonian rivers. *Boreal Environmental Research* 6: 191–203.

Environmental Engineering III – Pawłowski, Dudzińska & Pawłowski (eds)
© *2010 Taylor & Francis Group, London, ISBN 978-0-415-54882-3*

Mathematical model of sedimentation and flotation in a septic tank

M. Pawlak & R. Błażejewski
Rural Water Supply and Sanitation Section, University of life Science, Poznan, Poland

ABSTRACT: Septic tanks are commonly used for preliminary wastewater treatment in on-site plants. On-site wastewater treatment plants are popular in rural areas, where sewerage systems are economically and technically not justified. Three basic processes govern the operation of a septic tank; sedimentation, flotation, and digestion of the suspended solids which accumulate in the tank.

This paper presents a mathematical model describing the removal of suspended solids in a septic tank, except for their decomposition in the digestion process. This latter will be introduced as a second stage in developing the model.

Keywords: Flotation, mathematical modelling, OWTP, sedimentation, septic tank, wastewater treatment plant.

1 INTRODUCTION

Septic tanks are commonly used for domestic wastewater treatment. They are mostly popular in places without centralized sewerage systems; such installations are not feasible in sparsely populated areas.

According to official statistics, at the end of 2007 there were 42,000 on-site wastewater treatment plants (OWTPs) in Poland (COS008). This number will grow in the near future. Countries such as France, Italy, Spain, and Poland could reach a minimum of one million OWTPs, and the number throughout the whole of Europe could reach 10 million (IFAT, 2007). It can be assumed that OWTPs will serve about 10% of the Polish population, and a majority of them will include a septic tank. The largest application of septic tanks (STs) is observed in the USA, where 25% of the population is served by OWTS (Seabloom et al. 2004).

The septic tank was probably invented by the Frenchman Louis Mouras, who in 1860 built a tank of brick for the disposal of domestic wastewater. STs treat wastewater through sedimentation and flotation of the suspended solids. The heavier particles accumulate as a sludge, which is slowly digested. Gas, as a product of this digestion, rarely occurs (Seabloom et al. 2004), and the intensity of gas production depends on the phase of biochemical transformations in the ST. Particles lighter than water create a scum. On the scum's surface one can find a mould, which helps in the decomposition of the suspended solids in the scum (Jowett 2007).

The ST, despite its popularity, simplicity, and long history, has not been adequately investigated and there is no realistic, mathematical model for its operation.

This paper presents an attempt to create a mathematical model, which includes sedimentation, flotation, and suspended solids transport through a ST. The

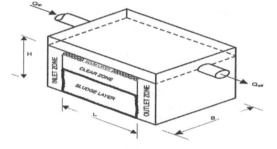

Figure 1. Scheme of a septic tank half full of sludge and scum – just before emptying.

model does not cover the biological processes occurring in the ST. These processes are to be addressed later.

There were several earlier trials to describe the processes in operation in a ST. The majority of them were merely empirical ones. The first scientifically based model was described by Schecher (1997); however, it does not seem to be realistic, due to an assumption that it is a completely mixed reactor operating under steady-state working conditions.

Septic tanks are typically designed for a two to three day hydraulic retention time (based on total tank volume) and one to seven years of sludge/scum accumulation (US EPA, 2002). It is recommended that the tank be emptied when the sludge or scum layer level is one-half the wastewater depth (H/2) in the ST (see Figure 1).

Investigating the scum in the ST, Pearson (1994) stated that it depends on the raw wastewater characteristics, inlet shape, and the application of any bio-additives. McKenzie (1999) found that the thickness of the scum layer varied between 4 cm and 8 cm, and that the impact of bio-additives is not significant. Field studies (Philip et al. 1993, McKenzie 1999) revealed that the scum growth rate is variable over time due to a random inflow of floating solids and their further digestion. This decreasing growth rate is reflected in the empirical formulas proposed by Bounds (1997), as well as by Weibel (Seabloom et al. 2004).

In spite of variations in the concentration of the suspended solids (SS) in the raw wastewater, the efficiency of its removal in a ST is fairly stable with a value ranging between 60% and 80% (Bounds 1997, US EPA 2002, Seabloom et al. 2004, Rothe & Lowe 2006). Rothe and Lowe (2006) estimate the concentration of total suspended solids (TSS) in raw domestic wastewater in the USA to be in a range of between 18 and 2232 mg/dm^3, while the US Environmental Protection Agency (US EPA 2002) projects values between 155 and 330 mg/dm^3. These sources also estimate the concentration of TSS in the ST effluent at 22 to 276 mg/dm^3 and 50 to 100 mg/dm^3, respectively.

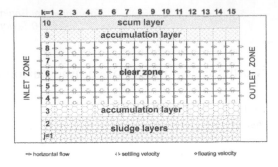

Figure 2. Division of the vertical section of the septic tank into zones.

$Xs_{j,k}$ – mean concentration of settling solids in the cell (j, k) [g/m^3]
$Xs_{i,j,k}$ – concentration of settling solids of the ith fraction in the cell (j, k) [g/m^3]

2 NOMENCLATURE

B – septic tank width [m]
cl – total number of layers in the clear zone [-]
d_i – mean size of a suspended particle of ith fraction [mm]
h – height of a single cell in the discretization net [m]
i – number of a suspended solids fraction of mean size d_i [-]
I – number of all fractions [-]
j – number of a layer (row) [-]
J – total number of layers [-]
k – number of a columns [-]
K – total number of columns [-]
l – length of a single cell in the discretization net [m]
L – total length of a septic tank, excluding inlet and outlet zones [m]
Q_j – flow rate through the jth row of cells [m^3/h]
Q_{in} – flow rate of incoming raw wastewater to the septic tank [m^3/h]
Q_{eff} – flow rate of treated wastewater at the septic tank outlet [m^3/h]
sc – number of scum layers [-]
sl – number of sludge layers [-]
V_i – settling velocity of a single particle of the ith fraction and size d_i [m/h]
W_i – floating velocity of a single particle of the ith fraction and size d_i [m/h]
X_{cr} – limiting concentration of total suspended solids (TSS), below which a given layer belongs to an accumulation layer or a clear zone [g/m^3]
X_{eff} – concentration of TSS in the effluent from a septic tank [g/m^3]
X_j – mean concentration of TSS in the jth layer (row) [g/m^3]
$Xf_{j,k}$ – mean concentration of floating solids in the cell (j, k) [g/m^3]
$Xf_{i,j,k}$ – concentration of floating solids ith fraction in the cell (j, k) [g/m^3]

3 MATERIALS AND METHODS

Our mathematical model concerns the sedimentation and flotation of the suspended solids particles in a rectangular parallelepiped shaped ST. It predicts the concentration of the suspended solids in the effluent as well as the volumes of the scum and sludge zones. The model is two-dimensional (Figure 2); that means that the concentration of the suspended solids is assumed to be distributed uniformly over the width of the ST.

3.1 Model assumptions

The model assumes that:

a) The wastewater level in the ST is constant and that it reaches the invert of the outlet pipe. Wastewater inflows can be variable over time, but mean hourly flows can be used for the calculations.
b) The total suspended solids flowing into the ST consist of floating (lighter than water) and settling (heavier than water) particles of size d_i. They are introduced uniformly through the inlet zone to the clear zone of the ST.
c) At the beginning there is no flow through the uppermost layer ($j = 10$ in Figure 2) which is treated as an accumulation layer for the scum and through the lowest (near-bottom) layer ($j = 1$ in Figure 2), which accumulates the settling sludge. After every time step of the simulation, the concentrations of solids in both accumulation layers are averaged over the layer volume. When the mean concentration of the floating or the settled solids is equal to or greater than a maximum allowable value (X_{cr}), then the layer is treated as fully fed and accumulation of solids starts in the next layer.
d) The scum and sludge accumulation layers are fed by solids vertically (Figure 3). A horizontal flow and transport of solid particles is possible through the clear zone, only.

498

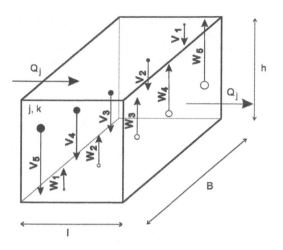

Figure 3. Elementary rectangular parallelepiped – cell (j, k) – with velocity vectors $\mathbf{V}$ for the settling and $\mathbf{W}$ for the floating suspended solids.

e) There is no effect of a flow in the clear zone on the accumulated scum or sludge, unless the flow rate is greater than a limiting (scouring) value.

f) There are no compactions and/or biochemical transformations in either the scum or the sludge layers. These phenomena will be taken into account in the further development of the model.

g) The ST effluent contains suspended solids flowing out from the last columns of the clear zone.

3.2 Model assumptions

The mass balances of the suspended solids in the clear zone can be expressed as follows:

– for the settling solids:

$$\frac{dXs_{i,j,k}}{dt} = \frac{Q_j\left(Xs_{i,j,k-1} - Xs_{i,j,k}\right)}{B \cdot l \cdot h} + \frac{V_i \cdot Xs_{i,j+1,k}}{h} - \frac{V_i \cdot Xs_{i,j,k}}{h} \tag{1}$$

and
– for the floating solids:

$$\frac{dXf_{i,j,k}}{dt} = \frac{Q_j\left(Xf_{i,j,k-1} - Xf_{i,j,k}\right)}{B \cdot l \cdot h} + \frac{W_i \cdot Xf_{i,j-1,k}}{h} - \frac{W_i \cdot Xf_{i,j,k}}{h} \tag{2}$$

The left hand side terms describe the accumulation of suspended solids of the ith fraction in a cell (j, k). The first terms on the right hand sides of Equations (1) and (2) describe the inflows and outflows of the suspended solids by a horizontal advection. The third and fourth terms show vertical inflows and outflows of the suspended solids due to sedimentation and flotation.

The concentrations of both the settling and floating SS in a given cell (j, k) are equal to the following sums, respectively:

$$Xs_{j,k} = \sum_{i=1}^{I} Xs_{i,j,k} \quad and \quad Xf_{j,k} = \sum_{i=1}^{I} Xf_{i,j,k} \tag{3}$$

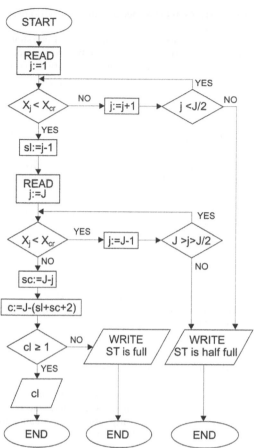

Figure 4. Flow chart to calculate the number of clear zone layers in a septic tank (ST).

The total suspended solids concentration in the jth layer can be calculated as:

$$X_j = \frac{\sum_{k=1}^{K}\left(Xs_{j,k} + Xf_{j,k}\right)}{K} \tag{4}$$

Wastewater flows through the clear zone only. Its flow rate through the jth layer (Q_j) is taken as:

$$Q_J = 0$$

$$Q_{j=J-1\,to\,2} = \begin{cases} \dfrac{Q_{in}}{cl} & if \quad X_{j-1} < X_{cr} \\ 0 & if \quad X_{j-1} \geq X_{cr} \end{cases}$$

$$Q_1 = 0 \tag{5}$$

The number of the clear zone layers (cl) can be calculated using the flow-chart shown in Figure 4.

The settling (floating) velocities V_i (W_i) depend mainly on particle sizes d_i and their densities. Unfortunately, there is a limited amount of data on these characteristics of the suspended solids in raw domestic wastewater.

To avoid re-suspension of sludge (scum), one should control the mean wastewater velocity across the clear zone, U. It should not be greater than a critical (souring) velocity, U_{cr}, i.e. it should comply with the following condition:

$$U = \frac{Q_{in}}{B \cdot cl \cdot h} \leq U_{cr} \qquad (6)$$

The concentration of TSS in the effluent from the simulated ST will be a sum of all fractions, as follows:

$$X_{eff} = \frac{\sum_{j=mincl}^{j=maxcl}\left(Xs_{j,K} + Xf_{j,K}\right) \cdot Q_j}{\sum_{j=mincl}^{j=maxcl} Q_j} = \frac{\sum_{j=mincl}^{j=maxcl}\left(Xs_{j,K} + Xf_{j,K}\right)}{cl} \qquad (7)$$

where: mincl = J – sc – cl; maxcl = J – sc – 1

The inflow rate (Q_{in}) is assumed to be equal to the outflow rate (Q_{eff}).

4 RESULTS AND DISCUSSION

The mathematical model presented above was implemented in Matlab/Simulink, a computer software program which is especially suited for calculations involving matrices. The first results were obtained for steady state flow $Q_{in} = Q_{eff}$ and two fractions (i = 2) of both settling and floating suspended solids. The results appear to be quite reasonable; however, there are no empirical data with which to check the model properly. Hence, there is a need to investigate the suspended solids in the raw wastewater entering different STs to determine their size distributions, densities, and other relevant characteristics.

5 CONCLUSIONS

The two-dimensional model presented above describes independently the sedimentation and flotation processes in a ST. Thus it can be applied to determine the concentrations of suspended solids in raw wastewater in an ST other than where there is induced hindered settling (floatation).

Because of the lack of input data there is a need for relevant empirical investigations on the characteristics of the suspended solids.

The next step in the model development will be its more detailed calibration and validation as well as the introduction of a component addressing the biochemical processes, similar to those described by the Anaerobic Digestion Model (ADM-1).

The model can also be utilized for the design of grease traps.

REFERENCES

Bounds, R. 1997. Design and performance of septic tanks. In M. S. Bedinger, A.I. Johnson & J.S. Fleming (ed.), *Site characterization and design of onsite septic systems* ASTM STP 901:217-233. American Society for Testing Materials: Philadelphia.

Central Statistical Office (CSO). 2008. Municipal infrastructure in 2007, Central Statistical Office. Warszawa.

IFAT. 2007. Millionen Kleinkläranlagen für Europa. *Abwasserbehandlung auf der IFAT 2008,* no. 04/d, http://media.nmm.de/36/04d_17462036.pdf, (Accessed 18.03.2009).

Jowett, E.C. 2007. Comparing the performance of prescribed septic tanks to long, narrow, flooded designs. Presented at *2007 WEFTEC Technical Program 1:* San Diego CA.

McKenzie, M.C. 1999. NC state produces landmark research on septic tank additives. *Small Flows* 13(3): 1–3.

Pearson. F. 1994. A study of scum control in septic tanks. *The Small Flows Journal* 1(1): 17–23.

Philip, H., Maunoir, S., Rambaud, A. & Philippi, L.S. 1993. Septic tank sludge. accumulation rate and biochemical characteristics. In: H. Odegaard, Tapir, Trondheim (ed.), *Small wastewater treatment plants:* 97–104.

Rothe, N.K. & Lowe, K.S. 2006. Wastewater composition and variability as obtained from literature source. In *NORWA 15th Annual Tech. Edu. Conf. & Exp.* Denver.

Schecher, W. 1997. *Septic tank treatment system (SepTTS) model for product chemicals.* Environmental Research Software. User's Manual Version 1.0. Hallowell, Maine. www.cleaning101.com/environment/SepTTS/Sda.pdf (Accessed 18.03.2009).

Seabloom, R.W., Bounds, T.R. & Loudon, T. 2004. *Septic tank.* University Curriculum Development for Decentralized Wastewater Management. London.

US EPA (US Environmental Protection Agency). 2002. *Onsite wastewater treatment systems manual.* Report No. EPA/625/R-00/008. US Environmental Protection Agency.

Environmental Engineering III – Pawłowski, Dudzińska & Pawłowski (eds)
© 2010 Taylor & Francis Group, London, ISBN 978-0-415-54882-3

Changeable character of both surface and retained water and its impact on the water treatment process

A. Rak

Opole University of Technology/ PML td Dublin branch in Wroclaw, Opole, Poland

ABSTRACT: This article focuses on the availability and quality of water from a catchment area, using 27 indicators. Tests were carried out in the reservoir and catchment area, and wastewater was sampled for the waste typical of the reservoir area. Tests carried out in the reservoir established its susceptibility to deterioration. Analysis of the water treatment process, on a semi-technical scale at a pilot station and on a technical scale, formed a basis for establishing border water quality parameters. Test results from the pilot station and under technical conditions were used to establish optimum water treatment systems so that the reservoir stock can be used year round.

Keywords: Surface water, quality indicators, pilot tests, technical tests, technology systems, unit processes.

1 INTRODUCTION

Considering water quality when examining water balance is not a new concept. However, it requires further investigation, especially at small mountain reservoirs. Surface water coming from mountain streams is contaminated to various degrees. Its quality is affected by both weather and water conditions in the reservoir area. Therefore, the water treatment process in such areas should provide for the removal of contaminants throughout the water year. This applies to indicators, such as suspension, turbidity, colour, pH, hardness and content of azonitrates, which vary with weather conditions in the reservoir area. When the water character is changeable, it is important to establish the optimum technology system and levels of chemical reagents required during the water treatment process.

Water quality and flow vary over time. It is important to identify the relationship between such variables as flow, indicator concentration and time, to prepare water and economy balances which consider water quality. Including real water quality parameters in water balances allows the acquisition of complete information on the available water stock. The classic approach to water stock (Z), i.e. without considering quality, is established from water indicators expressed as a function of flow (f) and time (t): $Z = f(Q,t)$. Water stock that takes account of water quality is the water volume and may be used provided any contamination (S) is at an acceptable level. Therefore, water stock is expressed as a function of additional contamination indicators $Z_D = f(Q,t,S)$ (Bartoszewski 1992). It is important when establishing water quality parameters to set an acceptable concentration S_d, which may be applied to the tested water. Other border (maximum) indicators might also be taken into account, depending

on the effectiveness of the system. The available stock $Z_D(S_d)$ is calculated form the water quality in a specific section of the reservoir. There may be three cases (Bartoszewski 1992):

- case 1: the average concentration of contaminants for the specific water treatment plant (WTP) is equal to or less than the acceptable level, i.e. $S \leq S_d$ all year round. In this case water quality does not affect stock availability, i.e. $Z_D(S_d) = Z_d$ when $Z_d > 0$;
- case 2: the average concentration of contaminants is less than S_d at certain times (t_{gr}) and more than S_d at other times during the year. In this case water quality does affect stock availability, i.e. $Z_D(S_d) < Z_d$ when $Z_D(S_d) > 0$; or
- case 3: the average concentration of contaminants is more than S_d, i.e. $S > S_d$, all year round. In this case, the water is unusable because of its high contamination level. This means that stock availability equals zero, i.e. $Z_D(S_d) = 0$ when $Z_d > 0$.

Establishing the function $Z_D = f(Q,t,S)$ is very complex because we are dealing with a mountain reservoir. In reality, we may have a number (k) of water contamination indicators, and consequently, k relations of the $S = f(Q,t)$ type. Once the stock availability is established for each indicator, we obtain k values for $Z_D(S_d)$. The minimum value, $Z_D(S_d)_{min}$, indicates the average stock availability of a particular reservoir.

Water quality prediction requires quick and effective sampling and analysis. The essential issue is to set up a water-quality control system that can predict contamination, based on water conditions at the reservoir.

Studies (Thorne & Fenner 2008) have indicated that it is important to take into account not only potential changes in weather conditions but also changes

Table 1. Basic characteristics of the reservoir and its catchment area.

Characteristic	Unit	Value	Susceptibility category
Area of indirect catchment area (F_1)	km^2	15.3	
Area of supplementary catchment area (F_2)	km^2	35.5	
Total area ($F_1 + F_2$)	km^2	50.8	
Max area of the reservoir (F_{res})	ha	178	
Water depth at the barrier	m	13.5	
Average depth of the water at the dam	m	8.15	II
Annual average flow of the indirect catchment area ($Q_{śR}$)	m^3/s	0.192	
Average low flow ($Q_{śN}$)	m^3/s	0.038	
Untouched flow (biological) ($Q_b = k*Q_{śN}$). ($k = 1.52$ for mountain streams)	m^3/s	0.577	
Annual average flow ($Q_{śR}$)	m^3/year	6,054,912	
Untouched annual outflow (V_b)	m^3/year	1,821,519	
Intake of water for consumption (Q_{WTP})	m^3/h	380.0	
(status for 2008)	m^3/year	3,328,800	
Planned intake of water for consumption	m^3/h	1000.0	
(Q_{WTP}) (projected status)	m^3/year	9,000,000	
Reservoir operating capacity ($V_{Capacity}$)	m^3	11,000,000	
Indicator of water exchange all over the year:			
without the water intake (W_{W1})	%	16.6	I
water intake (Q_{WTP}) (W_{W2})	%	46.8	II
projected water intake (W_{W3})	%	98.4	II
Schindler index W_S for F_1 (W_{S1})		1.39	I
Schindler index W_S for $F_1 + F_2$ (W_{S2})		4.62	II
Average Schindler index $W_{Sr} = (W_{S1} + W_{S2})/2$		3.00	II
Reservoir land development	% wood	65	I

in the use of soil and water exchange indicators due to the outlet or intake used for consumption purposes. The empiric models establishing water quality are based mainly on water data. They also consider any movement inside the reservoir. However, those models do not describe any effects on the fauna and flora of the reservoir or its catchment area (Lowrance 2007).

The surface waters of mountain reservoirs show very little contamination during stable weather conditions. However, contamination arises during heavy rain or snowfall. The contamination concentration parameters differ depending on different weather conditions in the reservoir area. A high level of variation in water quality indicates the need to include so-called 'treatment barriers' in order to stop and neutralize microorganisms, soluble organic matter and chemical micro-contamination (Sozański 1984). The most effective processes in such conditions are pre-ozonation, surface coagulation with flocculation on sand and anthracite filters, secondary ozonation and sorption on active carbon. It is important to carry out pilot tests because of changeable surface water characters, so that an optimal technical solution can be established as part of the water treatment process. The issue of pilot tests with different technical systems and filtration materials has been mentioned in the literature (Balcerzak & Łuszczek 2006, Johnson et al. 1995, Mołczan & Biłyk 2006, Sozański & Olanczuk-Neyman 2002).

2 CHARACTERISTICS OF THE RESERVOIR AREA AND TEST METHODS

The Sosnowska reservoir was constructed in the 1990s to supply water to the municipality of Jelenia Góra. It supplies a water treatment plant with a capacity of 25,000 m$^3 \cdot$ d^{-1}. The maximum capacity of the reservoir is 15.4 million m^3; the surface area is 1.78 km^2, the depth at the reservoir dam is 13.5 m and the average depth is 8.15 m (Table 1).

The wooded character of the reservoir area, low level of water exchange and lack of indirect contamination sources are very favourable for water quality. The average depth of the reservoir (8.5 m) contributes to its low rate of eutrophication. Moreover, at present, the reservoir's Schindler index puts it in category I, but this is likely to change to category II due to an increasing level of water intake used for consumption and the need to source additional water supplies.

Between 2005 and 2008, pilot tests were carried out and once the WTP started operations, more sampling was done. A series of samples were taken from the reservoir and its catchment area for water quality studies. At the same time, tests were carried out at the pilot station to establish the optimal technical treatment process for the water for consumption. Tests were done during the complete water year, from July 2005 to June 2006 (Rak 2008). The waters of both the Sosnówka reservoir and Podgórna river were sampled in cycles of a few days during each month of the study

Table 2. Technical systems during the pilot and technical tests.

Technical process	Technical systems								
	Pilot tests					Technical tests			
	WI	WII	VIIa	VIII	WIV	W1	W1A	W2	W3
Sieving with a 1 mm × 1 mm strainer	+	+	+	+	+	+	+	+	+
Pre-ozonation		+	+	+	+	+	+	+	+
Water pH correction with lime water	+	+	+	+	+	+	+		+
Aluminium sulphate coagulation	+	+	+			+	+		
Flocculation	+	+	+			+			
Accelerated anthracite and sand bed filtration	+	+	+	+	+	+	+	+	+
Secondary ozonation	+	+				+		+	+
Active carbon bed filtration	+	+	+	+	+	+	+	+	+
Final water quality correction						+	+	+	+
Disinfection						+	+	+	+

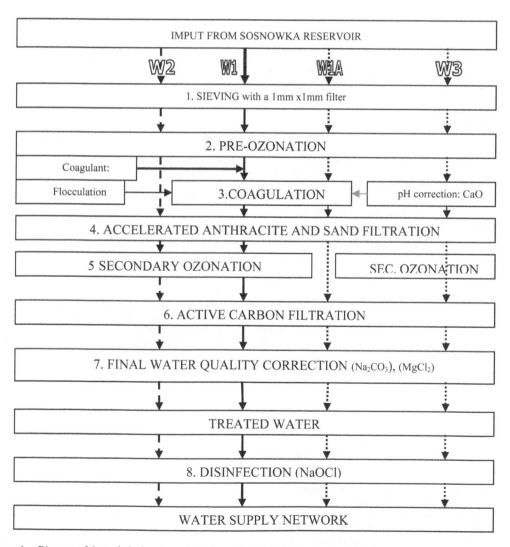

Figure 1. Diagram of the technical systems at the time of technical tests.

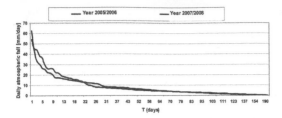

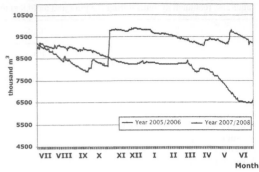

Figure 2. Curves of total daily atmospheric fall in the catchment area including the higher amounts in the study periods.

Figure 4. Water volume in the reservoir during the study periods.

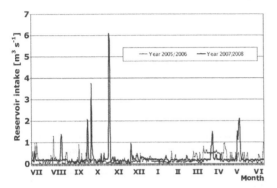

Figure 3. Amount of water input to the reservoir from the catchment area during the study periods.

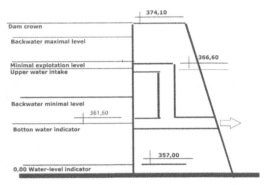

Figure 5. Main levels of backwater at Sosnówka reservoir.

year. There were 79 samples from the Sosnówka reservoir and 51 from the Podgórna river. Tests included air temperature and 26 contamination indicators. The C-200 multi-photometer was the basic instrument used for the measurements. The technical design of the pilot station allowed for variations in sampling, such as: sieving, pre-ozonation, coagulation with flocculation, water pH correction, filtration on anthracite and sand bed, secondary ozonation, and active carbon filtration. During this study, 20 of these variations were used.

The station allowed for technical testing in 4 main areas as shown in Table 2.

From November 2007 to October 2008, tests were carried out at the WTP. At the same time, technical tests were carried out under technical conditions. The tests included:

- water quality indices of raw water taken from the reservoir;
- water quality indices of treated water at various stages of the water treatment process;
- amount of reagents; and
- filtration process parameters (filtration rate, flow intensity).

The operating technology allows for a varying mix of technology systems, depending on the water quality at different periods of the water year. There are 3 main technical systems resulting from the analysis of the raw water quality and results of the technical tests show in Table 2. These technical systems are shown in Figure 1.

3 TEST RESULTS

The water catchment area of the Sosnówka reservoir is mountainous in character with widely variable weather conditions (temperature and rain/snowfall). The area has a short growing period (200–210 days), low temperatures (average in winter 1.4°C and in summer 12.3°C) and heavy snow/rainfall (800–1100 mm/year). There are no major sources of contamination. The area is heavily wooded. Hydrological conditions allow for the rapid flow of surface water. During the study periods, the annual precipitation was 983.8 mm and 1005.1 mm. Daily falls vary with the year. Daily values were 54 mm (23.09.2005) and 62.5 mm (29.10.2008). There was no atmospheric fall on 175 days of the study period during 2005/2006 and on 179 days during 2007/2008. There was an atmospheric fall of more than 20 mm/day on 9/10 days and of more than 10 mm/day on 24/28 days of the two study years (Figure 2).

During the study period, both water input to the reservoir and water status were noted at the reservoir (Figures 3 and 4).

The amount of water input to the reservoir depends on the amount of atmospheric fall occurring in the catchment area (Figures 3, 4). The heaviest atmospheric fall occurs in autumn (September – October) with lighter one during the summer (June – August).

Table 3. Typical values of the chosen water quality indicators during the technical testing period.

Indicators of tested water	Unit	Sosnówka reservoir			
		min	max	average	standard deviation
Water temperature	°C	3	15	9.2	1.620
Turbidity	mg SiO_2/dm^3	0.57	3.81	1.56	1.052
Colour	mg Pt/dm^3	5	14.9	8.21	2.791
pH	pH	7.2	8.4	7.63	0.366
Total hardness	$mval/dm^3$	0.62	1.03	0.85	0.072
Ammonium nitrogen	mg N/dm^3	0.01	0.11	0.05	0.004
Nitrate nitrogen	mg N/dm^3	2	5	2.44	1.300
Chlorides	mg Cl/dm^3	4	6.8	4.95	1.321
Oxidability	mg O_2/dm^3	1.0	5.22	3.458	1.532
Conductivity	μ s/cm	98.28	122.50	110.88	6.692
Dissolved oxygen	mg O_2/dm^3	9.4	12.4	11.18	1.172
Fe	mg Fe/dm^3	0.05	0.136	0.066	0.027
Mn	mg Mn/dm^3	0.012	0.036	0.021	0.007

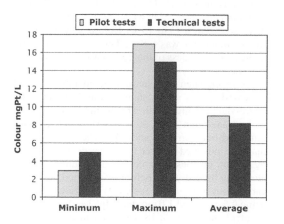

Figure 6. Minimum, maximum and average values for turbidity.

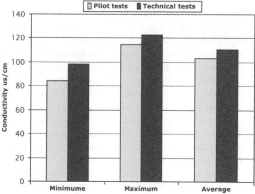

Figure 8. Minimum, maximum and average values for water pH.

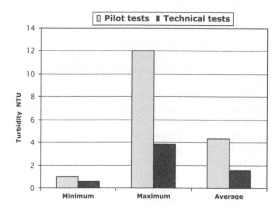

Figure 7. Minimum, maximum and average values for colour.

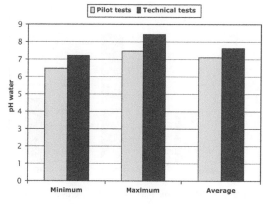

Figure 9. Minimum, maximum and average values for conductivity.

Most input to the reservoir occurs during the precipitation period, i.e. March – May. It is important to emphasize that water was not taken for consumption during the 2005/2006 study period. At that time, water management ensured the maintenance of a stable water level at all times and capacity reserves were secured in case of large outflows due to possible heavy atmospheric falls during the summer. During the second

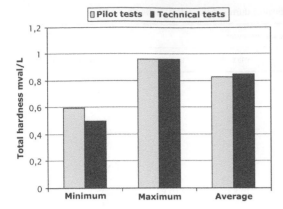

Figure 10. Minimum, maximum and average values for total hardness.

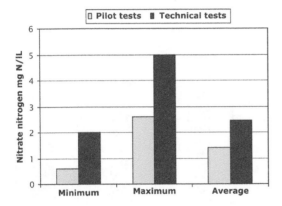

Figure 11. Minimum, maximum and average values for nitrogen (NO₃).

study period (2007/2008) when water was taken for consumption, the water level was stabilised by outflow control. The need to hold the water capacity level at 8 million m³ was essential due to the location of the pipes used for water intake for consumption. When the water was at that level, it allowed water to be supplied to the water treatment plant by gravitational flow. If the water were to fall below this level, it would be necessary to pump the water to the WTP (Figure 5).

In October 2007, the WTP started operating. Table 3 shows the minimum, maximum and average values of atmospheric fall.

The results of the tests carried out at the pilot station and at the water treatment plant (1st period of operation) are shown in Figures 6–11. During the technical testing, the water quality was much better, in particular with regard to turbidity and colour. The total hardness remained at the same level during the two study periods with a value of 0.5 mval/dm³ to 0.95 mval/dm³. The pH of the water was 8.4 during technical testing, whereas it was 7.2 during the pilot tests. The same trends were noted in ammonium nitrogen and conductivity.

4 DISCUSSION

Water quality was predicted from calculations based on the results of water quality tests at the Sosnówka reservoir and the Podgórna river. These results are shown in Table 4. They refer to the main methods for calculating dispersion in the tested water.

The predicted quality of the reservoir water was calculated using the formula below. It includes the impact the Podgórna river had on the quality.

$$S_{min} = S_{average} \cdot (1 - S_{rw}) \qquad (1)$$

$$S_{max} = S_{average} \cdot (1 + S_{rw}) \qquad (2)$$

where: S_{min} = the predicted minimum value of the contamination indicator; S_{max} = the predicted maximum value of the contamination indicator; $S_{average}$ = the average value of the contamination indicator at the reservoir based on the tests; S_{rw} = the equivalent diversification indicator, $[S_{rw} = (S_{r(Rz)} + S_{r(Z)})/2 \cdot 0.01]$; $S_{r(Z)}$ = the coefficient of water indicator for the reservoir water [%]; $S_{r(Rz)}$ = the coefficient of water indicator for the Podgórna river [%].

These results are shown in Table 5. The results for pH and colour are compared in graphs (Figures 12 and 13).

Comparing the results of the filtration tests reveals that anthracite and sand filtration is less effective under technical conditions (Figure 14). However, the same final goal was achieved in both tests, reducing the colour by 80% (Figure 15).

The stability and aggressiveness of water treated to reduce hardness and pH was studied – water pH and both Langelier and Rezner indices (I_L, I_R) were calculated when pH was satisfactory (pH$_S$). Water temperature, pH and pH$_S$, and I_L and I_R indices in technical system W2, showed that both raw and treated water was aggressive. These indicators were particularly marked during May when the water temperature increased from 8°C to 15°C. Before May, when the water temperature was 4–8°C, there was little increase in the aggressiveness of the treated water compared to raw water.

However, when both the water temperature and pH increased (pH$_{max}$ = 8.4) stability was lost and water aggressiveness increased. The I_L index was minus, with a maximum value of −1.6, while the I_R index remained between 10.5 and 10.0. The test results at that time indicated the need to begin water stabilisation by adding magnesium chloride and sodium carbonate to the treated water before disinfection. The tests carried out in technical system W2 were carried out using 0.5–0.75 and 1.0 mg/dm³ MgCl and Na₂CO₃, at a water temperature of 18–19°C and an average water pH of 7.65. The results indicated an increase in water stability due to increased doses of both magnesium chloride and sodium carbonate. Using 0.75 mg/dm³ of both reagents resulted in a hardness of more than 1,2 mval/dm³ in the treated water. The Langelier and Rezner indices were −0.3 and 8.3. Increasing the dose of both reagents to 1.0 mg/dm³ changed the I_L and I_R

Table 4.　Methods of testing dispersion in water from the reservoir and the river.

Indicators of tested water	Unit	Sosnówka reservoir			Podgórna river		
		S_X	(R)	S_r %	S_X.	(R)	S_r %
pH	pH	0.25	1.0	3.52	0.17	0.65	2.42
Turbidity	mg SiO$_2$/dm^3	2.22	11	48.9	6.77	26	105.1
Colour	mg Pt//dm^3	4.15	27	42.4	7.12	34.5	72.8
Total organic carbon	mg C/dm^3	0.53	1.81	21.2	0.56	2.4	27.86
Dissolved substances	mg/dm^3	5.4	19.0	9.02	12.00	34	40.97
Total hardness	mgCaCO$_3$/dm^3	3.62	18.0	8.78	8.57	25.2	44.75
Nitrate nitrogen	mg N/dm^3	0.4	2.0	32.8	0.37	1.72	38.14
Ca	mg Ca•/dm^3	0.5	1.32	7.38	1.32	3.70	31.35
Mg	mg Mg•/dm^3	0.14	0.47	17.72	0.13	0.43	37.14
P	mg Na•/dm^3	0.24	1.16	10.0	0.12	0.39	7.31
N	mg K•/dm^3	0.15	0.53	15.0	0.06	0.20	7.59
Chlorides	mg Cl•/dm^3	1.37	8.05	12.36	1.04	3.95	13.81
Sulphates	mg SO$_4$/dm^3	2.31	7.10	16.77	0.86	2.60	9.66

S_X = standard deviation; R = Spread = $S_{max} - S_{min}$; S_r = Diversification index = $S_X / S_{average}*100\%$.

Table 5.　Prognosis of the reservoir water quality.

Indicators of tested water	Unit	Sosnówka reservoir acc to tests				Prognosis	
		$S_{average}$	$S_{r(Z)}$ %	$S_{r(Rz)}$ %	S_{rw}	S_{min}	S_{max} (R)
pH	pH	7.1	3.52	9.56	0.065	6.64	7.56	0.92
Turbidity	mg SiO$_2$/dm^3	4.54	48.9	105.1	0.770	1.04	8.04	7.00
Colour	mg Pt/dm^3	9.78	42.4	72.8	0.576	4.15	15.4	11.25
Total organic carbon	mg C/dm^3	2.5	21.2	27.86	0.245	1.89	3.11	1.22
Dissolved sustances	mg/dm^3	59.84	9.02	40.97	0.249	44.94	74.74	29.80
Total hardness	mgCaCO$_3$/dm^3	41.23	8.78	44.75	0.267	30.22	52.24	22.02
Nitrate nitrogen	mg N /dm^3	1.33	32.8	38.14	0.355	0.86	1.80	0.94
Ca	mg Ca/dm^3	6.77	7.38	31.35	0.196	5.44	8.10	2.66
Mg	mg Mg/dm^3	0.79	17.72	37.14	0.274	0.57	1.00	0.43
P	mg Na /dm^3	2.04	10.0	7.31	0.086	1.86	2.22	0.36
N	mg K/dm^3	1.00	15.0	7.59	0.113	0.89	1.11	0.22
Chlorides	mg Cl/dm^3	11.08	12.36	13.81	0.131	9.63	12.53	2.90
Sulphates	mg SO$_4$/dm^3	13.77	16.77	9.66	0.132	11.95	15.59	3.64

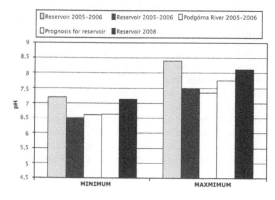

Figure 12.　Predicted minimum and maximum values for pH based on the test results.

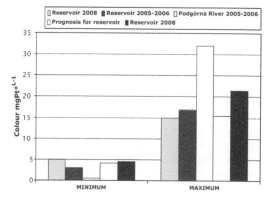

Figure 13.　Predicted minimum and maximum values for colour in the reservoir based on the test results.

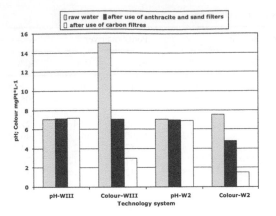

Figure 14. Effectiveness of colour reduction in technical systems WIII and W2 including change of pH.

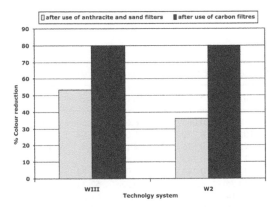

Figure 15. Effectiveness of colour reduction in technical systems WIII and W2 after implementation of filtration stages.

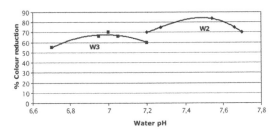

Figure 16. Effect of colour reduction subject to both water pH and technical system type.

indices to 0.1 and 7.2. Tests of the treated water in technical system W2 concluded that treatment with MgCl and Na$_2$CO$_3$ stabilised the water. It was not aggressive nor did not produce any CaCO$_3$ sludge.

5 CONCLUSIONS

This comparative analysis indicates a trend in the relationship between the pH of treated water and the effects

on colour reduction. The following examples show results obtained for similar weather conditions (water temperature 4.4–5.2°C):

1. In technical tests (W2) of raw water with a pH of 7.2–7.7, colour of 10–12 mg Pt/dm^3, and turbidity of 0.8–3.8 mg SiO$_2$/dm^3:

- the maximum colour reduction was 83.5%, with a pH of 7.54;
- the colour reduction curve was close to a parabola – at the same time as the water pH decreased from 7.7 to 7.54, the colour reduction increased from 75% to 83.5%;
- a further decrease in water pH to 7.22 caused a decrease in colour reduction effectiveness to 75%.

2. In pilot tests (WIII) of raw water with a pH of 6.9–7.05, colour of 10–12 mg Pt/dm^3, and muddiness of 5 mg SiO$_2$/dm^3:

- the maximum colour reduction was 75%, with a pH of 8.5;
- the colour reduction curve was close to a parabola – at the same time as the water pH decreased from 9.3 to 8.5, the colour reduction increased from 65% to 75%;
- a decrease in water pH to 8.22 caused a decrease in colour reduction effectiveness to 65%; and
- a further decrease in water pH to 7.5 caused a decrease in colour reduction to 20%.

Both the pilot station and technical tests indicated that there were a few technical systems suitable for the water treatment plant to use during operation (Figure 16):

- technology system W2, including ozone, 2 mg O$_3$/dm^3, in the pre-ozonation process and 1 mgO$_3$/dm^3 for secondary ozonation when the water used has an alkaline pH of 7.2–7.7, colour <15 mg Pt/dm^3, and turbidity <5 mg SiO$_2$/dm^3.
- technology system W3, including correction of pH with lime water, 3–5 mg CaO·/dm^{3}and ozone, 1,2 mgO$_3$/dm^3, in the pre-ozonation process and 0.8 mg O$_3$/dm^{3}for secondary ozonation when the water used has a pH of 6.65–7.2, colour <17 mg Pt/dm^3, and turbidity <10 mg SiO$_2$/dm^3.

In the light of the predicted quality of the reservoir water (Table 5), we can assume that the concentration of contaminants in the water (pH, turbidity, colour) will be less than the upper permissible limit set by the WTP, i.e. S ≤ S$_d$. The WTP's operating technology allows for effective water treatment of raw water. At the same time, the predicted water quality indicators have no effect on available stock, so case 1 is fulfilled, which guarantees year-round water intake for consumption purposes.

REFERENCES

Balcerzak, W. & Łuszczek, B. 2006. Pilot research as a way for selection of activated carbon in water treatment process; in Active carbon within environment protection and industry. Czêstochowa University of Technology. V Science Conference: 213.

Bartoszewski, K. 1992. Indicating the water stock including the water quality. Ochrona Środowiska 2–3/46–47: 37–40.

Johnson, B.A., Gong, B., Bellamy, W. & Tran, T. 1995. Pilot plant testing of dissolved air flotation for treating Boston's low-turbidity surface water supply. Water Science and Technology 31(3–4): 83–92.

Lowrance, R. 2007. Water Quality: Modeling, Water Science, Secondo Editio II: 1327–1329.

Mołczan, M. & Biłyk, A. 2006. Removal of organic substances of water in processes such as: ionic exchange, coagulation and adsorption. In: *Active carbon in environment protection and industry*: 204. Czêstochowa University of Technology. V Science Conference.

Rak, A. 2008. Treatment of water taken from a reservoir. *Polish Journal of Environmental Studies* 17(2C): 9–14.

Sozański, M.M. 1984. *Research issues in designing and operating of water treatment plants. Water supply to cities, towns and villages*. PZITS: Poznań.

Sozański, M.M. & Olanczuk-Neyman, K. 2002. *Status and perspectives of the water treatment process as a contemporary science discipline*. Environment Engineering Committee Monographs of the Committee for Environmental Engineering 9: Lublin.

Thorne, O.M. & Fenner, R.A. 2008. Modelling the impacts of climate change on a water treatment plant in South Australia, *Water Science & Technology*: Water Supply—WSTWS 8(3): 305–312. IWA Publishing.

Environmental Engineering III – Pawłowski, Dudzińska & Pawłowski (eds)
© *2010 Taylor & Francis Group, London, ISBN 978-0-415-54882-3*

Changes in the environments of water reservoirs in Warsaw resulting from transformations in their surrounding areas

T. Stańczyk & J. Jeznach

Department of Environmental Improvement, Warsaw University of Life Sciences-SGGW, Warsaw, Poland

ABSTRACT: This article presents the results of an investigation of the environmental states of 50 small water reservoirs in the area of Warsaw on the left-bank of the Vistula river. The investigations included measurement of the electrolytic conductance and transparency of the water as well as an assessment of certain aspects of environmental functioning. The results were analysed in the context of the anthropogenic transformations of the reservoir surroundings using cluster analysis. The results demonstrated that degeneration of a reservoir can be connected to the degree of transformation of its surroundings and also with the naturalness of the surrounding terrain.

Keywords: Water reservoir, pond, environmental assessment, cluster analysis.

1 INTRODUCTION

Small water reservoirs, often referred to as ponds or shallow lakes, are one of the last wildlife refuges that have not been altered by man (Bajkiewicz-Grabowska & Mikulski 1996, Oertli et al, 2002, Søndergaard et al. 2005). They create valuable biocoenoses, increasing the biological diversity of agricultural and forested lands (Nicolet et al. 2002, Ruggiero et al. 2008). Additionally, in urbanised areas, they may significantly improve the living conditions for humans, beautifying the environment in which they live (Szulczewska & Kaftan, 1996, Vermonden et al., 2009). Small water bodies positively influence urban climate and are recreation spots for the local inhabitants who visit them to enjoy the beauties of nature (Krzymowska-Kostrowicka 1997).

However, anthropogenic disturbances in the biological equilibrium and regulatory processes have the result that the positive effect of these water bodies is frequently limited. Recent studies have shown many threats to the existence and proper functioning of water reservoirs in the areas managed by man (Gibbs 2000, Stoianow et al. 2000).

In spite of the great importance of small water reservoirs as natural habitats and places of recreation and relaxation for city inhabitants, their role often goes unnoticed in land-use planning. This is reflected in the often environmentally unfriendly provisions of the planning documents (e.g. allowing construction of settlements on lands adjoining the reservoir or locating traffic routes too close, lack of buffer zones between built-up areas and reservoirs, lack of convenient access to the bank zone for inhabitants).

These unfavourable trends in land-use planning practice justify undertaking studies on the state of the environment of reservoirs surrounded by lands in varying states of transformation. Therefore, based on a survey and analysis of the characteristics of the land around reservoirs and the water in them, studies have been launched to assess the value of natural habitats and the usefulness of the reservoirs for recreational purposes.

2 MATERIALS AND METHODS

This research included small water bodies located in that area of Warsaw on the left bank of the Vistula River (left bank Warsaw), where more than 300 such habitats were identified. Fifty water reservoirs were selected during field studies carried out in the second half of 2008. Selection was based on the typology proposed by Stańczyk and Jeznach (2007) who divided the water reservoirs of left-bank Warsaw into 16 types. At least two reservoirs of each type were selected to be studied.

The objective of the first stage of the research was to analyse and identify, using a geographic information system (GIS), the land-use types of the surrounding land closest to the reservoirs.

Using the results of the topographic analysis, the boundary limits of the terrain depressions of the reservoirs were first marked on the map and then a 50-metre wide buffer zone was marked around them using the ArcMap® programme. Analysis of the topographic maps, interpretation of aerial photographs and site inventories made it possible to distinguish the predominant and concurrent types of land use in these 50 m zones around the reservoirs.

The land use types identified formed the basis for determining index indicators of the transformations of the areas around the reservoirs. The transformation of a surrounding area was assessed by assigning points to individual land-use types. The point values were then used in classifying the transformations into predominant and concurrent types. The values of the proposed transformation index adopted for different land use types are given in Table 1. The final assessment of the

Table 1. Indices of transformation of the water reservoir surroundings.

Transformation index	Land-use type in the surroundings of the water reservoirs
1	Forest, park with old stand, reedbeds, water (other water reservoirs)
2	Afforestation (e.g. Mid-field), thickets (woody species and shrubs), young-growth stand, wasteland
3	Park with young stand, meadows
4	Park, predominantly lawns, waste grasslands
5	Fallow lands, allotment gardens
6	Fields, farm buildings, orchards
7	Low buildings, tall buildings
8	Office buildings, cemeteries, storehouses
9	Industrial, traffic areas (roads, car parks, track-ways)

transformation is the sum of 60% of the value of the transformation for the predominant land-use type and 40% of this value for the concurrent type.

Further studies of the reservoir surroundings identified the existing recreational facilities on the bank area and the terrain depression of the reservoir. The inventories of the recreational infrastructure and facilities (benches, trails, platforms, beaches, etc.) were made during on-site inspections.

The states of the water reservoir environments, as well as the usefulness of their basins and terrain depressions for recreation and leisure purposes, were assessed using the index method of diagnosing the state of the environment (developed by Stańczyk in 2006). This approach is similar to methods widely used in pond, lakes, rivers and environment assessment (Pesce et al. 2000, Zhang et al. 2007)

The assessment was supplemented with measurements of the main quality parameters of the reservoir water. Measurements of the electrolytic conductivity of the water were carried out in the field with a mobile electrochemical meter and the measurements of water transparency, with the Secchi disc. For a better interpretation of the results, the values of the index assessments of the state of the environments and transformations of the surrounding, as well as the values of the measured electrolytic conductivity were subjected to a hierarchical cluster analysis. Such an analysis is widely applied in the natural sciences as an approach to effectively search for hidden regularities in even very large datasets (Kardaetz et al. 2008, Spruill et

Figure 1. Example of the marking out of the terrain depressions of reservoirs and their closest surroundings 1 – reservoir surrounding zones; 2 – terrain depression zones; 3 – grasslands and afforestation; 4 – industrial and transport; 5 – arable fields; 6 – reedbeds; 7 – settlements; 8 – water reservoirs.

al. 2002, Toth et al. 2008). The analysis was performed by the Ward's method using Euclidean distance inSTATISTICA® software. The cluster analysis made it possible to distinguish homogeneous groups in the 50 water reservoirs analysed.

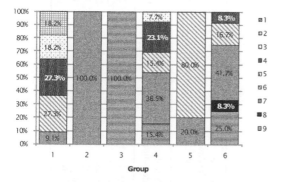

Figure 2. Differentiation of the predominant land-use types of the reservoir surroundings by groups 1 – traffic and industrial area; 2 – tall buildings; 3 – low buildings; 4 – fields and farm buildings; 5 – wasteland; 6 – lawns and meadows; 7 – park; 8 – afforestation; 9 – forest, old park, reedbeds.

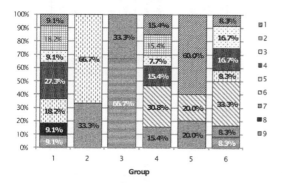

Figure 3. Differentiation of the concurrent land-use types of the reservoir surroundings by groups 1 – traffic and industrial area; 2 – tall buildings; 3 – low buildings; 4 – fields and farm buildings; 5 – wasteland; 6 – lawns and meadows; 7 – park; 8 – afforestation; 9 – forest, old park, reedbeds.

3 RESULTS AND DISCUSSION

The land-use analysis of the surroundings of the water reservoirs enabled identification of the units constituting a homogeneous form of land use of the depressions of the reservoirs and the 50 m wide zones adjoining them (Figure 1).

The predominant and concurrent land-use types were marked out for each reservoir. This made it possible to illustrate differences in the land use of the surroundings of the individual reservoir groups identified in the analysis (Figures 2 and 3).

The inventories of recreational infrastructure and facilities on the bank area and terrain depressions of the reservoirs were performed during on-site inspections. The most frequently used methods for adapting the reservoir surroundings for recreational use included setting out trails (17 reservoirs); installing a lighting system in the area (17); establishing a park (12); setting out benches (12); providing litter baskets (7); constructing playgrounds (6); and creating walkways (4).

Less frequently used approaches included developing grass beaches and constructing jetties, private gardens, footbridges, information boards, campfire sites and tourist routes.

The reservoir surroundings best equipped with tourist infrastructure facilities were those located in city parks. There is no doubt that the attractiveness of a permanent water feature is widely appreciated by users and landscape designers. However, even in the degraded or unkempt areas around the reservoirs there were many signs of a human presence (paths, provisional jetties built by anglers and campfire sites). Also a human presence can also be determined in the areas around the reservoirs that are lacking recreational infrastructure or are degraded. The lack of adaptation of these sites and facilities for recreational functions, combined with their intensive use by the inhabitants of the near-by settlements, causes damage to the reservoirs' environments.

The damage has usually occurred in the riparian vegetation on the banks and the slopes of the reservoirs, which are eroded.

There is also contamination of the general area.

Table 2. Measured values and results of the assessment of the general transformation of surroundings of the water reservoirs studied.

Group number	Number of reservoirs in a group	Water transparency [cm]		Electrolytic conductance [µS/cm]		General transformation of surroundings	
		Mean	Standard deviation	Mean	Standard deviation	Mean	Standard deviation
1	11	51	23	778	417	5.66	1.26
2	3	50	20	602	111	3.53	0.46
3	6	33	19	528	205	1.27	0.41
4	13	51	22	1016	1082	4.58	1.50
5	5	59	11	8483	868	4.48	0.83
6	12	117	41	1164	550	3.73	1.17

The last stage of the field studies was to assess the general state of the environment, as well as the usefulness of the reservoir basin and terrain depression for recreation and leisure purposes using the index method of assessing the state of the environment.

The results of these studies were used as the input data for a cluster analysis which made it possible to distinguish 6 groups in the set of 50 reservoirs. Table 2 lists the mean values and standard deviations of the analysed parameters as calculated for each group.

Reservoirs from Group 3 had the lowest degree of transformation of their surrounding environments, i.e. the highest degree of landscape naturalness. In spite of the very good state of the surrounding environment, the condition of the water in the reservoirs was assessed as definitely unsatisfactory. Two reservoirs from this group are located in the 'Skarpa Ursynowska' nature reserve. One effect of this passive form of nature protection is a lack of maintenances. This, together with advanced paludification processes, intense vegetation succession and eutrophication, increases the degradation of reservoirs.

The best environmental conditions were found in the reservoirs of Group 6. The degree of transformation of their surroundings, which were usually city parks, was moderate. The maintenance of park ponds exerts a favourable impact on their condition, reducing the threat of disappearance and increasing recreational attractiveness. These ponds are a valuable habitat for different species of flora and fauna, given the surrounding city conditions. The ability of parks to act as buffer zones reduces the adverse impact of adjacent built-up and traffic areas.

Water reservoirs from Group 2 were found to be in the worst general environmental state. However, this was not a consequence of an excessive transformation of the reservoir surroundings, which were generally insignificant.

Besides, the surroundings themselves were mostly managed as city parks. All the water reservoirs from this group were concrete constructions, having decorative roles in the city parks during the warmer months of the year. They provide good insulation against the effects of the surrounding environment. However the value of the reservoirs as natural habitats is, as a result, regrettably low.

It should, however, be remembered that this method of functioning and the construction of the reservoirs derive from their purposeful design. Any attempts at improving the assessment using the criteria discussed would entail a total redesign and reconstruction of the reservoir basin. The recreational functions of these reservoirs are quite good because of the broad range of facilities in their park surroundings; they provide much improved conditions for recreation in the area.

4 CONCLUSIONS

The research findings suggest the following conclusions and recommendations:

Table 3. Results of the assessment of the state of the environment of the water reservoirs studied.

Group number	Water conditions state		Resistance to water quality deterioration		Vegetation quality		Habitat quality		Resistance of the reservoir to the threat of disappearance		Recreational attractiveness		General state of the environment	
	Mean	Standard deviation	Mean	Standard deviation	Mean	Standard deviation	Mean	Standard deviation	Mean	Standard deviation	Mean	Standard deviation	Mean	Standard deviation
1	3.03	0.43	2.94	0.28	2.77	0.51	2.71	0.32	3.11	0.41	2.76	0.40	2.91	0.32
2	2.02	0.13	3.46	0.07	2.08	0.17	2.05	0.09	2.99	0.22	3.46	0.05	2.58	0.14
3	2.78	0.49	3.32	0.20	3.32	0.17	2.89	0.15	2.78	0.48	2.77	0.35	2.92	0.27
4	3.69	0.34	3.25	0.30	2.63	0.26	2.92	0.15	4.06	0.21	3.54	0.36	3.43	0.14
5	3.83	0.18	3.21	0.32	2.96	0.18	3.08	0.08	4.04	0.21	3.46	0.04	3.53	0.09
6	4.07	0.37	3.63	0.27	3.47	0.36	3.43	0.21	4.26	0.25	3.93	0.42	3.87	0.21

1) The diagnostic method used can also be applied as a support tool in studies on the conditions and directions of spatial land-use planning for communes and in identifying those water reservoirs that need to be placed under legal protection.

2) The data obtained during the inventory and diagnostic studies are valuable. They can be used for monitoring environmental changes and for other studies on the environment of water reservoirs in Warsaw.

3) It is recommended that those water reservoirs which were given the highest general rating (were assessed highest) and showed the most valuable environmental features should be placed under legal protection with the status of an ecological utility area as defined in the Act on nature conservation of 16 April 2004. This will reduce the urban pressure and, simultaneously, will make it possible to pursue active protection, which is more suitable for urbanised areas than reserve protection.

4) With a lack of active protection, the high vulnerability of small water reservoirs to eutrophication and paludification processes may cause them to degrade quickly, even when their surroundings represent high natural values and demonstrate a high degree of naturalness.

5) The reservoirs whose state was assessed as unsatisfactory should be subject to reclamation or improvement works with the objective of enhancing the areas' environmental and recreational functions.

6) For those reservoirs whose surroundings were transformed to a large degree, the improvement works should include proper adaptation of the bank areas to perform the function of a buffer zone. This would mitigate the negative impacts of the surroundings, prevent the inflow of pollutants, reduce the noise level, enhance biological diversity and coordinate tourist traffic.

REFERENCES

Bajkiewicz-Grabowska, E. & Mikulski Z. 1996. *General hydrology*. Warsaw: PWN.

Gibbs, J.P. 2000. Wetland loss and biodiversity conservation. *Conservation Biology* 13: 314–317.

Kardaetz, S., Strube, T., Bruggemann, R., & Nutzmann, G. 2008. Ecological scenarios analyzed and evaluated by a shallow lake model. *Journal of Environmental Management* 88: 120–135.

Krzymowska-Kostrowicka, A. 1997. *Geoecology of tourism and rest*. Warsaw: PWN.

Nicole, P., Biggs, J., Fox, J., Hodson, M.J., Reynolds, C., Whitfield, M. & Williams, P. 2002. The wetland plant and macroinvertebrate assemblages of temporary ponds in England and Wales. *Biological Conservation* 120: 261–278.

Oertli, B., Auderset Joye, D., Castella, E., Juge, R., Cambin, D. & Lachavanne, J-B. 2002. Does size matter? The relationship between pond area and biodiversity. *Biological Conservation* 104: 59–70.

Pesce, S. & Wunderlin, D.A. 2000. Use of water quality indices to verify the impact of Cordoba City (Argentina) on Suquia River. *Water Resources* 34(11): 2915–2926.

Ruggiero, A., Céréghino, R., Figuerola, J., Marty, P. & Angélibert, S. 2008. Farm ponds make a contribution to the biodiversity of aquatic insects in a French agricultural landscape. *Comptes Rendus Biologies* 331: 298–308.

Søndergaard, M., Jeppesen, E. & Jensen, J.P. 2005. Pond or lake: does it make any difference? *Archiv für Hydrobiologie* 162: 143–165.

Spruill, B.T., Showers, J.W. & Howe, S.S. 2002. Ground water quality application of classification-tree methods to identify nitrate sources in ground water. *Journal of Environmental Quality* 31(5): 15–38.

Stańczyk, T. & Jeznach, J. 2007. An attempt to establish a typology of small water reservoirs in the Left-Bank Warsaw Area. *Electronic Journal of Polish Agricultural Universities* 10(4). http://www.ejpau.media.pl/.

Stoianov, I., Chapra, S & Maksimovic, C. 2000. A framework linking urban park land use with pond water quality. *Urban Water* 2: 47–62.

Szulczewska, B. & Kaftan, J. 1996. *Urban natural system planning*. Warsaw: IGPiK.

Toto, L.G., Poikane, S., Penning, W.S., Free, G., Maemets, H., Kolada, A. & Hanganu, J. 2008. First steps in the Central-Baltic intercalibration exercise on lake macrophytes: where do we start? *Aquatic Ecology* 42: 265–275.

Vermonden, K, Leuven, R., Van Der Velde, G., Van Katwijk, M.M., Roelofs, J.G.M. & Hendriks, A.J. 2009. Urban drainage systems: An undervalued habitat for aquatic macroinvertebrates. *Biological Conservation* 142: 1105–1115.

Zhang, Y., Zhang, H., Gao, X. & Peng, B. 2007. Improved AHP method and its application in lake environmental comprehensive quality evaluation—a case study of Xuanwu Lake, Nanjing, China. *Chinese Journal of Oceanology and Limnology* 25(4): 427–433.

Environmental Engineering III – Pawłowski, Dudzińska & Pawłowski (eds)
© *2010 Taylor & Francis Group, London, ISBN 978-0-415-54882-3*

Removal of microcystin-LR from water by ozonation

D. Szczukocki, B. Macioszek & J. Dziegieć
Department of General and Inorganic Chemistry, University of Lodz, Lodz, Poland

ABSTRACT: In temperate climates, cyanobacteria (blue-green algae) occur most frequently in summer, when the demand for recreational water is highest. Blue-green algae can produce toxins and are also responsible for the taste and odour of water, which significantly impairs its quality. Cyanotoxins are very dangerous substances which can intoxicate hepatocytes and the nervous system in humans and animals. To prevent such situations, it is very important to remove cyanotoxins from water effectively during the pretreatment process.

In the present study, the ozonation of water containing microcystin-LR was tested. We performed this research at the laboratory scale as well as in a pretreatment plant near the Sulejow artificial lake during several seasons.

Keywords: Cyanobacterial toxins, microcystin, ozonation, water pre-treatment.

1   INTRODUCTION

Cyanobacteria (blue-green algae) are organisms that have some characteristics of bacteria and some of algae. They are similar to algae in size and, unlike other bacteria, they contain blue-green and green pigments and can perform photosynthesis. Pollution related to human activities, e.g. from agricultural runoff and inadequate sewage treatment, has led to excessive eutrophication (fertilization) of many water bodies. As a result, there is excessive proliferation of cyanobacteria in fresh water which has a considerable impact upon recreational water quality (as they are responsible for the taste and odour of water). In temperate climates, cyanobacterial dominance is most pronounced during the summer months, which coincides with the period when the demand for recreational water is the highest. The formation of cyanobacterial blooms is not a new phenomenon. The earliest reliable account of such blooms was at the end of the twelfth century (Ressom et al. 1994).

Blue-green algae produce several toxins, including neurotoxins, hepatotoxins, cylindrospermopsin and lipopolysaccharide endotoxins (Carmichael 1997). Toxic cyanobacteria are cosmopolitan. They have been recorded on every continent and about 50–75% of tested cyanobacterial blooms have been toxic (Codd 1995). Cyanobacteria have been implicated in various episodes of human and animal illnesses in Europe (Turner et al., 1990), North and South America (Billings 1981, Jochimsen et al. 1998), Asia (Yu 1989, Ueno et al. 1996), Africa (Zillberg 1966) and Australia (Falconer et al. 1983).

The type of cyanobacterial toxins most frequently found in fresh waters is microcystin-LR. This is a hepatotoxic peptide produced by a number of cyanobacterial genera, the most notable of which is the widespread

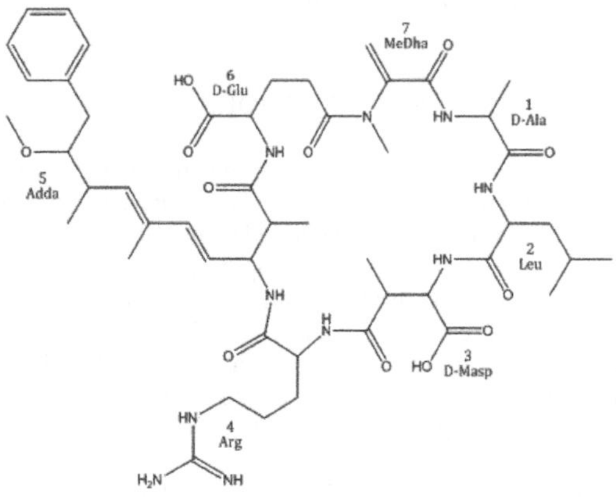

Figure 1.   The structure of microcystin-LR.

Microcystis, from which the toxins take their name. The microcystin-LR is a cyclic heptapeptide (Figure 1). The molecule consists of a seven-membered peptide ring which is made up of five non-protein amino acids and two protein amino acids: Leucine (L) and Arginine (R) in positions 2 and 4. The Adda side chain (position 5) is a key structural element necessary for biological activity. Separation of the Adda component from the cyclic peptide renders both components non-toxic (Carmichael 1992).

Many strategies for the removal of microcystin-LR from water have been investigated. Conventional methods of water treatment (sedimentation, filtration, coagulation) have been reported to be ineffective (Hoffman 1976, Himberg et al. 1989).

Activated carbon can remove microcystin-LR, but doses have to be higher than those generally used in water treatment (Falconer et al. 1989). Nicholson et al. (1994) have reported that microcystin-LR can be removed by chlorination, but unfortunately chlorination of organic compounds can also cause adverse health effects in humans. Chlorine dioxide is not effective at those doses used in drinking water treatment. Hydrogen peroxide was found to be ineffective in toxin removal; used alone or with UV radiation, it can remove only about 50% of microcystin-LR after 30 minutes (Rositano & Nicholson 1994). Ozone has been found to be most effective in oxidation of cell-bound microcystin, if it is applied at a sufficiently high dose and with a sufficiently long contact time. Dissolved air flotation has been proposed in which the recycled water is saturated with ozone-rich air (Baron et al. 1997). Ozone-rich air has also been proposed for dispersed air flotation. These approaches might result in the reduction of extracellular toxins as well as enhanced cell removal (Chorus & Bartram 1999).

This paper presents the results obtained by the ozonation of microcystin-LR in a water treatment plant near the Sulejow reservoir in Poland and at the laboratory scale. The studies concerned with the processes of water production were conducted during three seasons with different ozone concentrations. This is the first research work on such a scale.

2 MATERIALS AND METHODS

To remove mechanical, organic and inorganic pollutants, the water samples were first filtered using a membrane and then concentrated using the solid phase extraction method. After sample preconcentration, microcolumns were rinsed with methanol. The alcohol fraction was evaporated at inert gas flow at room temperature. After evaporation the samples were dissolved in acetonitrile–ammonium acetate buffer (74:26 v/v) filtered through a 0.45 mm filter and separated by RP–HPLC (Meriluoto 1997).

The concentration of ozone was determined by iodometric titration (Rakness et al. 1996).

3 RESULTS AND DISCUSSION

The changes in microcystin-LR concentrations (from an initial concentration of 16 mg/dm^3) during the ozonation process at the laboratory scale can be found in Table 1.

The total toxin content was oxidized after the addition of about 0.01 mg O_3 per 1 μg of microcystin-LR. This result suggests that about 220 ozone molecules combine with one molecule of microcystin-LR. It is highly probable that the peptide ring of the toxin molecule is broken, but the mechanism has not yet been identified. The changes in microcystin-LR concentrations in relation to different ozone levels are given in Figure 2.

The amount of oxidizing microcystin-LR per unit of time, along with the given speed of ozone addition, does not correlate with the microcystin-LR

Table 1. Changes in microcystin-LR concentrations (from an initial concentration of 16 mg/dm^3) during the ozonation process (laboratory scale).

Time of ozonation [s]	Ozone concentration in sample [mg/dm^3]	Mean yield of microcystin-LR residue [%]
0	0.00	100
5	0.06	78
10	0.11	38
11	0.12	32
12	0.13	25
13	0.14	17
14	0.15	11
15	0.16	5
16	0.17	0

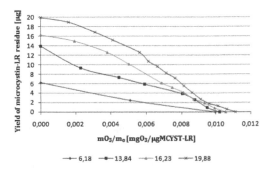

Figure 2. Changes in microcystin-LR concentrations in relation to ozone levels for different concentrations.

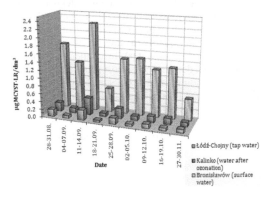

Figure 3. Concentrations of microcystin-LR in surface water, after ozonation in the Kalinko treatment plant, and in the Lodz-Chojny pumping station in the 2001 season.

concentrations in the solution. The microcystin-LR ozonation process is useful when cyanobacterial cells are removed.

The concentrations of microcystin-LR in surface water from the Sulejow artificial lake, after ozonation in a water treatment plant in Kalinko and at a final step in the Lodz-Chojny pumping station, are given in Figures 3, 4 and 5.

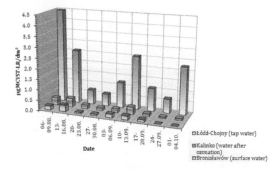

Figure 4. The concentrations of microcystin-LR in surface water, after ozonation in the Kalinko treatment plant, and in the Lodz-Chojny pumping station in the 2002 season.

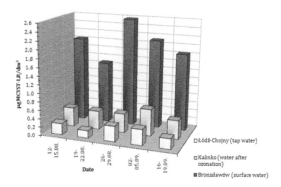

Figure 5. Concentrations of microcystin-LR in surface water, after ozonation in the Kalinko treatment plant, and in the Lodz-Chojny pumping station in the 2003 season.

Table 2. Doses of ozone used in the treatment system in Kalinko.

| Season | Ozone doses [g/m³] | | Mean yield of ozonation [%] |
	Min.	Max.	
2001	1.5	3.0	90.8
2002	1.4	3.5	92.1
2003	1.3	1.7	85.7

The ozone doses used in the treatment system in the Kalinko station can be found in Table 2.

Toxic cyanobacterial blooms were observed between 2001 and 2003. Microcystin-LR was found in surface water samples from the Sulejow reservoir even two months after the blooms had decayed.

4 CONCLUSIONS

The final efficiency of microcystin-LR removal in the ozonation process ranges between 82% and 100%; this results in concentrations lower than the maximum recommended by WHO ($1\,\mu g/dm^3$). This efficiency can be obtained using ozone doses of $1.3–3.5\,g\,O_3/m^3$ and a contact time of 35–48 mins.

After the ozonization process, water did not exhibit toxic properties. In spite of that, the products formed in the ozonolysis reaction may have adverse health effects. The toxicity of such substances is still under investigation.

Using a combination of ozone and UV light is possible. This method may enhance the efficiency of microcystin-LR removal but it has some disadvantages, e.g. it is a highly energy-consuming process. The application of such a technique may result in a higher cost for water production.

ACKNOWLEDGEMENTS

This study was supported by a grant from the Ministry of Faculty Science, Poland (project number N523 036 32/1348).

REFERENCES

Baron, J., Ionesco, N.M. & Bacquet, G 1997. Combining flotation and ozonation the Flottazone process. In: *Dissolved air flotation. Proceedings of an international conference.* Chartered Institution of Water and Environment Management. London.

Billings, W.H. 1981. Water-associated human illness in northeast Pennsylvania and its suspected association with blue-green algae blooms. In W.W. Carmichael (ed.), *The water environment: Algal toxins and health:* 243–255. Plenum Press: New York.

Carmichael, W.W. 1997. The cyanotoxins. In Callow J.A. (ed.), *Advances in botanical research.* Academic Press Inc.: San Diego 27: 211–256.

Carmichael W.W. 1992. Cyanobacteria secondary metabolites – the cyanotoxins. *J. Appl. Bacteriol.,* 72: 445–459.

Chorus, I. & Bartram, J. 1999. *Toxic cyanobacteria in water: A guide to their public health consequences, monitoring and management.,* WHO. London and New York.

Codd, G.A. 1995. Cyanobacterial toxins: Occurrence, properties and biological significance. *Water Sci. Technol.* 32(4): 149–156.

Falconer, I.R., Beresford, A.M. & Runnegar, M.T.C. 1983. Evidence of liver damage by toxin from a bloom of the blue-green alga. *Microcystis aeruginosa, Med. J. Aust.* 1: 511–514.

Falconer I.R., Runnegar M.T.C., Buckley T., Huyn V.L., Bradshaw P., 1989, Using activated carbon to remove toxicity from drinking water containing cyanobacterial blooms, *J. Am. Water Works Assoc.,* vol. 81, no. 2, pp. 102–105.

Himberg, K., Keijola, A-M., Hiisvirta, L., Pyysalo, H. & Sivonen, K. 1989. The effect of water treatment processes on the removal of hepatotoxins from Microcystis and Oscillatoria cyanobacteria: a laboratory study. *Water Res.* 23(8): 979–984.

Hoffmann, J.R.H. 1976. Removal of Microcystis toxins in water purification processes. *Water S. Afr.* 2(2): 58–60.

Jochimsen, E.M., Carmichael, W.W., An, J.S., Cardo, D.M., Cookson, S.T., Holmes, C.E.M., Antunes, M.B.D., De Melo, D.A., Lyra, T.M., Baretto, V.S.T., Azevedo, S.M.F.O. & Jarvis, W.R. 1998. Liver failure and death after exposure to microcystins at a hemodialysis center in Brazil. *N. Engl. J. Med.* 338: 873–878.

Meriluoto J. 1997. Chromatography of microcystins. *Anal. Chim. Acta* 352: 277–298.

Nicholson, B.C., Rositano, J. & Burch, M.D. 1994. Destruction of cyanobacterial peptide hepatotoxins by chlorine and chloramines. *Water Res.* 28(6): 1297–1303.

Rakness, K., Gordon, G., Langlais, B., Masschelein, W., Matsumoto, N., Richard, Y., Robson, M. & Somiya, I. 1996. Guideline for measurement of ozone concentration in the process gas from an ozone integrator. *Ozone Sci. Eng.* 18: 209–229.

Ressom R., Soong F.S., Fitzgerald J., Turczynowicz L., El Saadi O., Roder D.M., Maynard T. & Falconer I.R. 1994. *Health effects of toxic cyanobacteria (blue-green algae).* Australian Government Publishing Service: Canberra.

Rositano, J. & Nicholson, B.C. 1994. *Water treatment techniques for removal of cyanobacterial toxins from water.* Australian Centre for Water Quality Research, Salisbury.

Turner P.C., Gammie A.J., Hollinrake K., Codd G.A., 1990, Pneumonia associated with contact with cyanobacteria, *Br. Med. J.*, vol. 300, pp. 1440–1441.

Ueno, Y., Nagata, S., Tsutsumi, T., Hasegawa, A., Watanabe, M.F., Park, H-D., Chen, G-C., Chen, G. & Yu, S-Z., 1996, Detection of microcystins, a blue-green algal hepatotoxin, in drinking water sampled in Haimen and Fusui, endemic areas of primary liver cancer in China, by highly sensitive immunoassay. *Carcinogenesis* 17(6): 1317–1321.

Yu, S-Z. 1989. Drinking water and primary liver cancer. In Z-Y. Tang, M-C.Wu, S-S.Xia (eds), *Primary liver cancer: 30–37.* Springer-Verlag: Beijing.

Zillberg, B. 1966. Gastroenteritis in Salisbury European children – a five year study. *Centr. Afr. J. Med.* 12: 164–168.

Environmental Engineering III – Pawłowski, Dudzińska & Pawłowski (eds)
© *2010 Taylor & Francis Group, London, ISBN 978-0-415-54882-3*

Method of identification of operational states of water supply system

B. Tchórzewska-Cieślak & J. Rak
Faculty of Civil and Environmental Engineering, Rzeszow University of Technology, Rzeszow, Poland

ABSTRACT: A water supply system (WSS) is one of the most basic technological underground systems and is highly important for the livelihood and health of humans. In WSS operations, we deal with events that can cause breaks in water supply or water pollution. This system is characterised by continuous operation during which fluctuations in water supply can occur. This study aims to present a method for identifying the operational states of a WSS. This paper contains a description of system identification on the basis of either a single feature or many features. Application examples of the presented methods have been developed.

Keywords: Water supply system, dependability state, method of identification.

1 INTRODUCTION

The subject and main purpose of this study is to present methods of identification of operational states of a water supply system (WSS). The events of September 11, 2001 have significantly changed the approach to water management utilities. Previously, the consideration of terrorist threats to our nation's drinking water supply was minimal. Now, however, new trends in the technical management of WSSs are heavily focused on security issues related to the design, operation and management of these systems. Of particular interest in the future will be the evaluation of risk and reliability issues of the various components, subsystems and the systems as a whole, from the viewpoint of their susceptibility to terrorism (Ozger & Mays 2004).

A WSS is characterised by its spatial extensiveness and complexity to fulfil its main task to supply consumers with drinking water. It has many operational states during its continuous operation. WSS operations are conducted to process and service the drinking water system, and to connect the water supply to municipal water-pipe line users (consumers). The aim of WSS operations is to fulfil the expectations of the involved parties: waterworks companies and water consumers. The measure of the fulfilment of these expectations is water quality and quantity. Practice shows that most WSS failures are accompanied by a transient state, which is a partial system fault. A set of internal features of the WSS, described by interconnections of the processes taking place in its subsystems, is defined as a state of the system at time *t* (Shinstine et al. 2002, Tanyimboh et al. 1999). The following dependability

states can be identified (Rak & Tchórzewska-Cieslak 2007, Tchórzewska-Cieslak 2007):

– up state (dependability)
– inefficiency (failure) and an unreliability state:
– down state (fault)

For the purposes of this paper, the following notions have been defined:

• Up state: A set of features describing the system's ability to fulfil operational tasks. The system is in its up state if a random variable X describing its state takes value $X = 1$, and a set of features (C) is included in a set of requirements (S), which is written as $C \in S$. System operational dependability is $R = P(X = 1)$.

• A state of system inefficiency is characterised by the destructive processes that occur during operation. In this state, changes in operating characteristics take place, despite the fact that the operational potential of the system is maintained.

• A down state is characterised by a lack of possibilities to fulfil dependability requirements – that is, to supply a suitable amount of drinking water, with a required pressure and quality, at any time convenient to a customer, and at an acceptable price. The system is in a down state if a random variable X takes the value $X = 0$, and at least one feature does not fulfil the requirements, which is written as $C \notin S$. Unreliability of system functioning is $Q = P(X = 0)$.

- The first type of error means that the system is in an up state and an operator qualifies it as in a down state.
- The second type of error means that the system is in a down state and an operator qualifies it as in an up state.

2 MATERIALS AND METHODS

2.1 A method of identification of the state of a water supply system

To determine the state of the system, the values of the features compared with standard requirements need to be known (Fiok et al. 1998). The values of features are obtained by means of system monitoring. For the WSS, the features are water quality and quantity, which are understood as a supply. As a rule, monitoring is managed by a system operator who also qualifies the measurement data. It is assumed that: $R = P(Y = 1)$ – system dependability qualified by an operator $Q = P(Y = 0)$ – system unreliability qualified by an operator A measure of the first type of error is the probability of an event:

$$\alpha = P(Y = 0/X = 1) \tag{1}$$

A measure of the second type of error is the probability of an event:

$$\beta = P(Y = 1/X = 0) \tag{2}$$

Let D be an experience with the probable results $C_1, C_2, \ldots C_n$, which is executed with the probabilities $P_1, P_2 \ldots P_n$, then the information received as a result of experience D is a random variable that takes values $-\log P_i$, $i = 1, 2 \ldots n$ on a set C_i. An expected value of this information is called the experience entropy H(D) and is:

$$H(D) = -\sum_{i=1}^{n} P_j \log P_j \tag{3}$$

In some sense, the experience entropy is a measure of its indefiniteness and it can be used to control the states of the system. The more equal the probabilities of states, the bigger the entropy; the higher the difference in the probabilities of states, the lower the entropy (Bishop 2006). If the given event occurs with a probability of 1, then, as results from a simple substitution, the entropy is 0, because it is known in advance what will happen – uncertainty does not exist. The entropy (H), which is a measure of indefiniteness of system states, is calculated from the formula:

$$H = \sum_{i=1}^{n} P_i \cdot \ln \frac{1}{P_i} \tag{4}$$

It should be noticed that if $P_i = 1$, then $H = 0$, and if the probabilities P_i are equal and are $1/n$, then $H = \ln n$, obtaining the maximal value.

For the *a priori* probability P_i of a state of system, the following formula is valid:

$$\sum_{i=1}^{n} P_i = 1 \tag{5}$$

where: n = a number of system states.

A measure of quantity of information is the initial and final difference of entropy, which is written as:

$$I = H_0 - H \tag{6}$$

where: H_0 = entropy of a state of a given parameter until the system is controlled: H = entropy of a state of a given parameter after the system is controlled. The effectiveness of controlling the system state is:

$$E = \frac{H_0 - H}{H_0} \tag{7}$$

The control effectiveness index is in a range: $0 \leq E \geq 1$. For an ideal control, $H = 0$, which implicates $E = 1$; however, if the entropy does not change during the process of control, $H_0 = H_n$, then $E = 0$.

A system is characterised by a feature C, for which requirements are formulated. A random variable X that is controlled, was defined as:

$$X = \begin{cases} 1, \text{ when } C \in W \\ 0, \text{ when } C \notin W \end{cases} \tag{8}$$

As a result of control, C is identified as C^*, and a corresponding random value Y is defined in the following way (Newby 2008):

$$Y = \begin{cases} 1, \text{ when } C^* \in W \\ 0, \text{ when } C^* \notin W \end{cases} \tag{9}$$

As a result of control of C, the following events described by the conditional probabilities can occur:

$$P(Y = 1/X = 1) = P(C^* \in W/C \in W) = 1 - \alpha. \tag{10}$$

$$P(Y = 0/X = 1) = P(C^* \notin W/C \in W) = \alpha \tag{11}$$

$$P(Y = 1/X = 0) = P(C^* \in W/C \notin W) = \beta \tag{12}$$

$$P(Y = 0/X = 0) = P(C^* \notin W/C \in W) = 1 - \beta \tag{13}$$

α = the first type of error, a feature that fulfils requirements (true) was qualified as a feature that does not fulfil requirements (false) – a true hypothesis was rejected.

β = the second type of error, a feature that does not fulfil requirements (false) was qualified as a feature that fulfils requirements (true) – a false hypothesis was taken. Assuming the following notations of the probabilities: $P(X = 1) = R$ and $P(X = 0) = 1 - R$, we obtain:

$$P(X = 1/Y = 1) = P_{11} = R(1 - \alpha) \tag{14}$$

$$P(X = 1/Y = 0) = P_{10} = R \cdot \alpha \qquad (15)$$

$$P(X = 0/Y = 1) = P_{01} = (1-R) \cdot \beta \qquad (16)$$

$$P(X = 0/Y = 0) = P_{00} = (1-R) \cdot (1-\beta) \qquad (17)$$

The probabilities P(Y) are determined from the formulas:

$$r = P(Y = 1) = P_{11} + P_{01} = R(1-\alpha) + (1-R) \cdot \beta = \\ = \beta + R(1-\alpha-\beta) \qquad (18)$$

$$q = P(Y = 0) = P_{10} + P_{00} = R\alpha + (1-R) \cdot (1-\beta) = \\ = 1 - \beta - R(1-\alpha-\beta) \qquad (19)$$

The formulas (18) and (19) allow determination of the control effectiveness index E when the first type and the second type of errors occur. The initial entropy of a feature is:

$$H(X) = -[R \cdot \ln \cdot R + (1-R)\ln(1-R)] \qquad (20)$$

$$H(Y) = -[r \cdot \ln r + (1-r)\ln(1-r)] \qquad (21)$$

$r = P(Y = 1)$ – is calculated from the formula (18) Next it was calculated as:

$$H(X/Y) = -[P_{11} \cdot \ln P_{11} + P_{10} \ln P_{10} + P_{01} \ln P_{01} + P_{00} \ln P_{00}] \qquad (22)$$

The increase of entropy conditioned by the first type and the second type of errors is:

$$\Delta H = H(X/Y) - H(Y) \qquad (23)$$

The control effectiveness index is:

$$E = \frac{H(X) - \Delta H}{H_0(X)} \qquad (24)$$

Assuming the probabilities of the first and the second type of errors $\alpha = 0$ and $\beta = 0$, we obtain: $H(Y) = H(X)$ and $H(X/Y) = H(X)$, hence: $\Delta H = 0$, and the control effectiveness index $E = 1$.

The larger the first type of error, the smaller the function of control effectiveness.

2.2 Single system identification

A method concerns a single system identification for one feature (Fiok et al., 1998). The system can be found in two states:

$$P(C \in S) = P(X=1) = R \qquad (25)$$

$$P(C \notin S) = P(X=0) = Q \qquad (26)$$

$$Q + R = 1 \qquad (27)$$

The *a posteriori* probability is determined in the following way:

$$P(Y=1/X=1) = 1 - \alpha \qquad (28)$$

$$P(Y=1/X=0) = \beta \qquad (29)$$

$$P(Y=0/X=1) = \alpha \qquad (30)$$

$$P(Y=0/X=0) = 1 - \beta \qquad (31)$$

$$P_{11} = P(X=1/Y=1) = R(1 - \alpha) \qquad (32)$$

$$P_{01} = P(X=0/Y=1) = Q \cdot \beta \qquad (33)$$

$$P_{10} = P(X=1/Y=0) = R \cdot \alpha \qquad (34)$$

$$P_{00} = P(X=0/Y=0) = Q(1 - \beta) \qquad (35)$$

$$r = P(Y=1) = P_{11} + P_{01} = R(1 - \alpha) + (1 - R)\beta \qquad (36)$$

$$q = P(Y=0) = P_{10} + P_{00} = R\alpha + (1 - R)(1 - \beta) \qquad (37)$$

Finally, we obtained:

$$r_{11} = \frac{P_{11}}{r} = \frac{R \cdot (1-\alpha)}{R(1-\alpha) + (1-R) \cdot \beta} \qquad (38)$$

$$r_{00} = \frac{P_{00}}{q} = \frac{(1-R) \cdot (1-\beta)}{R \cdot \alpha + (1-R) \cdot (1-\beta)} \qquad (39)$$

$$q_{01} = \frac{P_{01}}{r} = \frac{(1-R) \cdot \beta}{R(1-\alpha) + (1-R)\beta} \qquad (40)$$

$$q_{10} = \frac{P_{10}}{q} = \frac{R \cdot \alpha}{R \cdot \alpha + (1-R) \cdot (1-\beta)} \qquad (41)$$

2.3 System identification of a state of system based on many features

A system is controlled by many qualitative features that were marked as $X_1, X_2, \ldots X_{n-1}, X_n$. For each feature, the system is qualified as being in an up state or in a down state. We assumed that the system is in an up state if it is in an up state for every separate feature X_i. When the features are independent, the probability that the system is in an up state is:

$$R = \prod_{i=1}^{n} P(X_i = 1) = \prod_{i=1}^{n} R_i \qquad (42)$$

The first type of error (αs) and the second type of error (β_s), for many features, are determined from dependences:

$$\alpha_s = 1 - \prod_{i=1}^{n}(1-\alpha_i) \qquad (43)$$

$$\beta_s = \frac{r_s - R_s(1-\alpha_s)}{1 - R_s} \qquad (44)$$

where:

$$r_s = \prod_{i=1}^{n} r_i = \prod_{i=1}^{n}[R_i(1-\alpha_i) + \beta_i(1-R_i)] \qquad (45)$$

$$r_i = R_i(1-\alpha_i)+\beta_i(1-R_i) \qquad (46)$$

When an operator qualifies a system as being in an up state (dependability), it is written as $r_s = P(Y_1 = 1, Y_2 = 1, ..., Y_n = 1)$.

For many features, a new random variable was assumed:

$$Z = \prod_{i=1}^{n} X_i \qquad (47)$$

which is defined as:

$$Z = \begin{cases} 1 \text{ - when system is in up \quad state} \\ 0 \text{ - when system is in down state} \end{cases}$$

The following dependences are valid:

$$R_s = P(Z = 1) \qquad (48)$$

$$Q_s = 1 - R_s = P(Z = 0) \qquad (49)$$

Analogically to Y for a single feature, a new random variable W for many features was defined:

$$W = \begin{cases} 1 \text{ - when an operator qualifies system as in up state} \\ 0 \text{ - when an operator qualifies system as in down state} \end{cases}$$

After appropriate transformations, the following final probabilities were obtained:

- the probability that the system qualified by an operator as being in an up state is, in reality, in an up state :

$$r_{11} = P(Z = 1/W = 1) = \frac{R_s(1-\alpha_s)}{R_s(1-\alpha_s)+(1-R_s)\beta_s} \qquad (50)$$

- the probability that the system qualified by an operator as being in an up state is, in reality, in a down state:

$$q_{01} = P(Z = 0/W = 1) = \frac{(1-R_s)\beta_s}{R_s(1-\alpha_s)+(1-R_s)\beta_s} \qquad (51)$$

- the probability that the system qualified by an operator as being in a down state is, in reality, in a down state:

$$r_{00} = P(Z = 0/W = 0) = \frac{(1-R_s)\cdot(1-\beta_s)}{R_s\cdot\alpha_s+(1-R_s)\cdot(1-\beta_s)} \qquad (52)$$

- the probability that the system qualified by an operator as being in a down state is, in reality, in an up state:

$$q_{10} = P(Z = 1/W = 0) = \frac{R_s\cdot\alpha_s}{R_s\cdot\alpha_s+(1-R_s)\cdot(1-\beta_s)} \qquad (53)$$

- On the basis of historical data, it was estimated that the probability to qualify treated water that fulfils standard requirements as water that is not suitable for use, based on a single water quality index monitoring, is $\alpha = 0.005$, and the probability that bad quality water will be, by mistake, qualified as good water is $\beta = 0.008$.

The availability index that characterises dependability of water treatment plant (WTP) operation is $R = K = 0.955$. If it was found, based on monitoring, that water fulfils standard requirements, then the probability that treated water, in reality, is suitable for consumption is:

$$r_{11} = \frac{0.955\cdot(1-0,005)}{0.008\cdot(1-0.955)+0.955\cdot(1-0,005)} = $$
$$= 0,9996212$$

The probability that treated water, in reality, is not suitable for consumption is:

$$q_{01} = \frac{(1-0.955)0.008}{0.008(1-0.955)+0.955(1-0.005)} = $$
$$= 3,78714\cdot10^{-4}$$

If it was found, based on monitoring, that water does not fulfil the standard requirements, then the probability that, in reality, water is not suitable for consumption is:

$$r_{00} = \frac{(1-0.955)\cdot(1-0.008)}{0.955\cdot0.005+(1-0.955)\cdot(1-0.008)} = $$
$$= 0.9033694$$

The probability that, in reality, water is suitable for consumption is

$$q_{10} = \frac{0,955\cdot0,005}{0,955\cdot0,005+(1-0,955)\cdot(1-0,008)} = $$
$$= 0,0966305$$

When the *a priori* probability K (the availability index) is not known, we assume that $Q = K = 0.5$ (dependability = unreliability). Then, the particular probabilities take the following values:
$r_{11} = 0.9920239$
$q_{01} = 7.97607\cdot10^{-3}$
$r_{00} = 5.01504\cdot10^{-3}$
$q_{10} = 0.005$

Comparing the probabilities with known K with the probabilities for K = 0.5, significant differences are found. More reliable are the probability values when the availability index is known. The calculations of control effectiveness are as follows:
For data R = 0.955, $\alpha = 0.005$, $\beta = 0.008$

According to the formulas (32), (33), (34), (35), (36) and (37), the particular probability values were calculated as:
$P_{11} = 0.950225$,
$P_{10} = 0.004775$,

Table 1. The results of calculations for the example.

α_1	α_2	β_1	β_2	R_1	R_2	R_s	r_1	r_2	r_s	αs	βs	r_{11}	q_{01}	r_{00}	q_{10}
0.05	0.10	0.05	0.05	0.95	0.9	0.855	0.907	0.815	0.74	0.145	0.059	0.988	0.012	0.524	0.476
0.10	0.05	0.02	0.1	0.9	0.8	0.72	0.815	0.764	0.623	0.145	0.0252	0.989	0.011	0.723	0.276
0.15	0.1	0.05	0.15	0.99	0.9	0.891	0.843	0.815	0.687	0.235	0.0461	0.993	0.007	0.332	0.668
0.2	0.1	0.08	0.20	0.9	0.9	0.810	0.730	0.818	0.597	0.28	0.0734	0.977	0.023	0.4302	0.562

Table 2. Calculations of control effectiveness.

αs	βs	R_s	$1-R_s$	P_{11}	P_{10}	P_{01}	P_{00}	r_s	q
0.145	0.059	0.855	0.145	0.731	0.124	0.009	0.136	0.740	0.260

H(X)	H(Y)	H(X/Y)	ΔH	E
0.413937	0.573496	0.800363	0.226868	0.451927

$P_{01} = 0.00036$,
$P_{00} = 0.04464$,
$r = 0.950585$,
$q = 0.049415$

Finally, the characteristics of entropy and effectiveness, according to the formulas, are:

H(X) = 0.183521,
H(Y) = 0.196789,
H(X/Y) = 0.21568,
$\Delta H = 0.018891$,
E = 0.897062

The control effectiveness value about 90% can be regarded as sufficient.

- The effectiveness of WTP operation, when underground water is a source, is identified by means of two features (index of iron and manganese content). The analyses from 5 years has shown that iron concentration in treated water fulfils the standard requirements, with a probability of $K_1 = P(X_1 = 1) = 0.95$.

The probability that manganese concentration in treated water fulfils the standard requirements is $R_z = P(X_2 = 1) = 0.9$.

The probability that WTP fulfils the requirements concerning both indices, according to formula (48), is:

$R_s = 0.95 \cdot 0.90 = 0.855$

Estimations of the first type and the second type of errors gave the following results:

$\alpha_1 = 0.05$, $\alpha_2 = 0.1$, $\beta_1 = 0.05$, $\beta_2 = 0.05$

The probabilities that, in the process of water quality identification, an operator qualifies water, for the given index, as drinkable are:

$r_1 = P(Y_1 = 1) = 0.95 \, (1-0.05) + 0.1(1-0.95) = 0.9075$
$r_2 = P(Y_2 = 1) = 0.9(1-0.1) + 0.05(1-0.9) = 0.815$

According to the formula (45), we obtain:
$r_s = $
$0.9075 \cdot 0.815 = 0.7396$

Using the formulas (43) and (44), the first type and the second type of errors were determined as:

$\alpha_s = 1 - (1-0.05) \cdot (1-0.1) = 0.145$

$\beta_s = \dfrac{0.7396 - 0.855 \cdot (1-0.145)}{1-0.855} = 0.059$

Finally, according to the formulas (50), (51), (52) and (53), the following numerical values of the probabilities were determined:

$r_{11} = \dfrac{0.855 \cdot (1-0.145)}{0.855(1-0.145) + (1-0.855) \cdot 0.059} = 0.9884$

$q_{01} = \dfrac{(1-0.855) \cdot 0.059}{0.855(1-0.145) + (1-0.855) \cdot 0.059} = 0.0116$

$r_{00} = \dfrac{(1-0.855) \cdot (1-0.059)}{0.855 \cdot 0.145 + (1-0.855) \cdot (1-0.059)} = 0.5239$

$q_{10} = \dfrac{0.855 \cdot 0.145}{0.855 \cdot 0.145 + (1-0.855) \cdot (1-0.059)} = 0.4761$

The results of calculations for various values ($\alpha_1, \alpha_2, \beta_1, \beta_2, R_1, R_2$) are given in Table 1.

In Table 2, the results of calculations of control effectiveness are given.

3 CONCLUSIONS

- Water supply systems belong to the critical infrastructure, and the dependability and safety of such systems have a direct influence on the quality of life for water consumers. During operation, the system should be continuously controlled, by means of different monitoring systems, in order to protect water consumers from the consequences of its unreliability.

- Water supply systems operate in a continuous way for a long period of time. Subsystems and elements that build it also operate intensively. The damage of an element usually results in it being eliminated it from the system, in order to repair the failure. This is why the correct identification of the state of the particular elements, subsystems and the WSS as a whole is so important. It is especially important in the decision-making process of a system operator.
- In many cases, making the second type of error results in much bigger losses than making the first type of error. When treated water is not allowed to flow into the distribution subsystem, despite the fact that it fulfils standard requirements, it results in financial losses for waterworks, caused by a lack of sales. However, when bad-quality water is allowed to flow into the distribution subsystem as a result of the operator's decision to consider it as good water, it can lead to mass public poisoning and even a threat to lives, which would be catastrophic.

REFERENCES

Bishop, Ch. 2006. *Pattern recognition and machine learning.* Springer: New York 2006.

Fiok, A., Grabsk, F. & Jaźwiński, J. 1998. Bayesian classification in the identification of the technical object state. *Proceeding of the 10th International Symposium on Development in Digital Measuring Instrumentation – ISDDMP'98.* Naples, September 17–18(I): 22–25.

Newby, M. 2008. Monitoring and maintenance of spares and one shot devices. *Reliability Engineering & System Safety* 93(4): 588–589.

TchórzewskA-Cieślak, B. 2007. Method of assessing risk of failure in water supply system. *Proceedings of the European safety and reliability conference ESREL, Norway, Stavanger. Risk, reliability and societal safety.* Taylor & Francis 2: 1535–1539.

Rak, J. & Tchórzewska-Cieślak, B. 2007. Czynniki ryzyka w eksploatacji systemów zaopatrzenia w wodę. *Oficyna Wydawnicza Politechniki Rzeszowskiej.*

Ozger, S. & Mays, L.W. 2004. Optimal location of isolation valves in water distribution systems a reliability/optimization approach. www.public.asu.edu/~lwmays/Ch07_Mays_144381-9.pdf.

Shinstine, D. S., Ahmed, I. & Lansey, K. 2002. Reliability/availability analysis of municipal water distribution networks: case studies. *Journal of Water Resources Planning and Management.* ASCE 128(2): 140–151.

Tanyimboh, T. T., Burd, R., Burrows, R. & Tabesh, M. 1999. Modelling and reliability analysis of water distribution systems. *Water Science Technology* 39(2): 4249–255.

Environmental Engineering III – Pawłowski, Dudzińska & Pawłowski (eds)
© 2010 Taylor & Francis Group, London, ISBN 978-0-415-54882-3

Water consumer safety in water distribution system

B. Tchórzewska-Cieślak
Department of Water Supply and Sewage Sludge, Rzeszow University of Technology, Rzeszow, Poland

ABSTRACT: In water distribution system operations, events such as failures in the water-pipe network can cause breaks in water supply or water pollution. People's notion of 'safety' is related to a guarantee that threat will be eliminated or minimised. For drinking water consumers, safety means the probability of avoiding threats resulting from consuming water that does not meet quality standards, or from lack of water. This paper presents the problem of relating water consumer safety to water distribution systems. It gives a definition of water consumer safety, and a method for its assessment on the basis of risk analysis.

Keywords: Water consumer safety, risk, water distribution system.

1 INTRODUCTION

In today's culture, there is an interest in problems connected with municipal systems' safety management, especially when a critical situation occurs. A water distribution system (WDS) affects the quality of water, which can undergo significant changes as it travels through the water distribution system from the point of supply and/or treatment to the point of delivery (Mays 2005). A water distribution system is one of the most important technical systems which belong to the so-called critical infrastructure of cities. Its aim is to supply consumers with a required amount of water, with a specific pressure and a specific quality, according to binding standards, and at an acceptable price. The programmes of the World Health Organization (WHO) concerning the sanitary quality of drinking water have been executed since the 1950s. In 1958, the WHO presented the first publication on that subject entitled "International standards for drinking water", which was then updated in 1963 and 1971. Since the 1980s, the WHO has not proposed any specific standard values but has presented and documented estimates of risk associated with drinking water contamination from microbiological, chemical and radiochemical elements, recommending values which do not create health threats or for which threats are marginal (WHO/SDE/WSH 2002).

For purposes of this paper, operational reliability of the WDS is defined as the ability to supply a constant flow of water for various groups of consumers, with a specific quality and a specific pressure, according to consumer demands, in specific operational conditions, at any or at a specific time. Safety of the WDS means the ability of the system to safely execute its functions in a given environment. The measure of WDS safety is risk. The size of risk (r) is a function of the probability that the undesirable event (failure) will occur and of the magnitude of the resulting consequences. This basic dependence describing risk can be expanded by the third parameter, the so-called 'protection against failure'. Failure is defined as the event in which the system fails to function with respect to its desired objectives. Failure can be grouped into either structural failure or performance failure. A structural failure, such as pipe breakage or a pump failure, can cause demands not to be met (Mays 2005, Tanyimboh et al. 1999).

Failure in a water-pipe network is a complex problem; every time it occurs, the primary reasons behind it must be analysed carefully. Consider the example of failure connected with pipeline corrosion. Possible reasons could be the following: ground corrosivity, a lack of anticorrosion protection (passive and active), or water corrosivity. The basic risk factors in the design process are as follows: an incorrect route for the water pipeline; errors in network hydraulic calculations (e.g. suitable water flow speeds are not ensured in network redimensioning); errors in construction such as a deviation from the design, a low quality of workmanship, or the network being constructed in a manner that is inconsistent with technological directives; errors in operation such as a lack of water pipeline operation monitoring, incomplete records of data about failures, or lack of a programme of risk management (Gardner 2008, Vreeburg et al. 1997). The factors that create the probability that negative consequences will occur are the following: the probability that the undesirable event will occur, frequency and degree of exposure, and the possibility of avoiding or minimising the negative consequences. Consequences connected with the occurrence of undesirable events can be divided into two groups (Shinstine et al. 2002):

• consequences for the waterworks, connected with breaks or lack of water supply, costs of restoring the WDS (e.g. failure repair, network disinfections, compensation payments) etc.

- consequences for the consumer and environment, such as the possibility of loss of the health or lives of water consumers, hygienic and sanitary inconvenience, and environmental damage.

The aim of the analysis of WDS safety is to create a basis of substantial criteria and information, which is needed in the decision-making process, in process optimisation, in system operation and control, and when protective actions preventing the occurrence of unfavourable consequences are undertaken.

The main objective of this paper is to present the issue of water consumer safety in the WDS. The paper explores the basic concepts related to water consumer safety and presents a method for its analysis, on the basis of risk connected with WDS operation.

2 MATERIALS AND METHODS

2.1 Water consumer safety

As a result of various types of undesirable events in the WDS, consumers can suffer a lack of running drinkable water for a few hours, days, or even weeks. The most common undesirable events that occur in the WDS are failures in the water-pipe network (Hipel et al. 2003). Such failures can be caused by errors made during design, construction and operation. In a water-pipe network, failures can occur in water-pipe connections, and in distributional pipes and water mains; secondary water contamination in the water-pipe network can also occur. The basic causes of such events are actions of the forces of nature; deliberate or incidental actions of a third party; defects in materials; industrial and building catastrophes; influences of the ground and water environment; changes in temperature; a low speed of water flow in the water-pipe network; the age of pipes; and human error made by the system operator (Pollard et al. 2004). For an accurate analysis of risk connected with a WDS operation, appropriate amounts of different information, and data recording and processing are necessary.

Risk analysis connected with the WDS operation is often performed in so-called 'uncertain information conditions', which are connected with uncertain (incomplete, imprecise or unreliable) data about subsystem operation. The uncertainty of such data is composed of many elements. Some of these are determined on the basis of the data's distribution, characterised by a standard deviation. The remaining elements are estimated on the basis of the assumed probability distribution, known from experience, or from other information. The following data, among others, are necessary to perform risk analysis in the WDS:

- data identifying the analysed object (e.g. water treatment station, distribution pipeline), the name and type of the object and its basic technical data,
- data about failures (undesirable events), repairs and other breaks in the WDS's operation (information

about the date, time and duration of failure, and a description of the failure),
- data relating to the reasons behind the occurrence of undesirable events,
- data relating to the consequences of these events.

The sources of the data needed to analyse risk are:

- data collected from waterworks about the WDS's operation,
- measurement data (e.g. measurements of pressure and water flow in the water-pipe network, measurements of water leaks in the water-pipe network),
- data collected from experts.

For the needs of consumer safety analysis, the concept of 'Current Level of Safety' (CLS) is introduced. This means the level of water consumer safety under the given WDS operating conditions. The following levels of consumer safety have been distinguished (Rak 2005):

- Tolerable Level of Safety (TLS) – a level of water consumer safety in which there are no threats to water consumers' health or lives, as a result of WDS operation.
- Controlled Level of Safety (CoLS) – a level of water consumer safety in which threats to water consumers' health can occur, as a result of a WSS operation, but there are sufficient protective barriers for drinking water consumers.
- Unacceptable Level of Safety (ULS) – a level of water consumer safety which, if exceeded, is a threat to consumers' health or lives, as a result of WSS operation.

In order to analyse water consumer safety, it is necessary to define the WDS safety conditions. The following WDS safety conditions are proposed:

- Full Safety Condition (FSC) – the WDS condition which is characterised by a failure-free operation. The system functions according to the binding legal regulations and consumer expectations. For water quantity, the nominal water production capacity Q_{dn} is higher than the maximum water daily consumption Q_{dmax}: $Q_{dn} \geq Q_{dmax}$, and quality (drinking water fulfils the requirements of binding regulations). The WSS does not threaten water consumers' health or lives: $CLS \geq TLS$.
- Limited Safety Condition (LiSC) – the WDS condition which is characterised by short disturbances in the WSS operation, $0.3Q_{dmax} \leq Q_{dn} < 0.7Q_{dmax}$, or breaks in water supply of up to 24 hours. In LiSC, there is the potential for undesirable events to escalate (the so-called 'domino effect' can occur). If pollution can be removed within 30 days, the appropriate Sanitary Inspector certifies drinking water usefulness under an accepted deviation (certifies water as acceptable to drink, under certain conditions (e.g. if it is boiled first). $CoLS \leq CLS < TLS$.
- Loss of Safety Condition (LSC) – the WDS is faulty, $0 < Q_{dn} < 0.3\, Q_{dmax}$, or breaks in water supply are

Table 1. Risk value according to Equation 1.

$P_{Si} \cdot C_{Si}$	O_{Si}		
	1	2	3
	r_k		
1	1	0.5	0.33
2	2	1	0.67
3	3	1.5	1
4	4	2	1.33
6	6	3	2
9	9	4.5	3

longer than 24 hours. Consumers are exposed to poor quality water. ULS ≤ CLS < CoLS.

• Disaster Safety Condition (DSC) – the WSS condition in which the system stops working and consumers cannot access acceptable quality drinking water. Lack of consumer protection against threats (e.g. as a result of flood or drought). CLS < ULS.

2.2 Method of water consumer safety analysis

As a result of the occurrence of the so-called representative emergency scenario, denoted by S_i, water consumers are exposed to the possibility of reduced levels of safety. The measure of this loss of safety for drinking water consumers is the risk connected with the possibility of consuming bad quality water or a lack of water (Rak, 2005). Water consumer risk r_k is calculated from the equation:

$$r_k = P_{Si} \cdot C_{Si} \cdot O_{Si}^{-1} \tag{1}$$

where: P_{Si} = the probability of S_i; C_{Si} = the degree, or point weight, of consequences connected with S_i for water consumers; O_{Si} = the level, or point weight, of protection of water consumers against S_i

For every situation, a score is assigned to the parameters P_{Si}, C_{Si} and O_{Si}, according to the following point scale: low = 1, medium = 2, high = 3. In this way, we obtain a point scale to measure risk: tolerable, controlled, unacceptable (Diamantidis, 2002), in a numerical form, within the range [0.33 ÷ 9], according to Equation 1, Table 1. Risk values are presented in Table 1, according to the three-parameter method (Rak, 2005).

For the probability of S_i, we propose to utilise the so-called Bayes networks, which are an acyclic directed graph with nodes representing random variables and edges determining cause-to-effect dependencies. Let us consider a configuration of random events mutually connected by cause and effect relationships. The occurrence of event X has some influence on the occurrence of event Y. However, this influence is not 'certain' and can be defined only by means of its likelihood. Such a configuration of events and their mutual relations

can be modelled by means of the directed graph D. Every event is interpreted as the graph node which is identified by a point on the plane. Relationships between events are represented by edges. If the occurrence of event X has an influence on the occurrence of event Y (Y depends on X), then in the graph model there is an edge (X, Y) going from X into Y (the direction is marked by an arrow). The node X will be called the parent of node Y. $\pi(X)$ means the set of all node X parents. To simplify the notation, every event X_i will be identified with the corresponding random variable with the same name. In our considerations, all the random variables corresponding to events are bivalent. The dependence between nodes (events) is expressed by means of the conditional likelihood. For the node X whose parents are in the set $\pi(X)$, these dependencies are represented by the conditional likelihood tables (CLT). The CLT for the variable X must define the likelihood values $P(X|\pi(X))$ for all possible combinations of variable values from the set $\pi(X)$. The table for a node without parents contains the likelihood that the random variable X will take its particular values. Formally the definition of a Bayesian network can be written as follows: Bayesian network B(D,CLT) is a pair (D, CLT), where D is an acyclic directed graph on whose nodes the function CLT has been defined, which assigns the conditional likelihood tables CLT to each node X. The table CLT for node X contains $P(X|\pi(X))$ (Teemu et. al. 2005).

If the network has n nodes: $X_1, \ldots, X_n$, then the joint probability distribution of all random variables can be shown as the equation (Bishop 2006):

$$P(X_1, \ldots, X_n) = \prod_{n=1}^{n} P(X_i \mid \pi(X_i)) \tag{2}$$

After calculating the value of the probability, we assume the following point weights for the parameter P_{Si}:

• low probability – once in 10–100 years – $P_{Si} = 1$;
• medium probability – once in 1–10 years – $P_{Si} = 2$;
• high probability – 1–10 times a year or more often $P_{Si} = 3$.

The values of point weights for the parameter of consequences resulting from S_i were assumed according to the following rule:

• small – perceptible organoleptic changes in water, isolated consumer complaints, financial losses are: $< (5 \cdot 10^3)$ EUR – $C_{Si} = 1$;
• medium – considerable organoleptic problems (odour, changed colour and turbidity), consumer health problems, numerous complaints, information in local public media, financial losses are: $(5 \cdot 10^3) \div 10^5$ EUR – $C_{Si} = 2$;
• large – endangered people require hospitalisation, professional rescue teams involved, serious toxic effects in test organisms, information in nationwide media, financial losses over 10^5 EUR – $C_{Si} = 3$.

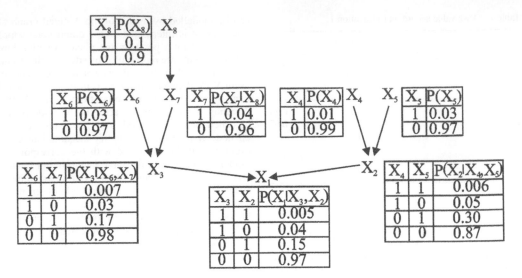

Figure 1. Bayes network for the example presented.

The values of point weights for the parameter of protection connected with S_i were assumed according to the following scale:

- low protection level – $O_{Si} = 1$, municipal water quality standard monitoring, according to regulations;
- medium protection level – $O_{Si} = 2$, over standard (greater than normal) monitoring, (e.g. utilising SCADA software, a detailed emergency plan is developed, including alternative water sources localisation);
- high protection level – $O_{Si} = 3$, special monitoring (e.g. within the framework of the multi-barrier system, including raw water biomonitoring based on test organisms, and use of industrial television cameras with movement detectors, as well as a well-developed complex plan for response to failure, including informing consumers about the threat).

The level of consumer safety r_k (Equation 1 and Table 1) was established as:

$r_k = [0.33 \div 2[$ – tolerable level of safety (TLS);
$r_k = (2 \div 3]$ – controlled level of safety (CoLS);
$r_k = (3 \div 9]$ – unacceptable level of safety (ULS).

3 RESULT AND DISCUSSION

In determining a level of water consumer safety for the selected S_i, the following notations were assumed for particular S_i events: X_1 – the possibility that consumer safety will be reduced; X_2 – lack of water supply; X_3 – pollution in water pipes; X_4 – failure of water mains; X_5 – power supply failure; X_6 – secondary water pollution in water-pipe network; X_7 – water treatment process failed; and X_8 – incidental pollution in water intake.

The cause and effect relationship is shown by means of the Bayesian network in Figure 1.

For the particular network nodes, the appropriate conditional likelihood tables are developed, assuming the following: $X = 1$ means that the event takes place; $X = 0$ means that the event does not take place. The joint probability distribution for the example presented in Figure 1 is:

$P(X_1, X_2, X_3, X_4, X_5, X_6, X_7, X_8) = P(X_1|X_3, X_2) \cdot P(X_2|X_4, X_5) \cdot$
$\cdot P(X_3|X_6, X_7) \cdot P(X_7|X_8) \cdot P(X_6) \cdot P(X_8)$

The S_i which resulted in the level of water consumer safety becoming lower was selected: lack of water supply took place (event $X_2 = 1$), water main failure took place (event $X_4 = 1$), water in the water-pipe network was not contaminated (event $X_3 = 0$), power supply failure did not take place (event $X_5 = 0$), secondary water pollution in the water-pipe network did not take place (event $X_6 = 0$), the water treatment process did not fail (event $X_7 = 0$), and incidental pollution in water intake did not take place (event $X_8 = 0$). We must determine the likelihood that the level of water consumer safety will become lower (event $X_1 = 1$). The following values for the particular probabilities were assumed according to Figure 1:

$P(X_1 = 1, X_2 = 1, X_3 = 0, X_4 = 1, X_5 = 0, X_6 = 0, X_7 = 0, X_8 = 0) =$
$= P(X_1|X_3, X_2) \cdot P(X_2|X_4, X_5) \cdot P(X_4) \cdot P(X_5) \cdot P(X_3|X_6, X_7) \cdot$
$\cdot P(X_7|X_8) \cdot P(X_6) \cdot P(X_8) =$
$= 0.15 \cdot 0.05 \cdot 0.01 \cdot 0.97 \cdot 0.98 \cdot 0.96 \cdot 0.97 \cdot 0.9 = 0.000066$

In order to determine the value of water consumer risk according to Equation 1, the following point weights for the particular parameters determining risk were assumed:

- for $P_{Si} = 0.000066 \Rightarrow$ point weight = 1,
- for consequences: considerable organoleptic problems $\Rightarrow$ point weight = 2,

- for protection: over standard (greater than normal) monitoring $\Rightarrow$ point weight $= 2$.

The risk value is $r_k = 1$, which, in the assumed scale of levels of water consumer safety, corresponds to the tolerable level of safety (TLS).

4 CONCLUSIONS

The WDS is a complex technological system, the reliable operation of which ensures water consumer safety. To determine the likelihood of different undesirable events scenarios in the WDS and to make a cause and effect analysis of these events, we can use Bayesian networks. One of the advantages of using a Bayesian network is a lower number of necessary likelihood values; instead of 2^n we need only $LN = \sum_{i=1}^{n} 2^{|\pi(X_i)|}$. The right Bayesian network construction requires a base of historical data in order to assess conditional likelihoods, and the knowledge of experts in the given field. An important challenge is also establishing the criteria values for risk levels (Diamantidis 2002), which should be made by cooperating teams of experts in the field of risk assessment, and experienced engineers, using up-to-date scientific and technological knowledge, as well as statistical data about the WDS's operation. By means of risk analysis, on the basis of detailed analysis of WDS operations, we can also determine a level of water consumer safety. The complete risk assessment for risk management (Rosen et al. 2008) should contain:

- establishment of a ranking of failures,
- determination of a level of risk,
- proposal of the activities aimed at risk minimisation,
- establishment of time after which risk can obtain its critical value, as a result of different processes, e.g. materials ageing.

If the calculated risk values (according to Equation 1) indicate the category is tolerable, then one can assume that the WDS fulfils its functions satisfactorily, with regard to both operational reliability and safety. If the risk values indicate the category is controlled then an improvement in the work of some elements of the system (e.g. network monitoring, protective stations) or repair of some sections of the water-pipe network should be considered. If the risk values indicate the category is unacceptable, this means that the WDS does not fulfil its functions, with regard to both operational reliability and safety. A thorough analysis of the main risk factors should be made, and the WDS should undergo complete modernisation or should even be redesigned. An important challenge is to define the tolerable risk level, the so-called ALARP (As Low As is Reasonably Practicable). The ALARP principle was first introduced in Great Britain, where the unacceptable (impermissible) value of the risk of death for the individual worker was determined to be $r = 0.001$ and the risk of death for the public was determined to be $r = 0.0001$ (HSE, 2001).

Depending on the distance between the ALARP level and the unacceptable risk level, we can formulate rules for choosing a risk assessment method:

- the smaller the difference between the unacceptable risk and ALARP, the more accurate should be the chosen method,
- in the case of a high risk level, the use of quantitative methods is recommended, e.g. fault tree analysis (FTA),
- if there is a large distance between ALARP and the unacceptable risk, the use of matrix methods is recommended,
- if the threat is not large, use of qualitative methods is recommended.

Risk-reducing processes should take into account a cost–benefit analysis. Risk levels should be determined at a level at which the costs of lowering it further are disproportionally high. Health and Safety Executive (HSE) directives introduce the notion, 'the cost for preventing a fatality', which is estimated, according to the mentioned above directives, at about 1 million GBP. Those directives also introduce a factor of proportionality (PF) and outline the following principles:

- if the PF is higher than required, it can be admitted that further risk reducing is not reasonably practicable,
- the higher the risk level, the more money it is worth spending on its reduction,
- at the unacceptable risk level, it is necessary to pay whatever sum of money is needed to reduce it to the tolerable level; if this is impossible, the system must be shut down.

The most important factor for water consumer safety is the security of a water supply system. Modern control and monitoring systems are increasingly becoming a standard element in the modernisation of water supply systems. Visualisation of water treatment (online process control on an ongoing basis, using software such as SCADA), pump station automation, and data transmitting by means of telemetry, are becoming standard.

ACKNOWLEDGEMENTS

Scientific work was financed from the measures (Ministry of Science and Higher Education) for research in the years 2007–2010, as a research project Nr N N523 3765 33 entitled, "Development of the methodology of analysis and assessment of risk of failure in water supply systems, with regard to water consumer safety".

REFERENCES

Bishop, C. 2006. *Pattern recognition and machine learning*. Springer. New York.

Diamantidis, D. 2002. Risk acceptance criteria for long tunnels. Paper presented at the International Conference PSAM 6. Puerto Rico.

Gardner, G. 2008. Implementing risk management for a water supply. A catalyst and incentive for change. *The Rangeland Journal CSIRO Publishing* 30: 149–156.

Hipel, K.W., Kilgour, D.M. & Zhao, N.Z. 2003. Risk analysis of the Walkerton drinking water crisis. *Canadian Water Resources Journal* 3: 395–397.

HSE (Health and Safety Executive). 2001. *Reducing risk – decision making process*, Health and Safety Executive. Sudbury. UK.

Mays L.W. 2005. The role of risk analysis in water resources engineering. Department of Civil and Environmental Engineering, Arizona State University. www.public.asu.edu/lwmays: 8–12.

Pollard, S.J.T., Strutt, J.E., Macgillivray, B.H., Hamilton, P.D. & Hrudey, S.E. 2004. Risk analysis and management in the water utility sector – a review of drivers, tools and techniques. *Process Safety and Environmental Protection* vol. 82(B6): 1–10.

Rak J. 2005. *Podstawy bezpieczeństwa systemów zaopatrzenia w wodę*. Polska Akademia Nauk (PAN) – Monografie Inżynierii Środowiska 28: 1–215.

Rosen, L., Linde, A., Hokstad, S., Sklet, J., Røstum, J. & Pettersson, T.J.R. 2008. Generic framework for integrated risk management in water safety plans. In *Proceedings of 6th Nordic Drinking Water Conference*. 9–11 June Oslo. The Nordic Network of the International Water Association – IWA: 193–203.

Shinstine D.S., Ahmed I. & Lansey K. 2002. Reliability/availability analysis of municipal water distribution networks: Case Studies. *Journal of Water Resources Planning and Management. ASCE* 128(2): 140–151.

Tanyimboh, T.T., Burd, R., Burrows, R. & Tabesh, M. 1999. Modelling and reliability analysis of water distribution systems. *Water Science Technology* 39(4): 249–255.

Teemu, R.,Wettig, H., Grunwald, P., Myllymaki, P. & Tirri, H. 2005. On discriminative Bayesian network classifiers and logistic regression. *Machine Learning* 59: 267–296.

Vreeburg, J.H.G. & Hoogsteen, K.J. 1997. Risk management: integral approach in the Netherlands. Paper presented at IWSA World Congress, Madrid, Spain.

WHO. 2004. *Guidelines for drinking-water quality*. WHO, Geneva.

WHO/SDE/WSH. 2002. *Water safety plans (revised draft)*. Report publication 02.09.2002.

Environmental Engineering III – Pawłowski, Dudzińska & Pawłowski (eds)
© 2010 Taylor & Francis Group, London, ISBN 978-0-415-54882-3

Formation of biofilm in tap water supply networks

T.M. Traczewska, M. Sitarska & A. Trusz-Zdybek

The Institute of the Faculty of Environmental Engineering, Wroclaw University of Technology, Wroclaw, Poland

ABSTRACT: The sanitary quality of water is affected by microbiological stability. A large number of microorganisms, along with water stagnation, results in the growth of biofilms on the inside surfaces of pipes and in the secondary microbiological contamination of water. The research looked at the susceptibility of plastic waterpipes to the development of bacteria in tap water, the influence of biofilm on the microbiological quality of water, and damage to synthetic materials caused by biofilm. It demonstrated that microorganisms present in tap water form biofilms on the surface of PE and PVC; these materials show a varied susceptibility to the development of biofilm.

Keywords: Biofilm, contamination of water, secondary microbiological, synthetic materials tap water network.

1 INTRODUCTION

During the last few years the use of synthetic materials developed in the 1930s have become popular in various aspects of human life. They are presently used in water distribution systems and their per cent share in tap water supply networks is steadily growing. This is due, in part, to their ready availability in the market, their relatively low price, and their ease of assembly. Furthermore, plastic pipes are installed to decrease the risk of the development of microbiological films in tap water networks which would, in turn, result in secondary microbiological contamination of water. Studies conducted in the 1960s demonstrated that such synthetic polymers as polyethylene or polyvinyl chloride show low surface roughness which considerably lowers their susceptibility to microbiological film creation. However, it does not totally eliminate it (Kuś & Ścieranka 2005, Zyska & Żakowska 2005).

The latest research proves that the smooth surface of polymers only slows the development of the first layer of biofilm. The next stages of its development can take place faster than in the case of other materials as the substances used in the pipe production, such as fixers, stabilisers or hardeners, can be a source of food, stimulating the growth of the microorganisms in the biofilm. Depending on the substrate properties, a variety of microflora was observed (Zyska 2000).

The rate of growth of biofilms in distribution systems depends significantly upon input water quality. The technologies applied in the purification process do not assure maintenance of the biological stability of the water during its flow. This can result in the water being received having worse parameters than the water initially supplied, constituting a threat to health (Hallam et al. 2001, Ollos et al. 2003).

The biofilms which develop in the distribution systems are a problem to many suppliers of drinking water. Oversized piping, old and corroded sections, as well as a constantly decreasing water intake causing stagnation, provide ideal conditions for the development of microorganisms on piping systems and fitting elements which in effect cause the growth of mature structures of biofilm. The bacteria which are washed out from biofilms quickly form new biofilms in further sections of the networks because of their greatly increased numbers in the water transported. This makes for an extremely adverse situation, especially where old and new pipes are connected (Kerneis et al. 1995, Zhang et al. 2002, Heidrich & Jędrzejkiewicz 2007).

The objective of this research was to evaluate the susceptibility of the synthetic materials commonly used in water distribution systems in contact with tap water to the development of microbiological biofilms, to determine the variety of species present in such biofilms, and to assess the influence of their metabolic processes on the structure of the material.

2 MATERIAL AND METHODS

Tests on the growth of microbiological films on the surfaces of polyvinyl chloride (PVC) and polyethylene (PE) were carried out at a test station located in the Municipal Water and Wastewater Company, Wrocław. The test station is located in the technology line before the water disinfection process. The test samples of the materials studied (20×25 mm and 3 mm thick) were placed collinearly in a custom-designed microstat, made of acid-resistant stainless steel, for a period of about 140 days. Given the construction of the

Figure 1. The microstat.

Figure 2. The surface of a new pipeline made of PE (a, magnification 4000×) and one of PVC (b, magnification 4000×).

microstat the tests for the different materials were conducted at different time of the year. The polyethylene samples were tested during the fall and winter, while the polyvinyl chloride samples were tested during the spring and summer.

The reactors were fed with purified infiltration water which had undergone deironing and demanganisation, filtration through sand deposits and sorption in active carbon. The speed of the water flow in the microstat was 0.2 m/s. Samples of water were taken every three to four days from the inlet and from the outlet of the test unit in order to verify the rate of the biofilm growth. The quantitative, microbiological analysis to calculate the number of mesophilic and psychrophilic bacteria was conducted using the Koch plate method on agar-rich substrate. The material samples were also subjected to microscopic analysis using a scanning electron microscope. This permitted a comparison of the structure of the polymer surface and the surfaces of the synthetic materials both with microorganisms and after their removal using ultrasound. The microorganisms received from the washouts of the polymers were quantitatively analyzed. A preliminary qualitative evaluation was conducted of gram-negative bacteria by direct isolation in an Analytical Profile Index (API) test. A similar evaluation was conducted for gram-positive bacteria i.e. Bacillus and Corynebacterium. A test for the presence of fungi was performed as well.

3 RESULTS AND DISCUSSION

The physical structure of a material surface plays a significant role in the rate of development of a biofilm. Increased surface roughness is matched by an increased number of deposited microorganisms which, when multiplying, begin to create colonies and in time form a mature biofilm (Lappin-Scott & Costerton 2003, Świderska-Bróż & Wolska 2003).

The pictures taken with the scanning electron microscope of fragments of new pipelines made from the plastics tested show the differences in their structure.

Polyethylene has numerous, different size corrugation resembling 'runs' which are irregularly located. This is why the structure of its surface seems less regular than that of the polyvinyl chloride which demonstrates a rather regular surface resembling a 'truss.' In PVC the strands of the material form oblique stripes similar to natural fibres with visible small pits of different sizes and shapes between them.

Under favourable conditions, the first stage of a biofilm can develop in three weeks. However, sudden changes in the water flow rate not only result in a washing out of individual bacteria cells, but in the displacement of whole fragments of biofilm as well. This in turn prolongs the state of development of the 'mature biofilm', increases the number of microorganisms in the water, and promotes the development of new biofilms in further sections of the pipeline (Świderska-Bróż 2003).

This was confirmed in our own tests. Polyethylene was tested during November and April. During the whole test period, the number of bacteria in the feedwater showed significant variations. The maximum number of mesophilic bacteria – 4 CFU/cm^3– was reached on day 98. In the analysis of the whole period of operation of the unit, it was significant that after about 90 days of testing mesophilic bacteria were not observed in the feedwater. The total number of the bacteria incubated at a temperature of $20°C \pm 2°$ fluctuated from 0 to 498 CFU/cm^3, with a mean value of 67 CFU/cm^3. In over 80% of the analyses conducted, the number of psychrophilic bacteria at the inlet did not exceed 100 CFU/cm^3.

During the study of the susceptibility of polyvinyl chloride to microbiological film, conducted between April and September, the total number of mesophilic bacteria at the inlet ranged from 0 to 25 CFU/cm^3, which testifies to their increased number in the water collected for municipal purposes in the summer months. Because the presence of these bacteria was not detected in a considerable number of samples, their mean level for the whole test period was 3 CFU/cm^3. The total number of psychrophilic bacteria in the inflow water during the first 52 days of operation of the unit varied from 35 to 381 CFU/cm^3. The mean value for that period was 197 CFU/cm^3. The admissible values for this type of contamination of tap water were exceeded in 64% of the tests conducted. However, it should be remembered that the water fed to the microstat was not disinfected. In the period from day 56 until the conclusion of the study, the number of bacteria did not exceed 250 CFU/cm^3 and the mean value was 59 CFU/cm^3. The total admissible number of bacteria was exceeded only in about 10% of the samples.

On day 52 of testing, the number of bacteria introduced to the microstat increased dramatically. The number of mesophilic bacteria at the inlet was measured at 349×10^2 CFU/cm^3 while the number of psychrophilic bacteria registered 385×10^2 CFU/cm^3. Because it was a unique event these values were not taken into account when calculating the mean values. It should be noted, however, that there must have been a significant impact on the developing biofilm because of their being deposited.

The quantitative microbiological analyses of the water fed to the microstat showed changes in their numbers probably because of the growth of the biofilms on the materials under test. Comparing the results of the quantitative tests for microorganisms in the water flowing from the microstat after contact with the PE and PVC it can be stated that these materials stimulated a growth in the number of bacteria in different ways. On the one hand, this phenomenon could be caused by the susceptibility of the materials to the growth of biofilms. On the other hand, the water flowing into the unit showed considerable fluctuations in the number of bacteria it contained. Most likely this was caused by the time of year when the observations were made. For the water which was in contact with the polyethylene, the total number of mesophilic bacteria increased over the first 95 days of testing. A comparable increase was not observed in the feedwater over the same period. In the period of March/April a slight increase in the mesophilic bacteria count of the feedwater took place. The maximum value occurred on day 134, when their number increased from 2 CFU/cm^3 to 31 CFU/cm^3. Over about 40 days of contact between the water and the PVC, the number of these microorganisms was from 3 to 20 times higher that that in the inflowing water. The maximum number of mesophilic bacteria (79 CFU/cm^3) was observed at the outlet on day 21.

The higher number of psychrophilic bacteria at the outlets from the reactors was observed after just one week of contact with water for both polymers. For the polyethylene, the initial slight increase remained approximately constant for about 10 days of operation of the unit and then, until day 35 of the experiment, the presence of psychrophilic bacteria was not observed at the inlet to or at the outlet from the reactor. This does not mean, however, that they are not present, only that their number was too small to be determined with the Koch plate method, which was used for the tests. On day 35 the total number of microorganisms incubated at a temperature of $20°C \pm 2°$ in the water flowing into the reactor reached a higher values than for the water flowing from the reactor.

The maximum value (413 CFU/cm^3) was observed on day 88 when the greatest increase occurred (as high as 328 CFU/cm^3) as compared to the water fed to the microstat. This increase remained the same until the conclusion of testing. The mean number of psychrophilic bacteria for the whole period of testing with PE was 110 CFU/cm^3.

With PVC in the microstat, the increase in the number of microorganisms at the outlets compared to the number in the feedwater remained the same until day 63 day of operation. The values ranged between 20 and 452 CFU/cm^3, having the lowest value at the beginning of the tests and the highest on day 21. Subsequently, for over 4 days the number of bacteria in the feedwater and at the outlet was similar – 15 to 80 CFU/cm^3. After that period, the number of psychrophilic bacteria at the outlet compared to the that at the inlet again grew and the trend remained the same until the unit ceased operation. The maximum value of the total number of psychrophilic bacteria for the PVC samples was 620 CFU/cm^3 on day 105 and day 136. The minimum number (7 CFU/cm^3) occurred on day 133. The mean number of the microorganisms incubated at a temperature of $20°C \pm 2°$ in the water flowing from the microstat for the whole test period was 240 CFU/cm^3.

The temporary decrease in the number of microorganisms observed in the water would suggest the growth of a mature biofilm in which structures have been created which closely adhere to one another and which, in turn, prevent the washing out of too large a number of microorganisms. During the tests, there were sudden temporary increases in the number of bacteria at the outlets from the reactors that were most probably caused by occasional detachments of fragments of biofilm that had developed on the synthetic materials under test, followed by periods of growth of the biofilm.

Figure 3 shows the total number of psychrophilic bacteria detected at various times throughout the experiment. Within the operating temperature range of 18 to 22°C there were significant variations in the

(a)

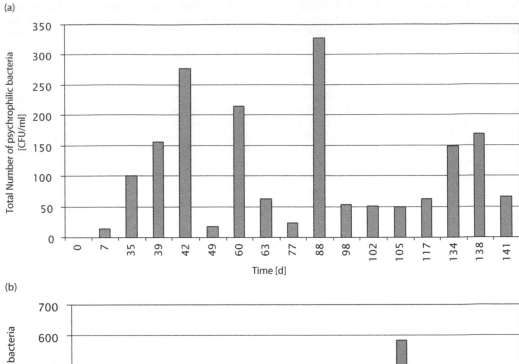

(b)

Figure 3. Total number of psychrophilic bacteria in the water in contact with samples of PE (a) and PVC materials (b) as compared to that in the feedwater.

number of microorganisms present. For about 80% of the time of operation of the unit, an increased number of bacteria was observed at the water outlets as compared to the inlets. In the case of the polyethylene samples, these changes demonstrated greater fluctuations than for the polyvinyl chloride ones, with values in the range of 2 to 328 CFU/cm^3. Maximum growth occurred on day 88. Growth of these mean values was observed between days 35 and 42 (178 CFU/cm^3), days 98 and 117 (54 CFU/cm^3); and days 134 and 141 (128 CFU/cm^3). The growth of the number of psychrophilic bacteria in the water flowing from the microstat in contact with the PVC was between 11 and 584 CFU/cm^3 which is almost twice that for the polyethylene. The growth of 11 CFU/cm^3 was reached

on day 31 and that of 584 CFU/cm^3, on day 105 of the operation of the unit.

At the end of each operation of the unit, the samples were subject to microscopic analysis using a scanning electron microscope. Figure 4 presents pictures of the biofilms which formed on the materials.

The tests conducted demonstrated the varied susceptibility of synthetic materials to microbiological film, both in respect of the number of organisms forming the biofilm and their biodiversity. Observations of the biological films showed clear differences in their structures depending on the substrate on which they developed. In the case of polyvinyl chloride, it seems more accumulated more uniformly, with a number of visible morphological forms of the bacteria. On the

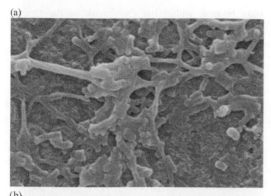

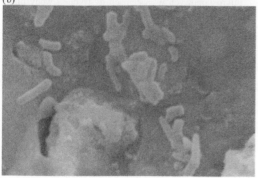

Figure 4. Biofilm formed on PE (a, magnification 4000×) and PVC (b, magnification 6000×) in purified water.

polyethylene, the biofilm looks more spatially diverse, with more fungi than on the polyvinyl chloride.

The film was removed from the surfaces of the polymers using ultrasound. Quantitative analysis of the two suspensions of microorganisms demonstrated a higher susceptibility of polyethylene to the formation of biofilms as compared to polyvinyl chloride. The number of mesophilic bacteria present in the film formed on the PE was 4×10^3 CFU/cm^2 of polymer surface while for the polyvinyl chloride the number was only 14 CFU/cm^2. Additionally, the number of psychrophilic bacteria for the PE was a little higher, at 84×10^3 CFU/cm^2, as compared to 73×10^3 CFU/cm^2 for the PVC. Slight differences were observed for the number of fungi. In the suspension derived from the washout from the polyethylene their number was 21×10^3 CFU/cm^2 and in the case of PVC it was 18×10^3 CFU/cm^2 of the polymer surface.

The initial identification of the organisms isolated from the biofilms which formed on the materials confirmed the data in the literature regarding the presence of bacteria and fungi, including provisionally pathogenic ones, in tap water (Evans et al. 2003, Olańczuk-Neyman & Sokołowska 2004, Grabińska-Łoniewska 2006).

The isolated bacteria included such genera as *Raoultella* sp., *Sarcina* sp., *Corynebacterium*, and Bacillus, as well as such species as *Escherichia coli*,

Aeromonas hydrophila, and *Vibrio parahaemoliticus*. The fungi included, primarily, *Candida* imperfect fungi.

The presence of such microorganisms can pose an epidemiological threat in as much as they decrease the effectiveness of disinfective agents, substantially reducing their penetration into the structure of the biofilm. The presence of bacteria from the Bacillus genus in the biofilm and their ability to sporalisation, due to their resistance to disinfecting agents, can result in a loss of the microbiological stability of the water in the distribution network. There is also the potential possibility of formation of other microbiological films in further sections of the distribution system (Evan et al. 2003).

Furthermore, it was demonstrated that the bacteria and fungi, which grow on the surface of plastics as a result of metabolic processes, cause their microbiological corrosion. This is demonstrated by the change in the structure of the surface of the materials tested. An analysis was made on the basis of the pictures of the material samples taken with the use of the scanning electron microscope after removing the biofilms from their surfaces.

Visible changes occurred to the polyethylene. Numerous pits and irregularities, which considerably increased its roughness, developed on the surface. The number of microorganisms was similar to that on PVC, which suggests that the substances covering the external surface of this material include chemical compounds which are easily assimilated by the microorganisms which then use them in their metabolic processes. Slight damage was observed on the surface of polyvinyl chloride.

The visible increase in surface roughness contributes to further growth of biofilm with an excess of microorganisms with poorer adhesive capabilities compared to the original deposit.

4 CONCLUSIONS

- The type of substrate significantly affects the formation of biofilms and their spatial structure, which is determined, primarily, by the number and kind of microorganisms deposited;
- Despite the higher number of bacteria introduced into the microstat, particularly for the PVC, their number in the biofilm was lower (73×10^3 CFU/cm^2) than for PE (88×10^3 CFU/cm^2). This would suggest a lower susceptibility of this polymer to the development of microorganisms caused by, for example, surface roughness or the presence of other chemical compounds used in production of the product;
- Opportunistic, pathogenic bacteria can be present in the biofilm which forms as a result of contact between the material and the treated water;
- Occasionally, fragments of biofilm can be displaced, as a result of changes in the flow rate of the water, for example, which results in a dramatic increase

Figure 5. Surfaces of PE (a, magnification 3000×) and PVC (b, magnification 4000×) polymers after removal of biofilm with ultrasound.

in the number of microorganisms in the water and consequently causes the formation of new biofilms in down stream sections of the tap water networks;

• As a result of the metabolic activity of the microorganisms, particularly the fungi present in the biofilms, the external layer of the polymer surface suffers damage which consequently increases its roughness.

ACKNOWLEDGEMENTS

This paper has been written as a results of realization of the project entitled: 'Detectors and sensors for measuring factors hazardous to environment – modeling and monitoring of threats'.

The project by the European Union via the European Regional Development Fund and the Polish state budget, within the framework of the Operational Programme Innovative Economy 2007÷2013. The contract for refinancing No. POIG.01.03.01-02-002/08-00.

REFERENCES

Besner, M-C., Gauthier, V., Barbeau, B., Millette, R., Chapleau, R. & Prevost, M. 2001. Understanding distribution system water quality. *Journal AWWA* 93(7): 101–114.

Evans, L.V. 2000. *Biofilms: recent advances in their study and control.* Harwood Academic Publishers. Amsterdam.

Grabińska-Łoniewska, A. 2006. Występowanie grzybów mikroskopowych w systemach dystrybucji wody pitnej i związane z tym zagrożenia zdrowotne. *Gaz, Woda i Technika Sanitarna,* lipiec-sierpień: 72-76.

Hallam, N.B., West, J.R., Forster, C.F. & Simms, J. 2001. The potential for biofilm growth in water distribution systems. *Water Research* 35(17): 4063–4071.

Heinrich, Z. & Jędrzejkiewicz, J. 2007. Analiza zużycia wody w miastach polskich w latach 1995-2005. *Ochrona Środowiska* 4: 29–34.

Kerneis, A., Nakache, F., Deguin, A. & Feinberg, M. 1995. The effects of water residence time on the biological quality in a distribution network. *Water Research* 29(7): 1719–1727.

Kuś, K. & Ścieranka, G. 2005. Wpływ materiału i parametrów eksploatacyjnych sieci wodociągowej na jakośæ wody na przykładzie Chorzowsko-Świętochłowickiego Przedsiębiorstwa Wodociągów i Kanalizacji w Chorzowie. *Ochrona Środowiska* 4: 31–33.

Lappin-Scott, H.M. & Costerton, J.W. 1995. *Microbial biofilms,.* Cambridge University Press: Cambridge.

Olańczuk-Neyman, K. & Sokołowska, A. 2004. Bakterie i wirusy w wodzie wodociągowej. *Inżynieria i Ochrona Środowiska* 7(3–4): 259–276.

Ollos, P., Huck, P. & Slawson, R. 2003. Factors affecting biofilm accumulation in Model Distribution Systems. *Journal AWWA* 95(1): 87–97.

Świderska-Bróż, M. 2003. Skutki braku stabilności biologicznej wody wodociągowej. *Ochrona Środowiska* 4: 7–12.

Świderska-Bróż, M. & Wolska, M. 2003. Korozyjność wody wodociągowej a zjawiska zachodzące w systemie jej dystrybucji. *Gaz, Woda i Technika Sanitarna* 1: 10–15.

Zhang, M., Semmens, M., Schuler, D. & Hozalski, R. 2002. Biostability and microbiological quality in a chloraminated distribution system. *Journal AWWA* 94(9): 112–122.

Zyska, B. 2000. *Mikrobiologiczny rozkład i korozja materiałów technicznych.* Politechnika Łódzka.

Zyska, B. & Żakowska, Z. 2005. *Mikrobiologia materiałów.* Politechnika Łódzka.

Environmental Engineering III – Pawłowski, Dudzińska & Pawłowski (eds)
© *2010 Taylor & Francis Group, London, ISBN 978-0-415-54882-3*

Calculating usable resources at surface water intakes using low-flow stochastic properties

B. Więzik

Hydrological and Meteorological Education Centre, Institute of Meteorology and Water Management, Warsaw, Poland

ABSTRACT: This paper presents modern methods, based on low-flow stochastic properties, for calculating the available and usable resources at surface water intake cross-sections. Until recently, such flows were calculated using the uniform water balance methodology. This methodology is based on a flow duration curve (FDC) that is determined from a long-term series of (decadal, daily) flows at the water intake cross-section. Decision analysis, based on stochastic processes of annual minimum flows, can now be the basic tool used for planning and implementing water management tasks, including the design of new water intakes or efficiency assessment of already existing water intakes.

Keywords: Water intake, available flow, usable flow, water abstraction, reliability, water deficit, Bayesian probability.

1 INTRODUCTION

The problems of water supply reliability and safety – taking into account whole systems and their particular elements, such as water intakes, water purification and distribution plants – often neglect the cases of extreme hydrological phenomena that can significantly influence the quality and quantity of water abstracted from rivers and streams. A significant problem for users is minimal water resources in watercourses where intakes were built and the usable flows not being correctly determined. During droughts that are occurring more and more frequently in our climatic zone – usually in summer but also in autumn and winter – serious problems exist in meeting water quantity demands if the necessary in-stream flow is preserved downstream of the intake (e.g., Smakhtin 2001, Environment Agency 2008). Building a new storage reservoir is often the only solution to increase the reliability of water abstraction.

When designing the water intake on mountain rivers and streams, the basic problem is the correct determination of the available and usable resources as a result of high flow variability. Available surface water resources are most frequently calculated using a long-term flow duration curve (FDC).

The amount of water to be abstracted is based on the usable resources of a fixed reliability that may be used for economic purposes (water supply for people, industry, agriculture) at the cross-section of the existing or planned water intake with the in-stream flow value preserved (Tallaksen & van Lanen 2004,

Environment Agency 2008). The available and usable flows are related according to the formula:

$$Q_d = Q_e + Q_{nn} \qquad (1)$$

where Q_d = available flow; m³·s⁻¹; Q_e = usable flow; m³·s⁻¹; Q_{nn} = in-stream flow, m³·s⁻¹.

The calculation of usable flow is thus based on a previous determination of the available flow and the fixed in-stream flow downstream of the water intake cross-section. Methods for determining usable resources and randomly occurring water deficits have been analysed for the Lecœnianka stream catchment at the cross-section of the existing water intake.

2 MATERIALS AND METHODS

2.1 *The catchment*

Systematic observations of water level and flow have been carried out for many years in the Lecœnianka stream catchment in southern Poland, where a water intake is located at the km 1.900 from the mouth. The data were used to estimate usable resources and to assess the level, which would meet users' needs during low-flow periods.

The upper part of the catchment of the Lecœnianka stream is located in the Beskid Śląski Mountains on the slopes of the Skrzyczne Peak; the lower part is located in the area of the large Żywiecka Valley (Kotlina Żywiecka). The area of the catchment is 37.1 km² and the stream is 14.6 km long.

The catchment boundary to the intake cross-section runs through the highest mountain peaks in the region: Skrzyczne (1257.0 m a.s.l.), Małe Skrzyczne (1210.8 m a.s.l.), Kopiec (1153.8 m a.s.l.), Magurka Wiślańska (1132.7 m a.s.l.), Cebula (1035.8 m a.s.l.) and Jaworzynka (951.5 m a.s.l.).

The springs of the Leśnianka stream are situated on the slopes of the Malinowska Skała at an elevation of 1050.0 m a.s.l.; the outlet of the stream to the Soła river is at an elevation of 362.5 m a.s.l. Forests cover more than 70% of the catchment area; the remaining part is covered by meadows, grazing land, arable land and dispersed settlements of the gmina (commune) of Lipowa.

A significant part of the total outflow is generated in the subcatchments of two streams – Leśnianka and Malinowski – which are situated in the upper compacted and forested part of the catchment. The total area of the Malinowski stream and the Leśnianka headwater reach is 18.8 km^2, which is 49.9% of the total area of the catchment, and the flow at this cross-section is as large as 72% of the flow at the outlet cross-section. The remaining part of the runoff comes from the flat valley in the middle and lower part of the catchment.

During periods of no rain, the watercourses are fed by ground water, the resources of which depend mainly on the geological structure of the catchment.

Based on direct measurements of the flow rate at certain characteristic cross-sections, it was determined that the flow increase at low water levels is not proportional to the increase in the catchment area.

2.2 Methods

According to the uniform water balance methodology, the available and usable flows are calculated using the FDC (Figure 1).

Using this method, which is based on a long-term FDC, standardisation of duration defines the reliability of the available flow and, if the in-stream flow is constant, also the reliability of water abstraction.

Taking into account the seasonal variability of the flow at a water intake cross-section on the Leśnianka stream, it was found that the water abstraction amount equals $Q_e = 0.070$ m$^3 \cdot$s^{-1} according to the water-abstraction permit and has very low reliability – $g = 0.68$ only, if the in-stream flow rate $Q_{nn} = 0.106$ m$^3 \cdot$s^{-1} is considered.

This means that there are serious limitations to water abstraction for the current users during hydrological drought, and there is no possibility of building a new water intake for the residents of the gmina.

Much more information on flow variability can be gained if FDCs are developed separately for each year. A set of FDCs obtained this way shows the variability of the flows that occurred during the long-term period under analysis.

In order to supply consumers with drinking water, when no alternatives for supplying water from other sources exist, it is necessary to determine the reliability of the duration of the usable flow with the in-stream

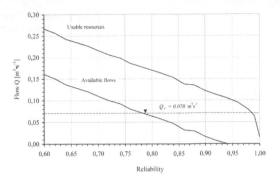

Figure 1. Usable water resources at the water intake cross-section. Q_e: usable flow.

flow unchanged. Basing calculations on the annual FDCs – apart from flow variability assessment – it is also possible to obtain the potential limiting values defined through the (upper and lower) envelopes of the FDCs.

If flows at the water intake cross-section for a fixed duration are considered as a random variable, the probability distribution of the flow can be developed. To this end, normal, log-normal and Gumbel one-dimensional distributions were used with parameters estimated by means of the maximum likelihood method. To choose the best distribution, the Kolmogorov–Smirnov, Pearson χ^2 and ORSTOM goodness-of-fit tests were applied.

In Figure 2, theoretical and empirical cumulative distribution functions (CDFs) are presented. A probabilistic description of flows of given duration at the water intake cross-section is often a side purpose. In engineering practice, this information is important as it is used for determining water abstraction reliability.

The probability of a k-fold exceedance of flow at the water intake cross-section during t years is usually described by the Poisson distribution, which represents random events in a large sample of small probabilities of success (Kaczmarek 1977):

$$P_{k/t} = \frac{(t \cdot p)^k}{k!} e^{-(t \cdot p)} \qquad (2)$$

where p = probability of exceedance of flow Q of fixed duration T; t = period length in years; and k = number of times the flow is exceeded.

Flow reliability is defined as the probability of non-exceedance during t years of flow Q of fixed duration T, i.e.:

$$g_{Q/T} = P_{0/t} \qquad (3)$$

Substituting equation (3) to (2) for $k = 0$, we finally get:

$$g_{Q/T} = e^{-p \cdot t} \qquad (4)$$

where g = reliability ranging from 0 to 1; t = time of the calculated reliability in years.

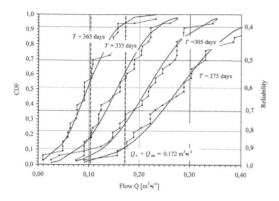

Figure 2. Reliability of the available flows. CDF: cumulative distribution function; Q_e: usable flow; Q_{nn}: in-stream flow.

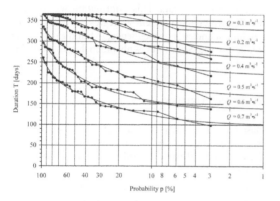

Figure 3. Probability of non-exceedance of duration of low flows at the water intake cross-section. Q: flow.

For small water intakes, when no alternatives of supplying water from another source exist, usually the value of $T = 1$ year is accepted, and the reliability of the available flows is determined for a fixed flow duration (Figure 2).

3 RESULTS AND DISCUSSION

3.1 *Estimation of water deficit at the water intake cross-section*

The problem of water deficits at water intake cross-sections requires the definition of basic terms, including the definition of the limit of low flows. To this end, one of the characteristic flows was adopted; the highest flow of the annual minimum flows (WNQ).

Using the annual FDCs for the water intake cross-section on the Leśnianka stream, empirical probabilities of non-exceedance of duration of selected flows were calculated (Figure 3).

For selected probability values, durations were read from the smoothed empirical curves. In this way a set of curves were developed that were used to determine the deficit duration and its probability.

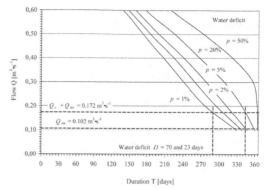

Figure 4. Water deficit at the water intake cross-section. Q_e: usable flow; Q_{nn}: in-stream flow.

At the water intake cross-section on the Leśnianka stream, a deficit of duration as high as almost 70 days occurs, on average, once per 100 years ($p = 1\%$). A 32-day deficit conditional on the water abstraction limit (because of the need to maintain in-stream flow) occurs, on average, once per 5 years ($p = 20\%$). In many cases, when a decision has to be made about building a new water intake, it is not the length of the deficit period, but the probability that interests a potential investor.

To answer the question of whether a water deficit period of given duration will occur once per year or once per many years, a Bayesian decision-making theory was applied.

3.2 *Bayesian probability*

In water resources management, the probability of a water deficit period occurring in a previously given time often dictates the building of a new water intake or determines the necessity of storing water in the catchment to equalise the flow. When using probabilistic analysis, more rational models often exist with various parameter estimation methods, and the conclusions from significance tests depend on the subjective selection of the acceptable error probabilities. However, the results obtained do not provide information that is so significant that, based on the results, decision algorithms can be formulated.

In recent years, another practical approach to probability calculus has been promoted. More and more often, along with a classical probabilistic model, Bayesian decision-making theory is applied that provides information that is necessary to carry out actions often of a strategic nature under uncertainty conditions (Benjamin & Cornell 1970).

Decision-making theory considers situations when the consequences of an action depend on certain factors of known or unknown random characteristics. These factors, called process state, may concern the occurrence of once or more per year of flows that are smaller than the limit – for example, the in-stream flow at the water intake cross-section.

The analysis of flow variability that is relevant to the study discussed hereafter refers to the conditions that arise when the consequences of decision-making depend on the future results of the random variable generating process, i.e., the process generating an annual minimum flow of given duration. The random state of the process was defined in terms of the values of its parameters and the information of their variability comes from historical observations.

The decision-making principle is defined as one variant of the available possible solutions. In practice, this is reduced to a limited number of potentially optimal solutions.

When designing a water intake or the further use of water resources at a given cross-section, the probability of events should be calculated based on a series of annual minimum flows of given duration (deficit period). In particular, the probability should be estimated such that, in successive m years, a low flow smaller than the in-stream flow will occur if, in the n years under analysis, such flow has occurred s times.

Let, in this case, $X_i = 1$ mean that a low flow that is smaller than the in-stream flow value (which in the water intake cross-section on the Leśnianka stream equals $Q_{nn} = 0.102\,\mathrm{m^3 \cdot s^{-1}}$) is a limit value. Assuming that the distribution of variable $Y = X_{n+1} + X_{n+2} + \ldots X_{n+m}$ is a binomial distribution $B(m, p)$, and p is known, then the conditional probability has the form:

$$p_Y(y \mid p) = \binom{m}{y} p^y (1-p)^{m-y} \qquad (5)$$

where $p =$ probability; $m =$ number of years with y deficit periods. We assume that p is a beta-distributed

$$f(p) = \frac{(n+1)!}{s!(n-1)!} p^s (1-p)^{n-s} \qquad \text{for} \quad 0 \le p \le 1 \qquad (6)$$

where $n =$ size of the series taken for analysis, in years; $s =$ number of occurrences of a deficit period during n years.

The Bayesian distribution of the random variable Y has the following form:

$$p_Y(y) = \frac{n+1}{s+y+1} \binom{m}{y} \binom{n}{s} \left[\binom{m+n+1}{s+y+1} \right]^{-1} \qquad (7)$$

where the variables are defined as in equations (5) and (6).

In the analysis, an $n = 36$-year series of annual minimum flows at the water intake cross-section on the Leśnianka stream was used and, during this period, a flow smaller than the limit occurred $s = 9$ times. In Figure 5, the Bayesian probability that the deficit period of duration 7, 14, 21 and 30 days will occur once per 2, 3, ..., 10 years is presented.

The calculation made shows that a 7-day deficit occurs once per year with a probability $p = 0.43$; with

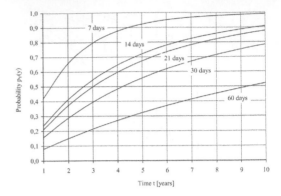

Figure 5. Probability of occurrence of deficit period.

the same probability, a 2-month deficit occurs, on average, once per 7 years. This information was used to develop water management strategies in the Leśnianka catchment and, as a consequence, to build a new water intake on the So³a river with much larger available resources.

4 CONCLUSIONS

Based on the analysis made of the available and usable resources at the water intake cross-section on the Leśnianka stream, the following conclusions were drawn:

1. Usable resources calculated according to the uniform water balance methodology, which have previously been the basis for issuing a permit for water abstraction from the Leśnianka stream, are of very low reliability.
2. A probabilistic analysis of available and usable flows at the water intake cross-section allowed an assessment of the probability and reliability of water abstraction for a given duration.
3. Application of Bayesian decision-making theory generated additional possibilities of assessing alternative solutions to the water supply for current and potential users.

REFERENCES

Benjamin, J.R. & Cornell, C.A. 1970. Probability, statistics and decision for civil engineers. McGraw-Hill: New York.
Environment Agency. 2008. Managing water abstraction interim update. EA publication GEH0D508150AH-E-E. UK Environment Agency: Bristol.
Kaczmarek, Z. 1977. Statistical methods in hydrology and meteorology, National Center for Scientific, Technical and Economic Information, available from U.S. Dept. of Commerce, National Technical Information Service.
Smakhtin, V.U. 2001. Low flow hydrology: a review. Journal of Hydrology 240: 147–186.
Tallaksen, L.M. & Van Lanen, H.A.J. 2004. Hydrological drought – processes and estimation methods for streamflow and groundwater. Developments in Water Science 48. Elsevier Science: Amsterdam.

Environmental Engineering III – Pawłowski, Dudzińska & Pawłowski (eds)
© 2010 Taylor & Francis Group, London, ISBN 978-0-415-54882-3

Trihalomethanes formation potential in water chlorinated with disinfectant produced in electrolyzers

A. Włodyka-Bergier & T. Bergier

AGH University of Science and Technology (AGH-UST); Faculty of Mining Surveying and Environmental Engineering, Department of Management and Protection of Environment, Cracow, Poland

ABSTRACT: This paper determines the amount of trihalomethanes (THMs) formed in water disinfected with a disinfectant produced in an electrolyzer and compares this with that produced in water disinfected with sodium hypochlorite. Research shows that using the disinfectant produced in the electrolyzer results in a potentially higher formation of THMs (approximately 30%) than when using sodium hypochlorite. The linear relation between the total THM concentration and the free chlorine content was observed in samples chlorinated with both disinfectants. The THM distribution was not different in samples chlorinated with both disinfectants, and mainly trichloromethane was formed.

1 INTRODUCTION

Chlorination is a cheap, efficient, and easy way to disinfect water. It is widely used in the treatment of water. However to eliminate the problems associated with the transport and storage of environmentally hazardous disinfectants, water treatment plants are frequently using the technology in which sodium hypochlorite is produced electrolytically on-site from sodium chloride (Brandt 1995, Destouches & Langlais 1996, Bashtan et al. 1999, Wasowski & Wasciszewska 2003). In recent years, new types of electrolyzers have been developed which are equipped with special cells to separate the products which are formed at the electrolyzer's anode and cathode. The disinfectant being formed through the reactions at the electrodes is a mixture of pure hypochlorous acid (approx. 95.9%), chlorine dioxide (approx. 2.3%) and ozone (approx. 1.8%). Such an oxidation mixture does not result in the specific smell and taste of chlorine in the water and it is chemically stable for 7 days (Wasowski & Wasciszewska 2003). Additionally, electro-chlorine is characterized by faster reactivity and higher disinfection effectiveness (Wasowski & Wasciszewska 2003, Bergmann et al. 2008). It also has the potential to generate lower amounts of trihalomethanes (THMs) when compared to sodium hypochlorite (Wasowski & Wasciszewska 2003).

Trihalomethanes are the main by-products of disinfection. They are generated during the reaction of chlorine with natural organic matter, such as the humic substances that naturally occur in untreated (raw) water. The total trihalomethanes content (TTHM) is the sum of four compounds; trichloromethane (TCM), bromodichloromethane (BDCM), dibromochloromethane (DBCM) and tribromomethane (TBM).

Several authors have reported that the presence of THMs in potable water raises the risk of urinary, bladder, and colon cancer. They also cause reproductive defects (WHO 2005, Wang et al. 2007), and low birth weight, and cleft, cardiac and central nervous system defects (Lee et al. 2006). The concentration of trihalomethanes in potable water depends on the natural organic matter concentration in the disinfected water, the chlorine dose, pH, temperature, and contact time. THMs formation is a long process, this is why THMs concentration cannot be measured just after disinfection. It is necessary to observe the THMs formation after some time has elapsed.

Even though solutions produced in electrolyzers are more frequently used for disinfecting water, there are relatively few research studies on the formation of their potential disinfection by-products. The authors conducted laboratory experiments to compare the THMs formation potential (THMFP) in water disinfected with commonly used sodium hypochlorite and with the disinfectant solution from an electrolyzer where the anode and cathode products are separated. This kind of solution is referred to as electro-chlorine in this article.

2 MATERIALS AND METHODS

2.1 Raw water

Water used for the research was taken from the 'Rudawa' Water Treatment Plant, which supplies the northern part of Cracow with potable water. The plant takes water from the Rudawa river. It is treated with following processes; oxidation, and coagulation, sand and active carbon rapid filtration, and, finally, disinfection. For the research, the water was collected from the

Table 1. Raw water quality.

Parameter	Value	Unit
pH	7.59	–
Turbidity	0.1	N.T.U.
Colour	3	$mgPt/dm^3$
Fe(II)+Fe(III)	0.02	mg/dm^3
Conductivity	661	$\mu S/dm$
Alkalinity	4.6	mg/dm^3
Alum	0.021	mg/dm^3
COD	1	mg/dm^3
UV_{254}	0.033	cm^{-1}
NH_4^+	<0.02	mg/dm^3
NO_3^-	19.4	mg/dm^3
NO_2^-	0.01	mg/dm^3

plant prior to the disinfection stage, but following all the other treatment processes. The authors decided to use water treated in the actual water treatment plant, so that the concentrations of THMs were as similar as possible to those observed in the output potable water. The plant chosen for the research has high natural organic matter (NOM) concentrations both in the raw and the treated water when compared to other Cracow water treatment plants. Table 1 presents the chemical analysis of the water samples. The analysis was performed by the certificated laboratory of MPWiK (the Municipal Waterworks and Sewers) in Cracow, and for this reason the authors have decided not to describe the analytical methods.

2.2 Disinfectants

Electro-chlorine, produced in an Annox generator, was used in the research. It was not generated on-site, but was transported to the laboratory prior to the research being conducted and kept at 4°C for 48 h. For this particular producer, the disinfectant could have been stored for 12 months under these conditions without loss of effectiveness. The disinfectant used for the research was NEUTHOX, a mixture of ATHOX (Cl_2, O_2, HOCl, HCl, $HClO_3$, ClO^-, ClO_2, O_2, H^+) and CATHOX (NaOH, H_2, (NaCl), H_2O_2). It has a pH in the range 7 to 7.5.

For comparison purposes, all experiments were repeated with a traditional solution of sodium hypochlorite purchased from POCH S.A. (Polish Chemical Reagents).

2.3 Experimental procedures

The experiments on the THMFP were conducted under laboratory conditions, in 0.5 litre, orange-glass bottles. The research was conducted at a constant temperature and constant pH. The water samples were incubated at 20°C and their pH adjusted to 7 by adding 0.1 normal sulphuric acid or sodium hydroxide and a phosphate buffer. The samples were then chlorinated using either sodium hypochlorite solution or electro-chlorine. The amount of disinfectant

was calculated from the chlorine demand curve (according to the Polish Standard PN-89/C-04600/10). THMFP was determined relative to the residual free chlorine ($0.03 \, mg/dm^3$, $0.05 \, mg/dm^3$, $0.10 \, mg/dm^3$, $0.25 \, mg/dm^3$, $0.50 \, mg/dm^3$) and reaction time (24 h, 48 h, 72 h, 96 h). The samples were kept in the incubator at 20°C, and, after given reaction times, the TCM, BDCM, DBCM, and TBM concentrations were analyzed.

2.4 Analytical methods

The free chlorine was analyzed using the DPD (N, N-diethyl-p-phenylendiamine) method (according to Polish Standard PN-ISO 7393-2). The free chlorine concentration was measured using an Auris 2021 UV-VIS spectrophotometer (Cecil Instruments). The detection limit was $0.03 \, mg/dm^3$ using this method.

The THMs concentrations were analyzed using a gas chromatograph with a Trace Ultra DSQII GC-MS mass spectrometer (Thermo Scientific). Helium was used as the carrier gas. The Rxi™-1ms capillary column (Restek) was used (100% polydimethylsiloxane; film thickness $0.25 \, \mu m$; column length 30 m; column diameter 0.25 mm). The THMs were extracted using the Head-Space method and analyzed on the GS-MS following Polish Standard PN-EN ISO 10301. The samples were incubated at 90°C for 40 minutes, the column was heated from 31°C (0 min) to 200°C (0 min) with the temperature increase rate of 16°C/min. The method detection limit was $0.01 \, \mu g/dm^3$ for TCM, BDCM, DBCM and $0.03 \, \mu g/dm^3$ for TBM.

3 RESULTS AND DISCUSSION

3.1 THMFP and reaction time

The time of the chlorine and organic matter reaction has a significant influence on THMs concentration in water disinfected with sodium hypochlorite and with electro-chlorine (Figures 1, 2).

As can been seen in Figure 1, in water chlorinated with electro-chlorine and at higher residual free chlorine concentrations ($0.25 \, mg/dm^3$ and $0.50 \, mg/dm^3$), the highest total trihalomethane (TTHM) increase was observed in the first 48 hours of the reaction.

After this time the TTHM concentration stabilized, and only a slight further increase was observed (maximum 3 to $4 \, \mu g/dm^3$ in 24 h). For the samples chlorinated with sodium hypochlorite, the TTHM concentration stabilized after 24 h. With lower residual free chlorine concentrations, $0.03 \, mg/dm^3$, $0.05 \, mg/dm^3$ and $0.10 \, mg/dm^3$, (Figure 2), TTHM concentration had generally stabilized after 24 h. After this time an insignificant increase was observed. The experiments were concluded after 96 h – after this time THMs concentrations in all samples had stabilized.

The TTHM concentration in the samples chlorinated with electro-chlorine were approximately 30% higher than those in the samples treated with sodium

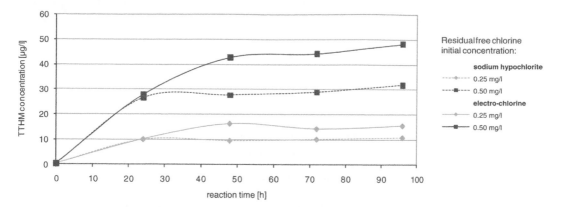

Figure 1.   TTHM formation kinetics in water disinfected with sodium hypochlorite and electro-chlorine (residual free chlorine concentrations: 0.25 mg/dm^3 and 0.50 mg/dm^3).

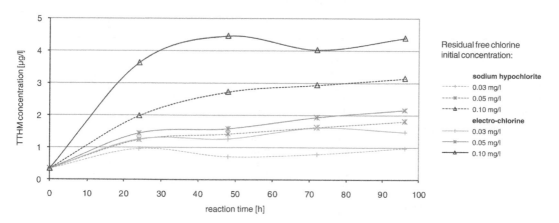

Figure 2.   TTHM formation kinetics in water disinfected with sodium hypochlorite and electro-chlorine (residual free chlorine concentrations: 0.03 mg/dm^3 and 0.05 mg/dm^3 and 0.10 mg/dm^3).

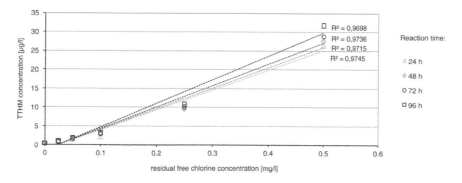

Figure 3.   TTHM concentration in water disinfected with sodium hypochlorite in relation to residual free chlorine.

hypochlorite. The TTHM concentration in any sample did not exceed $100\,\mu g/dm^3$, which is the Polish standard for TTHM in potable water (Rozp. Min. Zdrowia 2007). However the TCM concentration in the sample of water disinfected with electro-chlorine (for $0.5\,mg/dm^3$ residual free chlorine) after 96 h exceeded $30\,\mu g/dm^3$– the Polish standard for TCM in potable water (Rozp. Min. Zdrowia 2007).

3.2   *THMFP and residual free chlorine concentration*

The THMFP was significantly dependent on the residual free chlorine content. As can be observed in Figure 3 (for sodium hypochlorite) and Figure 4 (for electro-chlorine) the relationship between the TTHM concentration and the residual free chlorine for all the disinfectants examined is linear (R^2 between 0.9635

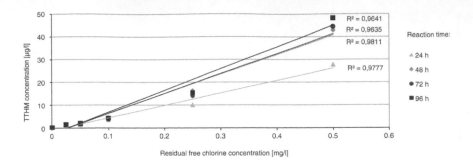

Figure 4. TTHM concentration in water disinfected with electro-chlorine in relation to residual free chlorine.

Table 2. Trihalomethane concentrations in water disinfected with sodium hypochlorite in relation to residual free chlorine and reaction time.

Residual free chlorine [mg/dm³]	Reaction time [h]				
	0	24	48	72	96
THM concentration [μg/l]					
0.03	0.24	0.66	0.56	0.65	0.65
0.05	0.24	0.98	1.08	1.20	1.26
0.10	0.24	1.42	2.21	2.40	2.55
0.25	0.24	8.53	7.64	8.39	8.91
0.50	0.24	20.82	21.64	22.85	24.45
BDCM concentration [μg/dm³]					
0.03	0.07	0.26	0.08	0.12	0.25
0.05	0.07	0.18	0.29	0.35	0.41
0.10	0.07	0.47	0.46	0.42	0.41
0.25	0.07	1.19	1.52	1.40	1.52
0.50	0.07	5.02	5.25	5.33	6.42
DBCM concentration [μg/dm³]					
0.03	0.03	0.03	0.07	0.07	0.08
0.05	0.03	0.09	0.06	0.09	0.12
0.10	0.03	0.10	0.06	0.13	0.15
0.25	0.03	0.16	0.31	0.29	0.26
0.50	0.03	0.60	0.77	0.68	0.81
TBM concentration [μg/dm³]					
0.03	nd	nd	nd	nd	nd
0.05	nd	nd	nd	nd	0.03
0.10	nd	nd	nd	nd	0.03
0.25	nd	nd	nd	nd	0.05
0.50	nd	nd	nd	nd	0.11
TTHMs concentration [μg/dm³]					
0.03	0.34	0.95	0.71	0.79	0.97
0.05	0.34	1.25	1.43	1.64	1.82
0.10	0.34	1.99	2.73	2.95	3.15
0.25	0.34	9.89	9.48	10.08	10.74
0.50	0.34	26.44	27.65	28.86	31.78

nd – not detected

and 0.9811). The linear correlation equations for each reaction time of the organic matter with chlorine are as follows:

for sodium hypochlorite:

- reaction time 24 h: $y = 52.875x - 1.3425$ ($R^2 = 0.9745$);
- reaction time 48 h: $y = 54.808x - 1.3921$ ($R^2 = 0.9715$);
- reaction time 72 h: $y = 57.160x - 1.3703$ ($R^2 = 0.9736$);
- reaction time 96 h: $y = 62.736x - 1.5361$ ($R^2 = 0.9698$);

and for electro-chlorine:

- reaction time 24 h: $y = 54.691x - 1.0159$ ($R^2 = 0.9777$);

Table 3. Trihalomethane concentrations in water disinfected with electro-chlorine in relation to residual free chlorine and reaction time.

Residual free chlorine [mg/l]	Reaction time [h]				
	0	24	48	72	96
THM concentration [μg/dm^3]					
0.03	0.24	0.87	0.97	1.09	0.97
0.05	0.24	1.12	1.17	1.47	1.57
0.10	0.24	2.47	3.14	2.95	3.24
0.25	0.24	6.67	11.60	12.29	10.93
0.50	0.24	22.42	32.65	34.20	36.62
BDCM concentration [μg/dm^3]					
0.03	0.07	0.19	0.23	0.36	0.45
0.05	0.07	0.23	0.32	0.37	0.45
0.10	0.07	1.01	1.06	0.87	0.95
0.25	0.07	2.97	3.88	1.52	3.78
0.50	0.07	4.62	8.70	9.23	9.82
DBCM concentration [μg/dm^3]					
0.03	0.03	0.20	0.07	0.16	0.06
0.05	0.03	0.09	0.10	0.10	0.10
0.10	0.03	0.15	0.27	0.21	0.16
0.25	0.03	0.40	0.67	0.37	0.64
0.50	0.03	0.76	1.41	0.90	1.64
TBM concentration [μg/dm^3]					
0.03	nd	nd	nd	nd	nd
0.05	nd	nd	nd	nd	0.04
0.10	nd	nd	nd	nd	0.04
0.25	nd	nd	nd	nd	0.05
0.50	nd	nd	nd	nd	0.13
TTHMs concentration [μg/dm^3]					
0.03	0.34	1.26	1.27	1.61	1.48
0.05	0.34	1.44	1.59	1.94	2.17
0.10	0.34	3.63	4.47	4.04	4.40
0.25	0.34	10.03	16.15	14.18	15.39
0.50	0.34	27.79	42.76	44.32	48.20

nd – not detected

- reaction time 48 h: $y = 85.852x - 2.1394$ ($R^2 = 0.9811$);
- reaction time 72 h: $y = 87.458x - 2.4114$ ($R^2 = 0.9635$);
- reaction time 96 h: $y = 95.284x - 2.6925$ ($R^2 = 0.9641$).

where x = residual free chlorine concentration in mg/dm^3; and y = TTHM concentration in μg/dm^3.

3.3 THMs species

As is shown in Tables 2 and 3, the distribution of THMs in the samples chlorinated with sodium hypochlorite and electro-chlorine is similar. In all samples, it is predominantly TCM that has been formed (between 67% and 86% in samples chlorinated with electro-chlorine and between 66% and 87% in the samples chlorinated with sodium hypochlorite). The next most abundant is BDCM (between 11% and 20% for electro-chlorine and between 11% and 30% for sodium hypochlorite). The third most abundant THM is DBCM (between 1% and 10% for electro-chlorine and between 2% and 16%

for sodium hypochlorite). TBM was not detectable in most samples. A small amount (from 0% to 2%) was detected for the longest reaction time (96 h) in samples disinfected with sodium hypochlorite and those treated with electro-chlorine.

4 CONCLUSIONS

The research shows that disinfection of water with the chlorine produced in the electrolyzers (with the separation of the products formed at the anode and cathode) resulted in the higher trihalomethanes formation potential than resulted from disinfecting with the traditional sodium hypochlorite. With the same residual free chlorine concentrations in the samples, disinfection with electro-chlorine resulted in approximately 30% more TTHM being generated.

For both disinfectants, the samples demonstrated a linear relationship between the TTHM concentrations and the free chlorine content. The distribution of THMs did not differ between the samples chlorinated with electro-chlorine and those chlorinated with sodium hypochlorite. TCM was the most abundant THM, representing between 67% and 86% of the THMs formed in samples chlorinated with electro-chlorine, and between 66% and 87% in those chlorinated with sodium hypochlorite.

During the electrolysis of brine, chlorine is produced along with some ozone and chlorine dioxide (Iriarte-Velasco et al. 2007). These latter products are powerful oxidants and, in comparison with chlorine alone, these mixed oxidants are more efficient at removing bacteria. The presence of such strong oxidants as ozone and chlorine dioxide causes the conversion of the hydrophobic fraction of the NOM to a hydrophilic one (Swietlik et al. 2004, Iriarte-Velasco et al. 2007). According to Marhaba and Van (2000) the hydrophilic acid fraction contains the most reactive precursors of THMs formation. Other research papers indicate that the hydrophilic base fraction of the NOM was one of the most inactive organic precursors of THMs formation (Panyapinyopol et al. 2005). The conversion of the NOM hydrophobic fraction to a hydrophilic one could be a reason for the higher THMs concentration in samples disinfected with electro-chlorine as compared to those chlorinated with sodium hypochlorite.

In this particular research only the relationship with the residual free chlorine content has been tested when comparing the higher efficiency of the electro-chlorine disinfection versus that of the sodium hypochlorite (Wasowski & Wasciszewska 2003, Iriarte-Velasco et al. 2007). Given the lower doses of electro-chlorine that are necessary for the same disinfection effect, the real TTHM concentrations would probably be lower in water treated this way than in water disinfected with sodium hypochlorite.

To study this phenomenon, further research, in which the relationship between THMFP and other factors (for instance the disinfection effects) would need to be investigated.

ACKNOWLEDGEMENTS

The work was completed under AGH-UST statutory research No. 11.11.150.008 for the Department of Management and Protection of Environment. The authors gratefully acknowledge the assistance of the Water System company and MPWiK, Cracow, in the preparations for the experiments.

REFERENCES

Bashtan, S., Goncharuk, V., Chebotareva, R., Belyakov, V. & Linkov, V. 1999. Production of sodium hypochlorite in an electrolyser equipped with a ceramic membrane. *Desalination* 126: 77–82.

Bergmann, H., Koparal, A.T., Koparal, A.S. & Ehrig F. 2008. The influence of products and by-products obtained by drinking water electrolysis on microorganisms. *Microchemical Journal* 89: 98–107.

Brandt, D. 1995. An electrolytic chlorination system for pretreatment and post-treatment in desalination systems. *Desalination* 102: 321–324.

Destouches, P. & Langlais, B. 1996. Electrochlorination – a prospective metod of water disinfection. *Ochrona Srodowiska* 60: 11–13.

Iriarte-Velasco, U., Alvarez-Uriarte, J. & Gonzalez-Velasco, J. 2007. Removal and structural changes in natural organic matter in a Spanish water treatment plant using nascent chlorine. *Separation and Purification Technology* 57: 152–160.

Lee, H-K., Yeh, Y-Y. & Chen, W-M. 2006. Cancer risk analysis and assessment of trihalomethanes in drinking water. *Stochastic Environmental Research and Risk Assessment* 21: 1–13.

Marhaba, T. & Van, D. 2000. The variation of mass and disinfection by-product formation potential of dissolved organic matter fractions along a conventional surface water treatment plant. *Journal of Hazardous Materials* A74: 133–147.

Panyapinyopol, B., Marhaba, T., Kanokkantapong, V. & Pavasant P. 2005. Characterization of precursors to trihalomethanes formation in Bangkok source water. *Journal of Hazardous Materials* B120: 229–236.

Rozp. Min. Zdrowia (Rozporzadzenie Ministra Zdrowia) 2007. Rozporzadzenie w sprawie jakosci wody przeznaczonej do spozycia przez ludzi. *Dz.U.* nr 61, poz. 417.

Swietlik, J., Dabrowska, A., Raczyk-Stanislawiak, U. & Nawrocki J. 2004. Reactivity of natural organic matter fractions with chlorine dioxide and ozone. *Water Research* 38: 547–558.

Wang, G-S., Deng, Y-C. & Lin T-F. 2007. Cancer risk assessment from trihalomethanes in drinking water. *Science of the Total Environment* 387: 86–95.

Wasowski, J. & Wasciszewska, M. 2003. Technological and economical aspects of water disinfection with the compounds produced in electrolyzers. *Ochrona Srodowiska* 25(4): 23–26.

World Health Organization (WHO) 2005. *Trihalomethanes in drinking-water: background document for development of WHO guidelines for drinking-water quality.* World Health Organization, Geneva (WHO/SDE/WSH/05.08/64).

Environmental Engineering III – Pawłowski, Dudzińska & Pawłowski (eds)
© 2010 Taylor & Francis Group, London, ISBN 978-0-415-54882-3

Advanced oxidation techniques in elimination of endocrine disrupting compounds present in water

E. Zacharska, H. Zatorska, B. Rut & J. Ozonek

Faculty of Environmental Engineering, Lublin University of Technology, Lublin, Poland

ABSTRACT: Many industries as well as agriculture, medicine and other human activities contribute to the growth of endocrine disrupting compounds (EDs) in the environment. This paper provides an overview of EDs, their impact on the environment and living organisms, their sources and migration pathways, and their degradation methods. The paper also presents the results of research into the use of ozonation to degrade selected pharmaceuticals: clofibric acid (CA) and 17 α-ethinyloestradiol (EE2). These results indicate that ozonation can be used for EE2 degradation, but it is not sufficient to significantly degrade clofibric acid. The efficiency for CA degradation was about 10%, while it was almost 100% for EE2 after 30 minutes ozonation. Further research on increasing ozonation efficiency by combining different methods and on by-products identification is required.

Keywords: Endocrine disrupting compounds (EDs), ozonation, advanced oxidation processes (AOPs), clofibric acid (CA), 17 α-ethinyloestradiol (EE2).

1 INTRODUCTION

The intensive industrial, scientific and human development in recent decades has generated new and complex chemical compounds. These substances are not neutral and they leave a permanent trace in the environment (Jngerslev et al. 2003). Scientists in several fields, including ecology, epidemiology, endocrinology and toxicology, have noted a group of chemical substances which can disturb endocrine functions when penetrating an organism, posing a serious danger to the health and life of people and animals. The compounds are called endocrine disruptors (EDs). According to the Environmental Protection Agency's (EPA) definition (of May 1997), EDs are exogenous agents that interfere with the synthesis, secretion, transport, binding, action and elimination of natural hormones in the body that are responsible for the maintenance of homeostasis, reproduction, development and/or behaviour. They disturb almost all the main vital functions of the living organisms, including growth and reproduction (Biłyk et al. 2003, Campbell et al 2006, Lintelman et al. 2003). Groups of substances that are classified as EDs include polycyclic aromatic hydrocarbons (PAHs), polychlorinated biphenyls (PCBs), pesticides, surfactants, personal hygiene products, pharmaceutical residues and some heavy metals (such as cadmium and mercury). As well as the anthropogenic compounds now present in the environment, some natural substances also show estrogenic activity. These natural substances are usually present in living organisms and plants, and consist of natural hormones produced by the living organisms as well as natural environmental estrogens (phytoestrogens), the compounds

occurring in plants (Kraszewska et al. 2007). Phytoestrogens show structural similarity to the natural female hormone – estradiol – and have the ability to combine with hormonal receptors, causing altered responses. High doses of environmental estrogens may influence the neural system and reproduction abilities by imitating or antagonising endogenous estrogens. Nevertheless, phytoestrogens do not show the ability for bioaccumulation in the tissues of living organisms, and they are easily degradable in comparison to synthetic estrogens. Endocrine disruptors constitute a serious problem, not only because of their negative influence on organisms but also because of their environmental persistence – conventional methods used in water and wastewater management do not provide a sufficient degree of degradation. Therefore, studies on endocrine disruptors' degradation constitute one of the main technological research areas in water, wastewater and landfill leachate management. Chemical oxidation, using ozone, seems to be promising for permanent organic contaminants degradation. Ozone, with its ability to disinfect bacteria and viruses, and its high oxidative potential, is more and more frequently used in water conditioning – most often for disinfection, iron and manganese oxidation and precipitation, taste improvement and organic substances oxidation. The aim of this work is to present the results from research trials on the degradation of selected endocrine disruptors through the use of ozone oxidation.

1.1 Sources of endocrine disruptors in the environment

The endocrine disruptors present in different parts of the environment originate both in large industrial

Table 1. Examples of EDs and their sources in the environment.

The class of EDs	Example compounds	Source
Phthalates	dibutyl phthalate, butylobenzyl phthalate	Used in plastics production and in plasticisers
Pesticides	DDT, atrazine, lindane, hexsachlorobenzene,aldryna	Widely used in agriculture; insecticides, herbicides, fungicides
Dioxins and furans	polychlorinated dibenzo-p-dioxins and dibenzofurans	Produced during the incineration of chlorinated organic compounds and during the production of plastics
Polycyclic aromatic hydrocarbons (PAHs)	fluorene, phenanthrene, antracene, naphtalene	The products of incomplete combustion of organic compounds
Heavy metals	cadmium, mercury, lead	Steel industry, mining
Phytoestrogens	genistein, daidzein, kumestrol	Natural substances present in plants
Bisphenols	bisphenol A	Used in polymers production
Pharmaceuticals	carbamazepine, ibuprofen, 17-α-ethinylestradiol, klofibrat, diclofenac	Hormonal drugs, anti-inflammatory drugs, cholesterol-lowering medicines, beta blockers

plants and small households. Nevertheless, the main sources include the chemical industry, which manufactures different products using plastics and polycarbonate resins, and agriculture, which uses pesticides on arable land and growth stimulating factors, including synthetic hormones, on large stock farms. The pharmaceutical industry also plays a significant role in the introduction of endocrine disruptors to the environment. Moreover, some medicines are not fully by organisms and some excreted metabolites show endocrine activity. Table 1 lists examples of different EDs and their main sources.

Some EDs are transported by air, especially combustion products such as dioxins, PAHs and polychlorinated biphenyls (PCBs) (Nakui et al. 2007). They can spread over very long distances from their original point of release into the environment. PCBs have been found in both Arctic and Antarctic environments, mainly as a result of air transport from Europe, Asia and North America (Choi et al. 2008). Some compounds migrate with the run-off from industrial and municipal wastewater, and landfill leachate, and as a result they penetrate to surface and groundwater, from where they can be easily absorbed by water organisms (Boyd et al. 2003, Buser et al. 1998, Cambell et al. 2006, Carballa et al. 2004, Dudziak et al. 2004, Farre et al. 2001, Kaleta et al. 2004, Kot-Wasik et al. 2003, Lintelman 2003, Weigel et al. 2002). These compounds are accumulated in the tissues of living organisms, and fish are especially endangered as they show the ability to accumulate different compounds in their adipose tissue. Transfer to other organisms is not difficult, as the compounds start to appear in the food chain.

Some substances are no longer produced because of their toxicity, but they are still present in the environment. This illustrates the persistence of some compounds and their tendency for bioaccumulation. Polychlorinated biphenyls (PCBs) were widely used as additives in anticorrosion paints and the cooling systems of transformers and heat exchangers because of their physico-chemical properties (Nakui et al. 2007, Choi et al. 2008).

About 1.5 million tonnes of polychlorinated biphenyls were produced worldwide, and two-thirds were still present in the environment at the end of the twentieth century (UNEP report 1999). PCBs are not produced today, so their migration is a consequence of redistribution in the environment. Similarly, pesticides belonging to the group of chlorinated hydrocarbons have been withdrawn from the use in North America and Europe due to their high durability and toxicity (Kaleta et al. 2004), nevertheless their residues are still found in the environment. These examples prove the high persistence and resistance to biodegradation of endocrine disruptors. Some EDs undergo slow transformations as a consequence of different biological, chemical and physical factors, however in natural conditions these processes rarely decompose the compound completely to carbon dioxide and water. Usually these transformations lead to new intermediates, which may reveal different properties that sometimes can be more dangerous than the parent substance.

The presence of EDs is also connected with the translocation of substances between different parts of the environment and their bioaccumulation in living organisms, and with the increase of their concentrations in the successive trophic levels, so called biomagnification. Therefore, not only are the organisms at the beginning of the food chain endangered, but also human life and health are threatened (Biłyk et al. 2003, Campbell et al. 2006, Dudziak et al. 2004, Farre et al. 2001, Jngerslev et al. 2003, Kraszewska et al. 2007, Lintelman 2003).

Pharmaceuticals constitute a large group of endocrine disruptors. They reach the environment from different sources, including households, hospitals and drug manufacturers. A significant number of medicines that are used frequently by people, such as antibiotics, hormones, sedatives, analgesics and contraceptives, are not entirely metabolised in their organisms, and they are excreted in urine or in faeces. Contemporary wastewater treatment technologies do not cause the complete degradation of pharmaceuticals metabolites, thus some harmful substances may reach the environment.

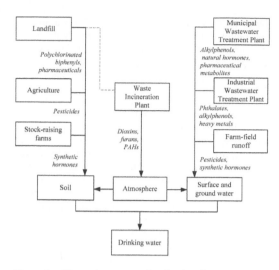

Figure 1. The main sources of endocrine disruptors in the environment.

1.2 Endocrine disruptors effects on living organisms

The work of the whole endocrine system responsible for the main mechanisms controlling different functions in the organisms is disrupted by EDs. Hormones, produced by endocrine glands, are the main information transmitters in this system. They are transported within the blood system to the appropriate internal organs, where they are bind with the receptors inducing the organism's biological response. Any disturbance to this mechanism may cause serious and permanent damage. The actions of EDs are complicated, and they depend on the kind of compound and its dose. Some EDs act as antagonists, blocking the receptor and preventing the organism from giving any response, others act like agonists affecting the appropriate biological response. Some substances may even influence the synthesis and release of hormones, altering their concentration in blood or haemolymph and changing various metabolic processes. The various impacts of EDs observed in living organisms are widely described in the literature. The effects most often mentioned include thyroid dysfunctions, changes in the immune system of sea mammals, anomalies in reproductive systems and metabolic disturbances in fish, reptiles, birds and mammals, a reduction in the number of hatching birds, and feminisation in males and masculinisation in females (Biłyk et al. 2003, Campbell et al. 2006, Dudziak et al. 2004, Kot-Wasik et al. 2003, Lintelman et al. 2003). In some cases, this can lead to population decrease. Moreover, there are also distressing reports about human exposure to different compounds. Endocrine disruptors are supposed to be responsible for increasing number of cancer cases, a decrease in sperm quantity and quality, and endometriosis. Today, reduced fertility and problems connected with conception and reproduction are serious issues. Not only has our way of life changed, but so has the environment in which we live. One substance

Table 2. Concentrations of ethinyloestradiol and estradiol measured in different countries (Ying et al. 2002).

Country	Ethinyloestradiol $\mu g/m^3$	Estradiol $\mu g/m^3$
Italy	<LOD[a]– 1.7	0.4 – 3.3
Netherlands	<LOD – 7.5	0.1 – 5
Germany	<LOD – 15	<LOD – 3
Canada	<LOD - 42	<LOD – 64

[a]LOD- limit of detection

which has been proved to be harmful for reproduction is dietylstilbestrol (DES). This synthetic estrogen was used as a medicine in the 1950s and 1960s, and its harmful effects were observed in the offspring of women who used this drug.

Contraceptives and hormones used in substitutive therapy are the most common pharmaceuticals influencing endocrine system. Contraceptives usually contain on average 20–40 µg of 17-α-ethinylestradiol (EE2). It is a synthetic equivalent for the natural hormone estradiol, and its endocrine potential is similar to the natural one, though its persistence in the environment is higher. Common usage has resulted in the presence of these substances (and their metabolites) in lakes, rivers and, even, inland seas. The conventional technologies used in wastewater treatment plants do not eliminate estrogens in sufficient levels. The highest measured concentrations of this synthetic hormone were found in treated sewage in Germany and in Canada (Table 2).

It is not just hormonal pharmaceuticals that show endocrine activity. Some compounds show similarities to others known as endocrinically active. For example, clofibric acid has a chemical similarity to a known herbicide mecoprop (methylchlorophenoxypropionic acid, MCPP) (Buser et al. 1998). Clofibric acid is a metabolite of klofibrat, a medicine that lowers the concentration of triglycerides and cholesterol in blood. Its migration and persistence in the environment has been examined by many researchers, and it is found in sewage effluent, surface water reservoirs, and even in ground water. The concentrations measured in the environment are on the level of ng/dm³ (Andreozzi et al. 2003, Buser et al. 1998, Carballa et al. 2004, Farre et al. 2001, Kot-Wasik 2003, Ternes et al. 2002, Weigel et al. 2002, Wenzel et al. 2003, Zwiener et al. 2000). The fact that clofibric acid is present even in groundwater and it is harmful for living organisms proves that it constitutes a serious ecological problem. Structural formulas of selected compounds are shown in Figure 2.

1.3 The methods of endocrine disruptors degradation

Conventional methods used in wastewater treatment plants and in water treatment stations do not eliminate or degrade endocrine disruptors completely. The research indicates that 60–90% of these compounds are removed from municipal sewage, depending on the technology used. In systems which use biological

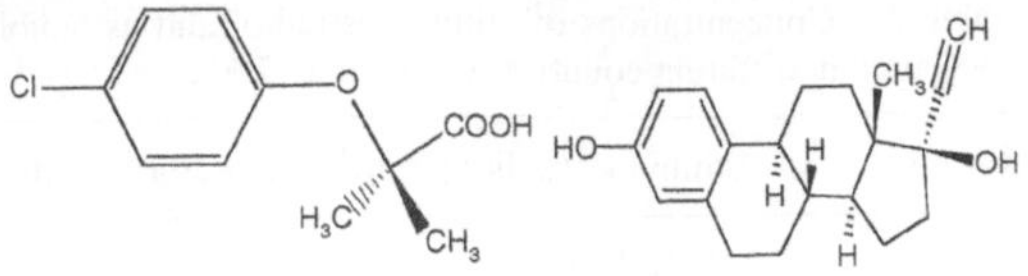

Clofibric acid ($C_{10}H_{11}ClO_3$)   17α-etynyloestardiol
($C_{20}H_{24}O_2$)

Figure 2. Structural formulas of clofibric acid and 17-α-etynyloestardiol.

reactors, some EDs may adsorb on the surface of sewage sludge and thus penetrate into the environment easily as a consequence of sewage sludge management. The efficiency of removing pharmaceuticals from municipal wastewater using a system with a primary settler, biological reactor and secondary settler has been examined. The results were 40–65% for anti-inflammatory drugs, 60% for antibiotics and 65% for synthetic hormones (Carballa et al. 2004). Vieno at al. examined chemical coagulation as a method for EDs elimination, however they concluded that by using this method alone it is not possible to remove pharmaceuticals from water entirely (Vieno et al. 2006). Snyder et. al examined many other methods, such as microfiltration, ultrafiltration, nanofiltration, reverse osmosis, reversed electrodialysis, membrane bioreactors and granular activated carbon and, despite the fact that their trials gave very good results in reducing the amount of EDs, several compounds were still detectable in the membrane permeate and carbon effluent (Snyder et al. 2007). Chemical oxidation seems to be considerably more efficient in eliminating EDs. Ozone appears to be one of the most effective oxidants; its oxidizing potential is 2.07 V, while chlorine is only 1.36 V. Solutions involving the ozonation of water may run either as reactions with molecular ozone or as free radicals processes. The reactions that use molecular ozone are very selective, and their effectiveness is limited to the elimination of unsaturated aliphatic compounds, and aromatic and amino compounds (Perkowski et al. 2005), while reactions with free radicals are less selective and they proceed in alkaline environment. Ozonation seems to be an efficient method for organic contaminant degradation, allowing a reduction in the amount of toxic, mutagenic and carcinogenic by-products generated during the process.

Nevertheless, this method has limitations; the main ones appear to be low solubility in water, low stability and the high costs of ozone production. Therefore there are a number of research studies that are attempting to adapt and increase the efficiency of ozonation in the elimination of toxic pollutants from the environment (Andreozzi et al. 2003, Esplugas et al. 2007, Ijeplaar et al. 2000). One of the alternatives to the conventional techniques of ozone application in water and wastewater treatment is the use of advanced oxidation processes (AOPs) (Andreozzi et al. 2003, Esplugas et al. 2007, Ijeplaar et al. 2000, Schulte et al. 1995). They lead to the generation of the OH radical, which

by its high oxidative potential (of 2.8 V) is a promising factor to use in the removal of persistent organic contaminants (Perkowski et al. 2005). There are several processes which generate OH radicals, including ozone oxidation, hydrodynamic and ultrasound cavitation, Fenton reactions, and thermal plasma reactions as well as a combination of these methods. The efficiency of the AOP method depends mainly on the nature of the pollutants present in the treated solution, but also on the type and dose of reagents used in the process, the proportion of oxidants to the compounds being oxidised, the reaction time, and how and where the method is introduced into the technological system. Despite the variety of advanced oxidation techniques, there are some general factors that support OH radicals formation such as hydrogen peroxide, UV radiation and chemical catalysts (Biń et al. 1998). Table 3 presents selected results for compound degradation obtained by various research groups using different methods.

1.4 Degradation of selected endocrine disruptors in laboratory conditions

Clofibric acid and 17-α-ethinyloestradiol are endocrine disrupting compounds belonging to the group of pharmaceuticals. They are found in many different parts of the environment and therefore their elimination is important. Many research groups have indicated that the levels of these EDs in the environment fluctuate with the levels of ng/L and μg/L (Alum et al. 2004, Boyd et al. 2003, Buser et al.1998, Campbell et al. 2006, Dudziak et al. 2004, Farre et al. 2001, Kot-Wasik 2003, Snyder et al. 2003, Weigel et al. 2002).

These compounds were selected for this study due to the fact that they are commonly used and their presence in the environment causes various changes in organisms exposed to their action. The main dysfunctions are connected with fertility, growth, metabolism and other vital processes.

2 MATERIALS AND METHODS

The trails of methods to degrade the two selected endocrine disruptors were carried out in the Environmental Analysis Laboratory of the Environmental Engineering Faculty of Lublin University of Technology. Clofibric acid and 17-α-etihnyloestradiol were prepared by dissolving in methanol. From the prepared samples 1 cm³ was diluted in 2 dm³ of HPLC-super gradient water for the sample of clofibric acid, and 1.92 dm³ of water from the Milipore system (Millipore Poland) and 0.8 dm³ of methanol for 17-α-ethinyloestradiol. Adding methanol to the EE2 solution was needed because of the limited solubility of the compound in water. The ozonation was carried out on a laboratory test stand, described in an earlier publication (Szkutnik et al. 2008). For this study, several chemical determinations were run for each solution. The prepared solution was introduced each time to a glass reactor, which had a diffuser in its bottom section allowing the introduction of air for mixing and ozone

Table 3. Selected results for compound degradation using different methods.

Compound	Type of tested solution	Process used for degradation	Process parameters	Research results	References
Clofibric acid	Drinking water	O_3/H_2O_2	$O_3/H_2O_2 = 2$; $C_{O3} = 5\,mg/dm^3$; $\tau_r = 10\,min$; $C_0 = 2\,\mu g/dm^3$	Removal degree 97.9%	Zwiener et al.
Clofibric acid	Drinking water	ozonation	$C_0 = 1\,\mu g/dm^3$; $C_{O3} = 0.5$–$3\,mg/dm^3$; $\tau_r = 20\,min$	For $C_{O3} = 0,5\,mg/dm^3$ Removal degree 10–15% For $C_{O3} = 3\,mg/dm^3$ Removal degree 40%	Ternes et al.
17-α-ethinyloestradiol	Distilled water	Ozonation	$C_0 = 100\,nmol/dm^3$ $t = 20°C$ $C_{O3} = 1.5\,mg/dm^3$	Removal degree 100%	Alum et al.
Atrazine	Ground water	O_3/H_2O_2	$pH = 7$ $H_2O_2/O_3 = 3.7$ $C_{H2O2} = 8,8\,mg/dm^3$	Removal degree 99%	Ijeplaar et al.
		Fe^{2+}/H_2O_2	$pH = 5.5$ $C_{H2O2} = 10\,mg/dm^3$ $C_{Fe^{2+}} = 5.1\,mg/dm^3$	Removal degree 75%	
Bisphenol A	Distilled water	ozonation	$C_0 = 100\,nmol/dm^3$ $t = 20°C$ $C_{O3} = 1.5\,mg/dm^3$	Removal degree 100%	Alum et al.

into the ozonation process. The first phase consisted of mixing the solution with air for at least 10 minutes and taking the 0 sample. In the next phase, the ozone generator was turned on and ozone was introduced to the reactor.

2.1 Clofibric acid ozonation

For clofibric acid, the process was examined under different conditions. In the first trail samples were taken every 5 minutes during ozonation (till the ozonation time reached 30 minutes). The flow of ozone was $0.9\,dm^3/min$ and ozone concentration $1.8\,g/Nm^3$.

The other trial lasted longer, and the ozonation lasted 95 minutes. The first sample was taken after first 5 minutes of ozonation, and subsequent samples were taken at 10 minutes time intervals. The flow of ozone was the same as in the first trial, but the ozone concentration was higher at $2.2\,g/Nm^3$.

2.2 17-α-ethinyloestradiol ozonation

The trial lasted for 60 minutes. The first sample was taken after 5 minutes of ozonation and next samples were taken at 5-minute time intervals. The flow of ozone was $0.9\,dm^3/min$ and the ozone concentration was $2.2\,g/Nm^3$.

Chromatographic analysis was carried out with the use of a liquid chromatograph Agilent 1200 series, consisting of a binary pump, automatic thermostatic samples batcher, columns thermostat and using DAD detectors and the MS/MS Q-Trap 4000 system from Applied Biosystems (Canada).

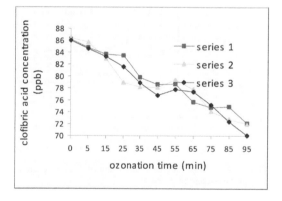

Figure 3. Clofibric acid concentration changes during the ozonation process.

3 RESULTS AND DISCUSSION

The results obtained during the first trial of CA ozonation, which lasted for 30 minutes, gave an average degradation efficiency of 8.9%. In the next trial, the ozonation process was longer and the concentration of ozone was higher to check if these factors change the degradation efficiency significantly. There was indeed an increase of the degradation efficiency but it was not significant, amounting on average 17.2%. The changes in CA concentration during the three experimental replications (marked as a series in Figure 4) of the ozonation process lasting for 95 minutes are shown in Figure 3.

These results suggest that unassisted ozonation is not a sufficient method for clofibric acid degradation. Therefore further studies are needed to increase the effectiveness of ozonation by combining it with other factors, such as UV radiation or hydrogen peroxide addition.

The tests of 17-α-ethinyloestradiol ozonation gave satisfying results. After 15 minutes of ozonation the degradation efficiency amounted on average to 99.94%. Longer period of ozonation gave 100% EE2 reduction. EE2 concentration changes during the two experimental runs (series 1 and 2) lasting for 60 minutes of ozonation are shown in Figure 4.

4 CONCLUSIONS

The results suggest that unassisted ozonation is not a sufficient method for clofibric acid degradation. Further studies are needed to increase the effectiveness of ozonation by combining it with other factors, such as UV radiation or hydrogen peroxide addition.

The results for EE2 degradation seem to be promising as the degradation occurs in a satisfying percentage, nevertheless the originating oxidation products should not be omitted in an evaluation of the process. It is necessary to recognise and identify the factors affecting the process and the products of the degradation process.

The concentration of the examined substance decreased significantly, nevertheless chromatographic analysis revealed the presence of an important intermediate of this degradation. The chromatogram showing gradual degradation of EE2 and an increasing amount of the intermediate is shown in Figure 5. The mass of the intermediate compound suggests that this is a close relative of EE2.

The results for 17-α-ethinyloestradiol seem to be promising as the degradation occurs in a satisfying percentage, nevertheless the originating oxidation products should not be omitted in an evaluation of the process. It is necessary to recognise and identify the factors affecting the process and the products of degradation process.

The concentration of the examined substance decreased significantly, nevertheless chromatographic analysis revealed the presence of an important intermediate of this degradation. The chromatogram showing gradual degradation of EE2 and an increasing amount of the intermediate is shown in Figure 5. The mass of the intermediate compound suggests that this is a close relative of EE2.

The results for 17-α-ethinyloestradiol seem to be promising as the degradation occurs in a satisfying percentage, nevertheless the originating oxidation products should not be omitted in an evaluation of the process. It is necessary to recognise and identify the factors affecting the process and the products of degradation process.

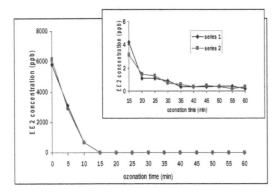

Figure 4. 17-α-ethinyloestradiol concentration changes during the ozonation process.

ACKNOWLEDGEMENTS

The research described in this paper were done thanks to the Ministry of Science and Higher Education within the framework of the project " Degradation of

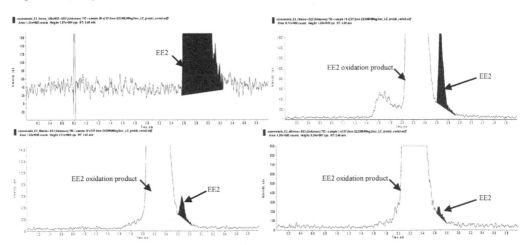

Figure 5. The chromatogram of gradual degradation of EE2.

bisphenols in water solutions with the use of ozone",
No. N N523 495134.

REFERENCES

Alum, A., Yoon, Y., Westerhopff, P. & Abbaszadegan, M. 2004. Oxidation of bisphenol A, 17β-estradiol and 17α-ethynyloestradiol and byproduct estrogenicity. *Environment Toxicology* 19: 257–264.

Andreozzi, R., Caprio, V., Marotta, R. & Radovnikovic, A. 2003. Ozonation and H2O2/UV treatment of clofibric acid in water: a kinetic investigation. *Journal of Hazardous Material* 103: 233–246.

Biłyk, A. & Nowak-Piechota, G. 2003. Polution of the environment with the substances causing disruption of the organism endocrine functions. (in polish) *Ochrona Środowiska* 3: 29–35.

Biń, A.K. 1998. Application of advanced oxidation processes for water treatment. (in polish) *Ochrona Środowiska* 1: 4–6.

Boyd, G.R., Reemsta, H., Grimm, D.A. & Mitras, S. 2003. Pharmaceuticals and personal care products (PPCPs) in surface water and treated waters in Louisiana, USA and Ontario, Canada. *The Science of the Total Environment* 311: 135–149.

Buser, H., Muller, M. & Theobald, N. 1998. Occurrence of the pharmaceuticals drug clofibric acid and the herbicide mecoprop in various Swiss Lake and in the North See. *Environmental Science and Technology* 32: 188–192.

Campbell, C.G., Borglin, S.E., Green, F.B., Grayson, A., Wozei, E. & Stringfellow, W.T. 2006. Biologically direct environmental monitoring, fate and transport of estrogenic compounds in water: A review. *Chemosphere* 65: 1265–1280.

Carballa, M., Omil, F., Lema, J.M., Llompart, M., Garcia-Jere,s C., Rodriguez, I., Gomez, M. & Ternes, T. 2004. Behavior of pharmaceutical, cosmetics and hormones in a sewage treatment plant. *Water Research* 38: 2918–2926.

Choi, Sung-Deuk; Baek, Song-Yee; Chang, Yoon-Seok; Wania, Frank; Ikonomou & Michael G. 2008. Passive Air Sampling of Polychlorinated Biphenyls and Organochlorine Pesticides at the Korean Arctic and Antarctic Research Stations: Implications for Long-Range Transport and Local Pollution. *Environmental Science & Technology* 42(19): 7125–7131.

Dudziak, M. & Luks-Betlej, K. 2004. Estimation of estrogens presence - steroid hormones In selected river waters in Poland. (in polish) *Ochrona Środowiska* 1: 21–24.

Esplugas, S., Bila, D., M., Krause, L. & Dezotti, M. 2007. Ozonation and advanced oxidation technologies to remove endocrine disrupting chemicals (EDCs) and pharmaceuticals and personal care products (PPCPs) in water effluent. *Journal of Hazardous Materials* 149: 631–642.

Farre, M., Ferrer, I., Ginebreda, A., Figuera,s M., Olivella, T., Tirapu, L., Vilanova, M. & Barcelo, D. 2001. Determination of drugs in surface water and wastewater samples by liquid chromatography-mass spectrometry: methods and preliminary results including toxicity studies with Vibrio fischer. *Journal of Chromatography A* 938: 187–197.

Ijeplaar, G.F., Meijers, R.T. & Hopman, R. 2000. Oxidation of herbicides in ground water by the Fenton process: a realistic alternative for O3/H2O2 treatment? *Ozone: Science and Engineering* 22: 607–616.

Jngerslev, F., Vaclavik, E. & Halling-Sorensen, B. 2003. Pharmaceuticals and personal care products: A source of endocrine disruption in the environment? IUPAC. *Pure and Applied Chemistry* 76: 1881–1893.

Kaleta J. 2004. Pesticides in water environment, (in polish) Zeszyty Naukowe Politechniki Rzeszowskiej Nr 218. *Budownictwo i Inżynieria Środowiska* 38: 23–37

Kot-Wasik, A., Dębska, J. & Namieśnik, J. 2003. The changes in concentration and determination of pharmaceutical residues in the environment. New horizons and challenges in environmental analysis and monitoring. Chapter 34 (in polish), Centrum Doskonałości Analityki i Monitoringu Środowiska (CEEAM). Wydział Chemiczny. Politechnika Gdańska: 722–744.

Kraszewska, O., Nynca, A., Kamińska, B. & Ciereszka, R. 2007. Phitoestrogens. I. Occurence, metabolizm and biological activity in females. (in polish) *Postępy biologii komórki* 34: 189–205.

Lintelman, J. 2003. Endocrine disruptors in the environment. *Pure and Applied Chemistry* 75: 631–681.

Nakui, H., Koyama, H., Watanabe, G., Sasaki, Y. & Suzuki, S. 2007. Simultaneous measurement of dioxins and organic halogenated compounds in working environment at waste incinerator. *Organohalogen Compounds* 69: 247/1–247/4.

Perkowski, J. & Zarzycki, R. 2005. The use of ozone. (in polish) Polska Akademia Nauk Oddział w Łodzi: Łódź.

Schulte, P., Bayer, A., Kuhn, F., Luy, Th. & Volkmer, M. 1995. H2O2/O3, H2O2/UV and H2O2/Fe2+ processes for the oxidation of hazardous waste. *Ozone: Science & Engineering* 17: 119–134.

Snyder, S., Yoon, Y., Wert, E., Westerhoff, P., Rexing, D. & Zegers, R. 2003. *The role of ozone in removing endocrine disruptors and pharmaceuticals from water.* International Ozone Association Conference.

Snyder, S.A., Adham, S., Redding, A.M., Cannon, F.S., DeCarolis, J., Oppenheimer, J., Wert, E.C. & Yoon, Y. 2007. Role of membranes and activated carbon in the removal of endocrine disruptors and pharmaceuticals. *Desalination* 202: 156–181.

Szkutnik, E. & Ozonek, J. 2008. Degradation products of tetrabromobisphenol during the ozonation process. *Polish Journal of Environmental Studies* 17(3A): 560–563.

Ternes, T.A., Meisenheimer, M., McDowell, D., Sacher, F., Brauch, H.J., Haist-Gulde, B., Preuss, G., Wilme, U. & Zulei-Siebert, N. 2002. Removal of pharmaceuticals during drinking water treatment. *Environmental Science and Technology* 36: 3855–3863.

UNEP (United Nations Environment Programme): UNEP 1999 report.

Vieno N., Tuhkanen T. & Kronberg L. 2006. Removal of pharmaceutical in drinking water treatment: Effect of chemical coagulation. *Environmental Technology* 27(2): 183–192.

Weigel, S., Kuhlmann, J. & Huhnerfun, H. 2002. Drugs and personal care products as ubiquitous pollutants: occurrence and distribution of clofibric acid, caffeine and DEET in the North See. *The Science of the Total Environment* 295: 131–141.

Wenzel, A., Muller, J. & Ternes, T. 2003. Study on endocrine disrupters in drinking water, Final Raport, Schmallenberg and Wiesbaden.

Ying, G.G., Kookana, R. & Ying-Jun, R. 2002. Occurrence and fate of hormone steroide in the environment. *Environmental International* 28: 545–551.

Zwiener, C. & Frimmel, F.H. 2000. Oxidative treatment of pharmaceuticals in water. *Water Research* 34: 1881–1885.

Environmental Engineering III – Pawłowski, Dudzińska & Pawłowski (eds)
© 2010 Taylor & Francis Group, London, ISBN 978-0-415-54882-3

Estimated changes of pluvio-thermal coefficient in Poland in the light of climatic changes

A. Ziernicka-Wojtaszek, P. Zawora, T. Sarna & T. Zawora
Department of Metrology and Agricultural Climatology, University of Agriculture in Krakow, Krakow, Poland

ABSTRACT: Images depicting spatial diversity of pluvio-thermal regions in Poland, based on 1971–2000 meteorological data, were compared; they were also contrasted with 2000 and 2007 estimates. Thermal index was characterised using active temperature sums ($\geq 10.0°C$), and humidity index using hydro-thermal coefficient (from June–August precipitation sum and average air temperature). Growing-season temperature increased by 0.48°C/decade, with no significant precipitation trends. Temperate–warm and warm regions (temperature sums $\geq 10.0°C$ of 2400–3200°C) covered 62% of Poland during 1971–2000, reaching 88 and 97% in 2000 and 2007, respectively. Temperate–dry area increased from 20% to 36% by 2000, and 45% in 2007.

Keywords: Global warming, Poland, thermal regions, humid regions.

1 INTRODUCTION

Water resources in Poland are limited compared to neighbouring countries. This fact is indicated by a rather low drainage coefficient of Polish rivers (drainage to precipitation ratio), which is significantly lower than in western Europe. In central Poland the drainage is equal to 60 mm, which is comparable to the Black Sea region and the Hungarian Lowland. Evaporation is the most significant index in the water balance equation, with most water evaporating during summer months due to very high temperatures. In addition, the coefficient of water resources has been evaluated at 1000–1700 m^3/person/y, making it among the lowest in Europe (Engelman & Leroy 1993).

During the last decades of the twentieth century a significant increase in global temperature has been detected, with similar temperature increases also detected in Poland (Hansen 2002, Trenberth et al. 2007). Climatic forecasts for Europe anticipate higher mean annual temperatures, in the years 2080–2099 compared to 1980–1999, of 2.3–5.3°C (Christensen et al. 2007). For Poland the anticipated temperature increase in this period is about 3.0–3.5°C.

The changes in precipitation observed around the world are not clear. The analysis of changes in atmospheric precipitation in Poland during the global warming period does not indicate homogenous trends. During the twentieth century there was a slight decreasing trend of 1.49 mm per decade, while in the second half of the century there was an increasing trend of 2.85 mm per decade (Kozuchowski 1996).

We cannot disregard the positive aspects of the observed and projected increase in temperature, such as prolongation of the growing and vegetation periods (Holden & Brereton 2004, Qian et al. 2005), earlier dates of heading of cultivated plants (Hu et al. 2005) and increasing coverage of thermophilic vegetables (Kenny & Harrison 1992). However, a remaining major problem will be the interaction of temperature increase on water resources in agricultural regions. Recent perennial studies indicate that in the near future we can anticipate considerable changes in water management. These changes will result in decreasing runoff and soil moisture, which subsequently will require water storage in watersheds. In addition the frequency of severe drought will increase, which will lower crop yields (Leathers et al. 2000, Pidgeon et al. 2001, Bootsma et al. 2005, Tao et al. 2008, Hlavinka et al. 2009).

Previous agriculture–climate and meteorological studies have usually been based on standard climatology data, obtained from 30-y periods. Most often the World Meteorological Organization has recommended two such periods: 1961–1990 and 1971–2000. In the latter period, and more precisely in the last two decades, climatologists in Poland have observed the highest increase in temperature thus far. The thermal and pluvio-thermal conditions at the end of this period (accounting for significant temperature increase) have deviated from the 30-y average. In light of further increases in temperature during the beginning of the twenty-first century some researchers claim that the 1971–2000 data cannot be used for analysis in subsequent years. Therefore, we postulate that any agrometeorological calculation should be based on the new climate features, and not previous climate standards, which are referred to as mean values from past years (Górski 2002, Milly et al. 2008).

The aim of this article is the assessment of recent impacts of climate change, which have occurred in the last three decades of the twentieth century and in the beginning of the twenty-first century, by comparing the images of pluvio-thermal regions in Poland. The abovementioned regions were determined on the basis of mean temperature data from 1971–2000, as well as from estimations made at the end of the period and in 2007. This work has both cognitive and methodological aspects. Comparative analysis was aimed at depicting differences in the images of thermal regions that were determined with the proposed method, particularly for the humid regions.

2 MATERIALS AND METHODS

We used monthly mean temperature data and monthly precipitation sums from 1971–2007. The meteorological data were published in the following: Monthly Agrometeorological Review, Decadal Agrometeorological Bulletin, National Hydrological and Meteorological Service Bulletin and Daily Meteorological Bulletin. The meteorological data was obtained from 21 meteorological stations that were evenly distributed in Poland. Due to insufficient weather stations in mountainous regions, there was practically no analysis for these regions.

Comparison of pluvio-thermal regions has been performed for mean temperature (1971–2000), mean precipitation sums (1971–2000), as well as for estimated temperatures and precipitation sums at the end of 1971–2007, respectively. The method of Daubenmire (1967), as modified by Cherszkowicz (1971), was employed to designate pluvio-thermal regions. This method assumes that thermal conditions in the growing period were determined on the basis of active temperature sums $\geq 10.0°C$. Humid conditions were determined by hydro-thermal coefficient, which expresses precipitation efficiency $K = 10P/t$ in the period June–August, where P denotes the sum of precipitation (mm) and t the daily temperatures sums (°C) in a given month. A linear function of temperature increase over time was established.

3 RESULTS AND DISCUSSION

During 1971–2000, the largest region (36% of the country) was that of temperate–warm-type with optimal humidity. This region encompasses the following geographic areas: the Mazowsze Lowland (except central and eastern portions), Silesian Lowland, Silesian Upland, Malopolska Upland and Sandomierz Basin (Table 1 and Figure 1).

Central and particularly central-western Poland was occupied by temperate–warm and temperate–dry types; encompassing the Wielkopolska Lakeland, Wielkopolska Lowland and small pockets of the central and eastern Mazowsze Lowland. In total these two types covered 19% of the country.

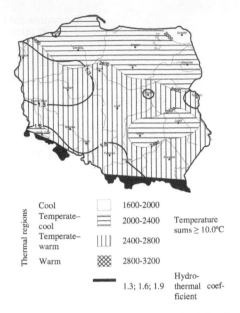

Thermal regions			Temperature sums $\geq 10.0°C$
	Cool	☐	1600–2000
	Temperate–cool	≡	2000–2400
	Temperate–warm	⫼	2400–2800
	Warm	▨	2800–3200
	—	1.3; 1.6; 1.9	Hydro-thermal coefficient

Figure 1. Pluvio-thermal regions in Poland based on 1971–2000 data.

Most of northern Poland (33%) was situated in the temperate–cool region with optimal humidity. These regions encompass coastland and lakeland geographical districts, with the exception of its most western portions as well as the Suwalki Lakeland. These regions also comprise the Podlasie Lowland, Lublin Uplands, Swietokrzyskie Mountains and Roztocze Region.

The temperate–warm and humid region covered nearly 7% of the country and was situated in southern part of the country, comprising such geographical regions as the Sudetes Foothills and Carpathian Foothills, with adjacent terrain in northern sections. The higher parts of Polish mountains are within the temperate–cool and humid region.

One of the characteristic features of values estimated for 2000 was the substantial decrease in size of the temperate–cool region (from 37 to 12%) and appearance of a warm region with temperature sums $\geq 10.0°C$, situated in the central Odra River valley and in the vicinity of the two major cities Wroclaw and Opole. This warm region covered 3% of the country. Most of the remainder of Poland (85%) belongs to temperate–warm regions with temperature sums $\geq 10.0°C$, i.e. in range 2400–2800°C. In terms of humidity regions, the temperate–dry region covered central Poland and ran parallel from west to east covering 36% of the country. North and south from the temperate–warm and temperate–dry regions was the temperate–warm region with optimal humidity, covering 44% of the country. The southern part of the Malopolska Upland, Carpathian Foothills with lower parts of Carpathian Mountains, as well as the Sudetes Foothills belonged to the temperate–warm and cool regions (Table 1 and Figure 2).

Table 1. Share of pluvio-thermal types in Poland during 1971–2000, and estimates for 2000 and 2007.

Period	Humidity region (hydro-thermal coefficient)	Thermal region (active temperature sums $\geq 10.0°C$)			
		Cool (1600–2000)	Temperate–cool (2000–2400)	Temperate–warm (2400–2800)	Warm (2800–3200)
1971–2000	Temperate–dry (1.0–1.3)		1	19	
	Optimal humid (1.3–1.6)	1	33	36	
	Humid (>1.6)		3	7	
Estimates for 2000	Temperate–dry (1.0–1.3)			36	
	Optimal humid (1.3–1.6)		11	44	3
	Humid (> 1.6)		1	5	
Estimates for 2007	Temperate–dry (1.0–1.3)			30	15
	Optimal humid (1.3–1.6)		3	38	8
	Humid (> 1.6)			3	3

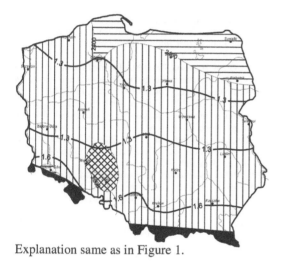

Explanation same as in Figure 1.

Figure 2. Pluvio-thermal regions in Poland using estimates for 2000.

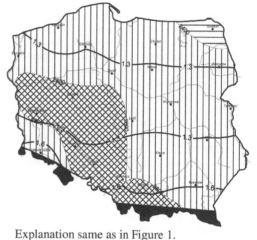

Explanation same as in Figure 1.

Figure 3. Pluvio-thermal regions in Poland from estimates for 2007.

One characteristic feature of the estimates for 2007 was a further decrease in size of the temperate–cool region, down to only 3% of the country. This region in particular comprises the city of Suwalki and environs. Other features included a substantial increase in size of the warm region (particularly in south-western Poland; from 3 to 26%) and an extension of up to 45% of the parallel temperate–dry belt, which had the lowest values of hydrothermal coefficient, of 1.1–1.3 (Table 1 and Figure 3).

The period 1971–2007 was characterised by a systematic increase ($P < 0.05$) in temperature at all stations used in this study. During the growing season (April–October) temperature increase was estimated at 0.48°C per decade. The largest increase was in April (0.71°C per decade). There were no significant trends for June, July or August (1971–2000).

In summary, 1971–2007 may be best characterised by systematic increase in air temperature and stable atmospheric precipitation. The images of pluvio-thermal regions, determined on the basis of standard mean values of 1971–2000, and estimated values for 2000 (derived from linear regression) and for 2007 (taking into account steady temperature increase) indicated substantial differences. If the calculated increase in temperature during the growing season (0.48°C per decade) is assumed we can speculate that temperature will rise by 1°C in 21 years, counting from the mid 1980s this 1°C increase was reached in 2006. Thus, the present coverage of pluvio-thermal regions is the fulfilled scenario of a 1°C increase in reference to the 1971–2000 mean.

Until recently, images of spatial diversity of hydrological and meteorological phenomena, as well as realisation of water management infrastructure projects have been made under assumptions of stationary hydro-climatological phenomena. When considering rapid climate change scenarios and forecast or ongoing temperature increase it is justifiable to reconsider this principle of stationary weather phenomena. Thus, we cannot simply determine the risk of water management economy on the basis of the abovementioned principle, because extreme values of meteorological and hydrological factors are now occurring more often, and in unprecedented fashion.

4 CONCLUSIONS

The results allow us to formulate the following remarks and observations:

1. The observed trend in air temperature during the growing season (1971–2007) was 0.48°C per decade, while there was no significant trend for atmospheric precipitation. For the purpose of this study we did not include the spatial diversity as well as differences in temperature during the growing season.
2. The image of pluvio-thermal regions based on mean values for 1971–2000 (accounting for temperature increase) differed from the estimates for 2000. The favourable increase in size of warm regions was accompanied by similar increase in the size of dry regions, which in this case was unprofitable. This particular tendency was exacerbated for 2007. Analysis of the temperature trend showed that a 1°C increase scenario for the vegetation period (April–October) compared to the 1971–2000 mean was fulfilled in 2006.
3. Hydro-climate and agro-climatic elaborations during the period of rapid climate change should not be based on assumptions of stationary characteristics of climate phenomena. Subsequently, this assumption should not be employed in determining the risk of water management economy, since extreme and unprecedented meteorological values are becoming more common.

REFERENCES

Bootsma, A., Gameda, S. & Mckenney, D.W. 2005. Impacts of potential climate change on selected agroclimatic indices in Atlantic Canada. *Canadian Journal of Soil Science* 16: 329–343.

Cherszkowicz, E. 1971. *Agro and climate related region taxonomy for fundamental crops cultivation in the socialistic European countries*. Sofia: Bulgarian Academy of Sciences, in Russian.

Christensen, J.H., Hewitson, B., Busuioc, A., Chen, A., Gao, X., Held, I., Jones, R., Kolli, R.K., Kwon, W.-T., Laprise, R., Magaña Rueda, V., Mearns, L., Menéndez, C.G., Räisänen, J., Rinke, A., Sarr, A. & Whetton, P. 2007. Regional Climate Projections. In Solomon S., Qin D., Manning M., Chen Z., Marquis M., Averyt K.B., Tignor M., Miller H.L. (eds), *Climate Change 2007: The Physical Science Basis. Contribution of Working Group I to the Fourth Assessment Report of the Intergovernmental Panel on Climate Change*: 847–940. Cambridge, UK, New York, NY, USA: Cambridge University Press.

Daubenmire, R.F. 1967. *Plants and environment*. New York: Wiley.

Engelman, R. & Leroy, P. 1993. *Sustaining water: Population and the future of renewable water supplies*. Washington DC: Population Action International.

Górski, T. 2002. Contemporary changes of Poland's agroclimate. *Pamietnik Pulawski* 130: 241–250, in Polish.

Hansen, J.W. 2002. Realizing the potential benefits of climate prediction to agriculture: issues, approaches, challenges. *Agricultural Systems* 74: 309–330.

Hlavinka P., Trnka M., Semerádová D., Dubrovský M., Žalud Z. & Možný M. 2009. Effect of drought on field variability of key crops in Czech Republic. *Agricultural and Forest Meteorology* 149: 431–442.

Holden, N.M. & Brereton, A.J., 2004. Definition of agroclimatic regions in Ireland using hydro-thermal and crop yield data. *Agricultural and Forest Meteorology* 122: 175–191.

Hu, Q., Weiss, A., Feng, S. & Baenziger, P.S. 2005. Earlier winter wheat heading dates and warmer spring in U.S. Great Plains. *Agricultural and Forest Meteorology* 135: 284–290.

Kenny, G.J. & Harrison, P.A. 1992. Thermal and moisture limits of grain maize in Europe: model testing and sensitivity to climate change. *Climate Research* 2: 113–129.

Kożuchowski, K. 1996. The present climatic changes in Poland against the background of the global changes. *Geographical Review* 68: 79–97, in Polish.

Leathers, D.J., Grundstein, A.J. & Ellis, A.W. 2000. Growing season moisture deficits across the northeastern US. *Climate Research* 14: 43–55.

Milly, P.C.D., Betancourt, J., Falkenmark, M., Hirsch, R.M., Kundzewicz, Z.W., Lettenmaier, D.P. & Stouffer, R.J. 2008. Stationarity is dead: whither water management? *Science* 319: 573–574.

Pidgeon, J.D., Werker, A.R., Jaggard, K.W., Richter, G.M., Lister, D.H. & Jones, P.D. 2001. Climatic impact on the productivity of sugar beet in Europe, 1961–1995. *Agricultural and Forest Meteorology* 109: 27–37.

Qian, B., Hayhoe, H. & Gameda, S. 2005. Developing daily climate scenarios for agricultural impact studies. In *16TH Conference On Climate Variability And Change*; *American Meteorological Society Conference, San Diego, California USA 9–13 January 2005*.

Tao, F., Yokozawa, M., Liu, J. & Zhang, Z. 2008. Climate-crop yield relationships at provincial scales in China and the impact of recent climate trends. *Climate Research* 38: 83–94.

Trenberth, K.E., Jones, P.D., Ambenje, P., Bojariu, R., Easterling, D., Klein Tank, A., Parker, D., Rahimzadeh, F., Renwick, J.A., Rusticucci, M., Soden, B. & Zhai, P. 2007. Observations: surface and atmospheric climate change. In Solomon S., Qin D., Manning M., Chen Z., Marquis M., Averyt K.B., Tignor M., Miller H.L. (eds), *Climate Change 2007: The Physical Science Basis. Contribution of Working Group I to the Fourth Assessment Report of the Intergovernmental Panel on Climate Change*: 235–336. Cambridge, UK, New York, NY, USA: Cambridge University Press.

Environmental Engineering III – Pawłowski, Dudzińska & Pawłowski (eds)
© *2010 Taylor & Francis Group, London, ISBN 978-0-415-54882-3*

Reliability analysis of water-pipe networks in Cracow, Poland

I. Zimoch

Silesian University of Technology, Institute of Water and Sewage Engineering, Gliwice, Poland

ABSTRACT: This paper presents a method of comprehensively estimating the reliability of water-pipe networks, which evaluates: failure and repair rates, mean time to failure, mean recovery time and probability of failure-free operation at any time. Following this, diameters, materials and exploitation times of water pipes were studied in a multifactorial regression analysis of unit failure intensity factors. Sophisticated statistical analysis methods were used to assess the reliability of water supply systems (WSS). The formula of unit failure intensity factor is a statistically significant model of reliability ($R = 0.886636$ and $R^2 = 0.76123$) and can be used not only for complex exploitation analyses of WSS, but also in the development of water company strategies.

Keywords: Water supply system, water distribution subsystem, reliability analysis, failure, failure intensity.

1 INTRODUCTION

In recent years, significant changes in management and in the attitude to the modernization of contemporary water supply systems (WSS) have been observed. Intensive research development, often conditioned by market needs, has caused a rise in new technologies associated with applied materials, as well as principles, of WSS operations. Technical progress and research development have led to WSS being treated as a complex process. It is necessary to assess the WSS operation rules based on an interdisciplinary approach to its operation analysis. This problem is difficult to solve due to the degree of complexity of WSS. The operation of modern water-pipe networks involves a combination of different fields of knowledge, such as: system engineering, reliability and safety theory, stores and queuing theory, administration of human and material resources, and economics (Denczew & Królikowski 2003, Pollard 2008). The proper realization of the exploitation process should include organization and modernization works, which ensure that the required technical condition of items is met, as well as detailed technical, economical and reliability analyses.

The reliability of water-pipe networks is a multifunctional operation process. This process requires continuous provision of water to consumers, in the required quantities and quality, at any moment, under suitable pressure, and at a price that is acceptable to the consumer. A literature review (Kirchsteiger 1994, Kwietniewski 1999, Ulanicka et al. 2000, Wieczysty et al. 2001, Wu & Simpson 2001, Gale 2002, Hotloś 2003, Rak 2005, Tchórzewska-Cieślak 2007, Pollard 2008, Rak 2008, Zimoch 2008) revealed that the reliability analysis of water-pipe networks is difficult and complicated. This is due to complexity and an extensive area of water distribution subsystem (WDS) operations. Other difficulties in reliability research follow on from the WDS structure (closed systems) and there is a diversity of items that form the subsystem (materials, diameter, pipe length, fittings, etc.).

Most of the technological objects of the water-pipe network are classified as so-called renewable elements, which undergo processes of operation and renovation. The water-pipe system parameters have a random distribution, and their analysis enables diagnosis of exploitation states, which, when ordered in time, give a full picture of the exploitation. Determination of the exploitation conditions, defined as reliability states, forms the basis for selection and estimation of suitable reliability factors of these objects. The two following reliability states are distinguished: state of operation, which means full ability; and state of partial or complete unavailability. Taking into consideration the above exploitation states and specification of objects and water-pipe devices in the reliability analysis of their operation, the following are defined: probability of being in a given state, intensity of state appearance and mean time of being in a given state. From the points of view of waterworks companies and water customers, it is necessary to achieve full efficiency of water-pipe networks (which is the exploitation state when the pipes do not need any immediate repairs), scheduling maintenance and delivery of water to customers in the required quantities, and of the required quality and pressure.

There are many variables that limit the capability of the water-pipe network to non-failure operation. Some of these depend on the manner and conditions of network operation – that is, operating time, pressure, flow, quality of transported water, maintenance intensity and repairs. However, the WDS failures are also connected with factors independently of the manner of operation. Such factors include: pipe materials, the technology and method of WDS building, soil condition of pipe location and exterior loads. Usually, reliability analysis of the water-pipe networks treats only one of the above listed variables. There is no information on the comprehensive influences of all these factors on the non-failure operation of WDS. This paper presents multifactorial reliability analysis methodology of water-pipe networks based on multiple regression procedures. The new multifactorial analysis was set to obtain a mathematical formula of failure intensity as a function of pipe diameter and age, as well as time to failure.

2 MATERIALS AND METHODS

The two-stage exploitation model was most often assumed in WDS reliability analyses. The following random events were distinguished in the accepted reliability model: (i) a pipe can pass through one state to another, (ii) unavailability state is removed when all broken-down elements are repaired, (iii) as a result of repair, pipe passes from any unavailability state (partial or complete) into an operational one, i.e. full ability.

To describe the exploitation model presented above, the following are considered:

- probability of failure-free operation $R(t)$ and renovation probability $R_o(t)$,
- mean time to failure T_p [h] and mean recovery time T_o [h],
- parameter of failure intensity ω [$a^{-1}km^{-1}$] and renovation intensity μ [h−1].

Knowing the probability distribution of the random variable – time to failure T'_p – allows the above-mentioned parameters to be calculated. The flux of the water-pipe element failures is a flux without consequences, singular and steady, and the renovation process is a Poisson process, for which the time is exponentially distributed (Kwietniewski 1999, Wieczysty et al. 2001, Pollard 2008). The above fact allows for the estimation of these variables according to the following formulas, as presented in Table 1.

In the water-pipe network reliability analysis, due to the linearity of the network components, the parameters presented above have to be related to the lengths of the pipes. A parameter that adequately describes the ability of failure-free operation to occur is the unit failure intensity parameter, which is calculated from the following relationship (Kwietniewski 1999, Wieczysty et al. 2001, Pollard 2008):

$$\omega^*(t) = \frac{n(t, t + \Delta t)}{L \cdot \Delta t} \qquad (1)$$

where $n(t, t + \Delta t)$ – failure number in period Δt; Δt – length of time period that observation period was divided to [a]; L – length of tested pipes [km].

Practical use of the formulas (Table 1) in reliability estimation of operation of WDS is based on

Table 1. Reliability index.

Reliability index	Mathematical formula	
Mean time to failure [h]	$T_p = \frac{1}{k+z} \left(\sum_{i=1}^{k} t_{pi} + z \cdot t \right)$	(1)
Mean recovery time [h]	$T_o = \frac{1}{n_o} \sum_{i=1}^{n_o} t_{oi}$	(2)
Parameter of failure intensity [h^{-1}]	$\omega = \frac{1}{T_p}$	(3)
Renovation intensity parameter [h^{-1}]	$\mu = \frac{1}{T_o}$	(4)
Probability of failure-free operation	$R(t) = P(T'_p \geq t)$ $= \exp(-\omega \cdot t)$	(5)
Renovation probability	$R_o(t) = 1 - \exp(-\omega \cdot t)$	(6)

k – number of operational periods of failing objects; t_{pi} – value of ith operational period [h];
t – observation period [h]; z – number of operational periods of unfailing objects;
n_o – number of failures during tested period; t_{oi} – time of lasting i^{th} renewal [h].

indispensable information obtained from water-pipe exploitation.

The methodology of research presented in this paper concerns the influence of exploitation factors on water-pipe network reliability, including: (i) the determination and classification of items and research time, as well as qualification of analysis range; (ii) the arrangement of the reliability model of studied items; (iii) carrying out the exploitation research; (iv) verification of the obtained results and calculation of fixed reliability factors (and their functional form); (v) the description of the influence of selected exploitation parameters on the reliability of water-pipe networks.

Operational research on the reliability of WDS in Cracow was conducted in 2004–2006 according to plan (l, W, t) (PN-84/N-04041.05). The plan is applied to items that are repaired after their failure occurred during research – W, the considered length of studied pipes is at least equal to l, and the time research is t. Moreover, the analysis was based on archival data (1996–2006) that was obtained from exploitation cards of water- pipe networks in Cracow. These data were made available by Cracow Water Company. The research was carried out on pipes with diameters of Ø25–Ø600 mm.

Conditions of research were based on an accuracy of $\delta = 0.1$ (relative error) of estimated reliability measures, with a confidence level of $\beta = 0.95$. The following data were collected for the research: the date and hour of failure occurrence and repair, the time of failure notification, repair beginning and end, information on the localization and type of failure as well as the method of repair, the diameter and material of pipe, and consequences of this failure for WSS or its distribution subsystem. Unfortunately, some archival data were not complete, which caused difficulties for detailed analysis.

3 OBJECT OF SCIENTIFIC RESEARCH – WATER DISTRIBUTION SUBSYSTEM OF CRACOW

Currently, the WDS consists of the complex water-pipe network with many storage tanks. Cracow is divided into separate water-pipe zones of supply, which are provided from independent sources: the Raba River, the Rudawa River, the Dłubnia River and the Sanka River, and one underground intake placed in Mistrzejowice. Localization and technical solutions of supply systems (in normal conditions of exploitation) deliver, on average, 169 000 m^3 of water daily to the distribution network. This guarantees a reliable operation of the WSS to the city. In Cracow, water is delivered to the consumers through a complex system of transit (Ø1400 mm, length 18 km), main networks (Ø1200–Ø330 mm, length 247.9 km) and distribution networks (Ø280–Ø80 mm, length 1131.1 km), as well as house waterworks terminals (Ø100–Ø25 mm, length 472.9 km). The total length of the network in the area serviced by the City of Cracow Water and Sewage Utility Company is 1869.9 km. This is characterized by substantial differentiation of age and materials. Most water pipes were built after 1975 (56% of total length). In addition, a substantial percentage of the water pipes were built in the 1990s (26% of total length), using modern materials and technology, which significantly increases their technical value. The most significant part of the material structure is steel pipes, which constitute 32% of the total length of the network; pipes made of cast iron constitute 26% and of PVC constitute 23%. There are 11 storage tanks, of total capacity equal to 276 200 m^3, which are inseparable elements of the WDS.

4 RESULTS AND DISCUSSION OF RELIABILITY ANALYSIS OF WDS IN CRACOW

Data collected within the confines of the research project PB no. 5T07E 044 25 (Zimoch 2007) were grouped in several ways. The first was grouped according to type and reason of failure, the second was grouped as treated material of which the network is built. The data were also split into three groups, according to the type of network (main, distribution network or waterworks terminals). The last division took into consideration the diameters of the pipes. Additionally, every sample was characterized by the city district at which a failure occurred and by the types of surface and soil where the pipe was located. For every pipe (with specified diameter, made of concrete material), realizations of the following random variables were determined: time to failure T_p [h], recovery time T_o [h], repair time T_n [h], and time of waiting for repair T_z [h]. All statistical analyses were carried out using the STATISTICA PL software package. The Kruskal–Wallis Test (Anderson et al. 1991, Sobczyk 2005) was applied to verify whether or not the ordered samples contained homogeneous elements from the same population.

It was assumed that the city district is the variable that uniquely identifies the sample membership in the general population. However, some values of the statistics H do not satisfy the condition $H < \chi^2_{(\alpha,v-1)}$, so this null hypothesis was rejected. For a new null hypothesis, H_0, it was assumed that the type of surface determines the sample membership in the general population and the Kruskal–Wallis Test gave results that satisfy $H < \chi^2_{(\alpha,v-1)}$. Thus, there are no sufficient grounds for the null hypothesis rejection on the accepted significance level, $\alpha = 0.05$. Random variables verified in this way are inputs for the evaluation of empirical reliability characteristics (Table 2). The estimator of time to failure T_p for steel pipes in house waterworks terminals equals 453.13 hours. As is presented in Table 2, for pipes with diameters between Ø25 and Ø80 mm, time to failure has interval values from 239.66 h to 1365.39 h. In distribution networks, mean time to failure is 444.19 h, which is similar to that in house waterworks terminals. The range of time to failure for steel distribution network pipes with diameters of Ø100–Ø250 mm varies between 324.59 h and 890 h. The Cracow steel main network is marked by the longest time to failure (1431.82 h). For cast-iron pipes, the mean time to failure is equal to 448.92 h for house waterworks terminals (it takes values from 191.3 h to 620.97 h for diameters of Ø25–Ø100 mm), 42.86 h for the distribution network (variation of the time to failure from 86.27 h to 430.56 h for diameters Ø80–Ø300 mm) and 543.58 h for the main network (range variation from 492.86 h to 669.23 h for diameters Ø200–Ø600 mm).

Causes of the water-pipe network failures were estimated by multifactorial analysis, in which the pipe diameter and age, as well as time to failure, were taken into consideration. The unit parameter of failure intensity per pipe length unit is an indicator of the ability of the water network to operate without failure of WDS. Therefore, the regression analysis was carried out for this reliability indicator. The probability density function of time to failure is exponential, so formula (3) allows the determination of the unit parameter of failure intensity ω' for all studied water pipes compared with their lengths. For the random variable ω', the Kolmogorov–Smirnov Test (at a significance level of $\alpha = 0.05$) was applied, but this did not allow acceptance of the null hypothesis about the normal distribution. In the next step, the new transformed random variable log $\left(\frac{\omega'}{\overline{\omega'}}\right)$, where $\overline{\omega'} = 0,037702$ [failure year^{-1} · km^{-1}] was the mean of unit parameter of failure intensity, was analyzed. In this case, there are no grounds for rejection of the null hypothesis – the value of statistics λ (d) is less than the critical value $\lambda_{crit}(d)$. Therefore, the function form of the water-pipe network reliability can be estimated based on the procedures of multiple regression. The results of the multiple regression show that diameter DN, the exploitation time (the age W) of a pipe, and time to failure T_p, significantly affect variable ω'. Standard statistical procedures confirm that the formula

Table 2. Empirical reliability indexes of water distribution subsystem in Cracow.

Diameter	N	Mean time to failure T_p [h]	Unit parameter of failure intensity ω^* [$a^{-1}km^{-1}$]	Mean recovery time T_o [h]	Mean repair time T_n [h]	Mean time of waiting for repair T_z[h]	Renovation intensity μ [h^-]
Steel water pipes							
House waterworks terminals	192	453.13	0.292000	26.05	7.41	18.65	0.038385
Ø25 and Ø 32	26	1365.39	0.158295	20.90	5.33	15.58	0.047838
Ø40	87	239.66	0.194825	36.34	6.74	29.60	0.027519
Ø50	32	506.25	0.529680	23.69	8.53	15.16	0.042216
Ø80	47	430.85	0.286149	26.52	8.01	18.51	0.037706
Distribution network	155	444.19	0.341993	29.05	9.24	19.81	0.034425
Ø100	42	435.71	0.370677	40.26	6.93	33.33	0.024837
Ø150	61	324.59	0.538364	23.04	6.81	16.23	0.043401
Ø200	37	482.43	0.326549	26.67	6.42	20.25	0.037497
Ø250	15	890.00	0.132385	30.10	6.77	23.33	0.033223
Main network	11	1431.82	0.028000	47.41	8.77	38.64	0.021000
Cast-iron water pipes							
House waterworks terminals	186	448.92	1.698630	25.49	6.56	18.92	0.039232
Ø25	5	420.00	0.292237	68.10	6.10	62.00	0.014684
Ø40	27	485.19	0.164384	28.43	5.83	22.59	0.035179
Ø50	31	604.84	0.188737	19.10	6.68	12.42	0.052365
Ø80	92	191.30	0.560122	24.09	6.91	17.17	0.041516
Ø100	31	620.97	0.188737	22.56	5.95	16.61	0.044317
Distribution network	875	42.86	1.052980	38.82	6.85	31.97	0.025760
Ø80	110	212.73	0.794253	55.76	6.22	49.55	0.017933
Ø100	310	256.45	2.238348	60.59	6.72	53.87	0.016505
Ø150	255	86.27	1.841222	34.68	6.74	27.94	0.028835
Ø200	118	266.95	0.852016	27.53	7.19	20.34	0.036319
Ø250	46	309.78	0.332142	21.04	6.70	14.35	0.047521
Ø300	36	430.56	0.259937	30.53	6.08	24.44	0.032757
Main network	109	543.58	0.422153	52.40	8.13	44.27	0.019084
Ø200.300	14	492.86	0.216886	15.21	7.71	7.50	0.065728
Ø400	26	669.23	0.402789	52.31	8.08	44.23	0.019118
Ø500	33	586.36	0.511232	51.32	8.14	43.18	0.019486
Ø600	36	433.33	0.557707	61.11	8.33	52.78	0.016364

presented below is a statistically significant model of reliability:

$$\log\left(\frac{\omega'}{\omega}\right) = -0.0211501 \cdot W - 0.002205 \cdot DN - 0.000607 \cdot T_p$$

It is worth stating that, for the above model, there are high coefficients of correlation and determination: $R = 0.886636$ and $R^2 = 0.76123$, respectively. Moreover, a hypothesis test for significance of the obtained model was carried out at the $\alpha = 0.05$ significance level. Verification results confirm the statistical significance of the mathematical formula of a unit failure intensity parameter, depending on the considered independent variables for $p = 0.000106$.

5 CONCLUSIONS

The complex method of reliability estimation, as well as the determination of factors that directly influence

water–pipe network failure, presented here, contains: (i) carrying out exploitation research based on the (l, W, t) plan and desirable aims; (ii) data handling and its verification; (iii) the identification of random variable distribution (i.e. time to failure T_p'); (iv) the determination of selected reliability estimators and their function models for verified distribution of random variables; (v) the application of multifactorial regression to the determination of factors that have the most significant influence on water-pipe network reliability, defined by the unit parameter of failure intensity ω'.

Results obtained for the Cracow water-pipe network indicate that the method presented in the paper can be also applied in order to estimate the reliability of other WDS.

The recommended value of the unit parameter of failure intensity is equal to: 0.3 [failure $\cdot$ year$^{-1} \cdot$ km^{-1}] for the main network, 0.5 [failure $\cdot$ year$^{-1} \cdot$ km^{-1}] for the distribution network and 1.0 [failure $\cdot$ year$^{-1} \cdot$km^{-1}]

for house waterworks terminals. These criteria were not satisfied only by the cast-iron distribution network and the cast-iron waterworks terminals.

Detailed reliability analysis allows us to determine the function form of the probability of failure-free operation $R(t) = \exp(-\omega \cdot t)$ and renovation probability $R_o(t) = 1 - \exp(-\omega \cdot t)$, where failure intensities ω are equal to: 0.002207 [h^{-1}] and 0.02228 [h^{-1}] for the steel and cast-iron house waterworks terminals, 0.002251 [h^{-1}] and 0.023333 [h^{-1}] for the steel and cast-iron distribution network, and 0.000698 [h^{-1}] and 0.00184 [h^{-1}] for the steel and cast-iron main network.

The quality and quantity analysis of negative exploitation events in water networks allows the precise and rational statement of conclusions, as well as the application of technical operations that will eliminate negative events. Hence, reliability analysis of WDS (with reference to its technical structure) not only has scientific meaning, but it can also be an important practical tool that is useful in the rational management of WSS.

REFERENCES

Anderson, D.R. Sweeney, D.J. & Williams, T.A. 1991. *Introduction to statistics: concepts and applications*. St. Paul, MN, USA: West Publication Company.

Denczew, S. & Królikowski, A. 2003. *Modern operation bases of water and sewage systems*. Warszawa: Arkady.

Gale, P. 2002. Using risk assessment to identify future research requirements. *Journal of American Water Works Association* 94(9): 30–38.

Hotloś, H. 2003. Reliability level of municipal water pip networks. *Environmental Protection Engineering* 29 (32): 141–151.

Kirchsteiger, CH. 1994. Nonparametric estimation of time-dependent failure rates probabilistic risk assessment. *Reliability Engineering and System Safety* 44: 1–9.

Kwietniewski, M. 1999. *Methods of water-pipe networks operations research in the aspect of consumer water delivery reliability*. Warszawa: Research Works of Environmental Engineering z.28, Warszawa University Publishing House.

PN-84/N-04041.05 Technology reliability. Assurance of technology objects reliability. General rules of test.

Pollard, S.J.T. 2008. *Risk Management for Water and Wastewater Utilities*. London: IWA Publishing.

Rak, J. 2005. *Safety bases of water supply systems*. 28: 121–132. Lublin: KIŚ PAN.

Rak, J. 2008. *Essentials problems in water supply reliability and safety*. Rzeszów: Rzeszów University Publishing House.

Sobczyk, M. 2005. Statistics. Warszawa: PWN. (in Polish)

Tchórzewska–Cieślak, B. 2007. Method of assessing of risk of failure in water supply system. *Reliability and Societal Safety* 2: 1535–1539. Norway: Taylor & Francis.

Ulanicka, K. Bounds, P.L.M. Ulanicki, B. & Rance, J.P. 2000. Experience with pressure control of a very large scale water distribution network. In Sozański, M. (ed.), *Water Supply and Water Quality; Conference Proceedings*: 1215-1226. Cracow: PZITS o/wielkopolski.

Wieczysty, A. Budziło, B. Iwanejko, R. Bajer, J. Lubowiecka, T. Rak, J. Zimoch, I. Głód, K. Knapik, K. Wierzbicki, R. Jarecka, U. & Kapcia, J. 2001. *Increase methods of water supply operating reliability* 2. Kraków: KIŚ PAN.

Wu, Z.Y. & Simpson, A. 2001. Competed genetic algorithm optimization of water distribution systems. *Journal of Computing in Civil Engineering ASCE* 15(2): 89–101.

Zimoch, I. 2008. Reliability analysis of water distribution subsystem. *Journal of KONBiN* 4(7): 307–316.

Zimoch, I. 2007. *Determination of reliability operation model of water supply system (WSS) in the aspect of secondary water contamination in a water pipe network*. Final report of research project KBN nr 5T07E 044 25, Gliwice.

Energy saving and recovery

Environmental Engineering III – Pawłowski, Dudzińska & Pawłowski (eds)
© 2010 Taylor & Francis Group, London, ISBN 978-0-415-54882-3

Comparison of heat losses in channel and preinsulated district heating networks

T. Cholewa & A. Siuta-Olcha

Faculty of Environmental Engineering, Lublin University of Technology, Lublin, Poland

ABSTRACT: This paper presents the results of a study into heat losses from channel and preinsulated district heating networks located in the city of Lublin during the 2006/2007 heating season. Special attention was paid to the percentage of delivered heat from the heating plant that was lost, which amounted to 5,6% for the channel network and 0,5% for the preinsulated network. A methodology for calculating the heat losses from channel and preinsulated networks was introduced and the real (measured) values were compared to the computational ones for both heating networks. It was found that replacing existing channel networks by preinsulated networks might be recommended in the near future. For that reason, a theoretical analysis of the reduction in heat loss that would be achieved by changing the existing channel network to a preinsulated network was carried out. The capital cost required to make this switch and achieve the savings was calculated.

Keywords: Heat losses, channel heat distribution network, preinsulated heat distribution network.

1 INTRODUCTION

Heat distribution networks were first designed and introduced in 1877 in Lockport, New York, USA (Comakli et al. 2004) and they have since been introduced to many countries. Very little attention has been paid to the heat losses from these systems over the years as they had clear benefits. The main advantage was to distribute heat to consumers living a long distance from the heating plant.

This situation has changed as energy saving policies have become popular due to increasing energy costs and because consumers are more environmentally aware. This is especially true in those countries which have signed international agreements (such as Kyoto and European Union decrees) to achieve long-term sustainable development and to meet international targets for reducing greenhouse gas emissions, such as Swedish and Finnish (described in Lygnerud & Peltola-Ojala 2010).

These concerns are also reflected in the scientific literature (Adamo et al. 1997, Benonysson et al. 1995, Bojic et al. 2000, Bohn 2000; Poredos & Kitanovski 2002), where there has been some research on how to cut heat losses from a heat distribution network to a minimum. Poredos & Kitonovski (2002) noted that heat losses in district heating network amount to 8–10%, and one of the most substantial factors influencing the size of this loss is the temperature of the heating medium in the pipes. Kalinci et al. (2008) determined optimum pipe diameters in a network and presented an energetic and exergetic assessment of a geothermal district heating system in Turkey. Popescu et al. (2009) showed the differences between district heating systems from Western European countries, designed to be demand-driven, and those from Central and Eastern European countries, designed to be production-driven. They developed an original simulation and prediction model for the space heating consumption of buildings connected to a partially-controlled system. Stevanovic et al. (2009) presented a model and numerical method for the computer prediction of thermal transients in a district heating systems, as well as the results of measurements and computer simulations of two thermal transients caused by the rapid increase and decrease of the heating power plant load. This computational tool provides reliable information about time periods of temperature front propagation and heat distribution from the heating source to consumers within the whole district heating network. Larsen et al. (2004) compared aggregated models for the simulation and operational optimisation of district heating networks.

A review of a sparse district heating research programme undertaken in Sweden between 2002 and 2006 was presented by Nilsson et al. (2008). They indicated that it is possible to enhance the profitability of sparse district heating investments, but the key is achieving higher productivity through more efficient construction routines and more suitable ways for customer communication, rather than more efficient district heating technology. Reidhav & Werner (2008) also analysed the profitability of sparse district heating projects in 74 areas with 3227 one-family houses connected to district heating in Göteborg, Sweden. They concluded that sparse district heating is possible with low investment costs for the local distribution network and low marginal costs for the heat generation.

Table 1.	Characteristics of Network A.				
D_n(m)	D_e(m)	D_{ins}(m)	D_{inr}(m)	L(m)
0.050	0.063	0.113	0.103	66.95
0.065	0.076	0.136	0.116	477.59
0.080	0.089	0.159	0.139	263.88
0.100	0.114	0.194	0.164	111.73
0.125	0.140	0.230	0.200	160.24
0.150	0.168	0.258	0.238	167.88

Table 2.	Characteristics of Network B.				
D_n(m)	D_e(m)	D_{ins}(m)	D_{inr}(m)	L(m)
0.065	0.076	0.140	0.140	48.00
0.080	0.089	0.160	0.160	123.48
0.100	0.114	0.200	0.200	232.90
0.125	0.140	0.225	0.225	55.84
0.150	0.168	0.250	0.250	148.79
0.250	0.273	0.400	0.400	8.10

According to recent Polish data (Stanny 2007, Niemyjski 2007), average annual losses from a channel distribution network could rise to 14% of the heat generated in the plant, and reach even 30% in the summer season. This is caused, among other things, by the large extent of the channel heat distribution network, which in terms of length comprises about 60% of the heat distribution network in Poland, and also by the dimensions (diameters) of heating network pipes in relation to present needs. This is why further research and analysis of heat distribution networks is necessary for the purpose of optimising heat distribution network parameters work to minimise heat losses, especially given that the annual amount of heat delivered through district heating systems worldwide is about 11 EJ (Werner 2004).

Therefore this paper pays special attention to the real heat losses from channel and preinsulated heating networks. It also considers the capital investment cost of district heating networks, combined with the payback time of these systems.

2 MATERIALS AND METHODS

The objects of analysis in this paper are two parts of a district heating network suppling heat to two settlements in Lublin. Lublin, a city of 400,000 inhabitants in eastern Poland, has a heating system operated by a central power plant. In different districts of the city one can find different types of heating networks.

Network A is channel two-pipe heat distribution network. Table 1 shows how the pipes are put together and the characteristics of the network. This network distributes heat in the "Pogodna" residential area. Pipes in this system are surrounded by glass wool in protective shell to act as thermal insulation. This has a thermal conductivity coefficient of $\lambda_{in} = 0.04$ W/mK. The pipes are arranged in a rectangular channel, measuring 1.0×0.6 m with sides 0.1 m thick.

The distance between the axis of network pipes and the ground level is 1.3 m on average. The heating medium is transported through these pipes from the thermal substation, where the parameters are transformed from 95/70°C to parameters $\tau_s/\tau_r = 54.2/41.9$°C, to consumers.

Measurements for network A were performed in February and March 2007 and the following parameters were examined:

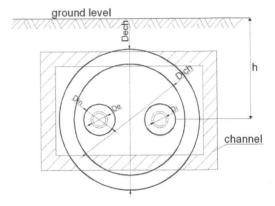

Figure 1. Model of a cross-section of a rectangular channel.

– thermal energy consumption
– temperature of supply heating medium (τ_s)
– temperature of return heating medium (τ_r)
– temporary power.

These readouts were made on the same time of the day in the thermal substation and in 15 individual thermal stations in the "Pogodna" residential area.

Network B, for which a similar set of measurements were taken, is located in the "Moniuszko" settlement. This is a two pipe preinsulated heat distribution network, arranged directly in the ground to a depth of 1.3 m. The system was modernised in 1999. Pipes are lagged with insulation with a thermal conductivity coefficient of $\lambda_{in} = 0.03$ W/mK and a protective shell of HDPE. The characteristics of network B are presented in Table 2.

Measurements for network B were made in February and March, and comprised the same set of readings that were made for network A. The measurements were taken at the same time of the day in the thermal substation and in 11 individual thermal stations. Additionally, as well as measuring the actual heat losses from the heating networks, an estimate of the heat losses from networks A and B were obtained though calculation. The heat losses of the channel network (network A) were calculated by using formulas 3–18 (Kamler 1979), which are presented in Appendix 1, and assuming a supplementary model of a cross-section of the rectangular channel as shown in Figure 1.

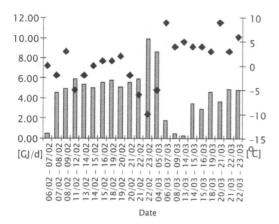

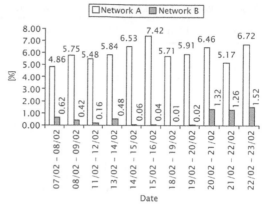

Figure 2. Daily heat losses for network A as a function of the average external air temperature.

Figure 4. Percentage of total heat supply lost by networks A and B.

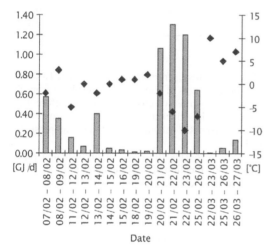

Figure 3. Daily heat losses for network B as a function of the average external air temperature.

3 RESULTS AND DISCUSSION

The measured daily heat losses and percentage of heat lost from network A as a function of the average external air temperature are presented in Figure 2. The same results for network B are presented in Figure 3.

It can be seen from Figure 2 that the daily heat losses from network A in the study period are in the range of 0.29–9.93 GJ/d and have a mean value of 4.5 GJ/d. The percentage of the total heat supplied from the plant that is lost by network A is in the range 0.5–8.9% and the average value is 5.6%.

The percentage of the total heat from plant that is lost by network B is between 0.01% and 1.32%, and the average value is 0.502%. The daily heat losses of network B are in range of 0.01–1.3 GJ/d and their average value is 0.379 GJ/d.

It can be noticed from Figures 2 and 3 that the external air temperature has a great impact on the heat losses, because a lower external temperature means

greater energy consumption, greater heating medium flow and higher heating medium temperature in the supply and return pipes and, as a result, higher energy losses.

Figure 4 Compares the percentage of total heat supply lost by networks A and B.

It appears from Figure 4 that network A has loses a considerably greater percentage of heat delivered from the plant in comparison to network B.

For the purpose of making a comparison of the real and computational heat loss values, an analysis of networks A and B through the utilization of equations 3–18 was made, and these are presented in Tables 3 and 4, respectively.

The average values of the daily heat losses obtained from measurements and calculations are put together in Table 5.

In the case of network A, the real (measured) value of the daily heat losses is greater than would be expected from the computational model. This might be caused by a reduction in the thermal insulating capacity of the glass wool, which loses this property over time. Therefore one can state that channel networks require complete reconditioning or should be replaced by a preinsulated network, where (in this study) the real heat losses are considerably smaller than calculated values. The relatively small real heat losses from the preinsulated heat distribution network could result from the comparatively warm winter experienced during the measuring period.

As a pipeline network is an essential feature of a district heating system – for hot water transport and distribution from heat sources to consumers – it should be seen as a necessary part of the urban infrastructure, as with sewage, water and roads (Stevanovic et al. 2007), and ways should be sought to optimise the network.

For the purpose of checking the advantages of replacing the channel network A with a preinsulated network, calculations for the operating conditions of network A ($\tau_s = 54.2$; $\tau_r = 41.9$) were made by using

Table 3. Results of heat loss calculations for network A.

D_n(m)	L(m)	q_s(W/m)	q_r(W/m)	Q_s(W)	Q_r(W)	Q_{js}(GJ/d)	Q_{jr}(GJ/d)
0.050	66.95	15.12	12.31	1012	824	0.087	0.071
0.065	477.59	15.38	14.00	7347	6685	0.635	0.578
0.080	263.88	15.58	13.67	4112	3608	0.355	0.312
0.100	111.73	16.95	16.22	1894	1812	0.164	0.157
0.125	160.24	18.10	16.73	2901	2680	0.251	0.232
0.150	167.88	20.51	17.24	3442	2894	0.297	0.250

Table 4. Results of heat loss calculations for network B.

D_n(m)	L(m)	q_s(W/m)	q_r(W/m)	Q_s(W)	Q_r(W)	Q_{js}(GJ/d)	Q_{jr}(GJ/d)
0.065	48.00	14.67	11.70	704	562	0.061	0.049
0.080	123.48	15.24	12.14	1882	1499	0.163	0.129
0.100	232.90	15.86	12.61	3694	2937	0.319	0.254
0.125	55.84	18.57	14.65	1037	818	0.090	0.071
0.150	148.79	21.86	17.06	3252	2539	0.281	0.219
0.250	8.10	22.67	17.65	184	143	0.016	0.012

Table 5. Average daily heat losses (Q_{ja}).

Kind of network	Q_{ja} on measurement basis	Q_{ja} on calculations basis
Network A	4.5	3.39
Network B	0.379	1.66

Table 6. Composition of daily heat losses for two kinds of heat distribution network A.

D_n (m)	L (m)	Channel network Q_{js} (GJ/d)	Q_{jr} (GJ/d)	Preinsulated network Q_{js} (GJ/d)	Q_{jr} (GJ/d)
0.050	66.95	0.087	0.071	0.065	0.048
0.065	477.59	0.635	0.578	0.499	0.403
0.080	263.88	0.355	0.312	0.255	0.206
0.100	111.73	0.164	0.157	0.127	0.103
0.125	160.24	0.251	0.232	0.212	0.170
0.150	167.88	0.297	0.250	0.152	0.123

equations 3, 4, 9, 10 and 16–18). Results of these calculations are presented in Table 6.

By exchanging channel pipes for preinsulated pipes, the average daily heat losses drop from 3.39 GJ/d to 2.36 GJ/d. This is very important, especially given the CO_2 emission limit in Poland (the Directive 2003/87/EC), and the associated charge for purchasing emission rights. The value of this charge was assumed to be 27 €/Mg CO_2 (Agency of energy market in Poland) for further calculations. As the result of simple calculations it can be noted that a decrease

of daily heat losses from network A of about 1.02 GJ/d generates savings of about 3.2 €/d, resulting from the reduced CO_2 emission charges. However before these savings can be achieved, the costs of investment must be included in the calculation, and these are presented in Table 7.

First, we define the costs of preinsulated pipes, which are a function of their diameter following Equation 1 (Dobersek & Goricanec 2009).

$$C_1 = A + B \cdot D_n + C \cdot D_n^2 \tag{1}$$

where A, B, C = constants and are equal to €18, 291 €/m, and 229 €/m², respectively.

Also the construction costs should be considered, and these depend on the pipe diameters in defined pipe sectors and on the construction site environment of the pipe network. The construction costs might be calculated by using Equation 2 (Dobersek & Goricanec 2009).

$$C_2 = D + E \cdot D_n + F \cdot D_n^2 \tag{2}$$

where D, E, F = constants and are equal to €287, 310 €/m, and 1275 €/m², respectively.

It can be seen that the total investment costs for replaing the channel network with a preinsulated network are very high – at €479,000 – in comparison to the savings achieved by the reduction in CO_2 emissions. Naturally there are also savings from the reductions in fuel consumption and pumping costs, which depend on the electricity energy price, but this is not sufficient to give a satisfactory pay-back time for this solution. However, as the lifetime of warm water pipe networks is limited, let us suppose that when network A has to be replaced, the new network should be a more efficient preinsulated one.

Table 7. Investment costs of replacing the channel network A with a preinsulated network.

D_n(m)	L(m)	C_1(€)	C_2(€)
0.050	66.95	2493	20,861
0.065	477.59	19,791	151,838
0.080	263.88	12,063	85,679
0.100	111.73	6050	37,866
0.125	160.24	10,132	56,948
0.150	167.88	12,314	62,966
Total		62,843	416,158

Another, but more radical, solution to limit heat losses in district heating networks is decentralisation of energy production, with the means of energy generation close to or at the consumption site. Decentralised power production has already reached the level of an individual building, where a combined heat and power generation plant – what one researcher terms a "micro-CHP" – is installed in a building. For example, Nystedt et al. (2006) examined whether a building's community could be self sufficient with respect to heat by using small CHP plants, and they explored the consequences of such an energy system. They showed that heat trading could be a functional way to develop decentralised energy systems, but further research is needed in respect of the economic aspects in order to understand better the possibilities of heat trading.

4 CONCLUSIONS

Measurements and calculations of heat losses from channel and preinsulated heat distribution networks showed that higher losses (up to 5.6% of total supplied energy) occurred for the channel network (network A), while the preinsulated network (network B) incurred only a 0.502% loss. Taking into consideration the significant operational advantages of a preinsulated network, substitution of existing (operating) channel networks by preinsulated networks could be recommended in the near future. This should generate direct profits through reductions in charges for CO_2 emissions and long-term benefits for the environment from better power management. But the pay-back time for this solution is very long and it is more reasonable to switch systems when a channel network actually needs to be replaced. Decentralisation of the heat source and greater usage of local heat sources, which would eliminate heat losses incurred during distribution, is another option.

APPENDIX 1

Formulas for heat losses calculations of channel and preinsulated heat distribution network.

$$Q_s = q_s \cdot L_s \tag{3}$$

$$Q_r = q_r \cdot L_r \tag{4}$$

$$q_s = \frac{\tau_s - t_{ach}}{\sum R_s} \tag{5}$$

for $\tau_s = 54.2$; $t_{ach} = 10.77$

$$q_r = \frac{\tau_r - t_{ach}}{\sum R_r} \tag{6}$$

for $\tau_r = 41.9$; $t_{ach} = 10.77$

$$\sum R_s = R_{ins} + R_{insa} + R_{ch} + R_{ach} + R_{soil} \tag{7}$$

$$\sum R_r = R_{inr} + R_{inra} + R_{ch} + R_{ach} + R_{soil} \tag{8}$$

$$R_{ins} = \frac{1}{2 \cdot \pi \cdot \lambda_{in}} \cdot \ln \frac{D_{ins}}{D_e} \tag{9}$$

for $\lambda_{in} = 0.04$.

$$R_{inr} = \frac{1}{2 \cdot \pi \cdot \lambda_{in}} \cdot \ln \frac{D_{inr}}{D_e} \tag{10}$$

for $\lambda_{in} = 0.04$.

$$R_{insa} = \frac{1}{\pi \cdot D_{ins} \cdot \alpha_{ina}} \tag{11}$$

for $\alpha_{ina} = 11.6$.

$$R_{inra} = \frac{1}{\pi \cdot D_{inr} \cdot \alpha_{ina}} \tag{12}$$

for $\alpha_{ina} = 11.6$.

$$R_{ch} = \frac{1}{2 \cdot \pi \cdot \lambda_{ch}} \cdot \ln \frac{D_{ech}}{D_{ich}} \tag{13}$$

for $\lambda_{ch} = 1.1$; $D_{ech} = 0.95$; $D_{ich} = 0.75$.

$$R_{ach} = \frac{1}{\pi \cdot D_{ich} \cdot \alpha_{ach}} \tag{14}$$

for $\alpha_{ach} = 11.6$; $D_{ich} = 0.75$.

$$R_{soil} = \frac{1}{2 \cdot \pi \cdot \lambda_{soil}} \cdot \ln \frac{4 \cdot h}{D_{ech}} \tag{15}$$

for $\lambda_{soil} = 1.16$; $D_{ech} = 0.95$; $h = 1.3$.

For heat losses calculations for the preinsulated heat distribution network (network B) Formulas (3), (4) and the following equations were used (Kamler 1979).

$$q_s = \frac{(\tau_s - t_{soil}) \cdot R_{inr} - (\tau_r - t_{soil}) \cdot R_{ad}}{R_{ins} \cdot R_{inr} - R_{ad}^2} \tag{16}$$

for $\tau_s = 54.74$; $\tau_r = 45.92$; $t_{soil} = 4$.

$$q_r = \frac{(\tau_r - t_{soil}) \cdot R_{ins} - (\tau_s - t_{soil}) \cdot R_{ad}}{R_{ins} \cdot R_{inr} - R_{ad}^2} \tag{17}$$

for $\tau_s = 54.74$; $\tau_r = 45.92$; $t_{soil} = 4$.

$$R_{ad} = \frac{1}{2 \cdot \pi \cdot \lambda_{soil}} \cdot \ln \sqrt{1 + \left(\frac{2 \cdot h}{a}\right)^2} \tag{18}$$

for $\lambda_{soil} = 1.2$; $a = 0.34$; $h = 1.3$.

For calculations of the insulation resistance of supply pipe (R_{ins}) and of return pipe (R_{inr}) Formulas 9–10 were used with the assumption $\lambda_{in} = 0.03$ W/mK.

NOMENCLATURE

a	distance between the axis of the supply and return pipe (m)
D_{inr}	diameter of return pipe insulation (m)
D_{ins}	diameter of supply pipe insulation (m)
D_n	nominal diameter of pipe (m)
D_i	internal diameter of pipe (m)
D_{ich}	internal substitute diameter of channel (m)
D_e	external diameter of pipe (m)
D_{ech}	external substitute diameter of channel (m)
h	distance from the axis of network pipes to the ground level (m)
L	lenght of pipe (m)
L_r	lenght of return pipe (m)
L_s	lenght of supply pipe (m)
q_r	individual heat loss of return pipe (W/m)
q_s	individual heat loss of supply pipe (W/m)
R_{ad}	additional thermal resistance (mK/W)
R_{soil}	thermal resistance of soil (mK/W)
R_{inr}	thermal resistance of insulated return pipe (mK/W)
R_{ins}	thermal resistance of insulated supply pipe (mK/W)
R_{ch}	thermal resistance of channel wall (mK/W)
R_{ach}	thermal resistance of taking over heat from air in the channel to the internal channel side (mK/W)
R_{inra}	thermal resistance of taking over heat from return pipe insulation to air in the channel (mK/W)
R_{insa}	thermal resistance of taking over heat from supply pipe insulation to air in the channel (mK/W)
ΣR_r	sum of thermal resistance for return pipe (mK/W)
ΣR_s	sum of thermal resistance for supply pipe (mK/W)
t_{soil}	temperature of soil (°C)
t_{ach}	temperature of air in the channel (°C)
Q_{ja}	average daily heat losses of heat distribution network (GJ/d)
Q_{jr}	average daily heat losses of return pipe (GJ/d)
Q_{js}	average daily heat losses of supply pipe (GJ/d)
Q_r	heat losses of return pipe (W)
Q_s	heat losses of supply pipe (W)
α_{ach}	coefficient of taking over heat from air in the channel to the internal channel side (W/m²K)
α_{ina}	coefficient of taking over heat from pipe insulation to air in the channel (W/m²K)
λ_{ch}	thermal conductivity coefficient of channel wall (W/mK)
λ_{in}	thermal conductivity coefficient of pipe insulation (W/mK)
λ_{soil}	thermal conductivity coefficient of soil (W/mK)
τ_r	temperature of heating medium in return pipe (°C)
τ_s	temperature of heating medium in supply pipe (°C)

REFERENCES

Adamo, G., Cammarata, A., Fichera A. & Marletta L. 1997. Improvement of a district network through thermoeconomic approach. *Renewable Energy* 10: 213–216.

Benonysson, A., Bohn, B. & Ravn H.F. 1995. Operational optimization in a district heating system. *Energy Convers. Mgmt.* 36: 297–314.

Bohn, B. 2000. On transient heat losses from buried district heating pipes. *Int. J. Energy Res.* 24: 1311–1334.

Bojic, M., Trifunovic, N. & Gustafsson S.I. 2000. Mixed 0–1 sequential linear programming optimization of heat distribution in a district-heating system. *Energy and Buildings* 32: 309–317.

Comakli, K., Yuksel, B. & Comakli, O. 2004. Evaluation of energy and exergy losses in district heating network. *Applied Thermal Engineering* 24: 1009–1017.

Dobersek, D. & Goricanec, D. 2009. Optimisation of tree path pipe network with nonlinear optimisation method. *Applied Thermal Engineering* 29: 1584–1591.

Kalinci, Y., Hepbasli, A. & Tavman, I. 2008. Determination of optimum pipe diameter along with energetic and exergetic evaluation of geothermal district heating systems: Modeling and application. *Energy and Buildings* 40: 742–755.

Kamler, W. 1979. *District Heating*. Warsaw: PWN.

Larsen, H.V., Bohm, B. & Wigbels, M. 2004. A comparison of aggregated models for simulation and operational optimisation of district heating networks. *Energy Conversion and Management* 45: 1119–1139.

Lygnerud, K. & Peltola-Ojala, P. 2010. Factors impacting district heating companies' decision to provide small house customers with heat. *Applied Energy* 87: 185–190.

Niemyjski, O. 2007. Heat losses in district heating networks and possibilities of their limitation. *District Heating, Heating, Ventilation* 7/8: 9–13.

Nilsson, S.F., Reidhav, Ch., Lygnerud, K. & Werner, S. 2008. Sparse district-heating in Sweden. *Applied Energy* 85: 555–564.

Nystedt, A., Shemeikka, J. & Klobut, K. 2006. Case analyses of heat trading between buildings connected by a district heating network. *Energy Conversion and Management* 47: 3652–3658.

Popescu, D., Ungureanu, F. & Hernandez-Guerrero, A. 2009. Simulation models for the analysis of space heat consumption of buildings. *Energy* 34: 1447–1453.

Poredos, A. & Kitanovski, A. 2002. Exergy loss as a basis for the price of thermal energy. *Energy Conversion and Management* 43: 2163–2173.

Reidhav, Ch. & Werner, S. 2008. Profitability of sparse district heating. *Applied Energy* 85: 867–877.

Stanny, A. 2007. Limitation of heat losses with the method ConduFill and their calculation by use of temperature recorders AS1922G and AS1922T. *District Heating, Heating, Ventilation* 6: 8–9.

Stevanovic, V.D., Prica, S., Maslovaric, B., Zivkovic, B. & Nikodijevic S. 2007. Efficient numerical method for district heating system hydraulics. *Energy Conversion and Management*, 48: 1536–1543.

Stevanovic, V.D., Zivkovic, B., Prica, S., Maslovaric, B., Karamarkovic, V. & Trkulja, V. 2009. Prediction of thermal transients in district heating systems. *Energy Conversion and Management* 50: 2167–2173.

Werner, S. 2004. District heating and cooling. *Encyclopedia of Energy*. Elsevier 1: 841–7.

Agency of energy market in Poland (www.cire.pl).

Environmental Engineering III – Pawłowski, Dudzińska & Pawłowski (eds)
© 2010 Taylor & Francis Group, London, ISBN 978-0-415-54882-3

Assessment of the combustion of biomass and pulverized coal by combining principal component analysis and image processing techniques

A. Kotyra & W. Wójcik
Department of Electronics, Lublin University of Technology, Lublin, Poland

T. Golec
Thermal Division, Lublin University of Technology, Lublin, Poland

ABSTRACT: Image processing is o many techniques for diagnosing combustion. Flame shape and thus the combustion process, can be analyzed through many different shape parameters. In this research, principal component analysis was used to assess selected geometric parameters of a flame shape during a combustion test when a coal-biomass mixture was burned.

Keywords: Biomass combustion, image processing, principal component analysis.

1 INTRODUCTION

Coal is still the main fuel used in electricity generation around the world. Solid fuels, such as coal, often contain impurities such as nitrogen and sulfur that can increase pollutant emissions signi?cantly. It has led to the development of new combustion techniques, such as air staging, reburning, and flue-gas recirculation (Li et al. 2003). What is more, fossil fuel depletion forces the use of renewable fuels, such as biomass. In the existing coal-fired power stations, biomass is milled and burned simultaneously with coal. However, low-emission combustion techniques as well as biomass co-combustion have negative side effects on combustion installations – increased corrosion, boiler slagging – which lower the efficiency and stability of the combustion process (Hein & Bemtgen 1998, Shaohua et al. 2007). To minimize these effects, a proper combustion monitoring system needs to be applied (Adamson & Cumming 2004).

The way in which a pulverized fuel is burned largely depends on the degree of its granularity, among other parameters. Coal particles of between 5 and $400\,\mu m$ in diameter are burned together in a swirling turbulent flame (Williams et al. 2001). The commonly applied, low-emission techniques of pulverized coal combustion use recirculation vortexes that lengthen the paths of the coal grains passing through the flame to minimize generation of thermal oxides of nitrogen (NO_x).

In order to make combustion of pulverized coal more efficient and clean, it is necessary to measure its key parameters. The information taken at the output (exhaust gas collector) is delayed and averaged. Thus, it cannot be used to assess the combustion quality for an individual burner. There are many combustion diagnostic techniques (Fristrom 1995), that can be directly used for the characterization and optimization of a combustion process, but most of them are expensive or impossible to utilize under industrial conditions. Optical methods, although suffering difficulties of access and the threat of dustiness do not require external illumination. The availability of rugged, low-cost charge coupled device (CCD) cameras and microcontroller systems allows the use of image processing techniques.

2 MATERIALS AND METHODS

2.1 Extracting the features of the flame areas

If a flame accompanies a combustion process, it reflects the chemical and physical phenomena that accompany the combustion process being examined. Flame properties, such as the geometry (size, position), radiation properties (the emission spectrum, irradiation distribution), and flicker frequency depend on the conditions in the location where the combustion process is proceeding. These conditions include the combustion chamber geometry, the burner type, fuel flow, air to fuel ratio, fuel granularity, etc (Fristrom, 1995). Analyzing the changes in the shapes of the flames that accompany variable burner input parameters, makes it possible to detect the different states of the combustion process, both in the laboratory (Lu et al. 2005) and under industrial conditions (Marques & Jorge 2000, Beak et al. 2001). Tens of parameters, resulting in a large feature vector, can be used to define shape. The problem is which shape features are most sensitive to variations in the burner input parameters and which can be neglected in the assessment of the combustion of biomass and pulverized coal. Size reductions of the feature vector, as well

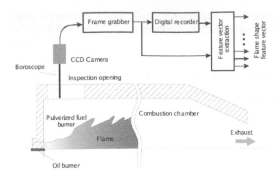

Figure 1. Scheme of the viewing system for capturing flame features.

as finding the most adequate shape features, can be obtained using principal component analysis (PCA). In this paper, only some of the geometric features of the flame image were taken into consideration.

The flame shape was analyzed using a vision system as shown in Figure 1. It consists of a CCD camera that was placed perpendicular to the burner outlet. To avoid any influence of the high temperature on the functioning of the camera it was equipped with a 0.7 m boroscope for transmitting the image. The digital video recording system captured 8-bit gray images with a resolution of 640 × 480 pixels at 15 frames per second. Not the whole image area was the subject of investigation, a region of interest (300 × 100 pixels) was chosen for further processing.

Estimating flame shape features requires extracting a flame area from every captured image. As the captured images were 8-bit gray bitmaps, image pixel amplitude was contained within the range of from 0 to 255. As the flame was the only luminous object, its boundary could be easily determined through pixel amplitude. However, as the flame boundary was not sharp, the Canny edge detection algorithm, with double thresholding, was chosen (Canny 1986). All points in the given image having a gray value larger than or equal to a high threshold ('secure' points) are immediately accepted. While all points with gray values less than a low threshold are immediately rejected. The other ('potential') points with gray values between both thresholds are accepted if they are connected to 'secure' points by a path of 'potential' points having a length of at most a certain number of points. A maximum distance between 'potential' and 'secure' points was arbitrarily assumed, taking into consideration the flame image resolution. Both the high and low thresholds were determined through an image histogram.

In the current analysis, the only featured vector elements are geometric flame shape parameters. These include:

- flame area, A, defined as a sum of all the pixels contained within the flame region,
- coordinates of the flame area center (AreaCent_X, AreaCent_Y), calculated as the mean values of the line or column coordinates of all flame area pixels,

- length of flame boundary, L, defined as a sum of all boundary pixels, assuming that the distance between two neighboring contour points parallel to the coordinate axes is rated 1, while the distance on the diagonal is rated $\sqrt{2}$,
- the maximum distance (Diameter) between two boundary points of a region and their coordinates - (Diam_Col1, Diam_Row1), (Diam_Col2, Diam_Row2),
- the number of holes (NumHoles) in a flame region,
- the radii (R_A, R_B) and the orientation, ϕ, of the ellipse having the 'same orientation' and the 'same side relation' as the flame region, defined as:

$$R_A = \frac{1}{2}\sqrt{8\left(M_{20} + M_{02} + \sqrt{(M_{20} - M_{02})^2 + 4M_{11}^2}\right)} \quad (1)$$

$$R_B = \frac{1}{2}\sqrt{8\left(M_{20} + M_{02} - \sqrt{(M_{20} - M_{02})^2 + 4M_{11}^2}\right)} \quad (2)$$

$$\varphi = -\frac{1}{2}\operatorname{atan2}(2M_{11}, M_{02} - M_{20}) \quad (3)$$

where $M_{i,j}$ denotes the geometric moment:

$$M_{i,j} = \sum_{(Z,S)\in R}(Z_0 - Z)^i(Z_0 - Z)^j \quad (4)$$

Z_0 and S_0 are the coordinates of the center of a region, where Z and S run through all the pixels of the region being considered,

- compactness, C, defined as:

$$C = L^2/4\pi A, \quad (5)$$

where $L =$ the length of the contour and A the area of the flame shape.

Compactness responds to the regularity of the contour (roughness) and to holes. If the region is long or has holes, then C is larger than 1;

- convexity is defined as the quotient of the area of the convex hull and the original area of the considered region. For convex shapes, such as triangles and rectangles, it equals 1, if a shape has indentation or holes, convexity is < 1,
- rectangularity is determined in the following way. First a rectangle is computed that has the same first and second order moments as the input region. The computation of the rectangularity measure is finally based on the difference in area between the computed rectangle and the input region normalized with respect to the area of the rectangle (Rosin 1999),
- circularity, calculates the similarity of the input region area with that of a circle. If A denotes the area of the flame and max is the maximum distance from the center to all contour pixels, the shape factor, Circularity, is defined as:

$$Circularity = A/(max^2 \cdot \pi), \quad (6)$$

Figure 2. Views of the laboratory combustion stand; an arrow indicates the place the camera is mounted.

- the central moments of flame area, I_1, I_2, I_3, I_4, are defined as:

$$I_1 = M_{2,0}M_{0,2} - M_{1,1}^2, \tag{7}$$

$$I_2 = \left(M_{3,0}M_{0,3} - M_{2,1}M_{1,2}\right)^2 -$$
$$4\left(M_{3,0}M_{1,2} - M_{2,1}^2\right)\left(M_{2,1}M_{0,3} - M_{1,2}^2\right) \tag{8}$$

$$I_3 = M_{2,0}\left(M_{2,1}M_{0,3} - M_{1,2}^2\right) - M_{1,1}\left(M_{3,0}M_{0,3} - M_{2,1}M_{1,2}\right)$$
$$+ M_{0,2}\left(M_{3,0}M_{1,2} - M_{2,1}^2\right), \tag{9}$$

$$I_4 = M_{3,0}^2 M_{0,2}^3 - 6M_{3,0}M_{2,1}M_{1,1}M_{0,2}^2 + 6M_{3,0}M_{1,2}M_{0,2}$$
$$\left(2M_{1,1}^2 - M_{2,0}M_{0,2}\right) + M_{3,0}M_{0,3}\left(6M_{2,1}M_{1,1}M_{0,2} - 8M_{1,1}^3\right)$$
$$+ 9M_{2,1}^2 M_{2,0}M_{0,2}^2 - 18M_{2,1}M_{1,2}M_{2,0}M_{1,1}M_{0,2}$$
$$+ 6M_{2,1}M_{0,3}M_{2,0}\left(2M_{1,1}^2 - M_{2,0}M_{0,2}\right) + 9M_{1,2}^2 M_{2,0}^2 M_{0,2}$$
$$- 6M_{1,2}M_{0,3}M_{1,1}M_{2,0}^2 + M_{0,3}^2 M_{2,0}^3. \tag{10}$$

These flame shape features can be determined in real time during combustion tests or from the stored image sequences and form a feature vector, that consisted of 21 elements for the case being considered. The flame shape feature vector was obtained for every frame captured during combustion tests.

2.2 Combustion tests

Combustion tests of coal-biomass mixture were done in a 0.5 MW_{th} (thermal megawatt) laboratory stand at the Institute of Power Engineering. This unit simulates the scaled down (10:1) combustion conditions of a full-scale swirl burner fired with pulverized coal with biomass added. The test stand comprises a horizontal layout consisting of a cylindrical combustion chamber 0.7 m in diameter and 2.5 m long, as shown in Figure 2.

A model of a low-NO_x swirl burner about 0.1 m in diameter is mounted at the front wall. The stand is equipped with all necessary supply systems; primary and secondary air, coal, and oil. A mixture of pulverized coal and biomass is prepared in advance and dumped into the coal feeder bunker.

A combustion test consisted of the following steps. First, the combustion chamber was warmed up by burning oil. When the temperature had risen sufficiently, the feeding device was started and the air-fuel mixture was delivered to the burner, simultaneously with the oil. After reaching the proper temperature level, the oil supply was switched off.

Primary air is used mainly for delivering pulverized coal to the burner nozzle, while secondary air is used for regulation purposes. Input parameters, such as the coal-biomass mixture and air flows, were changed several times during the tests, to create various combustion states.

2.3 Principal components analysis (PCA) overwiev

Principal component analysis (PCA) is a statistical technique that maps large number of independent variables into a set of principal components having less dimensions than those of the original variables. The PCA reduces the dimensionality of a set of variables while retaining the maximum variability in terms of the variance-covariance structure. Given a set, X, of m variables, a principal component model transforms these variables into a new set having fewer dimension, i.e., $c < m$, while still capturing most of the variability in the original data set. Each coordinate in the new transformed system is known as a principal component (Krzanowski, 2000).

Application of PCA requires a preprocessing stage in which the original variables are transformed in such a way that the general assumptions about the data set will hold best. There are several ways to preprocess a data set; usually unit variance scaling is applied. A transformed variable is obtained according to:

$$\bar{x}_i = \frac{x_i^n - \mu_i}{\sigma_i} \qquad (11)$$

where x_i, $i = 1, \ldots, n$, is a typical observation of the variable X_i, with mean μ_i and standard deviation σ_i^2.

Scaling transforms the variables of the data set in such a way that the variables have comparable values, thus preventing larger value variables from the smaller. This eliminates over representation of some variables merely on the basis of size.

The next step is to find a number of principal components that best represents the original variables, X, in the least squares sense. It is equivalent to finding a new coordinate system that predicts a matrix of observations, $\widehat{X}$, as an approximation of X. This step is carried out incrementally. Initially, the first principal component that captures the largest portion of variability in the initial data matrix X is extracted. Then the next components are found until a point is reached where the extracted principal components capture not less than a certain percentage of the variability in the data set.

The set of principal components forms an orthogonal system with point of origin given by the mean of the original data and orientation defined by the loading factors,

$$\mathbf{P}^T = \begin{bmatrix} p_{11}, \cdots p_{m1} \\ \vdots \ddots \vdots \\ p_{1c}, \cdots p_{mc} \end{bmatrix} \qquad (12)$$

where c is number of principal components, m is number of variables in the data matrix X. If matrix T defines the representation of the original observations in the new coordinate system, the predictions, $\widehat{X}$, of the principal component model for the preprocessed data set $\overline{X}$ is given by:

$$\widehat{X} = T \times P^T \qquad (13)$$

The difference between $\overline{X}$ and $\widehat{X}$ is a matrix E, that gives residuals. Combining it with equation 13 yields:

$$X = T \times P^T + E \qquad (14)$$

The loading factors, P^T, are arranged in order of the largest eigenvalues of the covariance matrix XX^T.

In order to assess the principal component model obtained, the fractions of the explained variation, R^2X,

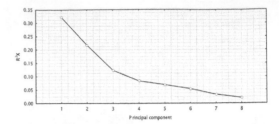

Figure 3. Plot of R^2X for the principal components obtained.

the predicted variation, Q^2X, are applied. R^2X is measured on the entire training sample, according to:

$$R^2X = 1 - \frac{\sum_{i,j}(x_{i,j} - \hat{x}_{i,j})^2}{\sum_{i,j} x_{i,j}^2} \qquad (15)$$

The matrix E of residuals can be used to find variable importance, that is, how well it is represented by the model obtained. It could be assessed by the modeling power, defined as:

$$Power = 1 - \frac{SV_j}{SV_j^0} \qquad (16)$$

where SV_j denotes the residual standard deviation of the jth variable and SV_j^0 its initial standard deviation.

3 RESULTS AND DISCUSSION

During the combustion experiment, a data set of $m = 21$ variables (flame shape parameters) was collected in $n = 77.600$ observations. The data were processed using the Statistica PCA module. First, the data were preprocessed with unit variance scaling (equation 11). Determination of the number of principal components was done by computing their significance using the cross-validation method. The number of significant principal components obtained was 8.

The plot of the fraction of the explained variation, R^2X, is shown in Figure 3. The more significant a principal component (PC) is, the higher the value of the explained variation to which it corresponds. Adding together all the R^2X for each PC yields 0.92 (explained variation), which indicates that the PC model obtained is adequate.

A plot showing the importance of the flame parameters analyzed is shown in Figure 4. It can be seen that almost all of the flame shape parameters are approximately equally represented in the PC model. As the power of most of the flame parameters is near 1, it means they are important in creating the PC model. The exception is the *NumHoles* parameter.

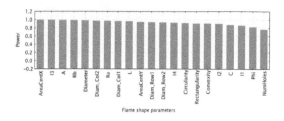

Figure 4. Modeling power of the flame shape parameters considered.

4 CONCLUSIONS

The flame shape parameters chosen were ones that could be determined in real time. The PCA analysis shows that the flame shape parameters obtained during a biomass-coal combustion test are nearly equally significant. The *NumHoles* parameter has the least power in sense of equation 16. Its value is slightly less than 0.8. (It may be relevant to add, that a variable is only completely relevant if its modeling power is equal to 1.) However, obtaining a PC model which has explained variation exceeding 0.9 for 21 initial variables requires up to 8 principal components. Thus, the set of shape parameters analyzed represent flame variability that occurs during variable combustion states, practically without redundancy.

REFERENCES

Adamson, E.J. & Cumming, R.G. 1997. Boiler flame monitoring systems for low NO$_x$ applications – an update. *Proceedings of the American Power Conference* 59–1: 340–344.

Baek, W.B., Lee, S.J., Baeg, S.Y. & Cho, Ch.H. 2001. Flame image processing and analysis for optimal coal firing of thermal power plant. *Proceedings ISIE 2001, IEEE International Symposium on Industrial Electronic* 2: 928–931.

Canny, J. 1986. A computational approach to edge detection, *IEEE Transactions on Pattern Analysis and Machine Intelligence. PAMI-8(6):* 679–698.

Fristrom, R.M. 1995. *Flame structure and processes.* Oxford University Press: London.

Hein K.R.G. & Bemtgen J.M. 1998. EU clean coal technology – co-combustion of coal and biomass. *Fuel Processing Technology:* 159–169.

Krzanowski, W.J. 2000. *Principles of multivariate analysis: a user's perspective.* Oxford University Press: London.

Li, Z.Q., Wei, F. & Jin, Y. 2003. Numerical simulation of pulverized coal combustion and NO formation. *Chemical Engineering Science* 58: 5161–5171.

Lu G., Gilbert G. & Yan Y. 2005. Vision based monitoring and characterization of combustion flames. *Journal of Physics: Conference Series* 15: 194–200.

Marques J.S. & Jorge M.P. 2000. Visual inspection of a combustion process in a thermoelectric plant. *Signal Processing* 80: 1577–1589.

Rosin P.L. 1999. Measuring rectangularity. *Machine Vision and Applications* 11: 191–196.

Shaohua, M., Zhiyuan, C., Ying, H., Xiaobai, L. & Yangyang, G. 2007. An approach of combustion diagnosis in boiler furnace based on phase space reconstruction. In D-S. Huang, L. Heutte, M. Loog (ed.), *Advanced intelligent computing theories and applications with aspects of contemporary intelligent computing techniques. Proceedings of the International Conference on Intelligent Computing ICIC 2007:* 528–535. Springer Verlag: Berlin.

Williams A., Pourkashanian M. & Jones J.M. 2001. Combustion of pulverized coal and biomass. *Progress in Energy and Combustion Science* 27: 587–610.

Environmental Engineering III – Pawłowski, Dudzińska & Pawłowski (eds)
© *2010 Taylor & Francis Group, London, ISBN 978-0-415-54882-3*

Energy demand for space heating of residential buildings in Poland

A. Siuta-Olcha & T. Cholewa
Faculty of Environmental Engineering, Lublin University of Technology, Lublin, Poland

ABSTRACT: Over the past few years in Poland, there have been aims to decrease the energy requirements for space heating of residential buildings, and this is now presented in this paper. Methods of heat demand calculations included in Polish standards and in European standards are compared. The results of calculations are presented for a typical multi-family residential building located in Lublin. As a result of thermo-modernization works, the design heat load of the analyzed residential building is decreased by 46% and the annual energy demand for space heating is reduced to about 90 kWhm^{-2} per year.

Keywords: Energy consumption, building shape factor, utilization factor of heat gains, building time constant, calculation methods.

1 INTRODUCTION

In 2006, the gross inland consumption in the European Union (EU-27) Member States was 1825.2 Mtoe and the EU final energy consumption was 1177.4 Mtoe, of which 483.92 Mtoe was in the buildings sector (41.1%) (Figure 1) (EC 2009 European Union Energy and Transport in Figures). Energy consumption in buildings in developed countries comprises 20–40% of the world's total final energy consumption (Pérez-Lombard et al. 2008). The residential sector consumed 25.9% of the total final energy consumption in the EU-27 in 2006. Energy used in residential buildings constitutes about 63% of the total energy consumption in the European buildings sector (Balaras et al. 2007, Poel et al. 2007). The EU-27 residential sector contributes 9.4% of the total greenhouse gases (GHG) emissions. In 2006, total GHG emission was 5142.8 million tonnes CO_2 equivalent in the EU-27 countries. The structure of GHG emissions by sectors in the EU and in Poland is illustrated in Figure 2. In 2006, the total annual CO_2 emissions from the EU-27 buildings stock was 731.4 million tonnes, of which 64% was from the residential sector and 36% was from the tertiary sector (non-residential buildings and agriculture) (EC 2009 European Union Energy and Transport in Figures).

In European residential buildings, about 57% of the total final energy consumption is used for space heating, 25% for domestic hot water and 11% for electricity (Chwieduk 2003, Balaras et al. 2007). The energy demands of the residential sector depend mainly on the climatic zone, the age of households, the characteristics and the quality of the building, and (particularly) the area of the building, the quality of the building envelope, the number of inhabitants, the efficiency

of the heating and cooling system, and the type of electrical appliances (Santamouris et al. 2007).

In order to improve the efficiency of energy used for heating and cooling of buildings, the Energy Performance of Buildings Directive (EPBD) 2002/91/EC was elaborated. According to the new trends in energy policy presented by the European Commission on January 10, 2007, the European Union countries should make the following changes by 2020: (i) a reduction in GHG emissions by 20%, (ii) an increase in renewable energy fractions in the fuel–energy balance by 20%, and (iii) a reduction in energy consumption by 20%.

In Poland, energy consumption in the buildings sector is excessive (Balaras et al. 2005, Wichowski 2007). There are approximately 11.8 million housing units, with a total usable area of about 700 million m^2 (CSO 2007 Energy Efficiency Policies and Measures in Poland 2006). The energy efficiency improvement of the buildings is closely connected with environmental protection. The governmental institution responsible for the operational implementation of energy efficiency policies is the Polish National Energy Conservation Agency. In 1998, a law on supporting thermo-modernization projects was passed. Prior to entering into the European Union, Poland accepted the Kyoto Protocol dated 1997 and assumed an obligation to decrease GHG emissions by 6% in relation to the year 1988 in the period 2008–2012. In Poland, the European Directive on the energy performance of buildings (EPBD) was implemented on 1 January 2009. On 16 July 2009, the Polish parliament passed an Act to change the Construction Law to redefine regulations concerning energy certificates for buildings. The novelization was adopted, with the aim of correcting the transposition of the Directive 2002/91/EC.

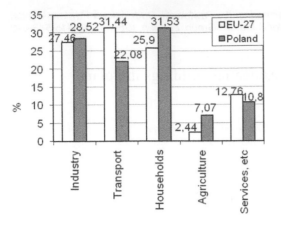

Figure 1. Structure of final energy consumption by sector in the EU-27 and Poland in 2006.

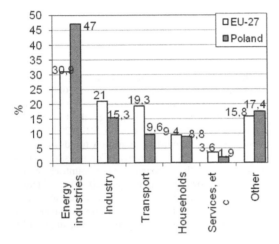

Figure 2. Structure of GHG emissions by sector in percentages for the EU-27 and Poland in 2006.

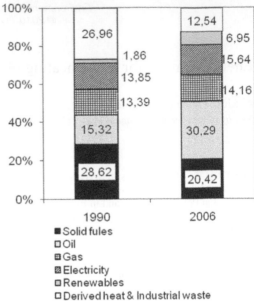

Solid fules
Oil
Gas
Electricity
Renewables
Derived heat & Industrial waste

Figure 3. Structure of final energy consumption by fuels in Poland in 1990–2006.

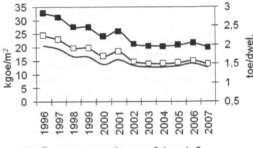

—■— Energy consumption per m2, kgoe/m2
—□— Energy consumption for heating per m2, kgoe/m2
—— Energy consumption per dwelling, toe/dwel.

Figure 4. Trends in energy consumption in the Polish residential sector in the period 1996–2007.

Changes in the structure of the final energy consumption by energy carrier between 1990 and 2006 are presented in Figure 3. In 2006, the total final energy consumption was 60.82 Mtoe. Energy demands in Polish building stock represented about 49.4% of the final energy consumption, of which 31.53% is from the residential sector, 10.8% is from the non-residential sector and 7.07% is from the agriculture sector (Figure 1) (EC 2009 European Union Energy and Transport in Figures). The structure of energy consumption by end use in Polish households for 2002 was: 71.2% for space heating, 15.1% for water heating, 6.6% for cooking, 4.5% for electrical equipment and 2.3% for lighting. Trends of energy consumption per m^2 in households in years 1996–2007 are showed in Figure 4. Energy demands for heating per m^2 were reduced from 24.5 kgoem^{-2} to 13.8 kgoem^{-2} between 1996 and 2007. Since 1996, energy consumption in dwellings has been decreased by 29%. In household sectors, the value of the energy efficiency indicator (ODEX) decreased from 111.2 in 1996 to 80.6 in

2007. The average annual improvement was 2.3% in this period (CSO 2009, Energy Efficiency in Poland in years 1997–2007).

In 2006, the total GHG emissions in Poland were 402.7 million tonnes CO_2 equivalent, of which 54.9 million tonnes CO_2 equivalent came from the buildings sector. CO_2 emissions by residential sector in Poland accounted for 9.9% of the total emissions, which was equal to 333 million tonnes (EC, European Union Energy and Transport in Figures 2009).

The average annual energy demand for space heating of residential buildings in Poland, over various periods of time, was as follows: 240–380 kWhm^{-2} before 1985, 160–200 kWhm^{-2} in 1985–1992, 120–160 kWhm^{-2} in 1993–1997, 90–120 kWhm^{-2} after 1998 (Chwieduk 2003). The main cause of large

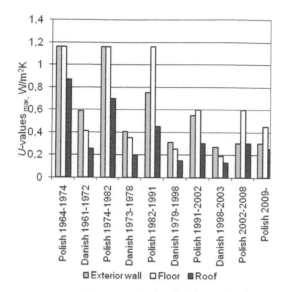

Figure 5. The maximum U-values of building's components according to Polish and Danish standards.

Table 1. Parameters of the analyzed building.

Parameter	Description	
	B1[a]	B2[b]
Length × width × height (m)	31.93 × 10.97 × 16.1	
A_n (m²)	1316	
V_n (m³)	3636	
North glazing surface (m²)	93.3	
East glazing surface (m²)	1.96	
South glazing surface (m²)	112.61	
West glazing surface (m²)	19.15	
A/V_n (m²m⁻³)	0.53	
U_{wall} (Wm⁻²K⁻¹)	1.74	0.32
U_{window} (Wm⁻²K⁻¹)	2.10	1.60
U_{roof} (Wm⁻²K⁻¹)	1.06	0.23
$U_{floorofcellarage}$ (Wm⁻²K⁻¹)	1.14	0.47
n_h (h⁻¹)	0.6	
Design heat load (kW)[c]	144.641	78.477

[a]B1: the building before thermal modernization,
[b]B2: the building after thermal modernization,
[c]according to PS-EN 12831:2006.

energy consumption is a lack of thermal insulation for buildings. At present, requirements concerning thermal protection of new buildings have become more strict in Poland. Polish standards and regulations were adapted to the European standards and directives. The annual energy demand for the space heating of a building is dependent on the U-values of different envelope elements.

Figure 5 illustrates the comparison of maximal values of the overall heat-transfer coefficient of buildings components, having been valid in Poland and Denmark over the past several decades. Currently, the required U-value of the window is equal to 1.7–1.8 Wm⁻²K⁻¹ for Poland and 1.5 W/m⁻²K⁻¹ for Denmark (Dyrbøl et al. 2005, Koczyk 2004, Tommerup & Svendsen 2006).

Multi-family residential buildings in Poland will be classified as energy-saving buildings if their annual energy consumption does not exceed 80 kWm⁻² (Pogorzelski 2007).

The purpose of this paper is to investigate the energy consumption for space heating in a typical residential building in Poland on the basis of Polish standards and regulations.

2 MATERIALS AND METHODS

2.1 Building description

The object of this analysis is the multi-family residential building located in Lublin. It is a five-storey building that consists of 30 apartments. It was designed based on traditional technology in the 1960s. The building was renovated in 2001. Thermo-modernization works consisted of insulating the external walls, roof and floor above non-heated cellars with foamed polystyrene of thermal conductivity

coefficient $\lambda = 0.045$ Wm⁻¹K⁻¹, as well as replacing some of the windows. On the basis of the weight criterion of the room envelope (Laskowski 2005), with consideration of the final envelope weight related to 1 m² of floor surface, the building has been classified as a medium-heavy class of building (426.1 kgm⁻²). The building is naturally ventilated. There are 110 inhabitants in the building. Parameters describing the building are presented in Table 1.

2.2 Weather data

The climate of Lublin is continental – moderately warm with an annual average temperature of 7.6°C. The average temperature of the season requiring use of heating is 2.2°C. The value of heating degree days for Lublin is equal to 3957.4 Kday per year. Basic climatic data for Lublin are collated in Table 2.

2.3 Comparison of calculation procedures for heat requirements, according to PS B-02025 and EN ISO 13790

The total annual (seasonal) heat energy requirements for heating are the sum of all heat demands in particular months in which they are higher than zero:

$$Q_h = \sum_m Q_{h,m} \qquad (1)$$

The heat demand for space heating of the building Q_h is calculated for each building zone and for each month as the difference between the heat loss of the building and the total heat gains with consideration of the utilization factor of heat gains:

$$Q_{h,m} = Q_{L,m} - \eta_m Q_{g,m} \qquad (2)$$

Table 2. Climate data for Lublin (Poland) (PS B-02025).

Month	Average external temperature, °C	Solar irradiance on vertical surface (MJm^{-2})				Heating period duration (h)
		south	west	north	east	
January	−3.9	142	78	56	72	744
February	−2.9	223	140	89	121	672
March	0.9	305	217	145	214	744
April	7.5	303	249	158	264	720
May	12.9	324	305	204	351	120
June	16.8	308	324	228	350	0
July	17.9	316	332	225	337	0
August	16.9	327	284	171	300	0
September	12.7	282	189	117	210	120
October	7.9	238	120	75	142	744
November	3.1	148	65	47	75	720
December	−1.1	99	40	29	45	744

When defining the heat losses of the building (Equations (3–6)), the following heat losses by transmission through building elements should be considered: external walls, windows, roof, floor on the ground, floor above the non-heated ceiling, as well as heat losses by ventilation. When determining heat losses by transmission in both procedures, the linear heat-loss coefficient of building thermal bridges must be considered.

$$Q_{L,m} = H(\theta_i - \theta_{e,m})t \qquad (3)$$

$$H = H_T + H_V \qquad (4)$$

$$H_T = \sum AU + l\Psi \qquad (5)$$

$$H_V = 0.34 n_h V_n \qquad (6)$$

According to the Polish Standard B-02025, the outdoor airflow rate should be calculated in compliance with the national standard B-03430. For a typical apartment in a multi-family building, this airflow rate amounts to 150 m^{3}h^{-1}. According to EN ISO 13790, air changes per hour (ACH) are given as 0.5 h^{-1}–1.2 h^{-1} depending on the class of envelope and leakproofness of the building. However, as there are no data on this particular aspect, it is recommended that the air change rate be 0.3 h^{-1}, which is equivalent to minimal ventilation. German regulations are an air change rate of 0.8 h^{-1}.

Total heat gains include internal and solar heat gains:

$$Q_{g,m} = (\phi_i + \phi_{s,m})t \qquad (7)$$

According to the Polish Standard B-02025, internal heat gains are calculated on the basis of the following formula:

$$\phi_i = 80NP + (230 + \phi_{light})NF \qquad (8)$$

Equation (8) makes allowances for heat gains arising from people, lighting, electrical appliances, domestic hot water and cooking. The European standard ISO 13790:2004 gives an average value of 4 Wm^{-2} for the heating floor area in order to define heat gains obtained from people, lighting and devices for residential buildings. On the basis of Polish regulations (Decree of the Ministry of Infrastructure 2008), unit values of internal gains can be taken within 3.2–6 Wm^{-2}.

German regulations for residential buildings state that internal heat gains expressed in kWh per year should not exceed the values calculated according to the following equation (Dilmac & Kesen 2003, Fayaz & Kari 2009):

$$Q_i = 8V \quad \text{or} \quad Q_i = 25A_n \qquad (9)$$

Solar heat gain for each direction j is described by the following formulae:

$$\phi_{s,m} = \sum_j \left(I_{j,m} \sum_n A_{s,nj} \right) \qquad (10)$$

$$A_s = A_w(1 - F_F)F_s g \qquad (11)$$

$$g = F_w g_n \qquad (12)$$

According to Polish Standard B-02025, the glass surface fraction in the window surface can be assumed to be equal to 0.6. To follow Polish regulations (Decree of the Ministry of Infrastructure 2008), this fraction totals 0.7, on average, and the shading factor should be taken as equal to 0.9, 0.95, 0.96 or 1.0, depending on the building location and types of shading elements. In Germany, a fixed shading factor value is 0.46 and is taken for computations for all types of buildings (Dilmac & Kesen 2003, Fayaz & Kari 2009). Values of the solar transmission factor of glazing are presented in Table 3.

According to Polish Standard B-02025, the utilization factor of heat gains depends only on the ratio

584

Table 3. Typical values of total solar energy transmittance for common types of glazing.

Glazing type	g in PS B-02025	g_n^a in PS-EN ISO 13790
Single glazing	0.82	0.85
Double glazing	0.70	0.75
Triple glazing	0.64	0.70
Double glazing with selective low-emissivity coating	0.64	0.67

$^a g = F_w g_n = 0.9 g_n$

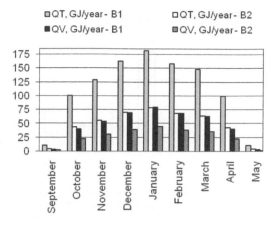

Figure 6. Monthly heat-loss balances for the analyzed building (B1: the building before thermal modernization; B2: the building after thermal modernization).

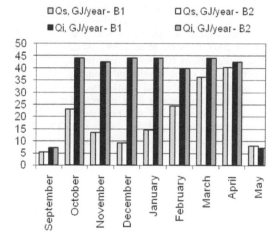

Figure 7. Monthly heat-gains balance for the analyzed building (B1: the building before thermal modernization, B2: the building after thermal modernization).

of total heat gains to the total heat losses and it is calculated on the basis of the following equation:

$$\eta_m = 1 - e^{-1/GLR} \tag{13}$$

$$GLR = \frac{Q_g}{Q_L} \tag{14}$$

In the case of simplified computations, $\eta = 0.9$.

In EN ISO 13790, while calculating the utilization factor of heat gains, it is necessary to allow for not only the ratio of total heat gains to total heat losses but also the building time constant:

$$\eta_m = \frac{1 - \gamma^a}{1 - \gamma^{a+1}} \quad \text{for} \quad \gamma \neq 1 \tag{15}$$

$$\eta_m = \frac{a}{a+1} \quad \text{for} \quad \gamma = 1 \tag{16}$$

$$a = a_0 + \frac{\tau}{\tau_0} \tag{17}$$

$$\tau = \frac{C}{H} \tag{18}$$

The building time constant characterizes the internal thermal inertia of heated areas. Considering the monthly calculation method, $a_0 = 1$ and $\tau_0 = 15\,h$ for residential buildings.

As a result of research conducted for climatic conditions of Finland, new numerical parameter values have been defined as being adequate for residential buildings $a_0 = 6$ and $\tau_0 = 7\,h$ (Jokisalo & Kurnitski 2007).

3 RESULTS AND DISCUSSION

The annual energy demand for space heating of the analyzed residential building is determined on the basis of its structure before and after thermal modernization. The computational results are obtained by using the computer programme PURMO OZC 4.0 (according to Polish Standard B-02025). Monthly heat balances divided into particular components are presented in Figures 6–7 for both cases. The most heat losses by transmission occur in January. The use of new windows and a thermal insulation – 10 cm thick for the external walls, 15 cm thick for the roof and 10 cm thick for the ceiling under the loft without heating – caused a reduction in heat losses by transmission from 1000.69 GJ per year to 433.68 GJ per year (about 56%). In a heating season, the heat losses by ventilation for the building before thermal modernization were 422.72 GJ per year and after modernization were 240.89 GJ per year. The total solar heat gains in the heating period were 174.82 GJ per year and the internal heat gains were 315.54 GJ per year.

Figure 8 shows the calculation results of the annual energy demands for space heating, expressed in kWh per square meter of heated floor area for the analyzed building. For building B1, the total heat losses are two times more and the total heat gains are 68% higher than those for building B2. The energy consumption in building B1 was decreased from 206.4 kWhm^{-2}y^{-1} to 88.8 kWhm^{-2}y^{-1} as a result of thermal modernization.

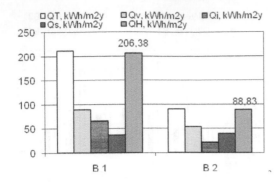

Figure 8. Comparison of annual energy consumption ($kWhm^{-2}$) in the analyzed building (B1: the building in Lublin before thermal modernization – PS B-02025; B2: the building in Lublin after thermal modernization – PS-EN ISO 13790).

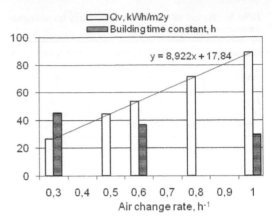

Figure 9. The influence of air change rate on annual heat demands of incoming air and on the time constant of the building.

Research results of interdependence between building shape and energy consumption are included in various publications (Depecker et al. 2001, Dilmac & Kesen 2003, Catalina et al. 2008). According to the Polish regulations (Decree of the Ministry of Infrastructure 2002), the residential building meets the requirements of thermal protection, if the energy demand rate for space heating, expressed in $kWhm^{-3}$ per year, does not exceed a maximal value dependent on the building shape coefficient. The shape coefficient has been defined as follows:

$$C_f = \frac{S_e}{V} \qquad (19)$$

Based on the results of calculations presented in Table 4, it was found that the heat energy demand for the analyzed building does not exceed maximal values presented in Polish regulations.

In Polish regulations, the windows area with an overall heat-transfer coefficient not less than $1.5\,Wm^{-2}K^{-1}$ in a residential building cannot exceed the value defined by means of the following formula:

$$A_w = 0.15A_e + 0.03A_i \qquad (20)$$

The French Thermal Standard (FTS) provides optimal values of the window to floor area ratio within 16.5–22% (Catalina et al. 2008). For the analyzed building, the value of the window to floor area ratio equals 24.6%. Regarding this case, the risk of overheating can appear during the summer season.

Heat losses by air infiltration and ventilation are a significant component in the energy balance of the building. Therefore, it is highly important to correctly determine the outdoor airflow rate value based on the type of the building, its operation and climatic conditions. The air change rate is an important parameter that affects the annual total energy consumption. The analysis with regard to the influence of air change rate on ventilation heat losses and the building time constant is presented in Figure 9.

Internal heat gains arising from people, lighting and electrical equipment enable us to reduce the heat demands for space heating. For the analyzed building, internal heat gains were determined according to Polish standard B-02025, EN ISO 13790, Polish and German regulations. The calculation results achieved are presented in Figure 10.

Table 4. Comparison of the calculated Q_H-values with the maximum permitted values given in German and Polish regulations (Dilmac & Kesen 2003, Koczyk 2004).

Description	B1[a]	B2[b]
According to calculations:	PS B-02025	PS B-02025 (PS EN ISO 13790)
$Q_{H,A}$ ($kWhm^{-2}y^{-1}$)	206.4	68.4 (88.8)
$Q_{H,V}$ ($kWhm^{-3}y^{-1}$)	74.7	24.8 (32.1)
According to Polish regulations:		
$Q_{H,Amax}$ ($kWhm^{-2}y^{-1}$)		90–120
$Q_{H,Vmax}$ ($kWhm^{-3}y^{-1}$)		32.6
According to German regulations:		
$Q_{H,Amax}$ ($kWhm^{-2}y^{-1}$)		61.7
$Q_{H,Vmax}$ ($kWhm^{-3}y^{-1}$)		17.1

[a]B1: the building before thermal modernization, [b]B2: the building after thermal modernization

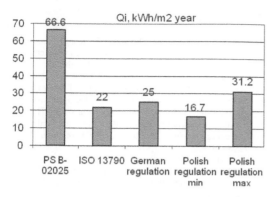

Figure 10. Annual internal heat gains of the analyzed building calculated with Polish standards and German regulations.

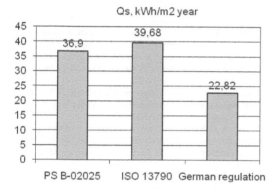

Figure 11. Annual solar heat gains of the analyzed building calculated with PS B-02025, EN ISO 13790 and German regulations.

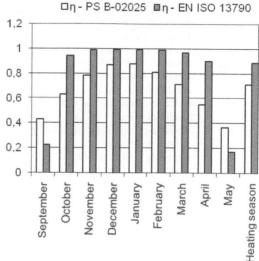

Figure 12. Comparison of monthly utilization factor values.

Annual internal heat gains per unit floor area $(kWhm^{-2}y^{-1})$ obtained by using the equations in Polish Standard B-02025 are different from those calculated by German regulation and Polish regulations.

Solar heat gains depend on solar irradiance and the effective solar collecting area of the glazed envelope element. The effect of the shading by different external obstacles (overhangs, fins) should be taken into consideration in the calculations. Figure 11 presents solar heat gains in the analyzed building, related to 1 m² of the heated area, calculated on the basis of Polish Standard B-02025, EN ISO 13790, Polish and German regulations. Annual solar heat gains per unit floor area $(kWhm^{-2}y^{-1})$ calculated by using the formulae in EN ISO 13790 are 73.9% higher than those calculated by the German regulations.

The building time constant defines the heat accumulation capacity of building construction. Catalina et al. (2008) reported that the building thermal inertia has a significant influence on the building energy consumption. The internal heat capacity of the analyzed residential building in Lublin amounts to 262.856 MJK^{-1}. Figure 12 demonstrates the values of the heat gains utilization factor for particular months of the heating season, calculated according to Polish Standard B-02025 and EN ISO 13790.

4 CONCLUSIONS

The reduction of energy consumption used for space heating by up to 30% constitutes a fundamental issue for environmental protection. Architects should be obliged to use energy-saving building technologies, engineers should design heating and ventilation systems of high efficiency, giving consideration to renewable energy sources. The shape of a building has an important impact on energy consumption. Low values of the overall heat transfer coefficient of building elements, as well as modern insulation techniques, enable a reduction in heat losses by transmission. Heat losses by ventilation are dependent on the air change rate. It is acceptable to calculate outdoor airflow rates based on national standards or on the basis of European standards. Comparing Polish standards with European standards and regulations, it is observed that there are distinct differences between calculation procedures in relation to internal heat gains and to solar heat gains. There is a need to conduct further research to define the influence of the internal thermal inertia of a building on the heat demands for space heating in Polish climatic conditions and with respect to different types of building construction.

NOMENCLATURE

a =numerical parameter in utilization factor;
A =heat loss surface (m²);
A_e =total amount of floor projection areas of all above-ground floors for a 5-meter-long space along the external walls (m²);

A_i =total amount of the remaining floor projection areas of all above-ground floors minus A_e (m^2);

A_n =net heating floor area (m^2);

A_{snj} =solar effective collecting area of the surface n having orientation j (m^2);

A_w =window area (m^2);

C =effective internal heat capacity of a conditional space (JK^{-1});

C_f =building shape coefficient $(\text{m}^2\text{m}^{-3})$;

F_F =frame area factor;

F_s =shading reduction factor;

F_w =correction factor for non-scattering glazing;

g =total solar energy transmittance of glazing;

g_n =total solar energy transmittance of glazing for the normal incidence;

H =total heat loss coefficient of the building (WK^{-1});

H_T =heat loss coefficient by transmission (WK^{-1});

H_V =heat loss coefficient by ventilation (WK^{-1});

I_j =solar radiation on vertical surface having orientation j (Wm^{-2});

l =length of thermal bridge (m);

n_h =air change rate (h^{-1});

NF =number of apartments;

NP =number of inhabitants;

Q_g =total heat gains, including internal and solar heat gains (J);

Q_L =heat loss of the building (J);

Q_h =heat demand of the building (J);

S_e =envelope surface of the building (m^2);

t =time (s);

U =thermal transmittance coefficient $(\text{Wm}^{-2}\text{K}^{-1})$;

V =volume of the building (m^3);

V_n =ventilated volume (m^3);

Greek letters

γ =heat gain and loss ratio (GLR);

η =utilization factor for heat gains;

θ_e =external air temperature (°C);

θ_i =internal air temperature (°C);

τ =building time constant (h);

Φ_i =average heat flow rate from internal heat source (W);

Φ_s =average heat flow rate from solar heat source (W);

Φ_{light} =heat flow rate from lighting (W per an apartment);

Ψ =linear thermal transmittance of thermal bridge $(\text{Wm}^{-1}\text{K}^{-1})$;

Subscripts

i =internal;

m =monthly;

s =solar.

Note: toe (tonne of oil equivalent) $= 41.868\,\text{MJkg}^{-1}$

REFERENCES

Balaras, C.A., Droutsa, K., Dascalaki, E., Kontoyiannidis, S. 2005. Heating energy consumption and resulting environmental impact of European apartment buildings. *Energy and Building* 37(5): 429–442.

Balaras, C.A., Gaglia, A.G., Georgopoulou, E., Mirasgedis, S., Sarafidis, Y., Lalas, D.P. 2007. European residential buildings and empirical assessment of the Hellenic building stock, energy consumption, emissions and potential energy savings. *Building and Environment* 42(3): 1298–1314.

Catalina, T., Virgone, J., Blanco, E. 2008. Development and validation of regression models to predict monthly heating demand for residential buildings. *Energy and Buildings* 40(10): 1825–1832.

CSO 2009. *Energy Efficiency in Poland in years 1997–2007*. Warsaw: Central Statistical Office Industry Division The Polish National Energy Conservation Agency.

CSO 2007. *Energy Efficiency Policies and Measures in Poland 2006*. Warsaw: Evaluation and Monitoring of Energy Efficiency in the New EU Member Countries and the EU-25 (EEE-NMC). Central Statistical Office The Polish National Energy Conservation Agency.

Chwieduk, D. 2003. Towards sustainable-energy buildings. *Applied Energy* 76(1–3): 211–217.

Depecker, P., Menezo, C., Virgone, J., Lepers, S. 2001. Design of buildings shape and energetic consumption. *Building and Environment* 36(5): 627–635.

Dilmac, S. & Kesen, N. 2003. A comparison of new Turkish thermal insulation standard (TS 825), ISO 9164, EN 832 and German regulation. *Energy and Buildings* 35(2): 161–174.

Dyrbøl, S., Tommerup, H., Svendsen, S. 2005. Savings potential in existing Danish building stock and new constructions. In: *ECEEE 2005 Summer study – what works & who delivers?*: 319–324.

EC 2009. *European Union Energy and Transport in Figures, 2009, ed. Part 2. Energy*. Brussels: European Commission. Directorate General for Energy and Transport.

Fayaz, R. & Kari, B.M. 2009. Comparison of energy conservation building codes of Iran, Turkey, Germany, China, ISO 9164 and EN 832. *Applied Energy* 86(10): 1949–1955.

Jokisalo, J. & Kurnitski, J. 2007. Performance of EN ISO 13790 utilisation factor heat demand calculation method in a cold climate. *Energy and Buildings* 39(2): 236–247.

Koczyk, H. 2004. Heat loads change of recipients in the aspect of the improvement of the thermal insulating power of buildings (in Polish). In Tomasz Mroz & Edward Szczechowiak (eds), *The technical progress in the district heating - Modern system-solutions*: 67-79. Poznan: PZITS.

Laskowski, L. 2005. *The thermal protection and the energy-characteristic of the building* (in Polish). Warsaw: Warsaw University of Technology.

Pérez-Lombard, L., Ortiz, J., Pout, C. 2008. A review on buildings energy consumption information. *Energy and Buildings* 40(3): 394–398.

Poel, B., Cruchten, G., Balaras, C.A. 2007, Energy performance assessment of existing dwellings. *Energy and Buildings* 39(4): 393–403.

Pogorzelski, J.A. 2007. The proposal of the initiating program of the Directive 2006/32/EC in the matter of the efficiency of the final energy utilization (in Polish). *Energy and Building* 10(10): 23–27.

Polish Standard PS B-03430. 1983. *Ventilation in dwelling and public utility buildings. Specification.*

Polish Standard PS B-02025. 2001. *Calculation of annual space heating requirements for residential and collective residential buildings.*

Polish Standard PS-EN ISO 13790. 2006. *Thermal performance of buildings – Calculation of energy use for space heating.*

Polish regulation: Decree of the Ministry of Infrastructure Dz.U. No 75 item 690, 12.04.2002.

Polish regulation: Decree of the Ministry of Infrastructure Dz.U. No 201 item 1238 and item 1240, 6.11.2008.

Santamouris, M., Kapsis, K., Korres, D., Livada, I., Pavlou, C., Assimakopoulos, M.N. 2007. On the relation between the energy and social characteristics of the residential sector. *Energy and Buildings* 39(8): 893–905.

Tommerup, H. & Svendsen, S. 2006. Energy savings in Danish residential building stock, *Energy and Buildings* 38(6): 618–626.

Wichowski, R. 2007. Energy consumption in residential buildings for selected European countries (in Polish). *District Heating, Heating, Ventilation* 11 (452): 70–71.

Environmental Engineering III – Pawłowski, Dudzińska & Pawłowski (eds)
© 2010 Taylor & Francis Group, London, ISBN 978-0-415-54882-3

Co-combustion of syngas obtained from air – biomass gasification with coal in small scale boilers

R.K. Wilk & P. Plis

Institute of Thermal Technology, Silesian University of Technology, Gliwice, Poland

ABSTRACT: The objective of this paper was an experimental investigation of syngas co-combustion with coal in a water boiler. The boiler used in the experimental system was a boiler with a 'retort' burner, with nominal heat power of 50 kW. The results showed that syngas combustion in the boiler provided 43.6–80.1% of actual boiler heat power. The experiments showed that biomass gasification and co-combustion of the syngas in a boiler was beneficial in reducing fossil fuel consumption in combustion systems.

Keywords: Gasification, biomass, retort boiler, syngas co-combustion.

1 INTRODUCTION

Biomass is widely known to be an alternative to fossil fuel combustion and can be used for many energy applications, such as heat and electricity generation. Choice of the method of utilization depends on the purpose of the combustion system. There are three basic concepts of biomass combustion: direct, indirect and parallel co-combustion (Brown et al. 2005). Of these, indirect co-combustion seems the most promising solution for using biomass of variable properties, but it is also more expensive than direct co-combustion. One of the chief disadvantages of biomass is the variation in properties, especially moisture content and elemental composition. This may require the adaptation of combustion installations for defined kinds of biomass. In indirect co-combustion systems this problem may be easily solved. Combustible syngas created in the gasification process is mainly composed of H_2, CO, CO_2, CH_4 and higher hydrocarbons, and may be further combusted in the power boiler, avoiding risks to burner and boiler operation associated with direct combustion. Use of syngas can reduce fossil fuel consumption, and since it contains hydrocarbons and other combustible species, it may be used as a reburning fuel to reduce NO_x emissions.

A review of the gasification technologies concerning feedstock properties and types of gasifiers (2002) confirmed that updraft gasification is a very simple and reliable technology and very importantly is suitable for wet biomass conversion. Updraft gasifiers can use biomass feedstock of differing sizes. For these reasons, updraft gasifiers are suitable for indirect biomass co-combustion, but the syngas should not be cooled because of the high concentration of condensable tars it contains.

2 MATERIALS AND METHODS

A laboratory system was designed and built for this experimental investigation (Figures 1, 2). The stand consisted of a water boiler and updraft gasifier connected by a pipe, thus the syngas obtained in the gasification process was supplied into the combustion zone of the boiler and was co-combusted with coal. Biomass was fed into the gasifier from the top, while air was supplied by blower from the bottom. The biomass feedstock moved in the opposite direction to gas flow and passed through the drying, pyrolysis, reduction and combustion zones. In the drying zone, moisture was evaporated; in the pyrolysis zone, biomass was thermally decomposed to volatiles and solid char; and in the reduction zone carbon was

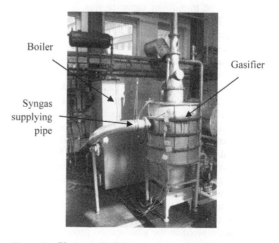

Figure 1. Photograph of the experimental system.

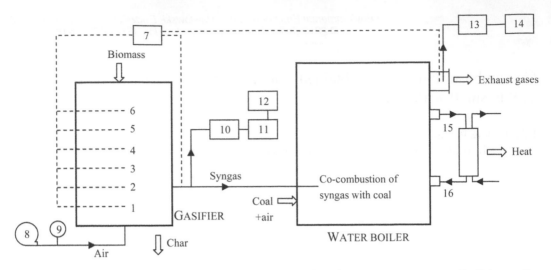

Figure 2. Schematic diagram of the experimental system: 1–6 – Points of temperature measurement, 7 – Data recording system, 8 – Air blower, 9 – Flow meter, 10 – Cleaning & cooling system, 11 – CO&H_2 analysers, 12 – Gas sample, 13 – Cleaning & cooling system, 14 – Exhaust gases analyser, 15 – Hot water, 16 – Feedwater.

converted, and CO and H_2 produced as the main combustible components of the syngas. In the combustion zone of the gasifier, the remaining char was combusted thus providing the heat for endothermic reactions in the upper zones. The gasifier was initially designed to ensure about 10% of the nominal heat power of the boiler.

The boiler used in the experiment was a water boiler with a 'retort' burner, commonly referred to as a 'retort boiler'. These types of boilers are very popular low-power boilers in Poland. The boiler was supplied with coal and had a nominal heat power of 50 kW, while its actual heat power was measured by an electronic heat meter. To balance the system, the temperature of exhaust gases was measured by a K-type thermocouple and their composition measured by an analyser.

The gasification chamber was constructed from four cylindrical segments lined with refractory rings, with an internal diameter of 0.25 m and height of 0.60 m. The maximum capacity was 20 kg of wood pellets. The syngas generator was placed on a scale to measure the mass decrement of gasified fuel. Temperature of the gasifier interior was measured by six N-type and K-type thermocouples located along the gasifier's vertical axis at different heights. The temperature of the syngas at the gasifier outlet was also measured. The air flow rate into the gasifier was measured by flow meter. At the gasifier outlet, there was a sampling point, where syngas samples were collected and cleaned by a system of filters, and then supplied to CO and H_2 analysers. Molar fractions of these two combustible species were measured on-line at the experimental system. For a few specified experimental points, full analysis of syngas composition was investigated by chromatographic analysis.

The coal parameters used in the boiler, and biomass parameters used in the gasification process are given in Table 1.

Table 1. Proximate and ultimate analyses of fuels used (on weight, as received basis).

		Coal	Wood pellets
Proximate analysis			
LHV	kJ/kg	28 903	19 190
Ash	%	2.72	1.07
Moisture content	%	6.77	7.0
Volatile matter	%	29.83	77.70
Ultimate analysis			
C	%	74.88	48.41
H	%	4.23	5.41
O	%	11.88	37.99
N	%	1.06	0.26
S	%	0.28	0.01

The experimental procedure started with firing up the boiler. When the boiler parameters reached a steady state (heat power), then the gasifier was fired up. The blower was switched on and biomass was introduced through the port in the upper part of the gasifier, and ignited through a port in the bottom. After approximately 2 h, the gasifier was heated and the experimental measurements started. The scale on which the gasifier was placed was turned on and the gasifier was filled with a fresh quantity of biomass. Once syngas production began, measurements of key variables started. The parameters measured were as follows:

- For the gasifier
  - Mass flow rate of biomass,
  - Air flow rate supplied into the gasifier,
  - Composition and temperature of the syngas produced by gasification and supplied into the boiler.

Table 2. Water boiler operating parameters for coal combustion and for syngas co-combustion with coal.

Case		Coal combustion	Syngas co-combustion with coal			
			No. 1	No. 2	No. 3	No. 4
Air supplied into the gasifier	m^3/h		6	8	10	12
Syngas composition						
CO_2	%		8.71	9.26	9.08	16.51
C_2H_4	%		0.02	0.10	0.02	0.42
C_2H_6	%		0.12	0.12	0.12	0.23
H_2	%		10.77	1.02	8.93	9.97
O_2	%		1.18	1.21	1.35	1.09
N_2	%		51.46	49.92	52.52	48.37
CH_4	%		2.58	2.42	2.27	4.00
CO	%		25.10	25.2	25.65	19.24
LHV	MJ/m_n^3		5.39	5.41	5.16	5.47
Exhaust gases						
CO_2	%	8.5	13.92	14.06	14.9	15.26
CO	%	0.02	0.73	0.99	2.5	3.72
O_2	%	11.6	5.94	5.58	3.92	3.14
N_2^*	%	79.9	79.41	797	78.8	77.88
NO_x	ppm	202	146	135	98	82
Temperature	°C	255	325	348	364	350
Heat power	kW	36.2	52	58.6	61.2	65.2
Gasification efficiency	%	–	67.9	65.9	64.9	59.7
Boiler efficiency	%	76.2	73.5	72.9	69.0	69.0
System efficiency	%	–	63.1	60.7	55.2	51.7
Coal feeding rate	kg/h	5.92	5.92	5.92	5.92	5.92

*by difference.

- For the boiler
  - Mass flow rate of coal,
  - Actual heat power of the boiler,
  - Composition and temperature of exhaust gases.

Energy balances used in the investigation were based on the assumption of semi-ideal gases, the reference temperature was set to 25° C; and combustible fraction in the slag from the boiler, and char from the gasifier were taken into account. Efficiency of the gasifier (Hot Gas Efficiency or HGE), boiler efficiency (η_b) and boiler–gasifier system efficiency (η_s) are defined as follows:

$$HGE = \frac{LHV\ of\ syngas + sensible\ heat\ of\ hot\ syngas}{Biomass\ flow\ rate \cdot LHV\ of\ biomass} \quad (1)$$

$$\eta_b = \frac{Heat\ power\ of\ the\ boiler}{Coal\ flow\ rate \cdot LHV\ of\ coal + syngas\ flow\ rate \cdot LHV\ of\ syngas} \quad (2)$$

$$\eta_s = \frac{Heat\ power\ of\ the\ boiler}{Coal\ flow\ rate \cdot LHV\ of\ coal + biomass\ flow\ rate \cdot LHV\ of\ biomass} \quad (3)$$

where LHV = the Lower Heating Value.

3 RESULTS AND DISCUSSION

The experimental results are presented in Table 2. Syngas co-combustion (supplementary fuel) in the power boiler changed the composition and amount of exhaust gases at the boiler outlet.

For syngas co-combustion with coal, exhaust gas composition from the boiler mainly depends on syngas composition and syngas flow rate. Syngas composition strongly depended on the air flow rate into the gasifier. Any change in air flow rate caused variation of gasification rate, process temperature, and composition and amount of syngas produced.

This influenced the boiler operating parameters and efficiency of co-combustion. Despite some differences in syngas composition, there were similar lower heating values (LHVs) of the syngas (Table 1). The main difference between the cases was syngas flow rate, which was highest for case No. 4. For that reason, the amount of CO and other combustible species supplied into the boiler was also the highest for case No. 4.

The composition of gases in the exhaust from the boiler for five cases is shown in Figure 3. The first case is coal combustion only, with an air flow rate equal to zero. Syngas co-combustion with coal decreased the O_2 molar fraction in exhaust gases, from 11.6 to 3.14%. A higher air flow into the gasifier caused more syngas to be created in the gasification process, thus there were higher amounts of combustible species to be combusted in the boiler.

Higher amounts of air supplied into the gasification process decreased the NO_x molar fraction in exhaust gases (Figure 3). For the maximum air flow rate of $12\,m^3/h$, NO_x was reduced by 59%, but there was a

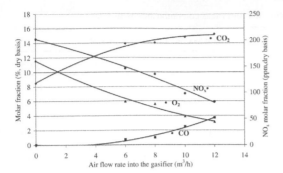

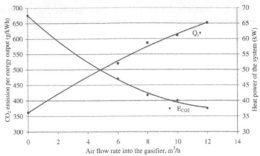

Figure 3. Composition of the exhaust gases from a boiler for coal combustion and syngas co-combustion.

Figure 4. CO$_2$ emission from the boiler compared with different amounts of syngas supplied to the boiler.

high CO molar fraction in the exhaust gases. A compromise has to be reached between NO$_x$ reduction and combustible species molar fraction in the exhaust gases; for the air flow of 6 m^3/h, the CO molar fraction in the syngas was much lower than for the air flow of 12 m^3/h and NO$_x$ was reduced by 28% (Figure 3).

The CO molar fraction in the exhaust gases increased markedly, from 0.02 to 3.72%. This phenomenon was not due to insufficient O$_2$ in the boiler combustion chamber. The air supplied to the combustion chamber was sufficient for combustion, because not all O$_2$ was used in the co-combustion process; in case No. 4, the boiler exhaust gases were 3.14% O$_2$. This indicates inadequate mixing of syngas with O$_2$ in the combustion chamber. At the experimental system, syngas was supplied into the coal combustion zone (into the flame), where temperature was expected to be highest. This should ensure burning of combustible species; however, the CO molar fraction in boiler exhaust gases indicated that not all CO was burned. This was due to a further important issue – the residence time of combustible species in the high temperature zone. Outside of the combustion zone there was a significant drop in temperature, to a point insufficient for complete combustion of CO. Experimental data showed that the CO molar fraction in the exhaust gases was unacceptably high. It may be that air supplied into the coal combustion chamber forced the syngas away from the high temperature zone, so it did not react completely with O$_2$ in the flue gases.

The CO$_2$ molar fraction in the exhaust gases increased with increased air flow into the gasifier. For coal combustion, the CO$_2$ molar fraction was 8.5%, while in the case No. 4 it was 15.26%. CO$_2$ emission from the coupled boiler–gasifier system was calculated as a CO$_2$ mass flow rate from the boiler divided by the system heat power, according to Equation (4):

$$E_{CO_2} = \frac{[CO_2]_{6\%} \cdot n_{eg} \cdot M_{CO_2} \cdot 3600 \cdot 1000}{Q_s}, \, g/kWh \quad (4)$$

where $[CO_2]_{6\%}$ (kmol CO$_2$/kmol dry exhaust gases) = the CO$_2$ concentration (6% O$_2$) in the exhaust gases; n_{eg} (kmol/s) = the exhaust gases molar flow

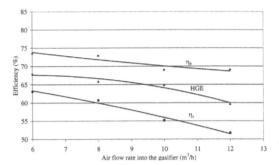

Figure 5. Efficiencies of the gasifier (HGE), boiler (η_b) and coupled boiler–gasifier system (η_s).

rate; M_{CO_2} (kg/kmol) = molar mass of CO$_2$; Q_s (kW) = a heat power of the system.

This approach allows comparison of the amounts of CO$_2$ released to the atmosphere per amount of heat produced in the boiler (Figure 4). The CO$_2$ emission per energy output decreased with increased syngas flow rate into the boiler.

A greater amount of air supplied into the gasifier increased the system heat power (Figure 4), but also reduced the boiler–gasifier system efficiency (Figure 5).

Gasifier efficiency decreased slightly with increased air flow into the gasifier, due to higher fuel consumption in the gasifier – a higher amount of air causes an increase in the biomass burnout rate. Compared to coal combustion, the efficiency of the boiler in co-combustion mode decreased because of increased CO molar fraction in the exhaust gases and increased exhaust gas temperature.

Efficiency of the boiler–gasifier system depends on the boiler and gasifier efficiencies. The highest system efficiencies were for case No. 1, where air flow into the gasifier was lowest. This was due to the lower amount of additional fuel supplied to the boiler.

To avoid significant losses of system efficiency, the gasifier should be operated at lower air flow rates; however, a major drawback is the lower heating value of the system. Increased air flow into the gasifier and thus increased syngas flow into the boiler, increased the system heat power.

4 CONCLUSIONS

The experimental investigation confirmed that syngas co-combustion with coal in a small-scale boiler was an interesting method for replacing fossil fuel combustion with alternative energy sources. Co-combustion of syngas supplied into the boiler increased the flue gas temperature at the boiler outlet, and this reduced boiler efficiency.

The boiler–gasifier system should be flexible in terms of the biomass feedstock used in the gasification process. Only one feedstock was investigated in this experiment, but a principal advantage of major updraft gasification is the possibility of using different sources of biomass, especially of high moisture content. Thus, this system should be suitable for a wide variety of biomass types since there is no problem with biomass pre-treatment compared to direct biomass combustion in the boiler.

Additionally, the syngas supply-line is of simple construction, and no special co-combustion nozzles were needed. Even with high amounts of tar in the syngas there were no problems of clogged nozzles. Moreover, tar content in syngas increases its heating value; however, the syngas supply-line should be insulated to prevent tar condensation on internal surfaces.

Supplying syngas into the coal combustion zone in the boiler increased the heat power of the system by 43.6–80.1% of the boiler actual heat power. Additionally, there was no need to shut down the boiler if the gasifier was out of operation.

ACKNOWLEDGEMENTS

This paper was prepared as part of the project financed by measures assigned for science in 2006–2009 as R&D by the Ministry of Science and Higher Education of Poland (No. R06 014 01).

REFERENCES

Brown, G., Hawkes, A.D., Bauen, A. & Leach, M.A. 2006. EUSUSTEL Report- Work Package 3: Electricity generation technologies and system integration – Biomass Application.
EUSUSTEL. Stuttgart, 2006. <http://www.eusustel.be/wp.php>
McKendry, P. 2002. Energy production from biomass (part 3): gasification technologies. *Bioresource Technology* 83: 55–63.

Environmental Engineering III – Pawłowski, Dudzińska & Pawłowski (eds)
© *2010 Taylor & Francis Group, London, ISBN 978-0-415-54882-3*

Narrow-band spectral models for diagnostic of gases produced during the biomass production

W. Wójcik & S. Cięszyk
Department of Electronics, Lublin University of Technology, Lublin, Poland

T. Golec
Thermal Divisions, Lublin University of Technology, Lublin, Poland

ABSTRACT: This paper presents methods for the use of biomass in the power industry, including biomass gasification and the subsequent combustion of the gases produced. During gasification a mixture of gases with spectral features in the mid-infrared region is formed. Spectral models used in spectroscopy and radiative transfer are presented and the Statistical Narrow-Band (SNB) model described in detail. Spectra of gases produced during biomass gasification are presented. Statistical models are used to calibrate and predict the high content of carbon monoxide. The method described is suitable for diagnostics in biomass gasification and the combustion of the gases produced.

Keywords: Biomass gasification, gas statistical narrow-band model, spectroscopy.

1   INTRODUCTION

There has recently been great interest in the use of biomass in power engineering. Biomass can be burned directly or in combination with coal or pre-processed. During biomass gasification a specific combination of gases is formed. The main energetic gases are carbon monoxide (CO), hydrogen (H_2) and methane (CH_4), together with large amounts of carbon dioxide (CO_2). So how can we determine the concentration of these gases? One method that can be used is spectroscopy, which relies on the determination of the absorption of the gas sample. The difficulty when working with high gas concentrations is the nonlinear relationship that exists between absorption and gas concentration. The gases produced during biomass gasification are strongly absorptive because of their high concentrations and high absorption coefficients, so high-resolution methods are not required. Assuming low spectral resolution, the approximation of spectral transmissivity of the gas may be used. This is mainly used in radiative transfer calculations.

biomass with another fuel is the usual process. Pyrolysis is a method of thermal decomposition of biomass in the absence of oxygen. As a result, gaseous fuel together with solid and liquid residues is produced. Pyrolysis is used to produce various types of fuels. Usually, the process is carried out at temperatures ranging from 400°C to 600°C. Biomass gasification may be based on the processes of pyrolysis, partial oxidation and reforming. Partial oxidation occurs when a substoichiometric mixture of fuel and oxygen is partially combusted. Reforming is a gasification process in the presence of an additional reagent. During gasification, CO and H_2 are the main products formed. Hydrocarbons are also present, mainly CH_4, as well as CO_2. A general characteristic is the increasing content of CO and H_2 with increasing temperature and the simultaneous decrease in levels of CH_4, CO_2 and H_2O.

Figure 1 shows changes over time in the levels of the main gases produced during biomass gasification. Measurements were carried out at the Institute of Power Engineering in Warsaw. As we can see, the main gas is CO at a level of above 25%.

2   MATERIALS AND METHODS

2.1   *Biomass utilizations in the power industry*

The basic way of using biomass is by direct combustion. The combustion of cellulose results in energy, CO_2 and steam (H_2O). Temperature and combustion efficiency depend mainly on water content and the availability of air. In power systems, co-combustion of

2.2   *Features of gas spectral modelling*

Gas molecules absorb and emit infrared radiation in discrete spectral bands. Each gas has spectral windows during which this phenomenon does not occur. This phenomenon of selective emission and absorption is utilised in laboratory spectroscopy. Another application is spectral remote sensing in astronomy, atmospheric research and non-invasive combustion

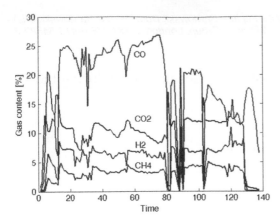

Figure 1. Example time plot of the content of the main gases from biomass gasification.

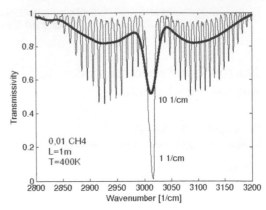

Figure 2. Methane transmissivity for 1 and 10 cm^{-1} resolution (Gaussian apodization).

diagnostics (Wójcik 2008). The gas molecule undergoes rotation and oscillation that affects the emission and absorption of radiation. Since the phenomenon is governed by the laws of quantum physics, the molecular energy change is quantized. Very narrow rotational lines with characteristic intensities and shapes appear in the spectrum. A set of these lines creates a vibration band. Gases can be identified based on their vibration-rotation bands. Models to calculate spectra based on physical laws are known as line-by-line methods. As the number of rotational lines for a particular gas is huge, a number of related parameters required for spectrum calculation are also important. Unfortunately, this means that a large amount of computer time is required for the whole band. Methods required for this type of gas modelling demand high spectral resolution capable of distinguishing rotational lines. Such high resolution methods are used in laboratory spectroscopic research. They are useful for distinguishing mixture of gases with similar spectral features and overlapping bands. In atmospheric spectral remote sensing, information from one or a few rotational lines is used. In radiative transfer calculations less exact but simpler spectral models are used, allowing faster calculation. Narrow-band models apply mainly to CO_2, H_2O and CO because of their dominant role in radiative transfer. These models do not correspond to the physical nature of gases but to their statistical properties. The main advantage of such models is their speed in the analysis and design of complex thermal objects. Statistical models are not suitable for general spectral analysis. For such purposes, more exact models that better reflect spectral features are required. In the gasification process, H_2O, CO_2, CO and CH_4 are the dominant molecules. For their analysis, narrow-band models are sufficient, especially for diagnostic purposes since high concentrations of these gases are present during gasification.

There are many databases for gas spectral calculations. They differ mainly in the temperature range to which they apply. HITRAN (Rothman et al. 2005) and its high temperature counterpart HITEMP are the most often used. They are line-by-line models containing rotational line data.

2.3 HITRAN database

HITRAN contains data for every particular rotational line – position, shape and other parameters. The HITEMP database contains additional lines active above 600 K, but these data are calculated theoretically. There are more than one million lines for H_2O and CO_2 in the HITEMP database. Absorption coefficient α, is a function of the line strength line S_i [cm^{-2} atm^{-1}], line shape function around the centre wavenumber ϕ [cm], pressure P [atm], molar fraction x_j and path length L (Rothman et al. 2005):

$$\alpha = S_i \phi P x_j L \qquad (1)$$

Line strength is a function of temperature (Rothman et al. 2005):

$$S(T) = S(T_0) \cdot \frac{Q(T_0)}{Q(T)} e^{\frac{hcE''}{k}\left(\frac{1}{T} - \frac{1}{T_0}\right)} \cdot \frac{1 - e^{-\frac{hcv_0}{kT}}}{1 - e^{-\frac{hcv_0}{kT_0}}} \qquad (2)$$

where: c = the speed of light [cm s^{-1}]; h = Planck's constant [J s]; k = Boltzmann constant [J K^{-1}]; E'' = lower state energy level [J]; v_0 = central wavenumber [cm^{-1}]; and Q = the partition function for a species.

The shape of the rotational line is determined by two main physical phenomena. At the temperature range we are concerned with, the main broadening mechanisms are effects of temperature and pressure. Temperature broadening is caused by the Doppler phenomenon depending on the velocity of the molecule relative to the observer. The higher the temperature of the gas, the wider the distribution of velocities in the gas. The distribution of velocities is described by the Boltzmann distribution. Pressure broadening is attributed to interaction between molecules. It is also called collision broadening because it increases

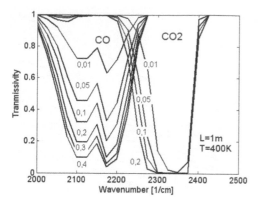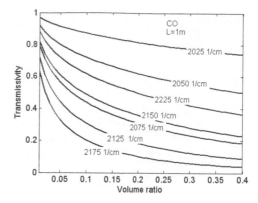

Figure 3. Comparison of CO and CO_2 spectral absorption (T = 400 K, L = 1 m) and transmissivity against CO content (volume ratio).

with the collision frequency, which is highly dependent on pressure. For typical applications, both types of broadening must be taken into account.

3 RESULTS AND DISCUSSION

The most often used narrowband gas models are RADCAL (Grosshandler 1993) and the SNB (Statistical Narrow-Band) model (Liu & Smallwood 2004, Soufiani & Taine 1997). They are used with a spectral resolution of 25 cm^{-1}, or in rare cases 10 or 5 cm^{-1}. The spectral line is defined by centre wavenumber, intensity, shape and half width. A rotation-vibration band consists of many lines close together. Mean transmission may be evaluated by adding together individual lines in a selected band taking into account their location. However, the changing behaviour of spectral line parameters with temperature, pressure and concentration of every line in the selected band must also be taken into account. In the literature, many types of statistical band parameter calculations are presented. These models differ in their way of estimating line distribution. The SNB model assumes a statistical distribution of line intensities. It is a good representation of line distribution for complex molecules, for which lines overlap in an irregular way. The Malkmus model is considered the most precise model for calculating gaseous radiation in the combustion process. It is statistical model with an exponential tailed distribution of line intensities. Transmissivity for this model may be written as follows (Caliot et al. 2009, Daszykowski et al. 2007, Liu & Smallwood 2004, Soufiani & Taine 1997):

$$\bar{\tau} = exp\left[-2\frac{\gamma}{\delta}\left(\sqrt{1 + xplk\frac{\delta}{\gamma}} - 1\right)\right] \quad (3)$$

where:k [cm^{-1} atm^{-1}] = mean line intensity to spacing ratio, and $1/\delta$ [cm] equivalent line spacing or typical distance between lines are parameters of the model, and γ [cm^{-1}] = typical collision half width of gas line.

As we can see (Equation 3) in the SNB model, transmissivity is calculated not absorption (absorption coefficient) as in spectroscopic models. Formulas for mean line broadening of combustion gases can be found in the literature. One of the most popular formulas for spectral line width calculation of CO, CO_2, H_2O and CH_4 is as follows (Perrin & Soufiani 2007, Soufiani & Taine 1997):

$$\bar{\gamma}_{CO} = \frac{p}{p_s}\left[0.075X_{CO_2}\left(\frac{T_s}{T}\right)^{0.6} + 0.12X_{H_2O}\left(\frac{T_s}{T}\right)^{0.82} \cdots \right.$$
$$\left. + 0.06 \cdot \left(\frac{T_s}{T}\right)^{0.7}\left(1 - X_{CO_2} - X_{H_2O}\right)\right] \quad (4)$$

$$\bar{\gamma}_{CO_2} = \frac{p}{p_s}\left(\frac{T_s}{T}\right)^{0.7}\left[0.07 \cdot X_{CO_2} + 0.1 \cdot X_{H_2O} + \cdots \right.$$
$$\left. + 0.058 \cdot \left(1 - X_{CO_2} - X_{H_2O}\right)\right] \quad (5)$$

$$\bar{\gamma}_{H_2O} = \frac{p}{p_s}\left\{0.462 \cdot X_{CO_2}\frac{T_s}{T} + \left(\frac{T_s}{T}\right)^{0.5} \cdots \right.$$
$$\left. \cdot \left[0.08\left(1 - X_{CO_2} - X_{O_2}\right) + 0.106X_{CO_2} + 0.036X_{H_2O}\right]\right\} \quad (6)$$

$$\bar{\gamma}_{CH_4} = 0,051\left(\frac{T_s}{T}\right)^{0.75} \quad (7)$$

where $p_s = 1$ atm and $T_s = 296$ K.

The procedure for evaluating parameters consists of multiple calculations of transmissivity for many values of the optical depth and a subsequent adjusting of parameters using the least squares method. The path length and gas concentration are selected in such a way that transmissivity changes in the range 0.05 to 0.95 in 0.05 steps. In this way, parameters are properly evaluated for various optical depths.

Example spectra for CO and CO_2 calculated using the SNB model are shown in Figures 3 and 4. Two sets of data, for learning and testing, were calculated using the SNB model. Next, the regression model

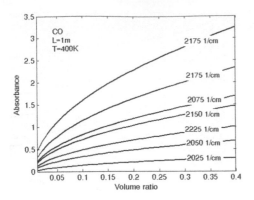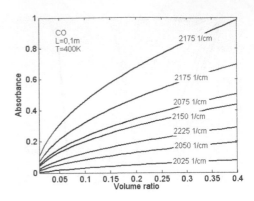

Figure 4. Absorbance dependence of CO (volume ratio) for a particular spectral band (T = 400 K, L = 1 m and L = 0.1 m).

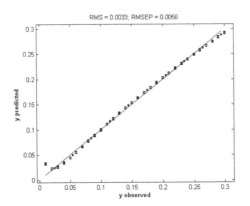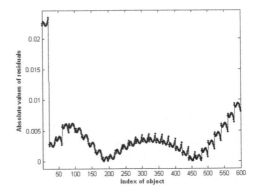

Figure 5. Relation between observed CO content and predicted CO content for the PLS model (T = 300 K, L = 0.1 m) and absolute values of residuals for the test set.

was created using the partial least squares procedure and the learning set. The MATLAB toolbox TOM-CAT (Daszykowski et al. 2007) was used to create the model. Its input is absorbance spectra between wavenumber 2000 and 2400 cm^{-1} and with 25 cm^{-1} resolution. The learning set consisted of 260 points of CO content changing from 0.01 to 0.3 volume ratio. Simultaneously, the content of CO_2 for the learning set was changing from 0.01 to 0.02. The test set consists of 600 similarly created data points (30 points for CO content from 0.01 to 0.3 volume ratio for each of 25 points of CO_2 content). Simulation results are shown in Figure 5. The root mean square error for the learning set was 0.0033. The root mean square error of prediction for the learning set was 0.0056. As shown in Figure 5b, test data error only exceeds 0.01 for the first 25 samples, that is for the least CO content of 0.01 volume ratio.

4 CONCLUSIONS

A mixture with a high content of energetic gases is produced during the biomass gasification process. Most of them (CO, CO_2, CH_4) have spectral features in the mid-infrared region. These gases belong to the

type that can be modelled using narrow-band models such as SNB. Since the levels produced are so high, these statistical models can be used for spectroscopic quantitative analysis. For example, calibration and prediction of CO content using partial least squares regression (PLS) is shown. The typical high concentrations of the gases do not require high spectral resolution measurements. Spectral analysis methods at 25 cm^{-1} resolution are sufficient for diagnostic purposes of biomass gasification. Calibration using the partial least squares method can be carried out by using RADCAL and the SNB model.

REFERENCES

Caliot, C., Abanades, S., Soufiani, A. & Flamand G. 2009. Effects of non-gray thermal radiation on the heating of a methane laminar flow at high temperature. *Fuel* 88: 617–624.

Daszykowski M., Serneels S., Kaczmarek K., Essen P., Croux C. & Walczak B. 2007. *TOMCAT:* A Matlab toolbox for multivariate calibration techniques. *Chemometrics and Intelligent Laboratory Systems* 85: 267–277.

Grosshandler, W.L. 1993. *RADCAL: A narrow-band model for radiation calculations in a combustion environment.* National Institute of Standards and Technology Technical Note 1402. April 1993.

Liu, F. & Smallwood, G.J. 2004. An efficient approach for the implementation of the SNB based correlated-k method and its evaluation. *Journal of Quantitative Spectroscopy & Radiative Transfer* 84: 465–475

Perrin, M.Y. & Soufiani, A. 2007. Approximate radiative properties of methane at high temperature. *Journal of Quantitative Spectroscopy & Radiative Transfer* 103: 3–13.

Rothman, L.S. et al. 2005. 2005. The HITRAN 2004 molecular spectroscopic database. *Journal of Quantitative Spectroscopy & Radiative Transfer* 96: 139–204.

Soufiani, A. & Taine, J. 1997. High temperature gas radiative property parameters of statistical narrow-band model for H_2O, CO_2 and CO, and correlated-K model for H_2O and CO_2. *International Journal of Heat Mass Transfer* 40(4): 987–991.

Wójcik, W., Cięszczyk, S., Komada, P. & Kisała, P. 2008. Pomiary widm procesów spalania z wykorzystaniem spektrometru FTIR. *Elektronika* 6: 230–232.

Author index

Printed and bound by CPI Group (UK) Ltd, Croydon, CR0 4YY

01/11/2024

01782599-0009